THE IMMUNE SYSTEM THIRD EDITION

THE IMMUNE SYSTEM THIRD EDITION
PETER PARHAM

The Immune System is adapted from *Immunobiology*
by Charles A. Janeway, Jr., Paul Travers, and Mark Walport;
also published by Garland Science.

GS **Garland Science**
Taylor & Francis Group

LONDON AND NEW YORK

Vice President: Denise Schanck
Senior Editor: Janet Foltin
Project Editor: Sigrid Masson
Assistant Editors: Katherine Ghezzi and Alex Engels
Text Editor: Eleanor Lawrence
Production Editor: Emma Jeffcock
Copyeditor: Bruce Goatly
Illustration and Design: Nigel Orme
Cover Photographer: Richard Denyer
Indexer: Liza Furnival

Peter Parham is on the faculty at Stanford University where he is Professor in
the Departments of Structural Biology, and Microbiology and Immunology.

Library of Congress Cataloging-in-Publication Data
Parham, P. (Peter), 1950-
  The immune system / Peter Parham. -- 3rd ed.
    p. ; cm.
  ISBN 978-0-8153-4146-8
1. Immune system. 2. Immunopathology. I. Title.
  [DNLM: 1. Immune System. 2. Immunity. QW 504 P229i 2009]
QR181.P335 2009
616.07'9--dc22

                                        2008034011

Published by Garland Science, Taylor & Francis Group, LLC, an informa business
270 Madison Avenue, New York, NY 10016, US, and
4 Park Square, Milton Park, Abingdon, OX14 4RN, UK

Printed in the United States of America
15 14 13 12 11 10 9 8 7 6 5 4 3 2 1

Taylor & Francis Group, an informa business

Visit our website at http://www.garlandscience.com

Preface

In the twentieth century, immunology was often taught as a form of epic saga, a progression of elegant experiments, breakthrough discoveries, and intrepid investigators. The merit of the approach was its simplicity, as in tracing history using the dates of monarchs, ministers, conflicts, and conquests. It also added spice and personality in an era when knowledge was hard won, incomplete and only tenuously related to the human condition. That situation has clearly changed, the direct consequence of technical innovations started in the latter part of the twentieth century. Today new information literally pours from the research pipeline, into the World Wide Web and onto the printed page. Elegant experiments, breakthrough discoveries, and intrepid investigators are now in such abundance that they in themselves can no longer provide a simplifying principle for explaining immunology. However, as a result of all this industry, immunologists now have a more coherent and intellectually more satisfying picture of the immune system and how it works, not always successfully, to protect the body from infection.

This book is aimed at students who are coming to immunology for the first time. Throughout the book the emphasis is upon the human immune system and how its successes, failures, and compromises affect the lives of each and every one of us. The first ten chapters of the book describe the cells and molecules of the immune system and how they work together in providing defenses against invading microorganisms. This part of the book is built around the biology of the B and T lymphocytes, the cells that adapt to infection and provide long-lasting protective immunity, but it also emphasizes the importance of other cells—macrophages, dendritic cells, granulocytes, mast cells, and natural killer cells—that provide the front-line defenses of innate immunity, the inflammation necessary to stimulate B and T cells and many of the weapons they deploy to destroy pathogenic microorganisms. The first chapter introduces the different types of cell that make up the immune system and the functions they perform. Two new chapters then follow: the first is devoted to innate immunity, and the second lays out the cellular and genetic principles underlying adaptive immunity. Antigen recognition, repertoire development, and effector functions of B and T cells are dealt with in turn in the next six chapters. Completing this part of the book is a chapter describing how innate and adaptive immunity work together to battle the common types of infection.

The next three chapters focus upon diseases that arise from inadequacies of the human immune system. The first describes situations in which infections fail to be controlled, often because the microorganism actively evades, exploits, or subverts the immune response. The following chapter examines conditions in which the immune system overreacts to innocuous substances in the environment and causes chronic inflammatory diseases such as allergies and asthma. The third of these chapters examines the autoimmune diseases, such as Graves' disease, insulin-dependent diabetes, multiple sclerosis, and rheumatoid arthritis, in which the immune system attacks healthy cells and causes tissue damage and loss of function. The final three chapters of the book illustrate how the immune system is being manipulated to improve human health. The first of these considers vaccination, a practice used for centuries and arguably the medical intervention that has saved the most human lives. The penultimate chapter of the book examines the transplantation of tissues and organs, life-saving and routine therapies that were principally developed in the twentieth century and which required a knowledge of the genetic differences between human beings and how their immune systems respond to them. The final chapter on cancer immunology looks to the future and to the long-felt hope for the more effective prevention and treatment of cancer through manipulation of the human immune response to malignant cells and oncogenic viruses.

This third edition of *The Immune System* represents a major revision involving a fundamental change of the order in which innate and adaptive immunity are presented, which was almost universally requested by the instructors using the book. Previously, innate immunity was introduced in Chapter 1 and only expanded upon in Chapter 8, where the complementary roles of innate and adaptive immunity were brought together in the context of infection. The cellular and molecular mechanisms of innate immunity have now been moved to the front of the book, being described in a new Chapter 2 that follows the introductory Chapter 1. Thus the reader becomes familiar with the breadth, depth, and importance of innate immunity before tackling the detailed account of adaptive immunity that is subsequently developed in Chapters 3–9. This reorganization has intrinsic biological logic, because innate immunity and adaptive immunity are dealt with in the same order in which they evolved and in which the human body uses them to respond to infection. This all makes for a much more coherent narrative, as is well illustrated by complement—many an instructor's nightmare—for which the fundamentals were previously, and somewhat arbitrarily, divided between Chapters 7 and 8 and are now logically consolidated in Chapter 2.

Removal of innate immunity from the chapter on the body's defenses against infection (now Chapter 10) provided the right opportunity to incorporate major new knowledge and understanding of the immune system. The section on immunological memory has been updated and new sections on mucosal immunity and on lymphocyte subsets that bridge innate and adaptive immunity have been added. At the request of many instructors, the last chapter on manipulation of the immune response (Chapter 12 in previous editions) has been divided into three separate chapters, each corresponding to one of its three parts: vaccination (Chapter 14), transplantation (Chapter 15) and cancer (Chapter 16). In addition to these major changes, all chapters have been subject to revision aimed at bringing the content up to date and improving its clarity. Exemplifying the extent of these changes, 147 (about 30% of the total) of the figures are new and they include new images generously provided by scientific colleagues, in particular Yasodha Natkunam.

I thank and acknowledge the authors of *Immunobiology* and of *Case Studies in Immunology* for giving me license with the text and figures of their books. Sherry Fuller-Espie (Cabrini College, Radnor, Pennsylvania) superbly composed the questions and answers for the end-of-chapter questions. Eleanor Lawrence expertly edited the text and figures, her steadfast sympathy for the reader ensuring that much complexity and other nonsense never made it to the printed page. Nigel Orme created many new illustrations for this edition, as well as the image for the front cover photographed by Richard Denyer. Emma Jeffcock was creative with the layout, and assiduous in proofing and production. I am indebted to Janet Foltin for orchestrating the entire operation and to Denise Schanck for her commitment to this project. Frances Brodsky has not only been a loyal user of the book but has generously given of her advice, suggestions, and much else to this third edition of *The Immune System*.

Reviewer Acknowledgments

The author and publisher would like to thank the following reviewers for their thoughtful comments and guidance: Adnane Achour, Karolinska Institute; Linda L. Baum, Rosalind Franklin University; Henry Beekhuizen, Leiden University Medical Center; Adam Benham, University of Durham; Nancy Bigley, Wright State University; Michelle Borrero, University of Puerto Rico; Kristen Brubaker, Bloomsburg University; Debra Burg, Grand Valley State University; Gerald N. Callahan, Colorado State University; Edmund Choi, University of Cincinnati College of Medicine; Michael Chorney, Pennsylvania State College of Medicine; Simon Fox, University of Plymouth; Alan B. Frey, New York University School of Medicine; David A. Fruman, University of California, Irvine; Ellen Fynan, Worcester State College; Jan Geliebter, New York Medical College; David Glick, King's College; Ananda Goldrath, University of California, San Diego; Gail Goodman-Snitkoff, Albany College of Pharmacy; Terri Hamrick, Campbell University School of Pharmacy; James X. Hartmann, Florida Atlantic University; Bethany Henderson-Dean, The University of Findlay; Evan Hermel, Touro University College of Osteopathic Medicine; Ann Hill, Oregon Health and Science University; Peter H. Koo, Northeastern Ohio Universities College of Medicine; J. Kwesi Kumi-Diaka, Florida Atlantic University; Janine LaSalle, University of California Davis School of Medicine; Cesar A. Lau-Cam, St. John's University; David A. Lawlor, Rochester Institute of Technology; Kristina Lejon, Umeå University; Osvaldo J. Lopez, Northern Michigan University; Kathleen L. McCoy, Virginia Commonwealth University; Mary Ann McDowell, University of Notre Dame; Judith Manning, University of Wisconsin–Madison; Glenn K. Matsushima, University of North Carolina; Mark A. Miller, University of Tennessee Health Sciences Center; Cathy Mitchelmore, Roskilde University; Philip F. Mixter, Washington State University; Karen G. Nakaoka, Weber State University; Stephen A. O'Barr, Western University of Health Sciences College of Pharmacy; Amy Obringer, University of St. Francis; Donald Ourth, University of Memphis; Kimberly J. Payne, University of Southern California, Keck School of Medicine; Leonard F. Peruski, Jr., Indiana University School of Medicine; Silvia S. Pierangeli, Morehouse School of Medicine; Joerg Reimann, University of Ulm; Michael R. Roner, The University of Texas at Arlington; John Samuel, University of Alberta; Virginia M. Sanders, The Ohio State University; James M. Sheil, West Virginia University; Rafael Solana, University of Cordoba; John Sternick, Mansfield University of Pennsylvania; David Ian Stott, University of Glasgow; John J. Taylor, University of Newcastle upon Tyne; W.J.M. Tax, University Medical Center Nijmegen; Julie Turner, University of Virginia; Fredric Volkert, SUNY Downstate Medical Center; Mary Warnock, Queen Margaret University College; Anthony P. Weetman, University of Sheffield; Jon Weidanz, Texas Tech University Health Sciences Center; Paul Whitley, University of Bath; and Robert V. Zackroff, Massachusetts College of Pharmacy and Health Sciences.

Instructor and Student Resources

Case Studies in Immunology, Fifth Edition by Raif Geha

Case Studies in Immunology reviews major topics of immunology as background to a selection of real clinical cases that serve to reinforce and extend the basic science. This new edition illustrates the importance of an understanding of immunology in diagnosis and therapy. This book can be used as a clinical companion together with *The Immune System,* Third Edition. A correlation guide matching relevant sections of the two books is packaged with this book.

The Art of The Immune System (IS3) (CD-ROM)

The Art of IS3 CD-ROM contains the figures from the book for presentation purposes. The figures are prepared in both PowerPoint® and JPEG formats.

Question Bank to Accompany The Immune System, Third Edition

The *Question Bank,* prepared by Sheryl L. Fuller-Espie, PhD, DIC (Cabrini College), features over 500 multiple choice and essay questions. The questions are organized by book chapter and designed for quiz and examination purposes.

Garland Science Classwire™

Available at: www.classwire.com/garlandscience, the Classwire course management system allows instructors to build websites for their courses easily. It also serves as an online archive for instructors' resources. After registering for Classwire, you will be able to download all of the figures from *The Immune System,* which are available in JPEG and PowerPoint formats. Additionally, instructors may download resources from other Garland Science textbooks. Garland Science Classwire is offered free of charge to all instructors who adopt *The Immune System.* Please contact science@garland.com for additional information on accessing the Classwire system. (Classwire™ is a trademark of Chalkfree, Inc.)

Contents

Detailed Contents

Chapter 16
**Cancer and Its Interactions with the Immune
System**   489

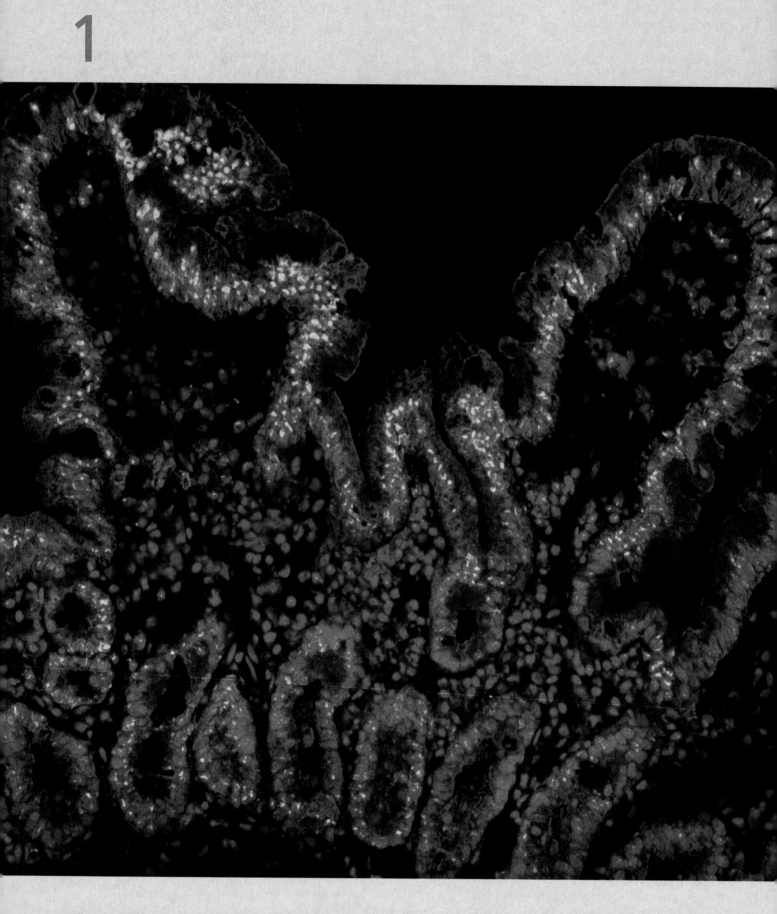

The small intestine is the major site in the human body that interacts with microorganisms.

Chapter 1

Elements of the Immune System and their Roles in Defense

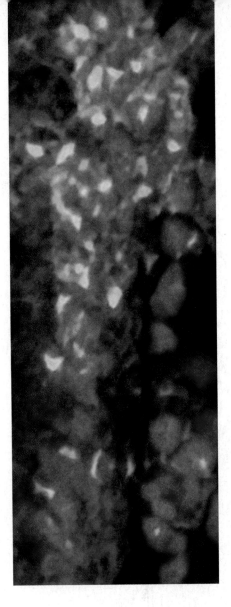

Immunology is the study of the physiological mechanisms that humans and other animals use to defend their bodies from invasion by other organisms. The origins of the subject lie in the practice of medicine and in historical observations that people who survived the ravages of epidemic disease were untouched when faced with that same disease again—they had become **immune** to infection. Infectious diseases are caused by microorganisms, which have the advantage of reproducing and evolving much more rapidly than their human hosts. During the course of an infection, the microorganism can pit enormous populations of its species against an individual *Homo sapiens*. In response, the human body invests heavily in cells dedicated to defense, which collectively form the **immune system**.

The immune system is crucial to human survival. In the absence of a working immune system, even minor infections can take hold and prove fatal. Without intensive treatment, children born without a functional immune system die in early childhood from the effects of common infections. However, in spite of their immune systems, all humans suffer from infectious diseases, especially when young. This is because the immune system takes time to build up its strongest response to an invading microorganism, time during which the invader can multiply and cause disease. To provide **immunity** that will provide protection from the disease in the future, the immune system must first do battle with the microorganism. This places people at highest risk during their first infection with a microorganism and, in the absence of modern medicine, leads to substantial child mortality, as witnessed in the developing world. When entire populations face a completely new infection, the outcome can be catastrophic, as experienced by indigenous Americans who were killed in large numbers by European diseases to which they were suddenly exposed after 1492. Today, infection with human immunodeficiency virus (HIV) and the acquired immune deficiency syndrome (AIDS) it causes are having a similarly tragic impact on the populations of several African countries.

In medicine the greatest triumph of immunology has been **vaccination**, or **immunization**, a procedure whereby severe disease is prevented by prior exposure to the infectious agent in a form that cannot cause disease. Vaccination provides the opportunity for the immune system to gain the experience needed to make a protective response with little risk to health or life. Vaccination was first used against smallpox, a viral scourge that once ravaged populations and disfigured the survivors. In Asia, small amounts of smallpox virus had been used to induce protective immunity for hundreds of years before 1721, when Lady Mary Wortley Montagu introduced the method into Western Europe. Subsequently, in 1796, Edward Jenner, a doctor in rural England, showed how inoculation with cowpox virus offered protection against the related smallpox virus with less risk than the earlier methods. Jenner called his procedure vaccination, after vaccinia, the name given to the mild disease produced by cowpox, and he is generally credited with its invention. Since his time, vaccination has dramatically reduced the incidence of smallpox worldwide, with the last cases being seen by physicians in the 1970s (Figure 1.1).

Effective vaccines have been made from only a fraction of the agents that cause disease and some are of limited availability because of their cost. Most of the widely used vaccines were first developed many years ago by a process of trial and error, before very much was known about the workings of the immune system. That approach is no longer so successful for developing new vaccines, perhaps because all the easily won vaccines have been made. But deeper understanding of the mechanisms of immunity is spawning new ideas for vaccines against infectious diseases and even against other types of disease such as cancer. Much is now known about the molecular and cellular components of the immune system and what they can do in the laboratory. Current research seeks to understand the contributions of these immune components to fighting infections in the world at large. The new knowledge is also being used to find better ways of manipulating the immune system to prevent the unwanted immune responses that cause allergies, autoimmune diseases, and rejection of organ transplants.

In this chapter we first consider the microorganisms that infect human beings and then the defenses they must overcome to start and propagate an infection. The individual cells and tissues of the immune system will be described, and how they integrate their functions with the rest of the human body. The first line of defense is innate immunity, which includes physical and chemical barriers to infection, and responses that are ready and waiting to halt infections before they can barely start. Most infections are stopped by these mechanisms, but when they fail, the more flexible and forceful defenses of the adaptive immune response are brought into play. The adaptive immune response is always targeted to the specific problem at hand and is made and refined during the course of the infection. When successful, it clears the infection and provides long-lasting immunity that prevents its recurrence.

1-1   Numerous commensal microorganisms inhabit healthy human bodies

The main purpose of the immune system is to protect the human body from infectious disease. Almost all infectious diseases suffered by humans are caused by microorganisms smaller than a single human cell. For both benign and dangerous microorganisms alike, the human body constitutes a vast resource-rich environment in which to live, feed, and reproduce. More than 500 microbial species live in the healthy adult human gut and contribute about two pounds to the body's weight; they are called **commensal species,** meaning they 'eat at the same table.' The community of microbial species that inhabits a particular niche in the human body—skin, mouth, gut, or vagina—is called the flora, for example the gut **flora.** Many of these species have not yet been studied properly because they cannot be propagated in the laboratory, growing only under the special conditions furnished by their human hosts.

Animals have evolved along with their commensal species and in so doing have become both tolerant of them and dependent upon them. Commensal organisms enhance human nutrition by processing digested food and making several vitamins. They also protect against disease, because their presence helps to prevent colonization by dangerous, disease-causing microorganisms. In addition to simple competition for space, *Escherichia coli,* a major bacterial component of the normal mammalian gut flora, secretes antibacterial

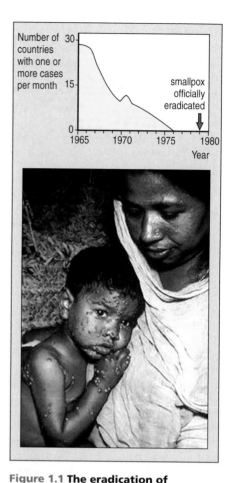

Figure 1.1 The eradication of smallpox by vaccination. Upper panel: smallpox vaccination was started in 1796. In 1979, after 3 years in which no case of smallpox was recorded, the World Health Organization announced that the virus had been eradicated. Since then the proportion of the human population that has been vaccinated against smallpox, or has acquired immunity from an infection, has steadily decreased. The result is that the human population has become increasingly vulnerable should the virus emerge again, either naturally or as a deliberate act of human malevolence. Lower panel: photograph of a child with smallpox and his immune mother. The distinctive rash of smallpox appears about 2 weeks after exposure to the virus. Photograph courtesy of the World Health Organization.

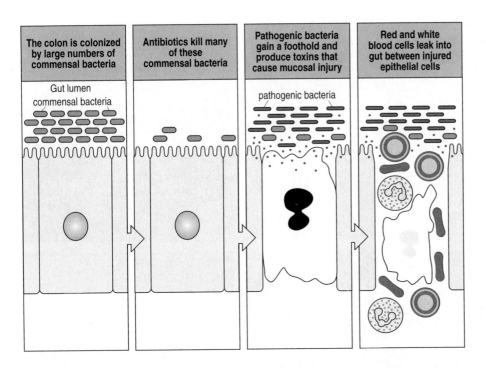

| The colon is colonized by large numbers of commensal bacteria | Antibiotics kill many of these commensal bacteria | Pathogenic bacteria gain a foothold and produce toxins that cause mucosal injury | Red and white blood cells leak into gut between injured epithelial cells |

Figure 1.2 Antibiotic treatments disrupt the natural ecology of the colon. When antibiotics are taken orally to counter a bacterial infection, beneficial populations of commensal bacteria in the colon are also decimated. This provides an opportunity for pathogenic strains of bacteria to populate the colon and cause further disease. *Clostridium difficile* is an example of such a bacterium; it produces a toxin that can cause severe diarrhea in patients treated with antibiotics. In hospitals, acquired *C. difficile* infections are an increasing cause of death for elderly patients.

proteins called colicins that incapacitate other bacteria and prevent them from colonizing the gut. When a patient with a bacterial infection takes a course of antibiotic drugs, much of the normal gut flora is killed along with the disease-causing bacteria. After such treatment the body is recolonized by a new population of microorganisms; in this situation, opportunistic disease-causing bacteria, such as *Clostridium difficile*, can sometimes establish themselves, causing further disease and sometimes death (Figure 1.2). *C. difficile* produces a toxin that can cause diarrhea and, in some cases, an even more serious gastrointestinal condition called pseudomembranous colitis.

1-2   Pathogens are infectious organisms that cause disease

Any organism with the potential to cause disease is known as a **pathogen**. This definition includes not only microorganisms like the influenza virus or the typhoid bacillus that habitually cause disease if they enter the body, but also ones that can colonize the human body to no ill effect for much of the time but cause illness if the body's defenses are weakened or if the microbe gets into the 'wrong' place. The latter kinds of pathogen are known as **opportunistic pathogens**.

Pathogens can be divided into four kinds: **bacteria**, **viruses**, and **fungi**, which are each a group of related microorganisms, and internal **parasites**, a less precise term used to embrace a heterogeneous collection of unicellular protozoa and multicellular invertebrates, mainly worms. In this book we consider the functions of the human immune system principally in the context of controlling infections. For some pathogens this necessitates their complete elimination, but for others it is sufficient to limit the size and location of the pathogen population within the human host. Figure 1.3 illustrates the variety in shape and form of the four kinds of pathogen. Figure 1.4 provides a list of common or well-known infectious diseases and the pathogens that cause them. Reference to many of these diseases and the problems they pose for the immune system will be made in the rest of this book.

Over evolutionary time, the relationship between a pathogen and its human hosts inevitably changes, affecting the severity of the disease produced. Most

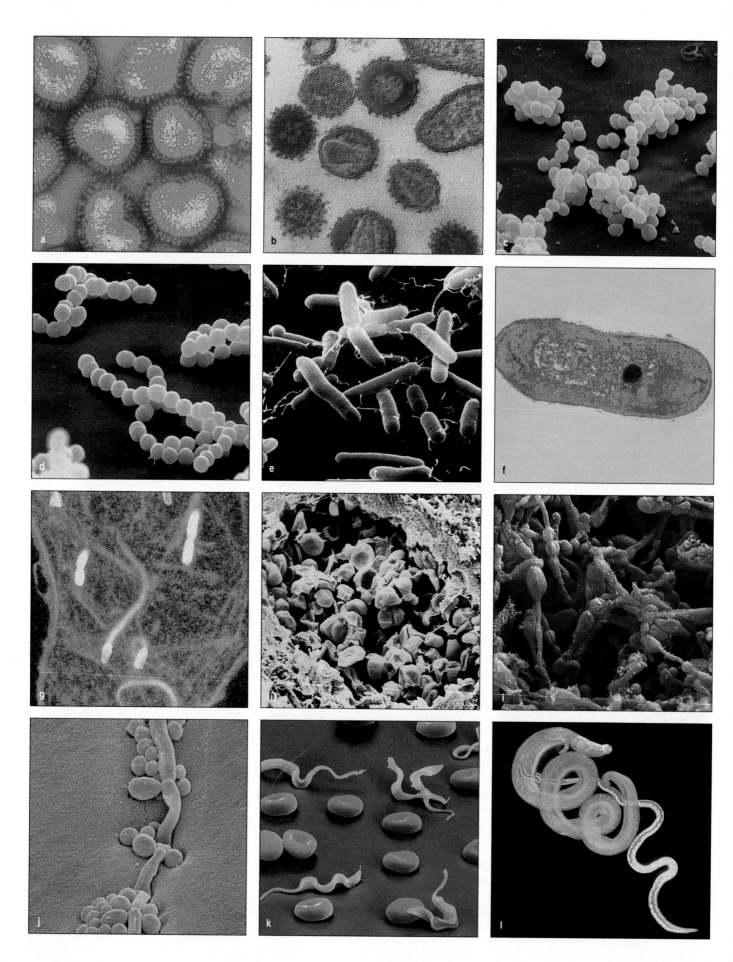

Figure 1.3 The diversity of microorganisms that are human pathogens. (a) Human immunodeficiency virus (HIV), the cause of AIDS. (b) Influenza virus. (c) *Staphylococcus aureus*, a bacterium that colonizes human skin, is the common cause of pimples and boils, and can also cause food poisoning. (d) *Streptococcus pyogenes*, the bacterium that is the principal cause of tonsillitis and scarlet fever and can also cause ear infections. (e) *Salmonella enteritidis*, the bacterium that commonly causes food poisoning. (f) *Mycobacterium tuberculosis*, the bacterium that causes tuberculosis. (g) A human cell (colored green) containing *Listeria monocytogenes* (colored yellow), a bacterium that can contaminate processed food, causing disease (listeriosis) in pregnant women and immunosuppressed individuals. (h) *Pneumocystis carinii*, an opportunistic fungus that infects patients with acquired immunodeficiency syndrome (AIDS) and other immunosuppressed individuals. The fungal cells (colored green) are in lung tissue. (i) *Epidermophyton floccosum*, the fungus that causes ringworm. (j) The fungus *Candida albicans*, a normal inhabitant of the human body that occasionally causes thrush and more severe systemic infections. (k) Red blood cells and *Trypanosoma brucei* (colored orange), a protozoan that causes African sleeping sickness. (l) *Schistosoma mansoni*, the helminth worm that causes schistosomiasis. The adult intestinal blood fluke forms are shown: the male is thick and bluish, the female thin and white. All the photos are false-colored electron micrographs, with the exception of (l), which is a light micrograph.

pathogenic organisms have evolved special adaptations that enable them to invade their hosts, replicate in them, and be transmitted. However, the rapid death of its host is rarely in a microbe's interest, because this destroys both its home and its source of food. Consequently, those organisms with the potential to cause severe and rapidly fatal disease often tend to evolve towards an accommodation with their hosts. In complementary fashion, human populations have evolved a degree of built-in genetic resistance to common disease-causing organisms, as well as acquiring lifetime immunity to endemic diseases. Endemic diseases are those, such as measles, chickenpox and malaria, that are ubiquitous in a given population and to which most people are exposed in childhood. Because of the interplay between host and pathogen, the nature and severity of infectious diseases in the human population are always changing.

Influenza is a good example of a common viral disease that, although severe in its symptoms, is usually overcome successfully by the immune system. The fever, aches, and lassitude that accompany infection can be overwhelming, and it is difficult to imagine overcoming foes or predators at the peak of a bout of influenza. Nevertheless, despite the severity of the symptoms, most strains of influenza pose no great danger to healthy people in populations in which influenza is endemic. Warm, well-nourished and otherwise healthy people usually recover in a couple of weeks and take it for granted that their immune system will accomplish this task. Pathogens new to the human population, in contrast, often cause high mortality in those infected—between 60% and 75% in the case of the Ebola virus.

1-3   The skin and mucosal surfaces form barriers against infection

The skin is the human body's first defense against infection. It forms a tough impenetrable barrier of **epithelium** protected by layers of keratinized cells. Epithelium is a general name for the layers of cells that line the outer surface and the inner cavities of the body. The skin can be breached by physical damage, such as wounds, burns, or surgical procedures, which exposes soft tissues and renders them vulnerable to infection. Until the adoption of antiseptic

Type	Disease	Pathogen	General classification*	Route of infection
Viruses	Severe acute respiratory syndrome	SARS virus	Coronaviruses	Oral/respiratory/ocular mucosa
	West Nile encephalitis	West Nile virus	Flaviviruses	Bite of an infected mosquito
	Yellow fever	Yellow fever virus	Flaviviruses	Bite of infected mosquito (*Aedes aegypti*)
	Hepatitis B	Hepatitis B virus	Hepadnaviruses	Sexual transmission; infected blood
	Chickenpox	Varicella-zoster	Herpes viruses	Oral/respiratory
	Mononucleosis	Epstein–Barr virus	Herpes viruses	Oral/respiratory
	Influenza	Influenza virus	Orthomyxoviruses	Oral/respiratory
	Measles	Measles virus	Paramyxoviruses	Oral/respiratory
	Mumps	Mumps virus	Paramyxoviruses	Oral/respiratory
	Poliomyelitis	Polio virus	Picornaviruses	Oral
	Jaundice	Hepatitis A virus	Picornaviruses	Oral
	Smallpox	Variola	Pox viruses	Oral/respiratory
	AIDS	Human immunodeficiency virus	Retroviruses	Sexual transmission, infected blood
	Rabies	Rabies virus	Rhabdoviruses	Bite of an infected animal
	Common cold	Rhinoviruses	Rhinoviruses	Nasal
	Diarrhea	Rotavirus	Rotaviruses	Oral
	Rubella	Rubella	Togaviruses	Oral/respiratory

procedures in the nineteenth century, surgery was a very risky business, principally because of the life-threatening infections that the procedures introduced. For the same reason, far more soldiers have died from infection acquired on the battlefield than from the direct effects of enemy action. Ironically, the need to conduct increasingly sophisticated and wide-ranging warfare has been the major force driving improvements in surgery and medicine. As an example from immunology, the burns suffered by fighter pilots during World War II stimulated studies on skin transplantation that led directly to the understanding of the cellular basis of the immune response.

Continuous with the skin are the epithelia lining the respiratory, gastrointestinal, and urogenital tracts (Figure 1.5). On these internal surfaces, the impermeable skin gives way to tissues that are specialized for communication with their environment and are more vulnerable to microbial invasion. Such surfaces are known as **mucosal surfaces** or **mucosae** as they are continually bathed in the **mucus** that they secrete. This thick fluid layer contains glycoproteins, proteoglycans, and enzymes that protect the epithelial cells from damage and help to limit infection. In the respiratory tract, mucus is continuously removed through the action of epithelial cells bearing beating cilia and is replenished by mucus-secreting goblet cells. The respiratory mucosa is thus continually cleansed of unwanted material, including infectious microorganisms that have been breathed in.

All epithelial surfaces also secrete antimicrobial substances. The sebum secreted by sebaceous glands associated with hair follicles contains fatty acids and lactic acids, both of which inhibit bacterial growth on the surface of the skin. All epithelia produce antimicrobial peptides called **defensins** that kill bacteria, fungi and enveloped viruses by perturbing their membranes. Tears and saliva contain lysozyme, an enzyme that kills bacteria by degrading their cell walls. Microorganisms are also deterred by the acidic environments of the stomach, the vagina, and the skin.

Type	Disease	Pathogen	General classification*	Route of infection
Bacteria	Trachoma	*Chlamydia trachomatis*	Chlamydias	Oral/respiratory/ocular mucosa
	Bacillary dysentery	*Shigella flexneri*	Gram-negative bacilli	Oral
	Food poisoning	*Salmonella enteritidis, S. typhimurium*	Gram-negative bacilli	Oral
	Plague	*Yersinia pestis*	Gram-negative bacilli	Infected flea bite, respiratory
	Tularemia	*Pasteurella tularensis*	Gram-negative bacilli	Handling infected animals
	Typhoid fever	*Salmonella typhi*	Gram-negative bacilli	Oral
	Gonorrhea	*Neisseria gonorrhoeae*	Gram-negative cocci	Sexually transmitted
	Meningococcal meningitis	*Neisseria meningitidis*	Gram-negative cocci	Oral/respiratory
	Meningitis, pneumonia	*Haemophilus influenzae*	Gram-negative coccobacilli	Oral/respiratory
	Legionnaire's disease	*Legionella pneumophila*	Gram-negative coccobacilli	Inhalation of contaminated aerosol
	Whooping cough	*Bordetella pertussis*	Gram-negative coccobacilli	Oral/respiratory
	Cholera	*Vibrio cholerae*	Gram-negative vibrios	Oral
	Anthrax	*Bacillus anthracis*	Gram-positive bacilli	Oral/respiratory by contact with spores
	Diphtheria	*Corynebacterium diphtheriae*	Gram-positive bacilli	Oral/respiratory
	Tetanus	*Clostridium tetani*	Gram-positive bacilli (anaerobic)	Infected wound
	Boils, wound infections	*Staphylococcus aureus*	Gram-positive cocci	Wounds; oral/respiratory
	Pneumonia, scarlet fever	*Streptococcus pneumoniae*	Gram-positive cocci	Oral/respiratory
	Tonsillitis	*Streptococcus pyogenes*	Gram-positive cocci	Oral/respiratory
	Leprosy	*Mycobacterium leprae*	Mycobacteria	Infected respiratory droplets
	Tuberculosis	*Mycobacterium tuberculosis*	Mycobacteria	Oral/respiratory
	Respiratory disease	*Mycoplasma pneumoniae*	Mycoplasmas	Oral/respiratory
	Typhus	*Rickettsia prowazekii*	Rickettsias	Bite of infected tick
	Lyme disease	*Borrelia burgdorferi*	Spirochetes	Bite of infected deer tick
	Syphilis	*Treponema pallidum*	Spirochetes	Sexual transmission
Fungi	Aspergillosis	*Aspergillus* species	Ascomycetes	Opportunistic pathogen, inhalation of spores
	Athlete's foot	*Tinea pedis*	Ascomycetes	Physical contact
	Candidiasis, thrush	*Candida albicans*	Ascomycetes (yeasts)	Opportunistic pathogen, resident flora
	Pneumonia	*Pneumocystis carinii*	Ascomycetes	Opportunistic pathogen, resident lung flora
Protozoan parasites	Leishmaniasis	*Leishmania major*	Protozoa	Bite of an infected sand fly
	Malaria	*Plasmodium falciparum*	Protozoa	Bite of an infected mosquito
	Toxoplasmosis	*Toxoplasma gondii*	Protozoa	Oral, from infected material
	Trypanosomiasis	*Trypanosoma brucei*	Protozoa	Bite of an infected tsetse fly
Helminth parasites (worms)	Common roundworm	*Ascaris lumbricoides*	Nematodes (roundworms)	Oral, from infected material
	Schistosomiasis	*Schistosoma mansoni*	Trematodes	Through skin by bathing in infected water

Figure 1.4 (opposite page and above) Diverse microorganisms cause human disease. Pathogenic organisms are of four main types—viruses, bacteria, fungi, and parasites, which are mostly protozoans or worms. Some important pathogens in each category are listed along with the diseases they cause. *The classifications given are intended as a guide only and are not taxonomically consistent; families are given for the viruses; general groupings often used in medical bacteriology for the bacteria; and higher taxonomic divisions for the fungi and parasites. The terms Gram-negative and Gram-positive refer to the staining properties of the bacteria; Gram-positive bacteria stain purple with the Gram stain, Gram-negative bacteria do not.

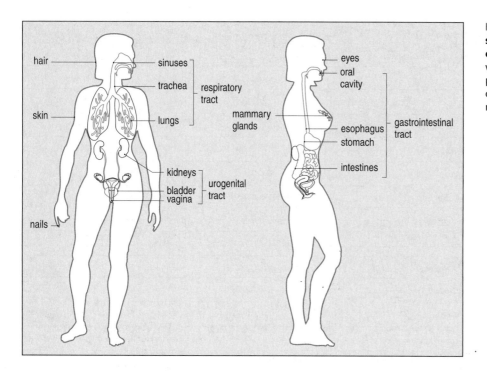

Figure 1.5 The physical barriers that separate the body from its external environment. In these images of a woman, the strong barriers to infection provided by the skin, hair, and nails are colored blue and the more vulnerable mucosal membranes are colored red.

With such defenses, skin and mucosa provide well-maintained mechanical, chemical, and microbiological barriers (see Section 1-1) that prevent most pathogens from gaining access to the cells and tissues of the body (Figure 1.6). When that barrier is breached and pathogens gain entry to the body's soft tissues, the fixed defenses of the innate immune system are brought into play.

1-4   The innate immune response causes inflammation at sites of infection

Cuts, abrasions, bites, and wounds provide routes for pathogens to get through the skin. Touching, rubbing, picking, and poking the eyes, nose, and mouth help pathogens to breach mucosal surfaces, as does breathing polluted air, eating contaminated food, and being around infected people. With very few exceptions, infections remain highly localized and are extinguished within a few days without illness or incapacitation. Such infections are controlled and terminated by the **innate immune response**, which is ready to react quickly.

	Skin	Gastrointestinal tract	Respiratory tract	Urogenital tract	Eyes
	Epithelial cells joined by tight junctions				
Mechanical	Flow of fluid, perspiration, sloughing off of skin	Flow of fluid, mucus, food, and saliva	Flow of fluid and mucus, e.g., by cilia Air flow	Flow of fluid, urine, mucus, sperm	Flow of fluid, tears
Chemical	Sebum (fatty acids, lactic acid, lysozyme)	Acidity, enzymes (proteases)	Lysozyme in nasal secretions	Acidity in vaginal secretions Spermine and zinc in semen	Lysozyme in tears
	Antimicrobial peptides (defensins)				
Microbiological	Normal flora of the skin	Normal flora of the gastrointestinal tract	Normal flora of the respiratory tract	Normal flora of the urogenital tract	Normal flora of the eyes

Figure 1.6 Various barriers prevent bacteria from crossing epithelia and colonizing tissues. Surface epithelia provide mechanical, chemical, and microbiological barriers to infection.

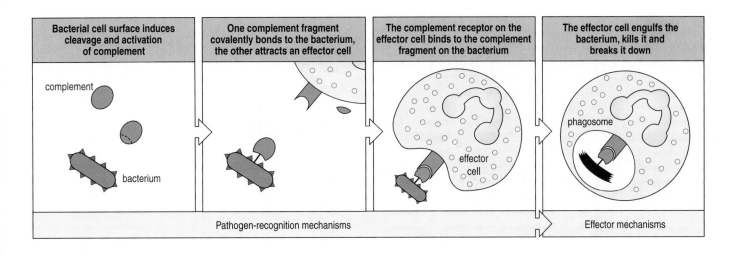

| Bacterial cell surface induces cleavage and activation of complement | One complement fragment covalently bonds to the bacterium, the other attracts an effector cell | The complement receptor on the effector cell binds to the complement fragment on the bacterium | The effector cell engulfs the bacterium, kills it and breaks it down |

Pathogen-recognition mechanisms　　　　　　　　　　　Effector mechanisms

This response consists of two parts (Figure 1.7). First is recognition that a pathogen is present. This involves soluble proteins and cell-surface receptors that bind either to the pathogen and its products or to human cells and serum proteins that become altered in the presence of the pathogen. Once the pathogen has been recognized, the second part of the response involves the recruitment of destructive **effector mechanisms** that kill and eliminate it. The effector mechanisms are provided by **effector cells** of various types that engulf bacteria, kill virus-infected cells, or attack protozoan parasites, and a battery of serum proteins called **complement** that help the effector cells by marking pathogens with molecular flags but also attack pathogens in their own right. Collectively, these defenses are called **innate immunity**. The word 'innate' refers to the fact that they are all determined entirely by the genes a person inherits from their parents. Many families of receptor proteins contribute to the recognition of pathogens in the innate immune response. They are of several different structural types and bind to chemically diverse ligands: peptides, proteins, glycoproteins, proteoglycans, peptidoglycan, carbohydrates, glycolipids, phospholipids, and nucleic acids.

An infection that would typically be cleared by innate immunity is of the sort experienced by skateboarders when they tumble onto a San Francisco sidewalk. On returning home the graze is washed, which removes most of the dirt and the associated pathogens of human, soil, pigeon, dog, cat, raccoon, skunk, and possum origin. Of the bacteria that remain, some begin to divide and set up an infection. Cells and proteins in the damaged tissue sense the presence of bacteria and the cells send out soluble proteins called **cytokines** that interact with other cells to trigger the innate immune response. The overall effect of the innate immune response is to induce a state of **inflammation** in the infected tissue. Inflammation is an ancient concept in medicine that has traditionally been defined by the Latin words *calor, dolor, rubor,* and *tumor;* for heat, pain, redness, and swelling, respectively. These symptoms, which are part of everyday human experience, are not due to the infection itself but to the immune system's response to it.

Cytokines induce the local dilation of blood capillaries which, by increasing the blood flow, causes the skin to warm and redden. Vascular dilation (vasodilation) introduces gaps between the cells of the **endothelium**, the thin layer of specialized epithelium that lines the interior of blood vessels. This makes the endothelium permeable and increases the leakage of blood plasma into the connective tissue. Expansion of the local fluid volume causes **edema** or swelling, putting pressure on nerve endings and causing pain. Cytokines also change the adhesive properties of the vascular endothelium, inviting white blood cells to attach to it and move from the blood into the inflamed tissue

Figure 1.7 Immune defense involves recognition of pathogens followed by their destruction. Almost all components of the immune system contribute to mechanisms for either recognizing pathogens or destroying pathogens, or to mechanisms for communicating between these two activities. This is illustrated here by a fundamental process used to get rid of pathogens. Serum proteins of the complement system (turquoise) are activated in the presence of a pathogen (red) to form a covalent bond between a fragment of complement protein and the pathogen. The attached piece of complement marks the pathogen as dangerous. The soluble complement fragment summons a phagocytic white blood cell to the site of complement activation. This effector cell has a surface receptor that binds to the complement fragment attached to the pathogen. The receptor and its bound ligand are taken up into the cell by phagocytosis, which delivers the pathogen to an intracellular vesicle called a phagosome, where it is destroyed. A phagocyte is a cell that eats, 'phago' being derived from the Greek word for eat.

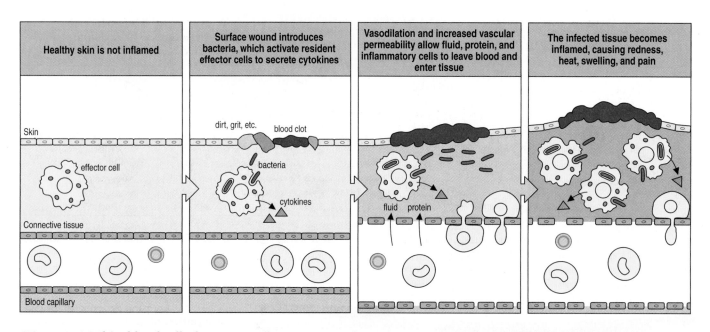

| Healthy skin is not inflamed | Surface wound introduces bacteria, which activate resident effector cells to secrete cytokines | Vasodilation and increased vascular permeability allow fluid, protein, and inflammatory cells to leave blood and enter tissue | The infected tissue becomes inflamed, causing redness, heat, swelling, and pain |

(Figure 1.8). White blood cells that are usually present in inflamed tissues and release substances that contribute to the inflammation are called **inflammatory cells**. Infiltration of cells into the inflamed tissue increases the swelling, and some of the molecules they release contribute to the pain. The benefit of the discomfort and disfigurement caused by inflammation is that it enables cells and molecules of the immune system to be brought rapidly and in large numbers into the infected tissue.

Figure 1.8 Innate immune mechanisms establish a state of inflammation at sites of infection. Illustrated here are the events following an abrasion of the skin. Bacteria invade the underlying connective tissue and stimulate the innate immune response.

1-5   The adaptive immune response adds to an ongoing innate immune response

Human beings are exposed to pathogens on a daily basis. The intensity of exposure and the diversity of the pathogens encountered increase with crowded city living and the daily exchange of people and pathogens from international airports. Despite this exposure, innate immunity keeps most people healthy for most of the time. Nevertheless, some infections outrun the innate immune response, an event more likely in people who are malnourished, poorly housed, deprived of sleep, or stressed in other ways. When this occurs, the innate immune response works to slow the spread of infection while it calls upon white blood cells called **lymphocytes** that increase the power and focus of the immune response. Their contribution to defense is the **adaptive immune response**. It is so called because it is organized around an ongoing infection and adapts to the nuances of the infecting pathogen. Consequently, the long-lasting **adaptive immunity** that develops against one pathogen provides a highly specialized defense that is of little use against infection by a different pathogen.

The effector mechanisms used in the adaptive immune response are similar to those used in the innate immune response; the important difference is in the cell-surface receptors used by lymphocytes to recognize pathogens (Figure 1.9). In contrast to the receptors of innate immunity, which are of many different types but are not specific for a particular pathogen, the receptors of adaptive immunity are all of the same molecular type and are highly pathogen-specific. They are not encoded by conventional genes but by genes that are cut, spliced, and modified to produce billions of variants of the basic receptor type, so that different lymphocytes bear different variants. During infection, only those lymphocytes bearing receptors that recognize the infecting pathogen are selected to participate in the adaptive response. These then proliferate and differentiate to produce large numbers of effector cells specific for

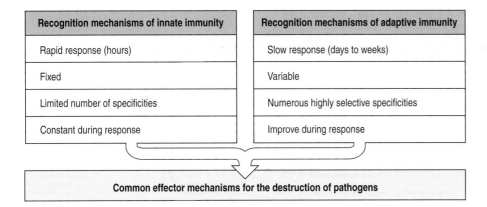

Recognition mechanisms of innate immunity	Recognition mechanisms of adaptive immunity
Rapid response (hours)	Slow response (days to weeks)
Fixed	Variable
Limited number of specificities	Numerous highly selective specificities
Constant during response	Improve during response

Common effector mechanisms for the destruction of pathogens

Figure 1.9 The principal characteristics of innate and adaptive immunity.

that pathogen (Figure 1.10). These processes, which select a small subset of lymphocytes for proliferation and differentiation into effector lymphocytes, are called **clonal selection** and **clonal expansion**, respectively. As these processes take time, the benefit of an adaptive immune response only begins to be felt about a week after the infection has started.

The value of the adaptive immune response is well illustrated in influenza, which is caused by infection of the epithelial cells in the lower respiratory tract with influenza virus. The debilitating symptoms start 3 or 4 days after the start of infection, when the virus has begun to outrun the innate immune response. The disease persists for 5–7 days while the adaptive immune response is being organized and put to work. As the adaptive immune response gains the upper hand, fever subsides and a gradual convalescence begins in the second week after infection.

Some of the lymphocytes selected during an adaptive immune response persist in the body and provide long-term **immunological memory** of the pathogen. These memory cells allow subsequent encounters with the same pathogen to elicit a stronger and faster adaptive immune response, which terminates infection with minimal illness. The adaptive immunity provided by immunological memory is also called **acquired immunity** or **protective immunity**. For some pathogens such as measles virus, one full-blown infection can provide immunity for decades, whereas for influenza the effect seems more short-lived. This is not because the immunological memory is faulty but because the influenza virus changes on a yearly basis to escape the immunity acquired by its human hosts.

The first time that an adaptive immune response is made to a given pathogen it is called the **primary immune response**. The second and subsequent times that an adaptive immune response is made, and when immunological memory applies, it is called a **secondary immune response**. The purpose of vaccination is to induce immunological memory to a pathogen so that subsequent

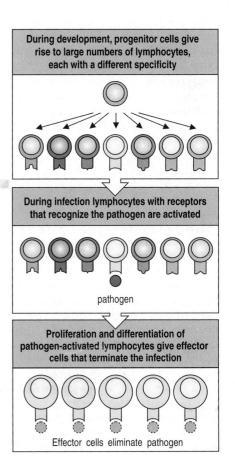

Figure 1.10 Selection of lymphocytes by a pathogen. Top panel: during its development from a progenitor cell (gray) a lymphocyte is programmed to make a single species of cell-surface receptor that recognizes a particular molecular structure. Each lymphocyte makes a receptor of different specificity, so that the population of circulating lymphocytes includes many millions of such receptors, all recognizing different structures, which enables all possible pathogens to be recognized. Lymphocytes with different receptor specificities are represented by different colors. Center panel: upon infection by a particular pathogen, only a small subset of lymphocytes (represented by the yellow cell) will have receptors that bind to the pathogen or its components. Bottom panel: these lymphocytes are stimulated to divide and differentiate, thereby producing an expanded population of effector cells from each pathogen-binding lymphocyte.

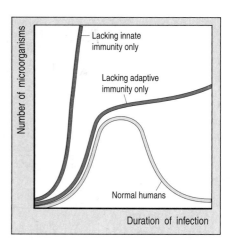

Figure 1.11 The benefits of having both innate and adaptive immunity. In normal individuals, a primary infection is cleared from the body by the combined effects of innate and adaptive immunity (yellow line). In a person who lacks innate immunity, uncontrolled infection occurs because the adaptive immune response cannot be deployed without the preceding innate response (red line). In a person who lacks adaptive immune responses, the infection is initially contained by innate immunity but cannot be cleared from the body (green line).

infection with the pathogen elicits a strong fast-acting adaptive response. Because all adaptive immune responses are contingent upon an innate immune response, vaccines must induce both innate and adaptive immune responses.

1-6 Adaptive immunity is better understood than innate immunity

The proportion of infections that are successfully eliminated by innate immunity is difficult to assess, mainly because such infections are overcome before they have caused symptoms severe enough to command the attention of those infected or of their physicians. Intuitively, it seems likely to be a high proportion, given the human body's capacity to sustain vast populations of resident microorganisms without these causing symptoms of disease. The importance of innate immunity is also implied by the rarity of inherited deficiencies in innate immune mechanisms and the considerable impairment of protection when these deficiencies do occur (Figure 1.11).

Much of medical practice is concerned with the small proportion of infections that innate immunity fails to terminate and in which the spread of the infection results in overt disease such as pneumonia, measles, or influenza and stimulates an adaptive immune response. In such situations the attending physicians and the adaptive immune response work together to effect a cure, a partnership that has historically favored the scientific investigation of adaptive immunity over innate immunity. Consequently, less has been learnt about innate immunity than adaptive immunity. Now that immunologists realize that innate immunity mechanisms are fundamental to every immune response, this gap in knowledge is being filled.

1-7 Immune system cells with different functions all derive from hematopoietic stem cells

The cells of the immune system are principally the white blood cells or **leukocytes**, and the tissue cells related to them. Along with the other blood cells, they are continually being generated by the body in the developmental process known as **hematopoiesis**. Leukocytes derive from a common progenitor called the **pluripotent hematopoietic stem cell**, which also gives rise to red blood cells (**erythrocytes**) and **megakaryocytes**, the source of **platelets**. All these cell types, together with their precursor cells, are collectively called **hematopoietic cells** (Figure 1.12). The anatomical site for hematopoiesis changes with age (Figure 1.13). In the early embryo, blood cells are first produced in the yolk sac and later in the fetal liver. From the third to the seventh month of fetal life the spleen is the major site of hematopoiesis. As the bones develop during the fourth and fifth months of fetal growth, hematopoiesis begins to shift to the **bone marrow** and by birth this is where practically all hematopoiesis takes place. In adults, hematopoiesis occurs mainly in the bone marrow of the skull, ribs, sternum, vertebral column, pelvis, and femurs. Because blood cells are short-lived, they have to be continually renewed, and hematopoiesis is active throughout life.

Figure 1.12 *(opposite page)* Types of hematopoietic cell. The different types of hematopoietic cell are depicted in schematic diagrams, which indicate their characteristic morphological features, and in accompanying light micrographs. Their main functions are indicated. We shall use these schematic representations for these cells throughout the book. Megakaryocytes (k) reside in bone marrow and release tiny non-nucleated, membrane-bound packets of cytoplasm, which circulate in the blood and are known as platelets. Red blood cells (erythrocytes) (l) are smaller than the white blood cells and have no nucleus. Original magnification × 15,000. Photographs courtesy of Yasodha Natkunam.

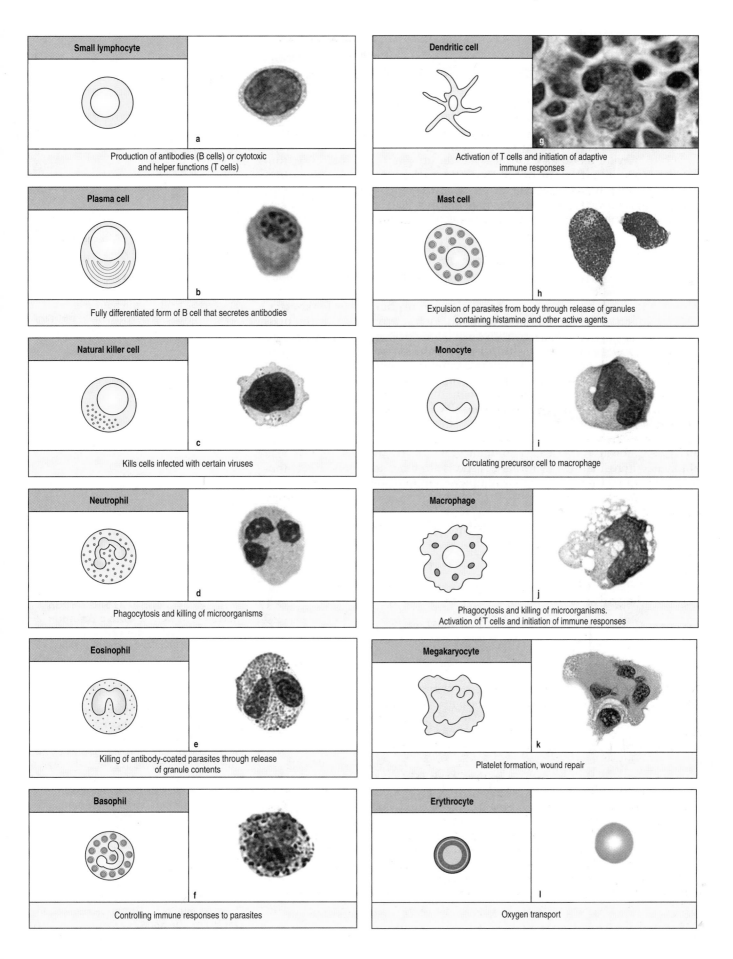

Small lymphocyte

a

Production of antibodies (B cells) or cytotoxic
and helper functions (T cells)

Plasma cell

b

Fully differentiated form of B cell that secretes antibodies

Natural killer cell

c

Kills cells infected with certain viruses

Neutrophil

d

Phagocytosis and killing of microorganisms

Eosinophil

e

Killing of antibody-coated parasites through release
of granule contents

Basophil

f

Controlling immune responses to parasites

Dendritic cell

g

Activation of T cells and initiation of adaptive
immune responses

Mast cell

h

Expulsion of parasites from body through release of granules
containing histamine and other active agents

Monocyte

i

Circulating precursor cell to macrophage

Macrophage

j

Phagocytosis and killing of microorganisms.
Activation of T cells and initiation of immune responses

Megakaryocyte

k

Platelet formation, wound repair

Erythrocyte

l

Oxygen transport

Hematopoietic stem cells can divide to give further hematopoietic stem cells, a process called **self renewal**; they can also become more mature stem cells that commit to one of three cell lineages: the erythroid, myeloid, and lymphoid lineages (Figure 1.14). The erythroid progenitor gives rise to the erythroid lineage of blood cells—the oxygen-carrying erythrocytes and the platelet-producing megakaryocytes. Platelets are small non-nucleated cell fragments of plate-like shape that maintain the integrity of blood vessels. They initiate and participate in the clotting reactions that block badly damaged blood vessels to prevent blood loss. Megakaryocytes are giant cells that arise from the fusion of multiple precursor cells and have nuclei containing multiple sets of chromosomes (megakaryocyte means cell with giant nucleus). Megakaryocytes are permanent residents of the bone marrow. Platelets are small packets of membrane-bounded cytoplasm that break off from these cells.

The **myeloid progenitor** gives rise to the **myeloid lineage** of cells. One group of myeloid cells consists of the **granulocytes**, which have prominent cytoplasmic granules containing reactive substances that kill microorganisms and enhance inflammation. Because granulocytes have irregularly shaped nuclei with two to five lobes, they are also called **polymorphonuclear leukocytes**. Most abundant of the granulocytes, and of all white blood cells, is the **neutrophil** (Figure 1.15), which is specialized in the capture, engulfment and killing of microorganisms. Cells with this function are called **phagocytes**, of which neutrophils are the most numerous and most lethal. Neutrophils are effector cells of innate immunity that are rapidly mobilized to enter sites of infection and can work in the anaerobic conditions that often prevail in damaged tissue. They are short-lived and die at the site of infection, forming **pus**, the stuff of pimples and boils (Figure 1.16). The second most abundant granulocyte is the **eosinophil**, which defends against helminth worms and other intestinal parasites. The least abundant granulocyte, the **basophil**, is also implicated in regulating the immune response to parasites but is so rare that relatively little is known of its contribution to immune defense. The names of

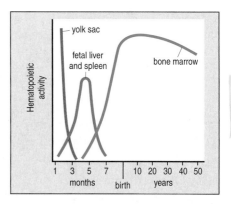

Figure 1.13 The site of hematopoiesis in humans changes during development. Blood cells are first made in the yolk sac of the embryo and subsequently in the embryonic liver. They start to be made in the bone marrow before birth, and by the time of birth this is the only tissue in which hematopoiesis occurs.

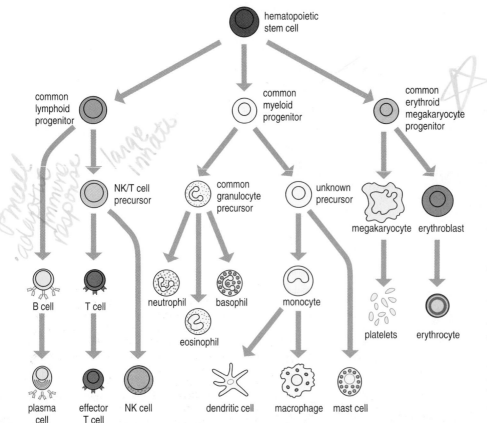

Figure 1.14 Blood cells and certain tissue cells derive from a common hematopoietic stem cell. The pluripotent stem cell (brown) divides and differentiates into more specialized progenitor cells that give rise to the lymphoid lineage, the myeloid lineage, and the erythroid lineage. The common lymphoid progenitor divides and differentiates to give B cells (yellow), T cells (blue), and NK cells (purple). On activation by infection, B cells divide and differentiate into plasma cells, whereas T cells differentiate into various types of effector T cell. The myeloid progenitor cell divides and differentiates to produce at least six cell types. These are: the three types of granulocyte—the neutrophil, the eosinophil, and the basophil; the mast cell, which takes up residence in connective and mucosal tissues; the circulating monocyte, which gives rise to the macrophages resident in tissues; and the dendritic cell. The word myeloid means 'of the bone marrow.'

the granulocytes refer to the staining of their cytoplasmic granules with commonly used histological stains: the eosinophil's granules contain basic substances that bind the acidic stain eosin, the basophil's granules contain acidic substances that bind basic stains such as hematoxylin, and the contents of the neutrophil's granules bind to neither acidic nor basic stains.

The second group of myeloid cells consists of monocytes, macrophages, and dendritic cells. **Monocytes** are leukocytes that circulate in the blood. They are distinguished from the granulocytes by being bigger, by having a distinctive indented nucleus, and by all looking the same: hence the name monocyte. Monocytes are the mobile progenitors of sedentary tissue cells called **macrophages**. They travel in the blood to tissues, where they mature into macrophages and take up residence. The name macrophage means 'large phagocyte,' and like the neutrophil, which was once called the microphage, the macrophage is well equipped for phagocytosis. Tissue macrophages are large, irregularly shaped cells characterized by an extensive cytoplasm with numerous vacuoles, often containing engulfed material (Figure 1.17). They are the general scavenger cells of the body, phagocytosing and disposing of dead cells and cell debris as well as invading microorganisms.

If neutrophils are the short-lived infantry of innate immunity, then macrophages are the long-lived commanders who provide a warning to other cells and orchestrate the local response to infection. Macrophages present in the infected tissues are generally the first phagocytic cell to sense an invading microorganism. As part of their response to the pathogen, macrophages secrete the cytokines that recruit neutrophils and other leukocytes into the infected area.

Dendritic cells are resident in the body's tissues and have a distinctive star-shaped morphology. Although they have many properties in common with macrophages, their unique function is to act as cellular messengers that are sent to call up an adaptive immune response when it is needed. At such times, dendritic cells that reside in the infected tissue will leave the tissue with a cargo of intact and degraded pathogens and take it to one of several lymphoid organs that specialize in making adaptive immune responses.

The last type of myeloid cell is the **mast cell**, which is resident in all connective tissues. It has granules like those of the basophil, but it is not developmentally derived from the basophil and the nature of its blood-borne progenitor is not yet known. The activation and degranulation of mast cells at sites of infection make a major contribution to inflammation.

Cell type	Proportion of leukocytes (%)
Neutrophil	40–75
Eosinophil	1–6
Basophil	<1
Monocyte	2–10
Lymphocyte	20–50

Figure 1.15 The relative abundance of the leukocyte cell types in human peripheral blood. The values given for each cell type are the normal ranges found in venous blood taken from healthy donors.

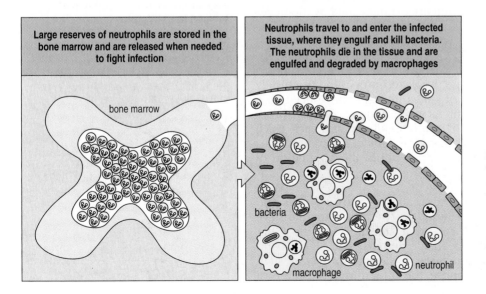

Figure 1.16 Neutrophils are stored in the bone marrow and move in large numbers to sites of infection, where they act and then die. After one round of ingestion and killing of bacteria, a neutrophil dies. The dead neutrophils are eventually mopped up by long-lived tissue macrophages, which break them down. The creamy material known as pus is composed of dead neutrophils.

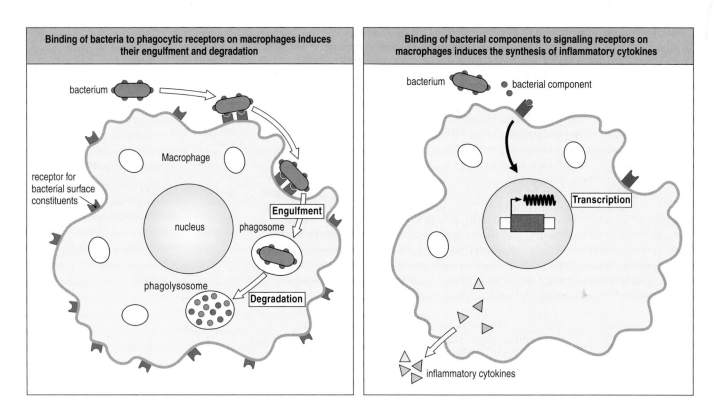

Binding of bacteria to phagocytic receptors on macrophages induces their engulfment and degradation

bacterium

Macrophage

receptor for bacterial surface constituents

nucleus   phagosome

Engulfment

phagolysosome

Degradation

Binding of bacterial components to signaling receptors on macrophages induces the synthesis of inflammatory cytokines

bacterium   bacterial component

Transcription

inflammatory cytokines

The **lymphoid progenitor** gives rise to the **lymphoid lineage** of white blood cells. Two populations of blood lymphocytes are distinguished morphologically: large lymphocytes with a granular cytoplasm and small lymphocytes with almost no cytoplasm. The **large granular lymphocytes** are effector cells of innate immunity called **natural killer cells** or **NK cells**. NK cells are important in the defense against viral infections. They enter infected tissues, where they prevent the spread of infection by killing virus-infected cells and secreting cytokines that impede viral replication in infected cells. The **small lymphocytes** are the cells responsible for the adaptive immune response. They are small because they circulate in a quiescent and immature form that is functionally inactive. Recognition of a pathogen by small lymphocytes drives a process of lymphocyte selection, growth and differentiation that after 1–2 weeks produces a powerful response tailored to the invading organism.

The small lymphocytes, although morphologically indistinguishable from each other, comprise several sublineages that are distinguished by their cell-surface receptors and the functions they are programmed to perform. The most important difference is between **B lymphocytes** and **T lymphocytes**, also called **B cells** and **T cells**, respectively. For B cells the cell-surface receptors for pathogens are **immunoglobulins**, whereas those of T cells are known as **T-cell receptors**. Immunoglobulins and T-cell receptors are structurally similar molecules that are the products of genes that are cut, spliced, and modified during lymphocyte development. As a consequence of these processes each B cell expresses a single type of immunoglobulin and each T cell expresses a single type of T-cell receptor. Many millions of different immunoglobulins and T-cell receptors are represented within the population of small lymphocytes in one human being.

T cells are further subdivided into two kinds, called **cytotoxic T cells** and **helper T cells** according to the effector functions they perform after they have become activated. Cytotoxic T cells kill cells that are infected with either viruses or bacteria that live and reproduce inside human cells. NK cells and cytotoxic T cells have similar effector functions, the former providing such functions during the innate immune response, the latter during the adaptive

Figure 1.17 Macrophages respond to pathogens by using different receptors to stimulate phagocytosis and cytokine secretion. The left panel shows receptor-mediated phagocytosis of bacteria by a macrophage. The bacterium (red) binds to cell-surface receptors (blue) on the macrophage, inducing engulfment of the bacterium into an internal vesicle called a phagosome within the macrophage cytoplasm. Fusion of the phagosome with lysosomes forms an acidic vesicle called a phagolysosome, which contains toxic small molecules and hydrolytic enzymes that kill and degrade the bacterium. The right panel shows how a bacterial component binding to a different type of cell-surface receptor sends a signal to the macrophage's nucleus that initiates the transcription of genes for inflammatory cytokines. The cytokines are synthesized in the cytoplasm and secreted into the extracellular fluid.

immune response. Helper T cells secrete cytokines that help other cells of the immune system become fully activated effector cells. For example, some subsets of helper T cells help activate B cells to become **plasma cells**. Plasma cells are effector cells that secrete soluble forms of immunoglobulin called **antibodies** that bind to pathogens and the toxic products they make.

1-8 Most lymphocytes are present in specialized lymphoid tissues

Although doctors and immunologists usually sample and study human lymphocytes from blood samples taken from their patients and voluntary donors, the vast majority of lymphocytes are to be found in specialized tissues known as **lymphoid tissues** or **lymphoid organs**. The major lymphoid organs are bone marrow, thymus, spleen, adenoids, tonsils, appendix, lymph nodes, and Peyer's patches (Figure 1.18). Less organized lymphoid tissue is also found lining the mucosal surfaces of the respiratory, gastrointestinal, and urogenital tracts. The lymphoid tissues are functionally divided into two types. **Primary** or **central lymphoid tissues** are where lymphocytes develop and mature to the stage at which they are able to respond to a pathogen. The bone marrow and the **thymus** are the primary lymphoid tissues; B and T lymphocytes both originate from lymphoid precursors in the bone marrow (see Section 1-7, p. 12), but whereas B cells complete their maturation in the bone marrow before entering the circulation, T cells leave the bone marrow at an immature stage and migrate in the blood to the thymus and mature there. Apart from the bone marrow and the thymus, all other lymphoid tissues are known as **secondary** or **peripheral lymphoid tissues**; they are the sites where mature lymphocytes become stimulated to respond to invading pathogens.

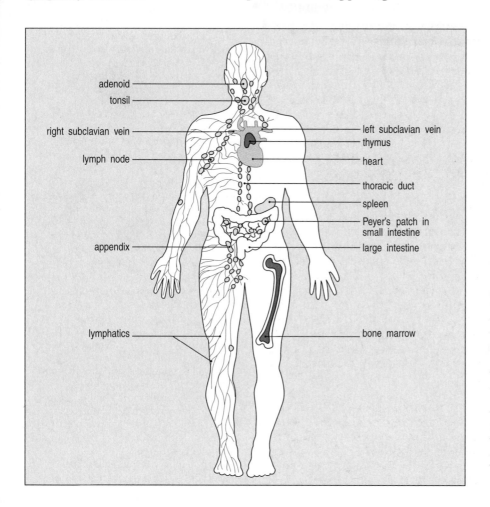

Figure 1.18 The sites of the principal lymphoid tissues within the human body. Lymphocytes arise from stem cells in the bone marrow. B cells complete their maturation in the bone marrow, whereas T cells leave at an immature stage and complete their development in the thymus. The bone marrow and the thymus are the primary lymphoid tissues and are shown in red. The secondary lymphoid tissues are shown in yellow and the thin black branching lines are the lymphatics. Plasma that has leaked from the blood is collected by the lymphatics as lymph and is returned to the blood via the thoracic duct, which empties into the left subclavian vein.

Lymph nodes lie at the junctions of an anastomosing network of **lymphatic vessels** called the **lymphatics**, which originate in the connective tissues throughout the body and collect the plasma that continually leaks out of blood vessels and forms the extracellular fluid. The lymphatics eventually return this fluid, called **lymph**, to the blood, chiefly via the thoracic duct, which empties into the left subclavian vein in the neck. Unlike the blood, the lymph is not driven by a dedicated pump and its flow is comparatively sluggish. One-way valves within lymphatic vessels, and the lymph nodes placed at their junctions, ensure that net movement of the lymph is always in a direction away from the peripheral tissues and towards the ducts in the upper body where the lymph empties into the blood. The flow of lymph is driven by the continual movements of one part of the body with respect to another. In the absence of such movement, as when a patient is confined to bed for a long time, lymph flow slows and fluid accumulates in tissues, causing the swelling known as edema.

A unique property of mature B and T cells, which distinguishes them from other blood cells, is that they move through the body in both blood and lymph. Lymphocytes are the only cell type present in lymph in any numbers; hence their name. When small lymphocytes leave the primary lymphoid tissues in which they have developed, they enter the bloodstream. When they reach the blood capillaries that invest a lymph node, or other secondary lymphoid tissue, small lymphocytes can leave the blood and enter the lymph node proper. If a lymphocyte becomes activated by a pathogen it remains in the lymph node; otherwise it will spend some time there and then leave in the efferent lymph and eventually be returned to the blood. This means that the population of lymphocytes within a node is in a continual state of flux, with new lymphocytes entering from the blood while others leave in the efferent lymph. This pattern of movement between blood and lymph is termed **lymphocyte recirculation** (Figure 1.19). It allows the lymphocyte population to continually survey the secondary lymphoid organs for evidence of infection. An

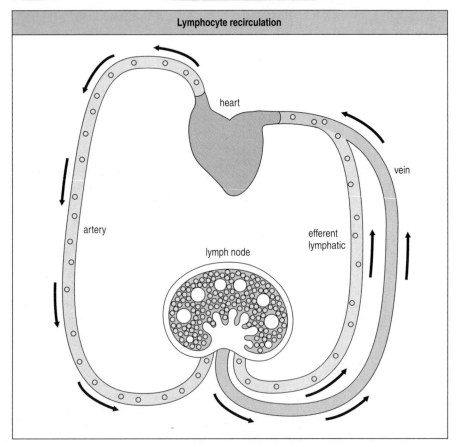

Lymphocyte recirculation

heart

vein

artery

efferent lymphatic

lymph node

Figure 1.19 Lymphocyte recirculation. Small lymphocytes are unique among blood cells in traveling through the body in the lymph as well as the blood. That is why they were named lymphocytes. Lymphocytes leave the blood through the walls of fine capillaries in secondary lymphoid organs. A lymph node is illustrated here. After spending some time in the lymph node, lymphocytes leave in the efferent lymph and return to the blood at the left subclavian vein. If a lymphocyte in a lymph node encounters a pathogen to which its cell-surface receptor binds, it stops recirculating.

exception to this pattern is the spleen, which has no connections to the lymphatic system. Lymphocytes both enter and leave the spleen in the blood.

The secondary lymphoid organs are dynamic tissues in which lymphocytes are constantly arriving from the blood and departing in the lymph. At any one time only a very small fraction of lymphocytes are in the blood and lymph; the majority are in lymphoid organs and tissues.

1-9   Adaptive immunity is initiated in secondary lymphoid tissues

Experiments involving deliberate infection of volunteers show that a large initial dose of pathogenic microorganisms is usually necessary to cause disease. To establish an infection a microorganism must colonize a tissue in sufficient numbers to overwhelm the cells and molecules of innate immunity that are promptly recruited from the blood to the site of invasion. Even in these circumstances the effects will usually be minor unless the infection can spread within the body. Frequent sites of infection are the connective tissues, which pathogens penetrate as a result of skin wounds. From such sites intact pathogens, components of pathogens and pathogen-infected dendritic cells are carried by the lymphatics to the nearest **lymph node**. The lymph node receiving the fluid collected at an infected site is called the **draining lymph node** (Figure 1.20).

The anatomy of the lymph node provides meeting places where lymphocytes coming from the blood encounter pathogens and their products brought from infected connective tissue (Figure 1.21). Arriving lymphocytes segregate

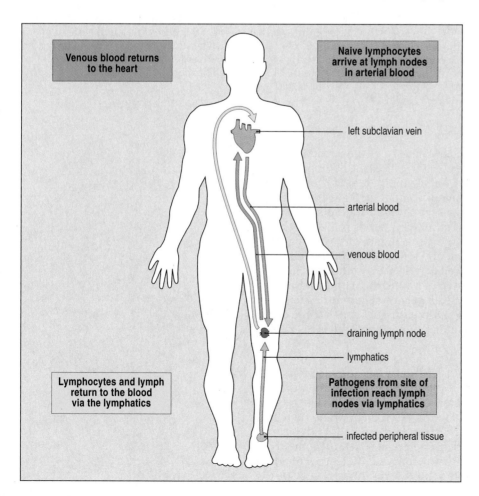

Figure 1.20 Circulating lymphocytes meet lymph-borne pathogens in draining lymph nodes. Lymphocytes leave the blood and enter lymph nodes, where they can be activated by pathogens in the afferent lymph draining from a site of infection. The circulation pertaining to a site of infection in the left foot is shown here. When activated by pathogens, lymphocytes stay in the node to divide and differentiate into effector cells. If lymphocytes are not activated, they leave the node in the efferent lymph and are carried by the lymphatics to the thoracic duct (see Figure 1.18), which empties into the blood at the left subclavian vein. Lymphocytes recirculate all the time, irrespective of infection. Every minute, 5×10^6 lymphocytes leave the blood and enter secondary lymphoid tissues.

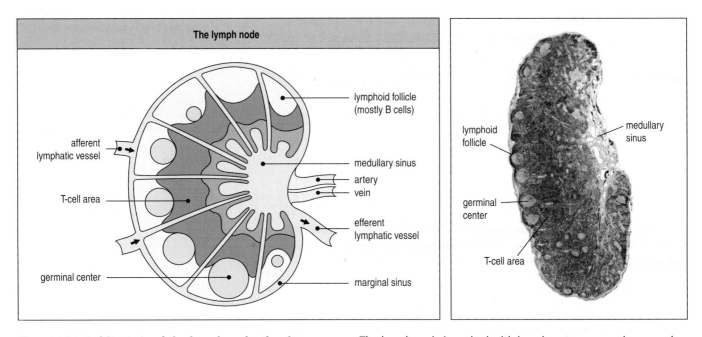

Figure 1.21 Architecture of the lymph node, the site where blood-borne lymphocytes respond to lymph-borne pathogens. Human lymph nodes are small kidney-shaped organs about 1.2 centimeters long and weighing 1 gram or less. They are located at junctions in the lymphatic system where a number of afferent lymph vessels bringing lymph from the tissues unite to form a single, larger efferent lymph vessel that takes the lymph away from the lymph node. A micrograph of a lymph node in longitudinal section is shown on the right and a lymph node is shown schematically in longitudinal section in the panel on the left. A lymph node is composed of a cortex (the yellow and blue areas on this panel) and a medulla (shown in dark pink). Lymph arrives via the afferent lymphatic vessels (pale pink), and during an infection pathogens and pathogen-laden dendritic cells arrive from infected tissue in the draining afferent lymph. The lymph node is packed with lymphocytes, macrophages and other cells of the immune system, between which the lymph percolates until it reaches the marginal sinus and leaves by the efferent lymphatic vessel. Lymphocytes arrive at lymph nodes in the arterial blood. They enter the node by passing between the endothelial cells that line the fine capillaries within the lymph node (not shown). Within the lymph node, B cells and T cells tend to congregate in anatomically discrete areas. T cells populate the inner cortex (paracortex) and B cells form lymphoid follicles in the outer cortex. During an infection, the expansion of pathogen-specific B cells forms a spherical structure called a germinal center within each follicle. Germinal centers are clearly visible in the photograph. Original magnification × 40. Photograph courtesy of Yasodha Natkunam.

to different regions of the lymph node: T cells to the T-cell areas and B cells to B-cell areas known as **lymphoid follicles**. Pathogens and pathogen-laden dendritic cells from the infected tissue arrive at a lymph node in **afferent lymphatic vessels**. Several of these unite at the node and then leave it as a single **efferent lymphatic vessel**. As the lymph passes through the node, the dendritic cells settle there and pathogens and other extraneous materials are filtered out by macrophages. This prevents infectious organisms from reaching the blood and provides a depot of pathogen and pathogen-carrying dendritic cells within the lymph node that can be used to activate lymphocytes. During an infection, pathogen-specific B cells that have bound the pathogen proliferate to form a dense spherical structure called a **germinal center** in each follicle. A lymph node draining a site of infection increases in size as a result of the proliferation of activated lymphocytes, a phenomenon sometimes referred to as "swollen glands."

In the lymph node the small fraction of B and T cells bearing receptors that bind to the pathogen or its products will be stimulated to divide and differentiate into effector cells. T cells are activated by dendritic cells, whereupon some of the T cells move to the associated lymphoid follicle where they help activate the B cells to become plasma cells. Other effector T cells and the antibodies secreted by plasma cells are carried by efferent lymph and blood to the infected tissues (Figure 1.22). There the effector cells and molecules of adaptive immunity work with their counterparts of innate immunity to subdue the

Figure 1.22 Activation of adaptive immunity in the draining lymph node. Pathogens, pathogen components, and dendritic cells carrying pathogens and molecules derived from them arrive in the afferent lymph draining the site of infection. Free pathogens and debris are removed by macrophages; the dendritic cells become resident in the lymph node and move to the T-cell areas, where they meet small lymphocytes that have entered the node from the blood (green). The dendritic cells specifically stimulate the division and differentiation of pathogen-specific small lymphocytes into effector lymphocytes (blue). Some helper T cells and cytotoxic T cells leave in the efferent lymph and travel to the infected tissue via the lymph and blood. Other helper T cells remain in the lymph node and stimulate the division and differentiation of pathogen-specific B cells into plasma cells (yellow). Plasma cells move to the medulla of the lymph node, where they secrete pathogen-specific antibodies, which are taken to the site of infection by the efferent lymph and subsequently the blood. Some plasma cells leave the lymph node and travel via the efferent lymph and the blood to the bone marrow, where they continue to secrete antibodies.

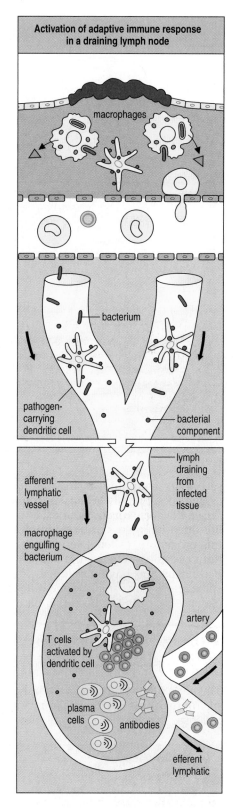

infection. Recovery from an infection involves clearance of infectious organisms from the body and repair of the damage caused by both infection and the immune response. A cure is not always possible. Infection can overwhelm the immune system, with death as the consequence. In the United States, some 36,000 deaths each year are associated with influenza. In intermediate situations the infection persists but its pathological effects are controlled by the adaptive immune response, as usually occurs with herpes viruses (see Figure 1.4).

1-10 The spleen provides adaptive immunity to blood infections

Pathogens can enter the blood directly, as occurs when blood-feeding insects transmit disease or when lymph nodes draining infected tissue have failed to remove all microorganisms from the lymph returned to the blood. The **spleen** is the lymphoid organ that serves as a filter for the blood. One purpose of the filtration is to remove damaged or senescent red cells; the second function of the spleen is that of a secondary lymphoid organ that defends the body against blood-borne pathogens. Any microorganism in the blood is a potential pathogen and source of dangerous systemic infection. Microorganisms and microbial products in the blood are taken up by splenic macrophages and dendritic cells, which then stimulate the B and T cells that arrive in the spleen from the blood. The spleen is made up of two different types of tissue: the red pulp, where red blood cells are monitored and removed, and the white pulp, where white blood cells gather to provide adaptive immunity. The organization and functions of splenic white pulp are similar to those of the lymph node, the main difference being that both pathogens and lymphocytes enter and leave the spleen in the blood (Figure 1.23).

Rare individuals have no spleen, a condition called asplenia (Figure 1.24). The underlying cause of asplenia is known to be genetic, because the condition runs in families, but the gene involved has yet to be defined. Children with asplenia are unusually susceptible to infections with so-called **encapsulated bacteria** such as *Streptococcus pneumoniae* (the pneumococcus) or *Haemophilus influenzae*, whose cells are surrounded by a thick polysaccharide capsule. A close relative of *S. pneumoniae*, *S. pyogenes*, which causes tonsillitis, can be seen in Figure 1.3d. Children with asplenia can be protected from these infections by immunization with vaccines incorporating the capsular polysaccharides of these bacteria. For good effect the vaccines are injected subcutaneously, into the connective tissue under the skin. From there they stimulate an immune response in the draining lymph nodes,

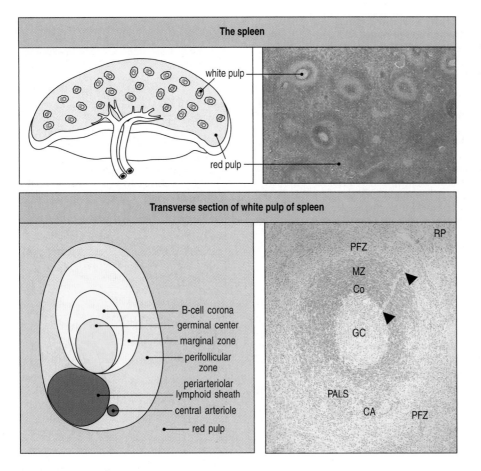

Figure 1.23 **The spleen has aggregations of lymphocytes similar to those in lymph nodes.** The human spleen is a large lymphoid organ in the upper left part of the abdomen, weighing about 150 grams. The upper diagram depicts a section of spleen in which nodules of white pulp are scattered within the more extensive red pulp. The red pulp is where old or damaged red cells are removed from the circulation; the white pulp is secondary lymphoid tissue, in which lymphocyte responses to blood-borne pathogens are made. The bottom diagram shows a nodule of white pulp in transverse section. It consists of a sheath of lymphocytes surrounding a central arteriole (CA). The sheath is called the periarteriolar lymphoid sheath (PALS). The lymphocytes closest to the arteriole are mostly T cells (blue region); B cells (yellow regions) are placed more peripherally. Lymphoid follicles each comprise a germinal center, a B-cell corona (Co) and a marginal zone (MZ), which contains differentiating B cells and macrophages. Both the follicle and the PALS are surrounded by a perifollicular zone (PFZ) abutting the red pulp and containing a variety of cell types, including erythrocytes, macrophages, T cells and B cells. Photographs courtesy of H.G. Burkitt and B. Young (top) and N.M. Milicevic (bottom).

secondary lymphoid organs that have normal immunological functions in asplenic individuals. When a person's spleen is damaged as a consequence of traumatic accidents or wounds, it will often be surgically removed to prevent life-threatening loss of blood into the abdominal cavity. Children who have this procedure, called splenectomy, can be as vulnerable to bacterial infections as asplenic children. For adults, who have already been infected by these pathogens and have developed protective immunity, the consequences of splenectomy are usually slight, but protective vaccination against some pathogens, especially *S. pneumoniae*, is advised.

1-11   Most secondary lymphoid tissue is associated with the gut

The parts of the body that harbor the largest and most diverse populations of microorganisms are the respiratory and gastrointestinal tracts. The extensive mucosal surfaces of these tissues make them particularly vulnerable to

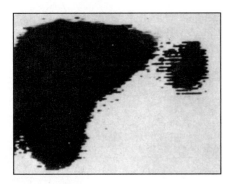
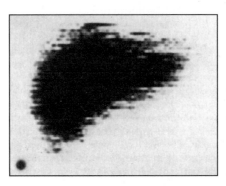

Figure 1.24 **Diagnosis of a child with asplenia.** The photos show scintillation scans of the abdomen of a mother (left panel) and child (right panel) after intravenous injection with radioactive colloidal gold. The large irregularly shaped organ to the left is the liver, which is seen in both panels. The smaller more rounded organ to the right in the mother is the spleen, which is not present in the child. Photographs courtesy of F. Rosen and R. Geha.

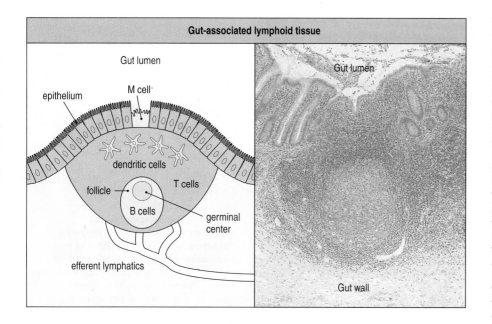

Gut-associated lymphoid tissue

Gut lumen

epithelium

M cell

dendritic cells

follicle

B cells

T cells

germinal center

efferent lymphatics

Gut lumen

Gut wall

Figure 1.25 A typical region of gut-associated lymphoid tissue (GALT). A schematic diagram (left panel) and a light micrograph (right panel) of a typical region of organized GALT such as a Peyer's patch. M cells of the gut epithelial wall deliver pathogens from the luminal side of the gut mucosa to the lymphoid tissue within the gut wall. The lymphoid tissue is organized similarly to the lymph node and the white pulp of the spleen, with distinctive B- and T-cell zones, lymphoid follicles, and germinal centers. White blood cells, including lymphocytes, are delivered from the blood through the walls of small blood capillaries, as in the lymph node. Lymphocytes activated in the GALT leave in the efferent lymphatics and are delivered via the mesenteric lymph nodes (not shown) into the thoracic duct and back into the blood, from which they specifically reenter the gut as effector lymphocytes. Photograph courtesy of N. Rooney.

infection and they are therefore heavily invested with secondary lymphoid tissue. The **gut-associated lymphoid tissues** (**GALT**) include the **tonsils**, **adenoids**, **appendix**, and the **Peyer's patches** which line the small intestine (see Figure 1.18). Similar but less organized aggregates of secondary lymphoid tissue line the respiratory epithelium, where they are called **bronchial-associated lymphoid tissue** (**BALT**), and other mucosal surfaces, including the gastrointestinal tract. The more diffuse mucosal lymphoid tissues are known generally as **mucosa-associated lymphoid tissue** (**MALT**).

Although different from the lymph nodes or spleen in outward appearance, the mucosal lymphoid tissues are similar to them in their microanatomy (Figure 1.25) and in their function of trapping pathogens to activate lymphocytes. The differences are chiefly in the route of pathogen entry and the migration patterns of their lymphocytes. Pathogens arrive at mucosa-associated lymphoid tissues by direct delivery across the mucosa, mediated by specialized cells of the mucosal epithelium called **M cells**. Lymphocytes first enter mucosal lymphoid tissue from the blood and, if not activated, leave via lymphatics that connect the mucosal tissues to draining lymph nodes. Lymphocytes activated in mucosal tissues tend to stay within the mucosal system, either moving directly out from the lymphoid tissue into the lamina propria and the mucosal epithelium, where they perform their effector actions, or reentering the mucosal tissues from the blood as effector cells after being recirculated.

1-12 Adaptive immune responses generally give rise to long-lived immunological memory and protective immunity

Clonal selection by pathogens (see Figure 1.10) is the guiding principle of adaptive immunity and explains the features of immunity that perplexed physicians in the past. The severity of a first encounter with an infectious disease arises because the primary immune response is developed from very few lymphocytes; the time taken to expand their number provides an opportunity for the pathogen to establish an infection to the point of causing disease. The clones of lymphocytes produced in a primary response include long-lived **memory cells**, which can respond more quickly and forcefully to subsequent encounters with the same pathogen.

The potency of such secondary immune responses can be sufficient to repel the pathogen before there is any detectable symptom of disease. The individual therefore appears immune to that disease. The striking differences between a primary and secondary response are illustrated in Figure 1.26. The immunity due to a secondary immune response is absolutely specific for the pathogen that provoked the primary response. It is the difference between the primary and the secondary response that has made vaccination so successful in the prevention of disease. Figure 1.27 shows how the introduction and availability of vaccines against diphtheria, polio, and measles have dramatically reduced the incidence of these diseases. In the case of measles there is also a corresponding reduction of subacute sclerosing panencephalitis, an infrequent but fatal spasticity caused by persisting measles virus that becomes manifest 7–10 years after measles infection and usually affects children who were infected as infants. When vaccination programs are successful to the point at which the disease is unfamiliar to physicians and public alike, then concerns can arise with real or perceived side effects of a vaccine that affect a very small minority of those vaccinated. Such concerns lead to fewer children being vaccinated and can lead to increasing occurrence of the disease.

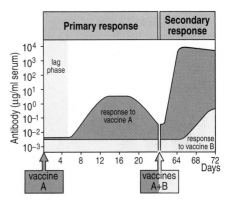

Figure 1.26 Comparison of primary and secondary immune responses. This diagram shows how the immune response develops during an experimental immunization of a laboratory animal. The response is measured in terms of the amount of pathogen-specific antibody present in the animal's blood serum, shown on the vertical axis, with time being shown on the horizontal axis. On the first day the animal is immunized with a vaccine against pathogen A. The levels of antibodies against pathogen A are shown in blue. The primary response reaches its maximum level 2 weeks after immunization. After the primary response has subsided, a second immunization with vaccine A on day 60 produces an immediate secondary response, which in 5 days is orders of magnitude greater than the primary response. In contrast, a vaccine against pathogen B, which was also given on day 60, produces a typical primary response to pathogen B as shown in yellow, demonstrating the specificity of the secondary response to vaccine A.

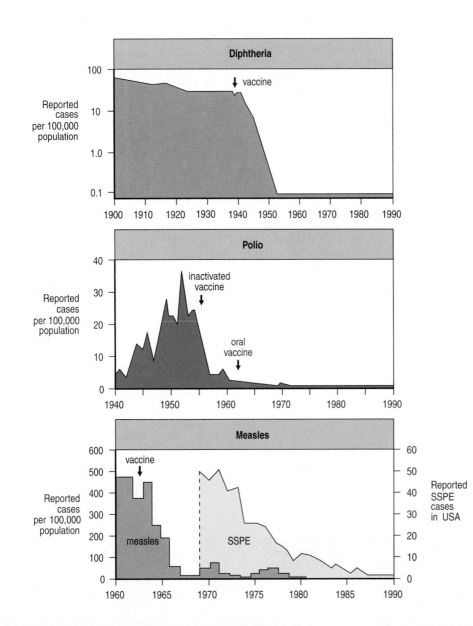

Figure 1.27 Successful vaccination campaigns. Diphtheria, poliomyelitis and measles have been virtually eliminated from the USA, as shown by these three graphs. The arrows indicate when the vaccination campaigns began. Subacute sclerosing panencephalitis (SSPE) is a brain disease that is a late consequence of measles infection for a minority of patients. Reduction of measles was paralleled by a reduction in SSPE 15 years later. Because these diseases have not been eradicated worldwide and the volume of international travel is so high, immunization must be maintained in much of the population to prevent recurrence of epidemic disease.

1-13 The immune system can be compromised by inherited immunodeficiencies or by the actions of certain pathogens

When components of the immune system are either missing or do not work properly, this generally leads to increased susceptibility to microbial infection. One cause of defective immune responses is inherited mutations in genes encoding proteins that contribute to immunity. Most people who carry a mutant gene are healthy because their other, normal, copy of the gene provides sufficient functional protein; the small and unfortunate minority who carry two mutant copies of the gene lack the function encoded by that gene. Such deficiencies lead to varying degrees of failure of the immune system and a wide range of **immunodeficiency diseases**, of which asplenia is one (see Section 1-10). In some of the immunodeficiency diseases, only one aspect of the immune response is affected, leading to susceptibility to particular kinds of infection; in others, adaptive immunity is completely absent, leading to a devastating vulnerability to all infections. These latter gene defects are rare, showing how vital is the protection normally afforded by the immune system. The discovery and study of immunodeficiency diseases has largely been the work of pediatricians, because such conditions usually show up early in childhood. Before the advent of antibiotics and, more recently, the possibility of bone marrow transplants and other replacement therapies, immunodeficiencies would usually have caused death in infancy.

Immunodeficiency states are caused not only by nonfunctional genes but also by pathogens that subvert the human immune system. An extreme example of an immunodeficiency due to disease is the **acquired immune deficiency syndrome** (**AIDS**), which is caused by infection with the **human immunodeficiency virus** (**HIV**). Although the disease has been recognized by clinicians only in the past 25 years, it is now at epidemic proportions, with around 33 million people infected worldwide (Figure 1.28). HIV infects helper

Figure 1.28 HIV infection is widespread on all the inhabited continents. Worldwide in 2007 there were about 33 million individuals infected with HIV, including about 2.5 million new cases, and about 2.1 million deaths from AIDS. Data are estimated numbers of adults and children living with HIV/AIDS at the end of 2007 (*AIDS Epidemic Update*, UNAIDS/World Health Organization; 2007).

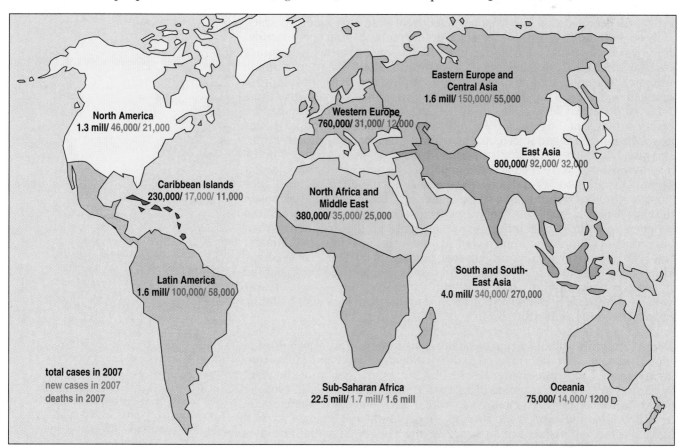

North America
1.3 mill/ 46,000/ 21,000

Eastern Europe and Central Asia
1.6 mill/ 150,000/ 55,000

Western Europe
760,000/ 31,000/ 12,000

East Asia
800,000/ 92,000/ 32,000

Caribbean Islands
230,000/ 17,000/ 11,000

North Africa and Middle East
380,000/ 35,000/ 25,000

Latin America
1.6 mill/ 100,000/ 58,000

South and South-East Asia
4.0 mill/ 340,000/ 270,000

total cases in 2007
new cases in 2007
deaths in 2007

Sub-Saharan Africa
22.5 mill/ 1.7 mill/ 1.6 mill

Oceania
75,000/ 14,000/ 1200

T cells, a cell type essential for adaptive immunity. During the course of an extended infection, which can last for up to 20 years, the population of helper T cells gradually diminishes, eventually leading to collapse of the immune system. Patients with AIDS become increasingly susceptible to a range of infectious microorganisms, many of which rarely trouble uninfected people. Death usually results from the effects of one or more of these opportunistic infections rather than from the direct effects of HIV infection.

Summary to Chapter 1

Throughout their evolutionary history, multicellular animals have been colonized and infected by microorganisms. To restrict the nature, size, and location of microbial infestation, animals have evolved a series of defenses, which humans still use today. The skin and contiguous mucosal surfaces provide physical and chemical barriers that confine microorganisms to the external surfaces of the body. When pathogens manage to breach the barriers and gain entry to the soft tissues, they are sought out and destroyed by the immune system. The cells of the immune system are principally the various types of leukocyte and allied tissue cells, such as dendritic cells, which all derive from a common stem cell in the bone marrow.

In responding to infection, the immune system starts with innate immune mechanisms that are fast, fixed in their mode of action, and effective in stopping most infections at an early stage. The cells and molecules of innate immunity identify common classes of pathogen and destroy them. Four key elements of innate immunity are: proteins such as mannose-binding lectin that noncovalently bind to the surfaces of pathogens; proteins such as complement that bond covalently to pathogen surfaces, forming ligands for receptors on phagocytes; phagocytic cells that engulf and kill pathogens; and cytotoxic cells that kill virus-infected cells. The cells and molecules of innate immunity have counterparts in vertebrates and invertebrates.

Vertebrates have evolved the additional defenses of adaptive immunity, which are brought into play when innate immunity fails to stop an infection. Although slow to start, the adaptive immune response eventually becomes powerful enough to terminate almost all of the infections that outrun innate immunity. The mechanisms of adaptive immunity are ones that improve pathogen recognition rather than pathogen destruction. They involve the T and B lymphocytes, which collectively have the ability to recognize the vast array of potential pathogens, and are initiated in specialized lymphoid tissues such as lymph nodes and spleen, to where infections that elude innate immunity spread. In these secondary lymphoid organs, small recirculating B and T lymphocytes with receptors that bind to pathogens or their macromolecular components are selected and activated. Because each individual B or T lymphocyte expresses receptors of a single and unique binding type, a pathogen stimulates only the small subset of lymphocytes that express receptors for the pathogen, focusing the adaptive immune response on that pathogen. When successful, an adaptive immune response terminates infection and provides long-lasting protective immunity against the pathogen that provoked the response. Failures to develop a successful response can arise from inherited deficiencies in the immune system or from the pathogen's ability to escape, avoid, or subvert the immune response. Such failures can lead to debilitating chronic infections or death.

Adaptive immunity builds on the mechanisms of innate immunity to provide a powerful response that is tailored to the pathogen at hand and can be rapidly reactivated on future challenge with that same pathogen, providing life-long immunity to many common diseases. Adaptive immunity is an evolving process within a person's lifetime, in which each infection changes the make-up of that individual's lymphocyte population. These changes are neither inherited nor passed on but, during the course of a lifetime, they determine a person's fitness and their susceptibility to disease. The strategy of vaccination aims at circumventing the risk of a first infection, and in the twentieth century successful campaigns of vaccination were waged against several diseases that were once both familiar and feared. Through the use of vaccination and antimicrobial drugs, as well as better sanitation and nutrition, infectious disease has become a less common cause of death in many countries.

Questions

1–1   Identify the four classes of pathogen that provoke immune responses in our bodies and give an example of each.

1–2   A bacterium that causes a common disease in a population that has been previously exposed to it is called:
  a. opportunistic
  b. resistant
  c. commensal
  d. endemic
  e. attenuated.

1–3
  A. Name three of the epithelia in the human body that act as barriers to infection.
  B. Describe the three main ways in which epithelia carry out this barrier function, giving details of the mechanisms employed.

1–4   An antimicrobial peptide that protects epithelial surfaces from pathogens is called:
  a. glycoprotein
  b. defensin
  c. proteoglycan
  d. lysozyme
  e. sebum.

1–5   How can antibiotics upset the barrier function of intestinal epithelia? Give a specific example.

1–6   Describe the characteristics commonly associated with inflammation and what causes them.

1–7   Which of the following are characteristics of innate immunity? (Select all that apply.)
  a. inflammation
  b. recognition of the pathogen improves during the response
  c. fast response
  d. highly specific for a particular pathogen
  e. cytokine production.

1–8   Which of the following statements regarding neutrophils is false?
  a. Neutrophils are mobilized from the bone marrow to sites of infection when needed.
  b. Neutrophils are active only in aerobic conditions.
  c. Neutrophils are phagocytic.
  d. Neutrophils form pus which is comprised of dead neutrophils.

  e. Dead neutrophils are cleared from sites of infection by macrophages.

1–9   What are the main differences between innate immunity and adaptive immunity?

1–10
  A. Identify the two major progenitor subsets of leukocytes.
  B. Where do they originate in adults?
  C. Name the white blood cells that differentiate from these two progenitor lineages.

1–11   Primary lymphoid tissues are the sites where lymphocytes _____, whereas secondary lymphoid tissues are the sites where lymphocytes _____.
  a. are stimulated; develop and mature
  b. encounter pathogens; undergo apoptosis
  c. develop and mature; become stimulated
  d. undergo clonal selection; differentiate from hematopoietic stem cells
  e. die; are phagocytosed after death.

1–12   The spleen differs from other secondary lymphoid organs in which of the following ways?
  a. It does not contain T cells.
  b. It filters blood as well as lymph.
  c. It is populated by specialized cells called M cells.
  d. It receives pathogens via afferent lymphatic vessels.
  e. It has no connection with the lymphatics.

1–13   What are clonal selection and clonal expansion in the context of an adaptive immune response, and describe how they shape the adaptive immune response.

1–14   What would be the consequence of a bioterrorist attack that released smallpox virus into a city?

1–15   Tim Schwartz, age 16, was hit by a car while riding his motorcycle. At the hospital he showed only minor abrasions and no bone fractures. He was discharged later that day. In the morning he experienced severe abdominal pain and returned to the hospital. Examination revealed tachycardia, low blood pressure, and a weak pulse. He received a blood transfusion without improvement. Laparoscopic surgery confirmed peritoneal hemorrhage due to a ruptured spleen. In addition to a splenectomy and a course of antibiotics, which of the following treatments would be administered?

a. plasmapheresis to remove autoantibodies (antibodies generated against self constituents)
b. regular intravenous injections of gamma globulin
c. vaccination and regular boosters with capsular polysaccharides from pathogenic pneumococcal strains
d. booster immunization with DPT (diphtheria toxoid, killed *Bordetella pertussis*, and tetanus toxoid)
e. regular blood transfusions.

1–16   Eileen Ratamacher is 83 years old and has been chronically ill with recurrent bacterial infections over the past year. Her physician prescribed broad-spectrum antibiotics which she had been taking for the past four months. This morning she suffered from severe abdominal cramping, vomiting, offensive smelling non-bloody diarrhea and a fever. Which of the following is not likely to be associated with her condition?

a. The gut flora has undergone a change in composition during the course of antibiotic treatment with displacement of commensal bacteria from the intestinal lumen.
b. There has been a reduction in colicin production by *Escherichia coli.*
c. There is food poisoning caused by enterohemorrhagic *Escherichia coli.*
d. There is toxin production by *Clostridium difficile* causing deterioration of the mucosal epithelium of the gastrointestinal tract.
e. The appearance of 'pseudomembranes' on the rectal surface seen on colonoscopy.

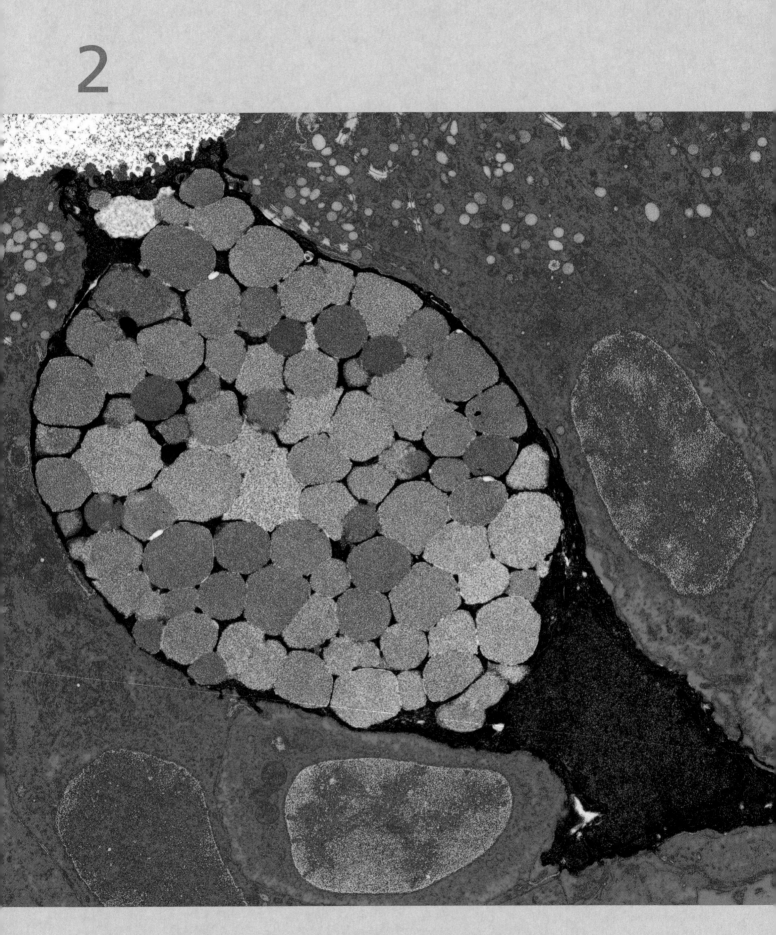

Goblet cells secrete the mucus that protects epithelial surfaces from invasion by microorganisms.

Chapter 2

Innate Immunity

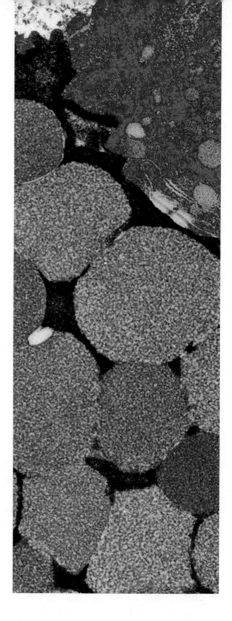

Although we become conscious of infectious agents only when we are suffering from the diseases they cause, microorganisms are always with us. Fortunately, the vast majority of microorganisms we come into contact with are prevented from ever causing an infection by barriers at the body's surface. The skin and the epithelia lining the gut and respiratory system provide an effective physical barrier against most organisms. These surfaces are also colonized by nonpathogenic resident microorganisms, which compete with the invading pathogen for nutrients and living space, as we saw in Chapter 1. If microorganisms penetrate these barriers and start an infection, most are eliminated within a few days by the innate immune response before they cause disease symptoms. The physical barriers and some of the mechanisms of innate immunity are ready for action at all times and function from the very beginning of an infection. Other innate immune mechanisms are mobilized after cells of the immune system detect the presence of infection and turn on the gene expression and protein synthesis needed to make the response. These responses are induced by the infection and need from a few hours to 4 days of development to become fully functional. The actual sequence of events induced by any particular infection is, however, dependent upon the particular type of pathogen and how it exploits the human body.

2-1 A variety of defense mechanisms have evolved to eliminate the different types of pathogen

Pathogens exploit and abuse human bodies in a variety of different ways. They also vary in the manner by which they live and replicate in the human body, and in the type of damage they cause (Figure 2.1). For the purposes of defense, a distinction can be made between pathogens that live and replicate in the spaces between human cells to produce **extracellular infections** and pathogens that replicate inside human cells to produce **intracellular infections**. Both extracellular and intracellular spaces can be further subdivided as shown in Figure 2.2, which also shows the mechanisms of innate immunity that are used against the pathogens that live in each space. Where a pathogen lives and replicates determines which form of immune mechanism will be more likely to succeed. Extracellular forms of pathogens are accessible to soluble molecules of the immune system, whereas intracellular forms are not. Intracellular pathogens that live in the nucleus or cytosol can be attacked by killing the infected cell. This interferes with the pathogen's life cycle and exposes pathogens released from the dead cells to the soluble molecules of the immune system. Pathogens that live in intracellular vesicles can be attacked by activating the infected cell to intensify its antimicrobial activity. And despite their differences, virtually all pathogens, whether viruses,

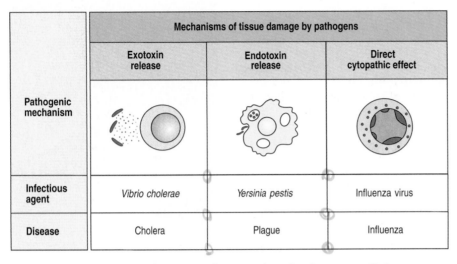

	Mechanisms of tissue damage by pathogens		
	Exotoxin release	Endotoxin release	Direct cytopathic effect
Pathogenic mechanism	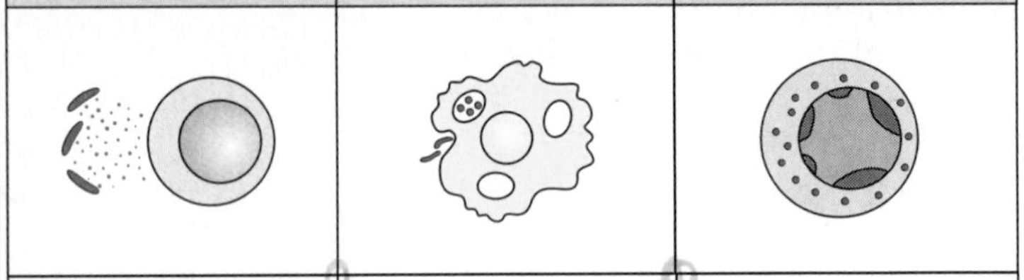		
Infectious agent	*Vibrio cholerae*	*Yersinia pestis*	Influenza virus
Disease	Cholera	Plague	Influenza

Figure 2.1 Pathogens damage tissues in different ways. Pathogens can kill cells and damage tissues in three ways. Exotoxins released by microorganisms act at the surfaces of host cells, usually via a cell-surface receptor (first column). When phagocytes degrade certain microorganisms, endotoxins are released that induce the phagocytes to secrete cytokines, causing local or systemic symptoms (second column). Cells infected by pathogens are usually killed or damaged in the process (third column).

bacteria, fungi, or parasites, spend some time in the extracellular spaces, where they can be attacked by soluble effector molecules of the immune system.

Most pathogens infect only a few related host species, and for this reason humans are rarely infected through transmission from another vertebrate species, such as the domesticated animals with which humans are often in contact, or wild animals that are hunted, butchered, and eaten. The vast majority of human infections result from transmission of the pathogen, either directly or indirectly, from another person who is already infected. Transmission can be directly from one person to another, or, as with many parasites, it requires an intermediate passage through a distantly related organism, for example an insect or mollusk, that is necessary for completing the pathogen's life cycle (see Figure 1.4, pp. 6–7).

The ability of different pathogens to persist outside the body varies considerably and determines the ease with which a particular disease is spread. The bacterial disease anthrax is spread by spores that are resistant to heat and desiccation and can therefore be passed over long distances from one person to another. It is these properties that make anthrax a 'hot topic' in discussions of germ warfare. In contrast, the human immunodeficiency virus (HIV) is very sensitive to changes in its environment and can be passed between individuals only by intimate contact and the exchange of infected body fluids and cells.

	Extracellular		Intracellular	
	Interstitial spaces, blood, lymph	Epithelial surfaces	Cytoplasmic	Vesicular
Site of infection	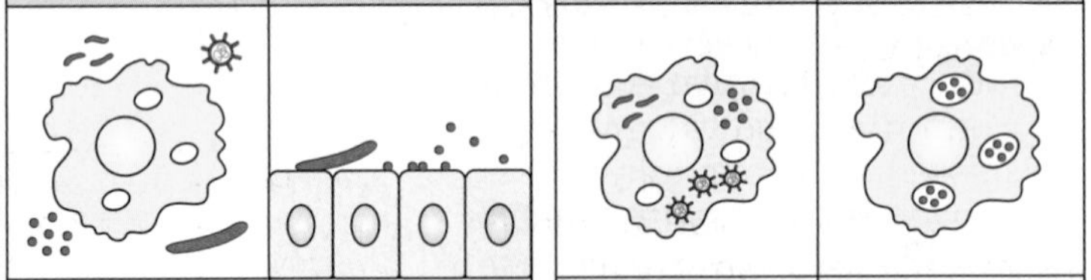			
Organisms	Viruses Bacteria Protozoa Fungi Worms	*Neisseria gonorrhoeae* *Candida albicans* Worms	Viruses *Listeria* Protozoa	Mycobacteria Trypanosomes *Cryptococcus neoformans*
Defense mechanism	Complement Macrophages Neutrophils	Antimicrobial peptides	NK cells	Activated macrophages

Figure 2.2 Pathogens exploit different compartments of the body that are defended in different ways by innate immunity. Virtually all pathogens have an extracellular stage in their life cycle. For the other compartments, a representative example of each type of pathogen that exploits the compartment is given. For some pathogens, all stages of their life cycle are extracellular, whereas others exploit intracellular sites as places to grow and replicate. Different components of the immune system contribute to defense against different types of microorganism in different locations. NK cells, natural killer cells.

2-2   Complement is a system of plasma proteins that marks pathogens for destruction

As soon as a pathogen penetrates an epithelial barrier and starts to live in a human tissue, the defense mechanisms of innate immunity are brought into play. One of the first weapons to fire is a system of soluble proteins that are made constitutively by the liver and are present in the blood, lymph, and extracellular fluids. These plasma proteins are collectively known as the **complement system** or just **complement**. Complement coats the surface of bacteria and extracellular virus particles and makes them more easily phagocytosed. Without such a protein coating, many bacteria resist phagocytosis, especially those that are enclosed in thick polysaccharide capsules.

Many complement components are proteolytic enzymes, or proteases, that circulate in functionally inactive forms known as zymogens. Infection triggers **complement activation**, which proceeds by a series, or cascade, of enzymatic reactions involving proteases, in which each protease cleaves and activates the next enzyme in the pathway. Each protease is highly specific for the complement component it cleaves, and cleavage is usually at a single site. Many of these enzymes belong to the large family of serine proteases, which also includes the digestive enzymes chymotrypsin and trypsin.

Although more than 30 proteins make up the complement system, **complement component 3 (C3)** is by far the most important. Although patients lacking other complement components suffer relatively minor immunodeficiencies, patients lacking C3 are prone to successive severe infections. Whenever complement is activated by infection it always leads to the cleavage of C3 into a small C3a fragment and a large C3b fragment. In the process, some of the C3b fragments become covalently bound to the pathogen's surface (Figure 2.3). This attachment of C3b to pathogen surfaces is the essential function of the complement system; it is called **complement fixation**, because C3b becomes firmly fixed to the pathogen. The bound C3b tags the pathogen for destruction by phagocytes and can also organize the formation of protein complexes that damage the pathogen's membrane. The soluble C3a fragment also contributes to the body's defenses by acting as a chemoattractant to recruit effector cells, including phagocytes, from the blood to the site of infection.

The unusual feature of C3 that underlies its unique and potent function is a high-energy thioester bond within the glycoprotein. C3 is made and enters the circulation in an inactive form, in which the thioester is sequestered and stabilized within the hydrophobic interior of the protein. When C3 is cleaved into C3a and C3b, the bond is exposed and becomes subject to nucleophilic attack by water molecules or by the amino and hydroxyl groups of proteins and carbohydrates on pathogen surfaces. This results in some of the C3b becoming covalently bonded to the pathogen (Figure 2.4). The thioester bonds of the vast majority of C3b molecules are attacked by water and so most C3b remains in solution in an inactive hydrolyzed form.

Three pathways of complement activation are defined. Although differing in how they are triggered and in the first few reactions in the cascade, they all lead to C3 activation, the deposition of C3b on the pathogen's surface and the recruitment of similar effector mechanisms for pathogen destruction (Figure 2.5). The pathway that works at the start of infection is the **alternative pathway of complement activation**. A second pathway, the **lectin pathway of complement activation**, is also a part of innate immunity but is induced by infection and requires some time before it gains strength. The third pathway, the **classical pathway of complement activation**, is a part of both innate and adaptive immunity and requires the binding of either antibody or an innate immune-system protein called C-reactive protein to the pathogen's surface. The names of the pathways reflect the order of their scientific discovery: the

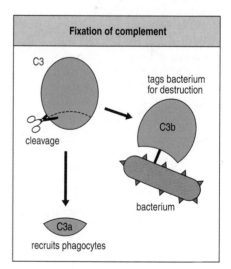

Figure 2.3 Complement activation achieves covalent attachment of C3b to a pathogen's surface. The key event in complement activation by a pathogen is the proteolytic cleavage of complement fragment C3. This cleavage produces a large C3b fragment and a small C3a fragment. C3b is chemically reactive and becomes covalently attached, or fixed, to the pathogen's surface, thereby marking the pathogen as dangerous. C3a recruits phagocytic cells to the site of infection.

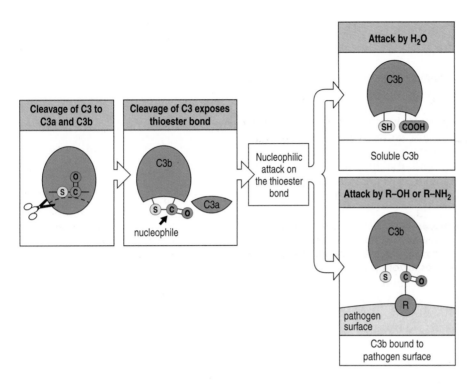

Figure 2.4 Cleavage of C3 exposes a reactive thioester bond that covalently attaches the C3b fragment to the pathogen surface. Circulating C3 is an inactive serine protease consisting of α and β polypeptide chains in which a thioester bond in the α chain is protected from hydrolysis within the hydrophobic interior of the protein. The thioester bond is denoted in the top two panels by the circled letters S, C, and O. The C3 molecule is activated by cleavage of the α chain to give fragments C3a and C3b. This exposes the thioester bond of C3b to the hydrophilic environment. The thioester bonds of most of the C3b fragments will be spontaneously hydrolyzed by water as shown in the bottom left panel, but a minority will react with hydroxyl and amino groups on molecules on the pathogen's surface, bonding C3b to the pathogen surface, as shown in the bottom right panel.

classical pathway was discovered first, then the alternative pathway and last the lectin pathway. The name complement was coined because the effector functions provided by these proteins were seen to 'complement' the pathogen-binding function of antibodies in the classical pathway of complement activation and pathogen destruction.

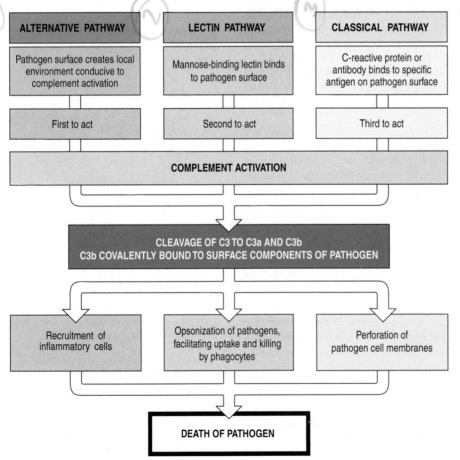

Figure 2.5 The three pathways of complement activation. The alternative pathway of complement activation is triggered by changes in the local physicochemical environment that are caused by the constituents of some bacterial surfaces. The alternative pathway acts at the earliest times during infection. The lectin-mediated pathway is initiated by the mannose-binding lectin of plasma, which binds to carbohydrates found on bacterial cells and other pathogens. The lectin-mediated pathway is induced by infection and contributes to innate immunity. The classical pathway is initiated in the innate immune response by the binding of C-reactive protein to bacterial surfaces, and in the adaptive immune response by the binding of antibodies to pathogen surfaces.

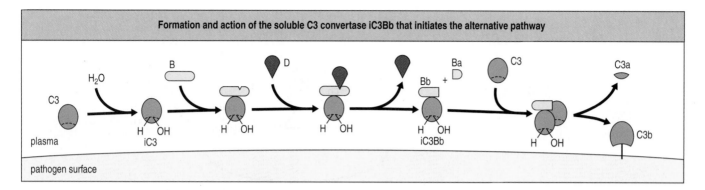

Figure 2.6 **Formation and action of the soluble C3 convertase that initiates the alternative pathway of complement activation.** In the plasma close to a microbial surface the thioester bond of C3 spontaneously hydrolyzes at low frequency. This activates the C3, which then binds factor B. Cleavage of B by the serine protease factor D produces a soluble C3 convertase, called iC3Bb, which then activates C3 molecules by cleavage into C3b and C3a. Some of the C3b fragments become covalently attached to the microbial surface.

2-3 At the start of an infection, complement activation proceeds by the alternative pathway

We shall start by describing the alternative pathway of complement activation, which is one of the first responses of the innate immune system, especially to bacterial infection. When C3 is first made in the liver, the thioester bond is sequestered inside the protein, but when C3 is secreted into the aqueous environment of the plasma, a conformational change occurs in the protein that makes the thioester bond available for hydrolysis. The first step in the alternative pathway of complement activation involves exposure and hydrolysis of the thioester bond of a small proportion of C3 molecules to give a form of C3 called **iC3** or **C3(H$_2$O)**, a reaction that does not involve cleavage of the C3. This reaction occurs spontaneously at a low rate in plasma but is catalyzed by the environment in the vicinity of certain pathogens, particularly bacteria. Also facilitating the spontaneous hydrolysis reaction is the high concentration of C3 in blood (about 1.2 mg/ml). iC3 binds to the inactive complement **factor B**, making factor B susceptible to cleavage by the protease **factor D**. This reaction produces a small fragment, Ba, which is released, and a large fragment, Bb, that has protease activity and remains bound to iC3. The iC3Bb complex binds intact C3 molecules and its protease activity cleaves them efficiently into C3a and C3b fragments, with the consequent activation of the thioester bond (see Figure 2.4) and some C3b becoming covalently bonded to the pathogen (Figure 2.6).

Proteases that cleave and activate C3 are called **C3 convertases**, iC3Bb being an example of a soluble C3 convertase. Like iC3, pathogen-bound C3b binds factor B and facilitates the cleavage of factor B by factor D. This reaction leads to the release of Ba and the formation of a C3bBb complex on the microbial surface. C3bBb is a potent C3 convertase, called the **alternative C3 convertase**, which works right at the surface of the pathogen (Figure 2.7). C3bBb binds C3 and cleaves it into C3b and Bb with activation of the thioester bond. Because this convertase is situated at the pathogen's surface, and is unable to diffuse away like iC3Bb, a larger proportion of the C3b fragments it produces become fixed to the pathogen. Once some C3 convertase molecules have been assembled, they cleave more C3 and fix more C3b at the microbial surface, leading to the assembly of yet more convertase. This positive feedback process, in which the C3b product of the enzymatic reaction can assemble more enzyme, is one of progressive amplification of C3 cleavage. From the initial deposition of a few molecules of C3b the pathogen rapidly becomes coated (Figure 2.8).

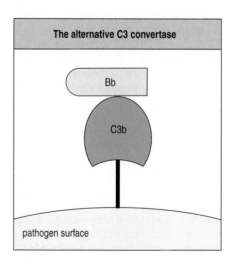

Figure 2.7 **The C3 convertase of the alternative pathway is a complex of C3b and Bb.** In this complex the Bb fragment of factor B provides the protease activity to cleave C3, and the C3b fragment of C3 locates the enzyme to the pathogen's surface.

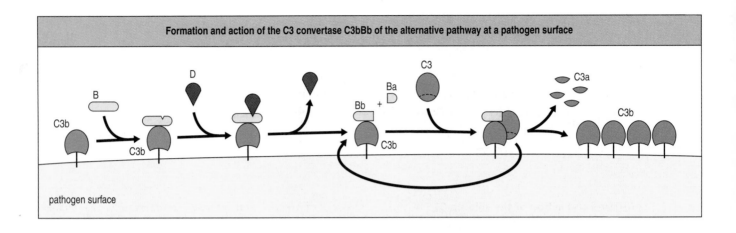

Formation and action of the C3 convertase C3bBb of the alternative pathway at a pathogen surface

pathogen surface

2-4   Regulatory proteins determine the extent and site of C3b deposition

As we have seen in the previous section, the alternative C3 convertase, C3bBb, is capable of rapid and explosive reactions because one molecule of C3bBb can make numerous additional molecules of C3bBb. Two broad categories of **complement control proteins** have evolved to regulate these reactions, which they do mainly by stabilizing or degrading C3b at cell surfaces. One class comprises plasma proteins that interact with C3b attached to human and microbial cell surfaces; the other includes membrane proteins on human cells that prevent complement fixation at the cell surface.

The plasma protein **properdin** (**factor P**) increases the speed and power of complement activation by binding to the C3 convertase C3bBb on microbial surfaces and preventing its degradation by proteases (Figure 2.9, upper panel). Countering the effect of properdin is plasma protein **factor H**, which binds to C3b and facilitates its further cleavage to a form called iC3b by the plasma serine protease **factor I** (Figure 2.9, middle panel). Fragment iC3b cannot assemble a C3 convertase, so the combined action of factors H and I is to reduce the number of C3 convertase molecules on the pathogen surface. The importance of the negative regulation by factors H and I is illustrated by the immunodeficiency suffered by patients who, for genetic reasons, lack factor I. In these people, formation of the C3 convertase C3bBb runs away unchecked until it depletes the reservoir of C3 in blood, extracellular fluid, and lymph. When faced with bacterial infections, people with factor I deficiency fix abnormally small amounts of C3b on bacterial surfaces, making for less efficient bacterial clearance by phagocytes. Consequently, these people are more susceptible than usual to ear infections and abscesses caused by encapsulated bacteria; that is, bacteria enclosed in a thick polysaccharide capsule (see Section 1-10), which are phagocytosed much more efficiently when they are coated with complement.

The second category of complement control proteins comprises membrane proteins of human cells that interfere with complement activation at human cell surfaces. The **decay-accelerating factor** (**DAF**) binds to the C3b component of the alternative C3 convertase, causing its dissociation and inactivation. **Membrane co-factor protein** (**MCP**) also has this function, but the binding of MCP to C3b makes it also susceptible to cleavage and inactivation by factor I (Figure 2.9, bottom panel). The functions of MCP are similar to those of the soluble complement regulator, factor H, which can also become membrane associated; factor H has a binding site for sialic acid, a component of human cell-surface carbohydrates but absent from most bacteria. As a strategy to evade the actions of complement, some species of bacteria, such as *Streptococcus pyogenes* and *Staphylococcus aureus*, cover their cell surfaces

Figure 2.8 Formation and action of the C3 convertase, C3bBb, of the alternative pathway at a microbial surface. Through the action of the soluble C3 convertase, iC3Bb, C3b fragments are bound to the microbial surface (see Figure 2.6). These bind factor B, which is then cleaved by factor D to produce C3bBb, the surface-bound convertase of the alternative pathway. This enzyme cleaves C3 to produce further C3b fragments bound to the microbe and small soluble C3a fragments. The C3b fragments can be used either to make more C3 convertase, which amplifies the activation of C3, or to provide ligands for the receptors of phagocytic cells. The small, soluble C3a fragments attract phagocytes to sites of complement fixation.

with sialic acid. By this means the bacteria mimic human cells. Consequently, when C3b becomes deposited on the surface of these bacteria it is readily inactivated by the factor H bound to the bacterial sialic acid.

Many of the diverse proteins that regulate complement, such as DAF, MCP, and factor H, are elongated structures built from varying numbers of structurally similar modules known as **complement control protein (CCP) modules**. Each module consists of about 60 amino acids that fold into a compact sandwich formed from two slices of β-pleated sheet stabilized by two conserved disulfide bonds. Proteins made up of CCP modules are also called **regulators of complement activation (RCA)**.

The combined effect of the reactions that promote and regulate C3 activation is to ensure that C3b is in practice deposited only on the surfaces of pathogenic microorganisms and not on human cells. In this manner the complement system provides a simple and effective way of distinguishing human cells from microbial cells, and for guiding mechanisms of death and destruction toward invading pathogens and away from healthy cells and tissues. In immunology, this type of distinction is called the discrimination of non-self from self.

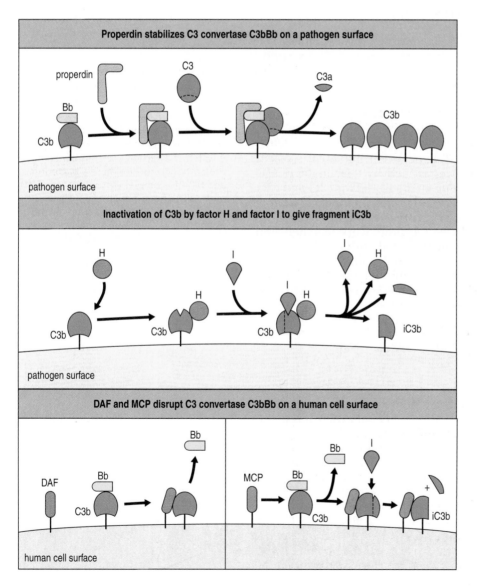

Figure 2.9 Formation and stability of the alternative C3 convertase on cell surfaces is determined by complement control proteins. Upper panel: the soluble protein properdin (factor P) binds to C3bBb and extends its lifetime on the microbial surface. Middle panel: factor H binds to C3b and changes its conformation to one that is susceptible to cleavage by factor I. The product of this cleavage is the iC3b fragment of C3, which remains attached to the pathogen surface but cannot form a C3 convertase. Lower panel: when C3bBb is formed on a human cell surface it is rapidly disrupted by the action of one of two membrane proteins: decay-accelerating factor (DAF) or membrane cofactor protein (MCP). In combination, these regulatory proteins ensure that much complement is fixed to pathogen surfaces and little is fixed to human cell surfaces.

2-5   Phagocytosis by macrophages provides a first line of cellular defense against invading microorganisms

When a pathogen invades a human tissue, the first effector cells of the immune system it encounters are the resident macrophages. **Macrophages** are the mature forms of circulating monocytes (see Figure 1.12, p. 13) that have left the blood and taken up residence in the tissues. They are prevalent in the connective tissues, the linings of the gastrointestinal and respiratory tracts, the alveoli of the lungs, and in the liver, where they are known as **Kupffer cells**. Macrophages are long-lived phagocytic cells that participate in both innate and adaptive immunity.

Although macrophages phagocytose bacteria and other microorganisms in a nonspecific fashion, the process is made more efficient by receptors on the macrophage surface that bind to specific ligands on microbial surfaces. One such receptor binds to C3b fragments that have been deposited at high density on the surface of a pathogen through activation of the alternative pathway of complement. This receptor is called **complement receptor 1** or **CR1**. The interaction of an array of C3b fragments on a pathogen with an array of CR1 molecules on the macrophage facilitates the engulfment and destruction of the pathogen. Bacteria coated with C3b are more efficiently phagocytosed than uncoated bacteria: the coating of a pathogen with a protein that facilitates phagocytosis is called **opsonization** (Figure 2.10).

CR1 also serves to protect the surface of cells on which it is expressed. Like MCP and factor H, CR1 disrupts the C3 convertase by making C3b susceptible to cleavage by factor I. During phagocytosis, some of a macrophage's CR1 molecules will have this protective role whereas others will engage the C3b fragments deposited on the pathogen's surface. Like MCP and factor H, CR1 is made up of CCP modules.

Two other macrophage receptors, **complement receptor 3** (**CR3**) and **complement receptor 4** (**CR4**) bind to iC3b fragments on microbial surfaces. Although the iC3b fragment has no C3 convertase activity, it facilitates phagocytosis and pathogen destruction by serving as the ligand for CR3 and CR4. These receptors are structurally unrelated to CR1, being members of a family of surface glycoproteins, the integrins, that contribute to adhesive interactions between cells. The CR1, CR3 and CR4 receptors work together more effectively in the phagocytosis of complement-coated pathogens than does each receptor on its own. The combination of opsonization by complement activated through the alternative pathway and subsequent phagocytosis by macrophages allows pathogens to be recognized and destroyed from the very beginning of an infection.

Figure 2.10 Complement receptors on phagocytes trigger the uptake and breakdown of C3b-coated pathogens. Covalently attached C3b fragments coat the pathogen surface, here a bacterium, and bind to complement receptor 1 (CR1) molecules on the phagocyte surface, thereby tethering the bacterium to the phagocyte. Intracellular signals generated by CR1 enhance the phagocytosis of the bacterium and the fusion of lysosomes containing degradative enzymes and toxic molecules with the phagosome. Ultimately, the bacterium is killed.

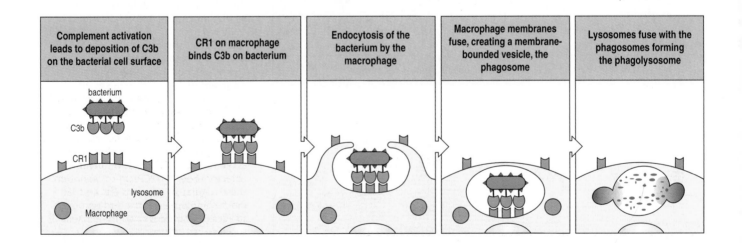

| Complement activation leads to deposition of C3b on the bacterial cell surface | CR1 on macrophage binds C3b on bacterium | Endocytosis of the bacterium by the macrophage | Macrophage membranes fuse, creating a membrane-bounded vesicle, the phagosome | Lysosomes fuse with the phagosomes forming the phagolysosome |

The terminal complement components that form the membrane-attack complex		
Protein	Concentration in serum (μg/ml)	Function
C5	85	On activation the soluble C4b fragment initiates assembly of the membrane-attack complex in solution
C6	60	Binds to and stabilizes C5b. Forms a binding site for C7
C7	55	Binds to C5b6 and exposes a hydrophobic region that permits attachment to the cell membrane
C8	55	Binds to C5b67 and exposes a hydrophobic region that inserts into the cell membrane
C9	60	Polymerization on the C5b678 complex to form a membrane-spanning channel that disrupts the cell's integrity and can result in cell death

Figure 2.11 The terminal components of the complement pathway.

2-6   The terminal complement proteins lyse pathogens by forming a membrane pore

As we have seen, the most important product of complement activation is C3b bonded to pathogen surfaces. However, the cascade of complement reactions does extend beyond this stage, involving five additional complement components (Figure 2.11). C3b binds to the alternative C3 convertase to produce an enzyme that acts on the C5 component of complement and is called the **alternative C5 convertase**; it consists of Bb plus two C3b fragments and is designated C3b$_2$Bb (Figure 2.12).

Complement component C5 is structurally similar to C3 but lacks the thioester bond and has a different function. It is cleaved by the C5 convertase into a smaller C5a fragment and a larger C5b fragment (see Figure 2.12). The function of C5b is to initiate the formation of a **membrane-attack complex**, which can make holes in the membranes of bacterial pathogens and eukaryotic cells. In succession, C6 and C7 bind to C5b—interactions that expose a hydrophobic site in C7, which inserts into the lipid bilayer. When C8 binds to C5b a hydrophobic site in C8 is exposed and, on insertion into the membrane, this part of C8 initiates the polymerization of C9, the component that forms the transmembrane pores (Figure 2.13). The components of the membrane-attack complex are listed and their activities summarized in Figure 2.11.

Although in the laboratory the perforation of membranes by the membrane-attack complex seems dramatic, clinical evidence demonstrating the importance of the C5–C9 components remains limited. The clearest effect of deficiency in any of these components is to increase susceptibility to infection

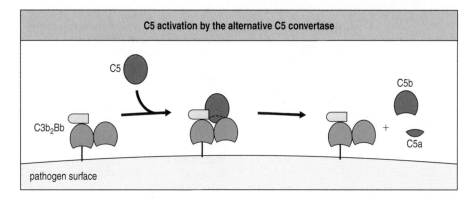

C5 activation by the alternative C5 convertase

C5   C3b$_2$Bb   C5b   C5a   pathogen surface

Figure 2.12 Complement component C5 is cleaved by C5 convertase to give a soluble active C5b fragment. The C5 convertase of the alternative pathway consists of two molecules of C3b and one of Bb (C3b$_2$Bb). C5 binds to the C3b component of the convertase and is cleaved into fragments C5a and C5b, of which C5b initiates the assembly of the terminal complement components to form the membrane-attack complex.

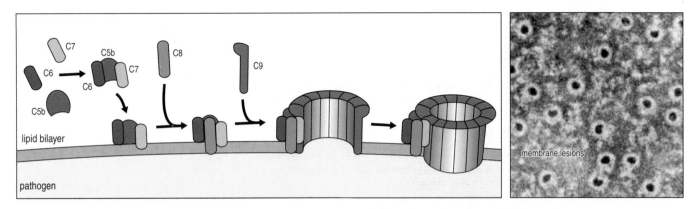

Figure 2.13 The membrane-attack complex assembles to generate a pore in the lipid bilayer membrane. The sequence of steps and their approximate appearance is shown here in schematic form. C5b is generated by the cleavage of C5 by the alternative C5 convertase C3b$_2$Bb. C5b then forms a complex by the successive binding of one molecule each of C6, C7, and C8. In forming the complex, C7 and C8 undergo a conformational change that exposes hydrophobic sites, which insert into the membrane. This complex causes some membrane damage and also induces the polymerization of C9. As each molecule of C9 is added to the polymer, it exposes a hydrophobic site and inserts into the membrane. Up to 16 molecules of C9 can be added to generate a transmembrane channel 100 Å in diameter. The channel disrupts the bacterial outer membrane, killing the bacterium. In the laboratory, the erythrocyte is a convenient cell with which to measure complement-mediated lysis. The electron micrograph shows erythrocyte membranes with membrane-attack complexes seen end on. Photograph courtesy of S. Bhakdi and J. Tranum-Jensen.

by bacteria of the genus *Neisseria*, different species of which cause the sexually transmitted disease gonorrhea and a common form of bacterial meningitis. Inherited deficiency for some complement components is not uncommon. For example, 1 in 40 Japanese people are heterozygous for C9 deficiency, predicting that 1 in 1600 of them will be completely deficient in C9.

The activity of the terminal complement components on human cells is regulated by soluble and surface-associated proteins. The soluble proteins called S protein, clusterin, and factor J prevent the soluble complex of C5b with C6 and C7 from associating with cell membranes. At the human cell surface, proteins called homologous restriction factor (HRF) and **CD59** (also called **protectin**) prevent the recruitment of C9 by the complex of C5b, C6, C7, and C8 (Figure 2.14). DAF, HRF, and CD59 are all linked to the plasma membrane by glycosylphosphatidylinositol lipid tails. Impaired synthesis of this tail is the common cause of **paroxysmal nocturnal hemoglobinuria**, a disease characterized by episodes of complement-mediated lysis of red blood cells that lack cell-surface DAF, HRF, or CD59.

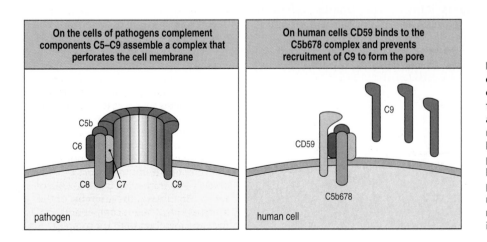

Figure 2.14 CD59 prevents assembly of the membrane attack complex on human cells. Left panel: the formation of a pore by the membrane attack complex (MAC) on a pathogenic microorganism. Right panel: how the human cell-surface protein CD59 prevents pore formation on human cells. By binding to the C5b678 complex, CD59 prevents the polymerization of C9 in the membrane to form a pore. Homologous restriction factor (HRF, not shown) works in the same way.

2-7   Small peptides released during complement activation induce local inflammation

During complement activation, C3 and C5 are each cleaved into two fragments, of which the larger (C3b and C5b) continue the pathway of complement activation. The smaller soluble C3a and C5a fragments are also physiologically active, increasing inflammation at the site of complement activation through binding to receptors on several cell types. Inflammation (see Section 1-4, p. 8) is a major consequence of the innate immune response to infection, which is also sometimes called the inflammatory response. In some circumstances, the C3a and C5a fragments induce anaphylactic shock, which is an acute inflammatory response that occurs simultaneously in tissues throughout the body; they are therefore referred to as **anaphylatoxins**. Of the anaphylatoxins, C5a is more stable and more potent than C3a. Phagocytes, endothelial cells, and mast cells have receptors specific for C5a and C3a. The two receptors are related and are of a type that is embedded in the cell membrane and signals through the activation of a guanine-nucleotide-binding protein.

The anaphylatoxins induce the contraction of smooth muscle and the degranulation of mast cells and basophils, with the consequent release of histamine and other vasoactive substances that increase capillary permeability. They also have direct vasoactive effects on local blood vessels, increasing blood flow and vascular permeability. These changes make it easier for plasma proteins and cells to pass out of the blood into the site of an infection (Figure 2.15).

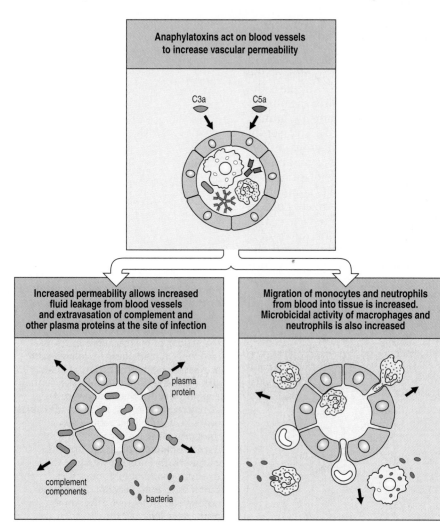

Figure 2.15 Local inflammatory responses can be induced by the small complement fragments C3a and C5a. These small anaphylatoxic peptides are produced by complement cleavage at the site of infection and cause local inflammatory responses by acting on local blood vessels. They cause increased blood flow, increased binding of phagocytes to endothelial cells, and increased vascular permeability, leading to the accumulation of fluid, plasma proteins, and cells in the local tissues. The complement and cells recruited by this inflammatory stimulus remove pathogen by enhancing the activity of phagocytes, which are themselves also directly stimulated by the anaphylatoxins. C5a is more potent than C3a.

C5a also acts directly on neutrophils and monocytes to increase their adherence to blood vessel walls, and acts as a chemoattractant to direct their migration toward sites where complement is being fixed. It also increases the capacity of these cells for phagocytosis, as well as raising the expression of CR1 and CR3 on their surfaces. In these ways, the anaphylatoxins act in concert with other complement components to speed the destruction of pathogens by phagocytes.

2-8   Several classes of plasma protein limit the spread of infection

In addition to complement, several other types of plasma protein impede the invasion and colonization of human tissues by microorganisms. Damage to blood vessels activates the **coagulation system**, a cascade of plasma enzymes that forms blood clots, which immobilize microorganisms and prevent them from entering the blood and lymph, as well as reducing the loss of blood and fluid. Platelets are a major component of blood clots, and during clot formation they release a variety of highly active substances from their storage granules. These include prostaglandins, hydrolytic enzymes, growth factors, and other mediators that stimulate various cell types to contribute to antimicrobial defense, wound healing, and inflammation. Further mediators, including the vasoactive peptide bradykinin, are produced by the **kinin system**, a second enzymatic cascade of plasma proteins that is triggered by tissue damage. By causing vasodilation, bradykinin increases the supply of the soluble and cellular materials of innate immunity to the infected site.

As part of their invasive mechanism, many pathogens carry proteases on their surface or secrete them. These proteases break down human tissues and aid the pathogen's dissemination, and can inactivate antimicrobial proteins. In some instances the proteases are made by the pathogen; in others the pathogen hijacks a human protease for its own purposes. One example is the bacterium *Streptococcus pyogenes* (see Figure 1.4), which acquires the human protease plasmin on its surface. To counter such invasive mechanisms, human secretions and plasma contain **protease inhibitors**. About 10% of serum proteins are protease inhibitors. Among these are the α_2-macroglobulins, glycoproteins with a molecular mass of 180 kDa that circulate as monomers, dimers, and trimers and are able to inhibit a broad range of proteases. α_2-Macroglobulins have structural similarities to complement component C3, including the presence of internal thioester bonds. The α_2-macroglobulin molecule lures a protease with a bait region that it is allowed to cleave. This activates the α_2-macroglobulin, producing two effects: first, the thioester is used to attach the protease covalently to the α_2-macroglobulin; second, the α_2-macroglobulin undergoes a conformational change by which it envelops the protease and prevents it attacking other substrates (Figure 2.16). The resulting complexes of protease and α_2-macroglobulin are rapidly cleared from the circulation by a receptor present on hepatocytes, fibroblasts, and macrophages.

Figure 2.16 α_2-Macroglobulin inhibits potentially damaging proteases. Microbial invasion and colonization of human tissues is often aided by the actions of microbial proteases. In response, human plasma is loaded with protease inhibitors of different kinds. One class, the α_2-macroglobulins, contain a highly reactive thioester bond. An α_2-macroglobulin first traps the microbial protease with a 'bait' region. When the protease cleaves the bait, the α_2-macroglobulin binds the protease covalently through activation of the thioester group. It enshrouds the protease so that it cannot access other protein substrates, even though the protease is still catalytically active.

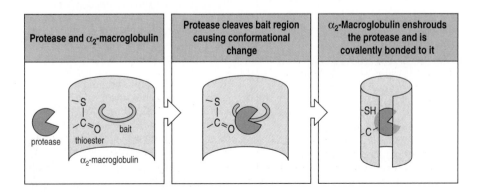

| Protease and α_2-macroglobulin | Protease cleaves bait region causing conformational change | α_2-Macroglobulin enshrouds the protease and is covalently bonded to it |

2-9   Defensins are a family of variable antimicrobial peptides

As touched on in Chapter 1 (see Figure 1.6, p. 8) antimicrobial peptides contribute to the innate immune response. The major family of human antimicrobial peptides comprises the **defensins**, peptides of 35–40 amino acids that are rich in positively charged arginine residues and which characteristically have three intra-chain disulfide bonds. They divide into two classes—the α-**defensins** and the β-**defensins**. The defensin molecule is amphipathic in character, meaning that its surface has both hydrophobic and hydrophilic regions. This property allows defensin molecules to penetrate microbial membranes and disrupt their integrity—the mechanism by which they destroy bacteria, fungi, and enveloped viruses.

The α-defensins are expressed mainly by neutrophils, the predominant phagocytes of innate immunity, and by Paneth cells, specialized epithelial cells of the small intestine that are situated at the base of the crypts between the intestinal villi (Figure 2.17). In addition to α-defensins HD5 and HD6 (also called **cryptdins**), Paneth cells secrete other antimicrobial agents, including lysozyme, that contribute to innate immunity. The β-defensins are expressed by a broad range of epithelial cells, in particular those of the skin, the respiratory tract and the urogenital tract. To prevent defensins from disrupting human cells they are synthesized as part of longer, inactive polypeptides that are then cleaved to release the active fragment. Even then, they function poorly under the physiological conditions in which they are actually produced, needing the lower ionic strength of sweat, tears, the gut lumen or the phagosome to become fully active. In neutrophils, the defensins kill pathogens that have been taken up by phagocytosis. In the gut, the defensins secreted by Paneth cells kill enteric pathogens and maintain the normal gut flora.

The set of defensins made varies from one individual to another. There are at least six α-defensins and four β-defensins (Figure 2.18). The regions of the human genome that encode the defensins are even more variable because individuals differ in their number of copies of a defensin gene: from 2 to 14 copies for α-defensin genes and from 2 to 12 copies for β-defensin genes. The gene copy number determines the amount of protein made, with the result that the arsenal of defensins that a neutrophil carries varies from one person to another. Variation in the amino acid sequences of defensins correlates with their different skills in killing microorganisms. For example, the β-defensin HBD2 specializes in killing Gram-negative bacteria, whereas its relative HBD3 kills both Gram-positive and Gram-negative bacteria. The terms Gram-positive and Gram-negative are traditionally used in bacteriology to distinguish between two large classes of medically important bacteria: one stains purple with the Gram stain; the other does not retain this stain (see Figure 1.4). The defensins can also differ in the epithelial surfaces they protect, the α-defensin HD5 being secreted in the female urogenital tract, and the β-defensin HBD1 being secreted in the respiratory tract as well as the urogenital tract.

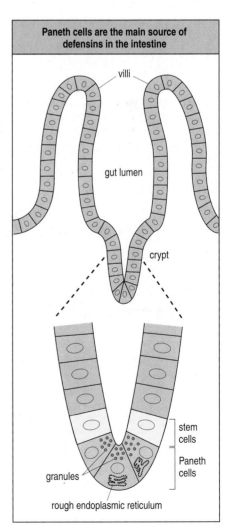

Paneth cells are the main source of defensins in the intestine

villi

gut lumen

crypt

stem cells

Paneth cells

granules

rough endoplasmic reticulum

Figure 2.17 Paneth cells are located in the crypts of the small intestine. The α-defensins HD5 and HD6, also known as cryptdins, are made only by Paneth cells. The upper part of the diagram shows the location of a crypt between two villi in the distal part of the small intestine (ileum). The lower part of the diagram shows the Paneth cells at the base of the crypt and the epithelial stem cells that give rise to them. Paneth cells also secrete other antimicrobial factors, including lysozyme and phospholipase A2. Although they are of epithelial, not hematopoietic, origin, Paneth cells can be considered cells of the immune system.

Defensin		Site of synthesis	Tissues defended	Regulation of synthesis
Class	Name			
α	HNP1	Neutrophils > monocytes, macrophages, NK cells, B cells, and some T cells	Intestinal epithelium, placenta, and cervical mucus plug	Constitutive
α	HNP2			
α	HNP3			
α	HNP4	Neutrophils	Not determined	Constitutive
α	HD5	Paneth cells > vaginal epithelial cells	Salivary glands, gastrointestinal tract, eyes, female genital tract, and breast milk	Constitutive and induced by sexually transmitted infection
α	HD6	Paneth cells	Salivary glands, gastrointestinal tract, eyes, and breast milk	
β	HBD1	Epithelial cells > monocytes, macrophages, dendritic cells, and keratinocytes	Gastrointestinal tract, respiratory tract, urogenital tract, skin, eyes, salivary glands, kidneys, and blood plasma	Constitutive and induced by infection
β	HBD2			
β	HBD3			
β	HBD4	Epithelial cells	Stomach (gastric antrum) and testes	

Figure 2.18 Human defensins are variable antimicrobial peptides. Defensins are small antimicrobial peptides that are found at epithelial surfaces and in the granules of neutrophils. They form two families: the α-defensins and the β-defensins. HNP, human neutrophil protein; HD, human defensin; HBD, human β-defensin. The gastric antrum is that part of the stomach nearer the outlet and does not secrete acid.

Under pressure from human innate immunity, pathogens evolve ways to escape from attack by defensins. In return, the pressures that those pathogens impose on the human immune system selects for new variants of human defensins that kill the pathogens more efficiently. This evolutionary game of cat-and-mouse never ends, and its consequence is the abundance and variability of the defensin genes accumulated by the human population. In comparison with most other genes in the genome, the defensin genes evolve rapidly. Such instability, or plasticity, is not unique to the defensin genes but also occurs in some other families of genes that encode pathogen-binding proteins of innate immunity.

2-10   Innate immune receptors distinguish features of microbial structure

In their structure and biochemistry, microorganisms differ from animal cells in ways that have allowed the evolution of receptors on mammalian cells that recognize these differences. Macrophages express many such receptors that work in concert with the complement receptors to phagocytose bacteria and other pathogens (Figure 2.19). Many of the microbial ligands for the receptors of innate immunity are carbohydrates and lipids. The carbohydrates present on the surfaces of microorganisms have components and structures that are not present on eukaryotic cells, and are targets for many different receptors on innate immune cells. As a group, the receptors and plasma proteins that recognize carbohydrates are called **lectins**. Examples of lectins on the macrophage surface are the **mannose receptor** and the **glucan receptor**. The aptly named **scavenger receptor** is a phagocytic receptor of macrophages that is not a lectin; it binds to an assortment of ligands that share the property of being negatively charged. Ligands for the scavenger receptor include sulfated polysaccharides, nucleic acids and the phosphate-containing lipoteichoic

Figure 2.19 Macrophages have many different cell-surface receptors by which they recognize pathogens. The mannose, glucan, and scavenger receptors are phagocytic receptors that bind microbial consitutents not found in human cells. Binding to such receptors results in the internalization of the pathogen by phagocytosis and its destruction in a phagolysosome. The Toll-like receptor (TLR) represents a class of signaling receptors that detect the presence of a wide variety of microbial components. CD14 is a lectin that binds the lipopolysaccharide of Gram-negative bacteria and becomes associated with one of the TLRs. CR3 is a receptor for the complement component iC3b.

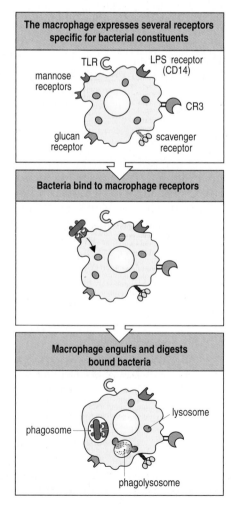

acids present in the cell walls of Gram-positive bacteria. The surfaces of Gram-positive and Gram-negative bacteria are composed of quite different types of macromolecule and are recognized by different macrophage receptors.

The complement receptors CR3 and CR4 recognize several microbial products in addition to their iC3b ligand (see Section 2-5), including **bacterial lipopolysaccharide** (**LPS**), a major component of the surface of Gram-negative bacteria, the lipophosphoglycan of the protozoan parasite *Leishmania*, the filamentous hemagglutinin, a protein on the surface of the bacterium *Bordetella pertussis*, and cell-surface structures on pathogenic yeasts such as *Candida* and *Histoplasma*. Many of the ligands for these receptors are present in regular arrays at the microbial surface, facilitating the simultaneous engagement of many receptors and an irreversible attachment of the pathogen to the macrophage surface. The ligands also tend to be molecules that are common to particular groups of pathogen and have been stable over evolutionary time. They therefore provided a common and constant target against which immune-system receptors could be selected and refined.

The binding of these macrophage receptors to their microbial ligands initiates a process of engulfment called **receptor-mediated endocytosis**, in which the receptor-bound pathogen is surrounded by the macrophage membrane and internalized into a membrane-bounded vesicle called an **endosome** or **phagosome**. Phagosomes then fuse with the cellular organelles called **lysosomes** to form **phagolysosomes**, vesicles that are loaded with degradative enzymes and toxic substances that destroy the pathogen (see Figure 2.19).

As well as the phagocytic receptors, the macrophage carries another class of receptors whose job is not to promote phagocytosis but to send signals into the interior of the cell when pathogens are detected. The signals activate the macrophage to make and secrete small biologically active proteins called **cytokines**, which recruit other immune-system cells into the infected tissue, where they work together with the macrophages to limit the spread of infection (see Section 1-4). These cytokines are not initially present as part of the macrophages's fixed defenses of innate immunity, but their synthesis is induced by the presence of pathogens. In this way additional defenses of innate immunity are mobilized as needed if the infection gains strength. Chief among the signaling receptors of innate immunity are the Toll-like receptors (TLRs), which are described in the next two sections.

2-11 Toll-like receptors sense the presence of infection

The **Toll-like receptors** (**TLRs**) are a family of signaling receptors, each of which is specific for a different set of microbial products. Toll-like receptors are expressed by different types of cell, allowing the type of innate immune response to be varied according to the type of pathogen and the site of infection. Macrophages express TLR4, which has specificity for the bacterial lipopolysaccharide (LPS) and related compounds present on the outside of Gram-negative bacteria. In the presence of bacterial infection TLR4 sends signals to the macrophage's nucleus that change the pattern of gene expression. In particular, the genes for cytokines that induce innate immune responses and inflammation at the site of infection are switched on. These cytokines are known as **inflammatory cytokines**. The stimulation of Toll-like receptors by microbial ligands at an early phase of infection is not only essential for the innate immune response but also provides the conditions necessary for the adaptive immune response should it be needed.

Toll-like receptors are transmembrane proteins composed of an extracellular domain that recognizes the pathogen and a cytoplasmic signaling domain that conveys that information to the inside of the cell. The pathogen-recognition domain consists of a repeated motif of 20–29 amino acid residues that is rich in the hydrophobic amino acid leucine and is termed a leucine-rich repeat region (LRR). The Toll-like receptor proteins vary in their number of LRRs, which together with other sequence differences account for the receptors' differing specificities. In three dimensions the pathogen-recognition domain forms a horseshoe-shaped structure (Figure 2.20). The cytoplasmic domain of the Toll-like receptor is called the Toll–interleukin receptor (TIR) domain because it is present in both Toll-like receptors and the receptor for interleukin 1, one of the inflammatory cytokines made by macrophages in response to signaling through TLR4.

Humans have ten *TLR* genes, distributed between five chromosomes, each encoding a different Toll-like receptor polypeptide. Some Toll-like receptors, such as TLR4, consist of homodimers of a single polypeptide, whereas others, such as TLR1:TLR2, are heterodimers consisting of two different polypeptides. The different receptors and the microbial products they recognize are listed in Figure 2.21. As well as responding to the LPS of Gram-negative bacteria, TLR4 also responds to components of other pathogens that are chemically or structurally related to LPS. Other members of the TLR family sense different microbial constituents; in each case they are molecules common to groups of pathogens and are not found in human cells. For example, TLR3 senses double-stranded RNA, which is present in many viral infections, TLR2:TLR6 detects zymosan, which is derived from yeast cell walls, and TLR9 detects the unmethylated CpG nucleotide motifs that are abundant in bacterial and viral genomes but not in human DNA (see Figure 2.21). So although the number of human Toll-like receptors is limited, they recognize features that are typical of all the different groups of pathogens, and thus can detect the presence of many different species of microorganisms.

Because the recognition of microbial components by Toll-like receptors can involve the participation of other cofactors, called co-receptors, Toll-like

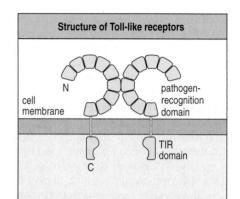

Structure of Toll-like receptors

Figure 2.20 Toll-like receptors sense infection with a horseshoe-shaped structure. A Toll-like receptor (TLR) protein is a transmembrane polypeptide with a Toll–interleukin receptor (TIR) signaling domain on the cytoplasmic side of the membrane and a horseshoe-shaped sensor domain on the other side. Functional receptors can be homodimers (as shown here) or heterodimers of TLR polypeptides.

Recognition of microbial products through Toll-like receptors				
Receptor	**Ligands**	**Microorganisms recognized**	**Cells carrying receptor**	**Cellular location of receptor**
TLR1:TLR2 heterodimer	Lipopeptides	Bacteria	Monocytes, dendritic cells, eosinophils, basophils, mast cells	Plasma membrane
	GPI	Parasites e.g., trypanosomes		
TLR2:TLR6 heterodimer	Lipoteichoic acid	Gram-positive bacteria		Plasma membrane
	Zymosan	Yeasts (fungi)		
TLR3	Double-stranded viral RNA	Viruses e.g., West Nile virus	NK cells	Endosomes
TLR4:TLR4 homodimer	Lipopolysaccharide	Gram-negative bacteria	Macrophages, dendritic cells, mast cells, eosinophils	Plasma membrane
TLR5	Flagellin	Motile bacteria having a flagellum	Intestinal epithelium	Plasma membrane
TLR7	Single-stranded viral RNAs	Viruses e.g., human immunodeficiency virus (HIV)	Plasmacytoid dendritic cells, NK cells, eosinophils, B cells	Endosomes
TLR8	Single-stranded viral RNAs	Viruses e.g., influenza	NK cells	Endosomes
TLR9	Unmethylated CpG-rich DNA	Bacteria Viruses e.g., herpes viruses	Plasmacytoid dendritic cells, B cells, eosinophils, basophils	Endosomes
TLR10 homodimer and heterodimers with TLR1 and 2	Unknown		Plasmacytoid dendritic cells, basophils, eosinophils, B cells	Unknown

Figure 2.21 The human Toll-like receptors allow the detection of many different types of infection. Each of the known Toll-like receptors (TLRs) seems to recognize one or more characteristic features of microbial macromolecules, but TLR5 is the only TLR so far for which a direct interaction with a microbial product, the bacterial protein flagellin, has been demonstrated. There are 10 TLR genes in humans, each encoding a distinct TLR polypeptide. Some TLRs are known to be heterodimers of these polypeptides; some, such as TLR4, are known to act only as homodimers. The Toll-like receptors take their name from their structural similarities to a receptor called Toll in the fruitfly *Drosophila melanogaster*, which is involved in the adult fly's defense against infection.

receptors are often described as 'sensing' or 'detecting' the presence of a microbial component, rather than directly binding to it. Families of receptors that each detect a different microbial ligand are characteristic of innate immune systems in animals and plants. Toll-like receptors are present in both vertebrates and invertebrates, telling us they are an ancient system for detecting infection.

Toll-like receptors such as TLR5, TLR4, TLR1:TLR2, and TLR2:TLR6, which sense proteins, carbohydrates, and lipids characteristic of microbial cell surfaces, are located on the surfaces of human cells (see Figure 2.21). They reside in the plasma membrane, where direct contact can be made with extracellular pathogens and their distinctive surface components. In contrast, TLR3, TLR7, TLR8 and TLR9, which detect the nucleic acids of pathogens, are not present on the surfaces of human cells but reside in the membranes of endosomes within the cytoplasm. In these vesicles, the DNA and RNA released from pathogens taken up from the extracellular environment and degraded are available for detection (Figure 2.22).

Responses to LPS mediated by TLR4 are important in the body's innate immune defenses against Gram-negative bacteria, many of which are potential pathogens, and TLR4 is the most thoroughly studied Toll-like receptor. When LPS is released from bacterial surfaces it is bound on the macrophage surface by a protein called CD14, which acts as a co-receptor to TLR4. Alternatively, LPS in the plasma can be picked up by a soluble LPS-binding protein and delivered to CD14 on the macrophage surface. The TLR4 dimer associates with a protein called MD2 and together they form a complex with CD14 and LPS (Figure 2.23). This complex then generates intracellular signals via the cytoplasmic signaling domain of TLR4. The result of these signals is that the macrophage switches on a set of genes encoding the inflammatory cytokines. The cytokine proteins are synthesized and secreted into the extracellular environment, where they work together to change the environment at the infected site in ways that help limit the spread of infection.

2-12 Signaling through Toll-like receptors leads to two different cytokine responses

Macrophages in a tissue infected with Gram-negative bacteria recognize extracellular LPS with the cell-surface complex of TLR4, MD2, and CD14 (see Figure 2.23). This event triggers intracellular reactions that lead to the macrophage's secreting inflammatory cytokines. The first part of this pathway occurs in the cytoplasm and leads to the activation of the transcription factor **nuclear factor κB (NFκB)**. This transcription factor has a major role in both innate and adaptive immune responses. When not required, NFκB is held in the cytoplasm in an inactive complex with the inhibitor of κB (IκB). Activation of NFκB requires its release from this complex and its translocation from the cytoplasm to the nucleus. The second part of the pathway takes place in the nucleus, where NFκB initiates the transcription of genes encoding inflammatory cytokines.

How TLR4 signaling leads to the mobilization of NFκB is shown in Figure 2.24. Extracellular recognition of LPS causes the TIR domain of TLR4 inside the cell to bind to a similar TIR domain in the protein MyD88. MyD88 is an example

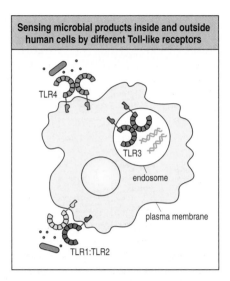

Sensing microbial products inside and outside human cells by different Toll-like receptors

Figure 2.22 Different Toll-like receptors sense bacterial infection outside the cell and viral infection inside the cell. The TLR4 homodimer and the TLR1:TLR2 heterodimer at the cell surface are shown sensing a bacterial infection, and the homodimer of TLR3 in an endosomal vesicle is detecting a viral infection.

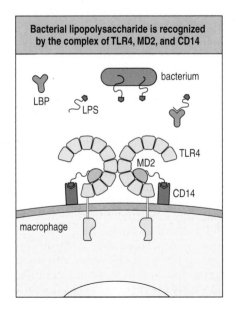

Bacterial lipopolysaccharide is recognized by the complex of TLR4, MD2, and CD14

Figure 2.23 TLR4 recognizes bacterial lipopolysaccharide with help from other proteins. Bacterial lipopolysaccharide (LPS) is recognized by a complex of the TLR4, MD2, and CD14 proteins at the cell surface. MD2 is a soluble protein that associates with the extracellular domains of TLR4, but not with other TLR family members, and confers sensitivity to LPS. The soluble lipopolysaccharide-binding protein (LBP) can also deliver LPS to this cell-surface complex.

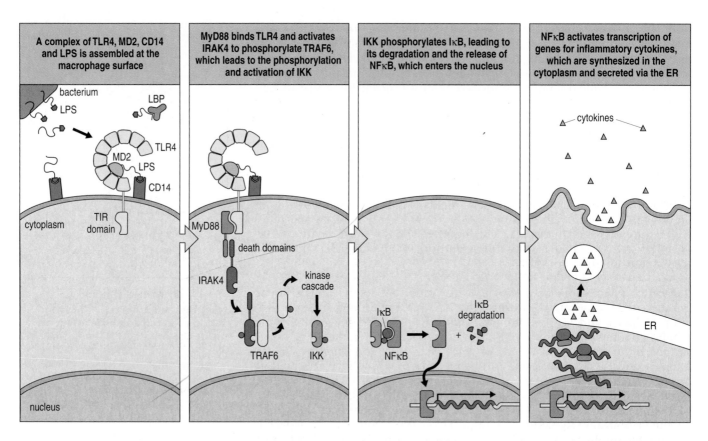

| A complex of TLR4, MD2, CD14 and LPS is assembled at the macrophage surface | MyD88 binds TLR4 and activates IRAK4 to phosphorylate TRAF6, which leads to the phosphorylation and activation of IKK | IKK phosphorylates IκB, leading to its degradation and the release of NFκB, which enters the nucleus | NFκB activates transcription of genes for inflammatory cytokines, which are synthesized in the cytoplasm and secreted via the ER |

Figure 2.24 Sensing of LPS by TLR4 on macrophages leads to activation of the transcription factor NFκB and the synthesis of inflammatory cytokines. First panel: LPS is detected by the complex of TLR4, CD14, and MD2 on the macrophage surface. Second panel: the activated receptor binds the adaptor protein MyD88, which binds the protein kinase IRAK4. IRAK4 binds and phosphorylates the adaptor TRAF6, which leads via a kinase cascade to the activation of IKK. Third panel: in the absence of a signal, the transcription factor NFκB is bound by its inhibitor, IκB, which prevents it from entering the nucleus. In the presence of a signal, activated IKK phosphorylates IκB, which induces the release of NFκB from the complex; IκB is degraded. NFκB then enters the nucleus where it activates genes encoding inflammatory cytokines. Fourth panel: cytokines are synthesized from cytokine mRNA in the cytoplasm and secreted via the endoplasmic reticulum (ER). This MyD88–NFκB pathway is also stimulated by the receptors for cytokines IL-1 and IL-18.

of an adaptor, a protein that acts as a bridge to bring other signaling proteins together. It does this by means of its TIR domain and a second domain, of a type known as a death domain, with which it recruits the next member of the pathway—a protein kinase called IRAK4. Death domains are so called because they were first identified in proteins involved in apoptosis, a normal process by which cells are killed in a tidy fashion that leaves no mess and does not stimulate the immune system.

Protein kinases are enzymes that phosphorylate other proteins. The phosphorylation can alter the activity of the target protein, or enable it to bind to specific proteins, or both. Because of this, protein kinases are key components of intracellular signaling pathways. IRAK4 has a death domain with which it binds the death domain of MyD88. The kinase is activated by this binding and phosphorylates itself, whereupon it dissociates from the complex and phosphorylates another adaptor protein called TRAF6. Additional steps in the pathway eventually lead to the activation of a kinase complex called the inhibitor of κB kinase (IKK). This phosphorylates IκB, causing its dissociation from the complex with NFκB and its eventual destruction. Once released from its inhibitor, NFκB moves into the nucleus, where it directs the activation of genes for cytokines, adhesion molecules and other proteins that expand and intensify the macrophage's effector functions.

Children with a rare genetic disease called **X-linked hypohydrotic ectodermal dysplasia and immunodeficiency** or **NEMO deficiency** lack one of the subunits of IKK and thus have impaired activation of NFκB. This makes them susceptible to bacterial infections because macrophage activation through TLR4 signaling is inefficient. The gene for the kinase subunit, called IKKγ or NEMO, is on the X chromosome, and so this syndrome is more frequent in boys, who inherit a single copy of the X chromosome, than in girls, who inherit two copies, both of which have to be defective for disease to be present. NFκB has functions in development as well as in immunity, and the other consequences of IKKγ deficiency are abnormalities in the development of tissues derived from embryonic ectoderm: the skin, teeth, and hair (Figure 2.25).

Most human Toll-like receptors signal through the pathway that is initiated by the binding of the MyD88 adaptor protein and activates NFκB. An exception is TLR3, which uses another signaling pathway that leads to the activation of the transcription factor interferon response factor 3 (IRF3) and to the production of antiviral cytokines called type I interferons. This pathway is specialized to sense and respond to viral infections. TLR4 also uses this second pathway and is the only human Toll-like receptor that can use both pathways (Figure 2.26). The second pathway does not involve MyD88; instead, it uses two alternative adaptor proteins called the Toll receptor-associated activator of interferon (TRIF) and the Toll receptor-associated molecule (TRAM). These adaptors form a complex with TLR3 or TLR4 after they have detected their ligands, and initiate a signaling pathway that involves TRAF3, which is related to TRAF6, and a kinase cascade (in which a series of protein kinases phosphorylate and activate each other) that leads to the phosphorylation of IRF3 in the cytoplasm (see Figure 2.26). Phosphorylated IRF3 enters the nucleus, where it directs the transcription of the genes for type I interferons, cytokines that are central to the innate immune response to infection with viruses and intracellular bacteria.

These signaling pathways are currently the best understood of the pathways used by TLRs. They illustrate how macrophages and other cells having TLRs can tailor the innate immune response to different types of infection. For an extracellular bacterial infection, the production of inflammatory cytokines is more likely to eliminate the infection, whereas for a viral infection it is the production of interferons that is likely to do so. This distinction is reflected in the disease susceptibilities of people deficient for the kinase IRAK4. Because they activate NFκB poorly, the ability to make inflammatory cytokines is impaired and these patients suffer from recurrent infections with encapsulated bacteria. In contrast, they maintain good responses to most common viral infections, presumably because the ability of their TLR3 and TLR4 to activate IRF3 and produce type I interferons is normal.

2-13 Activation of resident macrophages induces inflammation at sites of infection

On sensing the presence of pathogens through TLR4 and other receptors, macrophages are stimulated to secrete a battery of cytokines and other substances that recruit effector cells, prominently neutrophils, into the infected

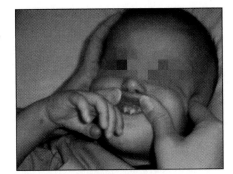

Figure 2.25 Infant with X-linked ectodermal dysplasia and immunodeficiency. This condition is caused by impairment of NFκB activation as a result of a lack of a functional IKKγ polypeptide. As well as immunological deficiencies, the lack of NFκB activation leads to developmental defects. The physical features of patients with this syndrome include deep-set eyes, fine or sparse hair, and conical or missing teeth. Photograph courtesy of F. Rosen and R. Geha.

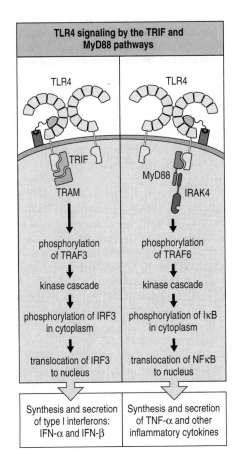

Figure 2.26 TLR4 activation can lead to the production of either inflammatory cytokines or antiviral type I interferons. TLR4 can stimulate two different intracellular signaling pathways, depending on whether the adaptor protein MyD88 or TRIF is recruited to the activated receptor. TLR4 signaling through TRIF leads to activation of the transcription factor interferon response factor 3 (IRF3) and the production of type I interferons. Signaling through MyD88 leads to activation of the transcription factor NFκB and the production of inflammatory cytokines such as IL-6 and TNF-α. TLR3 also uses the TRIF pathway.

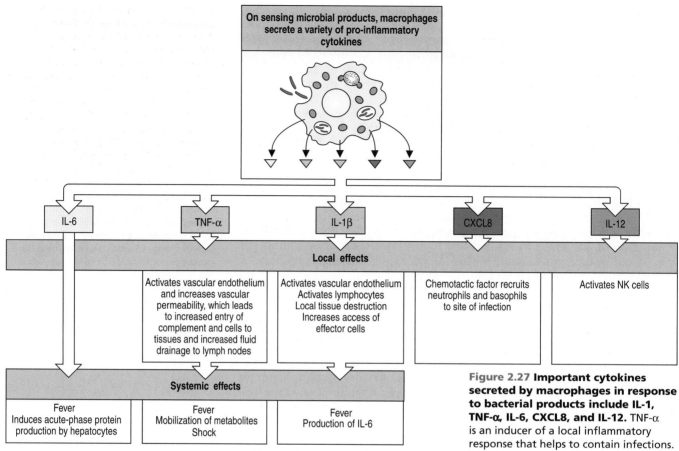

Figure 2.27 Important cytokines secreted by macrophages in response to bacterial products include IL-1, TNF-α, IL-6, CXCL8, and IL-12. TNF-α is an inducer of a local inflammatory response that helps to contain infections. It also has systemic effects, many of which are harmful. The chemokine CXCL8 is also involved in the local inflammatory response, helping to attract neutrophils to the site of infection. IL-1, IL-6, and TNF-α have a critical role in inducing the acute-phase response in the liver and induce fever, which favors effective host defense in various ways. IL-12 activates natural killer (NK) cells.

area. The infiltrating cells cause a state of **inflammation** to develop within the tissue. Inflammation describes the local accumulation of fluid accompanied by swelling, reddening, and pain. These effects stem from changes induced in the local blood capillaries that lead to an increase in their diameter (a process called dilation), reduction in the rate of blood flow, and increased permeability of the blood vessel wall. The increased supply of blood to the region causes the local redness and heat associated with inflammation. The increased permeability of blood vessels allows the movement of fluid, plasma proteins, and white blood cells from the blood capillaries into the adjoining connective tissues, causing the swelling and pain.

Translocation of NFκB to the macrophage nucleus (see Figure 2.24) initiates the transcription of various cytokine genes. Cytokines are small proteins with a molecular mass of about 25kDa that are made by a cell in response to an external stimulus and influence other cells by binding to a specific receptor on their surfaces. Prominent cytokines produced by activated macrophages are **IL-1**, **IL-6**, **CXCL8**, **IL-12**, and **tumor necrosis factor-α** (**TNF-α**). These inflammatory cytokines have powerful effects that can be localized to the infected tissue or can be manifested systemically throughout the body (Figure 2.27).

CXCL8 (previously called IL-8) is one of a large family of about 40 chemoattractant cytokines, or **chemokines**. Chemokines are messengers that direct the flow of leukocyte traffic; they differ in the type of cell or tissue that makes them and in the type of cell they attract. Some chemokines, including CXCL8, attract leukocytes into sites of tissue damage or infection. Others direct the traffic of leukocytes during their development and during their recirculation through lymphoid tissues. Chemokines are small, structurally similar proteins of about 60–140 amino acids. Two major subfamilies are defined on the basis

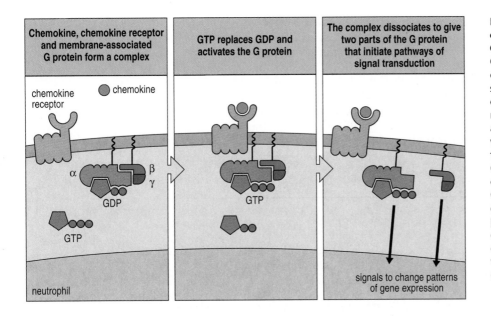

Chemokine, chemokine receptor and membrane-associated G protein form a complex	GTP replaces GDP and activates the G protein	The complex dissociates to give two parts of the G protein that initiate pathways of signal transduction

Figure 2.28 Chemokines bind to chemokine receptors that are G-protein-coupled receptors. Chemokine receptors are a family of seven-span receptors that have seven transmembrane helices. When a chemokine such as CXCL8 binds to its receptor, the receptor associates with an intracellular GTP-binding (G) protein, which in its inactive state consists of three polypeptides (α, β, and γ) and has GDP bound. On association with the chemokine receptor GDP is replaced by GTP which leads to dissociation of the α chain of the G protein from the β and γ chain. The α chain, and to a lesser extent the β and γ chain, bind to other cellular proteins that generate signals which change the cell's pattern of gene expression.

of pairs of cysteine residues, which are either adjacent (CC) or separated by another amino acid (CXC). Cells are attracted from the blood into infected tissue by following a concentration gradient of chemokine produced by cells within the infected site. Chemokines interact with their target cells by binding to specific cell-surface receptors, which in humans comprise a family of 16 seven-span transmembrane proteins that signal through associated GTP-binding proteins (Figure 2.28).

The principal function of CXCL8, a CXC chemokine, is to recruit neutrophils from the blood into infected areas. Circulating neutrophils express two chemokine receptors, CXCR1 and CXCR2, which will bind CXCL8 emanating from an infected tissue. Interaction with a chemokine has two distinct effects on the targeted leukocyte: first, the cell's adhesive properties are altered so that it can leave the blood and enter tissue; second, its movement is guided toward the center of infection along a gradient of the chemokine, present both in solution and attached to the extracellular matrix and endothelial cell surfaces. Chemokines have structural and functional similarities to the defensins (see Section 2-9); some chemokines have antimicrobial activity, whereas some defensins have chemoattractant properties and bind to chemokine receptors.

The cytokine IL-12 serves to activate a class of lymphocyte called **natural killer (NK) lymphocytes**, which enter infected sites soon after infection. NK cells are lymphocytes of innate immunity that specialize in defense against viral infections. The cytokines IL-1 and TNF-α facilitate the entry of neutrophils, NK cells, and other effector cells into infected areas by inducing changes in the endothelial cell walls of the local blood vessels. Other effector molecules released by macrophages are plasminogen activator, phospholipase, prostaglandins, oxygen radicals, peroxides, nitric oxide, leukotrienes, and platelet-activating factor (PAF), which all contribute to inflammation and tissue damage. In the course of complement activation, the soluble complement fragments C3a and C5a recruit neutrophils from the blood into infected tissues and stimulate mast cells to degranulate, releasing the inflammatory molecules histamine and TNF-α, among others. Molecules involved in the induction of inflammation are known generally as **inflammatory mediators**. The combined effect of all this activity is to produce a local state of inflammation with its characteristic symptoms.

The TNF-α released by macrophages as a result of Toll-like receptor stimulation can have both beneficial and harmful consequences. In response to

TNF-α, vascular endothelial cells make platelet-activating factor, which triggers blood clotting and blockage of the local blood vessels. This restricts the leakage of plasma from the blood and prevents pathogens from entering the blood and disseminating infection throughout the body, a condition known as **systemic infection**. If an infection does spread to the blood, as can occur in patients who have suffered severe burns and loss of the skin's protective barrier, bacterial endotoxins such as LPS provoke the widespread production of TNF-α, which then acts in ways that can become catastrophic (Figure 2.29). Infections of the blood are known as **sepsis** or **septicemia**.

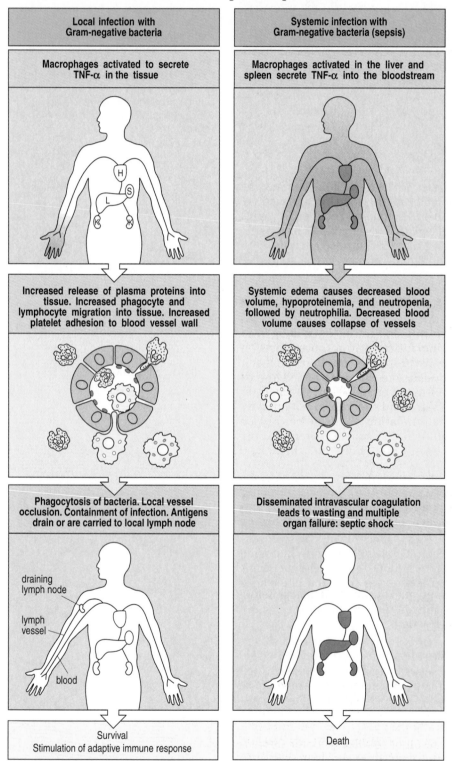

Figure 2.29 **TNF-α released by macrophages induces protection at the local level but can lead to catastrophe when released systemically.** The panels on the left describe the causes and consequences of the release of TNF-α within a local area of infection. In contrast, the panels on the right describe the causes and consequences of the release of TNF-α throughout the body. The initial effects of TNF-α are on the endothelium of blood vessels, especially venules. It causes increased blood flow, vascular permeability, and endothelial adhesiveness for white blood cells and platelets. These events cause the blood in the venules to clot, preventing the spread of infection and directing extracellular fluid to the lymphatics and lymph nodes, where the adaptive immune response is activated. When an infection develops in the blood, the systemic release of TNF-α and the effect it has on the venules in all tissues simultaneously induce a state of shock that can lead to organ failure and death. H, heart; K, kidney; L, liver; S, spleen.

A systemic bacterial infection induces macrophages in the liver, spleen, and other sites to release TNF-α, which causes the dilation of blood vessels and massive leakage of fluid into tissues throughout the body, leading to a profound state of shock called **septic shock**. One symptom of septic shock is widespread blood clotting in capillaries—disseminated intravascular coagulation (DAC)—which exhausts the supply of clotting proteins. More critically, septic shock frequently leads to the failure of vital organs such as the kidneys, liver, heart, and lungs, which are soon compromised by the lack of a normal blood supply. Consequently, septic shock has a high mortality rate. It causes the death of more than 100,000 people in the United States each year, with Gram-negative bacteria being the most common trigger.

The role of TLR4 in defense against infection with Gram-negative bacteria is shown by the association of a TLR4 variant with an increased risk of septic shock. People who carry one defective copy and one good copy of the *TLR4* gene (that is, they are heterozygous for the *TLR4* gene) are over-represented in patients suffering septic shock compared with the population overall. Indeed, the only person known to have been homozygous for this TLR4 variant (that is, carrying two copies of the defective gene) died in adolescence of septic shock following a kidney infection with the Gram-negative commensal bacterium *Escherichia coli*. The variant TLR4 protein has a glycine residue at position 299 in the amino acid sequence instead of the asparagine found in the common form of TLR4, and generates a weaker response to LPS, making it more likely that the bacterial infection will become systemic. Once the infection is systemic, TNF-α will be produced throughout the body at sufficient levels to induce shock.

2-14 Neutrophils are dedicated phagocytes that are summoned to sites of infection

By engulfing and killing microorganisms, phagocytic cells are the principal means by which the immune system destroys invading pathogens. The two kinds of phagocyte that serve this purpose—the macrophage and the neutrophil—have distinct and complementary properties. Macrophages are long-lived: they reside in the tissues, work from the very beginning of infection, raise the alarm, and have functions other than phagocytosis. **Neutrophils**, in contrast, are short-lived dedicated killers that circulate in the blood awaiting a call from a macrophage to enter infected tissue.

Neutrophils are a type of granulocyte, having numerous granules in the cytoplasm, and are also known as **polymorphonuclear leukocytes** because of the variable and irregular shapes of their nuclei (see Figure 1.12, p. 13). Neutrophils were historically called microphages because they are smaller than macrophages. What they lack in size they more than make up for in number: they are the most abundant white blood cells, a healthy adult having some 50 billion in circulation at any time. This abundance combined with the short life span of the circulating neutrophil—less than 2 days—means that about 60% of the hematopoietic activity of the bone marrow is devoted to neutrophil production. Mature neutrophils are kept in the bone marrow for about 5 days before being released into the circulation; this constitutes a large reserve of neutrophils that can be called on at times of infection.

Neutrophils are excluded from healthy tissue, but at infected sites the release of inflammatory mediators attracts neutrophils to leave the blood and enter the infected area in large numbers, where they soon become the dominant phagocytic cell. Every day, some 3×10^9 neutrophils enter the tissues of the mouth and throat, the most contaminated sites in the body. The arrival of neutrophils is the first of a series of reactions, called the **inflammatory response**, by which cells and molecules of innate immunity are recruited into sites of wounding or infection. Although neutrophils are specialized for working under the anaerobic conditions that prevail in damaged tissues, they still die within a few hours after entry. In doing so, they form the creamy **pus** that characteristically develops at infected wounds and other sites of infection. This is why extracellular bacteria such as *S. aureus*, which are responsible for the superficial infections and abscesses that neutrophils tackle in large numbers, are known as pus-forming or **pyogenic** bacteria.

2-15   The homing of neutrophils to inflamed tissues involves altered interactions with vascular endothelium

The movements of leukocytes between blood and tissues, which are crucial to all aspects of the immune response, are determined by interactions between complementary pairs of **adhesion molecules**, one of which is expressed on the leukocyte surface, the other on the surface of vascular endothelial cells or other tissue cells. The adhesion molecules of the immune system comprise four structural classes of protein: **selectins**, cell-surface **mucins** called **vascular addressins**, **integrins**, and members of the **immunoglobulin superfamily** (Figure 2.30). Selectins are carbohydrate-binding proteins, or lectins, which have specificity for the oligosaccharides of different vascular addressins. Integrins comprise a large family of adhesion molecules with a common structure of α-chain and β-chain polypeptides; the complement receptors CR3 and CR4 (see Section 2-5) are examples of integrins. The extracellular parts of adhesion molecules in the immunoglobulin superfamily have compact protein modules about 100 amino acids in length; these are called immunoglobulin-like domains because they were first discovered in immunoglobulins (see Section 1-7). Whereas the ligands for selectins are carbohydrates, the ligands for integrins are proteins, many of which are immunoglobulin superfamily members (see Figure 2.30).

Neutrophils have surface receptors for inflammatory mediators such as the chemokine CXCL8 secreted by activated macrophages and the C5a anaphylatoxin cleaved from C5 during complement activation (see Section 2-6). Another neutrophil receptor binds to chemoattractants produced only by bacterial infections. This receptor binds only peptides containing *N*-formylmethionine, a common component of bacterial, but not of human, proteins. During an infection, these ligand–receptor interactions induce the expression of adhesion molecules on the neutrophil surface. Correspondingly, inflammatory mediators also induce expression of the ligands for these adhesion molecules by the endothelium of blood capillaries in and around the infected site. Vascular endothelium that has undergone these changes is said to be activated. Together, these changes enable neutrophils in the blood to bind to activated vascular endothelium within an infected site, to squeeze between the endothelial cells and enter the infected tissue. The further release of inflammatory mediators by the increasing numbers of neutrophils steadily increases the inflammation within the tissue.

The process by which neutrophils migrate out of blood capillaries and into tissues is called **extravasation**, and it occurs in four steps. The first is an interaction between circulating leukocytes and blood vessel walls that slows down the neutrophils. This interaction is mediated by the carbohydrate-binding selectins (see Figure 2.30). In healthy tissue, vascular endothelial cells contain granules, known as **Weibel–Palade bodies**, which contain **P-selectin**. On exposure to inflammatory mediators, including leukotriene LTB$_4$, C5a, and histamine, the P-selectin in the Weibel–Palade bodies is transported to the cell surface. A second selectin, **E-selectin**, is also expressed on the endothelial cell surface a few hours after exposure to LPS or TNF-α. The two selectins bind to a carbohydrate side chain on glycolipids and glycoproteins on the leukocyte cell surface. This carbohydrate is rich in sialic acid and is known as sialyl-Lewis[x] because it is one of the antigens in the Lewis blood group system.

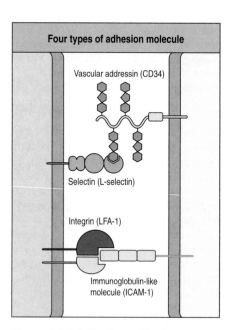

Four types of adhesion molecule

Vascular addressin (CD34)

Selectin (L-selectin)

Integrin (LFA-1)

Immunoglobulin-like molecule (ICAM-1)

Figure 2.30 Adhesion of leukocytes to vascular endothelium involves interactions between adhesion molecules of four structurally different types. These are the vascular addressins, the selectins, the integrins, and proteins containing immunoglobulin-like domains.

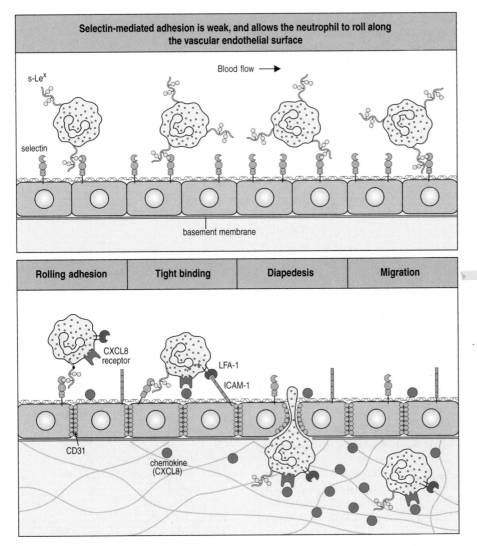

Selectin-mediated adhesion is weak, and allows the neutrophil to roll along the vascular endothelial surface

Blood flow →

s-Le^x

selectin

basement membrane

Rolling adhesion	Tight binding	Diapedesis	Migration

CXCL8 receptor

LFA-1

ICAM-1

CD31

chemokine (CXCL8)

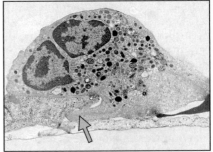

Figure 2.31 Neutrophils are directed to sites of infection through interactions between adhesion molecules. Inflammatory mediators and cytokines produced as the result of infection induce the expression of selectin on vascular endothelium, which enables it to bind leukocytes. The top panel shows the rolling interaction of a neutrophil with vascular endothelium as a result of transient interactions between selectin on the endothelium and sialyl-Lewis^x (s-Le^x) on the leukocyte. The bottom panel shows the conversion of rolling adhesion into tight binding and subsequent migration of the leukocyte into the infected tissue. The four stages of extravasation are shown. Rolling adhesion is converted into tight binding by interactions between integrins on the leukocyte (LFA-1 is shown here) and adhesion moleules on the endothelium (ICAM-1). Expression of these adhesion molecules is also induced by cytokines. A strong interaction is induced by the presence of chemoattractant cytokines (the chemokine CXCL8 is shown here) that have their source at the site of infection. They are held on proteoglycans of the extracellular matrix and cell surface to form a gradient along which the leukocyte can travel. Under the guidance of these chemokines, the neutrophil squeezes between the endothelial cells and penetrates the connective tissue (diapedesis). It then migrates to the center of infection along the CXCL8 gradient. The electron micrograph shows a neutrophil that has just started to migrate between adjacent endothelial cells but has yet to break through the basement membrane, which is at the bottom of the photograph. The blue arrow points to the pseudopod that the neutrophil is inserting between the endothelial cells. The dark mass in the bottom right-hand corner is an erythrocyte that has become trapped under the neutrophil. Photograph (× 5500) courtesy of I. Bird and J. Spragg.

These reversible interactions allow the neutrophils to adhere to the blood vessel walls and to 'roll' slowly along them by forming new adhesive interactions at the front of the cell while breaking them at the back (Figure 2.31, top panel).

The second step in extravasation depends on interactions between the integrins LFA-1 and CR3 on the neutrophil and adhesion molecules on the endothelium, for example ICAM-1, whose expression is also induced by TNF-α. Under normal conditions, LFA-1 and CR3 interact only weakly with endothelial adhesion molecules, but exposure to the CXCL8 coming from cells in the inflamed tissues induces conformational changes in the LFA-1 and CR3 on a rolling leukocyte that strengthen their adhesion. As a result, the neutrophil holds tightly to the endothelium and stops rolling (Figure 2.31, bottom panel).

In the third step, the neutrophil crosses the blood vessel wall. LFA-1 and CR3 contribute to this movement, as does adhesion involving the immunoglobulin superfamily protein CD31, which is expressed by both neutrophils and endothelial cells at their junctions with one another. The leukocyte squeezes between neighboring endothelial cells, a maneuver known as **diapedesis**, and reaches the basement membrane, a part of the extracellular matrix. It then crosses the basement membrane by secreting proteases that break down the membrane. The fourth and final step in extravasation is movement of the neutrophil toward the center of infection in the tissue. This migration is accomplished on the gradient of CXCL8, which originates within the infected site (see Figure 2.31).

At various stages in their maturation, activation, and execution of their effector functions, all types of white blood cell leave the blood and migrate to particular tissues, a process known as **homing**. All these migrations involve mechanisms analogous to those that control the entry of neutrophils into infected tissue. Cytokines and chemokines induce changes in adhesion molecules on white blood cells and vascular endothelium that determine where and when extravasation occurs.

2-16   Neutrophils are potent killers of pathogens and are themselves programmed to die

Neutrophils phagocytose microorganisms by mechanisms similar to those used by macrophages. Neutrophils have a range of phagocytic receptors that recognize microbial products as well as complement receptors that facilitate the phagocytosis of pathogens opsonized by complement fixation (Figure 2.32). The range of particulate material that neutrophils engulf is greater than that tackled by macrophages, as is the diversity of the microbicidal substances stored in their granules. Because mature neutrophils are programmed to die young, they devote more of their resources to the storage and delivery of antimicrobial weaponry than the longer-living macrophage.

Almost immediately after a pathogen has been engulfed by a neutrophil, a battery of degradative enzymes and other toxic substances is brought to bear upon it and death occurs quickly. Phagosomes containing recently captured microorganisms are fused with two types of preformed neutrophil granules: **azurophilic** (or **primary**) **granules**, and **specific** (or **secondary**) **granules** (Figure 2.33). The azurophilic granules are packed with proteins and peptides that can disrupt and digest microbes. These include lysozyme, defensins, myeloperoxidase, neutral proteases such as cathepsin G, elastase and proteinase 3, and a bactericidal/permeability-increasing protein that binds LPS and kills Gram-negative bacteria. Binding these proteins and peptides together in the granule is a negatively charged matrix of sulfated proteoglycans. The combination of this matrix and the acidity of the granule's interior sequesters the weaponry in a safe, inactive form until it is needed.

The specific granules contain unsaturated lactoferrin, which competes with pathogens for iron and copper by binding to proteins that contain these metals. They also contain lysozyme and several membrane proteins, including components of **NADPH oxidase**, an essential enzyme for neutrophil function that is assembled in the phagosome after fusion of the phagosome with the azurophilic and specific granules.

NADPH oxidase produces superoxide radicals that are converted into hydrogen peroxide by the enzyme superoxide dismutase (Figure 2.34). These reactions rapidly consume hydrogen ions and have the direct effect of raising the pH of the phagosome to 7.8–8.0 within 3 minutes of phagocytosis. At this pH the antimicrobial peptides and proteins become activated and attack the trapped pathogens. The pH of the phagosome then slowly goes down, reaching neutrality (pH 7.0) after 10–15 minutes. At this point some of the neutrophil's lysosomes fuse with the phagosome to form the phagolysosome. The lysosomes contribute a variety of degradative enzymes, collectively called acid hydrolases, that are active at the lower pH of the phagolysosome and ensure the continued and complete breakdown of the pathogen's macromolecules.

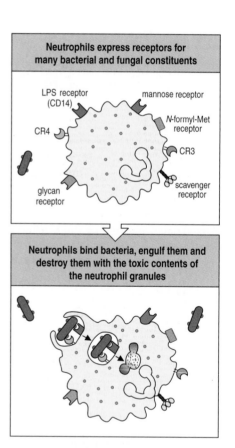

Figure 2.32 Bacteria binding to neutrophil receptors induce phagocytosis and microbial killing. Upper panel: the neutrophil has several different receptors for microbial products. Lower panel: the mechanism of phagocytosis for two such receptors, CD14 and CR4, which are specific for bacterial lipopolysaccharide (LPS). A bacterium binding to these receptors stimulates its phagocytosis and degradation.

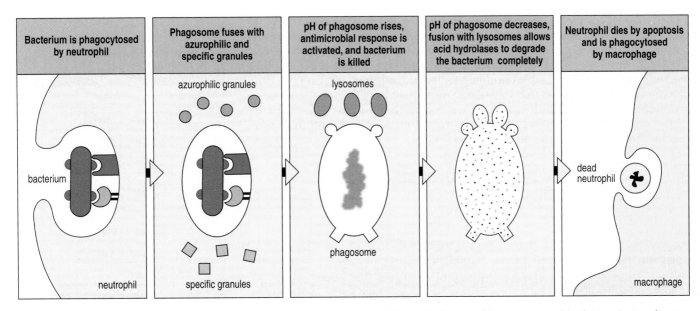

| Bacterium is phagocytosed by neutrophil | Phagosome fuses with azurophilic and specific granules | pH of phagosome rises, antimicrobial response is activated, and bacterium is killed | pH of phagosome decreases, fusion with lysosomes allows acid hydrolases to degrade the bacterium completely | Neutrophil dies by apoptosis and is phagocytosed by macrophage |

Figure 2.33 Killing of bacteria by neutrophils involves the fusion of two types of granule and lysosomes with the phagosome. After phagocytosis (first panel), the bacterium is held in a phagosome inside the neutrophil. The neutrophil's azurophilic granules and specific granules fuse with the phagosome, releasing their contents of antimicrobial proteins and peptides (second panel). NAPDH oxidase components contributed by the specific granules enable the respiratory burst to occur, which raises the pH of the phagosome. Antimicrobial proteins and peptides are activated and the bacterium is damaged and killed. A subsequent decrease in pH and the fusion of the phagosome with lysosomes containing acid hydrolases results in complete degradation of the bacterium. The neutrophil dies and is phagocytosed by a macrophage.

Powering the neutrophil's ferocious intracellular attack, which can kill both Gram-positive and Gram-negative bacteria, as well as fungi, is a transient increase in oxygen consumption called the **respiratory burst**. The products of the respiratory burst are several toxic oxygen species that can diffuse out of the cell and damage other host cells. To limit the damage, the respiratory burst is also accompanied by the synthesis of enzymes that inactivate these potent small molecules: one such enzyme is catalase, which degrades hydrogen peroxide to water and oxygen (see Figure 2.34).

The mature neutrophil cannot replenish its granule contents, so once they are used up the neutrophil dies by apoptosis and is ultimately phagocytosed by a macrophage (see Figure 2.33). The dependence of the body's defenses on neutrophils is well illustrated by **chronic granulomatous disease**, a genetic syndrome caused by defective forms of the genes encoding NADPH oxidase subunits. In the absence of functional NADPH oxidase, there is no respiratory burst after phagocytosis and the pH of the neutrophil's phagosome cannot be

Figure 2.34 Killing of bacteria by neutrophils is dependent on a respiratory burst. In the absence of infection the antimicrobial proteins and peptides in neutrophil granules are kept inactive at low pH. After the granules fuse with the phagosome the pH within the phagosome is raised through the first two reactions, involving the enzymes NADPH oxidase and superoxide dismutase. Each round of these reactions eliminates a hydrogen ion, thereby reducing the acidity of the phagosome. A product of the two reactions is hydrogen peroxide, which has the potential to damage human cells. (In hair salons and the manufacture of paper it is used as a powerful bleach.) The third reaction, involving catalase, the most efficient of all enzymes, promptly gets rid of the hydrogen peroxide produced during the neutrophil's respiratory burst, raising the pH of the phagosome and enabling activation of the antimicrobial peptides and proteins.

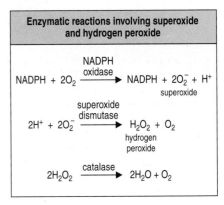

Enzymatic reactions involving superoxide and hydrogen peroxide

$$NADPH + 2O_2 \xrightarrow{\text{NADPH oxidase}} NADPH + 2O_2^- + H^+ \text{ (superoxide)}$$

$$2H^+ + 2O_2^- \xrightarrow{\text{superoxide dismutase}} H_2O_2 + O_2 \text{ (hydrogen peroxide)}$$

$$2H_2O_2 \xrightarrow{\text{catalase}} 2H_2O + O_2$$

raised to the level needed to activate a successful attack by antimicrobial peptides and proteins. Bacteria and fungi are not cleared and persist as chronic intracellular infections of neutrophils and macrophages. Because of the actions of other mechanisms of innate and adaptive immunity, the infections become contained in localized nodules, called **granulomas**, which imprison the infected macrophages that have eaten more than their fill of infected neutrophils. The bacteria and fungi that most commonly cause infections in chronic granulomatous disease include organisms such as *E. coli* that form part of the normal flora of healthy people (Figure 2.35).

2-17   Inflammatory cytokines raise body temperature and activate hepatocytes to make the acute-phase response

A systemic effect of the inflammatory cytokines IL-1, IL-6, and TNF-α is to cause the rise in body temperature called **fever**. The cytokines act on temperature-control sites in the hypothalamus, and on muscle and fat cells, altering energy mobilization to generate heat (Figure 2.36). Molecules that induce fever are called **pyrogens**. Some pathogen products also raise the body's temperature and generally do so through inducing the production of these cytokines. In this context, the bacterial products are called 'exogenous' pyrogens, because they originate outside the body, and the cytokines are called 'endogenous' pyrogens because they originate inside the body. On balance, a raised body temperature helps the immune system fight infection, because most bacterial and viral pathogens grow and replicate faster at temperatures lower than that of the human body, and adaptive immunity becomes more potent at higher temperatures. In addition, human cells become more resistant to the deleterious effects of TNF-α when experiencing fever.

A further systemic effect of IL-1, IL-6 and TNF-α is to change the spectrum of soluble plasma proteins secreted by hepatocytes in the liver, thus producing the **acute-phase response**. Those proteins whose synthesis and secretion is increased during the acute-phase response are called **acute-phase proteins**. Two of the acute-phase proteins—mannose-binding lectin and C-reactive protein—enhance the fixation of complement at pathogen surfaces.

Mannose-binding lectin (**MBL**) is a calcium-dependent lectin that binds to mannose-containing carbohydrates of bacteria, fungi, protozoans, and viruses. The structure of MBL resembles a bunch of flowers in which each stalk is a triple helix made from three identical polypeptides. These helices are just like those found in collagen molecules and fibers. Each polypeptide contributes a carbohydrate-recognition domain, the three together forming a

Fungus
Aspergillus fumigatus
Bacteria
Staphylococcus aureus
Chromobacterium violaceum
Burkholderia cepacia
Nocardia asteroides
Salmonella typhimurium
Serratia marcescens
Mycobacterium fortuitum
Several species of *Klebsiella*
Escherichia coli
Several species of *Actinomyces*
Legionella bosmanii
Clostridium difficile

Figure 2.35 The species of fungi and bacteria most commonly responsible for infections in chronic granulomatous disease.

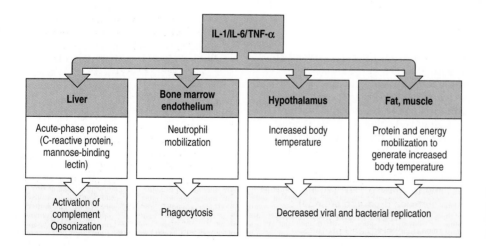

Figure 2.36 The macrophage-produced cytokines TNF-α, IL-1, and IL-6 have a spectrum of biological activity.

'flower' (Figure 2.37). Each molecule of mannose-binding lectin has five or six flowers, giving it either 15 or 18 potential sites for attachment to a pathogen's surface. Even relatively weak individual interactions with a carbohydrate structure can be developed into an overall strong binding through the use of multipoint attachments. Although some carbohydrates on human cells contain mannose, they do not bind mannose-binding lectin because their geometry does not permit multipoint attachment. When bound to the surface of a pathogen, mannose-binding lectin triggers the lectin pathway of complement activation; it also serves as an opsonin that facilitates the uptake of bacteria by monocytes in the blood (Figure 2.38). These cells lack the macrophage mannose receptor but have receptors that can bind to mannose-binding lectin coating a bacterial surface. Mannose-binding lectin is a member of a protein family that is called the **collectins** because its members combine the properties of collagen and lectins. The pulmonary surfactant proteins A and D (SP-A and SP-D) are also collectins; they defend the lungs by opsonizing pathogens such as *Pneumocystis carinii*.

C-reactive protein (**CRP**), a member of the **pentraxin** family of proteins, contains 5 identical subunits that form a pentamer (Figure 2.39). C-reactive protein binds to the phosphocholine component of lipopolysaccharides in bacterial and fungal cell walls, but not to the phosphorylcholine present in the phospholipids of human cell membranes. It was originally named for its propensity to bind the C polysaccharide of *Streptococcus pneumoniae*, which contains phosphocholine. In binding to bacteria, C-reactive protein acts as an opsonin, and triggers the classical pathway of complement fixation in the absence of specific antibody. In the absence of infection, C-reactive protein and mannose-binding lectin are present at low levels in plasma, but levels can increase by up to 1000-fold during the peak of the acute-phase response, about 2 days after its start. Because C-reactive protein and mannose-binding

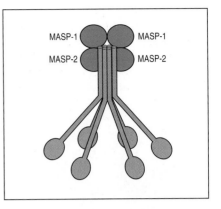

Figure 2.37 Structure of mannose-binding lectin. It resembles a bunch of flowers, with each flower composed of three identical polypeptides. The stalks are rigid triple helices like collagen, with a single bend; each flower comprises three carbohydrate-binding domains. Associated with the mannose-binding lectin (blue) are the mannose-binding lectin associated serine proteases (MASP) 1 and 2.

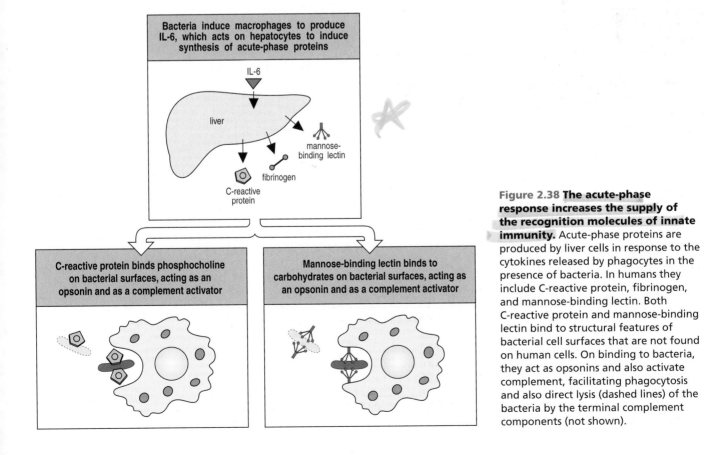

Figure 2.38 The acute-phase response increases the supply of the recognition molecules of innate immunity. Acute-phase proteins are produced by liver cells in response to the cytokines released by phagocytes in the presence of bacteria. In humans they include C-reactive protein, fibrinogen, and mannose-binding lectin. Both C-reactive protein and mannose-binding lectin bind to structural features of bacterial cell surfaces that are not found on human cells. On binding to bacteria, they act as opsonins and also activate complement, facilitating phagocytosis and also direct lysis (dashed lines) of the bacteria by the terminal complement components (not shown).

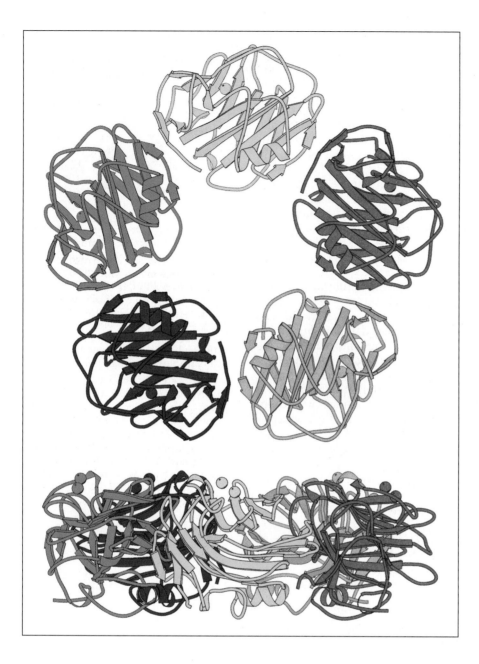

Figure 2.39 **The structure of C-reactive protein.** C-reactive protein belongs to the pentraxin family, so called because these proteins are composed of five identical subunits. The polypeptide backbones of the five subunits are traced by ribbons of different color. Overall, C-reactive protein resembles a pentagonal slab with a hole in the middle, as is seen by comparing a view from above (upper image) with one from the side (lower image). Images courtesy of Annette Shrive and Trevor Greenhough.

lectin both bind to distinct structures that are common features of pathogens but not of human cells, they thereby distinguish non-self from self. How complement activation by mannose-binding lectin and C-reactive protein differs from complement activation by the alternative pathway is examined in the next two sections.

2-18   The lectin pathway of complement activation is initiated by mannose-binding lectin

Mannose-binding lectin circulates in plasma as a complex with two serine protease zymogens: MBL-associated serine protease (MASP) 1 and 2. Two molecules each of MASP-1 and MASP-2 associate with the main stalk of the mannose-binding lectin (see Figure 2.37). When the MBL complex binds to mannose-containing macromolecules at a pathogen surface, one molecule of MASP-2 is induced to become enzymatically active and cut itself. It then cuts

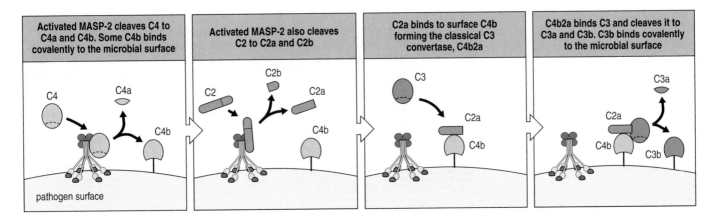

| Activated MASP-2 cleaves C4 to C4a and C4b. Some C4b binds covalently to the microbial surface | Activated MASP-2 also cleaves C2 to C2a and C2b | C2a binds to surface C4b forming the classical C3 convertase, C4b2a | C4b2a binds C3 and cleaves it to C3a and C3b. C3b binds covalently to the microbial surface |

the second MASP-2 molecule. It is not known whether MASP-1 has an enzymatic role in lectin-mediated complement activation. Substrates for the activated MASP-2 proteases are the **C4** and **C2** complement components. C4 is similar to C3 in its structure, function, and thioester bond, whereas C2 is a serine protease zymogen similar to factor B.

When a C4 molecule interacts with activated MASP-2, it is cleaved into a large C4b fragment and a small C4a fragment. This cleavage exposes the thioester bond of C4b, which is rapidly subjected to nucleophilic attack, leading to the covalent bonding, or fixation, of some C4b fragments to the pathogen surface (Figure 2.40). The soluble C4a fragment is an anaphylatoxin that can recruit leukocytes to the site of C4b fixation, but its activity is weaker than that of either C3a or C5a (see Section 2-7). When a C2 molecule interacts with activated MASP-2 it is cleaved into a larger enzymatically active fragment called C2a that binds to pathogen-bonded C4b and a small inactive fragment called C2b. (For historical reasons, the small cleavage product of C2 is called C2b and the larger product is called C2a, whereas for other complement components the larger fragment is called 'b' and the smaller 'a'.) The complex of C4b and C2a, designated as C4bC2a, is a C3 convertase. Although called the **classical C3 convertase**, it is actually a component of both the lectin and classical pathways of complement activation, for it is at this stage that the lectin and classical pathways converge. The unique aspects of the lectin pathway are the contribution of mannose-binding lectin to binding pathogens, and the activation of C4 and C2 by the MASP proteins.

The classical C3 convertase, C4bC2a, binds and cleaves C3 to yield C3b fragments attached to the pathogen surface. These in turn bind and activate factor B to assemble molecules of the alternative C3 convertase, C3bBb (Figure 2.41). It is at this stage that the lectin and classical pathways converge on the alternative pathway of complement activation. Because C3 is present at much higher concentrations in plasma than C4, the contribution of the alternative convertase to the fixation of complement far exceeds that of the classical convertase.

Alleles encoding nonfunctional variants of MBL are present at frequencies greater than 10% in human populations. Consequently, deficiency of MBL is common and causes increased susceptibility to infection. Individuals who carry two nonfunctional alleles are more likely to develop severe meningitis caused by *Neisseria meningitidis*, a bacterium that is carried as a harmless

Figure 2.40 The activated MBL complex cleaves C4 and C2 to produce C4b and C2a, which associate to form the classical C3 convertase. First panel: a complex of MBL and MASP-1 and MASP-2 binds to the pathogen surface. This activates MASP-2, which binds and cleaves C4 to reveal the thioester bond of the C4b fragment. C4b becomes covalently bound to the microbial surface. Second panel: C2 binds to the MBL complex and is cleaved by activated MASP-2. Third panel: the C2a fragment binds to C4b to form the classical C3 convertase, C4bC2a. Fourth panel: C3 is bound and cleaved by C4bC2a. The thioester bond of the C3b fragment is exposed and C3b becomes covalently bound to the microbial surface.

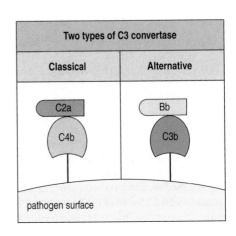

Figure 2.41 The two types of C3 convertase have similar structures and functions. In the C3 convertase produced by the classical pathway, C4bC2a, the activated protease C2a cleaves C3 to C3b and C3a (not shown). In the analogous C3 convertase of the alternative pathway, C3bBb, the activated protease Bb carries out exactly the same reaction.

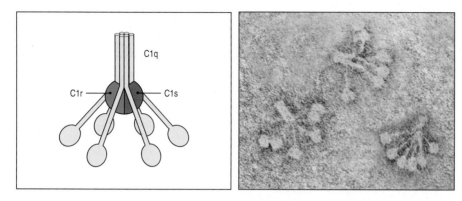

Figure 2.42 The complement component C1. The C1 molecule consists of a complex of C1q, C1r, and C1s. The C1q component consists of six identical subunits, each with one binding site for the Fc region of IgM or IgG and extended amino-terminal stalk regions that interact with each other and with two molecules each of the proteases C1r and C1s. The electron micrograph on the right contains images of three C1q molecules. Photograph courtesy of K.B.M. Reid.

commensal by about 1% of the population. Similar susceptibility is observed in people who are deficient for a terminal complement component, showing that complement-mediated killing of the bacteria is the mechanism by which healthy carriers keep their *N. meningitidis* in order.

2-19   C-reactive protein triggers the classical pathway of complement activation

Once C-reactive protein binds to a bacterium it can also interact with C1, the first component of the classical pathway of complement activation. C1 has an organization and structure like that of the complex of mannose-binding lectin with MASP-1 and 2 (Figure 2.42). In the C1 molecule, a bunch of six flowers is formed from 18 C1q polypeptides and two molecules each of C1r and C1s, which are inactive serine proteases similar to MASP-1 and 2. Each stalk is formed by a collagen-like triple helix of three C1q molecules. C-reactive protein binds to the C1q stalks and causes one molecule of C1r to cut itself, the other molecule of C1r and both molecules of C1s. In this manner C1s becomes an active protease. It cleaves C4, leading to the covalent attachment of C4b to the pathogen surface (Figure 2.43). It also cleaves C2, leading to the formation of the classical C3 convertase C4bC2a. At this stage the classical and lectin pathways converge; the unique aspects of the classical pathway of complement activation being the contribution of C1q to binding pathogens and of C1r and C1s in the activation of C4 and C2.

At the start of infection, complement activation is mainly by the alternative pathway. As the inflammatory response develops and acute-phase proteins are produced, mannose-binding lectin and C-reactive protein provide increased activation of complement via the lectin and classical pathways, respectively. All three pathways contribute to innate immunity and they work together to produce quantities of C3b fragments and C3 convertases at the pathogen surface.

2-20   Type I interferons inhibit viral replication and activate host defenses

The proteins of innate immunity discussed so far in this chapter act on pathogens in their extracellular phases. We shall now look at the specific defenses of the innate immune system against viruses once they have entered cells. When any human cell becomes infected with a virus it responds by making cytokines called **type I interferons**, or simply **interferon**. The immediate effects of type I interferon are to interfere with viral replication by the infected cell, and to signal neighboring uninfected cells that they too should prepare to resist a viral infection. Further effects of type I interferon are to alert cells of the immune system that an infection is about, and to make virus-infected cells more vulnerable to attack by killer lymphocytes. As almost all types of human cell are susceptible to viral infections, virtually all cells are equipped to make

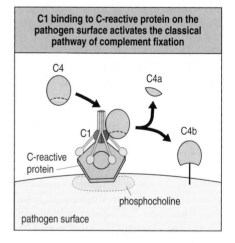

C1 binding to C-reactive protein on the pathogen surface activates the classical pathway of complement fixation

Figure 2.43 C-reactive protein can initiate the classical pathway of complement activation. C-reactive protein bound to phosphocholine on bacterial cell surfaces binds complement component C1, resulting in the cleavage of C4 and opsonization of the bacterial surface with C4b.

both type I interferons and their receptor. The receptor is always present on cell surfaces, ready to bind interferon newly made in response to infection. Although type I interferon is barely detectable in the blood of healthy people, upon infection it becomes abundant.

There are many different forms, or isotypes, of type I interferon. Humans have a single form of **interferon-β (IFN-β)**, multiple forms of **interferon-α (IFN-α)** and several additional isotypes: IFN-δ, -κ, -λ, -τ, and -ω. The isotypes have a similar structure, bind to the same cell-surface receptor, and are specified by a family of linked genes on human chromosome 9.

Type I interferon synthesis is induced by intracellular events that follow viral infection or the triggering of a signaling receptor, for example the sensing of double-stranded RNA by TLR3. Double-stranded RNA, a type of nucleic acid not found in healthy human cells, is a component of some viral genomes, and an intermediary nucleic acid in viral life cycles. Infection or ligand sensing triggers the phosphorylation of the transcription factor IRF3 in the cytoplasm (see Section 2-12), which dimerizes and enters the nucleus to help initiate transcription of the IFN-β gene, which also requires the transcription factors NFκB and AP-1. Once IFN-β is secreted, it acts both in an autocrine fashion, binding to receptors on the cell that made it, and in a paracrine fashion, binding to receptors on uninfected cells nearby (Figure 2.44).

When interferon binds to its receptor, the intracellular Jak1 and Tyk2 kinases associated with the receptor initiate reactions that change the expression of a variety of human genes, a process called the **interferon response** (Figure 2.45). Among the cellular proteins induced by interferon are some that interfere directly with viral genome replication. An example is the enzyme oligoadenylate synthetase, which polymerizes ATP by 2′–5′ linkages rather than the 3′–5′ linkages normally present in human nucleic acids. These unusual oligomers activate an endoribonuclease that degrades viral RNA. Also activated by IFN-α and IFN-β is a serine/threonine protein kinase called protein kinase R (PKR) that phosphorylates and inhibits the protein synthesis initiation factor eIF-2, thereby preventing viral protein synthesis and the production of new infectious virions.

Several of the interferon-induced proteins are transcription factors similar to IRF3, the only one of the group that is made constitutively. These other interferon response factors are instrumental in turning on the transcription of

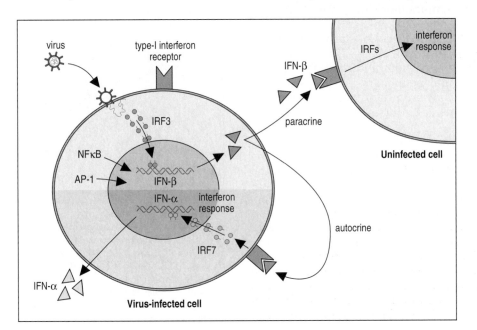

Figure 2.44 Virus-infected cells are stimulated to produce type I interferons. The cell on the left is infected with a virus that triggers signals that lead to the phosphorylation, dimerization, and passage to the nucleus of the transcription factor interferon-response factor 3 (IRF3). Transcription factors NFκB and AP-1 are also mobilized and coordinate with IRF3 to turn on transcription of the interferon (IFN)-β gene. These events are depicted in the upper half of the cell. Secreted IFN-β binds to the interferon receptor on the infected cell surface, acting in an autocrine fashion to mobilize other interferon-response factors and change patterns of gene expression to give the interferon response. These events are depicted in the lower half of the cell, being exemplified by IRF7 turning on transcription of the IFN-α gene, which it does without the need for AP-1 or NFκB. Secreted IFN-β will also bind to the interferon receptor expressed by nearby cells that are not infected by the virus, acting in a paracrine fashion to induce the interferon response that helps these cells to resist infection.

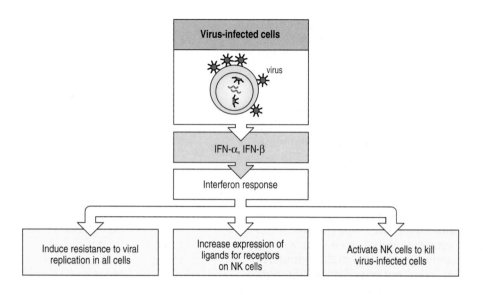

Figure 2.45 **Major functions of the type I interferons.** Interferon-α and interferon-β (IFN-α and IFN-β) have three major functions. First, they induce resistance to viral replication by activating cellular genes that destroy viral mRNA and inhibit the translation of viral proteins. Second, they increase the expression of ligands for NK cell receptors on virus-infected cells. Third, they activate NK cells to kill virus-infected cells.

many different genes, including those for interferons other than IFN-β. Interferon response factor 7 (IRF7) initiates the transcription of IFN-α, which does not require the participation of NFκB and AP-1. In this manner, a positive feedback loop develops, in which a small initial amount of interferon serves to increase both the size and range of future production.

As well as interfering with viral replication, interferon also induces cellular changes that make the infected cell more likely to be attacked by killer lymphocytes. NK cells are lymphocytes of innate immunity that provide defense against viral infections by secreting cytokines and killing infected cells. When IFN-α or IFN-β bind to the interferon receptors on circulating NK cells, these become activated and are drawn into infected tissues, where they attack virus-infected cells. Because of its power to boost the immune response, type I interferon has been explored as a treatment for human disease. It has been found to ameliorate several conditions: infections with hepatitis B or C viruses; the degenerative autoimmune disease multiple sclerosis, which affects the central nervous system; and certain leukemias and lymphomas.

Although almost all human cells can secrete some type I interferon, specialized cells called **interferon-producing cells** (**IPCs**) or **natural interferon-producing cells** (**NIPCs**) secrete up to 1000 times more interferon than other cells. These lymphocyte-like cells are present in the blood, making up less than 1% of the total leukocytes, and are distinguished by having cytoplasm resembling that of a plasma cell, another cell type engaged in the massive production of secreted protein (Figure 2.46). Interferon-producing cells express Toll-like receptors 6, 7, 9, and 10, making them responsive to a range of viral infections (see Figure 2.21). These receptors are thought to signal for interferon production using a pathway different from that of TLR3.

Within the first day after stimulation by viral infection, interferon-producing cells produce massive amounts of type I interferons. In the next 2 days the interferon-producing cell differentiates into a type of dendritic cell called the **plasmacytoid dendritic cell**, which retains the ability to produce interferon. During an infection, these cells congregate in the T-cell areas of draining lymph nodes, after having entered from the blood across the walls of high endothelial venules. Although there are some similarities between plasmacytoid dendritic cells and the myeloid dendritic cells described in Section 1-7, which are known as conventional dendritic cells, plasmacytoid dendritic cells are not thought to be much involved in the activation of T cells in adaptive immunity, which is the main function of conventional dendritic cells. In the context of innate immunity, conventional dendritic cells make relatively small amounts of type I interferons but produce large amounts of IL-12, a cytokine

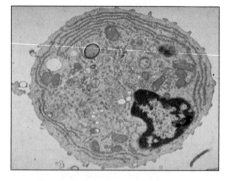

Figure 2.46 **Type-I-interferon-producing cell from human peripheral blood.** Note the extensive rough endoplasmic reticulum that is similar in appearance to that of a plasma cell and is due to the massive synthesis and secretion of interferon by these cells. Image courtesy of Dr Yong-Jun Liu.

that works with type I interferons to activate the NK-cell response to viral infection. Throughout this book, dendritic cells will mean conventional dendritic cells unless specified otherwise.

2-21   NK cells provide an early defense against intracellular infections

Natural killer cells (**NK cells**) are the killer lymphocytes of the innate immune response. They comprise 5–25% of the lymphocytes in the blood and are distinguished from circulating B cells and T cells by their larger size and well-developed cytoplasm containing cytotoxic granules. When first discovered, NK cells were called 'large granular lymphocytes;' they provide innate immunity against intracellular infections and migrate from the blood into infected tissues in response to inflammatory cytokines. Patients who lack NK cells suffer from persistent viral infections, particularly of herpes viruses, which these patients cannot clear without help from antiviral drugs despite making a normal adaptive immune response. These rare individuals demonstrate the importance of NK cells in managing virus infections and show how the NK-cell response complements that of the cytotoxic T cells of adaptive immunity. NK cells have two types of effector function—cell killing and the secretion of cytokines—that are used in different ways depending on the pathogen. To a rough approximation, NK cells perform functions in the innate immune response similar to those of cytotoxic T cells in the adaptive immune response.

Laboratory experiments have shown that NK cells freshly isolated from human blood will kill certain types of target cell in the absence of inflammatory cytokines. This base level of cytotoxicity is increased 20–100-fold on exposure to the IFN-α and IFN-β produced in response to viral infection. Type I interferons also induce the proliferation of NK cells. NK cells are also activated by IL-12, which especially targets them, and by TNF-α, both of which are produced by macrophages and dendritic cells early in many infections. The actions of these four cytokines produce a wave of activated NK cells during the early part of a virus infection that either terminates the infection or contains it during the time required to develop the cytotoxic T-cell response (Figure 2.47).

Stimulation of NK cells with IFN-α and IFN-β favors the development of the cells' killer functions, whereas stimulation with IL-12 favors the production of cytokines. The principal cytokine released by NK cells is **IFN-γ**, also called **type II interferon**, which is unrelated in structure and function to the type I interferons. A major function of IFN-γ is to activate macrophages. Macrophage secretion of IL-12 and NK-cell secretion of IFN-γ create a system of positive feedback that increases the activation of both types of cell within an infected tissue. Interactions between NK cells and dendritic cells can also lead either to mutual activation or to killing of dendritic cells, events that influence whether and when dendritic cells migrate to secondary lymphoid tissue and initiate the adaptive immune response. In the early stage of an infection, NK cells are the major producers of IFN-γ, which activates macrophages to secrete cytokines that help activate T cells, thus initiating the adaptive immune response. Once effector T cells have been produced and enter the infected

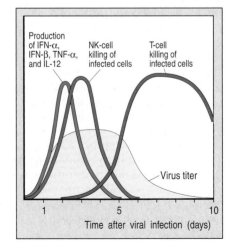

Figure 2.47 NK cells provide an early response to virus infection. The kinetics of the immune response to an experimental virus infection of mice are shown. As a result of infection, a burst of cytokines is secreted, including IFN-α, IFN-β, TNF-α, and IL-12 (green curve). These induce the proliferation and activation of NK cells (blue curve), which are seen as a wave emerging after cytokine production. NK cells control virus replication and the spread of infection while effector killer T cells (red curve) are developing. The level of virus (the virus titer) is given by the curve described by the yellow shading.

site, they become the major source of IFN-γ and of cell-mediated cytotoxicity. With the arrival of effector T cells, NK-cell functions are turned off by IL-10, an inhibitory cytokine made by cytotoxic T cells.

2-22   NK-cell receptors differ in the ligands they bind and the signals they generate

NK cells respond quickly to infection because they circulate in a partly activated state, as seen from their large size and their cytoplasmic granules loaded with toxic effector molecules. In contrast, B and T lymphocytes circulate in small, quiescent forms that require an extended period of stimulation and differentiation before they acquire effector functions. A further characteristic distinguishing NK cells from B and T cells is that NK cells do not express surface receptors produced from rearranging genes. Keeping NK cells in a state of readiness for infection, while curbing their potential to attack healthy tissue, is an extensive range of cell-surface receptors, some of which deliver activating signals and others inhibitory signals. Most NK-cell receptors fall into two broad structural types: immunoglobulin-like receptors and lectin-like receptors (Figure 2.48). For the **NK-cell immunoglobulin-like receptors** the extracellular ligand-binding site is composed of immunoglobulin domains. The second type of NK-cell receptors have extracellular ligand-binding sites that are structurally similar to the carbohydrate-recognition domain of mannose-binding lectin. Although it is convenient to call the latter group the **NK-cell lectin-like receptors**, many actually bind protein ligands rather than carbohydrates.

Although ligands for NK-cell receptors are as varied as the receptors, they are mainly cell-surface proteins whose expression is altered in response to infection, malignancy or other trauma. The alteration can involve changes in the abundance of the ligand, its intracellular distribution or its structure. When an NK cell interacts with a healthy cell, the combined signals it receives from its inhibitory and activating receptors binding to ligands on the healthy cell have the overall effect of preventing it from attacking. In contrast, when the NK cell interacts with a virus-infected cell, the balance of activating and inhibitory signals is altered to favor NK-cell attack on the virus-infected cell. In this manner NK cells are able to discriminate between healthy cells that should be protected and unhealthy cells that should be destroyed. By killing virus-infected cells, the NK cell impedes the production of new virions (virus particles) and the further infection of healthy human cells.

We will illustrate how NK cells can respond to infection by considering its activation via NKG2D, an activating lectin-like NK-cell receptor that binds to ligands called MIC-A and MIC-B, cell-surface proteins that are produced in response to stress (Figure 2.49). The only tissue that makes MIC-A and MIC-B constitutively is intestinal epithelium, and there the amount is small. When any epithelial cell becomes infected, damaged, or cancerous, however, expression of MIC-A and MIC-B is induced, and in intestinal epithelium their abundance increases. Once expressing MIC-A and MIC-B, epithelial cells become targets for NK-cell attack through the receptor NKG2D. Through the agency of adaptor proteins that associate with the cytoplasmic tail of NKG2D, protein kinases are activated whose actions lead to the release of the NK cell's cytotoxic granules and cytokines.

Although some receptors, such as NKG2D, are expressed by all NK cells, most are expressed only by subpopulations of NK cells. Consequently, individual NK cells express different combinations of receptors, imparting heterogeneity to a person's NK-cell population and providing a repertoire of responses to pathogens. Among other cell types involved in the innate immune response, such as macrophages, dendritic cells, and neutrophils, individual cells also express different combinations of receptors.

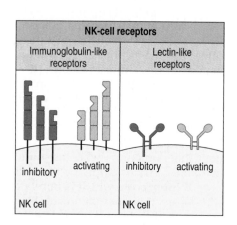

Figure 2.48 Immunoglobulin-like and lectin-like NK-cell receptors. Most NK-cell receptors have extracellular ligand-binding regions that are made up of immunoglobulin domains (left panel) or lectin-like domains resembling that of mannose-binding lectin (right panel). Activating receptors have short cytoplasmic tails and charged amino acid residues in the transmembrane domain that facilitate interaction with intracellular signaling proteins. Inhibitory receptors have long cytoplasmic tails that contain a short amino acid sequence motif called an immunoreceptor tyrosine-based inhibitory motif (ITIM), which binds protein phosphatases that act to inhibit the activating pathways.

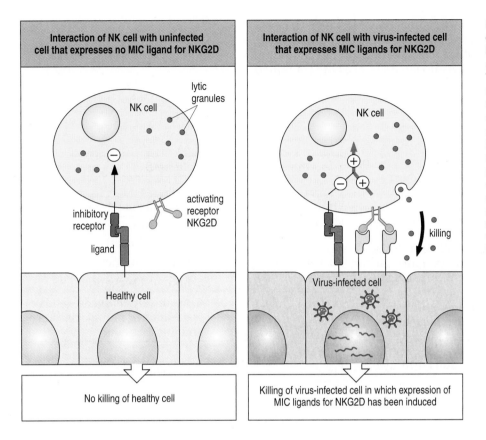

Figure 2.49 NK cell receptors distinguish unhealthy cells from healthy cells. NK cells have activating and inhibitory cell-surface receptors. The ligands for NKG2D, an activating receptor present on all human NK cells, are MIC-A and MIC-B, proteins that are not expressed by healthy cells but are expressed by cells stressed by virus infection or other trauma. Healthy cells resist attack by NK cells because signals generated from inhibitory receptors dominate those generated from activating receptors (left panel). NK cells attack the virus-infected cell because the signal generated by NKG2D interacting with MIC proteins tips the balance from inhibition to activation.

Summary to Chapter 2

The human body has several lines of defense, all of which must be overcome if a pathogen is to establish an infection and then exploit its human host for the remainder of that person's life. The first defense is the protective epithelial surfaces of the body and their commensal microorganisms, which successfully prevent most pathogens from ever gaining entry to the rich resources of the body's interior. Any pathogen that succeeds in penetrating an epithelial surface is immediately faced by the effector cells and molecules of the innate immune response. Innate immunity provides a variety of defenses that work immediately a pathogen is first confronted or soon after. These fixed defenses are always available and do not improve with repeated exposure to the same pathogen.

Inhibiting a pathogen's progress in colonizing tissues and spreading infection are the protease inhibitors, blood-clotting cascade, and kinin reactions. A host of plasma proteins and cell-surface molecules provide systems for identifying microbiological invaders and distinguishing them from self. Complement provides a general means to tag almost any component at a microbial surface; more specific receptors bind common chemical aspects of microbial macromolecules that are not a part of the human body. As well as helping resident macrophages to phagocytose pathogens, these interactions induce the macrophages to pour out inflammatory cytokines that summon neutrophils and NK cells to the site of infection. Interactions between these cells, and with resident macrophages and dendritic cells, produce mutual activation and cytokine secretion that heightens the state of inflammation in the infected tissue.

Bacterial infections are frequently overcome by the phagocytic powers and potent poisons of the abundant neutrophils. In viral infections, the production of type I interferons by infected cells and interferon-producing cells will often set the stage for NK cells to terminate the infection. Most infections are efficiently cleared by the innate immune response and lead to neither disease nor incapacitation. In the minority of infections that escape innate immunity and spread from their point of entry, the pathogen then faces the combined forces of innate and adaptive immunity.

Questions

2–1   Which of these pairs are mismatched?
a. cytosol: intracellular pathogen
b. surface of epithelium: extracellular pathogen
c. nucleus: intracellular pathogen
d. lymph: intracellular pathogen.

2–2   Although activation of the three different pathways of complement involves different components, the three pathways converge on a common enzymatic reaction referred to as complement fixation.
A. Describe this reaction.
B. Describe the enzyme responsible for this reaction in the alternative pathway.
C. Identify the three effector mechanisms of complement that are enabled by this common pathway.

2–3   Which of the following is the soluble form of C3 convertase of the alternative pathway of complement activation?
a. iC3
b. iC3b
c. C3b
d. iC3Bb
e. C3bBb.

2–4   Explain the steps that take place when a bacterium is opsonized via C3b:CR1 interaction between the bacterium and a resident macrophage in tissues.

2–5   In the early stages of the alternative pathway of complement activation there are complement control proteins that are soluble (factors H and I) and cell surface-associated (DAF and MCP). Identify the (i) soluble and (ii) cell surface-associated complement control proteins that operate in the terminal stages of the alternative pathway of complement activation, and describe their activities.

2–6
A. Review the differences between the three pathways of complement (alternative, lectin, and classical) in terms of how they are activated.
B. Distinguish which pathway(s) are considered part of an adaptive immune response and which are considered part of innate immunity, and explain why.

2–7   Match the innate immune receptor in column A with its ligand(s) in column B. More than one ligand may be used for each immune receptor.

Column A	Column B
a. lectin receptor	1. iC3b
b. scavenger receptor	2. lipophosphoglycan
c. CR3	3. carbohydrates (e.g., mannose and glucan)
d. CR4	4. filamentous hemagglutinin
e. CR1	5. lipopolysaccharide (LPS)
f. TLR4:TLR4	6. negatively charged ligands (e.g., sulfated polysaccharides and nucleic acids)
g. TLR5	7. C3b
h. TLR3	8. flagellin
	9. RNA

2–8   Other than their ligand specificity, what is a key difference between TLR5, TLR4, TLR1:TLR2, and TLR2:TLR6 compared with TLRs 3, 7, 8, and 9?

2–9   Explain why TLRs can detect many different species of microbes despite the limited number of different TLR proteins.

2–10   Explain the importance of NFκB in mediating signals through TLRs.

2–11   What is the name given to the earliest intracellular vesicle that contains material opsonized by macrophages?
a. opsonome
b. membrane-attack complex
c. lysosome
d. phagosome
e. phagolysosome.

2–12
A. What are the main (i) similarities and (ii) differences in the general properties and roles of macrophages and neutrophils?
B. How do they both destroy extracellular pathogens? Give details of the process.

2–13   In response to TNF-α, vascular endothelium produces _____, which induces localized blood clotting?
a. platelet-activating factor
b. IL-12
c. CXCL8

d. IL-1β

e. IL-6.

2–14

  A. What induces the production of type I interferon by virus-infected cells?

  B. Do normal cells produce this inducer? Why, or why not?

  C. Discuss the mechanisms by which type I interferons exert their antiviral effects.

2–15 Which of the following activities are most closely associated with natural killer cells? Select all correct answers.

  a. production of TNF-α

  b. lysis of virus-infected cells

  c. phagocytosis of bacteria

  d. release of reactive oxygen intermediates

  e. production of IFN-γ.

2–16 Jonathan Miller, age 6 years, was brought to emergency room by his parents presenting with fever, severe headache, a petechial rash, stiff neck and vomiting. Jonathan had a history of recurrent sinusitis and otitis media, all caused by pyogenic bacteria and treated successfully with antibiotics. Suspecting bacterial meningitis, the attending physician began an immediate course of intravenous antibiotics and requested a lumbar puncture. *Neisseria meningitidis* was grown from the cerebrospinal fluid. The physician was concerned about the recurrence of infections caused by pyogenic bacteria and suspected an immunodeficiency. He ordered blood tests and found the serum complement profiles to have low C3, factor B, and factor H, and undetectable factor I. Which of the following explains why a factor I deficiency is associated with infections caused by pyogenic bacteria?

  a. Elevated levels of C3 convertase C3bBb interfere with the activation of the classical pathway of complement activation.

  b. Rapid turnover and consumption of C3 in the serum cause inefficient fixation of C3b on the surface of pathogens, compromising opsonization and phagocytosis.

  c. Factor I is an opsonin that facilitates phagocytosis.

  d. Factor I is a chemokine and is important for the recruitment of phagocytes.

  e. Factor I is required for the assembly of the terminal components of the complement pathway.

2–17 Mary Hanson, a 2-year-old, was brought to the doctor's office by her mother after she discovered two swollen and painful lumps in Mary's groin. *Staphylococcus aureus* was cultured from phagocytic cells obtained from the affected lesions and granulomas were noted. Additional tests revealed that neutrophils from Mary do not activate a respiratory burst after phagocytosis. The most likely protein defect causing the inability to generate reactive oxygen intermediates during the respiratory burst in Mary's phagocytes would be:

  a. NADPH oxidase subunit

  b. IFN-β

  c. IL-6

  d. TNF-α

  e. mannose-binding lectin.

3

The internal structure of the human immunodeficiency virus which can slowly destroy the adaptive immune system.

Chapter 3

Principles of Adaptive Immunity

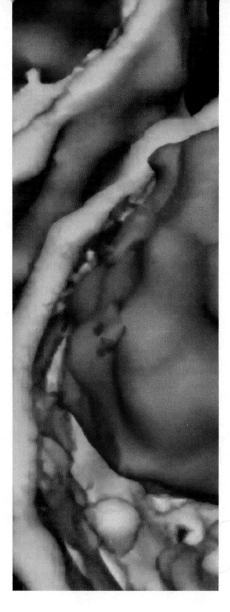

The **adaptive immune response**, or **adaptive immunity**, is the body's third line of defense, being brought into play only after physical barriers have been breached and the innate immune response has failed to vanquish the invading pathogen. Adaptive immunity is essentially due to two types of lymphocyte: the B lymphocyte, or B cell, and the T lymphocyte, or T cell. The effector mechanisms recruited by B and T cells to rid the body of pathogens are similar to those of innate immunity and involve many of the cells and molecules described in Chapter 2. The unique feature of adaptive immunity is the elaborate system by which B cells and T cells recognize pathogens and distinguish one microorganism from another. This enables the body to make a more focused and forceful response to any pathogen than is possible with the innate immune response alone. The rest of this book is largely concerned with adaptive immunity: its molecular and cellular mechanisms, its successes and failures in countering infection, and the chronic conditions caused by its misplaced actions. In this chapter these topics will be introduced in the context of the principles that distinguish adaptive immunity from innate immunity.

3-1 Innate and adaptive immunity differ in their strategies for pathogen recognition

In the innate immune response, pathogens are recognized by a fixed repertoire of cell-surface receptors and soluble effector molecules that has evolved by natural selection over hundreds of millions of years. Genes encoding these innate immune molecules are inherited from one generation to the next in a stable form. Very occasionally, new variants arise through mutation and meiotic recombination; a very small fraction of these will have novel pathogen-binding functions that confer an advantage and will be positively selected for. In particular, the evolution of gene families through successive gene duplications has been a common and successful strategy for diversifying innate immune functions, as is illustrated by the families of genes encoding the defensins and the Toll-like receptors (see Sections 2-9 and 2-11). In general, innate immune receptors recognize either structures shared by many different pathogens or alterations to human cells that are commonly induced by the presence of pathogens.

The adaptive immune response uses an entirely different strategy of pathogen recognition. B cells and T cells each recognize pathogens by using cell-surface receptors of just one molecular type—called **B-cell receptors** in the case of B cells and **T-cell receptors** in the case of T cells. These proteins can, however, be made in an almost infinite number of different versions, each with a binding site for a different ligand. The genes encoding the B-cell and T-cell receptors are unique among mammalian genes in that they 'evolve' into many different forms during the course of lymphocyte development, with the result that each person has at their disposal at any one time millions of B-cell and T-cell receptors with different binding sites, each expressed by a small subset of lymphocytes. On infection with a pathogen, only those lymphocytes with receptors that can bind to components of that particular pathogen are selected to divide, proliferate, and differentiate into effector lymphocytes (see Figure 1.10). This means that the immune system channels all its energy into the current crisis, thereby increasing the likelihood that the infection will be terminated in a timely fashion.

Adaptive immunity has evolved only in vertebrates, and it complements the mechanisms of innate immunity that vertebrates share with many other animals and even, in the case of the defensins, with plants. This restriction suggests that its evolution became feasible or advantageous only with the degree of cellular and anatomical complexity attained by the vertebrates. The adaptive immune system adds several powerful new features to immunological defenses. One is that a strong immune response can be precisely targeted against a particular pathogen by making use of the many small differences that distinguish one pathogen from another and from human cells. A second advantage is that some of the pathogen-specific B and T cells that proliferate during a first infection are retained long after the infection has been terminated, which enables subsequent infections by the same pathogen to be countered and terminated promptly and with less disease. This state of immunity, or immunological memory, is what vaccination seeks to induce (see Section 1-5). A third advantage of the adaptive immune system is that it gives the slowly evolving vertebrates a response that can keep up with the multiplicity of rapidly evolving microorganisms, because it has the capacity and versatility to recognize and attack any new target.

3-2 Immunoglobulins and T-cell receptors are the highly variable recognition molecules of adaptive immunity

B-cell receptors and T-cell receptors are the proteins on which adaptive immunity is based. B-cell receptors are also known as **immunoglobulins**. They are expressed on the surface of B cells, where they bind to pathogens and serve as the B cell's pathogen-recognition receptors. Effector B cells, called plasma cells, secrete soluble forms of these immunoglobulins, which are known as **antibodies**. In contrast, T-cell receptors are only ever expressed as cell-surface recognition receptors, never as soluble proteins.

Any molecule, macromolecule, virus particle, or cell that contains a structure recognized and bound by an immunoglobulin or T-cell receptor is called its corresponding **antigen**. Surface immunoglobulins and T-cell receptors are thus also referred to as the **antigen receptors** of lymphocytes. Immunoglobulins can bind to a vast variety of different chemical structures, whereas T-cell receptors recognize a more limited range of antigens. Immunoglobulins and T-cell receptors are said to be **specific**, or to have **specificity**, for the antigens they bind.

Immunoglobulins and T-cell receptors are structurally related molecules with a shared ancestry (Figure 3.1). Immunoglobulins are formed from two different polypeptides called the heavy and light chains; each Y-shaped immunoglobulin molecule consists of two identical heavy chains and two identical light chains. Both types of polypeptide have an amino-terminal **variable region** that differs in amino-acid sequence from one immunoglobulin to the next, and a **constant region** that is very similar in amino-acid sequence between immunoglobulins. The variable regions contain the sites that bind

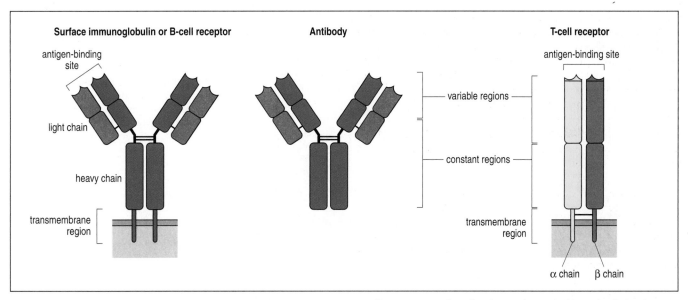

Figure 3.1 Comparison of the structures of surface immunoglobulin, antibody, and the T-cell receptor. The heavy chains of surface immunoglobulin, also known as the B-cell receptor, and antibody are shown in blue, the light chains in red. The α chain of the T-cell receptor is shown in yellow, the β chain is shown in green.

antigens. Surface immunoglobulin is anchored in the membrane by two transmembrane regions, one at the carboxyl end of each heavy chain. Antibodies are a secreted form of immunoglobulin that lack these transmembrane regions but are otherwise identical to surface immunoglobulins (see Figure 3.1).

A typical T-cell receptor consists of an α chain (TCRα) and a β chain (TCRβ), both anchored in the T-cell membrane. Like the light chains and heavy chains of immunoglobulins, the α and β chains of T-cell receptors each consist of a variable region and a constant region, with the variable regions forming an antigen-binding site (see Figure 3.1).

Differences in the amino-acid sequences of the variable regions of immunoglobulins and T-cell receptors create a vast variety of binding sites that are specific for different antigens and thus for different pathogens. A consequence of this specificity is that the adaptive immune response made against one pathogen does not provide immunity to another. For example, antibodies made in response to a measles infection bind to measles virus but not to influenza virus (Figure 3.2); conversely, antibodies specific for influenza virus do not bind to measles virus.

The constant regions of antibodies contain binding sites for cell-surface receptors on phagocytes and other inflammatory cells and also for complement proteins (see Chapter 2). The secreted antibody thus acts as a molecular adaptor or bridge. With one site it binds to pathogens and with another it binds to effector cells and molecules that can destroy the pathogen. For immunoglobulins, but not T-cell receptors, there are several different types of constant region, which confer different effector functions on the secreted antibodies and can also target them to different sites in the body.

3-3 The diversity of immunoglobulins and T-cell receptors is generated by gene rearrangement

Diversity in the antigen receptors of B and T cells is generated by genetic mechanisms that are unique to the immunoglobulin and T-cell receptor genes. Different parts of the variable regions of immunoglobulin and T-cell

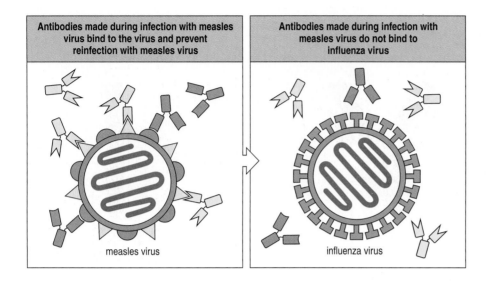

Figure 3.2 **The antibodies made against a pathogen are highly specific for that pathogen.** After recovering from infection with measles virus, a person's body fluids contain many different antibodies that can bind to the measles virus and prevent reinfection. None of these antibodies binds to an unrelated virus such as influenza.

receptor chains are encoded by separate gene segments called V, D, and J, each of which is present in the genome as a tandem array of variant forms. The genes for the immunoglobulin heavy chain and T-cell receptor β chain have arrays of V, D, and J segments, whereas the genes for the immunoglobulin light chain and the T-cell receptor α chain have only V and J. In these disconnected forms, the genes cannot be transcribed and translated into protein chains. For a functional gene to be made, one each of the different gene segments must be brought into register by breaking and splicing the DNA with elimination of the intervening regions. This enzyme-catalyzed process of recombination, called **gene rearrangement**, makes a variable-region sequence that can be transcribed and translated (Figure 3.3). The immunoglobulin and T-cell receptor genes are the only human genes requiring rearrangement in order to be functional, and they are jointly referred to as **rearranging genes**. Rearrangements at the heavy- and light-chain gene loci occur only in B cells, whereas rearrangements at the α- and β-chain gene loci occur only in T cells. Before gene rearrangement, the immunoglobulin and T-cell receptor genes are said to be in the **germline configuration** because that is how they are present in the germ cells—eggs and sperm. The process of gene rearrangement in B cells and T cells is called **somatic recombination** because it occurs in somatic cells—those cells of the body that are not germ cells.

The numerous combinations of V, D, and J (or V and J) segments that can be brought together by gene rearrangement are the principal source of variable region diversity. Further diversity arises from imprecision in the enzymatic machinery used to 'cut and paste' the DNA during gene rearrangement. Its effect is to introduce additional nucleotides into the joints between gene segments. Together, these processes create sequence diversity in the variable regions of the individual chains of the immunoglobulin and T-cell receptor polypeptides (Figure 3.4). Additional diversity in the antigen-binding sites arises from the combinatorial association of different heavy and light chains in immunoglobulins and of different α and β chains in T-cell receptors.

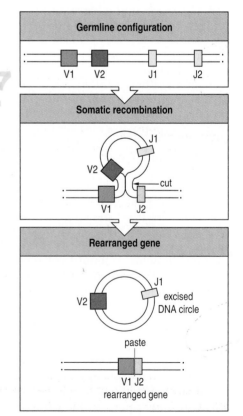

Figure 3.3 **The type of gene rearrangement that occurs in immunoglobulin and T-cell receptor genes.** In this simplified example the unrearranged DNA contains two alternative V segments and two alternative J segments. A functional exon encoding the variable region consists of one V segment joined to one J segment. This rearrangement is achieved by a process of 'cut and paste' in which the intervening DNA is removed as a circle.

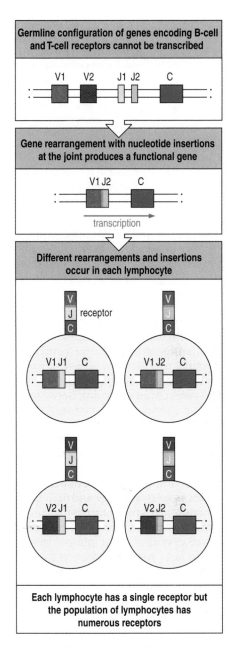

Figure 3.4 Gene rearrangement produces a diversity of antigen receptors in lymphocytes. Top panel: in this highly simplified germline gene for an antigen receptor chain, multiple V and J segments lie upstream of the sequence (C) encoding the constant region. Center panel: after gene rearrangement has juxtaposed one V and one J segment, the gene can be transcribed. Additional diversity is generated by random insertion of nucleotides at the joint between the V and J segments, as indicated by the dark orange line. Bottom panel: expression of the rearranged gene produces a cell-surface antigen receptor. Individual lymphocytes express different antigen receptors as a result of different combinations of V and J segments being brought together by the rearrangement process. The same principle of gene rearrangment applies to the genes for all the protein chains that make up B-cell receptors, antibodies and T-cell receptors.

3-4 Clonal selection of B and T lymphocytes is the guiding principle of the adaptive immune response

A direct consequence of the mechanism of immunoglobulin or T-cell receptor gene rearrangement is that each lymphocyte expresses immunoglobulins or T-cell receptors of a single specificity. As a population, however, the lymphocytes of the human immune system make millions of different immunoglobulins and T-cell receptors. This combination of uniformity at the level of the individual and diversity at the level of the population means that the lymphocyte response can be tailored toward any particular pathogen by selecting only those lymphocytes with the appropriate receptors. When an infection occurs, only a very small proportion of lymphocytes have receptors that recognize the particular pathogen. To increase their numbers, each lymphocyte stimulated by the pathogen first proliferates to give rise to a clone of cells all expressing an identical immunoglobulin or T-cell receptor. As they proliferate, the pathogen-selected T and B cells differentiate into effector cells that work together to eliminate the pathogen (Figure 3.5). The process by which pathogens select particular lymphocytes for expansion is called **clonal selection**, and the proliferation of the selected clones is called **clonal expansion**. The use of a minute fraction of the total lymphocyte repertoire to respond to each pathogen ensures that the adaptive response is highly specific for the infection at hand.

The adaptive immune response we outline in this chapter is the one that is stimulated when someone first becomes seriously infected by a pathogen— that is, the infection has overcome the defenses of innate immune mechanisms acting alone. This adaptive immune response is called the **primary immune response**. It takes several days to produce sufficient numbers of antigen-specific B and T cells to be effective, and during this period the pathogen is least restrained and most able to cause disease.

3-5 Adaptive immune responses are initiated in secondary lymphoid tissues by antigen-bearing dendritic cells and T cells

We saw in Chapter 2 how an innate immune response is initiated at the site of infection and later involves the liver, where proteins of the acute-phase response are made. In contrast, adaptive immune responses are initiated in specialized lymphoid tissues and organs that are dedicated to this purpose and are found throughout the body. These include the lymph nodes, the white pulp of the spleen, and the Peyer's patches of the gut (see Sections 1-9 to 1-11, pp. 19–22). These tissues all have a similar microanatomy and circulation that promote both the meeting of pathogen-derived antigens with the rare pathogen-specific T cells and B cells and the subsequent activation and differentiation of the antigen-activated lymphocytes into effector cells.

Figure 3.5 An adaptive immune response to a pathogen is due to the selection and expansion of the small fraction of B lymphocytes and T lymphocytes that carry surface receptors that recognize the pathogen. The rearranging genes that encode the antigen receptors of lymphocytes produce vast repertoires of B cells (B) and T cells (T) in which each cell expresses a single receptor and very few cells express identical receptors. Because of the breadth of the repertoires, an adaptive immune response can be made to any pathogen. The drawback is that only a small number of cells will be able to respond to a given pathogen. Upon infection these few cells are triggered to divide, forming expanded clones of pathogen-specific B cells and T cells. These differentiate into large numbers of effector T cells and B cells that work together to eliminate the infection.

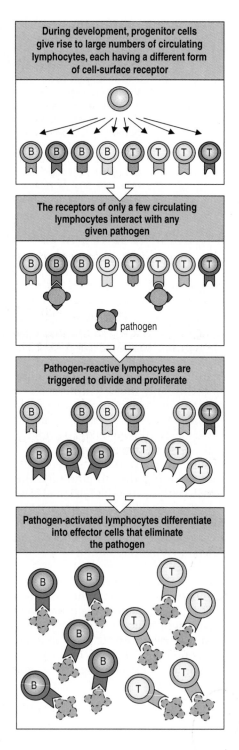

The first step in an adaptive immune response is for the pathogen to be carried by dendritic cells from the infected site to the nearest secondary lymphoid tissue. We will use an infected wound in the skin as our example here (Figure 3.6). **Dendritic cells** are a type of phagocytic leukocyte developmentally related to macrophages (see Figure 1.14) and, like macrophages, they are resident in tissues and are specialized in the uptake and breakdown of infectious agents. Unlike macrophages, however, the presence of infection causes dendritic cells to differentiate into mobile cells that transport bacteria and their antigens through the lymph to the secondary lymphoid tissue that drains the infected site. In the case of a wound to the skin, this lymphoid tissue will be the nearest lymph node (see Figure 1.20). Dendritic cells comprise the essential connecting link between innate immunity and the activation of an adaptive immune response. Dendritic cell differentiation and movement occur if the innate immune response is unable to contain the infection, a situation that is assessed by means of interactions between dendritic cells, natural killer cells (NK cells), and other cells of innate immunity.

The second step in the adaptive immune response is for the antigen-loaded dendritic cells to meet and activate the small proportion of T cells with receptors that recognize pathogen-specific antigens. Upon entering the secondary lymphoid tissue, dendritic cells settle in the T-cell areas. There they can test the receptors of circulating T cells that enter the lymph node from the blood. These cells have yet to encounter their antigen and are called **naive** T cells. The receptors of almost all the naive T cells will not recognize the bacterial antigens displayed by the dendritic cell, and those T cells will continue their circulation, leaving the node in the efferent lymph. The minute fraction of T cells with receptors specific for a bacterial antigen will bind their antigens on the dendritic cell surface and stay in contact with the dendritic cells, where they are activated to divide and differentiate into **effector T cells**. T-cell activation is the essential first step in most adaptive immune responses, because effector T cells are almost always needed to help activate B cells. Because of this dependence, we will discuss T-cell activation further before we look at B-cell activation.

3-6 T-cell receptors recognize degraded fragments of pathogen proteins

Unlike many of the receptors of innate immunity, such as the phagocytic receptors on macrophages and neutrophils, T-cell receptors do not recognize the native structures of the surface macromolecules of a pathogen. Consequently, they do not interact directly with the pathogen itself. The antigens recognized by T-cell receptors are short peptides, about 8–25 amino acids long, generated by the degradation of a pathogen's proteins. The strategy of peptide recognition simplifies the task of making a strong and specific T-cell response, first because it concentrates on one type of macromolecule—proteins—and second because it ignores the complexity of the proteins'

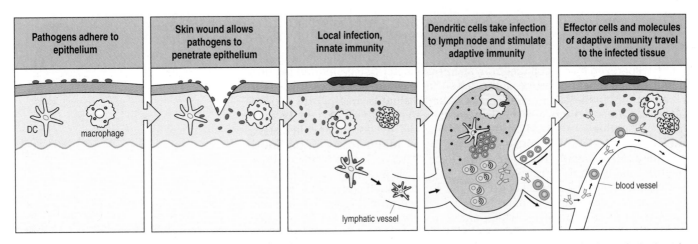

Figure 3.6 Dendritic cells carry pathogens to the draining lymph node, where they stimulate the adaptive immune response. The outer surface of the skin is always colonized by potential pathogens. A wound gains them entry to the connective tissue underlying the protective epithelium, a rich environment where they live and multiply. Here, pathogens encounter the defenses of innate immunity. If these do not eliminate the infection, dendritic cells (DCs) take up the pathogens and their components and migrate through the lymph to the draining lymph node. In the node, the infected dendritic cells settle in the T-cell areas, where they stimulate the small fraction of circulating T cells (small blue cells) that are specific for the pathogen. Stimulation of T cells by dendritic cells initiates the adaptive immune response, which eventually leads to antibodies and effector T cells (larger blue cells) traveling from the lymph node to the infected tissue via the lymph and blood.

three-dimensional structures and instead targets small, linear elements of primary structure. A variety of antigenic peptides can be generated from a typical protein, and the diversity of T-cell receptors in the T-cell population is such that a specific, adaptive T-cell response can be made to any pathogen.

Peptides destined to become T-cell antigens are produced in the dendritic cells that carry the pathogen from the infected tissue to the secondary lymphoid tissue. These cells degrade the pathogen's proteins, by using either the proteases that are normally used to remove dendritic cell proteins that are misfolded or no longer needed, or the proteases that break down proteins made by other cells of the body, and are taken up by dendritic cells. The production of antigenic peptides from pathogen proteins by these mechanisms is called **antigen processing** (Figure 3.7, first three panels).

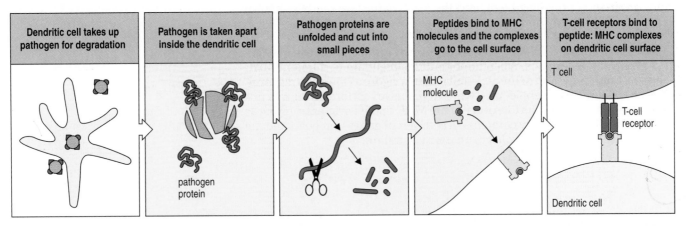

Figure 3.7 T-cell receptors recognize peptide antigens produced by the degradation of pathogen proteins.
Dendritic cells internalize pathogens and degrade their proteins into small peptides and amino acids. Some of the peptides are bound inside the cell by proteins called MHC molecules, which carry the peptides to the dendritic cell surface. Once at the surface, the complex of MHC molecule and peptide antigen is accessible to the receptors of circulating T cells. If there is a match between the antigen and the T-cell receptor, the T cell will stay in contact with the dendritic cell and be stimulated to divide and differentiate. Although other types of cell have MHC molecules and present peptide antigens, the dendritic cell is the most adept at this and is responsible for the initiation of a primary immune response—that is, the immune response to a pathogen when it infects a person and causes disease for the very first time.

3-7   T-cell receptors recognize peptide antigens bound to human cell-surface molecules

To be recognized by a T-cell receptor, an antigenic peptide generated inside a dendritic cell must be carried to the cell surface, where it is accessible to the receptors of circulating T cells. This is accomplished by having the peptide bind to a human protein also destined for the cell membrane. Inside the dendritic cell, pathogen-derived peptides are bound by glycoproteins called **MHC molecules**, which are encoded by genes in the **major histocompatibility complex** (MHC) region of the genome. Each MHC molecule binds a single peptide, and when this occurs the complex of MHC molecule and peptide (a peptide:MHC complex) is taken to the cell surface. Such complexes, and not the peptides alone, are the ligands for T-cell receptors (Figure 3.7, last two panels). In these complexes, which are very stable, the peptide is stretched out by the MHC molecule and is made linear and accessible to the peptide-binding sites in the T-cell receptor. MHC molecules are said to **present** antigens to T cells, and cells carrying peptide:MHC complexes on their surface are known as **antigen-presenting cells**. Although dendritic cells are not the only cells that can present antigens to naive T cells and activate them, they are the most specialized, and for starting off a primary T-cell response they are essential.

In contrast with T-cell receptors, which are specific for only one or a few closely related peptides, a given MHC molecule can bind many different peptides. Which peptides bind, however, is constrained by peptide length and certain features of the amino-acid sequence. Consequently, only a very small fraction of all the peptides produced by breakdown of a pathogen's proteins are presented by MHC molecules to T-cell receptors and thus act as antigens. To increase the efficiency of antigen presentation, dendritic cells express several different forms of MHC molecule, each with a different peptide-binding specificity. Unlike the variable immunoglobulins and T-cell receptors, each form of an MHC molecule is encoded by a separate gene. Adding further to MHC diversity, there are many different genetic variants, or **alleles**, for each of these genes within the human population. This means that most people are heterozygous for MHC genes (that is, they carry two different alleles of each MHC gene). This enables their dendritic cells to present a greater variety of pathogen-derived peptides than would be possible in homozygous individuals (who carry two identical alleles of a given gene). Another word for this allelic variation is **polymorphism**, and the MHC genes are the best-known examples of highly polymorphic genes.

MHC polymorphism is the basis of **tissue type** and is the chief cause of rejection of transplanted organs. When donor and recipient are of different MHC types, the immune system of the recipient makes a vigorous immune response against the donor's MHC molecules, which it perceives as 'foreign.' It was this phenomenon that led to the discovery of the MHC and its original naming as the major histocompatibility complex—a complex of genes that governed the compatibility of tissues (Greek *histo*) on transplantation.

3-8   Two classes of MHC molecule present peptide antigens to two types of T cell

Microorganisms that infect human tissues are of two broad kinds: extracellular pathogens, such as many bacteria, that live and replicate in the spaces between human cells, and intracellular pathogens, such as viruses, that live and replicate inside human cells. Because of this topological difference, proteins from extracellular pathogens are degraded within endocytic vesicles and lysosomes, whereas proteins from intracellular pathogens are degraded in the cytosol, also called cytoplasm. To aim T-cell responses at these different infections, two types of MHC molecule are used: **MHC class I** presents antigens from intracellular pathogens and **MHC class II** presents antigens from extracellular pathogens (Figure 3.8). Peptides produced by the cytosolic degradation of intracellular pathogens are delivered to the endoplasmic reticulum, where MHC class I molecules bind them and take them to the cell surface. In contrast, MHC class II molecules first travel to endosomal vesicles, where they bind peptides produced by lysosomal degradation of extracellular pathogens, before moving to the cell surface.

Corresponding to the two classes of MHC class I molecule are two specialized types of effector T cell: the **cytotoxic T cell** that defends against intracellular infection and the **helper T cell** that defends against extracellular infection. We will first examine how cytotoxic T cells work with MHC class I molecules and then how helper T cells work with MHC class II molecules.

Cytotoxic T cells are effective against intracellular infections because by killing infected cells they prevent the pathogen from replicating and spreading the infection to healthy cells. To kill an infected cell, the antigen receptor of a cytotoxic T cell must recognize a pathogen-derived peptide presented by an MHC class I molecule. To ensure that cytotoxic T cells recognize only antigens presented by MHC class I molecules they carry a cell-surface molecule called **CD8**, which binds to a conserved site on MHC class I molecules that is different from the site recognized by the T-cell receptor (Figure 3.9, left). The initial activation of a naive CD8 T cell by an infected dendritic cell in a secondary lymphoid tissue, and the mature cytotoxic T cell's killing of infected cells in a peripheral tissue, require interactions of MHC class I molecules with both the T-cell receptor and CD8. Because the CD8 molecule acts cooperatively with the T-cell receptor in activating cytotoxic T cells it is also called the **co-receptor** of cytotoxic T cells. Helper T cells cannot respond to antigens presented by MHC class I molecules because they lack CD8.

Helper T cells defend against extracellular infections by enhancing the phagocytosis of extracellular pathogens by macrophages and neutrophils. This is accomplished in several different ways, one of which is through direct contact with macrophages and the secretion of cytokines that drive the macrophages to a higher state of activation. Helper T cells also help B cells make antibodies that act as strong opsonizing agents, again aiding the phagocytosis of the pathogen (see Section 2-5, p. 38). When naive helper T cells are first activated in a secondary lymphoid tissue, their T-cell receptors interact with antigenic peptides presented by MHC class II molecules on the surfaces of dendritic cells. Such peptides are produced in the dendritic cell by the engulfment and lysosomal degradation of extracellular pathogens. When the activated helper T cells subsequently interact with B cells or macrophages, their T-cell receptors interact with the same peptide:MHC class II complexes on the surface of the B cell or macrophage. To ensure that helper T cells recognize only antigens presented by MHC class II molecules they express a co-receptor, called **CD4**, that binds to a conserved site on MHC class II molecules and functions analogously to the CD8 co-receptor of cytotoxic cells (see Figure 3.9, right).

All T cells in the circulation express either CD4 or CD8, but not both. The consequence of this mutually exclusive expression is that cytotoxic CD8 T cells respond only to antigens from intracellular infections presented by MHC class I molecules, and helper CD4 T cells respond only to antigens from extracellular infections presented by MHC class II molecules. Because all types of human cell are susceptible to infection, MHC class I molecules are expressed by almost all cells in the body. This allows cytotoxic T cells to attack the vast majority of infected cells. For helper T cells to attack the full range of extracellular infections they need to interact only with dendritic cells, macrophages and B cells, and ubiquitous expression of MHC class II molecules is limited to these three cell types.

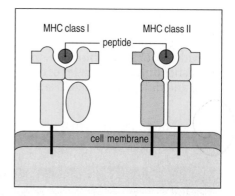

Figure 3.8 There are two types of MHC molecule, MHC class I and MHC class II. The two classes of MHC molecule have similar overall three-dimensional structures. Where they differ is in their constituent polypeptide chains. An MHC class II molecule is made of two similarly sized polypeptides that contain two extracellular domains and are anchored in the plasma membrane. In the MHC class I molecule, a larger polypeptide contains three extracellular domains and is anchored in the plasma membrane; a smaller polypeptide comprises the fourth extracellular domain and is not attached to the membrane.

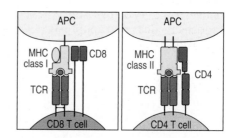

Figure 3.9 MHC class I and MHC class II molecules bind to different T-cell co-receptors. Left panel: the CD8 co-receptor of a CD8 T cells binds to an MHC class I molecule on an antigen-presenting cell (APC). Right panel: the CD4 co-receptor of a CD4 T cell binds to an MHC class II molecule on an APC. TCR, T-cell receptor. Although CD4 and CD8 perform an analogous function they have different polypeptide chains and three-dimensional structures.

In this book we will encounter many cell-surface proteins on immune system cells that are, like CD4 and CD8, numbered in the 'CD' series. More than 240 such molecules, or groups of molecules, have been numbered so far. Individually or in combination, the cell-surface expression of CD markers distinguishes different lineages of cell types and different stages in their development and differentiation. CD stands for 'clusters of differentiation.'

3-9 MHC class I molecules present antigens of intracellular origin to CD8 T cells

Cytotoxic CD8 T cells function in the adaptive immune response to intracellular infections, in which their role is to kill infected cells. Of the four main groups of pathogens (see Figure 1.4), viruses are the ones that must always get inside human cells to replicate and survive, using the cell's own machinery for protein synthesis to do so. Cytotoxic T cells are thus important in viral infections. Some types of bacteria also live and replicate inside human cells and stimulate cytotoxic T-cell responses.

At the beginning of the adaptive immune response to a virus, virus-infected dendritic cells travel from the infected site to a secondary lymphoid tissue. Viral proteins are synthesized on the ribosomes of the dendritic cells and are degraded by the cytosolic proteases that normally degrade damaged or faulty human proteins. The peptides produced are transported into the endoplasmic reticulum, where they can bind to MHC class I molecules. Once an MHC class I molecule has bound a peptide, the complex leaves the endoplasmic reticulum and is transported through the Golgi apparatus to the plasma membrane. A virus-infected dendritic cell thus displays complexes of MHC class I molecules and viral peptides on its surface. In the secondary lymphoid tissue, the infected dendritic cells encounter circulating naive CD8 T cells with receptors specific for the viral peptide:MHC class I complexes on the dendritic cell surface. This first encounter with their corresponding antigen promotes the activation and clonal expansion of these CD8 T cells and their differentiation into highly cytotoxic effector CD8 T cells.

At this point the cytotoxic CD8 T cells leave the secondary lymphoid tissue and travel via the lymph and the bloodstream to the infected tissue, where virus-infected cells will be displaying complexes of MHC class I molecules and viral peptides on their surface. When the T-cell receptors and CD8 co-receptors bind to these complexes, the interaction triggers the cytotoxic T cell to deliver a battery of toxic substances onto the surface of the infected cell, which induces its death by apoptosis (Figure 3.10). Destruction of the infected cell interrupts viral replication and prevents the further infection of healthy cells.

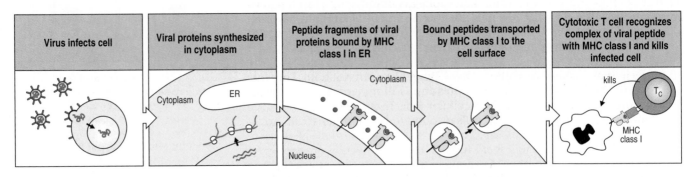

Figure 3.10 **The MHC class I pathway presents antigens derived from intracellular infections to cytotoxic CD8 T cells.** In virus-infected cells, new viral proteins are made on ribosomes in the cytoplasm. Some of these proteins are degraded in the cytoplasm and the resultant peptides are transported into the endoplasmic reticulum (ER). MHC class I molecules bind peptides in the ER and then transport them to the surface of the infected cell, where they can be recognized by a virus-specific effector cytotoxic T cells (T_C). The cytotoxic T cell kills the virus-infected cell.

Viral infections can target almost any cell in the human body, an exception being the red blood cell, which lacks a nucleus and cannot support viral replication. MHC class I molecules are made constitutively by most nucleated cell types, and at infected sites the cytokines produced by the innate immune response increase the levels of MHC class I expression. Such ubiquitous expression of MHC class I molecules allows effector cytotoxic CD8 T cells to recognize and kill any type of cell infected with a virus.

3-10 MHC class II molecules present antigens of extracellular origin to CD4 T cells

About a third of the T lymphocytes circulating in the bloodstream of a healthy individual are CD8 T cells that recognize peptide antigens presented by MHC class I molecules. The other two-thirds carry the CD4 co-receptor. When naive CD4 T cells are activated by antigen they differentiate into helper T cells, which are the **effector CD4 T cells** of the adaptive immune response. Helper CD4 T cells have the potential to secrete many different cytokines and are functionally more versatile than cytotoxic CD8 T cells. They exist in several different functional subtypes and contribute to the adaptive immune response against all types of pathogen. The characteristic feature of CD4 T cells is that they respond to pathogens present in the fluid and spaces that separate cells and tissues. All four types of pathogen—bacteria, viruses, fungi, and parasites—can infect these extracellular locations.

CD4 T-cell function is directed to extracellular infections by the presentation of pathogen-derived peptides by MHC class II molecules. At sites of extracellular infection dendritic cells use a variety of cell-surface receptors to take up pathogens from the extracellular milieu by endocytosis (as described in Section 2-10 for macrophages), and they then degrade the pathogen proteins into peptides in endocytic vesicles. MHC class II molecules are assembled in the endoplasmic reticulum but, unlike MHC class I molecules, this is not where they bind peptides. They are transported through the Golgi apparatus to the endocytic vesicles, which contain peptides derived from the breakdown of extracellular pathogens. Here the MHC class II molecules bind these peptides, after which they move to the plasma membrane.

Dendritic cells carrying complexes of MHC class II molecules and peptide antigens on their surface move from infected tissue to a secondary lymphoid organ. Interaction of these complexes with the antigen receptors on naive CD4 T cells stimulates these cells to proliferate into several different types of helper T cell. Some travel in the lymph and blood to the infected tissue, where they secrete cytokines that attract neutrophils and monocytes to the tissue. The latter mature into macrophages. In the tissue, helper T cells interact directly and specifically with those macrophages that have ingested pathogens for which the T cells are specific. Like dendritic cells, macrophages process the pathogens they take up, and display pathogen-derived peptides in combination with MHC class II molecules on the macrophage surface. Helper T cells specific for these peptide:MHC class II complexes interact with the macrophage and make it more effective at killing the pathogens it has engulfed (Figure 3.11). As described in the next section, another population of helper T cells remains in the secondary lymphoid organ. There they help stimulate B cells to produce antibodies, the main weapon of the adaptive immune response.

3-11 Effector CD4 T cells help B cells become antibody-producing plasma cells

Whereas T cells have a variety of functions in the adaptive immune response, B cells have a sole purpose, namely to make antibodies. These are soluble

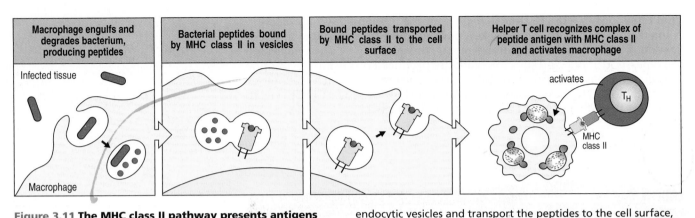

| Macrophage engulfs and degrades bacterium, producing peptides | Bacterial peptides bound by MHC class II in vesicles | Bound peptides transported by MHC class II to the cell surface | Helper T cell recognizes complex of peptide antigen with MHC class II and activates macrophage |

Figure 3.11 The MHC class II pathway presents antigens derived from extracellular infections to helper CD4 T cells. In a tissue infected with bacteria, a resident macrophage is shown phagocytosing extracellular bacteria and degrading their proteins in endocytic vesicles. MHC class II molecules bind peptides in endocytic vesicles and transport the peptides to the cell surface, where they are recognized by a helper CD4 T cell (T_H). Through cell contact and release of cytokines the helper T cell activates the macrophage, making it more effective in killing bacteria.

forms of the immunoglobulin molecules that B cells use as their antigen receptors (see Figure 3.1). Activation of B cells by a pathogen leads to proliferation and differentiation of the B cells into dedicated antibody factories called **plasma cells**. Depending on the site of infection, antibodies are poured into the circulation or onto mucosal surfaces, where they bind tightly to pathogens and prevent their exploitation of the human body. Sometimes the binding of antibody is sufficient to achieve this goal, as when antibody bound to the influenza virus prevents the virus from infecting human cells. More often, the antibody serves to target the pathogen for destruction by a phagocyte or other effector cell.

Although antibodies can be made against all manner of chemical structures, the most useful antibodies are those that bind tenaciously to the outsides of live pathogens. The antigens that these antibodies recognize are parts of the native structures of the macromolecules that make up the outer surface of the pathogen (Figure 3.12). This emphasizes the fundamental difference and complementarity of the antigens recognized by B-cell and T-cell receptors. Whereas T-cell receptors recognize amino-acid sequences in the degraded fragments of a pathogen's proteins, the useful antibodies are those that recognize surface elements in complete biological macromolecules.

When dendritic cells move from an infected tissue to a lymph node, they carry the infection with them. Bacteria and virus particles are also carried in the lymph to the draining lymph node. Intact pathogens or pathogen proteins within the lymph node can then interact with the antigen receptors of circulating B cells that enter the node from the blood. B cells that have yet to encounter their antigen are called **naive B cells**. Those naive B cells with receptors specific for antigens on the surface of a bacterium, for example, will bind to the bacterium and stay in the lymph node, whereas B cells with other receptors will leave the lymph node in the efferent lymph. Binding of a bacterium to the B-cell receptor induces a process called **receptor-mediated endocytosis**, which causes the bacterium to be engulfed and delivered to endocytic vesicles in the B-cell cytoplasm, in which it is killed and degraded. Peptides derived from the proteins of the internalized pathogen are bound by MHC class II molecules in the endocytic vesicles and delivered to the B-cell surface. B cells can therefore act as antigen-presenting cells to CD4 T cells, a feature that is essential for their differentiation into antibody-producing cells.

Recognition of antigen alone is not usually sufficient to drive naive B cells to differentiate and produce antibody. For this, they need the help of helper CD4 T cells that have been activated by the same antigen. In our example, the lymph node will already contain clones of helper CD4 T cells that have been

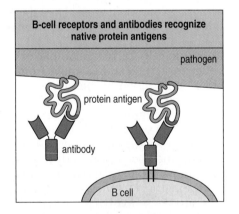

Figure 3.12 B cells recognize native macromolecules on the outside surfaces of pathogens. Whereas T-cell receptors recognize only short degraded fragments of a pathogen's proteins bound by MHC molecules, the immunoglobulin receptors and soluble antibodies of B cells directly recognize the native conformations of proteins, as well as the many other types of macromolecule present on pathogen surfaces, including carbohydrates, proteoglycans, glycoproteins, and glycolipids.

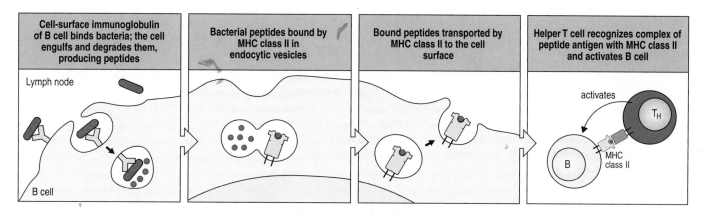

| Cell-surface immunoglobulin of B cell binds bacteria; the cell engulfs and degrades them, producing peptides | Bacterial peptides bound by MHC class II in endocytic vesicles | Bound peptides transported by MHC class II to the cell surface | Helper T cell recognizes complex of peptide antigen with MHC class II and activates B cell |

Figure 3.13 Helper CD4 T cells help activate B cells to make antibody. Pathogens are brought by the lymph from a site of infection to the draining lymph node. In the lymph node a pathogen-specific B cell binds the pathogen (illustrated here as a bacterium) via its cell-surface immunoglobulin. The bound bacterium is endocytosed and its proteins are degraded into peptides. Some of these are bound by MHC class II molecules in endocytic vesicles and transported to the cell surface, where they can be recognized by the antigen receptor of a helper CD4 T cell (T$_H$). Through cell contact and secretion of cytokines the helper T cell helps to activate the B cell fully. That then divides and differentiates into plasma cells secreting antibody against the bacterium (not shown).

activated by complexes of bacterial peptides and MHC class II molecules on dendritic cells. Identical peptide:MHC class II complexes are also present on the surface of a B cell that has engulfed and degraded the same bacteria (Figure 3.13). A helper CD4 T cell with receptors of the right specificity can thus bind to the B cell via these complexes. This mutual interaction causes the helper CD4 T cell to secrete cytokines that, together with signals from the antigen-bound B-cell receptor itself, help drive the B cell to proliferate and differentiate into plasma cells.

The important principle that governs cellular cooperation in the antibody response is that the interacting T cell and B cell must both be specific for the same or different parts of the same antigen—in this case a bacterial cell. In vaccine design, in which the goal is usually a strong antibody response, it is essential to incorporate protein sequences into the vaccine that will stimulate a strong response by the CD4 T cells that are needed to activate the appropriate B cells.

3-12 Extracellular pathogens and their toxins are eliminated by antibodies

The immunoglobulins produced by B cells and plasma cells are divided into five **classes** or **isotypes**, which differ in their heavy-chain constant regions and have specialized functions when secreted as antibodies. The isotypes are called **IgA**, **IgD**, **IgE**, **IgG**, and **IgM**, where Ig stands for immunoglobulin. Cell-surface IgM and IgD are the antigen receptors on circulating B cells that have yet to encounter antigen. IgM is always the first antibody to be secreted in the immune response (IgD antibody is also made, but in negligible amounts). As the immune response matures, antibodies of other isotypes are made. Immunity due to antibodies and their actions is often known as **humoral immunity**. This is because antibodies were first discovered circulating in body fluids (*humors*) such as blood and lymph.

IgM, IgA, and IgG are the main antibodies present in blood, lymph, and the intercellular fluid in connective tissues. Antibodies secreted into the blood from lymph nodes, spleen, and bone marrow circulate in the bloodstream. At sites of infection and inflammation, blood vessels dilate and become permeable, increasing the flow of fluid into the infected tissue and with it the supply of antibodies for binding to extracellular pathogens (bacteria and virus particles) and to the protein toxins secreted by many bacterial pathogens. IgA is also made in the lymphoid tissues underlying mucosa and then selectively transported across the mucosal epithelium to bind extracellular pathogens and their toxins on the mucosal surfaces.

One way in which antibodies reduce infection is by binding tightly to a site on a pathogen so as to inhibit pathogen growth, replication, or interaction with human cells. This mechanism is called **neutralization**. For example, certain antibodies when bound to an influenza virus particle prevent the virus from infecting human cells. Similarly, the lethal actions of bacterial toxins can be prevented by a bound antibody (Figure 3.14).

The most important function of IgG antibodies is to facilitate the engulfment and destruction of extracellular microorganisms and toxins by phagocytes. Neutrophils and macrophages have cell-surface receptors that bind to the constant regions of the IgG heavy chains. A bacterium coated with IgG is

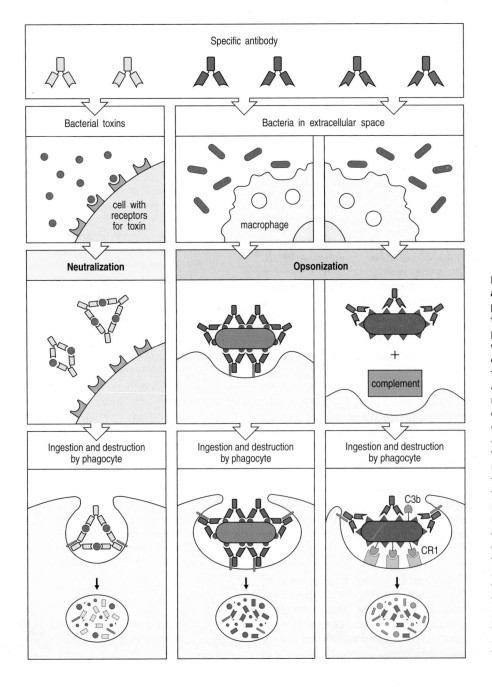

Figure 3.14 Mechanisms by which antibodies combat infection. Left panels: antibodies bind to a bacterial toxin and neutralize its toxic activity by preventing the toxin from interacting with its receptor on human cells. The complex of toxin and antibodies binds to macrophage receptors through the antibody's constant region. Finally the macrophage ingests and degrades the complex. Center panels: the opsonization of a bacterium by coating it with antibody. When the bacterium is coated with IgG antibodies, their constant regions point outward and can bind to the receptors on a phagocyte, which then ingests and degrades the bacterium. Right panels: opsonization of a bacterium by a combination of antibody and complement. The bacterium is first coated with IgG molecules, which trigger the classical pathway of complement activation. C3b fragments fixed to the bacterial surface provide ligands for the complement receptor CR1 on macrophages. The combined interaction of phagocyte receptors for complement and for the constant region of IgG makes for efficient phagocytosis.

therefore more efficiently phagocytosed than an uncoated bacterium, a phenomenon called **opsonization** (see Figure 3.14). Opsonization can also be achieved by a coating of complement (see Section 2-5). There are no receptors on phagocytes for the constant region of IgM antibodies, but IgM on the surface of a pathogen binds to a complement component that initiates the classical pathway of complement activation. As a consequence, the pathogen becomes coated with C3b (see Chapter 2), facilitating its uptake and phagocytosis by macrophages via their complement receptors. A combination of IgG antibody and complement produces the greatest stimulation of phagocytosis because macrophage receptors for IgG constant regions and complement both participate in the process (see Figure 3.14).

The constant regions of IgE antibodies bind tightly to receptors on the surface of mast cells. Most of the IgE antibody in the body is in this form, rather than circulating freely in blood or lymph like IgM and IgG. In response to worms and other parasitic infestations, the IgE bound to mast cells triggers strong inflammatory reactions that help expel or destroy the parasites. However, in developed countries, where parasitic infections are not a major health problem, IgE is most often encountered as the antibody involved in unwanted allergic reactions to innocuous substances such as plant pollens.

3-13   Antibody quality improves during the course of an adaptive immune response

Antibodies are adaptor molecules that bring together pathogens and the means for their destruction. Antigen-binding sites formed by the immunoglobulin variable regions bind to pathogens, whereas binding sites formed by the immunoglobulin constant regions trigger complement fixation on the pathogen surface and binding of the pathogen to specific receptors on phagocytes and other effector cells.

The antigen receptors of naive B cells are always of the IgM isotype, and the first antibody secreted by a clone of B cells that has been activated by specific antigen is IgM. As clones of activated B cells expand in the germinal centers of secondary lymphoid tissues (see Section 1-9, p. 19), two mutational mechanisms improve the effectiveness of the antibodies they will make. The first is a process called **somatic hypermutation**, which introduces nucleotide substitutions throughout the variable regions of the rearranged immunoglobulin heavy- and light-chain genes. Some of the variant immunoglobulins produced by somatic hypermutation bind the pathogen more tightly than the original B-cell receptors, and the cells producing them are preferentially chosen to become antibody-secreting plasma cells (Figure 3.15).

Figure 3.15 Somatic hypermutation and isotype switching during an adaptive immune response improve the quality of the antibody that is made. At the beginning of an adaptive immune response to infection, IgM is the only antibody made (left panel). In the population of B cells that are making pathogen-specific IgM, somatic hypermutation of the variable regions of the heavy- and light-chain genes produces some new variant antibodies that bind more tightly to the pathogen (center panel). Isotype switching to IgG changes the constant region of the heavy chain (right panel). In this case it provides the antibody with a binding site for a receptor on phagocytes that is specific for IgG and does not bind IgM. Binding of the IgG-coated pathogen to this IgG receptor ensures its phagocytosis and destruction. Whereas somatic hypermutation strengthens the antigen-binding properties of the antibodies, isotype switching improves their recruitment of effector functions. Together these two processes improve the ability of antibodies to remove pathogens.

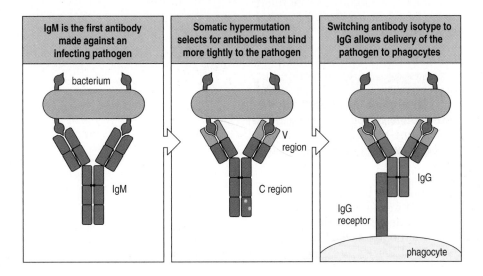

| IgM is the first antibody made against an infecting pathogen | Somatic hypermutation selects for antibodies that bind more tightly to the pathogen | Switching antibody isotype to IgG allows delivery of the pathogen to phagocytes |

The second mutational mechanism that acts in activated B cells is called **iso-type switching** or **class switch recombination** and changes the isotype of the immunoglobulin but not its antigen-binding site (see Figure 3.15). In B cells making IgM, an array of untranscribed gene segments encoding the constant regions of all the other immunoglobulin isotypes lies downstream of the transcribed gene segment encoding the μ chain constant region. Germinal center B cells are able to remove the gene segment encoding the μ constant region and bring the rearranged variable region into juxtaposition with a gene segment encoding a different constant region. The length of the DNA segment removed can vary, enabling the isotype to be switched from IgM to IgG, IgA, or IgE. In the lymph nodes and spleen, where the adaptive immune responses to infections in tissues and blood are made, the predominant switch is from IgM to IgG. In the Peyer's patches of gut epithelium and the lymphoid tissues that line the mucosa, the predominant switch is from IgM to IgA. In parasite infections, the switch of isotype from IgM to IgE enables eosinophils, basophils, and mast cells to be recruited to the adaptive immune response. These cells, which have specialized weapons for killing parasites, all have a cell-surface receptor that binds tightly to the IgE constant region.

Through the combination of somatic hypermutation and isotype switching, the adaptive immune response selects for antibodies that bind more tightly to the pathogen, are more efficiently delivered to the site of infection, and are better able to involve phagocytes and other inflammatory cells in pathogen elimination. All these improvements are retained by a person's B-cell population after the infection has ended, and they comprise one aspect of immunological memory. On a subsequent infection with the same pathogen, high-affinity antibodies of the most suitable isotype can be made from the very beginning of the infection.

3-14 Immunological memory is a consequence of clonal selection

In the course of the adaptive immune response to infection, clones of pathogen-specific T cells and B cells are expanded to provide the means for a targeted attack on the pathogen by effector T cells and antibodies. Once an infection has been successfully terminated, regulatory mechanisms come into play to stop the immune response, reduce inflammation and allow damaged tissues to be repaired. Effector T cells, whose cytokines often contribute to inflammation, are signaled to die off, but plasma cells are allowed to persist, and pathogen-specific antibodies can be maintained in the circulation for years. The benefit of this is that high-affinity antibodies that bind tightly to a pathogen's surface and target it for phagocytosis are the most effective way to halt a subsequent infection at a very early stage. Antibodies are the main means of preventing recurrent disease during the annual episodes of colds and other common viral diseases that affect humans.

The clonal expansion of pathogen-specific B cells and T cells not only provides effector cells for the short-term fight against an ongoing infection but also produces long-lived clones of **memory cells**. These cells underlie the **immunological memory** that ensures that all subsequent infections with the same pathogen will be met with a faster and stronger response than during the first infection. Second and subsequent immune responses to the same pathogen are known as **secondary adaptive immune responses** (see Figure 1.26). Immunological memory can last for life.

The greater power of a secondary immune response is the result of several factors. The specific memory cells that respond to the pathogen are more numerous than the naive lymphocytes that responded in the primary infection. Memory cells are more quickly activated than naive cells. Unlike naive T cells, memory T cells can patrol non-lymphoid tissues and thus detect infection at an earlier stage. Memory B cells make better immunoglobulins than naive B cells, because they benefit from the somatic hypermutations and isotype switching that occurred in the primary response. Because of the far superior response of memory cells, activation of naive pathogen-specific lymphocytes is completely shut down when pathogen-specific memory cells are available.

3-15 Clonal selection makes T cells and B cells tolerant of self and responsive to pathogens

The power of the adaptive immune system is its versatility—its capacity to make immunoglobulins and T-cell receptors with a virtually infinite number of different binding-site specificities. The drawback of this is that some fraction of developing B cells and T cells inevitably carry receptors that bind to normal components of the healthy human body. Such lymphocytes have the potential to attack and damage healthy tissues, and several mechanisms have evolved to prevent this. These mechanisms are said to induce a state of **immunological tolerance** towards **self**, where self refers to the entire molecular constituents of the healthy human body.

Tolerance to some important **self antigens** is set up during lymphocyte development. During T-cell development in the thymus, immature T cells called thymocytes are subject to two rounds of clonal selection. The first is called **positive selection** because it identifies and selects T cells bearing antigen receptors that work effectively with the particular subset of MHC class I and class II molecules that a person has (Figure 3.16, top three panels). T cells recognize their peptide antigens only in combination with the body's own MHC molecules (**self-MHC**), and so useful T cells must have receptors that recognize features of self-MHC molecules as well as peptide antigens. In the human population as a whole there are more than 1000 variant forms of MHC molecule, and each person has fewer than 20 of them. As a consequence, each person makes numerous T cells bearing receptors that bind poorly to their own MHC molecules (but would bind well to the different MHC molecules of other people). Those immature T cells that can bind to a self-MHC molecule are given a signal that allows them to survive, whereas the remaining T cells, which bind weakly to self-MHC, die by apoptosis.

T cells surviving positive selection are then subject to a process of **negative selection**. This identifies T cells bearing receptors that bind too strongly to self-MHC and are likely to be capable of attacking healthy cells expressing these proteins. Such cells are also signaled to die by apoptosis. So stringent are these selections that fewer than 1% of thymocytes mature to become circulating naive T cells (see Figure 3.16, bottom two panels). The negative selection imposed on potentially self-reactive T cells helps to render the circulating pool of T cells non-responsive to, or **tolerant of**, the normal components of the body. The T-cell repertoire is thus said to be **self-tolerant**.

B-cell activation depends on helper T cells that recognize the same antigen as the B cell (see Figure 3.13). Because of this requirement, the negative selection of T cells reactive to a self protein also means that B cells reactive to that

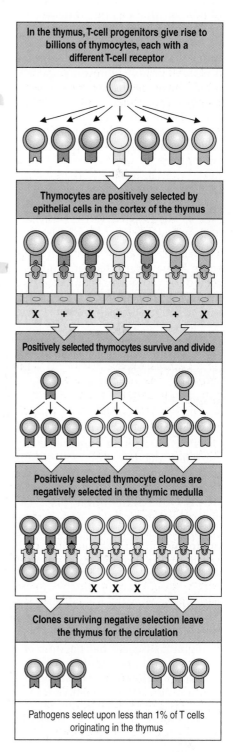

Figure 3.16 Stringent processes of positive and negative selection in the thymus determine the small fraction of T cells that mature and enter the peripheral circulation. Different clones of T cells are represented by different colors. Positive selection by peptide:MHC complexes on thymic cells ensures that the T cells that survive are able to interact with self-MHC molecules. Negative selection on macrophages in the thymic medulla ensures that T cells that react strongly against self-antigens presented by self-MHC molecules are eliminated. The resulting naive T-cell repertoire represents a very small fraction of the total diversity of T-cell receptors generated.

protein cannot be activated. Consequently, it is not so critical to eliminate all B cells with receptors that bind to components of the body in order to maintain self-tolerance. Nevertheless, during the maturation of B cells in the bone marrow, negative selection is used to prevent the emergence of some cells that could make potentially harmful antibodies. Clones of B cells with immunoglobulin receptors that bind strongly to the constituents of bone marrow tissue are negatively selected and die by apoptosis.

Apoptosis is a common feature of pathways of lymphocyte development, activation and effector function. Individual cells are induced to commit a tidy form of cell suicide, in which there is no generalized and messy tissue damage of the sort that cell death by necrosis produces. Membrane changes in cells undergoing apoptosis lead to their phagocytosis by macrophages.

So far, we have seen how the elimination of clones of self-reactive lymphocytes is used to produce immunological tolerance. This is sometimes called central tolerance because it is achieved in the primary lymphoid organs. Other mechanisms of tolerance, called peripheral tolerance, occur in the circulation, the secondary lymphoid organs and the rest of the body—the periphery. One of these involves a subset of effector CD4 T cells called **regulatory T cells** that suppress the responses of T cells reactive against healthy cells and tissues.

3-16   Unwanted effects of adaptive immunity cause autoimmune disease, transplant rejection and allergy

The sensitivity and force of the adaptive immune response, which are helpful traits in battling infection, can under some circumstances become misdirected toward normal components of the human body or innocuous materials in the environment. When this happens, various kinds of chronic and noninfectious disease emerge.

Despite the mechanisms for creating and maintaining self-tolerance, a wide range of diseases are caused by adaptive immune responses that become chronically misdirected towards self and gradually erode a target tissue. These conditions are called **autoimmune diseases** and result from **autoimmune responses** directed against normal components of healthy human cells. In insulin-dependent diabetes, for example, the insulin-secreting β cells of the islets of Langerhans in the pancreas are the target of an autoimmune response. Because the human pancreas can produce insulin well in excess of what is normally needed, disease symptoms do not appear until long after the destructive response has become established. Pancreatic β-cell destruction is believed to be due to B and T cells that respond to proteins made only in β cells. Other autoimmune diseases are rheumatoid arthritis, multiple sclerosis, myasthenia gravis, and Graves' disease. Autoimmune responses often seem to be by-products of a protective immune response made against an acute microbial infection. Once activated by pathogen-derived antigens, certain B or T cells cross-react with self components to which they were previously tolerant (Figure 3.17).

Over the past century, hygiene, vaccination, and antimicrobial drugs have all served to reduce mortality from infectious disease in the developed countries, particularly for children. In these populations the treatment of chronic and malignant diseases has become of growing importance in medicine. For an increasing number of these diseases, the replacement of tissues or organs by **transplantation** is now being used to restore health and prolong life. A major factor limiting the benefits of transplantation are the tissue incompatibilities caused by the extensive polymorphism of MHC class I and class II genes in the human population. MHC-compatible donors can be found within families, but only for some of the patients needing a transplant. For patients seeking a

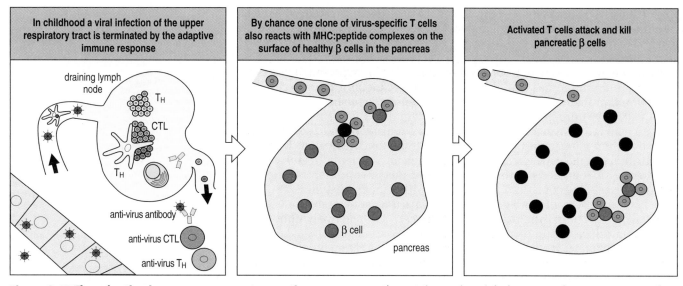

| In childhood a viral infection of the upper respiratory tract is terminated by the adaptive immune response | By chance one clone of virus-specific T cells also reacts with MHC:peptide complexes on the surface of healthy β cells in the pancreas | Activated T cells attack and kill pancreatic β cells |

Figure 3.17 The adaptive immune response to a pathogen can sometimes trigger autoimmune reactions that eventually cause disease. Insulin-dependent diabetes mellitus (type 1 diabetes mellitus) is an autoimmune disease in which the insulin-producing β cells of the pancreas are gradually destroyed. A history of viral infection, for example with the Coxsackie viruses, has been correlated with type 1 diabetes and in such cases the autoimmunity might be a secondary consequence of the adaptive immune response to the virus. This could happen if a clone of T cells activated by the virus has, by chance, a T-cell receptor that reacts with an peptide:MHC complex present on the surface of pancreatic β cells. This peptide:MHC complex will then continually re-stimulate the T cells, which in turn persistently attack the β cells.

transplant from an unrelated donor, the chances of finding one who is MHC identical are low. In practice, therefore, most clinical transplants are performed across MHC differences that have the potential to stimulate strong immune responses involving all the arms of adaptive immunity. Currently, the only means of preventing transplant rejection is the prolonged use of drugs that suppress adaptive immunity. Although facilitating graft acceptance, such drugs have toxic effects and also render the transplant patient more vulnerable to infection and cancer.

A state of **allergy** or **hypersensitivity** arises when the immune system reacts against innocuous substances in the environment, such as foods, grass pollens, house dust, or dander from pets. Any such non-infectious antigen that causes an allergy is known as an **allergen**. The most common types of allergy, such as hay fever and asthma, are caused when some people make antibodies of the IgE isotype on first encounter with the allergen. The IgE antibodies circulate and become bound by their constant regions to mast cells in tissues. On a repeat exposure to the allergen, it binds to the IgE on mast cells. The combination of IgE and allergen causes the mast cell to release inflammatory substances that produce symptoms very similar to those caused by some infections (Figure 3.18), and which sometimes lead to violent and life-threatening reactions such as those of asthma and systemic anaphylaxis. The damage and incapacitation caused by allergic reactions is greatly out of

Figure 3.18 Many allergies are caused when antibodies of the IgE isotype are made against innocuous substances. The example shown here is of an allergy to airborne pollen, such as ragweed pollen, that is breathed in and stimulates an allergic response in the mucosa of the respiratory system. For genetic and environmental reasons some people make an IgE antibody response to an antigen in the pollen. Once made, the IgE antibody becomes bound to mast cells in the mucosal tissues, where it acts as a receptor. On subsequent exposure to pollen, the pollen allergen binds to and cross-links the cell-bound IgE, sending a signal for the mast cell to release the granules it contains. These granules contain cytokines and other substances, such as histamine, that induce inflammation, sneezing, and other symptoms that are usually associated with infectious diseases.

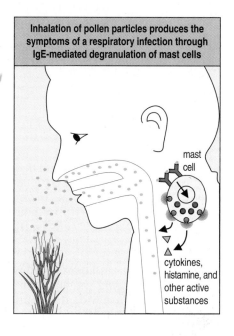

Inhalation of pollen particles produces the symptoms of a respiratory infection through IgE-mediated degranulation of mast cells

proportion to the threat posed by the allergen. Not all hypersensitivity diseases are caused by IgE; some are mediated by IgG, CD4 T cells, or CD8 cytotoxic T cells.

Many of the unwanted actions of adaptive immunity in hypersensitivity diseases, in autoimmune diseases, and in transplant recipients are due to chronic states of inflammation that facilitate a persistent response by inflammatory cells of adaptive immunity: CD4 T cells, CD8 cytotoxic T cells, and mast cells charged with IgE. The overall incidence of hypersensitivity and autoimmune diseases is increasing in the richer countries, a trend that the **hygiene hypothesis** attributes to the widespread practice of hygiene, vaccination, and antibiotic therapy. It is argued that in growing up in this environment, children's immune systems become less able to deal with proper infections and are more likely to charge after imaginary foes in ways that, at best, are not helpful and, at worst, are literally self-defeating.

Summary to Chapter 3

The mechanisms unique to adaptive immunity are ones that improve pathogen recognition rather than pathogen destruction. Adaptive immune responses are due to the lymphocytes of the immune system, which collectively have the ability to recognize a vast array of antigens. Somatic gene rearrangement and somatic mutation in the genes for antigen receptors provide the lymphocyte population with a set of highly diverse antigen receptors—immunoglobulins on B lymphocytes, and T-cell receptors on T lymphocytes. An individual lymphocyte expresses receptors of a single and unique antigen-binding specificity. A pathogen therefore stimulates only the small subset of lymphocytes that express receptors for the pathogen antigens, focusing the adaptive immune response on that pathogen.

One major difference between B and T cells is the type of antigens they recognize. Whereas the immunoglobulin receptors of B cells bind whole molecules and intact pathogens, T-cell receptors recognize only short peptide antigens that are bound to major histocompatibility complex molecules on cell surfaces. A second major difference is that all B cells have the same general function of making immunoglobulins, whereas T cells can have distinctive functions. CD8 cytotoxic T cells kill cells infected with viruses by recognizing viral peptide antigens presented by MHC class I molecules; CD4 T cells help macrophages and B cells become activated by recognizing peptides presented by MHC class II molecules. The CD4 T cells are further subdivided into ones that help activate macrophages, and others that help activate B cells. Before they can perform these functions, all T cells must themselves be activated within a secondary lymphoid tissue through recognition of their specific antigen on a professional antigen-presenting cell.

B cells differentiate into plasma cells that make antibody, a secreted form of the cell-surface immunoglobulin. The production of antibodies by a B cell usually requires help from a CD4 T cell that responds to the same antigen and activates the B cell. Antibodies bind to extracellular pathogens and their toxins and deliver them to phagocytes and other effector cells for destruction. Immunity due to antibodies and their actions is known as humoral immunity. Whereas B-cell activating CD4 T cells work within secondary lymphoid tissues, CD8 cytotoxic T cells and macrophage-activating CD4 T cells must travel to the infected site to perform their functions. Immunity that is due principally to CD8 cytotoxic T cells and/or macrophage-activating CD4 T cells is known as cell-mediated immunity.

When successful, an adaptive immune response terminates infection and provides long-lasting protective immunity against the pathogen that provoked the response. Adaptive immunity is an evolving process within a person's lifetime, in which each infection changes the make-up of that individual's lymphocyte population. These changes are neither inherited nor passed on but, during the course of a lifetime, they determine a person's fitness and their susceptibility to disease. Failures to develop a successful adaptive response can arise from inherited deficiencies in the immune system or from the pathogen's ability to escape, avoid, or subvert the immune response. Such failures can lead to debilitating chronic infections or death. Another category of failure of adaptive immunity is the chronic disease caused by a misdirected response. Allergy is the consequence of a strong response to a harmless substance, whereas autoimmune disease is caused when the destructive potential of the immune system is directed at one or more of the body's own tissues. A further medical context for unwanted adaptive immunity is transplantation. Here the differences in MHC type between donor and recipient trigger an immune response that, if left unchecked, will reject the transplanted tissue.

Questions

3–1   Give three advantageous features that distinguish vertebrate adaptive immune responses from the innate immune responses made by both vertebrates and invertebrates.

3–2   The process by which a pathogen stimulates only those lymphocytes with receptors specific for that pathogen is called
  a. germline recombination
  b. somatic recombination
  c. clonal selection
  d. antigen processing
  e. antigen presentation.

3–3   Explain why a pathogen is most capable of causing disease the first time it is encountered.

3–4   Which of the following statements characterizing B-cell receptors and T-cell receptors is incorrect?
  a. Antibodies are a secreted form of the B-cell receptor, whereas immunoglobulins are a transmembrane form.
  b. The diversity of B-cell and T-cell receptors is generated by the same molecular process.
  c. The constant region of immunoglobulins differs between immunoglobulins, whereas the variable region is conserved.
  d. B-cell receptors and T-cell receptors are expressed as transmembrane polypeptides on the cell surface.
  e. T-cell receptors are more restricted than B-cell receptors in the range of antigens to which they bind.

3–5   The process of _____ results in the recombination of gene segments of the T-cell receptor and immunoglobulin genes giving rise to lymphocytes with unique specificities for antigens:
  a. germline recombination
  b. somatic recombination
  c. clonal selection
  d. antigen processing
  e. antigen presentation.

3–6   During a typical infection, dendritic cells first encounter pathogen in _____ and then transport the pathogen to _____ where pathogen-derived peptide fragments are presented to _____:
  a. secondary lymphoid tissue; infected peripheral tissue; T cells
  b. secondary lymphoid tissue; infected peripheral tissue; B cells
  c. endocytic vesicles; cell surface; MHC molecules
  d. infected peripheral tissue; secondary lymphoid tissue; T cells
  e. infected peripheral tissue; secondary lymphoid tissue; B cells.

3–7   Explain the fundamental differences between CD4 and CD8 T cells in terms of (A) their effector function and (B) their interaction with MHC molecules.

3–8
  A. Identify the two major sources of the pathogen-derived antigens recognized by T cells.
  B. Explain why it is advantageous to the host to use different effector mechanisms to combat antigens encountered in these two distinct locations.

3–9   Describe how (A) MHC class I molecules and (B) MHC class II molecules encounter the peptides to which they will bind and ultimately present to T cells.

3–10   Which of the following is paired correctly?
  a. IgG: present in minute amounts in the blood
  b. IgA: first antibody secreted during a B-cell response
  c. IgG: transported across mucosal epithelium
  d. IgD: complement fixation
  e. IgE: mast cells and defense against parasites.

3–11   Discuss why a vaccine based solely on carbohydrate moieties (for example capsular polysaccharides) of a pathogen would be ill conceived if the goal was to stimulate not only antibody responses but also strong helper T-cell responses.

3–12   Discuss (A) the similarities and (B) the differences between neutralization and opsonization.

3–13   If there are no receptors on macrophages for the constant region of IgM, then how does IgM coating a pathogen promote its phagocytosis?

3–14   Describe briefly the two mutational mechanisms that drive the enhancement of antibody quality in activated B cells.

3–15   Describe three types of unwanted and potentially harmful immune response.

3–16   At 42 years old, Stephanie Goldstein developed occasional blurred and double vision, numbness and "pins and needles" in her arms and legs (paresthesia), and bladder incontinence. After a month of these symptoms she went to her doctor, who sent her to the neurology specialist. An MRI scan revealed areas of demyelination in the central nervous system (CNS) and Stephanie was diagnosed with the autoimmune disease multiple sclerosis (MS). Which of the following best explains why some people are susceptible to the development of MS?
    a. Negative selection of autoreactive T cells occurs during T-cell development.
    b. Apoptosis of autoreactive B cells occurs in the bone marrow during B-cell development.
    c. An inability to produce immunological tolerance toward CNS-derived constituents results in the generation of self-reactive lymphocytes.
    d. An immunodeficiency inhibiting somatic recombination of immunoglobulins and T-cell receptors results in impaired lymphocyte development.
    e. Regulatory T cells fail to activate autoreactive T cells in secondary lymphoid organs.

4

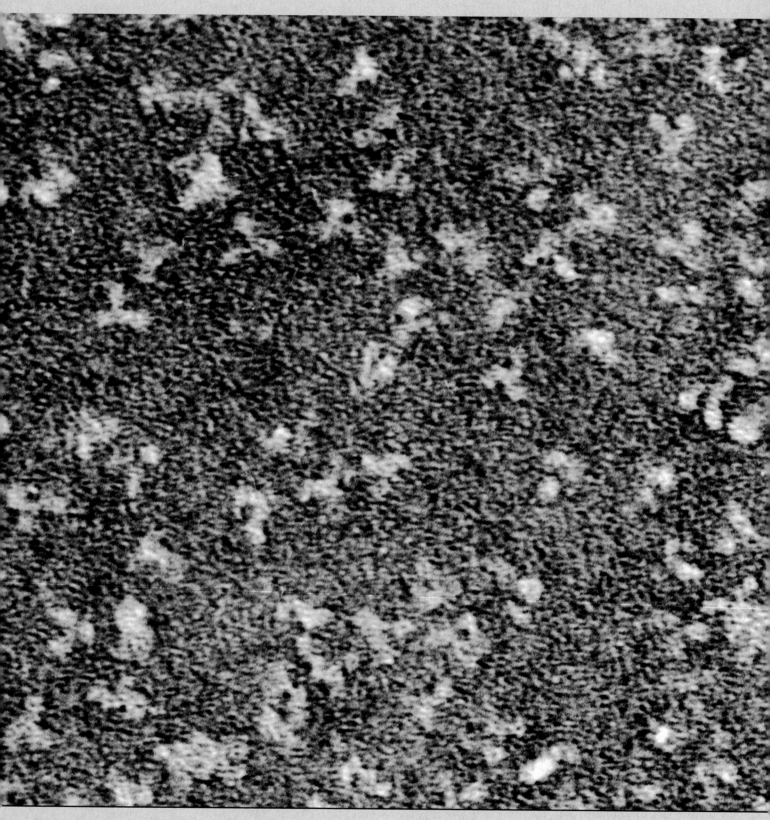

Individual molecules of human immunoglobulin G, the predominant antibody in blood and connective tissue.

Chapter 4

Antibody Structure and the Generation of B-Cell Diversity

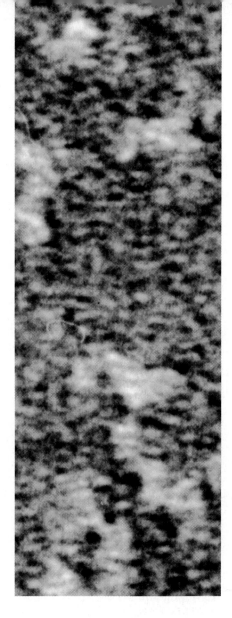

Pathogens and their products can be found either within infected cells or outside cells in the intercellular fluid and blood. In an adaptive immune response, the body is cleared of extracellular pathogens and their toxins by means of **antibodies**, the secreted form of the B-cell antigen receptor (see Section 3-2). Antibodies are produced by the effector B lymphocytes, or plasma cells, of the immune system in response to infection. They circulate as a major component of the plasma in blood and lymph and are always present at mucosal surfaces. Antibodies can recognize all types of biological macromolecule, but, in practice, proteins and carbohydrates are the antigens most commonly encountered. Binding of antibody to a bacterium or virus particle can disable the pathogen and also render it susceptible to destruction by other components of the immune system. Antibodies are the best source of protective immunity, and the most successful vaccines protect through stimulating the production of high-quality antibodies.

As we learned in Chapter 3, antibodies are very variable proteins. An individual antibody molecule is said to be **specific**; that is, it can bind to only one antigen or a very small number of different antigens. The immune system has the potential to make antibodies against a vast number of different substances, paralleling the many different foreign antigens that a person is likely to be exposed to during their lifetime. Collectively, therefore, antibodies are diverse in their antigen-binding specificities; the total number of different specific antibodies that can be made by an individual is known as the **antibody repertoire** and it might be as high as 10^{16}. In practice, the number of B cells limits the actual repertoire to closer to 10^9.

Antibodies are the secreted form of proteins known more generally as immunoglobulins. During its development, an individual B cell becomes committed to producing immunoglobulin of a unique antigen specificity; thus, mature B cells collectively produce immunoglobulins of many different specificities. Before it has encountered antigen, a mature B cell only expresses immunoglobulin in a membrane-bound form that serves as the cell's receptor for antigen. When antigen binds to this receptor, the B cell is stimulated to proliferate and to differentiate into **plasma cells**, which then secrete large amounts of antibodies with the same specificity as that of the membrane-bound immunoglobulin (Figure 4.1). This is an example of the principle of clonal selection, on which adaptive immune responses are based (see Section 3-4). Antibody production is the single effector function of the B lymphocytes of the immune system. An inability to make immunoglobulins, called agammaglobulinemia, renders those people affected highly susceptible to certain types of infection.

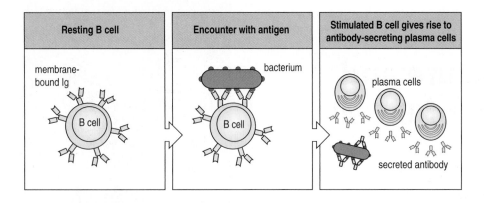

Figure 4.1 **Plasma cells secrete antibody of the same antigen specificity as the membrane-bound immunoglobulin expressed by their B-cell precursor.** A mature B cell expresses membrane-bound immunoglobulin (Ig) of a single antigen specificity. When a foreign antigen first binds to this immunoglobulin, the B cell is stimulated to proliferate. Its progeny differentiate into plasma cells that secrete antibody of the same specificity as the membrane-bound immunoglobulin.

The first part of this chapter describes the general structure of immunoglobulins, and the second part considers how immunoglobulin diversity is generated in developing B cells. After a mature B cell has encountered its specific antigen, changes are made in the specificity and effector functions of the antibodies that it produces; these are discussed in the final part of the chapter.

The structural basis of antibody diversity

The function of an antibody molecule in host defense is to recognize and bind its corresponding antigen, and to target the bound antigen to other components of the immune system. These then destroy the antigen or clear it from the body. Antigen binding and interaction with other immune system cells and molecules are carried out by different parts of the antibody molecule. One part is highly variable in that its amino acid sequence differs considerably from antibody to antibody. This variable part contains the site of antigen binding and confers specificity on the antibody. The rest of the molecule is far less variable in its amino acid sequence. This constant part of the antibody interacts with other immune-system components.

Immunoglobulins are split into five classes, or isotypes—IgA, IgD, IgE, IgG, and IgM. These are distinguished on the basis of structural differences in the constant part of the molecule and have different effector functions. We first consider the general structure of immunoglobulins and then look more closely at the structural features that confer antigen specificity. IgG, the most abundant antibody in blood and lymph, is used to illustrate the common structural features of immunoglobulins.

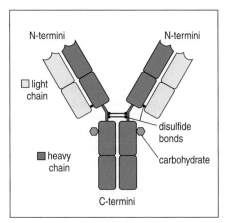

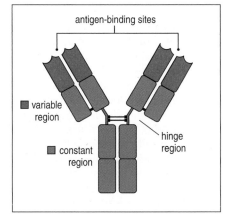

Figure 4.2 **The immunoglobulin G (IgG) molecule.** As shown in the top panel, each IgG molecule is made up of two identical heavy chains (green) and two identical light chains (yellow). Carbohydrate (turquoise) is attached to the heavy chains. The lower panel shows the location of the variable (V) and constant (C) regions in the IgG molecule. The amino-terminal regions (red) of the heavy and light chains are variable in sequence from one IgG molecule to another; the remaining regions are constant in sequence (blue). The carbohydrate is omitted from this panel and from most subsequent figures for simplicity. In IgG a flexible hinge region is located between the two arms and the stem of the Y.

4-1 Antibodies are composed of polypeptides with variable and constant regions

Antibodies are glycoproteins that are built from a basic unit of four polypeptide chains. This unit consists of two identical **heavy chains** (**H chains**) and two identical, smaller, **light chains** (**L chains**), which are assembled into a structure that looks like the letter Y (Figure 4.2, top panel). An IgG molecule has a molecular weight of about 150 kDa, to which each heavy chain contributes approximately 50 kDa and each light chain approximately 25 kDa. Each arm of the Y is made up of a complete light chain paired with the amino-terminal part of a heavy chain, covalently linked by a disulfide bond. The stem of the Y consists of the paired carboxy-terminal portions of the two heavy chains. The two heavy chains are also linked to each other by disulfide bonds.

The polypeptide chains of different antibodies vary greatly in amino acid sequence, and the sequence differences are concentrated in the amino-terminal region of each type of chain; this is known as the **variable region** or **V region**. This variability is the reason for the great diversity of antigen-binding

Figure 4.3 The Y-shaped immunoglobulin molecule can be dissected with a protease. Treatment of IgG with the enzyme papain results in proteolytic cleavage of the hinge of each heavy chain, as shown by the scissors, and reduction of the disulfide bonds that connect the two hinges. The IgG molecule is thus dissected into three pieces: two Fab fragments and one Fc fragment. Reduction of the disulfide bonds is not due to papain activity but is caused by free cysteine in the reaction mixture, which is needed to stabilize the enzyme.

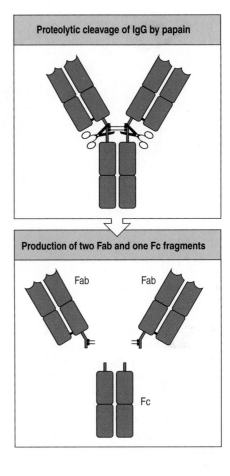

specificities among antibodies, because the paired V regions of a heavy and a light chain form the **antigen-binding site** (Figure 4.2, bottom panel). Every Y-shaped antibody molecule has two identical antigen-binding sites, one at the end of each arm. The remaining parts of the light chain and the heavy chain have much more limited variation in amino acid sequence between different antibodies and are therefore known as the **constant regions** or **C regions**.

In IgG, a relatively unstructured portion in the middle of the heavy chain forms a flexible hinge region at which the molecule can be cleaved to produce defined antibody fragments. These fragments were instrumental in providing information about antibody structure and function because they can be used for separate analysis of the antigen-binding and effector functions of the molecule. Digestion with the plant protease papain produces three fragments, corresponding to the two arms and the stem (Figure 4.3). The fragments corresponding to the arms are called **Fab** (Fragment antigen binding) because they bind antigen. The fragment corresponding to the stem is called **Fc** (Fragment crystallizable) because it was seen to crystallize in the first experiments of this sort. The stem of a complete antibody molecule is therefore often known as the **Fc region** or **Fc piece**, and the arms as Fab. The hinge of the IgG molecule allows the two Fabs to adopt many different orientations with respect to each other. This flexibility enables antigens spaced at different distances apart on the surfaces of pathogens to be bound tightly by both Fab arms of an IgG molecule (Figure 4.4).

Differences in the heavy-chain C regions define five main **isotypes** or **classes** of immunoglobulin, which have different functions in the immune response. They are **immunoglobulin G (IgG)**, **immunoglobulin M (IgM)**, **immunoglobulin D (IgD)**, **immunoglobulin A (IgA)**, and **immunoglobulin E (IgE)** (Figure 4.5). Their heavy chains are denoted by the corresponding lower-case Greek letter (γ, μ, δ, α, and ε, respectively).

The light chain has only two isotypes or classes, which are termed **kappa (κ)** and **lambda (λ)**. No functional difference has been found between antibodies carrying κ light chains and those carrying λ light chains; light chains of both isotypes are found associated with all the heavy-chain isotypes. Each antibody, however, contains either κ or λ light chains, not both. The relative abundances of κ and λ light chains vary with the species of animal. In humans, two-thirds of the antibody molecules contain κ chains and one-third have λ chains.

4-2 Immunoglobulin chains are folded into compact and stable protein domains

Antibodies function in extracellular environments in the presence of infection, where they can encounter variations in pH, salt concentration, proteolytic enzymes, and other potentially destabilizing factors. Their structure, however, helps them to withstand such harsh conditions. Heavy and light chains each consist of a series of similar sequence motifs; a single motif is about 100–110 amino acids long and folds up into a compact and exception-

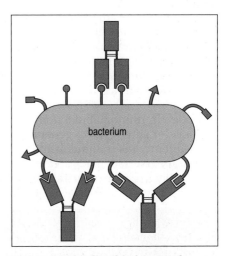

Figure 4.4 The flexible hinge of the IgG molecule allows it to bind with both arms to many different arrangements of antigens on the surfaces of pathogens. Three IgG molecules are shown binding with both arms to antigens that are located at different distances apart on the surface of a bacterium.

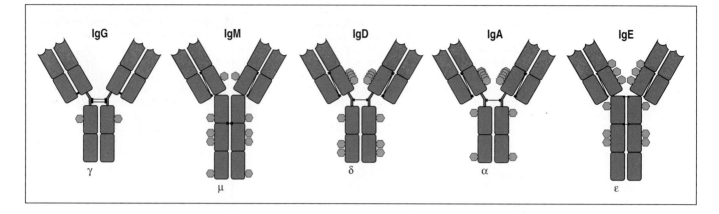

ally stable protein domain called an **immunoglobulin domain**. Each immunoglobulin chain is composed of a linear series of these domains.

The V region at the amino-terminal end of each heavy or light chain is composed of a single **variable domain** (**V domain**): V_H in the heavy chain and V_L in the light chain. A V_H and a V_L domain together form an antigen-binding site. The other domains have little or no sequence diversity within a particular class of antibody and are termed the **constant domains** (**C domains**), which make up the C regions. The constant region of a light chain is composed of a single C_L domain, whereas the constant region of a heavy chain is composed of three or four C domains, depending on the isotype. The γ heavy chains of IgG have three domains—C_H1, C_H2, and C_H3. Some other isotypes have four C domains (see Figure 4.5). In the complete IgG molecule the pairing of the four polypeptide chains produces three globular regions, corresponding to the two Fab arms and the Fc stem, respectively, each of which is composed of four immunoglobulin domains (Figure 4.6).

The structure of a single immunoglobulin domain can be compared to a bulging sandwich in which two β sheets (the slices of bread) are held together by strong hydrophobic interactions between their constituent amino acid side chains (the filling). The structure is stabilized by a disulfide bond between the two β sheets. The adjacent strands within the β sheets are connected by loops of the polypeptide chain. This arrangement of β sheets provides the stable structural framework of all immunoglobulin domains, whereas the amino acid sequence of the loops can be varied to confer different binding properties on the domain. The structural similarities and differences between the C and V domains are illustrated for the light chain in Figure 4.7).

Although immunoglobulin domains were first discovered in antibodies, very similar domains, known generally as **immunoglobulin-like domains**, have subsequently been found in many other proteins and are particularly common in cell-surface and secreted proteins of the immune system. Such proteins collectively form the **immunoglobulin superfamily**.

Figure 4.5 The structures of the human immunoglobulin classes. In particular, note the differences in length of the heavy-chain C regions, the locations of the disulfide bonds linking the chains, and the presence of a hinge region in IgG, IgA, and IgD but not in IgM and IgE. The heavy-chain isotype in each antibody is indicated by the Greek letter. The isotypes also differ in the distribution of *N*-linked carbohydrate groups (turquoise). All these immunoglobulins occur as monomers in their membrane-bound form. In their soluble, secreted form, IgD, IgE, and IgG are always monomers, IgA forms monomers and dimers, and IgM forms only pentamers.

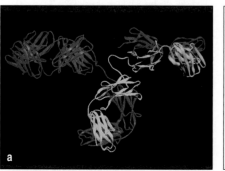

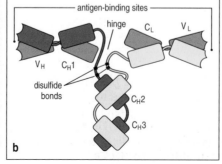

Figure 4.6 The heavy and light chains of an immunoglobulin molecule are made from a series of similar protein domains. Panel a shows a ribbon diagram tracing the course of the polypeptide backbones of chains in an IgG molecule. The heavy chains are shown in yellow and purple, the light chains in red. The molecule is composed of three similarly sized globular portions, which correspond to the Fc stem and the two Fab arms. Each of these globular portions is made up of four immunoglobulin domains. The schematic diagram in panel b shows the domain structure of an IgG molecule. The light chain (red) has one variable domain (V_L) and one constant domain (C_L). The γ heavy chain (one yellow and one purple) has a variable domain (V_H) and three constant domains (C_H1, C_H2, and C_H3). Panel a courtesy of L. Harris.

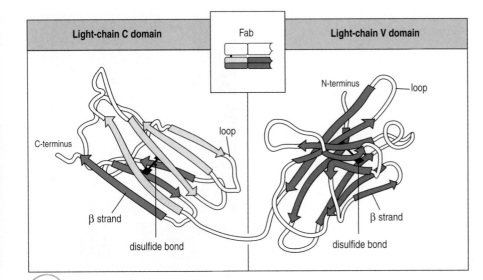

| Light-chain C domain | Fab | Light-chain V domain |

Figure 4.7 **The three-dimensional structure of immunoglobulin C and V domains.** An individual light chain is folded into a single C domain (left panel) and a single V domain (right panel). The inset shows the location of this light chain in the Fab part of the IgG molecule. The folding of the polypeptide chain backbone is depicted as ribbon diagrams in which the thick colored arrows correspond to the parts of the chain that form the β strands of the β sheet. The arrows point from the amino terminus of the chain to the carboxy terminus. Adjacent strands in each β sheet run in opposite directions (antiparallel), as shown by the arrows, and are connected by loops. In a C domain there are four β strands in the upper β sheet (yellow) and three in the lower sheet (green). In a V domain there are five strands in the upper sheet (blue) and four in the lower (red).

4-3 An antigen-binding site is formed from the hypervariable regions of a heavy-chain V domain and a light-chain V domain

A comparison of the V domains of heavy and light chains from different antibody molecules shows that the differences in amino acid sequence are concentrated within particular regions called **hypervariable regions** (**HV**), which are flanked by much less variable **framework regions** (Figure 4.8, top panel). Three hypervariable regions are found in each V domain. When mapped onto the folded structure of a V domain, the hypervariable regions locate to three of the loops that are exposed at the end of the domain farthest from the constant region. The framework regions correspond to the β strands and the remaining loops. This relationship is illustrated for the light-chain V domain in Figure 4.8 (center panel). Thus, the structure of the immunoglobulin domain provides the capacity for localized variability within a structurally constant framework.

The pairing of a heavy and a light chain in an antibody molecule brings together the hypervariable loops from each V domain to create a composite hypervariable surface, which forms the antigen-binding site at the tip of each Fab arm (Figure 4.8, bottom panel). The differences in the loops between different antibodies create both the specificity of antigen-binding sites and their

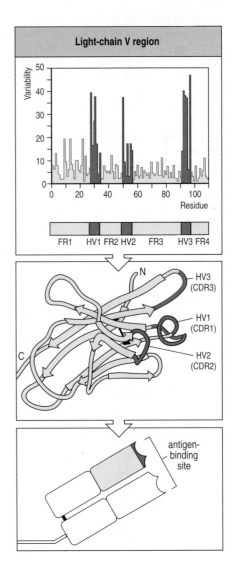

Figure 4.8 **The hypervariable regions of antibody V domains lie in discrete loops at one end of the domain structure.** The top panel shows the variability plot for the 110 positions within the amino acid sequence of a light-chain V domain. It is obtained from comparison of many light-chain sequences. Variability is the ratio of the number of different amino acids found at a position to the frequency of the most common amino acid at that position. The maximum value possible for the variability is 400, the square of 20, the number of different amino acids found in antibodies. The minimum value is 1. Three hypervariable regions (HV1, HV2, and HV3) can be discerned (red) flanked by four framework regions (FR1, FR2, FR3, and FR4) (yellow). The center panel shows the correspondence of the hypervariable regions to three loops at the end of the V domain farthest from the C region. The location of hypervariable regions in the heavy-chain V domain is similar (not shown). The hypervariable loops contribute much of the antigen specificity of the antigen-binding site located at the tip of each arm of the antibody molecule. Hypervariable regions are also known as complementarity-determining regions: CDR1, CDR2, and CDR3. The bottom panel shows the location of the light-chain V region in the Fab part of the IgG molecule.

diversity. The hypervariable loops are also called **complementarity-deter-mining regions** (**CDRs**) because they provide a binding surface that is complementary to that of the antigen.

4-4 Antigen-binding sites vary in shape and physical properties

The function of antibodies is to bind to microorganisms and facilitate their destruction or ejection from the body. The antibodies that are most effective against infection are generally those that bind to the exposed and accessible molecules making up the surface of a pathogen. The part of an antigen to which an antibody binds is called an **antigenic determinant** or **epitope** (Figure 4.9). In nature, these structures are usually either carbohydrate or protein, because the surface molecules of pathogens are commonly glycoproteins, polysaccharides, glycolipids, and peptidoglycans. Complex macromolecules such as these will usually contain several different epitopes, each of which can be bound by a different antibody. Individual epitopes are typically composed of a cluster of amino acids or a small part of a polysaccharide chain. Any antigen that contains more than one epitope, or more than one copy of the same epitope, is known as a **multivalent** antigen (Figure 4.10).

Antibodies can be made against a much wider range of chemical structures than proteins and carbohydrates. Such antibodies are, however, more commonly implicated in allergic reactions and autoimmune diseases than in protective immunity against infection. Antibodies specific for DNA, for example, are characteristic of the autoimmune disease systemic lupus erythematosus (SLE), and allergies can be caused by antibodies against small organic molecules, such as antibiotics and other drugs.

The antigen-binding sites of antibodies vary according to the size and shape of the epitope they recognize. Antibodies that bind to the end of a polysaccharide or polypeptide chain, for example, use a deep pocket formed between the heavy-chain and light-chain V domains (Figure 4.11, first panel). In such cases not all the CDRs make contact with the antigen. By contrast, antibodies that bind to several of adjacent sugars within a polysaccharide, or amino acids within a polypeptide, use longer, shallower clefts formed by all the opposing CDRs of the V_H and V_L domains (Figure 4.11, second panel). Epitopes of this kind, in which the antibody binds to parts of a molecule that are adjacent in the linear sequence, are called **linear epitopes** (Figure 4.12, left panel).

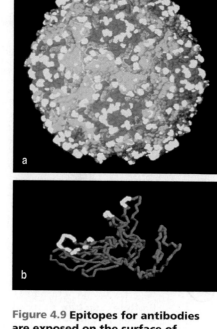

a

b

Figure 4.9 Epitopes for antibodies are exposed on the surface of pathogens. Panel a shows a spherical poliovirus particle. Its coat is an array of three different proteins (indicated in yellow, blue, and pink). In panel b a molecule of one of the viral coat proteins, VP1 (blue), is shown as folded in the virus particle. VP1 contains several epitopes (white). They are located at the surface of the protein and are exposed on the surface of the virus particle. Photograph courtesy of D. Filman and J.M. Hogle.

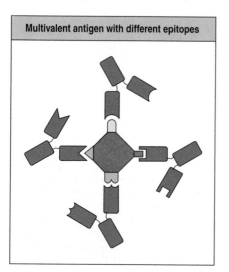

Multivalent antigen with different epitopes

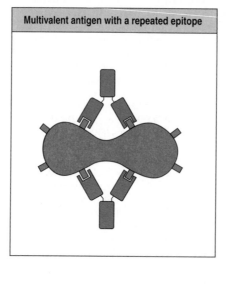

Multivalent antigen with a repeated epitope

Figure 4.10 Two kinds of multivalent antigen. Many soluble protein antigens have several different epitopes, but each is represented only once on the surface of the protein. This situation is depicted in the left panel, where four IgG molecules with different specificities all bind to the protein antigen using a single Fab arm. On pathogen surfaces there are numerous copies of the same epitope, as illustrated for poliovirus in Figure 4.9. This situation, depicted in the right panel, allows many IgG molecules with identical antigenic specificity to bind to the multivalent antigen with both Fab arms.

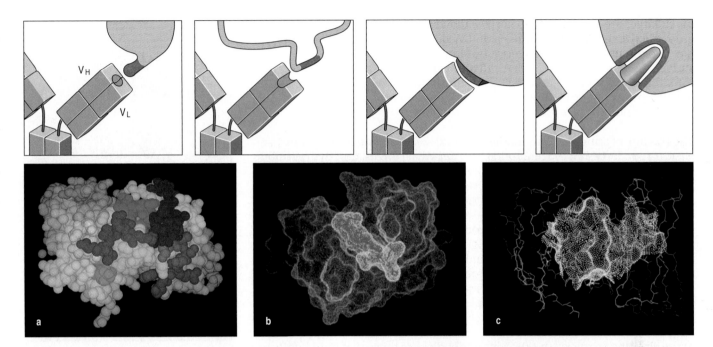

Figure 4.11 Epitopes can bind to pockets, grooves, extended surfaces, or knobs in antigen-binding sites. The top row shows schematic representations of antibodies (blue) binding to four different types of epitope (red). The first panel shows a small compact epitope binding to a pocket in the antigen-binding site; in the second panel an epitope consisting of a relatively unfolded part of a polypeptide chain binds to a shallow groove; the third panel shows an epitope with an extended surface binding to a similarly sized but complementary surface in the antibody; and in the fourth panel the epitope is a pocket into which the antigen-binding site intrudes. Only one Fab arm of each antibody is shown. The crystallographic images in the bottom row are of Fab fragments binding to antigens in ways corresponding to the first three panels in the top row. Panel a shows a peptide epitope (red) bound to a pocket formed by the CDR loops (colored), panel b shows a peptide epitope (red) bound to a groove between the two V domains (green), and panel c shows Fab binding to a surface epitope of the protein lysozyme. The surface contour of lysozyme (yellow dots) is superimposed on the antigen-binding site of the Fab. The peptide backbone of the Fab is shown in blue and the amino acids that contact lysozyme are in red. Photographs courtesy of I.A. Wilson, R.L. Stanfield, and S. Sheriff.

Another kind of epitope is formed by parts of a protein that are separated in the amino acid sequence but are brought together in the folded protein. These are called **conformational** or **discontinuous epitopes** (Figure 4.12, right panel). Antibodies that bind to complete proteins often make contact with a

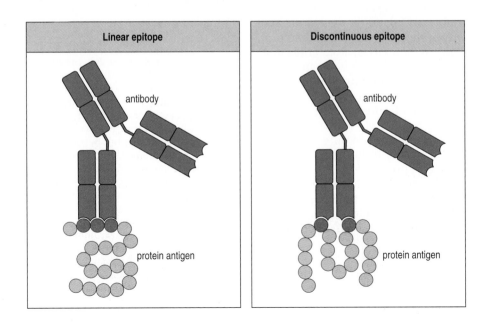

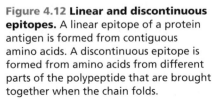

Figure 4.12 Linear and discontinuous epitopes. A linear epitope of a protein antigen is formed from contiguous amino acids. A discontinuous epitope is formed from amino acids from different parts of the polypeptide that are brought together when the chain folds.

much larger area of the protein surface than can be accommodated within a groove between the CDRs of the V_H and V_L domains. These antibodies might have no discernible binding groove but interact with the antigen by using a surface (in area usually 700–900 Å²) that is formed by the CDRs and can even extend beyond them (Figure 4.11, top and bottom rows, third panel). The fourth type of epitope is a pocket in the antigen, into which the antigen-binding site intrudes (Figure 4.11, top row, fourth panel).

The binding of antigens to antibodies is based solely on noncovalent forces—electrostatic forces, hydrogen bonds, van der Waals forces, and hydrophobic interactions. The antigen-binding sites of antibodies are unusually rich in aromatic amino acids, which can participate in many van der Waals and hydrophobic interactions. The better the general fit between the interacting surfaces of antigen and antibody, the stronger are the bonds formed by these short-range forces. Thus, small differences in shape and chemical properties of the binding site can give several antibodies specificity for the same epitope, but they bind to it with different binding strengths or **affinities**. Binding caused by van der Waals forces and hydrophobic interactions is complemented by the formation of electrostatic interactions (salt bridges) and hydrogen bonds between particular chemical groups on the antigen and particular amino acid residues of the antibody.

In the immune response, the effective antibodies are those that bind tightly to an antigen and do not let go. This behavior contrasts with that of enzymes, which bind a substrate, chemically change the substrate's structure, and then release it. However, the same set of 20 amino acids is used to make enzymes and antibodies and, after intensive searching, some antibodies have been found that catalyze chemical reactions involving the antigen they bind. The application of such **catalytic antibodies** in medicine is being explored. A possible use for a catalytic antibody would be to convert a toxic chemical within the body into an innocuous product. Toward this end, catalytic antibodies have been made that bind to cocaine or methamphetamine and convert these highly addictive drugs to derivatives that have no pyschostimulating activity.

4-5   Monoclonal antibodies are produced from a clone of antibody-producing cells

The specificity and strength of an antibody for its antigen make antibodies useful reagents for detecting and quantifying a particular antigen. They are used in this way in a range of clinical and laboratory tests and in biological research generally. The traditional method for making antibodies of desired specificity is to immunize animals with the appropriate antigen and then prepare **antisera** from their blood. The specificity and quality of such antisera is highly dependent upon the purity of the immunizing antigen preparation because antibodies will be made against all the foreign components it contains. Consequently, antisera specific for a single protein or carbohydrate can only be made when the antigen is already available in a purified form.

A more modern method for making antibodies does not require a purified antigen. In this method, B cells are isolated from the immunized animals, usually mice, and immortalized by fusion with a tumor cell to form **hybridoma** cell lines that grow and produce antibodies indefinitely. The individual hybridoma cells are then separated from each other, and the cells making antibodies of the desired specificity are identified and selected for further propagation. The antibodies made by a hybridoma cell line are all identical and are therefore called **monoclonal antibodies** (Figure 4.13).

Since the invention of hybridomas in the 1970s, monoclonal antibodies have replaced antisera in many clinical assays. They have also been used to identify numerous previously unknown surface proteins of cells of the immune

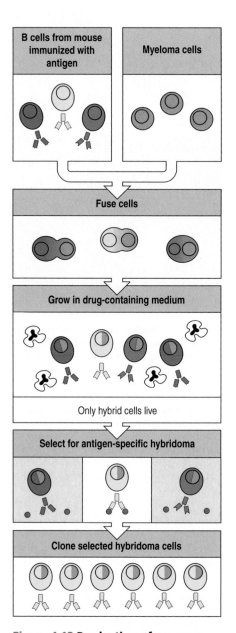

Figure 4.13 Production of a mouse monoclonal antibody. Lymphocytes from a mouse immunized with the antigen are fused with myeloma cells by using polyethyleneglycol. The cells are then grown in the presence of drugs that kill myeloma cells but permit the growth of hybridoma cells. Unfused lymphocytes also die. Individual cultures of hybridomas are tested to determine whether they make the desired antibody. The cells are then cloned to produce a homogeneous culture of cells making a monoclonal antibody. Myelomas are tumors of plasma cells; those used to make hybridomas were selected not to express heavy and light chains. Thus, hybridomas only express the antibody made by the B-cell partner.

system. All of the 250 or so cell-surface proteins in the CD series (for example, CD4 and CD8) were discovered by using monoclonal antibodies. The reactions of monoclonal antibodies with different types of cell are measured by the technique of flow cytometry (Figure 4.14).

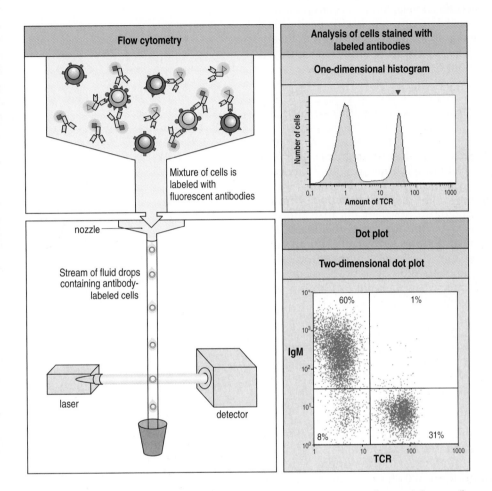

Figure 4.14 The flow cytometer allows individual cells to be identified by their cell-surface molecules. The left-hand panels illustrate the principles of flow cytometry. Human cells are labeled with mouse monoclonal antibodies specific for human cell-surface proteins, antibodies of different specificity being tagged with fluorescent dyes of different colors. The labeled cells pass through a nozzle, forming a stream of droplets each containing one cell. The stream of cells passes through a laser beam, which causes the fluorescent dyes to emit light of different wavelengths. The emitted signals are analyzed by a detector: cells with particular characteristics are counted and the abundance of each type of labeled cell-surface molecule measured. The right-hand panels show how the data obtained can be represented, as exemplified by the presence of IgM and T-cell receptor (TCR) on lymphocytes from the human spleen. The expression of IgM and TCR distinguishes between B cells and T cells, respectively. When the presence of one type of molecule is analyzed, the data are usually displayed as a histogram, as shown in the upper right panel for TCR. Two populations of cells are distinguished in the histogram. The larger peak to the left consists of lymphocytes (mostly B cells) that do not bind the anti-TCR monoclonal antibody; it comprises 58% of the total cell number. The smaller peak to the right corresponds to the T cells that bind the anti-TCR antibody and comprises 32% of the total. Two-dimensional plots

(lower right panel) are used to compare the expression of two cell-surface molecules. The horizontal axis represents the amount of fluorescent anti-TCR antibody bound by a cell; the vertical axis represents the amount of fluorescent anti-IgM antibody bound. Each dot represents the values obtained for a single cell. Many thousands of cells are usually analyzed, and the dots can blend into each other in those parts of the plot that are heavily populated. The dot plot is divided into four quadrants that roughly correspond to the four cell populations distinguished by analysis with two antibodies. In the top left quadrant are B cells; these bind anti-IgM antibody but not anti-TCR antibody and comprise 60% of the cells. In the bottom right quadrant are the T cells; these bind anti-TCR antibody but not anti-IgM antibody and comprise 31% of the cells. In the bottom left quadrant are cells that bind neither anti-IgM nor anti-TCR antibody; these comprise 8% of the cells and probably include NK cells and some contaminating non-lymphoid leukocytes. Theoretically, cells in the upper right quadrant would correspond to lymphocytes that bind both anti-IgM and anti-TCR. As no lymphocytes express both IgM and TCR at their surface, these double reactions (1% of the total) arise from imprecision in using quadrants to separate the cell population and from experimental artifact, for example the nonspecific sticking of antibody molecules to cells that do not express their specific antigen.

In medicine, flow cytometry is used to analyze the cell populations in peripheral blood and detect perturbations caused by disease. For example, children with a rare inherited genetic defect have no B cells and make no antibodies. For a period of around a year after birth and while being breastfed, these children are protected from infection by antibodies provided by their mothers, but this protection declines with age. They then become particularly susceptible to infections with encapsulated bacteria such as *Streptococcus pneumoniae* or *Haemophilus influenzae*, which, if left untreated with antibiotics, would eventually lead to death. Knowledge that a person is B-cell deficient allows infections to be prevented by regular injections of antibodies purified from the pooled sera of healthy blood donors.

4-6   Monoclonal antibodies are used as treatments for a variety of diseases

Monoclonal antibodies are used in therapy as well as in diagnosis. The first successful therapeutic application was of a mouse monoclonal antibody specific for the CD3 antigen on human T cells. This was used to block T-cell responses and thus prevent the imminent T cell-mediated rejection of transplanted kidneys. This antibody was of limited use, however, because after the first treatment, humans make antibodies specific for the C regions of the mouse antibody. The interaction of these antibodies with a subsequent dose of mouse monoclonal antibodies diminishes the latter's therapeutic effect and increases the likelihood of complications. To reduce this problem, genetic engineering has been used to make **chimeric monoclonal antibodies** that combine mouse V regions with human C regions (Figure 4.15).

One chimeric monoclonal antibody of this sort, called rituximab, is used to treat some types of non-Hodgkin B-cell lymphoma. Rituximab is specific for the cell-surface protein CD20, which is found on normal and malignant human B cells and influences B-cell proliferation and survival. Rituximab binds strongly to both normal and malignant B cells and targets their rapid destruction by NK cells, which have a surface receptor that binds to the Fc of IgG bound to the B cells by its Fab arms. Despite their temporary loss of normal B cells, treated patients continue to make antibodies because plasma cells lack CD20 and are unaffected by rituximab. Normal populations of circulating B cells are usually reconstituted within a year after treatment. Since its approval for medical use in 1997, more than 600,000 patients have been treated with rituximab.

Another approach to reducing the tendency for useful mouse monoclonal antibodies to provoke human antibodies against them is to **humanize** the monoclonal antibodies by genetic engineering. In this procedure, sequences encoding the CDR loops of the mouse monoclonal antibody's heavy and light chains are used to replace the corresponding CDR sequences of a human immunoglobulin (see Figure 4.15). Cells expressing the humanized heavy-chain and light-chain genes make an antibody in which only the CDR loops are of mouse origin and the rest is human. Omalizumab is a humanized

Figure 4.15 Monoclonal antibodies as treatments for disease. Mouse monoclonal antibodies were the first antibodies used therapeutically, but the human immune system perceives the mouse protein as foreign and makes antibodies that bind to the mouse antibodies and get rid of them. To reduce this problem, chimeric antibodies have been made in which the constant regions of mouse monoclonal antibodies (blue) are replaced with human constant regions (yellow). Taking this approach further are humanized antibodies, in which only the CDR loops are of mouse origin. Fully human antibodies can now be made from human hybridomas or by mice in which the mouse immunoglobulin genes are replaced by human immunoglobulin genes. All four types of antibody have been used therapeutically.

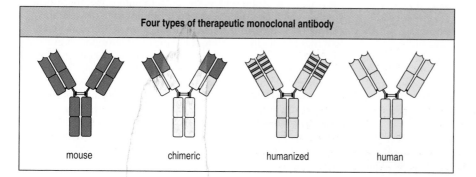

Four types of therapeutic monoclonal antibody

mouse   chimeric   humanized   human

antibody specific for human IgE that is being assessed as a treatment for allergic asthma. By binding to the constant region of IgE, the therapeutic antibody prevents the recruitment of mast cells, eosinoils and basophils to the allergic response. Two other methods produce all-human monoclonal antibodies for therapeutic use and avoid the problems associated with mouse immunoglobulins. In the first, the antibodies are made by mice in which the mouse immunoglobulin genes have been replaced with their human counterparts. The second is to produce human hybridomas and grow them in tissue culture. Adalimumab is a widely used all-human monoclonal antibody specific for the inflammatory cytokine TNF-α (see Section 2-13). It is used in the treatment of rheumatoid arthritis, a condition in which chronic inflammation of the joints is perpetuated by TNF-α.

Summary

IgG antibodies are made of four polypeptide chains—two identical heavy chains and two identical light chains. Each chain has a V region that contributes to the antigen-binding site and a C region that, in the heavy chain, determines the antibody isotype and its specialized effector functions. The IgG molecule is Y-shaped, in which the stem and the arms are of comparable sizes and can flex with respect to each other. Each arm contains an antigen-binding site, whereas the stem contains binding sites for cells and effector molecules that enable bound antigen to be cleared from the body. Immunoglobulin chains are built from a series of structurally related immunoglobulin domains. Within the V domain, sequence variability is localized to three hypervariable regions corresponding to the three loops that are clustered at one end of the domain. In the antibody molecule the hypervariable loops of the heavy and light chains form a variable surface that binds antigen. The type of antigen bound by an antibody depends on the shape of the antigen-binding site: antigens that are small molecules can be bound within deep pockets; linear epitopes from proteins or carbohydrates can be bound within clefts or grooves; and the binding of conformational epitopes of folded proteins takes place over an extended surface area. Monoclonal antibodies are antibodies of a single specificity that are derived from a clone of identical antibody-producing cells, which are used in diagnostic tests and as therapeutic agents.

Generation of immunoglobulin diversity in B cells before encounter with antigen

The number of different antibodies that can be produced by the human body seems to be virtually limitless. To achieve this, the immunoglobulin genes are organized differently from other genes. In all cells, except B cells, the immunoglobulin genes are in a fragmented form that cannot be expressed; instead of containing a single complete gene, immunoglobulin heavy-chain and light-chain loci consist of families of **gene segments**, sequentially arrayed along the chromosome, with each set of segments containing alternative versions of parts of the immunoglobulin V region. The immunoglobulin genes are inherited in this form through the germline (egg and sperm); this arrangement is therefore called the **germline form** or **germline configuration**.

For an immunoglobulin gene to be expressed, individual gene segments must first be rearranged to assemble a functional gene, a process that occurs only in developing B cells. Immunoglobulin-gene rearrangements occur during the development of B cells from B-cell precursors in the bone marrow. When gene rearrangements are complete, heavy and light chains can be produced, and membrane-bound immunoglobulin appears at the B-cell surface. The B cell can now recognize and respond to an antigen through this receptor. Much of the diversity within the mature antibody repertoire is generated during this process of gene rearrangement, as we shall see in this part of the chapter.

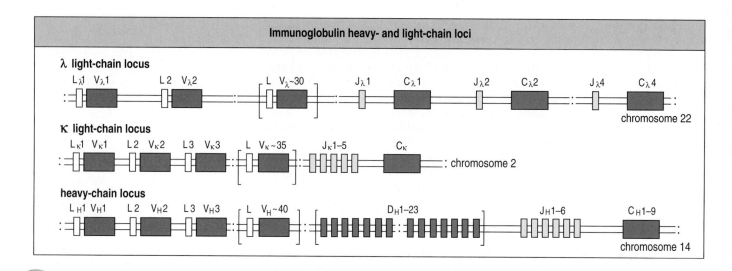

Immunoglobulin heavy- and light-chain loci

λ light-chain locus

chromosome 22

κ light-chain locus

: chromosome 2

heavy-chain locus

chromosome 14

4-7 The DNA sequence encoding a V region is assembled from two or three gene segments

In humans, the immunoglobulin genes are found at three chromosomal locations: the heavy-chain locus on chromosome 14, the κ light-chain locus on chromosome 2 and the λ light-chain locus on chromosome 22. Different gene segments encode the leader peptide (L), the V region (V) and the constant region (C) of the heavy and light chains. Gene segments encoding C regions are commonly called C genes (Figure 4.16). Within the heavy-chain locus are C genes for all the different heavy-chain isotypes.

The gene segments encoding the leader peptides and the C regions consist of exons and introns, like those found in other human genes, and they are ready to be transcribed. In contrast, the V regions are encoded by two (V_L) or three (V_H) gene segments that require rearrangement to produce an exon that can be transcribed. The two types of gene segment that encode the light-chain V region are called **variable (V)** and **joining (J) gene segments**. The heavy-chain locus includes an additional set of **diversity (D) gene segments** that lies between the arrays of V and J gene segments (see Figure 4.16).

The V region of a light chain is encoded by the combination of one V and one J segment, whereas the C region is encoded by a single C gene. The main differences between V gene segments are in the sequences that encode the first and second hypervariable regions of the V domain (see Section 4-3); the third hypervariable region is determined by the junction between the V and J segments. In an individual B cell, only one of the light-chain loci (κ or λ) gives rise to a functional light-chain gene. The V region of the heavy chain is encoded by one V, one D, and one J segment. The first two hypervariable regions are determined by differences in the sequences of the heavy-chain V gene segments; the third hypervariable region is determined by differences in the D gene segments and the junctions they make with the V and J gene segments. The C region of a heavy chain is encoded by one of the C genes.

4-8 Random recombination of gene segments produces diversity in the antigen-binding sites of immunoglobulins

During the development of B cells, the arrays of V, D, and J segments are cut and spliced by DNA recombination. This process is called **somatic recombination** because it occurs in cells of the soma, a term that embraces all of the body's cells except the germ cells. A single gene segment of each type is

Figure 4.16 The germline organization of the human immunoglobulin heavy-chain and light-chain loci. The upper row shows the λ light-chain locus, which has about 30 functional $V_λ$ gene segments and four pairs of functional $J_λ$ gene segments and $C_λ$ gene segments. The κ locus (center row) is organized in a similar way, with about 35 functional $V_κ$ gene segments accompanied by a cluster of five $J_κ$ gene segments but with a single $C_κ$ gene segment. In approximately half of the human population, the entire cluster of $V_κ$ gene segments is duplicated (not shown, for simplicity). The heavy-chain locus (bottom row) has about 40 functional V_H gene segments, a cluster of about 23 D segments and 6 J_H gene segments. For simplicity, a single C_H gene ($C_H1–9$) is shown in this diagram to represent the nine C genes. The diagram is not to scale: the total length of the heavy-chain locus is more than 2 megabases (2 million bases), whereas some of the D segments are only six bases long. L, leader sequence.

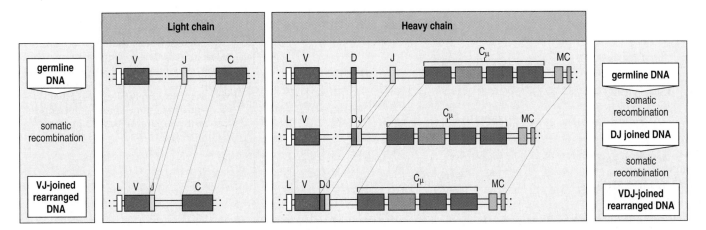

Figure 4.17 V-region sequences are constructed from gene segments. Light-chain V-region genes are constructed from two segments (left panel). A variable (V) and a joining (J) gene segment in the genomic DNA are joined to form a complete light-chain V-region (V_L) exon. After rearrangement, the light-chain gene consists of three exons, encoding the leader (L) peptide, the V region, and the C region, which are separated by introns. Heavy-chain V regions are constructed from three gene segments (right panels). First the diversity (D) and J gene segments join, then the V gene segment joins to the combined DJ sequence, forming a complete heavy-chain V-region (V_H) exon. For simplicity, only the first of the heavy-chain genes, C_μ, is shown here. Each immunoglobulin domain is encoded by a separate exon, and two additional membrane-coding exons (MC, colored light blue) specify the hydrophobic sequence that will anchor the heavy chain to the B-cell membrane.

brought together to form a DNA sequence encoding the V region of an immunoglobulin chain. For light chains, a single recombination occurs, between a V_L and a J_L segment (Figure 4.17, left panels) whereas for heavy chains, two recombinations are needed, the first to join a D and a J_H segment, and the second to join the combined DJ segment to a V_H segment (Figure 4.17, right panels). In each case, the particular V, D, and J gene segments that are joined together are selected at random. Because of the multiple gene segments of each type, numerous different combinations of V, D, and J gene segments are possible. Thus, the gene rearrangement process generates many different V-region sequences in the B-cell population. This is one of several factors contributing to the diversity of immunoglobulin V regions.

For the human κ light chain, approximately 35 V_κ gene segments and 5 J_κ gene segments (Figure 4.18) can be recombined in 175 (35 × 5) different ways. Similarly, some 30 V_λ gene segments and 4 J_λ gene segments can be recombined in 120 (30 × 4) different ways. At the heavy-chain locus, 40 V_H segments, some 23 D segments and 6 J_H segments can produce 5520 (40 × 23 × 6) combinations. If recombination between different gene segments were the only mechanism producing V-region diversity, then a maximum of 295 different light chains and 5520 different heavy chains could be made. With random combination of heavy and light chains this diversity could make 1.6 million different antibodies.

Somatic recombination is carried out by enzymes that cut and rejoin the DNA, and it exploits some of the mechanisms more universally used by cells for DNA recombination and repair. The recombination of V, J, and D gene segments is directed by sequences called **recombination signal sequences** (**RSSs**), which flank the 3' side of the V segment, both sides of the D segment, and the 5' side of the J segment (Figure 4.19). There are two types of RSS, and recombination can occur only between different types. One type of RSS comprises a defined heptamer [7 base pairs (bp); CACAGTG] sequence and a defined nonamer (9 bp; ACAAAAACC) separated by a 12-bp spacer; the other comprises the heptamer and the nonamer sequences separated by a 23-bp spacer. As well as providing recognition sites for the enzymes that cut and rejoin the DNA, RSSs ensure that the gene segments are joined in the correct order, as shown for light-chain V_J recombination in Figure 4.20. Because of

Number of gene segments in human immunoglobulin loci			
Segment	**Light chains**		**Heavy chain**
	κ	λ	H
Variable (V)	31–36	29–33	38–46
Diversity (D)	0	0	23
Joining (J)	5	4–5	6
Constant (C)	1	4–5	9

Figure 4.18 The numbers of functional gene segments available to construct the variable and constant regions of human immunoglobulin heavy chains and light chains.

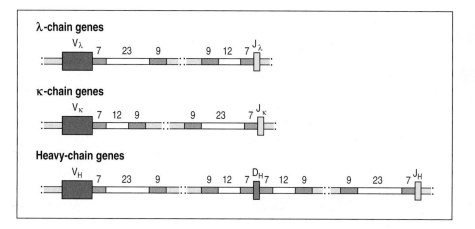

Figure 4.19 Each V, D, or J gene segment is flanked by recombination signal sequences (RSSs). There are two types of RSS. One consists of a nonamer (9 bp, shown in purple) and a heptamer (7 bp, shown in orange) separated by a spacer of 12 bp (white). The other consists of the same 9- and 7-nucleotide sequences separated by a 23-bp spacer (white).

this strict requirement, recombination in the heavy-chain DNA cannot join V$_H$ directly to J$_H$ without the involvement of D$_H$, because the V$_H$ and J$_H$ segments are flanked by the same type of RSS (see Figure 4.19).

4-9   Recombination enzymes produce additional diversity in the antigen-binding site

The set of enzymes needed to recombine V, D, and J segments is called the **V(D)J recombinase**. Two of the component proteins are made only in lymphocytes; they are specified by the **recombination-activating genes** (*RAG-1* and *RAG-2*). The other components are present in all nucleated cells and have

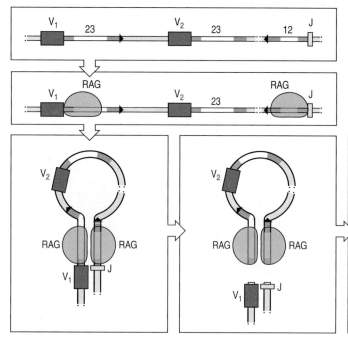

Figure 4.20 Gene segments encoding the variable region are joined by recombination at recombination signal sequences. The recombination between a V (red) and a J (yellow) segment of a light-chain gene is shown here. A RAG complex binds to the 23-bp spacer and another to the 12-bp spacer, so that a recombination signal sequence (RSS) containing a 12-bp spacer is brought together with that containing a 23-bp spacer. This is known as the 12/23 rule and ensures that gene segments are joined in the correct order. The DNA molecules are broken at the ends of the heptamer sequences (orange) and are then joined together with different topologies. The region of DNA that was originally between the V and J segments to be joined is excised as a small circle of DNA that has no function. The joint made in forming this circle is called the signal joint. Within the chromosomal DNA, the V and J segments are joined to form the coding joint. Formation of this joint involves the opening up of hairpins that were formed at the original point of cleavage at one end of each V and J segment and then repairing the DNA in a way that introduces additional variability into the nucleotide sequence around the joint. The additional enzymes involved in these processes are represented in blue. Nonamers are shown in purple, heptamers in orange, spacers in white.

activities that repair double-stranded DNA, bend DNA, or modify the ends of broken DNA strands. They include the enzymes DNA ligase IV, DNA-dependent protein kinase (DNA-PK), the nuclease Artemis, and the Ku protein associated with DNA-PK. The RAG-1 and RAG-2 proteins interact with each other and with other proteins, known as the high-mobility group of proteins, to form a RAG complex. In the first step in recombination, one RAG complex binds to one type of RSS and another complex binds the second type of RSS (see Figure 4.20). Interaction between the two RAG complexes aligns the two RSSs and then cleaves the DNA at the ends of the immunoglobulin gene segments in such a way as to create a DNA hairpin at the end of each gene segment and a clean break at the ends of the two heptamer sequences (see Figure 4.20). The DNA molecules are held in place by the RAG complexes while the broken ends are rejoined by DNA repair enzymes in a process called nonhomologous end-joining. This brings together the ends of two gene segments to form a so-called **coding joint** in the chromosome while joining the ends of the removed DNA in a **signal joint**.

The enzymes that open the hairpins and form the coding joint introduce additional sequence diversity into the third hypervariable region (CDR3) of immunoglobulin heavy and light chains. This diversity is not encoded in the sequence of the germline DNA and is generated in several ways (Figure 4.21). First, the nick that opens the hairpin can occur at any of several positions. It forms a single-stranded end in which bases that were complementary in the two DNA strands are now on the same strand. This creates a sequence that will be a palindrome in double-stranded DNA—a palindrome is a sequence that is identical when read from either end (see Figure 4.21). The added nucleotides are thus called **P nucleotides** (for **p**alindromic nucleotides). The ends of the opened hairpins can then be variably modified by exonucleases that remove germline-encoded nucleotides and by the enzyme **terminal deoxynucleotidyl transferase** (TdT), which randomly adds nucleotides. The added nucleotides need not correspond to the germline sequence in either order or number, and are called **N nucleotides** because they are **n**ontemplated (not encoded) in germline DNA. Once the single-stranded tails of the two gene segments are able to pair, the single-strand gaps are filled in with complementary nucleotides to complete the coding joint (see Figure 4.21). The same enzymes are used for gene rearrangement in T cells, and genetic deficiencies in some of these enzymes are among the causes of severe combined immunodeficiency, in which gene rearrangement fails and neither B cells nor T cells are produced.

The contribution of P nucleotides and N nucleotides to the resulting amino acid sequence diversity in the third hypervariable region is called **junctional diversity**. This is an important source of immunoglobulin variability, having the potential to increase the overall diversity by a factor of up to 3×10^7. The third hypervariable region of the light-chain V domain is encoded by the

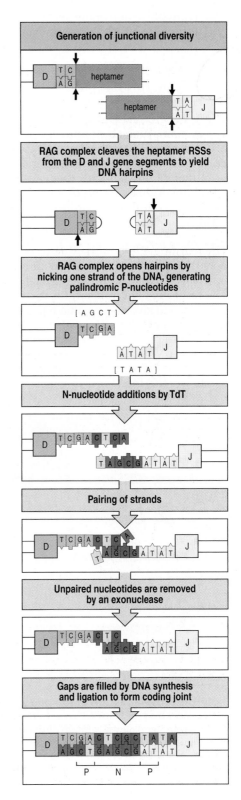

Figure 4.21 The generation of junctional diversity during gene rearrangement. The process is illustrated for a D to J rearrangement. The RSSs are brought together and the RAG complex cleaves (arrows) between the heptamer sequences and the gene segments (top panel). This leads to excision of the DNA that separates the D and J segments. The ends of the two DNA strands of the D and J segments are joined to form hairpins. Further cleavage (arrows) on one DNA strand of the D and J segments releases the hairpins and generates short single-stranded sequences at the ends of the D and J segments. The extra nucleotides are known as P nucleotides because they make a palindromic sequence in the final double-stranded DNA (as indicated on the diagram). Terminal deoxynucleotidyl transferase (TdT) adds nucleotides randomly to the ends of the single strands. These nucleotides, which are not encoded in the germline, are known as N nucleotides. The single strands pair, and through the action of exonuclease, DNA polymerase, and DNA ligase the double-stranded DNA molecule is repaired to give the coding joint.

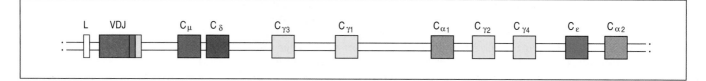

junction formed between the V and J segments (see Figure 4.8), and the third hypervariable region of the heavy-chain V domain is formed by the D segment and its junctions with the rearranged V and J segments.

4-10 Developing and naive B cells use alternative mRNA splicing to make both IgM and IgD

The isotype of an antibody is determined by its heavy chain, and the only heavy chains made by mature B cells before they encounter antigen are μ and δ. Circulating B cells that have yet to encounter antigen are known as **naive B cells**, and they express both IgM and IgD on their surfaces. These are the only immunoglobulin isotypes that can be produced simultaneously by a B cell. Simultaneous expression of both μ and δ chains from the same heavy-chain locus is accomplished by differential splicing of the same primary RNA transcript, a process that involves no rearrangement of genomic DNA.

The rearrangement of the V, D, and J segments of the heavy-chain locus that occurs during B-cell development brings a gene promoter and an enhancer into closer juxtaposition, which enables the rearranged gene to be transcribed. The resulting RNA transcript is then spliced and processed and the mRNA is translated to give a heavy-chain protein. In the rearranged heavy-chain locus, the exons encoding the leader peptide and the V region are on the 5′ side (upstream) of the DNA encoding the nine different C regions. Closest to the rearranged V-region gene is the μ gene, which is followed by the δ gene (Figure 4.22). In each C gene, separate exons encode each immunoglobulin domain, as shown for the μ and δ genes in Figure 4.23. In mature naive B cells,

Figure 4.22 Rearrangement of V, D, and J segments produces a functional heavy-chain gene. The assembled VDJ sequence lies some distance from the cluster of C genes. Only functional C genes are shown here. The four different γ genes specify four different subtypes of the γ heavy chain, whereas the two α genes specify two subtypes of the α heavy chain. For simplicity, individual exons in the C genes are not shown. The diagram is not to scale.

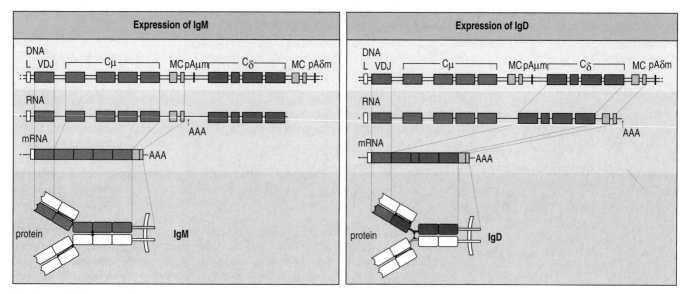

Figure 4.23 Coexpression of IgD and IgM is regulated by RNA processing. In mature B cells, transcription initiated at the V_H promoter extends through both the C_μ and C_δ genes. For simplicity we have not shown all the individual C-gene exons but only those of relevance to the production of IgM and IgD. The long primary transcript is then processed by cleavage, polyadenylation, and splicing. Cleavage and polyadenylation at the μ site (pAμm; the 'm' denotes that this site produces membrane-bound IgM) and splicing between C_μ exons yields an mRNA encoding the μ heavy chain (left panel). Cleavage and polyadenylation at the δ site (pAδm) and a different pattern of splicing that removes the C_μ exons yields mRNA encoding the δ heavy chain (right panel). AAA designates the poly A tail. MC, exons that encode the transmembrane region of the heavy chain.

transcription of the heavy-chain gene starts upstream of the exons encoding the leader peptide and the V region, continues through the μ and δ C genes and terminates downstream of the δ gene, before the γ3 C gene. This long primary RNA transcript is then spliced and processed in two different ways: one yields mRNA for the μ heavy chain (Figure 4.23, left panel) and one yields mRNA for the δ heavy chain (Figure 4.23, right panel). In making μ-chain mRNA from the primary transcript, the entire δ-gene RNA is removed along with the introns from the μ gene. Conversely, in making δ-chain mRNA the entire μ-gene RNA is removed as well as the δ-gene introns.

4-11 Each B cell produces immunoglobulin of a single antigen specificity

In a developing B cell, the process of immunoglobulin-gene rearrangement is tightly controlled so that only one heavy chain and one light chain are finally expressed, a phenomenon known as **allelic exclusion**. This ensures that each B cell produces IgM and IgD of a single antigen specificity. Although every B cell has two copies, or alleles, of the heavy-chain locus and two copies of each light-chain locus, only one heavy-chain locus and one light-chain locus are rearranged to produce functional genes.

In a population of B cells, the same functional light-chain gene rearrangement is found associated with different functional heavy-chain gene rearrangements. Conversely, the same functional heavy-chain gene rearrangement is found associated with different functional light-chain gene rearrangements. Because an antigen-binding site is formed by the association of a heavy chain and a light chain, the combinatorial association of heavy and light chains makes an important contribution to the overall diversity of immunoglobulins. B cells are free to produce any combination of light and heavy chains, and, thus, the potential number of antibodies of different antigen specificities that can be made is the product of the total numbers of different heavy and light chains (see Section 4-8).

The fact that B cells are monospecific means that an encounter with a given pathogen engages a subset of B cells that will make antibodies that bind only to the pathogen. This clonal selection is one of the central principles of adaptive immunity (see Chapter 3). It ensures the specificity of the antibody response against an infection and focuses the response on that pathogen. It is also the reason that vaccination against diphtheria, for example, provides protection against diphtheria but not against influenza.

The fact that the DNA sequence of the expressed immunoglobulin genes varies from one clone of B cells to the next can be used to detect the large clonal populations of cancer cells in patients with B-cell lymphoma or leukemia. Because the cancer cells are derived from a single clone of B cells, they can be readily distinguished from healthy B cells by comparing the immunoglobulin-gene rearrangements by DNA analysis (Figure 4.24). In the clinic, such analyses are used to determine the presence of cancer cells in blood samples or tissue biopsies and to monitor the response to therapy.

4-12 Immunoglobulin is first made in a membrane-bound form that is present on the B-cell surface

When a B cell first makes IgM and IgD, the heavy chains (but not the light chains) have a hydrophobic sequence near the carboxy terminus by which the immunoglobulins associate with cell membranes. Like all proteins destined for the cell surface, immunoglobulin chains enter the endoplasmic reticulum as soon as they are synthesized. There they associate with each other to form immunoglobulin molecules attached to the endoplasmic

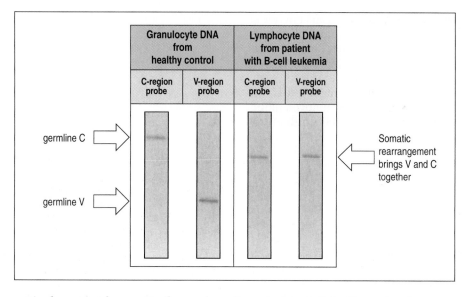

Granulocyte DNA from healthy control		Lymphocyte DNA from patient with B-cell leukemia	
C-region probe	V-region probe	C-region probe	V-region probe

germline C →

germline V →

← Somatic rearrangement brings V and C together

Figure 4.24 DNA analysis of immunoglobulin genes can help to diagnose B-cell tumors. The two photographs on the left show agarose gel electrophoresis of a restriction enzyme digest of DNA from peripheral blood granulocytes of a healthy person (healthy control). The state of the immunoglobulin genes is examined by hybridization with V-region and C-region DNA probes. In these cells, mostly neutrophils, the immunoglobulin genes are in the germline configuration and so the V and C regions are found in quite separate DNA fragments. The two photographs on the right are of a similar restriction digest of DNA from peripheral blood lymphocytes from a patient with chronic lymphocytic leukemia, in which a particular clone of B cells is greatly expanded. For this reason, a unique immunoglobulin gene rearrangement is visible—the V and C regions are found in the same fragment. Normal B lymphocytes in the patient's blood each have a different gene rearrangement, none of which is sufficiently well represented to be visible as a band. Photographs courtesy of S. Wagner and L. Luzzatto.

reticulum membrane. By themselves, these immunoglobulin molecules cannot be transported to the cell surface. For that to happen they must associate with two additional transmembrane proteins called Igα and Igβ (Figure 4.25). These proteins are invariant in sequence, unlike immunoglobulins, and travel to the B-cell surface in a complex with the immunoglobulin molecule. At the cell surface this complex of IgM with Igα and Igβ forms the B-cell receptor for antigen. The function of the less abundant complex of IgD with Igα and Igβ is not known and is one of the unsolved mysteries of immunology.

In responding to specific antigen, the immunoglobulin component of the B-cell receptor is responsible for binding to the specific antigen. However, this interaction alone cannot give the signal that tells the cell's interior when an antigen has bound. The cytoplasmic portions of immunoglobulin heavy chains are very short and do not interact with the intracellular proteins that signal B cells to divide and differentiate. That function is provided by the longer cytoplasmic tails of the Igα and Igβ components of the B-cell receptor. They contain amino acid motifs to which intracellular signaling proteins bind.

Summary

In the human genome, the immunoglobulin heavy-chain and light-chain genes are in a form that is incapable of being expressed. In developing B cells, however, the immunoglobulin genes undergo structural rearrangements that permit their expression. The V domains of immunoglobulin light and heavy chains are encoded in two (V and J) or three (V, D, and J) different kinds of gene segment, respectively, that are brought into juxtaposition by recombination reactions. One mechanism contributing to the diversity in V-region sequences is the random combination of different V and J segments in light-chain genes and of different V, D, and J segments in rearranged heavy-chain genes. A second mechanism is the introduction of additional nucleotides (P and N nucleotides) at the junctions between gene segments during the process of recombination. A third mechanism that creates diversity in the antigen-binding sites of antibodies is the association of heavy and light chains in different combinations. Gene rearrangement in an individual B cell is strictly controlled so that only one type of heavy chain and one type of light chain are expressed, resulting in an individual B cell and its progeny expressing only immunoglobulin of a single antigen specificity. A naive B cell that has never encountered antigen expresses only membrane-bound immunoglobulin of the IgM and IgD classes. The μ and δ heavy chains are produced from a single transcriptional unit that undergoes alternative RNA processing and splicing.

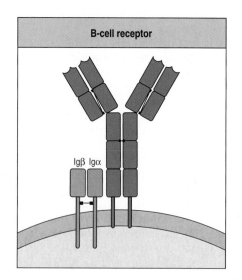

Figure 4.25 Membrane-bound immunoglobulins are associated with two other proteins, Igα and Igβ. Igα and Igβ are disulfide-linked. They have long cytoplasmic tails that can interact with intracellular signaling proteins, and the complex of immunoglobulin with Igα and Igβ serves as the functional B-cell receptor. The immunoglobulin shown here is IgM, but all isotypes can serve as B-cell receptors.

Diversification of antibodies after B cells encounter antigen

A mature B cell's first encounter with antigen marks a watershed in its development. Binding of antigen to the surface immunoglobulin of a mature naive B cell triggers the cell's proliferation and differentiation, and ultimately the secretion of antibodies. As the immune response progresses, antibodies with different properties are made. In this part of the chapter we look at the structural and functional changes that occur in immunoglobulins after antigen has been encountered, and how antibodies with different functions are made.

4-13   Secreted antibodies are produced by an alternative pattern of heavy-chain RNA processing

Gene rearrangement in an immature B cell leads to the expression of functional heavy and light chains and to the production of membrane-bound IgM and IgD on the mature B cell. After an encounter with antigen, these isotypes are produced as secreted antibodies. IgM antibodies are produced in large amounts and are important in protective immunity, whereas IgD antibodies are produced only in small amounts and have no known effector function.

All the isotypes, or classes, of immunoglobulin (IgA, IgD, IgE, IgG, and IgM) can be made in two forms: one that is bound to the cell membrane and serves as the B-cell receptor for antigen, and one, the antibody, that is secreted to bind to antigen and aid its destruction. During differentiation to antibody-secreting plasma cells, B cells change from making the membrane-bound form to making the secreted form; plasma cells only make secreted antibody. The difference between membrane-bound and secreted immunoglobulin lies at the carboxy terminus of the heavy chain: here, membrane-associated immunoglobulin has a hydrophobic anchor sequence that is inserted into the membrane, whereas antibody has a hydrophilic sequence. This difference is determined by different patterns of RNA splicing and processing of the same primary RNA transcript, and involves no rearrangement of the underlying genomic DNA. The alternative patterns of splicing for IgM are compared in Figure 4.26.

The hydrophilic carboxy terminus of the secreted μ chain is encoded at the 3' end of the exon encoding the fourth C-region domain, whereas the

Figure 4.26 The surface and secreted forms of an immunoglobulin are derived from the same heavy-chain gene by alternative RNA processing. Each heavy-chain C gene has two exons (membrane-coding (MC), light blue) encoding the transmembrane region and cytoplasmic tail of the surface form of that isotype, and a secretion-coding (SC) sequence (orange) encoding the carboxy terminus of the secreted form. The events that dictate whether a heavy-chain RNA will result in a secreted or transmembrane immunoglobulin occur during processing of the initial transcript and are shown here for IgM. Each heavy-chain C gene has two potential polyadenylation sites (shown as pAμs and pAμm). In the left panel, the transcript is cleaved and polyadenylated at the second site (pAμm). Splicing between a site located between the fourth C_μ exon and the SC sequence, and a second site at the 5' end of the MC exons, removes the SC sequence and joins the MC exons to the fourth C_μ exon. This generates the transmembrane form of the heavy chain. In the right panel, the primary transcript is cleaved and polyadenylated at the first site (pAμs), eliminating the MC exons and giving rise to the secreted form of the heavy chain. AAA designates the poly A tail.

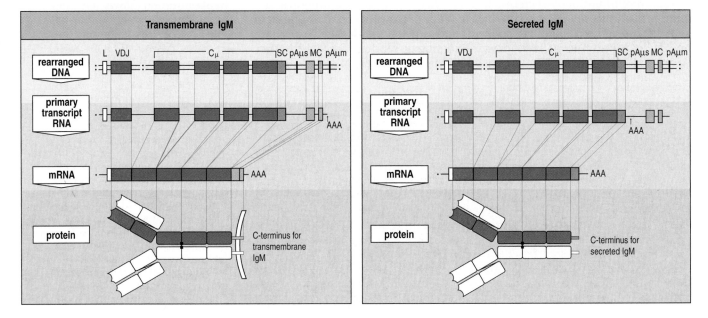

hydrophobic membrane anchor of the membrane-associated μ chain is encoded by two small, separate exons downstream. The splicing to give secreted μ chain is the simpler pattern: the sequence encoding the hydrophilic carboxy terminus is retained and the sequences 3′ of that, including the exons encoding the hydrophobic membrane anchor, are discarded (Figure 4.26, right panel). To produce mRNA encoding membrane-associated μ chain, alternative splicing in the exon encoding the fourth C-region domain removes the sequence encoding the hydrophilic anchor, whereas the exons encoding the hydrophobic carboxy terminus are retained and incorporated into the mRNA when the introns are spliced out (Figure 4.26, left panel).

4-14   Rearranged V-region sequences are further diversified by somatic hypermutation

The diversity generated during gene rearrangement is concentrated in the third CDR of the V_H and V_L domains. Once a B cell has been activated by antigen, however, further diversification of the whole of the V-domain coding sequences occurs through a process of **somatic hypermutation**. This almost randomly introduces single-nucleotide substitutions (point mutations) at a high rate throughout the rearranged V regions of heavy-chain and light-chain genes (Figure 4.27). The immunoglobulin constant regions are not affected, and neither are other B-cell genes. The mutations occur at a rate of about one mutation per V-region sequence per cell division, which is more than a million times greater than the ordinary mutation rate for a gene.

Somatic mutation is dependent on the enzyme **activation-induced cytidine deaminase** (**AID**), which is made only by proliferating B cells. AID converts cytosine in single-stranded DNA to uracil, a normal component of RNA but not DNA. During transcription, when the two DNA strands of the immunoglobulin gene become temporarily separated, AID converts cytosine bases to uracil. Other enzymes, which are not specific to B cells but are components of general pathways of DNA repair and modification, can then act to convert the uracil to any one of the four bases of normal DNA.

Somatic hypermutation gives rise to B cells bearing mutant immunoglobulin molecules on their surface. Some of these mutant immunoglobulin molecules have substitutions in the antigen-binding site that increase its affinity for the antigen. B cells bearing these mutant high-affinity immunoglobulin receptors compete most effectively for binding to antigen and are preferentially selected to mature into antibody-secreting plasma cells. The mutant antibodies that emerge from this selection do not have a random distribution of amino-acid substitutions. The changes are concentrated at positions in the heavy-chain and light-chain CDR loops that form the antigen-binding site and directly contact antigen (Figure 4.28). As the adaptive immune response to infection

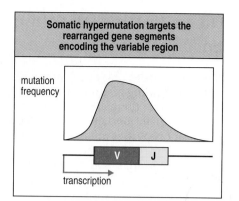

Figure 4.27 Somatic hypermutation is targeted to the rearranged gene segments that encode immunoglobulin V regions. The frequency of mutations at positions in and around the rearranged VJ sequence of an expressed light-chain gene is shown here.

Figure 4.28 The almost random variation produced by somatic hypermutation allows selection of variant immunoglobulins with improved antigen-binding sites. B cells were collected 1 and 2 weeks after immunization with the same epitope and used to make hybridomas secreting monoclonal antibodies. The amino acid sequences of the heavy and light chains expressed by the hybridoma cells were determined. Each line represents one antibody and the red bars represent amino acid positions that differ from the prototypic sequence. One week after primary immunization, most of the B cells make IgM, which shows some new sequence variation in the V region. This variation is confined to the CDRs, which form the antigen-binding site. Two weeks after immunization, both IgG- and IgM-producing B cells are present and their antibodies show increased variation that now involves all six CDRs of the antigen-binding site.

1 week after primary immunization				2 weeks after primary immunization			
IgM				IgM/IgG			
heavy-chain V region		light-chain V region		heavy-chain V region		light-chain V region	
CDR1 CDR2 CDR3		CDR1 CDR2 CDR3		CDR1 CDR2 CDR3		CDR1 CDR2 CDR3	

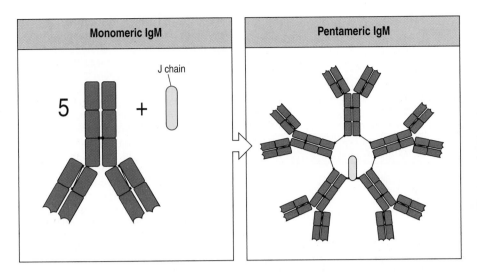

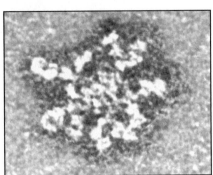

Figure 4.29 IgM is secreted as a pentamer of immunoglobulin monomers. The left two panels show schematic diagrams of the IgM monomer and pentamer. The IgM pentamer is held together by a polypeptide called the J chain, for joining chain (not to be confused with a J segment). The monomers are cross-linked by disulfide bonds to each other and to the J chain. The right panel shows an electron micrograph of an IgM pentamer, showing the arrangement of the monomers in a flat disc. The lack of a hinge region in the IgM monomer makes the molecule less flexible than, say, IgG, but this is compensated for by the pentamer having five times as many antigen binding sites as IgG. Faced with a pathogen having multiple identical epitopes on its surface, IgM can usually attach to it with several binding sites simultaneously. Photograph (× 900,000) courtesy of K.H. Roux and J.M. Schiff.

proceeds, antibodies of progressively higher affinity for the infecting pathogen are produced—a phenomenon called **affinity maturation**.

Affinity maturation is a process of evolution in which variant immunoglobulins generated in a random manner are subjected to selection for improved binding to a pathogen. It achieves in a few days what would require thousands, if not millions, of years of classical Darwinian evolution in a conventional gene. This capacity for extraordinarily rapid evolution in pathogen-binding immunoglobulins is a major factor in allowing the human immune system to keep up with the generally faster-evolving pathogens.

4-15 Isotype switching produces immunoglobulins with different C regions but identical antigen specificities

IgM is the first class of antibody made in a primary immune response. Whereas the surface IgM of the B-cell receptor is monomeric, secreted effector IgM consists of a circular pentamer of the Y-shaped immunoglobulin monomers (Figure 4.29). Because of its 10 antigen-binding sites, IgM binds strongly to the surface of pathogens with multiple repetitive epitopes, but it is limited in the effector mechanisms that it uses to clear antigen from the body. Antibodies with other effector functions are produced by the process of **isotype switching** or **class switching**, in which a further DNA recombination event enables the rearranged V-region coding sequence to be used with other heavy-chain C genes. Isotype switching, like somatic hypermutation, is dependent on AID and similarly occurs only in B cells proliferating in response to antigen.

Isotype switching is accomplished by a recombination within the cluster of C genes that excises the previously expressed C gene and brings a different one into juxtaposition with the assembled V-region sequence. Thus, the antigen specificity of the antibody remains unchanged, even though its isotype changes. Flanking the 5′ side of each C gene, with the exception of the δ gene, are highly repetitive sequences that mediate recombination. They are called **switch sequences** or **switch regions** (Figure 4.30).

An example of isotype switching from IgM (plus IgD) to IgG1 is shown in Figure 4.30. The first step is the initiation of transcription of the C-region gene to which the cell will be switched, in this case $C_{\gamma 1}$ and its flanking switch region, $S_{\gamma 1}$. AID then targets the cytosines in the $S_{\mu 1}$ and the $S_{\gamma 1}$ switch regions for deamination to uracil (see Figure 4.30). Next, uracil is removed by the enzyme uracil-DNA-glycosylase (UNG), leaving a nucleotide lacking a base. A specific endonuclease, APE1, then excises this abasic nucleotide, leaving a

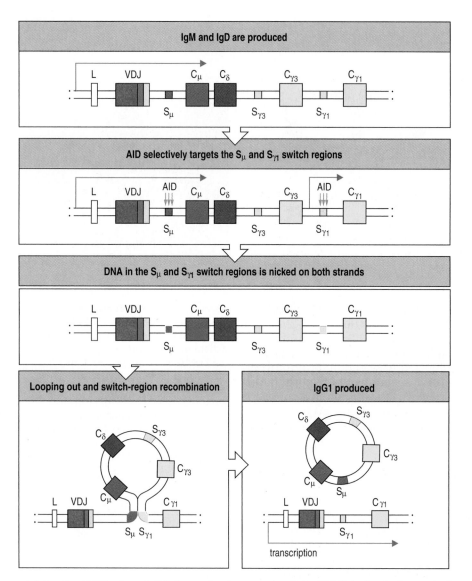

Figure 4.30 Isotype switching involves recombination between specific switch regions. Repetitive DNA sequences are found to the 5′ side of each of the heavy-chain C genes, with the exception of the δ gene. Immunoglobulin isotype switching occurs by recombination between these switch regions (S), with deletion of the intervening DNA. The switch regions are targeted by AID, which leads to nicks being made in both strands of the DNA. These nicks facilitate recombination between the switch regions, which leads to excision of the intervening DNA as a non-functional circle of DNA and brings the rearranged VDJ segments into juxtaposition with a different C gene. The first switch a clone of B cells makes is from the μ isotype to another isotype. A switch from μ to the γ1 isotype is shown here. Further switching to other isotypes can take place subsequently.

nick in the DNA strand. Nicks on both strands of the DNA in both the switch sequences facilitate their recombination, by which the μ, δ, and $C_{\gamma 3}$ genes are excised as a circular DNA molecule, bringing the V region into juxtaposition with the $C_{\gamma 1}$ gene. The mRNA for the new immunoglobulin is then produced as described in Section 4-9. Switching can take place between the μ switch region and that of any other isotype. Sequential switching can also occur, for example from μ to γ_1 to α_1. Isotype switching only occurs during an active immune response, and the patterns of isotype switching are regulated by cytokines secreted by antigen-activated T cells.

Knowledge of the molecular mechanisms underlying isotype switching and somatic hypermutation has been aided by the study of patients deficient in the various enzymes involved in the mutation and modification of DNA in the V-region gene and in the switch sequences of the C-region genes. In particular, the immunoglobulin genes in the B cells of patients lacking a functional AID gene do not undergo somatic hypermutation or isotype switching. The only antibodies that these patients make are low-affinity IgM, which are produced in greater amounts than in normal individuals. For this reason their condition is called **hyper IgM immunodeficiency**. The main consequence of this immunodeficiency is susceptibility to infection with pyogenic bacteria, particularly in the sinuses, ears, and lungs. These infections can usually be terminated with antibiotics and prevented by regular injections of immunoglobulin purified from the plasma of healthy blood donors.

4-16 Antibodies with different C regions have different effector functions

In humans and other mammals, the five classes of immunoglobulin are IgA, IgD, IgE, IgG, and IgM (see Figure 4.5). Certain classes are divided further into subclasses, which differ in both nomenclature and properties between species. In humans, IgA is divided into two subclasses (IgA1 and IgA2), whereas IgG is divided into four subclasses (IgG1, IgG2, IgG3, and IgG4), which are numbered according to their relative abundance in plasma, IgG1 being the most abundant. The heavy chains of the human IgA subclasses are designated α1 and α2 and the heavy chains of the human IgG subclasses by γ1, γ2, γ3, and γ4. The α, δ, and γ heavy-chain C regions are made up of three C domains, whereas the μ and ε heavy chains have four (see Figure 4.5). Each C domain is encoded by a separate exon in the relevant C gene. Additional exons are used to encode the hinge region and the carboxy terminus, depending on the isotype. The physical properties of the human immunoglobulin isotypes are given in Figure 4.31).

Antibodies aid in the clearance of pathogens from the body in various ways. **Neutralizing antibodies** directly inactivate a pathogen or a toxin and prevent it from interacting with human cells. Neutralizing antibodies against viruses, for example, bind to a site on the virus that is normally used to gain entry to cells. Another function of antibodies is opsonization, a term used to describe the coating of pathogens with an immune-system protein (see Section 2-5). The common **opsonins** are antibodies and complement proteins. Opsonized pathogens are more efficiently ingested by phagocytes, which have receptors for the Fc region of some antibodies and for certain complement proteins. Activation of complement by antibodies bound to a bacterial surface can also lead to the direct lysis of the bacterium.

IgM is the first antibody produced in an immune response against a pathogen. It is made principally by plasma cells resident in lymph nodes, spleen, and bone marrow, and it circulates in blood and lymph. On initiation of an immune response, most of the antibodies that bind the antigen will be of low affinity, and the multiple antigen-binding sites of IgM are needed if enough antibody is to bind sufficiently strongly to a microorganism to be of any use. The overall strength of binding at multiple sites is called the **avidity** of an antibody, in contrast to its affinity, which is the strength of binding at a single site. When bound to antigen, sites exposed in the constant region of IgM initiate reactions with complement that can kill microorganisms directly or facilitate their phagocytosis. Because somatic hypermutation leads to antibodies of increased affinity for the antigen, two antigen-binding sites are then sufficient

	Immunoglobulin class or subclass								
	IgM	IgD	IgG1	IgG2	IgG3	IgG4	IgA1	IgA2	IgE
Heavy chain	μ	δ	γ_1	γ_2	γ_3	γ_4	α_1	α_2	ε
Molecular weight (kDa)	970	184	146	146	165	146	160	160	188
Serum level (mean adult mg/ml)	1.5	0.03	9	3	1	0.5	2.0	0.5	5×10^{-5}
Half-life in serum (days)	10	3	21	20	7	21	6	6	2

Figure 4.31 **The physical properties of the human immunoglobulin isotypes.** The molecular weight given for IgM is that of the pentamer (see Figure 4.29), the predominant form in serum. The molecular weight given for IgA is that of the monomer. Large amounts of IgA are also produced in the form of dimers, which is the form found in secretions at mucosal surfaces.

Function	IgM	IgD	IgG1	IgG2	IgG3	IgG4	IgA	IgE
Neutralization	+	–	+++	+++	+++	+++	+++	–
Opsonization	–	–	+++	*	++	+	+	–
Sensitization for killing by NK cells	–	–	++	–	++	–	–	–
Sensitization of mast cells	–	–	+	–	+	–	–	+++
Activation of complement system	+++	–	++	+	+++	–	+	–

Property	IgM	IgD	IgG1	IgG2	IgG3	IgG4	IgA	IgE
Transport across epithelium	+	–	–	–	–	–	+++ (dimer)	–
Transport across placenta	–	–	+++	+	++	++	–	–
Diffusion into extravascular sites	+/–	–	+++	+++	+++	+++	++ (monomer)	+
Mean serum level (mg/ml)	1.5	0.03	9	3	1	0.5	2.5	5×10^{-5}

Figure 4.32 Each human immunoglobulin isotype has specialized functions and distinct properties. The major effector functions of each isotype (+++) are shaded in dark red; lesser functions (++) are shown in dark pink, and very minor functions (+) in pale pink. Other properties are similarly marked, with mean serum concentrations shown in the bottom row. Opsonization refers to the ability of the antibody itself to facilitate phagocytosis. Antibodies that activate the complement system indirectly cause opsonization via complement. The properties of IgA1 and IgA2 are similar and are given here under IgA. *IgG2 acts as an opsonin in the presence of a genetic variant of its phagocyte Fc receptor, which is found in about 50% of Caucasians. NK cells, natural killer cells.

to produce strong binding. By switching isotype, different effector functions can be brought into play while preserving antigen specificity: synthesis of IgM gives way to synthesis of IgG.

IgG is the most abundant antibody in the internal body fluids, including blood and lymph. Like IgM, it is made principally in the lymph nodes, spleen, and bone marrow and circulates in lymph and blood. IgG is smaller and more flexible than IgM, properties that give it easier access to antigens in the extracellular spaces of damaged and infected tissues. The flexibility of the hinge region in IgG allows the two Fab arms to move relative to each other (see Figure 4.4). This enables both of the antigen-binding sites to bind to repeated epitopes on the surfaces of pathogens. IgG antibodies can also implement more effector functions than IgM (Figure 4.32). Once they have bound an antigen, the IgG1 and IgG3 subclasses can directly recruit phagocytic cells to ingest the antigen:antibody complex, as well as activating the complement system. During pregnancy, IgG antibodies can be transferred across the placenta, providing the fetus with protective antibodies from the mother in advance of possible infection.

Monomeric IgA is made by plasma cells in lymph nodes, spleen, and bone marrow and is secreted into the bloodstream. IgA can also be made as a dimer, two monomers being joined by a J chain identical to that in pentameric IgM (Figure 4.33). Dimeric IgA is made principally in the lymphoid tissues underlying mucosal surfaces and is the antibody secreted into the lumen of the gut; it is also the main antibody in other secretions, including milk, saliva, sweat, and tears. The mucosal surface of the gastrointestinal tract provides an extensive surface of contact between the human body and the environment, and the transport processes involved in the uptake of food make it vulnerable to infection. In total, more IgA is made than any other isotype. Some of it is directed against the resident microorganisms that colonize mucosal surfaces, keeping their population in check.

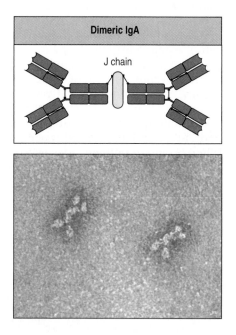

Figure 4.33 IgA molecules can form dimers. In mucosal lymphoid tissue, IgA is synthesized as a dimer in association with the same J chain as that found in pentameric IgM. In dimeric IgA, the monomers have disulfide bonds to the J chain but not to each other. The bottom panel shows an electron micrograph of dimeric IgA. Photograph (× 900,000) courtesy of K.H. Roux and J.M. Schiff.

The IgE class of antibodies is highly specialized toward recruiting the effector functions of mast cells in epithelium, activated eosinophils present at mucosal surfaces, and basophils in blood. These cell types carry a high-affinity receptor that binds IgE in the absence of antigen. The presence of antigen, which binds to the IgE, triggers strong physical and inflammatory reactions that can expel and kill infecting parasites. In countries where parasite infections are rare, the major impact of the IgE response is from the allergies and asthma caused when IgE is made against otherwise harmless antigens.

Summary

Membrane-bound immunoglobulins of the IgM and IgD classes serve as antigen receptors, or B-cell receptors, for mature, naive B cells. When a pathogen binds to a B-cell receptor, the cell receives signals that cause cell proliferation and the differentiation of some of the resulting B cells into plasma cells. Although they have no immunoglobulin on their surface, plasma cells secrete copious quantities of soluble IgM antibody and some IgD. The change from membrane-associated immunoglobulin to secreted antibody is caused by altered splicing of heavy-chain mRNA that eliminates the sequence encoding the hydrophobic transmembrane anchor. Other members of the clone of proliferating B cells are subject to somatic hypermutation, which introduces mutations throughout the heavy-chain and light-chain V regions. After hypermutation, the B cells carrying cell-surface immunoglobulins with the highest affinity for antigen are selected to become plasma cells. As antibody affinity increases, the dependence on pentameric IgM for strong binding to antigen relaxes. It then becomes advantageous to change the immunoglobulin from IgM/IgD to another isotype, either IgG, IgE, or IgA, that has effector functions more suited to the type of infecting pathogen and the anatomical site of infection. This process of isotype switching is accomplished by further somatic recombination of the expressed heavy-chain gene, in which the rearranged VDJ segment is moved next to a 'new' C region, with the excision of the μ and δ C genes. Both somatic hypermutation and isotype switching depend upon activation-induced cytidine deaminase (AID), a highly specific enzyme of adaptive immunity that is made only by B cells proliferating in response to antigen.

Summary to Chapter 4

The principal function of B lymphocytes is to produce antibodies, which are secreted immunoglobulins that bind tightly to infectious agents and tag them for destruction or elimination. Each antibody is highly specific for its corresponding antigen; the antibody repertoire of each person is enormous because it is composed of many millions of different antibodies that can bind a wide variety of different antigens. Antibodies can also be divided into five different effector classes—IgM, IgG, IgD, IgA, and IgE—that have different functions in the immune response. This chapter has provided an overview of the structure and function of the antibody molecule and of the unusual genetic mechanisms that create this diversity in specificity and effector function. Within an antibody molecule, the V regions that bind antigen are physically separated from the C region that interacts with effector molecules and cells of the immune system, such as complement, phagocytes, and other leukocytes. Antigen binding is the property of the paired V domains of the heavy and light chains, which can form an almost unlimited number of different binding sites with structural complementarity to a vast range of molecules. The immunoglobulin genes (heavy chain, κ light chain, and λ light chain) are expressed only in B cells, and their expression involves an unusual process of DNA rearrangement in which somatic DNA recombination assembles a V-region coding sequence from sets of gene segments that are present in the unrearranged gene. The random selection of gene segments for assembly creates much of the collective diversity of antigen-binding sites. The lymphocyte-specific proteins and ubiquitous enzymes of DNA repair and recombination are involved in the recombination machinery. Imprecision is inherent in some of their reactions, which creates additional diversity at the junctions between gene segments. In an individual B cell, only one rearranged heavy-chain gene and one rearranged light-chain gene become functional, ensuring that each B cell expresses immunoglobulin of a single specificity. The series of gene rearrangements that result in the production of membrane-bound IgM, the first immunoglobulin produced, is summarized in Figure 4.34.

On mature B cells, the membrane-bound immunoglobulin functions as the specific receptor for antigen; on encountering antigen, the B cell is stimulated to proliferate and differentiate into plasma cells that secrete antibody of the same specificity as the membrane-bound immunoglobulin. This ensures that an immune response is directed only against the invading pathogen or immunizing antigen. On the stimulation of a B cell by its specific antigen, a mechanism of somatic hypermutation introduces point mutations into the rearranged V-region DNA, diversifying the clone of proliferating B cells. Further selection of B cells by antigen increases the overall binding strength of the antibodies for antigen. Thus, the diversity of antibodies is due in part to inherited variation that is encoded in the genome and in part to non-inherited diversity that develops in B cells during an individual's lifetime. The first antibody produced after an encounter with antigen is always IgM. As the immune response proceeds, the process of isotype switching further rearranges the expressed heavy-chain gene so that the protein made has the same

Figure 4.34 Gene rearrangement and the synthesis of cell-surface IgM in B cells. Before immunoglobulin light-chain (center panel) and heavy-chain (right panel) genes can be expressed, rearrangements of gene segments are needed to produce exons encoding the V regions. Once this has been achieved, the genes are transcribed to give primary transcripts containing both exons and introns. The latter are spliced out to produce mRNAs that are translated to give κ or λ light chains and μ heavy chains that assemble inside the cell and are expressed as membrane-bound IgM at the cell surface. The main stages in the biosynthesis of the heavy and light chains are shown in the panel on the left.

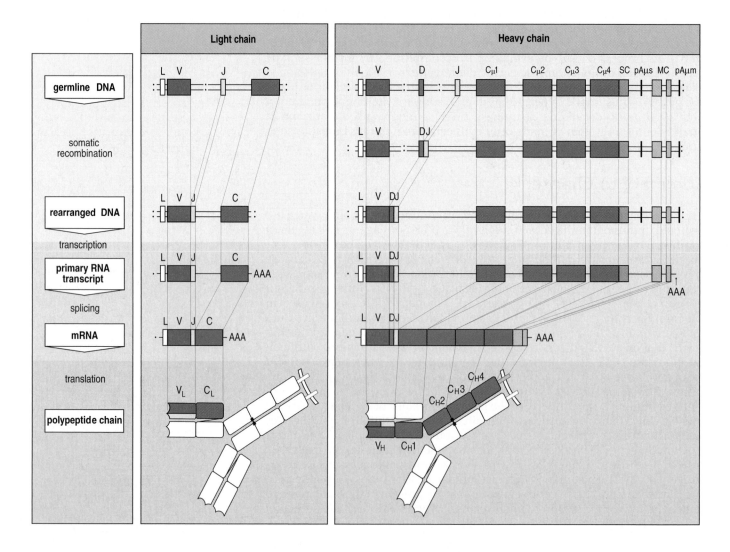

Changes in immunoglobulin genes during a B cell's life		
Event	**Mechanism**	**Permanence of change to the B cell's genome**
1 V-region assembly from gene fragments	Somatic recombination of genomic DNA	Irreversible
2 Generation of junctional diversity	Imprecision in joining rearranged DNA segments adds nongermline nucleotides (P and N) and deletes germline nucleotides	Irreversible
3 Assembly of transcriptional controlling elements	Promoter and enhancer are brought closer together by V-region assembly	Irreversible
4 Transcription activated with coexpression of surface IgM and IgD	Two patterns of splicing and processing RNA are used	Reversible and regulated
5 Synthesis changes from membrane Ig to secreted antibody	Two patterns of splicing and processing RNA are used	Reversible and regulated
6 Somatic hypermutation	Point mutation of genomic DNA	Irreversible
7 Isotype switch	Somatic recombination of genomic DNA	Irreversible

Figure 4.35 The changes in the immunoglobulin genes that occur over a B cell's lifetime.

V region but a different C region. The functional effect of isotype switching is to produce antibody molecules with the same antigen specificity but different effector functions—IgG, IgA, and IgE. The changes in the immunoglobulin genes that occur throughout the life of a B cell are summarized in Figure 4.35.

Questions

4–1
  A. What is the difference between antibodies and immunoglobulins?
  B. Which cell types produce each type of molecule?

4–2 Describe the structure of an antibody molecule and how this structure enables it to bind to a specific antigen. Include the following terms in your description: heavy chain (H chain), light chain (L chain), variable region, constant region, Fab, Fc, antigen-binding site, hypervariable region, and framework region.

4–3 The molecules to which antibodies bind are called
  a. constant regions
  b. framework regions
  c. complementarity-determining regions
  d. antigens
  e. gene segments.

4–4
  A. What is an epitope?
  B. Define the term multivalent antigen.
  C. How does a linear epitope differ from a conformational epitope?
  D. Do antibodies bind their antigens via noncovalent bonding or via covalent bonding?

4–5 The process of gene rearrangement in immunoglobulin and T-cell receptor genes is called
  a. somatic hypermutation
  b. isotype switching
  c. somatic recombination
  d. apoptosis
  e. clonal selection.

4–6
  A. Explain briefly how a vast number of immunoglobulins of different antigen specificities can be produced from the relatively small number of immunoglobulin genes present in the genome. Include the following terms in your explanation: somatic recombination; germline configuration; V, D, and J segments.
  B. What is the final arrangement of gene segments in the rearranged immunoglobulin heavy-chain gene V region, and in what order do these gene segment rearrangements occur?
  C. In what order do the various immunoglobulin loci rearrange?

4–7   Which of the following recombinations is not permitted during somatic recombination in the heavy-chain and light-chain immunoglobulin loci? (Select all correct answers.)
   a. $D_H:J_H$
   b. $V_\lambda:J_\lambda$
   c. $D_\kappa:V_H$
   d. $V_H:J_H$
   e. $V_H:D_H$.

4–8   Junctional diversity during gene rearrangement results from the addition of
   a. switch region nucleotides
   b. P and N nucleotides
   c. V, D, and J nucleotides
   d. recombination signal sequences
   e. mutations in complementarity-determining regions.

4–9   What would be the effect of a genetic defect that resulted in a lack of somatic recombination between V, D, and J segments?

4–10   Which of the following statements regarding immunoglobulins is correct?
   a. Immunoglobulins make up five classes (or isotypes) called IgA, IgD, IgE, IgG, and IgM.
   b. Regardless of their isotype, immunoglobulins all have the same effector function.
   c. Antibodies consist of four identical heavy chains and four identical light chains.
   d. Peptide bonds hold the heavy and light chains together.
   e. The constant regions make up the antigen-binding site.

4–11   Indicate which of the following statements are true (T) or and which are false (F) with reference to immunoglobulin structure.
   a. __The antibody secreted by a plasma cell has a different specificity for antigen than the immunoglobulin expressed by its B-cell precursor.
   b. __The amino-terminal regions of heavy and light chains of different immunoglobulins all differ in amino acid sequence.
   c. __A flexible hinge region holds the heavy chain and light chain together.
   d. __The heavy-chain constant region is responsible for the effector function of immunoglobulins.
   e. __λ and κ light chains have different functions.

4–12   Which of the following is mismatched
   a. surface immunoglobulin: B-cell antigen receptor
   b. affinity maturation: isotype switching
   c. constant region of antibodies: binding to complement proteins
   d. activation-induced cytidine deaminase: somatic hypermutation
   e. switch sequences: class switching.

4–13   Which of the following statements about the production and use of monoclonal antibodies is incorrect?
   a. Production of monoclonal antibodies requires a purified form of antigen.
   b. A monoclonal antibody has specificity for only one epitope of an antigen.
   c. B cells are fused with a tumor cell called a myeloma, to immortalize the resulting hybridoma.
   d. Monoclonal antibodies made in mice have limited therapeutic potential.
   e. Humanized monoclonal antibodies reduce complications associated with using mouse monoclonal antibodies.

4–14   The process used to produce either surface or secreted forms of the immunoglobulin heavy chain is called
   a. alternative RNA processing
   b. isotype switching
   c. somatic recombination
   d. somatic hypermutation
   e. opsonization.

4–15   Aliya Agassi, 3-year-old girl with pneumonia, a temperature of 40.8°C, respirations 42 per minute (normal 20), and blood oxygen saturation of 90% (normal is more than 98%) was admitted to the hospital. Her neck and armpit lymph nodes were enlarged, and X rays confirmed inflammation in the lower lobe of her right lung. Her medical history revealed two previous cases of pneumonia and six middle-ear infections that were treated successfully with antibiotics. A blood culture grew *H. influenzae*. Blood tests showed elevated levels of IgM above normal, while IgA and IgG were not detected. Her father had normal levels of serum IgA, IgG, and IgM. Which of the following is the most likely cause of her symptoms?
   a. acute lymphoblastic leukemia
   b. IgA deficiency
   c. X-linked agammaglobulinemia
   d. severe combined immunodeficiency
   e. X-linked hyper IgM syndrome
   f. AID deficiency
   g. myeloma.

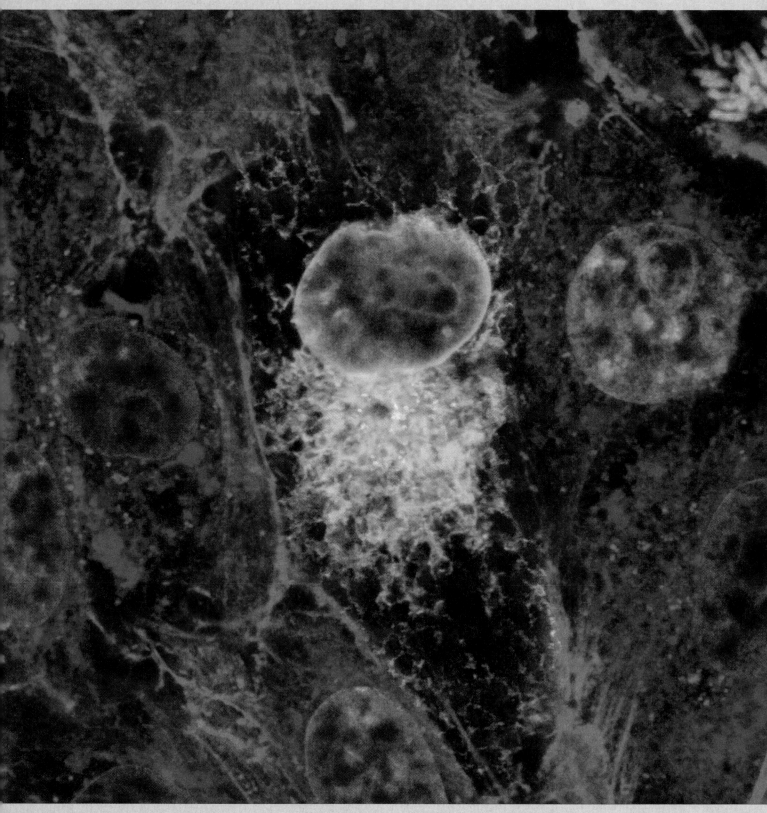

In cells infected with mumps virus the viral proteins are processed into peptides that enter the endoplasmic reticulum to be bound by MHC class I molecules.

Chapter 5

Antigen Recognition by T Lymphocytes

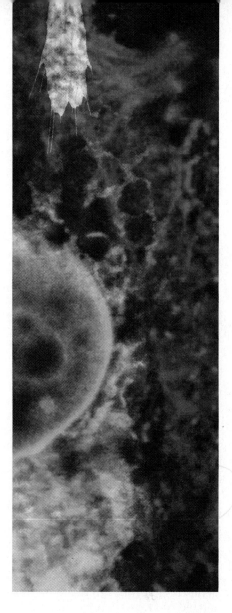

Like B cells, T lymphocytes—or T cells—recognize and bind antigen through highly variable antigen-specific receptors. Whereas the sole function of B cells is to produce secreted antibodies (see Chapter 4), T cells have more diverse roles, all of which involve interactions with other cells. The same basic principles underlie antigen recognition by B cells and T cells, as discussed in Chapter 3, but because of the distinct functions of B and T cells there are important differences between them, particularly in the type of antigen that they recognize.

In the first part of the chapter we describe the structure of the T-cell antigen receptor and the mechanisms that generate its antigen specificity and diversity. The antigen receptor on T cells is commonly referred to simply as the **T-cell receptor** or **TCR**. T-cell receptors have much in common with the immunoglobulins. They have a similar structure, are produced as a result of gene rearrangement, and are highly variable and diverse in their antigen specificity. Like B cells, each clone of T cells expresses a single species of antigen receptor, and thus different clones possess different and unique antigen specificities. This clonal distribution of diverse T-cell receptors is produced by genetic mechanisms similar to those that generate immunoglobulins during B-cell development.

As we learned in Chapter 3, the antigens recognized by T cells are quite distinct from those recognized by immunoglobulins. Immunoglobulins bind epitopes on a wide range of intact molecules, such as proteins, carbohydrates, and lipids. These kinds of epitope are present on the surfaces of bacteria, viruses, and parasites, and also on soluble protein toxins. T-cell receptors, in contrast, recognize and bind mainly to peptide antigens, which are derived from the pathogen's proteins.

The effector function of T cells, in contrast to that of B cells, is carried out via antigen-specific interactions with other cells of the body. This is reflected in the type of antigen that T-cell receptors recognize. As described in Chapter 3, they only recognize peptide antigens when the peptides are complexed with specialized glycoproteins called MHC molecules on the surface of another human cell. The ligand for a T-cell receptor is therefore not simply a peptide antigen but the combination of peptide and MHC molecule on a cell surface. In the second part of the chapter we trace the antigen-processing pathways by which peptides derived from the proteins of infecting microorganisms become bound to MHC molecules and are displayed on cell surfaces.

As noted in Chapter 3, the genes that encode the MHC molecules are clustered in the chromosomal region called the **major histocompatibility complex (MHC)**. For several of the MHC proteins there are hundreds of genetically determined variants in the human population and when organs are transplanted, the differences between MHC molecules of the donor and recipient are the major cause of tissue incompatibility and transplant rejection. The third part of the chapter deals with the MHC itself and with the immunological implications of the exceptional genetic polymorphism of the MHC genes.

T-cell receptor diversity

The T-cell receptor is a membrane-bound glycoprotein that closely resembles a single antigen-binding arm of an immunoglobulin molecule. It is composed of two different polypeptide chains and has one antigen-binding site. T-cell receptors are always membrane bound and there is no secreted form as there is for immunoglobulins. Like immunoglobulins, each chain has a variable region, which binds antigen, and a constant region. During T-cell development, gene rearrangement produces sequence variability in the variable regions of the T-cell receptor by the same mechanisms that are used by B cells to produce the variable regions of immunoglobulins. However, after the T cell is stimulated with antigen, there is no further change in the T-cell receptor and therefore no equivalent of somatic hypermutation of the antigen-binding site or switching of the constant-region isotype as occurs for immunoglobulins (see Chapter 4). These differences correlate with the fact that T-cell receptors are used only as receptors to recognize antigen, whereas immunoglobulins serve as both recognition and effector molecules.

5-1 The T-cell receptor resembles a membrane-associated Fab fragment of immunoglobulin

A T-cell receptor consists of two different polypeptide chains, termed the **T-cell receptor α chain (TCRα)** and the **T-cell receptor β chain (TCRβ)**. The genes encoding the α and β chains have a germline organization similar to that of the genes encoding immunoglobulin heavy-chain and light-chain genes in that they are made up of sets of gene segments that must be rearranged to form a functional gene. As a consequence of the gene rearrangements that occur as part of T-cell development, each mature T cell expresses one functional α chain and one functional β chain, which together define a unique T-cell receptor molecule. Within the population of T cells in a healthy human being there are many millions of different T-cell receptors, each of which defines a clone of T cells and a single antigen-binding specificity.

Comparison of the amino acid sequences of the α and β chains from different T-cell clones shows that each chain has a variable region (V region) and a constant region (C region) like those found in immunoglobulin chains. The α and β chains are folded into discrete protein domains resembling those in immunoglobulin chains. Each chain consists of an amino-terminal V domain, followed by a C domain, and then a membrane-anchoring domain. The antigen-recognition site of T-cell receptors is formed from the V_α and V_β domains and is the most variable part of the molecule, as in the immunoglobulins. The three-dimensional structure of the four extracellular domains of the T-cell receptor is very similar to that of the antigen-binding Fab fragment of IgG (Figure 5.1).

Comparison of the amino acid sequences of V domains from different clones of T cells shows that sequence variation in the α and β chains is clustered into regions of hypervariability, which correspond to loops of the polypeptide chain at the end of the domain farthest from the T-cell membrane. These loops form the binding site for antigen and are termed complementarity-determining regions (CDRs), as in the immunoglobulins. The T-cell receptor α-chain and β-chain V domains each have three CDR loops, called CDR1, CDR2, and CDR3 (Figure 5.2).

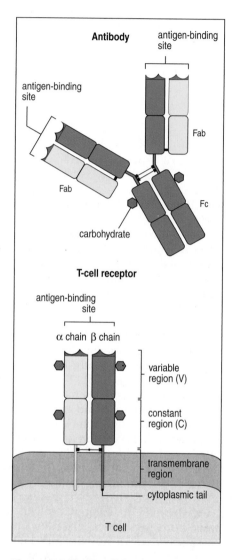

Figure 5.1 The T-cell receptor resembles a membrane-bound Fab fragment. Comparison of the T-cell receptor with an IgG antibody. The T-cell receptor is a membrane-bound heterodimer composed of an α chain of 40–50 kDa and a β chain of 35–46 kDa. The extracellular portion of each chain consists of two immunoglobulin-like domains: the one nearest to the membrane is a C domain and the domain farthest from the membrane is a V domain. The α and β chains both span the cell membrane and have very short cytoplasmic tails. The three-dimensional structure formed by the four immunoglobulin-like domains of the T-cell receptor resembles that of an antigen-binding Fab fragment of antibody.

Immunoglobulins possess two or more binding sites for antigen; this strengthens the interactions of soluble antibody with the repetitive antigens found on the surfaces of microorganisms. In contrast, T-cell receptors possess a single binding site for antigen and are used only as cell-surface receptors for antigen, never as soluble antigen-binding molecules. Antigen binding to T-cell receptors occurs always in the context of two opposing cell surfaces, where multiple copies of the T-cell receptor bind to multiple copies of the antigen:MHC complex on the opposing cell, thus achieving multipoint attachment.

5-2 T-cell receptor diversity is generated by gene rearrangement

In Chapter 4 we divided the mechanisms that generate immunoglobulin diversity into two categories: those operating before the B cell is stimulated with specific antigen, and those operating afterwards. In the first category were the gene rearrangements that generate the V-region sequence, whereas in the second category were changes in mRNA splicing that produce a secreted immunoglobulin, C-region DNA rearrangements that switch the heavy-chain isotype, and somatic hypermutation of the V-region gene to produce antibodies of higher affinity. In T cells the mechanisms that generate diversity before antigen stimulation are essentially the same as those in B cells, but after antigen stimulation the picture is quite different: whereas immunoglobulin genes continue to diversify, the genes encoding T-cell receptors remain unchanged. This fundamental difference reflects the fact that the T-cell receptor is used only for the recognition of antigen and has no role in effector functions, which are handled by other proteins produced by T cells.

The human T-cell α-chain locus is on chromosome 14 and the β-chain locus is on chromosome 7. The organization of the gene segments encoding T-cell receptor α and β chains is essentially like that of the immunoglobulin gene segments (Figure 5.3). The main difference is the simplicity of the T-cell

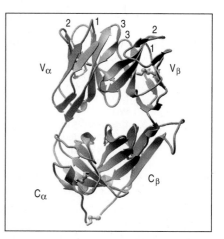

Figure 5.2 Three-dimensional structure of the T-cell receptor showing the antigen-binding CDR loops. The ribbon diagram shows the α chain (in magenta) and the β chain (in blue). The receptor is viewed from the side as it would sit on a cell surface, with the highly variable CDR loops, which bind the peptide:MHC molecule ligand, arrayed across its relatively flat top surface. The CDR loops are numbered 1–3 for each chain. Courtesy of I.A. Wilson.

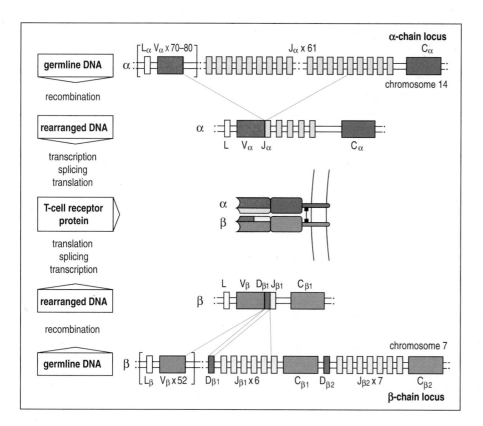

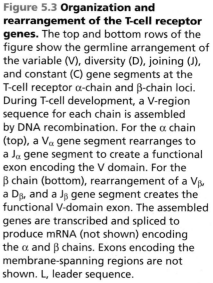

Figure 5.3 Organization and rearrangement of the T-cell receptor genes. The top and bottom rows of the figure show the germline arrangement of the variable (V), diversity (D), joining (J), and constant (C) gene segments at the T-cell receptor α-chain and β-chain loci. During T-cell development, a V-region sequence for each chain is assembled by DNA recombination. For the α chain (top), a V_α gene segment rearranges to a J_α gene segment to create a functional exon encoding the V domain. For the β chain (bottom), rearrangement of a V_β, a D_β, and a J_β gene segment creates the functional V-domain exon. The assembled genes are transcribed and spliced to produce mRNA (not shown) encoding the α and β chains. Exons encoding the membrane-spanning regions are not shown. L, leader sequence.

Figure 5.4 Severe combined immunodeficiency syndrome (SCID). SCID is characterized by a lack of functional T cells and B cells and the inability to make an adaptive immune response. Infants with SCID typically show infections with opportunistic pathogens. Panel a shows chronic *Candida albicans* infection in the mouth of an infant with SCID. SCID can be caused by various genetic defects, one of which is complete loss of RAG function. Panel b shows an infant with Omenn syndrome, a similar immunodeficiency which is due to a genetic defect that results in 80% loss of RAG activity. The bright red rash on the face and shoulders, which is due to chronic inflammation, is a characteristic of this condition. Unless an immune system can be reconstituted by bone marrow transplantation from a healthy donor, babies with SCID or Omenn syndrome die in infancy. Panel a courtesy of Fred Rosen; panel b courtesy of Luigi Notarangelo.

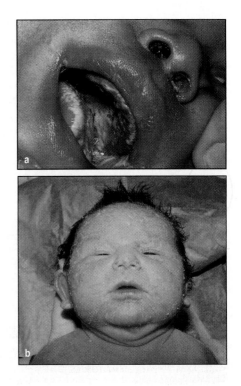

receptor C region: there is only one C_α gene, and, although there are two C_β genes, no functional distinction between them is known. The T-cell receptor α-chain locus is otherwise similar to an immunoglobulin light-chain locus, containing sets of V and J gene segments only; the β-chain locus is similar to an immunoglobulin heavy-chain locus, containing D gene segments in addition to V and J gene segments. The V domain of the T-cell receptor α chain is thus encoded by a V gene segment and a J gene segment; that of the β chain is encoded by a D gene segment in addition to a V and a J gene segment.

T-cell receptor gene rearrangement occurs during T-cell development in the thymus, and the mechanism is similar to that outlined in Chapter 4 for the immunoglobulin genes. In the α-chain gene, a V gene segment is joined to a J gene segment by somatic DNA recombination to make the V-region sequence; in the β-chain gene, recombination first joins a D and a J gene segment, which are then joined to a V gene segment (see Figure 5.3). The T-cell receptor gene segments are flanked by recombination signal sequences similar to those found in the immunoglobulin genes and the RAG complex, and the same DNA-modifying enzymes are involved in the recombination process (see Sections 4-8 and 4-9, pp. 106–110). During recombination, additional, non-templated P and N nucleotides are inserted in the junctions between the V, D, and J gene segments of the T-cell receptor β-chain coding sequence and between the V and J gene segments of the α-chain sequence. These mechanisms contribute junctional diversity in the CDR3 to T-cell receptor α and β chains.

Genetic defects that result in an absence of RAG proteins are one cause of a rare syndrome called **severe combined immunodeficiency disease** (**SCID**). The disease is called 'combined' because functional B and T lymphocytes are both absent, and it is 'severe' compared with immunodeficiencies in which only B cells are lacking. Without a bone marrow transplant or other medical intervention, children with SCID die in infancy from common infections (Figure 5.4, upper panel). Rare cases of missense mutations that produce RAG proteins with partial enzymatic activity have also been found in humans; these also cause a rapidly fatal immunodeficiency that differs from SCID in some of its symptoms and is known as **Omenn syndrome** (Figure 5.4, lower panel).

After gene rearrangement, functional α-chain and β-chain genes consist of exons encoding the leader peptide, the V region, the C region, and the membrane-spanning region. The exons are separated by introns, which in the case of the intron between the V-domain and the C-domain exons may contain unrearranged gene segments (see the α-chain gene in Figure 5.3). When the gene is transcribed, the primary RNA transcript is spliced to remove the introns and is processed to give mRNA. Translation of the α-chain and β-chain mRNA produces α and β chains, respectively. Like all proteins destined for the cell membrane, newly synthesized α and β chains enter the endoplasmic reticulum. There they pair to form the **α:β T-cell receptor** (see Figure 5.3).

5-3   The *RAG* genes were key elements in the origin of adaptive immunity

T cells and B cells use identical mechanisms of gene rearrangement, often called **V(D)J recombination,** to generate the clonal diversity in their antigen receptors. Key to this recombination are the two subunits of the RAG recombinase, which are made only by lymphocytes (see Section 4-9, p. 108). The RAG proteins are thus specific to adaptive immunity and are essential for its function, as shown by the detrimental effects of their absence or malfunction (see Figure 5.4). These properties are consistent with the appearance of the *RAG* genes in a common ancestor of the vertebrates being a formative event in the evolution of the adaptive immune system.

The *RAG* genes lack the introns that characterize eukaryotic genes. In this unusual feature they resemble the transposase gene of a transposon, a type of genetic element that can make and move copies of itself to different chromosomal locations. The essential components of a transposon are a transposase—an enzyme that cuts double-stranded DNA—and regions of repetitive DNA, called the terminal repeat sequences, that are recognized by the transposase (Figure 5.5). These two features allow the transposon to be excised from one location and inserted into another. The similarity of the RAG recombinase and transposase has led to the suggestion that the mechanism now used to rearrange immunoglobulin and T-cell receptor gene segments originated from the insertion of a transposon into some type of innate immune receptor gene in a vertebrate ancestor. The inserted transposase genes evolved to encode the RAG proteins, and the terminal repeat sequences evolved to became recombination signal sequences for the first rearranging gene segments. The transposase gene and the long terminal repeats of the transposon became separated to become components of different genes, both being specifically expressed in lymphocytes. Today, the human *RAG* genes are on chromosome 11 and the much-expanded sets of rearranging antigen-receptor genes are on four other chromosomes.

5-4   Expression of the T-cell receptor on the cell surface requires association with additional proteins

T-cell receptors are diverse and specific receptors for antigen. By themselves, however, heterodimers of α and β chains are unable to leave the endoplasmic reticulum and be expressed on the T-cell surface. In this respect the T-cell receptor resembles immunoglobulin. Before leaving the endoplasmic reticulum, an α:β heterodimer associates with four invariant membrane proteins.

Figure 5.5 Components of a transposon could have evolved to become the *RAG* genes and the recombination signal sequences of immunoglobulin and T-cell receptor genes. The first event in the evolution of rearranging antigen-receptor genes is thought to have occurred more than 400 million years ago, when a transposon integrated into a gene encoding an innate immune receptor protein (first panel). The transposon separated the gene into two segments, each flanked by a piece of repetitive transposon DNA (second panel). Subsequently, chromosomal rearrangements placed the transposase gene (or genes) onto a different chromosome from the primordial rearranging gene. The repetitive DNAs became the RSSs of a primordial rearranging gene and the transposase genes became the ancestral *RAG-1* and *RAG-2* genes (third panel). Over 400 million years of evolution, the family of rearranging genes has expanded and is now spread over five different human chromosomes (fourth panel).

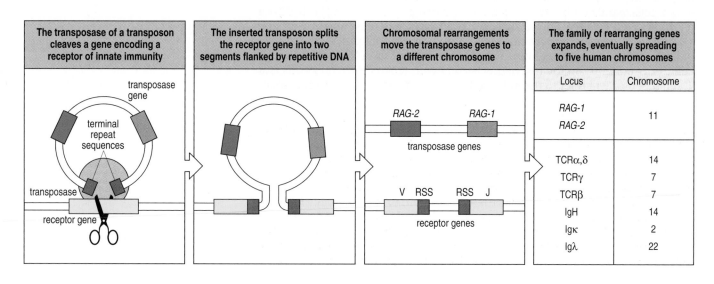

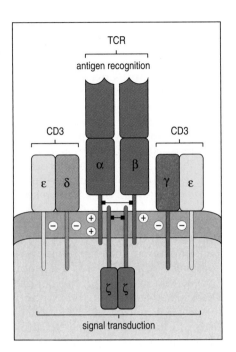

Figure 5.6 Polypeptide composition of the T-cell receptor complex. The functional antigen receptor on the surface of T cells is composed of eight polypeptides and is called the T-cell receptor complex. The α and β chains bind antigen and form the core T-cell receptor (TCR). They associate with one copy each of CD3γ and CD3δ and two copies each of CD3ε and the ζ chain. These associated invariant polypeptides are necessary for the transport of newly synthesized TCR to the cell surface and for the transduction of signals to the cell's interior after the TCR has bound antigen. The transmembrane domains of the α and β chains contain positively charged amino acids (+), which form strong electrostatic interactions with negatively charged amino acids (–) in the transmembrane regions of the CD3γ, δ, and ε chains.

Three of the proteins are encoded by closely linked genes on human chromosome 11 and are homologous to each other: these proteins are collectively termed the **CD3 complex** and are individually called CD3γ, CD3δ, and CD3ε. The fourth protein is known as the ζ chain and is encoded by a gene on human chromosome 1.

At the cell surface the CD3 proteins and the ζ chain remain in stable association with the T-cell receptor and form the functional **T-cell receptor complex** (Figure 5.6). In this complex the CD3 proteins and the ζ chain transmit signals to the cell's interior after antigen has been recognized by the α:β chain heterodimer. Like the Igα and Igβ proteins of the B-cell receptor complex, the cytoplasmic domains of the CD3 proteins and the ζ chain contain sequences that associate with intracellular signaling molecules. In contrast, the T-cell receptor α and β chains have very short cytoplasmic tails that lack signaling function. In people lacking functional CD3δ or CD3ε chains, transport of T-cell receptors to the cell surface is inefficient and their T cells have abnormally low numbers of receptors that do not signal effectively. As a consequence of both the low receptor numbers and impaired signal transduction these people suffer from immunodeficiency.

5-5   A distinct population of T cells expresses a second class of T-cell receptor with γ and δ chains

There is a second type of T-cell receptor that is similar in overall structure to the α:β receptor but is formed of two different protein chains called the T-cell receptor **γ chain** (not to be confused with CD3γ) and the T-cell receptor **δ chain**. TCRγ resembles the α chain, and TCRδ resembles the β chain (Figure 5.7). T cells express either α:β receptors or γ:δ receptors but never both. Those expressing α:β T-cell receptors are called **α:β T cells**, whereas those expressing γ:δ receptors are called **γ:δ T cells**. Cells with γ:δ receptors form a small subset of all T cells. Much more is known of the functions of α:β T cells than of γ:δ T cells. Consequently, in the rest of this book, T cells will refer to α:β T cells and T-cell receptor will refer to the α:β T-cell receptor, unless specified otherwise.

The organization of the γ and δ loci resembles that of the β and α loci, but there are some important differences (Figure 5.8). The δ gene segments are situated within the α-chain locus on chromosome 14, between the V_α and J_α gene segments. This location means that DNA rearrangement within the α-chain locus inevitably results in the deletion and inactivation of the δ-chain locus. The human γ-chain locus is on chromosome 7. The γ-chain and δ-chain loci contain fewer V gene segments than the α-chain or β-chain loci, and so, in theory, might produce less diverse receptors, but in the case of the δ chain this is compensated for by an increase in junctional diversity. Rearrangement at the γ and δ loci proceeds as for the α and β loci, with the exception that during δ-gene rearrangement two D segments can be incorporated into the final

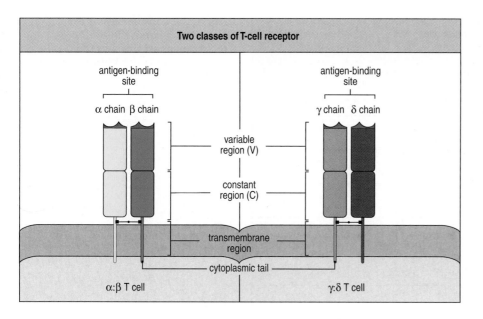

Two classes of T-cell receptor

Figure 5.7 There are two classes of T-cell receptor. The α:β T-cell receptor (left panel) and the γ:δ T-cell receptor (right panel) have similar structures, but they are encoded by different sets of rearranging gene segments and have different functions.

gene sequence. This increases the variability of the δ chain in two ways. First, the potential number of combinations of gene segments is increased. Second, extra N nucleotides can be added at the junction between the two D segments, as well as at the VD and DJ junctions.

T cells bearing γ:δ receptors comprise about 1–5% of the T cells found in the circulation, but they can be the dominant T-cell population in epithelial tissue. The immune function of γ:δ T cells is less well defined than that of α:β T cells, as are the antigens to which these cells respond and the ligands that their receptors engage. Unlike α:β T cells, the γ:δ T cells are not restricted to the recognition of peptide antigens associated with MHC molecules.

Summary

T cells recognize antigen through a cell-surface receptor known as the T-cell receptor, which has structural and functional similarities to the membrane-bound immunoglobulin that serves as the B-cell receptor for antigen. T-cell receptors are heterodimeric glycoproteins in which each polypeptide chain consists of a V domain and a C domain, similar to those found in immunoglobulin chains, and a membrane-spanning region. The three-dimensional structure of the extracellular domains of the T-cell receptor resembles that of the antigen-binding Fab fragment of IgG. There are two types of T-cell receptor: one made up of α and β chains and expressed by T cells whose function is understood, and another made up of γ and δ chains and carried by T cells

Figure 5.8 The organization of the human T-cell receptor γ-chain and δ-chain loci. The TCR γ and δ loci, like the α and β loci, contain sets of variable (V), diversity (D), joining (J), and constant (C) gene segments. The δ locus is located within the α-chain locus on chromosome 14, lying between the clusters of V_α and J_α gene segments. There are at least three V_δ gene segments, three D_δ gene segments, three J_δ gene segments, and a single C_δ gene segment. V_δ segments are interspersed among V_α and other gene segments. The γ locus, on chromosome 7, resembles the β locus, with a set of V segments and two C gene segments each with its own set of J segments.

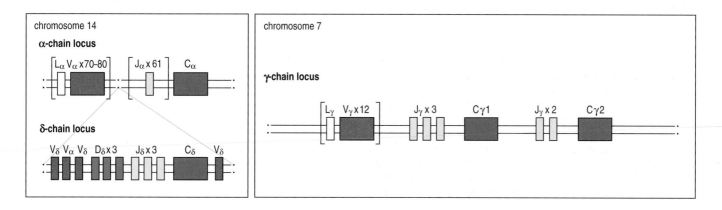

Element	Immunoglobulin		α:β T-cell receptors	
	H	κ+λ	β	α
Variable segments (V)	40	70	52	~70
Diversity segments (D)	25	0	2	0
D segments read in three frames	rarely	–	often	–
Joining segments (J)	6	5(κ) 4(λ)	13	61
Joints with N- and P-nucleotides	2	50% of joints	2	1
Number of V gene pairs	1.9×10^6		5.8×10^6	
Junctional diversity	$\sim3 \times 10^7$		$\sim2 \times 10^{11}$	
Total diversity	$\sim5 \times 10^{13}$		$\sim10^{18}$	

Figure 5.9 Comparison of the potential diversity in the T-cell receptor repertoire and the B-cell receptor repertoire before encounter with antigen.

whose function remains elusive. All four types of T-cell receptor chain are encoded by genes that resemble the immunoglobulin genes and require similar DNA rearrangements in order to be expressed. In their capacity to diversify the T-cell repertoire before encounter with antigen, the T-cell receptors exceed the immunoglobulins of B cells (Figure 5.9).

The critical difference between T-cell receptors and immunoglobulins is that T-cell receptors serve only as cell-surface receptors and are not secreted as soluble proteins with effector function; T cells use other molecules for effector function. This explains why the T-cell receptor has only a single binding site for antigen that does not change its affinity on encountering antigen, and a simple constant region that does not switch isotype. Expression of the T-cell receptor at the T-cell surface requires association with proteins of the CD3 complex. These proteins transmit signals to the interior of the cell when the T-cell receptor binds antigen.

Antigen processing and presentation

Unlike the immunoglobulins of B cells, which can recognize a wide range of molecules in their native form, a human T-cell receptor can recognize antigen only in the form of a peptide bound to an MHC molecule on the surface of another human cell (see Section 3-7, p. 78). This means that, to be recognized by T cells, pathogen-derived proteins must be degraded into peptides. This occurs inside the body's own cells and is called **antigen processing**. The peptides are then assembled into peptide:MHC molecule complexes for display on cell surfaces, where they can be recognized by T cells. The binding of a peptide antigen by an MHC molecule and its display at the cell surface is termed **antigen presentation** (Figure 5.10).

The microorganisms that infect the human body can be broadly divided into those that propagate within cells, such as viruses, and those, such as most bacteria, that live in the extracellular spaces. The α:β T-cell population that fights infection is composed of two subpopulations: one is specialized to fight intracellular infections, the other to fight extracellular sources of infection. The latter include bacteria that live in extracellular spaces, and virus particles

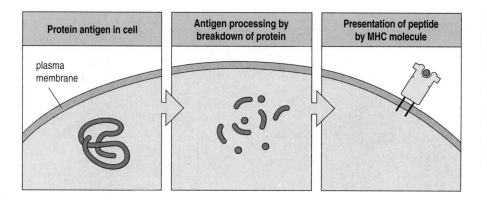

| Protein antigen in cell | Antigen processing by breakdown of protein | Presentation of peptide by MHC molecule |

plasma membrane

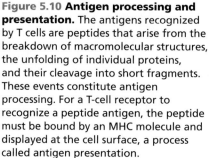

Figure 5.10 Antigen processing and presentation. The antigens recognized by T cells are peptides that arise from the breakdown of macromolecular structures, the unfolding of individual proteins, and their cleavage into short fragments. These events constitute antigen processing. For a T-cell receptor to recognize a peptide antigen, the peptide must be bound by an MHC molecule and displayed at the cell surface, a process called antigen presentation.

present in the extracellular fluid after their release from infected cells. In this part of the chapter we consider how these two sources of antigen are distinguished by the immune system and processed by intracellular pathways, and how the appropriate type of T cell is activated.

5-6   The two classes of MHC molecule present antigen to CD8 and CD4 T cells, respectively

As we saw in Chapter 3, circulating α:β T cells fall into one of two mutually exclusive classes, one defined by expression of the co-receptor CD4 on the cell surface and the other by expression of the co-receptor CD8 (Figure 5.11). CD8 T cells are cytotoxic and their main function is to kill cells that have become infected with a virus or some other intracellular pathogen. The general function of CD4 T cells, or helper T cells, is to help other cells of the immune system to respond to extracellular sources of infection. Helper T cells are involved in stimulating B cells to make antibodies, which bind to extracellular bacteria and virus particles. In addition, they activate tissue macrophages to phagocytose and kill extracellular pathogens, and to secrete cytokines and other biologically active molecules that affect the course of the immune response (Figure 5.12).

The human immunodeficiency virus (HIV), which causes acquired immunodeficiency syndrome (AIDS), selectively infects CD4 T cells by exploiting the CD4 molecule as its receptor. On binding to CD4 on a T-cell surface the virus gains entry to the cell where it will replicate. As the HIV infection progresses, the number of circulating CD4 T cells gradually declines and, in the absence of effective treatment, will eventually reach a level at which the adaptive immune response to other types of infection becomes fatally compromised.

The MHC molecules are crucial in ensuring that the appropriate class of T cells is activated in response to a particular source of infection. As introduced in Chapter 3, there are two different classes of MHC molecule—MHC class I and MHC class II—and each presents peptides from one kind of antigen source to one subpopulation of T cells. MHC class I molecules present antigens of intracellular origin to CD8 T cells, whereas MHC class II molecules present antigens of extracellular origin to CD4 T cells (see Sections 3-9 and 3-10, pp. 80–81).

Figure 5.11 The structures of the CD4 and CD8 glycoproteins. CD4 and CD8 are members of the immunoglobulin superfamily of proteins. CD4 has four extracellular immunoglobulin-like domains (D1–D4) with a hinge between the D2 and D3 domains. CD8 consists of an α and a β chain, which both have an immunoglobulin-like domain that is connected to the membrane-spanning region by an extended stalk. C denotes the carboxy terminus.

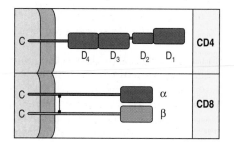

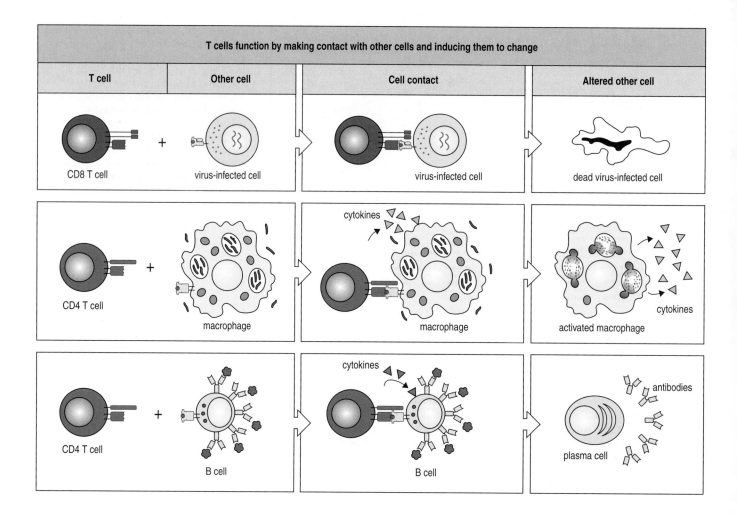

Figure 5.12 **T cells function by making contact with other cells.** Top panels: a cytotoxic CD8 T cell makes contact with a virus-infected cell, recognizes that it is infected, and kills it. Middle panels: a CD4 helper T cell contacts a macrophage that is engaged in the phagocytosis of bacteria and secretes cytokines that increase the microbicidal power of the macrophage and its secretion of inflammatory cytokines. Bottom panels: a CD4 helper T cell contacts a B cell that is binding its specific antigen and secretes cytokines that cause the B cell to differentiate into an antibody-secreting plasma cell.

Before we consider how the appropriate MHC molecules become associated with peptides from different sources, we shall look at their structure and general peptide-binding properties, and how they associate with CD4 and CD8.

5-7 The two classes of MHC molecule have similar three-dimensional structures

MHC class I and class II molecules are membrane glycoproteins whose function is to bind peptide antigens and present them to T cells. Underlying this common function is a similar three-dimensional structure, which is formed in different ways in the two MHC classes.

An MHC class I molecule is made up of a transmembrane heavy chain, or α chain, which is noncovalently complexed with the small protein **β2-microglobulin** (Figure 5.13). The heavy chain has three extracellular domains (α_1, α_2, and α_3). The peptide-binding site is formed by the folding of α_1 and α_2, the domains farthest from the membrane, and is supported by the α_3 domain and the β2-microglobulin. The MHC class I heavy chain is encoded by a gene in the MHC, whereas β2-microglobulin is not.

In contrast, an MHC class II molecule consists of two transmembrane chains (α and β), each of which contributes one domain to the peptide-binding site and one immunoglobulin-like supporting domain (see Figure 5.13). Both these chains are encoded by genes in the MHC. Thus, both classes of MHC molecule have similar three-dimensional structures consisting of two pairs of extracellular domains, with the paired domains farthest from the membrane

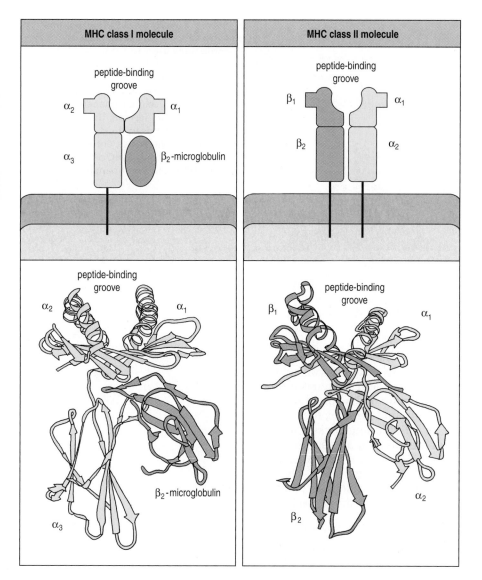

Figure 5.13 The structures of MHC class I and MHC class II molecules are variations on a theme. An MHC class I molecule (left panels) is composed of one membrane-bound heavy (or α) chain and noncovalently bonded β2-microglobulin. The heavy chain has three extracellular domains, of which the amino-terminal α1 and α2 domains resemble each other in structure and form the peptide-binding site. An MHC class II molecule (right panels) is composed of two membrane-bound chains, an α chain (which is a different protein from MHC class I α) and a β chain. These have two extracellular domains each, the amino-terminal two (α1 and β1) resembling each other in structure and forming the peptide-binding site. The β2 domain of MHC class II molecules should not be confused with the β2-microglobulin of MHC class I molecules. The ribbon diagrams in the lower panels trace the paths of the polypeptide backbone chains.

resembling each other and forming the peptide-binding site. In both MHC types, the domains supporting the peptide-binding domains are immunoglobulin-like domains: α3 and β2-microglobulin in MHC class I molecules, and α2 and β2 in MHC class II molecules.

The immunoglobulin-like domains of MHC class I and II molecules are not just a support for the peptide-binding site; they also provide specific binding sites for the CD4 and CD8 co-receptors. The sites on an MHC molecule that interact with the T-cell receptor and the co-receptor are separated, allowing the MHC molecule to engage both a T-cell receptor and a co-receptor simultaneously (Figure 5.14).

5-8   MHC molecules bind a variety of peptides

The peptide-binding site in an MHC molecule can bind peptides of many different amino acid sequences. This **promiscuous binding specificity** contrasts with the specificity of an immunoglobulin or a T-cell receptor for a single epitope. The peptide-binding site is a deep groove on the surface of the MHC molecule (Figure 5.15), within which a single peptide is held tightly by noncovalent bonds.

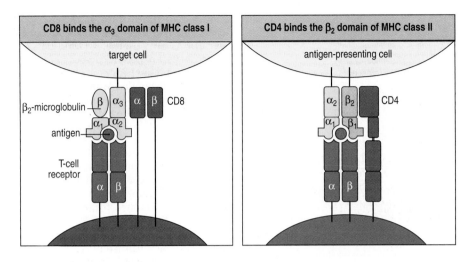

Figure 5.14 MHC class I molecules bind to CD8, and MHC class II molecules bind to CD4. The CD8 co-receptor binds to the α_3 domain of the MHC class I heavy chain, ensuring that MHC class I molecules present peptides only to CD8 T cells (left panel). In a complementary fashion, the CD4 co-receptor binds to the β_2 domain of MHC class II molecules, ensuring that peptides bound by MHC class II stimulate only CD4 T cells (right panel).

There are some constraints on the length and amino acid sequence of peptides that are bound by MHC molecules. These are dictated by the structure of the peptide-binding groove, which differs between MHC class I and class II. The length of the peptides bound by MHC class I molecules is limited because the two ends of the peptide are grasped by pockets situated at the ends of the peptide-binding groove (see Figure 5.15, left panel). The vast majority of peptides that bind MHC class I molecules are eight, nine, or ten amino acids in length; most are nine amino acids. The differences in length are accommodated by a slight kinking of the extended conformation of the bound peptide. Features that are common to all peptides—the amino terminus, the carboxy terminus, and the peptide backbone—interact with binding pockets present in all MHC class I molecules, and these form the basis for all peptide–MHC class I interactions. Most peptides bound by MHC class I molecules have either a hydrophobic or a basic residue at the carboxy terminus. The two groups of peptides bind to different forms of MHC class I that have a binding pocket complementary to either hydrophobic or basic amino acid side chains.

In MHC class II molecules, the two ends of the peptide are not pinned down into pockets at each end of the peptide-binding groove (see Figure 5.15, right panel). As a consequence, they can extend out at each end of the groove and so peptides bound by MHC class II molecules are both longer and more variable in length than peptides bound by MHC class I. Peptides that bind to MHC class II molecules are usually 13–25 amino acids long, and some are much longer.

Figure 5.15 The peptide-binding groove of MHC class I and MHC class II molecules. The T-cell receptor's view of the peptide-binding groove, with a peptide bound, is shown. In the MHC class I molecule (left panel) the groove is formed by the α_1 and α_2 domains of the MHC class I heavy chain; in the MHC class II molecule (right panel) it is formed by the α_1 domain of the class II α chain and the β_1 domain of the class II β chain. Amino acid side chains on the MHC molecule that are important for making interactions with the bound peptide are shown. The dotted blue lines indicate hydrogen bonds and ionic interactions made between the peptide and the MHC molecule. Peptides bind to MHC class I molecules by their ends (left panel), whereas in MHC class II molecules the peptide extends beyond the peptide-binding groove and is held by interactions along its length (right panel).

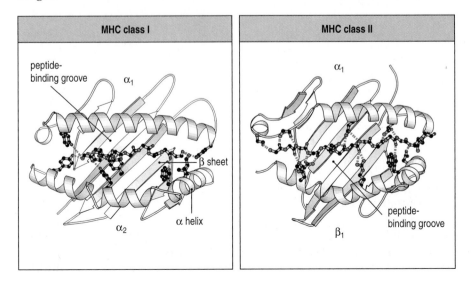

5-9 MHC class I and MHC class II molecules bind peptides in different intracellular compartments

The peptide antigens that are presented by MHC molecules are generated inside cells of the body by the breakdown of larger protein antigens. Proteins derived from 'intracellular' and 'extracellular' antigens are present in different intracellular compartments (Figure 5.16). They are processed into peptides by two intracellular pathways of degradation, and bind to the two classes of MHC molecule in separate intracellular compartments. Peptides derived from the degradation of intracellular pathogens are formed in the cytosol and delivered to the endoplasmic reticulum. This is where MHC class I molecules bind peptides. In contrast, extracellular microorganisms and proteins are taken up by cells via phagocytosis and endocytosis into phagosomes and endocytic vesicles, and they are degraded in the lysosomes and other vesicles of the endocytic pathways. It is in these cellular compartments that MHC class II molecules bind peptides. In this way, the class of the MHC molecule labels the peptide as being extracellular or intracellular in origin.

5-10 Peptides generated in the cytosol are transported into the endoplasmic reticulum, where they bind MHC class I molecules

When viruses infect human cells they exploit the cell's ribosomes to synthesize viral proteins, which are therefore present in the cytosol before being assembled into viral particles. In response, the infected cell uses and adapts its normal processes of breakdown and turnover of cellular proteins to degrade some of the viral proteins into peptides that can be bound by MHC class I molecules and presented to CD8 T cells.

Proteins in the cytosol are degraded by a large barrel-shaped protein complex called the **proteasome**, which has several different protease activities. It consists of 28 polypeptide subunits, each of 20–30 kDa molecular weight. In the cells of healthy tissue the proteasome is used to break down proteins that are damaged, poorly folded or no longer needed. In infected tissues, the NK cells of the innate immune system secrete the cytokine interferon-γ (IFN-γ), which causes the tissue cells to change their proteasomes to favor the production of peptides that bind to MHC class I molecules. Two proteasome subunits are replaced with alternative forms, and another protein, the PA28 proteasome activator, binds to both ends of the barrel. This modified proteasome, called the **immunoproteasome**, specializes in making peptides having a hydrophobic or a basic residue at the carboxy terminus, features that enable them to bind to MHC class I molecules.

Once formed, the antigenic peptides are transported out of the cytosol and into the endoplasmic reticulum (Figure 5.17). Transport of peptides across the endoplasmic reticulum membrane is accomplished by a protein, called the **transporter associated with antigen processing** (**TAP**), embedded in the membrane. TAP is a heterodimer consisting of two structurally related polypeptide chains, TAP-1 and TAP-2. Peptide transport by TAP is dependent on the binding and hydrolysis of ATP, properties shared with the other transporters in this family. The types of peptide that are preferentially transported

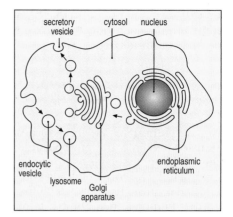

Figure 5.16 There are two major compartments within cells, separated by membranes. One compartment is the cytosol, which is contiguous with the nucleus via the pores in the nuclear membrane. The other compartment is the vesicular system, which consists of the endoplasmic reticulum, the Golgi apparatus, endocytic vesicles, lysosomes, and other intracellular vesicles. The vesicular system is effectively contiguous with the extracellular fluid. Secretory vesicles bud off from the endoplasmic reticulum and, by successive fusion and budding with the Golgi membranes, move vesicular contents out of the cell. In the reverse direction, endocytic vesicles formed from infolding of the plasma membrane take up extracellular material into the vesicular system.

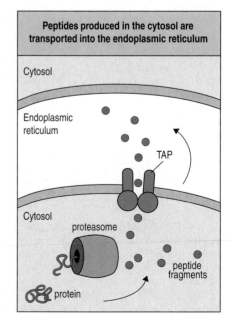

Figure 5.17 Formation and transport of peptides that bind to MHC class I molecules. In all cells, proteasomes degrade cellular proteins that are poorly folded, damaged, or unwanted. When a cell becomes infected, pathogen-derived proteins in the cytosol are also degraded by the proteasome. Peptides are transported from the cytosol into the lumen of the endoplasmic reticulum by the protein called transporter associated with antigen processing (TAP), which is embedded in the endoplasmic reticulum membrane.

by TAP are similar to those that bind MHC class I molecules: they are eight or more amino acids long and have either hydrophobic or basic residues at the carboxy terminus.

5-11   MHC class I molecules bind antigenic peptide as part of a peptide-loading complex

Newly synthesized MHC class I heavy chains and β_2-microglobulin are also translocated into the endoplasmic reticulum, where they partly complete their folding, associate, and finally bind peptide, at which point folding is completed. The correct folding and peptide loading of MHC class I molecules is aided by proteins known as chaperones. These are proteins that assist in the correct folding and subunit assembly of other proteins while keeping them out of harm's way until they are ready to enter cellular pathways and carry out their functions.

When MHC class I heavy chains (α chains) first enter the endoplasmic reticulum they bind the membrane protein known as **calnexin**, which retains the partly folded heavy chain in the endoplasmic reticulum (Figure 5.18). Calnexin is a calcium-dependent lectin, a carbohydrate-binding protein that retains many multisubunit glycoproteins, including T-cell receptors and immunoglobulins, in the endoplasmic reticulum until they have folded correctly. When the MHC class I heavy chain has bound β_2-microglobulin, calnexin is released and the heterodimer is then incorporated into the **peptide-loading complex**. This complex of at least six proteins stabilizes the partly folded heterodimer of heavy chain and β_2-microglobulin and positions it to sample peptides coming from the cytosol. This is achieved by the disulfide-bonded

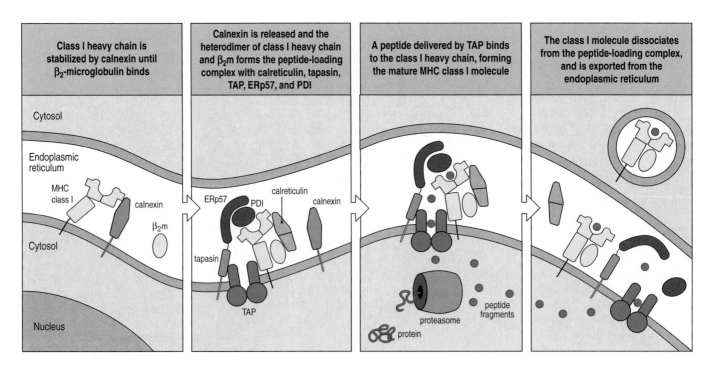

Figure 5.18 **Proteins of the peptide-loading complex aid the assembly and peptide loading of MHC class I molecules in the endoplasmic reticulum.** MHC class I heavy chains assemble in the endoplasmic reticulum with the membrane-bound protein calnexin. When this complex binds β_2-microglobulin (β_2m) the partly folded MHC class I molecule is released from calnexin and then associates with the TAP, tapasin, calreticulin, ERp57, and protein disulfide isomerase (PDI) to form the peptide-loading complex. The MHC class I molecule is retained in the endoplasmic reticulum until it binds a peptide, which completes the folding of the molecule. The peptide:MHC class I molecule is then released from the other proteins and leaves the endoplasmic reticulum for transport to the cell surface.

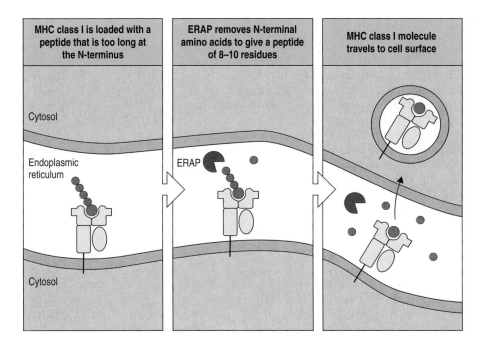

| MHC class I is loaded with a peptide that is too long at the N-terminus | ERAP removes N-terminal amino acids to give a peptide of 8–10 residues | MHC class I molecule travels to cell surface |

Figure 5.19 An aminopeptidase in the endoplasmic reticulum can trim peptides bound to MHC class I to improve their binding affinity. The endoplasmic reticulum aminopeptidase (ERAP) binds to MHC class I molecules in which the amino terminus of an overlong peptide hangs out of the binding site. It removes the accessible amino acid residues to leave a peptide of 8–10 amino acids with an improved fit to the heterodimer of the class I heavy chain and β2-microglobulin. It is not known whether ERAP acts on the class I molecule alone, as illustrated here, or when it is part of the peptide-loading complex, or both.

complex of the oxidoreductase enzyme **ERp57** and **tapasin**, a bridging protein that binds to both the MHC class I heavy chain and the TAP-1 subunit of the peptide transporter. Other components of the peptide-loading complex are protein disulfide isomerase (PDI), which stabilizes the disulfide bond of the α2 domain of the class I heavy chain, and calreticulin, which resembles a soluble form of calnexin and has a similar chaperone function.

While it is a part of the peptide-loading complex, the heterodimer of the class I heavy chain and β2-microglobulin will test various peptides for binding until it finds one that is compatible with its binding site. At this point the carboxy terminus of the peptide will fit well, but the peptide might be too long and overhang at the amino terminus (Figure 5.19). In that case an enzyme called **endoplasmic reticulum aminopeptidase** (**ERAP**) will remove amino acids from the amino-terminal end of the peptide to improve its fit.

On binding a peptide, the completely assembled MHC class I molecule is released from all chaperones and leaves the endoplasmic reticulum in a membrane-enclosed vesicle. It then makes its way through the Golgi stacks to the plasma membrane. Most of the peptides transported by TAP are not successful in binding an MHC class I molecule; they are transported out of the endoplasmic reticulum and back to the cytosol.

MHC class I molecules cannot leave the endoplasmic reticulum unless they have bound a peptide. In one form of a rare disease called **bare lymphocyte syndrome**, the TAP protein is nonfunctional and so no peptides enter the endoplasmic reticulum. Cells of patients with this defect have less than 1% of the normal level of MHC class I molecules on their surface. As a consequence of this deficiency, patients develop very poor CD8 T-cell responses to viruses and suffer from chronic respiratory infections from a young age.

Protein degradation and peptide transport occur continuously, not only when cells are infected. In the absence of infection, MHC class I molecules carry **self peptides** derived from normal human self proteins, as also do MHC class II molecules. These do not normally provoke an immune response because of mechanisms that act during T-cell development and eliminate or inactivate T cells reactive to these self peptides. Occasionally, however, this state of T-cell tolerance to self peptides breaks down, resulting in autoimmunity.

5-12   Peptides presented by MHC class II molecules are generated in acidified intracellular vesicles

Proteins of extracellular bacteria, extracellular virus particles, and soluble protein antigens are processed by a different intracellular pathway from that followed by cytosolic proteins, and their peptide fragments end up bound to MHC class II molecules. Most cells are continually internalizing extracellular fluid and material bound at their surface by the process of endocytosis or phagocytosis. Cells specialized for phagocytosis and defense against infection—dendritic cells, macrophages, and neutrophils—also have receptors on their surface that bind to pathogen surfaces and promote their phagocytosis (see Section 2-10, p. 44). In addition, phagocytes can engulf larger objects such as dead cells. All these uptake mechanisms produce intracellular vesicles known as endosomes or phagosomes, in which the vesicular membrane is derived from the plasma membrane and the lumen contains extracellular material.

These membrane-enclosed vesicles become part of an interconnected vesicle system that carries materials to and from the cell surface. As vesicles travel inwards from the plasma membrane, their interiors become acidified by the action of proton pumps in the vesicle membrane and they fuse with other vesicles, such as lysosomes, that contain proteases and hydrolases that are active in acid conditions. Within the phagolysosomes formed by this fusion, enzymes degrade the vesicle contents to produce, among other things, peptides from the proteins and glycoproteins of the internalized pathogens.

Microorganisms in the extracellular environment are taken up by macrophages and dendritic cells by phagocytosis, and are then degraded within phagolysosomes. B cells also bind specific antigens via their surface immunoglobulin, and then internalize these antigens by receptor-mediated endocytosis. These antigens are similarly degraded within the vesicular system. Peptides produced within phagolysosomes become bound to MHC class II molecules within the vesicular system, and the peptide:MHC class II complexes are carried to the cell surface by outward-going vesicles. Thus, the MHC class II pathway samples the extracellular environment, complementing the MHC class I pathway, which samples the intracellular environment (Figure 5.20).

Certain pathogens, for example the mycobacteria that cause leprosy and tuberculosis, actually exploit the vesicular system as a protected site for intracellular growth and replication. As their proteins do not enter the cytosol, they are not processed and presented to cytotoxic CD8 T cells by MHC class I molecules. They protect themselves from degradation by lysosomal enzymes by preventing fusion of the phagosome with a lysosome. This also prevents the presentation of mycobacterial antigens by MHC class II molecules.

5-13   MHC class II molecules are prevented from binding peptides in the endoplasmic reticulum by the invariant chain

Newly synthesized MHC class II α and β chains are translocated from the ribosomes into the membranes of the endoplasmic reticulum. There, an α chain and a β chain associate with a third chain, called the **invariant chain** (Figure 5.21). This is so called because it is identical in all individuals, whereas the α and β chains vary from one person to another. One function of the invariant chain is to prevent the peptide-binding site formed by the association of an α and a β chain from binding peptides present in the endoplasmic reticulum; these peptides are therefore targeted to MHC class I molecules only.

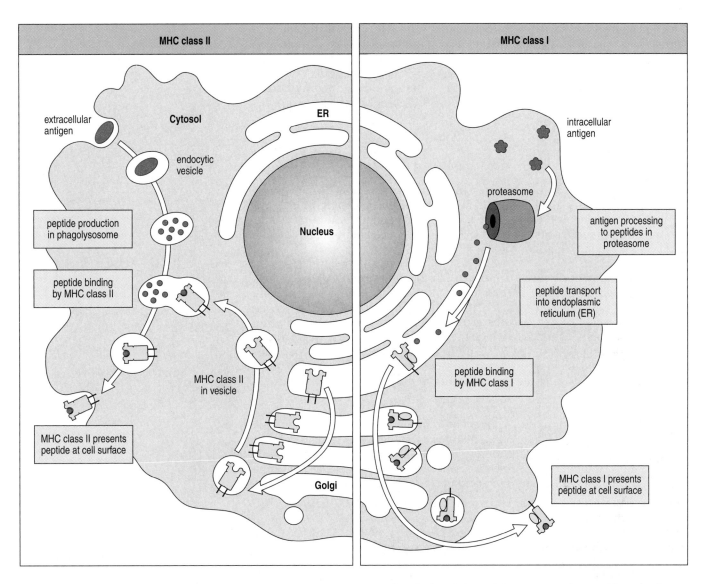

Figure 5.20 Processing of antigens for presentation by MHC class II or MHC class I molecules occurs in different cellular compartments. The left half of the figure shows the fate of peptides derived from extracellular antigens and pathogens. Extracellular material is taken up by endocytosis and phagocytosis into the vesicular system of the cell, in this case a macrophage. Proteases in these vesicles break down proteins to produce peptides that are bound by MHC class II molecules, which have been transported to the vesicles via the endoplasmic reticulum (ER) and the Golgi apparatus. The peptide:MHC class II complex is transported to the cell surface in outgoing vesicles. The right half of the figure shows the fate of peptides generated in the cytosol as a result of infection with viruses or intracytosolic bacteria. Proteins from such pathogens are broken down in the cytosol by the proteasome to peptides, which enter the ER. There the peptides are bound by MHC class I molecules. The peptide:MHC class I complex is transported to the cell surface via the Golgi apparatus.

A second function of the invariant chain is to deliver MHC class II molecules to endocytic vesicles, where they bind peptide. These vesicles, which have been called **MIIC**, for **MHC class II compartment**, contain proteases, for example cathepsin S, that selectively attack the invariant chain. A series of cleavages leaves just a small fragment of the invariant chain that covers up the MHC class II peptide-binding site. This fragment is called **class II-associated invariant-chain peptide** (**CLIP**). Removal of CLIP and binding of peptide is aided by the interaction of the MHC class II molecule with a glycoprotein in the vesicle membrane called **HLA-DM** (see Figure 5.21). HLA-DM resembles an MHC class II molecule in structure but does not bind peptides or appear on the cell surface. On binding to the MHC class II molecule, HLA-DM causes the displacement of CLIP, which then allows the MHC class II molecule to sample other peptides. By repeated rebinding to the MHC class II molecule, HLA-DM displaces weakly bound peptides and makes sure that the MHC class II molecule eventually ends up with a strongly bound peptide. When the MHC class II molecule has lost its invariant chain and has a tightly bound peptide it is carried to the cell surface by outward-going vesicles.

5-14 The T-cell receptor specifically recognizes both peptide and MHC molecule

Once a peptide:MHC complex appears on the cell surface it can be recognized by its corresponding T-cell receptor. When the T-cell receptor binds to a peptide:MHC molecule complex it makes contacts with both the peptide and the surrounding surface of the MHC molecule. Thus, each peptide:MHC complex forms a unique ligand for a T-cell receptor.

The floor of the peptide-binding groove of both classes of MHC molecule is formed by eight strands of antiparallel β-pleated sheet, on which lie two antiparallel α helices (see Figure 5.15). The peptide lies between the helices and parallel to them, such that the top surfaces of the helices and peptide form a roughly planar surface to which the T-cell receptor binds. The residues of the peptide that bind to the MHC molecule lie deep within the peptide-binding groove and are inaccessible to the T-cell receptor; side chains of other peptide amino acids stick out of the binding site and bind to the T-cell receptor.

In its overall organization the T-cell receptor antigen-binding site resembles that of an antibody (see Figure 5.2). The interaction between T-cell receptors and ligands comprising peptides bound to MHC molecules has been visualized by X-ray crystallography. Analysis of several complexes has revealed similar interactions for peptides bound to either MHC class I or class II molecules,

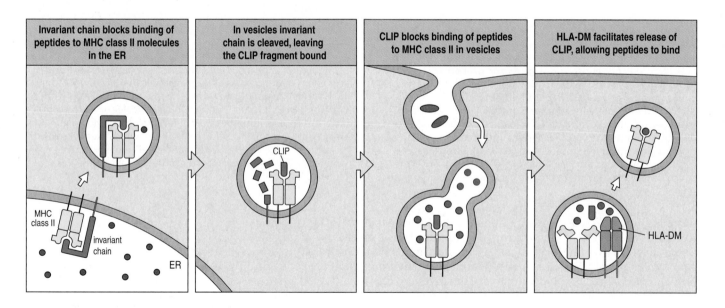

Figure 5.21 The invariant chain prevents peptides from binding to an MHC class II molecule until it reaches the site of extracellular protein breakdown. In the endoplasmic reticulum (ER), MHC class II α and β chains are assembled with an invariant chain that fills the peptide-binding groove; this complex is transported to the acidified vesicles of the endocytic system. The invariant chain is broken down, leaving a small fragment called class II-associated invariant-chain peptide (CLIP) attached in the peptide-binding site. The vesicle membrane protein HLA-DM catalyzes the release of the CLIP fragment and its replacement by a peptide derived from endocytosed antigen that has been degraded within the acidic interior of the vesicles.

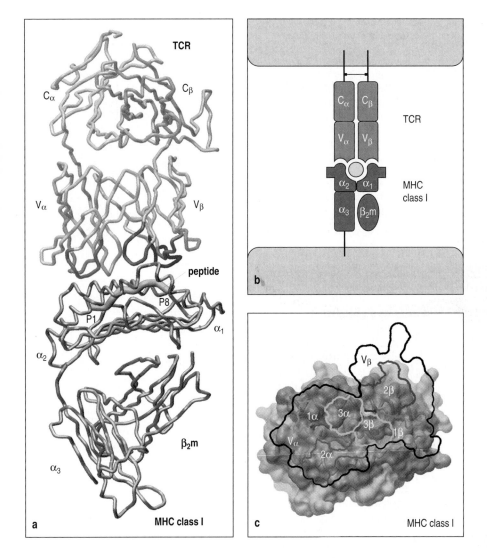

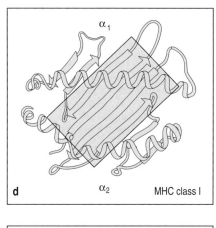

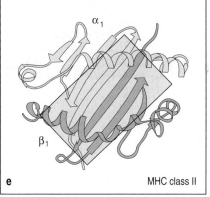

Figure 5.22 The peptide:MHC:T-cell receptor complex. Panel a shows a diagram of the polypeptide backbone of the complex of a T-cell receptor (TCR) bound to its peptide:MHC class I ligand. Panel b shows a schematic representation of this view of the receptor:ligand complex. In panel a the T-cell receptor's CDRs are colored: the α-chain CDR1 and CDR2 are light and dark blue respectively, while the β-chain CDR1 and CDR2 are light and dark purple respectively. The α chain CDR3 is yellow and the β-chain CDR3 is dark yellow. The eight-amino-acid peptide is colored yellow and the positions of the first (P1) and last (P8) amino acids are indicated. Panel c is a view rotated 90° from that of panel a and shows the surface of the peptide:MHC class I ligand and the footprint made on it by the T-cell receptor (outlined in black). Within this footprint the contributions of the CDRs are outlined in different colors and labeled. In panels d and e the diagonal orientations of the T-cell receptor with respect to the peptide-binding grooves of MHC class I and class II molecules, respectively, are shown in schematic diagrams. The T-cell receptor is represented by the black rectangle superimposed on the ribbon diagram of the peptide-binding domains of the MHC molecules. Panels a and c courtesy of I.A. Wilson.

and the interaction is principally illustrated here for MHC class I (Figure 5.22). The T-cell receptor binds to the peptide:MHC class I complex with the long axis of its binding site oriented diagonally across the peptide-binding groove of the MHC class I molecule (see Figure 5.22, panel d). The T-cell receptor binds to a peptide:MHC class II complex in a similar orientation (see Figure 5.22, panel e). The CDR3 loops of the T-cell receptor α and β chains form the central part of the binding site and they grasp the side chain of one of the amino acids in the middle of the peptide. In contrast, the CDR1 and CDR2 loops form the periphery of the binding site and contact the α helices of the MHC molecule. The CDR3 loops that directly contact the peptide are the most variable part of the T-cell receptor antigen-recognition site; the α-chain CDR3 includes the joint between the V and J sequences, and the β chain CDR3 includes the joints between V and D, the whole of the D segment, and the joint between D and J.

5-15   The two classes of MHC molecule are expressed differentially on cells

T-cell responses are guided in appropriate directions by differential expression of the two classes of MHC molecules on human cells (Figure 5.23). Virtually all the cells of the body express MHC class I molecules constitutively,

Figure 5.23 Most human cells express MHC class I, whereas only selected cell types express MHC class II. MHC class I molecules are expressed on almost all nucleated cells, although they are most abundant on hematopoietic cells. MHC class II molecules are normally expressed by only a subset of hematopoietic cells and by stromal epithelial cells in the thymus, although they can be produced by other cell types on exposure to the cytokine interferon-γ. *Activated T cells express MHC class II molecules, whereas resting T cells do not. †Most cell types in the brain are MHC class II-negative, but microglial cells, which are related to macrophages, are MHC class II-positive.

Tissue/cell	MHC	
	class I	class II
Hematopoietic		
T cells	+++	+*
B cells	+++	+++
Macrophages	+++	++
Dendritic cells	+++	+++
Neutrophils	+++	—
Erythrocytes	—	—
Non-hematopoietic		
Thymic epithelium	+	+++
Liver hepatocytes	+	—
Kidney epithelium	+	—
Brain	+	—†

and as all cell types are susceptible to infection by viruses, this enables comprehensive surveillance by CD8 T cells. The erythrocyte is one cell type that lacks MHC class I, a property that might facilitate its persistent infection by malaria parasites.

MHC class II molecules are, in contrast, constitutively expressed on only a few cell types, which are cells of the immune system specialized for the uptake, processing, and presentation of antigens from the extracellular environment. This distribution is consistent with MHC class II function, alerting CD4 T cells to the presence of extracellular infections. To be effective, this function need not be performed by every cell within a tissue or organ, but requires only a sufficient number of cells equipped to guard the extracellular territory. These professional antigen-presenting cells are: dendritic cells, which are supremely specialized for antigen presentation and the activation of T cells; macrophages, which take up antigens by phagocytosis and endocytosis; and B cells, which efficiently internalize specific antigens bound to their surface immunoglobulin.

During the course of an immune response, cells can increase the synthesis and cell-surface expression of MHC molecules beyond constitutive levels. This upregulation, which enhances antigen presentation and T-cell activation, is induced by several cytokines produced by activated immune system cells, in particular the interferons. In addition to increasing the levels of constitutively expressed MHC molecules, IFN-γ can induce the expression of MHC class II molecules on some cell types that do not normally produce them. In this manner, presentation of antigen to CD4 T cells by MHC class II molecules can be induced and increased in infected or inflamed tissues.

5-16   Cross-presentation allows extracellular antigens to be presented by MHC class I

Initiation of a cytotoxic CD8 T-cell response to a virus infection requires MHC class I-mediated presentation of virus-derived peptides by a professional antigen-presenting cell—a dendritic cell, macrophage, or B cell. If a virus does not infect a professional antigen-presenting cell, however, it cannot be presented as MHC class I-bound peptides derived from endogenously synthesized viral proteins. This is the case for the hepatitis C virus, for example, which infects only liver cells (hepatocytes). In such situations, a CD8 T-cell response can still be generated via a third pathway of antigen presentation. This involves the endocytosis or phagocytosis of extracellular material from virus-infected cells by professional antigen-presenting cells and its delivery to the MHC class I pathway of antigen presentation rather than to the MHC class II pathway. Because this process implies a connection between the class II and class I pathways it is called **cross-presentation**. When an immune response is initiated by cross-presentation this is known as **cross-priming** of the immune response.

Although the phenomena of cross-presentation and cross-priming are well established, the underlying mechanisms remain poorly defined. One proposal is that phagocytosis of dead, infected cells by the professional antigen-presenting cell allows complexes of MHC class I and viral peptides to be transferred from the dead cell to the membranes of intracellular vesicles inside the professional antigen-presenting cell, with subsequent transport to the cell surface where they can be engaged by the T-cell receptors of CD8 T cells. Another speculation is that viral components are delivered from endocytic vesicles into the cytosol for degradation by the proteasome and presentation by the usual MHC class I pathway. Such extracellular delivery of viral proteins to the cross-presentation pathway of dendritic cells, macrophages and B cells is being explored as a novel way to vaccinate people against viruses such as the human immunodeficiency virus, for which more conventional approaches to vaccine development have failed.

Summary

T cells expressing α:β T-cell receptors recognize peptides presented at cell surfaces by MHC molecules, the third type of antigen-binding molecule in the adaptive immune system. Unlike immunoglobulins and T-cell receptors, however, MHC molecules have promiscuous binding sites and each MHC molecule can bind peptides of many different amino acid sequences. The peptides are produced by the intracellular degradation of proteins of infectious agents and of self proteins. An α:β T cell either expresses the CD8 co-receptor and recognizes peptides presented by MHC class I molecules, or expresses the CD4 co-receptor and recognizes peptides presented by MHC class II molecules. The CD8 co-receptor interacts specifically with MHC class I molecules, and the CD4 co-receptor interacts specifically with MHC class II molecules. Protein antigens from intracellular and extracellular sources are processed into peptides by two different pathways. Peptides generated in the cytosol from viruses and other intracytosolic pathogens enter the endoplasmic reticulum, where they are bound by MHC class I molecules. Thus, these peptides are recognized by CD8 T cells, which are specialized to fight intracellular infections. As all human cells are susceptible to infection, MHC class I molecules are expressed by most cell types. Extracellular material that has been taken up by endocytosis is degraded into peptides in endocytic vesicles and these peptides are bound by MHC class II molecules within the vesicular system. The resulting complex is recognized by CD4 T cells that are specialized to fight extracellular sources of infection by mobilizing other cells of the immune system, such as B cells and macrophages. MHC class II molecules are expressed by a few cell types of the immune system that are specialized to take up extracellular antigens efficiently and activate CD4 T cells. Cross-presentation of extracellular antigens by MHC class I molecules enables professional antigen-presenting cells to stimulate cytotoxic CD8 T-cell responses against viruses that do not infect them directly.

The major histocompatibility complex

MHC molecules and other proteins involved in antigen processing and presentation are encoded in a cluster of closely linked genes, which in humans is located on chromosome 6. This region is called the major histocompatibility complex (MHC) because it was first recognized as the site of genes that cause T cells to reject tissues transplanted from unrelated donors to recipients (see Section 3-6). Now we know that these genes encode the MHC class I and II molecules that present antigens to T cells. For some MHC class I and class II molecules, numerous genetic variants are present in the human population. Each variant functions differently in the peptides it binds and the T cells it triggers, differences that, individually and collectively, have helped the human species to survive predation by diverse and numerous pathogens. Although the magnitude of MHC diversity is much smaller than that of immunoglobulins or T-cell receptors, it has a major impact on the immune response, disease susceptibility, and the practice of medicine. In this part of the chapter we consider the immunological and medical consequences of MHC diversity within the human population.

5-17 The diversity of MHC molecules in the human population is due to multigene families and genetic polymorphism

The human MHC is called the **human leukocyte antigen (HLA) complex** because the antibodies used to identify human MHC molecules react with the white cells of the blood—the leukocytes—but not with the red cells, which lack MHC molecules. This observation distinguished the MHC from the other known systems of cell-surface antigens, for example the ABO system that has to be matched in blood transfusion, which all involve antigens on the surface of red blood cells. Human MHC class I and II molecules are called **HLA class I molecules** and **HLA class II molecules**, respectively.

In contrast to immunoglobulins and T-cell receptors, the MHC class I and II molecules are encoded by conventionally stable genes that neither rearrange nor undergo any other developmental or somatic process of structural change. The inherited diversity of MHC molecules has two components. The first component is provided by **gene families**, consisting of multiple similar genes encoding the MHC class I heavy chains, MHC class II α chains and MHC class II β chains. The second component is **genetic polymorphism**, which is the presence within the population of multiple alternative forms of a gene.

The products of the different genes in a MHC class I or class II family are called isotypes. The different forms of any given gene are called alleles, and their encoded proteins are called **allotypes**. When considering the diversity of MHC class I or II molecules that arises from the combination of multiple genes and multiple alleles, the term **isoform** can be useful to denote any particular MHC protein. The numerous alleles of certain MHC class I and II genes, and the many differences that distinguish them, make these MHC genes stand out from other polymorphic genes, and they are therefore said to be **highly polymorphic**. MHC class I and II genes that have no polymorphism are described as **monomorphic** and genes having a few alleles are described as **oligomorphic**. An important consequence of high genetic polymorphism is that most children inherit different alleles of a gene from their parents, in which case each child is said to be **heterozygous**; children who inherit the same allele from both parents are said to be **homozygous**.

In humans there are six MHC class I isotypes, namely **HLA-A**, **HLA-B**, **HLA-C**, **HLA-E**, **HLA-F**, and **HLA-G**, and five MHC class II isotypes, namely **HLA-DM**, **HLA-DO**, **HLA-DP**, **HLA-DQ**, and **HLA-DR** (Figure 5.24). Of the class I isotypes, HLA-A, HLA-B, and HLA-C are highly polymorphic and their function is to present antigens to CD8 T cells and to form ligands for receptors on natural killer (NK) cells. HLA-E and HLA-G are oligomorphic and form ligands for NK-cell receptors. HLA-F seems to be monomorphic and remains intracellular; its function is unknown. The human MHC class II isotypes also display a range of properties. The three highly polymorphic molecules, **HLA-DP**, **HLA-DQ**, and **HLA-DR**, are those that present peptide antigens directly to CD4 T cells, whereas the oligomorphic **HLA-DM** and **HLA-DO** molecules have functions that regulate peptide loading of HLA-DP, HLA-DQ, and HLA-DR. The numbers of alleles currently known for each HLA locus are listed in Figure 5.25.

The polymorphism of the HLA-A, B, and C molecules is the property of the heavy chain, β$_2$-microglobulin being monomorphic. By contrast, the polymorphic HLA class II isotypes differ in the diversity contributed by their α and β chains. In HLA-DR the α chain contributes almost no diversity and

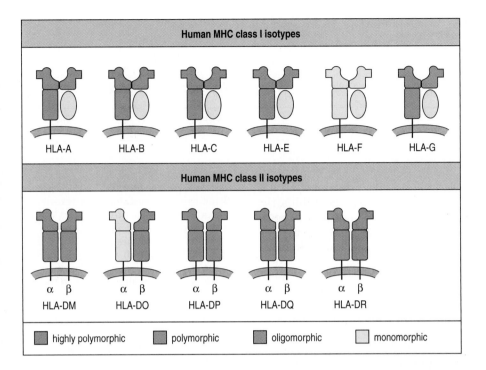

Figure 5.24 Human MHC class I and II isotypes differ in function and in the extent of their polymorphism. Of the human MHC class I isotypes, HLA-A, HLA-B, and HLA-C are highly polymorphic. They present peptide antigens to CD8 T cells and also interact with NK-cell receptors. HLA-E and HLA-G are oligomorphic and interact with NK-cell receptors. HLA-F is intracellular and of unknown function, and occurs as a single isotype. Of the human MHC class II isotypes, HLA-DP, HLA-DQ, and HLA-DR are polymorphic and present peptide antigens to CD4 T cells, whereas HLA-DM and HLA-DO occur in only a few isotypes, are intracellular, and regulate the loading of peptides onto HLA-DP, HLA-DQ, and HLA-DR.

the β chain is highly polymorphic, whereas the α and β chains of HLA-DP and HLA-DQ are both polymorphic. Overall, there is greater diversity in HLA class I than in HLA class II.

5-18 The HLA class I and class II genes occupy different regions of the HLA complex

The human MHC, the HLA complex, consists of about 4 million base pairs of DNA on the short arm of chromosome 6 and is divided into three regions (Figure 5.26). The **class I region**, at the end of the complex farthest from the chromosome's centromere, contains the six expressed HLA class I genes as well as several nonfunctional class I genes and gene fragments. At the other end of the complex is the **class II region**, which contains all the expressed class II genes and several nonfunctional class II genes. Separating the class I and II regions is a region of around 1 million base pairs called the **class III region** or the **central MHC**; although dense with other types of gene, it contains no class I or II genes. Notably absent from the HLA complex is the gene encoding β2-microglobulin, the invariant light chain of HLA class I molecules, which is located on human chromosome 15.

For each HLA class II isotype, the genes encoding the α and β genes are called A and B respectively, for example *HLA-DMA* and *HLA-DMB*. When there is more than one gene, including nonfunctional genes, a number in series is added, for example *HLA-DQA1* and *HLA-DQA2*. The genes for the α and β chains of the HLA-DM, HLA-DP, HLA-DQ, and HLA-DR class II isotypes cluster together in different subregions within the class II region of the MHC (see Figure 5.26). An exception is HLA-DO, for which the *HLA-DOA* and *HLA-DOB* genes are separated by *HLA-DM* and other genes. There are two pairs of genes for both HLA-DP and HLA-DQ, one pair being functional (*HLA-DPA1, DPB1* and *HLA-DQA1, DQB1*) and the other nonfunctional (*HLA-DPA2, DPB2* and *HLA-DQA2, DQB2*). For HLA-DR there is a single *HLA-DRA* gene, but four different genes encoding HLA-DR β chains (*DRB1, DRB3, DRB4,* and *DRB5*) and several nonfunctional genes (*DRB2, DRB6, DRB7, DRB8,* and *DRB9*). Only *DRB1* is present on all chromosomes 6, and for some people this is the only DRB gene expressed. Three other types of chromosome 6 carry either *DRB3, DRB4,* or *DRB5* in addition to *DRB1* (Figure 5.27).

HLA polymorphism		
MHC class	**HLA locus**	**Number of allotypes**
MHC class I	A	506
	B	872
	C	274
	E	3
	F	4
	G	10
MHC class II	DMA	4
	DMB	7
	DOA	3
	DOB	4
	DPA1	15
	DPB1	114
	DQA1	25
	DQB1	66
	DRA	2
	DRB1	466
	DRB3	37
	DRB4	7
	DRB5	15

Figure 5.25 Some HLA class I and class II genes are highly polymorphic. The number of known functional alleles in the human population for each gene is shown. Data is from the website http://www.ebi.ac.uk/imgt/hla/

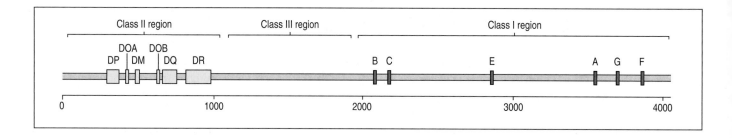

The particular combination of HLA alleles found on a given chromosome 6 is known as the **haplotype**. Within the HLA complex, meiotic recombination occurs at a frequency of about 2%. In the population this mechanism reassorts alleles of polymorphic HLA genes into new haplotypes, although in most families the parental HLA haplotypes are inherited intact. Although there are no more than a few hundred alleles for any one HLA gene (see Figure 5.25), in the course of human history they have been recombined into many thousands of different haplotypes. The combination of two HLA haplotypes in each individual thus means that millions of different HLA isoform combinations are represented in the human population. Individuals homozygous for the HLA complex are rare, but they are usually healthy. They express three class I (HLA-A, B, and C) and three class II (HLA-DP, DQ, and DR) isoforms that present antigens to their T cells. HLA heterozygous individuals can express up to six class I and eight class II isoforms; the maximum number is when each HLA haplotype contains two functional DRB genes and contributes a different allele for all of the polymorphic HLA class I and II genes.

5-19   Other proteins involved in antigen processing and presentation are encoded in the HLA class II region

The HLA complex as a whole contains more than 200 genes, of which the HLA class I and II genes are a minority. The other genes embrace a variety of functions including several that are important for the immune system. Particularly striking is the fact that the class II region of the HLA is almost entirely dedicated to genes involved in the processing of antigens and their presentation to T cells. In addition to genes encoding the α and β chains of the five HLA class II isotypes, the class II region contains genes encoding the two polypeptides of the TAP peptide transporter, the gene for tapasin, and genes encoding two subunits of the proteasome called LMP2 and LMP7 (Figure 5.28). The gene encoding the invariant chain is an exception; it is on chromosome 5.

Genes encoding proteins that work together in antigen processing and presentation are coordinately regulated by the cytokines IFN-α, -β, and -γ, which are produced at sites of infection at an early stage in the immune response. These cytokines stimulate cells in the vicinity to increase their expression of HLA class I heavy chains, β₂-microglobulin, TAP, and the LMP2 and LMP7 proteasome subunits. LMP2 and LMP7 are not constitutive components of the proteasome in healthy cells but are components of the immunoproteasome made specifically in response to interferon (see Section 5-10). When they replace the corresponding constitutive proteasome subunits they bias the proteasome toward the production of peptides compatible with MHC class I binding requirements.

Figure 5.26 The MHC is divided into three regions containing different types of gene. The positions within the HLA complex (the human MHC) of the HLA class I and II genes are shown here. The class I genes (red) are all contained in the class I region, and the class II genes (yellow) are all contained in the class II region. Separating the class I and class II regions is the class III region, which contains a variety of genes (not shown), none of which contributes to antigen processing and presentation. For HLA-DM, HLA-DP, HLA-DQ, and HLA-DR, the α-chain and β-chain genes are close together and are shown as a single yellow block; for HLA-DO the α and β genes (DOA and DOB, respectively) are separated by the DM genes and are therefore shown separately. Approximate distances are given in thousands of base pairs (kb).

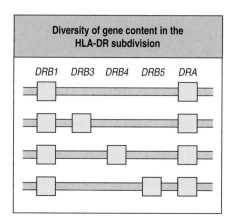

Figure 5.27 Human MHC regions differ in their number of DR genes. The MHC on every human chromosome 6 contains one gene (DRA) for the HLA class II DR α chain and one gene (DRB1) for the DR β chain. In addition, some MHCs have either DRB3 or DRB4 or DRB5. Any DRβ chain can pair with the DRα chain to form a class II molecule.

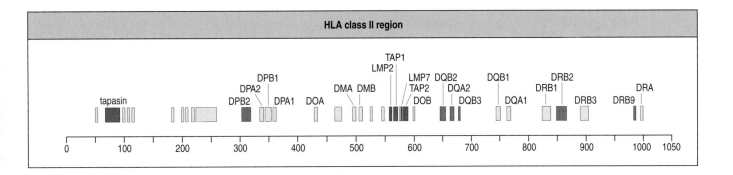

Figure 5.28 Almost all of the genes in the HLA class II region are involved in the processing and presentation of antigens to T cells. A detailed map of the HLA class II region is shown. Genes shown in dark gray are pseudogenes that are related to functional genes but are not expressed. Unnamed genes in light gray are not involved in immune system function. In addition to genes encoding the MHC class II isoforms, the class II region includes genes for the peptide transporter (TAP), proteasome components (LMP) and tapasin. Approximate distances are given in thousands of base pairs (kb).

Expression of the HLA-DM, HLA-DP, HLA-DQ, HLA-DR, and invariant-chain genes is coordinated by the cytokine IFN-γ. These genes are turned on by a transcriptional activator known as **MHC class II transactivator** (**CIITA**), which is itself induced by IFN-γ. Inherited impaired CIITA function leads to a form of bare lymphocyte syndrome in which HLA class II molecules are not made and CD4 T cells cannot function.

The majority of genes in the class I region are not involved in the immune system; neither do the class I genes form as compact a cluster as the class II genes (see Figure 5.26). Whereas genes encoding class II molecules are only present in the class II region of the MHC, genes encoding class I molecules and related class I-like molecules are found on several different chromosomes. A further difference is that MHC class II molecules are dedicated components of adaptive immunity that serve only to present antigen to T cells, whereas MHC class I molecules encompass a broader range of functions, including uptake of IgG in the gut, regulation of iron metabolism, and regulation of NK-cell function in the innate immune response. Together, these genetic and functional differences show that MHC class I is the older form of MHC molecule and that MHC class II evolved more recently from MHC class I.

5-20   MHC polymorphism affects the binding and presentation of peptide antigens to T cells

The alleles of highly polymorphic MHC genes encode proteins that differ by 1–50 amino acid substitutions. The substitutions are not randomly distributed within the sequence but are mainly in the domains that bind peptide and interact with the T-cell receptor: the α_1 and α_2 domains of MHC class I, and the α_1 and β_1 domains of MHC class II. More specifically, the substitutions are focused at positions within those domains that contact either bound peptide or the T-cell receptor (Figure 5.29). Not all the contact residues vary, as is apparent from the HLA-DR molecule in which the α_1 domain is invariant. In contrast, variability occurs in both the α_1 and β_1 domains of HLA-DP and HLA-DQ molecules.

Figure 5.29 Variation between MHC allotypes is concentrated in the sites that bind peptide and T-cell receptor. In the HLA class I molecule (left), allotype variability is clustered in specific sites (shown in red) within the α_1 and α_2 domains. These sites line the peptide-binding groove, lying either in the floor of the groove, where they influence peptide binding, or in the α helices that form the walls, which are also involved in binding the T-cell receptors. In the HLA class II molecule illustrated (right), which is a DR molecule, variability is found only in the β_1 domain because the α chain is monomorphic.

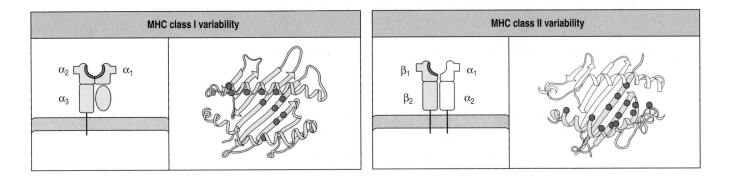

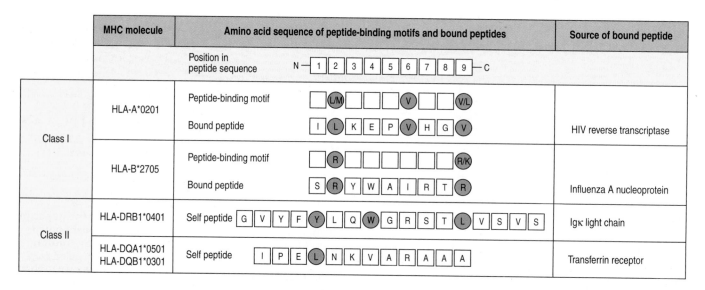

MHC molecule		Amino acid sequence of peptide-binding motifs and bound peptides	Source of bound peptide
		Position in peptide sequence — N — 1 2 3 4 5 6 7 8 9 — C	
Class I	HLA-A*0201	**Peptide-binding motif** (L/M at 2, V at 6, V/L at 9) **Bound peptide** I L K E P V H G V	HIV reverse transcriptase
	HLA-B*2705	**Peptide-binding motif** (R at 2, R/K at 9) **Bound peptide** S R Y W A I R T R	Influenza A nucleoprotein
Class II	HLA-DRB1*0401	**Self peptide** G V Y F Y L Q W G R S T L V S V S	Igκ light chain
	HLA-DQA1*0501 HLA-DQB1*0301	**Self peptide** I P E L N K V A R A A A	Transferrin receptor

Figure 5.30 Peptide-binding motifs of some MHC isoforms and the sequences of peptides bound. For the HLA-A and HLA-B isoforms, both the peptide-binding motif of the MHC molecule and the complete amino acid sequence of one peptide presented by that isoform are given. Blank boxes in the peptide-binding motifs are positions at which the identity of the amino acid can vary. For the HLA-DR and HLA-DQ isoforms, only the sequence of a self peptide that is bound by the isoform is shown. Anchor residues are in green circles. Peptide-binding motifs for MHC class II molecules are not readily defined. The one-letter code for amino acids is used here.

Variation in the peptide-contact residues on the floor and sides of the peptide-binding groove determines the types of peptide that each isoform binds. At certain positions within the peptide sequence, most of the peptides that bind an MHC isoform have the same amino acid, or one of a few chemically similar amino acids. The preferences arise because the side chains of the amino acids at these positions are bound by complementary pockets within the binding groove. These amino acids are called **anchor residues** because they anchor the peptide to the MHC molecule. The combination of anchor residues that binds to a particular MHC isoform is called its **peptide-binding motif**. For MHC class I molecules, which bind mostly nonamer peptides, positions 2 and 9 are the usual anchor residues. For MHC class II the anchor residues are less clearly defined, in part because of the heterogeneity in length of the bound peptides (Figure 5.30).

The number of peptide-binding motifs is limited and so MHC allotypes that differ by only a few amino acids often bind overlapping populations of peptides. To a rough approximation, the greater the difference in sequence between two MHC allotypes, the more disparate will be the populations of peptides they bind.

In the complex of peptide bound to an MHC molecule, the anchor residues are buried and inaccessible to the T-cell receptor. In contrast, the peptide's other positions, which are occupied by a much greater diversity of amino acids, are available for contact with T-cell receptors. These form part of the planar surface that interacts with the T-cell receptor and include variable residues on the upper surfaces of the α helices of the MHC molecule. Any given T-cell receptor is therefore specific for the complex of a particular peptide bound to a particular MHC molecule. This basic principle of T-cell biology is known as **MHC restriction** because the antigen-specific T-cell response is restricted by the MHC type. As a consequence of MHC restriction, a T cell that responds to a peptide presented by one MHC allotype will not respond to another peptide bound by that same MHC allotype or to the same peptide bound to another MHC allotype (Figure 5.31).

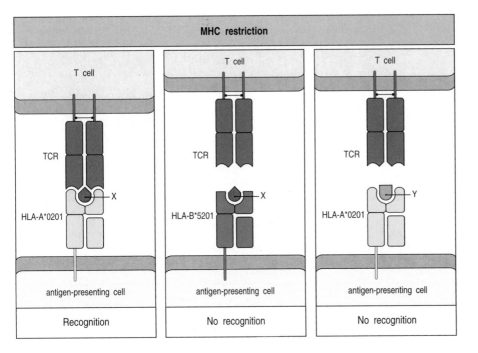

Figure 5.31 T-cell recognition of antigens is MHC restricted. The receptor of the CD8 T cell shown in the left panel is specific for the complex of peptide X with the class I molecule HLA-A*0201. Because of this co-recognition, which is called MHC restriction, the T-cell receptor (TCR) does not recognize the same peptide when it is bound to a different class I molecule, HLA-B*5201 (middle panel). Nor does the T-cell receptor recognize the complex of HLA-A*0201 with a different peptide, Y (right panel). X is HIV-1 Nef residues 190–198, AFHHVAR. Y is influenza A matrix protein residues 58–68, GILGFVFTL.

5-21 MHC diversity results from selection by infectious disease

We have just seen that the amino acid substitutions that distinguish between MHC isoforms are concentrated at sites that affect peptide binding and presentation. Further nonrandomness is present in the gene sequences, where the frequency of nucleotide substitutions that result in a change of amino acid is much greater than would be generated by chance. The inescapable conclusion is that MHC diversity is due to natural selection, and because of the immunological functions of MHC molecules the likely sources of the selection are the infections caused by pathogens.

For an individual the advantage of having multiple MHC class I and class II genes is that they contribute different peptide-binding specificities, allowing a greater number of pathogen-derived peptides to be presented during any infection. This improves the strength of the immune response against the pathogen by increasing the number of activated pathogen-specific T cells. The same argument applies to polymorphism at any given MHC locus, the advantage to the heterozygote being that two different peptide-binding specificities can be brought into play compared with one for the homozygote (Figure 5.32). Moreover, the high degree of polymorphism of the antigen-

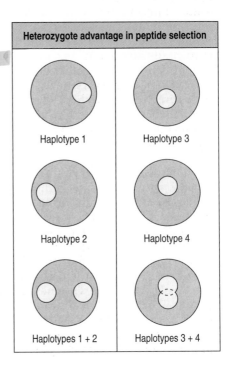

Heterozygote advantage in peptide selection

Haplotype 1

Haplotype 3

Haplotype 2

Haplotype 4

Haplotypes 1 + 2

Haplotypes 3 + 4

Figure 5.32 The advantage of being heterozygous for the MHC. The large circles represent the total number of antigenic peptides derived from a pathogen that can be presented by human MHC class I and II molecules. The small circles represent the subpopulation of peptides that can be presented by the MHC class I and II molecules encoded by the genes of particular MHC haplotypes. These subpopulations differ between the haplotypes. In general, heterozygous individuals (that is, those with haplotypes 1 + 2 and haplotypes 3 + 4 in our example), will have a set of MHC class I and II molecules able to present a wider range of pathogen-derived peptides than a homozygote. The extent of this benefit varies, however. A person who has the divergent haplotypes 1 and 2 will, on average, present a larger number of different peptides from any pathogen than a person who has the more closely related haplotypes 3 and 4.

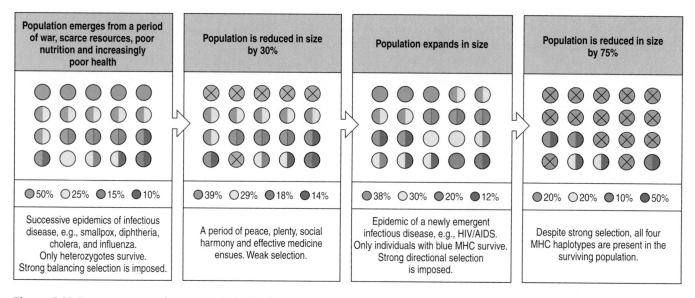

Population emerges from a period of war, scarce resources, poor nutrition and increasingly poor health	Population is reduced in size by 30%	Population expands in size	Population is reduced in size by 75%
● 50% ○ 25% ● 15% ● 10%	● 39% ○ 29% ● 18% ● 14%	● 38% ○ 30% ● 20% ● 12%	● 20% ○ 20% ● 10% ● 50%
Successive epidemics of infectious disease, e.g., smallpox, diphtheria, cholera, and influenza. Only heterozygotes survive. Strong balancing selection is imposed.	A period of peace, plenty, social harmony and effective medicine ensues. Weak selection.	Epidemic of a newly emergent infectious disease, e.g., HIV/AIDS. Only individuals with blue MHC survive. Strong directional selection is imposed.	Despite strong selection, all four MHC haplotypes are present in the surviving population.

Figure 5.33 Exposure to pathogens selects for MHC polymorphism. In our example population there are four different MHC haplotypes, each represented by a different color. The frequencies of different genotypes are represented by the 20 circles in each panel, the frequencies of the four haplotypes being given underneath. First panel: the population experiences a period characterized by balancing selection arising from successive epidemic infections, after which only heterozygotes survive (second panel) and in which 30% of the population die, as indicated by circles containing an X. After recovery of the population during a period of relative calm and health (third panel) it becomes subject to directional selection by a new and particularly nasty infection. Only individuals with the blue MHC haplotype survive and 75% of the population dies (fourth panel). As a result of these selections the frequencies of the MHC haplotypes change considerably, but all four MHC haplotypes are retained within the population.

presenting HLA isotypes ensures that most members of the population are heterozygotes (the extent of the heterozygote advantage will vary, however, depending on the difference in the peptide-binding specificities of the two allotypes). All these selective processes act to maintain a variety of MHC isoforms in the population and are known as **balancing selection** (Figure 5.33, first two panels).

A different mode of selection favors certain MHC alleles, or combinations of alleles, at the expense of others and is imposed by specific, epidemic disease. Here, presentation of particular pathogen-derived peptides by particular MHC allotypes is advantageous and in the extreme case makes the difference between life and death. As a consequence, the selected alleles are driven to higher frequency while other alleles decrease in frequency (Figure 5.33, third and fourth panels). This type of selection disrupts the balance and is therefore called **directional selection**. The numerous HLA differences between human populations of different ethnicity and geographical origins are evidence for directional selection. Only a minority of HLA alleles are common to all human populations, the majority being of recent origin and specific to particular ethnic groups.

Because pathogens adapt to the MHC of their host populations, it has been argued that rare, recently formed MHC alleles to which the pathogen has not adapted will more probably confer advantage on the host and be selected during disease epidemics. New variants of HLA class I and II alleles arise through point mutation and several types of recombination, which can involve alleles of the same gene or alleles of two different genes in the same family (Figure 5.34). Particularly favored seem to be new HLA alleles in which a small segment of one allele has been replaced by the homologous section of another, with the introduction of several amino acid substitutions that change contact residues in the peptide-binding groove (see Figure 5.29). The recombination mechanism that produces such variants has been termed **interallelic conversion** or **segmental exchange** (see Figure 5.34, left panel). Selection

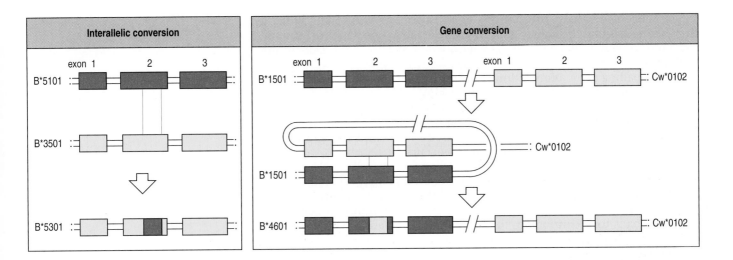

Interallelic conversion	Gene conversion

for new alleles of this type has been particularly strong upon the HLA-B locus in the indigenous populations of South and Central America, because today a majority of their HLA-B alleles are 'new variants' specific to these populations.

In industrialized countries the current epidemic of HIV infection provides a unique opportunity to study the effects of HLA polymorphism on an infectious disease and vice versa. In addition to a general advantage of HLA heterozygosity (Figure 5.35), certain families of alleles (HLA-B14, B27, B57, HLA-C8, C14) are associated with slow progression of the disease, whereas others are associated with rapid progression (HLA-A29, HLA-B22, B35, HLA-C16, and HLA-DR11). Almost all the correlations are with HLA class I; this is consistent with the principal mechanism for controlling the infection being killing of virus-infected cells by cytotoxic CD8 T cells.

5-22   MHC polymorphism triggers T-cell reactions that can reject transplanted organs

During T-cell development, any cells having T-cell receptors that respond to complexes of self peptide and MHC class I and II molecules at healthy cell surfaces are eliminated. This quality-control mechanism, which prevents a person's T cells from attacking their own healthy tissue and causing disease, encompasses only the MHC isoforms expressed by that person and not other MHC isoforms. In this context, the **self-MHC** isoforms are described as **autologous**, and all other MHC isoforms are described as **allogeneic**. Accordingly, in every person's circulation there are T cells that can respond to complexes of peptide and allogeneic MHC class I and II molecules that are present on healthy cells of other individuals. T cells with this property are called **alloreactive T cells**, and those reactive against any given allogeneic cell constitute between 1% and 10% of circulating T cells.

Figure 5.34 New MHC alleles are generated by interallelic conversion or gene conversion. Recombination between alleles of the same gene (HLA-B*5101 and HLA-B*3501), as shown on the left, and between alleles of different genes (HLA-B*1501 and HLA-Cw*0102), as shown on the right, can both result in the formation of a new allele in which a small block of DNA sequence has been replaced. B*5301 is characteristic of African populations and has been associated with resistance to severe malaria. B*4601 is found in southeast Asian populations and has been associated with susceptibility to nasopharyngeal carcinoma.

Figure 5.35 MHC heterozygosity delays the progression to AIDS in people infected with HIV-1. When people who have been infected with HIV-1 start to make detectable antibodies to the virus they are said to have undergone seroconversion. The onset of overt symptoms of AIDS occurs years after seroconversion. The rate of progress to AIDS decreases with the extent of HLA heterozygosity, as shown here by a comparison of individuals who are heterozygous for all the highly polymorphic HLA class I and II loci (red) with those who are homozygous for one locus (yellow) or for two or three loci (blue).

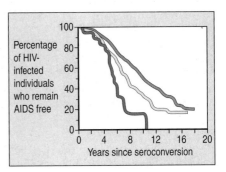

When allogeneic kidneys are transplanted to patients with renal failure, a major danger is that the kidney graft will be rejected by the patient's immune system. One way in which this happens is when alloreactive T cells in the patient's circulation are activated by allogeneic HLA molecules expressed by the graft, leading to an **alloreaction**, a potent T-cell response that attacks the graft. To reduce the probability of kidney graft rejection, donors may be selected who have identical or similar combinations of HLA alleles to the patient's. The combination of HLA alleles that a person has is called their **HLA type**. Immunosuppressive drugs are also used to preempt the alloreactive T-cell response and to treat rejection when it occurs.

A natural situation in which alloreactions occur is pregnancy, when the mother's immune system can be stimulated by the HLA molecules of the fetus that derive from the father but are not expressed on the mother's cells. This response leads to alloantibodies in the mother's circulation with specificity for paternal MHC molecules. **Alloantibody** is the name given to any antibody raised in one member of a species against an allotypic protein from another member of the same species. Although of no harm to the fetus, which is protected, the alloantibodies produced by pregnancy can have disastrous effects should the mother need a kidney transplant in the future. If these preexisting alloantibodies react with the allogeneic MHC class I of the transplanted kidney cells, they cause a type of graft rejection that is almost impossible to treat. To avoid this outcome the patient's serum is tested for reactivity with the potential donor's leukocytes and transplantation is undertaken only when the reaction is acceptably low. Before molecular genetic methods were available, the HLA types of transplant recipients and donors were determined using the anti-HLA class I and class II alloantibodies present in sera obtained from women who had had several children.

Summary

In humans, the highly polymorphic MHC class I and II genes are closely linked in the HLA region on chromosome 6, which comprises the human MHC. In contrast to immunoglobulin and T-cell receptor genes, MHC class I and II genes have a conventional organization and do not rearrange. In humans, the MHC class I genes encode the heavy (α) chains of three different class I molecules—HLA-A, HLA-B, and HLA-C—whereas the MHC class II genes encode the α and β chains of three different MHC class II molecules—HLA-DP, HLA-DQ, and HLA-DR. β_2-Microglobulin, the light chain of MHC class I molecules, is encoded outside the MHC, on chromosome 15. Certain MHC class I and class II genes are highly polymorphic; some have several hundred alleles. Genes encoding other proteins involved in antigen presentation are located in the MHC and, like the MHC class I and II genes, their expression is regulated by the interferons produced during an immune response. The strategy used by MHC molecules to bind diverse antigens contrasts with that of the T-cell receptors. MHC molecules have highly promiscuous binding sites for peptides; an MHC molecule is therefore usually able to present a diversity of peptide antigens to a large number of T-cell receptors with highly specific binding sites. Polymorphism in families of MHC class I heavy-chain genes and MHC class II α-chain and β-chain genes is a secondary strategy that serves to increase the breadth and strength of T-cell immunity. It also diversifies T-cell immunity within human populations, a strategy that helps them survive epidemic disease but has also created the main immunological barrier to clinical transplantation.

Summary to Chapter 5

The general structure of T-cell antigen receptors resembles that of the membrane-bound immunoglobulins of B cells and they are encoded by similarly organized genes that undergo gene rearrangement before they are expressed.

As with B cells, this gives rise to a population of T cells each expressing a unique receptor. Differences between immunoglobulins and T-cell receptors reflect the fact that the T-cell receptor is used only as a membrane-bound receptor, whereas immunoglobulins are also used as secreted effector molecules.

T-cell receptors are more limited than immunoglobulins in the antigens that they bind, recognizing only short peptides bound to MHC molecules on a cell surface. The two main classes of T cell—CD8 and CD4—are specialized to respond to intracellular and extracellular pathogens, respectively. They are activated in response to an antigen from the appropriate source by specific interactions between the MHC molecules displaying the peptide antigen and the CD4 or CD8 glycoproteins on the T-cell surface. Processing of extracellular and intracellular pathogen antigens into peptides, and their binding to MHC molecules, occur inside the cells of the infected host. Cytotoxic CD8 T cells are activated by peptides presented by MHC class I molecules and are directed to destroy the antigen-presenting cell. Most cells express MHC class I molecules and so can present pathogen-derived peptides to CD8 T cells if infected with a virus or other pathogen that penetrates the cytosol. The function of CD4 T cells is the activation of other types of effector cell and their recruitment to sites of infection through cell–cell interactions and the production of cytokines. They are activated by peptides presented by MHC class II molecules, which are usually expressed only on specialized antigen-presenting cells that take up and process material from the extracellular environment and can activate CD4 T cells.

MHC molecules have promiscuous peptide-binding sites, a feature that enables the relatively small number of different MHC molecules present in each individual to bind peptides of many different sequences. The diversity of peptides that can be presented by the human population as a whole is further increased by the highly polymorphic nature of MHC class I and II genes. Each person differs from almost all other individuals in some or all of the MHC alleles that they possess. Thus, the population is able to respond to the pathogens it encounters with a wide diversity of individual immune responses. Because of the polymorphism of the MHC, organs or tissues transplanted between individuals of different MHC type provoke strong T-cell responses directed against the 'foreign' MHC molecules.

Questions

5–1   Describe (A) five ways in which T-cell receptors are similar to immunoglobulins, and (B) five ways in which they are different (other than the way in which they recognize antigen).

5–2   Compare the organization of T-cell receptor α and β genes (the TCRα and TCRβ loci) with the organization of immunoglobulin heavy-chain and light-chain genes.

5–3   T-cell receptors do not undergo isotype switching. Suggest a possible reason for this.

5–4   The role of the CD3 proteins and ζ chain on the surface of the cell is to
   a. transduce signals to the interior of the T cell
   b. bind to antigen associated with MHC molecules
   c. bind to MHC molecules
   d. bind to CD4 or CD8 molecules
   e. facilitate antigen processing of antigens that bind to the surface of T cells.

5–5   Which of the following accurately completes this statement: "The function of _____ T cells is to make contact with _____ and _____"? (Select all that apply.)
   a. CD8; virus-infected cells; kill virus-infected cells
   b. CD8; B cells; stimulate B cells to differentiate into plasma cells
   c. CD4; macrophages; enhance microbicidal powers of macrophages
   d. CD4; B cells; stimulate B cells to differentiate into plasma cells
   e. All of the above are accurate.

5–6   The immunological consequence of severe combined immunodeficiency disease (SCID) caused by a genetic defect in either RAG-1 or RAG-2 genes is
   a. lack of somatic recombination in T-cell receptor and immunoglobulin gene loci
   b. lack of somatic recombination in T-cell receptor loci

c. lack of somatic recombination in immunoglobulin loci
d. lack of somatic hypermutation in T-cell receptor and immunoglobulin loci
e. lack of somatic hypermutation in T-cell receptor loci.

5–7

A. (i) Describe the structure of an MHC class I molecule, identifying the different polypeptide chains and domains. (ii) What are the names of the MHC class I molecules produced by humans? Which part of the molecule is encoded within the MHC region of the genome? (iii) Which domains or parts of domains participate in the following: antigen binding; binding the T-cell receptor; and binding the T-cell co-receptor? (iv) Which domains are the most polymorphic?
B. Repeat this for an MHC class II molecule.

5–8   Amino acid variation among MHC class II allotypes that present antigens to CD4 T cells is concentrated
a. where MHC contacts the co-receptors CD4 or CD8
b. in the β chain, because the α chain is monomorphic
c. where the MHC molecule contacts peptide and the T-cell receptor
d. in the α chain, because the β chain is monomorphic
e. throughout both the α and β chains in all domains.

5–9   What is meant by the terms (a) antigen processing and (b) antigen presentation? (c) Why are these processes required before T cells can be activated?

5–10

A. Describe in chronological order the steps of the endogenous antigen-processing pathway for intracellular, cytosolic pathogens.
B. (i) What would be the outcome if a mutant MHC class I α chain could not associate with β_2-microglobulin, and (ii) what would happen if the TAP transporter were lacking as a result of mutation? Explain your answers.

5–11   Which of the following removes CLIP from MHC class II molecules?
a. HLA-DM
b. HLA-DO
c. HLA-DP
d. HLA-DQ
e. HLA-DR.

5–12

A. Describe in chronological order the steps of the antigen-processing pathway for extracellular pathogens.
B. What would be the outcome (i) if invariant chain were defective or missing, or (ii) if HLA-DM were not expressed?

5–13

A. What is the difference between MHC variation due to multigene family and allelic polymorphism influence the antigens that a person's T cells can recognize?
B. How do MHC variation due to multigene family and allelic polymorphism influence the antigens that a person's T cells can recognize?

5–14   What evidence supports the proposal that MHC diversity evolved by natural selection caused by infectious pathogens rather than exclusively by random DNA mutations?

5–15   Brittany Hudson, 16 years of age, was seen by her physician after the development of a small pustule around her nostrils that had expanded and was now showing signs of ulceration typical of chronic granulomatous inflammation. In the past year, Jennifer had experienced a similar lesion on her left thigh that healed slowly, leaving a hyperpigmented scar. She also had a history of chronic bacterial infections of the upper and lower respiratory tract. Flow cytometric analysis of peripheral blood revealed abnormally low numbers of MHC class I molecules on cell surfaces and abnormally low numbers of CD8 T cells. A diagnosis of type I bare lymphocyte syndrome was made. A deficiency in which of the following would explain this etiology?
a. HLA-DM
b. Invariant chain
c. Class II-associated invariant chain peptide (CLIP)
d. TAP1 or TAP2
e. MHC class II transactivator (CIITA).

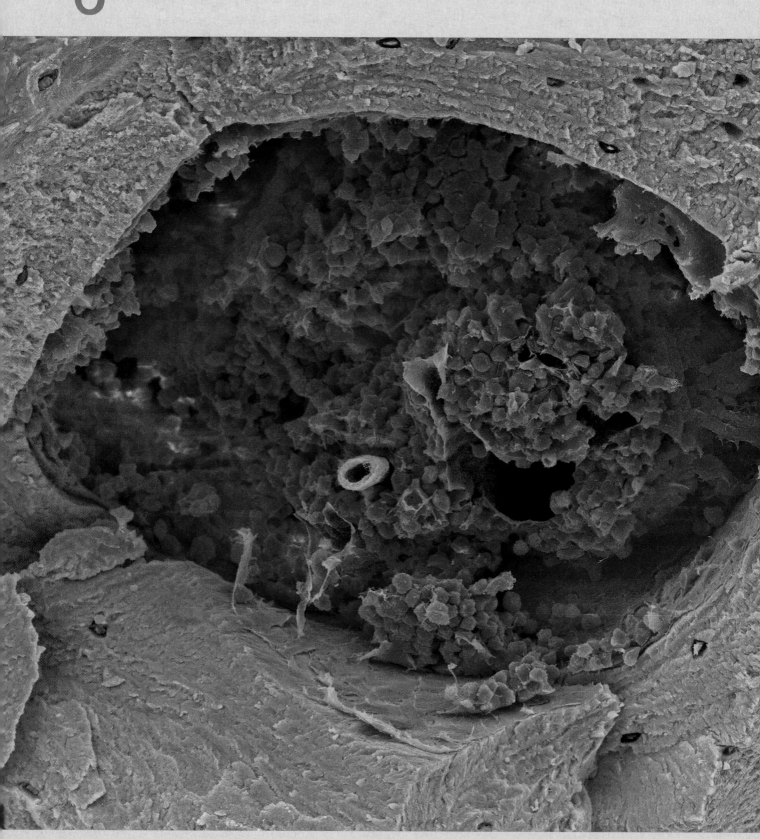

The marrow cavities of the bones in which B-cell development occurs.

Chapter 6

The Development of B Lymphocytes

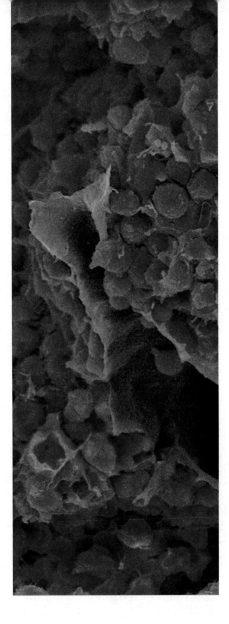

The B cells of the human immune system have the capacity to make immunoglobulins specific for almost every nuance of chemical structure, which gives each person the potential to make antibodies against all the infectious microorganisms that could possibly be encountered in a lifetime. The body does not, however, stockpile all the B cells needed to do this. That would probably mean devoting the vast majority of the body's resources to the immune system, leaving little for the immune system to protect. Instead, the body carries a less complete inventory of B cells, but one that expands and contracts its individual clones according to need and circumstance. Fueling this system are stem cells in the bone marrow, which generate more than sixty billion new B cells every day of your life.

The development of B cells from bone marrow stem cells through to antibody-producing plasma cells is the subject of this chapter. B-cell development can be divided into six functionally distinct phases (Figure 6.1). In the first phase of development, which is described in the first part of the chapter, B-cell precursors in the bone marrow acquire functional antigen receptors through the immunoglobulin gene rearrangements described in Chapter 4. Although each mature B cell expresses immunoglobulin of just one antigen specificity as its B-cell receptor, the B-cell population as a whole represents a vast repertoire of immunoglobulins with different binding specificities. The second part of this chapter describes how this repertoire is modified in various ways as B cells mature and travel from the bone marrow to the secondary lymphoid organs, where they can be activated by antigen to function in the body's defense.

The second phase in B-cell development is one of negative selection, which prevents the emergence of mature B cells bearing receptors that bind to normal constituents of the human body. Such cells are potentially dangerous because of their potential to attack healthy tissue and cause autoimmune disease. Negative selection begins in the bone marrow and continues as the immature B cells leave the bone marrow and travel to the secondary lymphoid

Figure 6.1 The development of B cells can be divided into six functionally distinct phases.

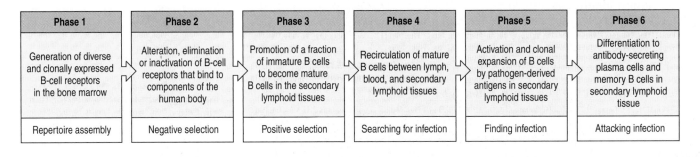

Phase 1	Phase 2	Phase 3	Phase 4	Phase 5	Phase 6
Generation of diverse and clonally expressed B-cell receptors in the bone marrow	Alteration, elimination or inactivation of B-cell receptors that bind to components of the human body	Promotion of a fraction of immature B cells to become mature B cells in the secondary lymphoid tissues	Recirculation of mature B cells between lymph, blood, and secondary lymphoid tissues	Activation and clonal expansion of B cells by pathogen-derived antigens in secondary lymphoid tissues	Differentiation to antibody-secreting plasma cells and memory B cells in secondary lymphoid tissue
Repertoire assembly	Negative selection	Positive selection	Searching for infection	Finding infection	Attacking infection

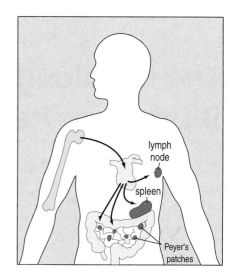

Figure 6.2 **B cells develop in bone marrow and then migrate to secondary lymphoid tissues.** B cells leaving the bone marrow (yellow) are carried in the blood to lymph nodes, the spleen, Peyer's patches (all shown in green), and other secondary lymphoid tissues such as those lining the respiratory tract (not shown).

organs. The third phase is one of positive selection, in which the immature B cells compete for the limited number of sites in the follicles of the secondary lymphoid tissues where they must go to complete their maturation (Figure 6.2). In the fourth phase, the mature B cells travel between secondary lymphoid tissues in the lymph and blood, patrolling for infections and the pathogen-derived antigens to which the B-cell receptors bind. Activation of B cells by antigen in the fifth phase leads to the proliferation and clonal expansion of the antigen-specific B cells. In the sixth phase, differentiation and diversification of B cells within each expanded clone gives rise to plasma cells that provide antibodies to attack ongoing infection and memory B cells for the speedy elimination of any future infections by the same pathogen. At every phase in B-cell development, significant numbers of cells are, for various reasons, prevented from advancing to the next stage. Overall, the efficiency of B-cell development is very low, but that is the price paid for having an almost infinite capacity to make antibodies against any possible pathogen.

The development of B cells in the bone marrow

The sole purpose of a B cell is to make an immunoglobulin. Consequently, B-cell development in the bone marrow can be divided into stages that correspond to successive steps in the rearrangement and expression of the immunoglobulin genes. The quality of the gene rearrangements and the proteins they make are assessed at two key checkpoints, and further development is halted if they are found wanting. In this part of the chapter we shall see how gene rearrangements are ordered and controlled to produce an immature B cell that makes only one type of heavy chain and one type of light chain, and thus expresses immunoglobulin of a single antigen specificity on its surface.

6-1   B-cell development in the bone marrow proceeds through several stages

In the bone marrow, pluripotent hematopoietic stem cells give rise to the common lymphoid progenitor cells, which have the potential to produce both B cells and T cells (see Figure 1.14). Some of these cells develop into precursor cells that are committed to becoming B cells (Figure 6.3). All these undifferentiated precursor cells can be distinguished by cell-surface markers. One of these is the protein CD34, which is present on all human hematopoietic stem cells and is exploited medically, through the use of anti-CD34 monoclonal antibodies, to separate hematopoietic stem cells from the other bone marrow cells for use in therapeutic transplantation.

The earliest identifiable cells of the B-cell lineage are called **pro-B cells** (see Figure 6.3). These progenitor cells retain a limited capacity for self-renewal, dividing to produce both more pro-B cells and cells that will go on to develop further. The main event in the pro-B-cell stage is the rearrangement of the heavy-chain genes, which always precedes rearrangement of the light-chain genes. Joining of D_H and J_H gene segments occurs at the **early pro-B-cell** stage, followed at the **late pro-B-cell** stage by joining of a V_H segment to the rearranged DJ_H. The rearranged gene is transcribed through to the μ C-region gene, the nearest C gene to the rearranged V region (see Figure 4.22). The RNA transcript is spliced to produce mRNA for the μ heavy chain, the first type of

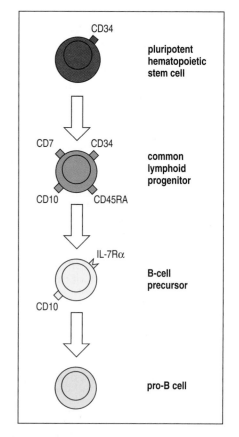

Figure 6.3 **Pro-B cells develop from the pluripotent hematopoietic stem cell.** Cells at different stages can be identified by protein markers on their surface.

	Stem cell	Early pro-B cell	Late pro-B cell	Large pre-B cell	Small pre-B cell	Immature B cell
H-chain genes	Germline	D–J rearrangement	V–DJ rearrangement	VDJ rearranged	VDJ rearranged	VDJ rearranged
L-chain genes	Germline	Germline	Germline	Germline	V–J rearranging	VJ rearranged
Ig status	None	None	None	μ heavy chain is made	μ chain in endoplasmic reticulum	μ heavy chain. λ or κ light chain. IgM on surface.

Figure 6.4 The development of B cells in the bone marrow proceeds through stages defined by the rearrangement and expression of the immunoglobulin genes. In the stem cell, the immunoglobulin (Ig) genes are in the germline configuration. The first rearrangements are of the heavy-chain (H chain) genes. Joining D_H to J_H defines the early pro-B cell, which becomes a late pro-B cell on joining V_H to DJ_H. Expression of a functional μ chain defines the large pre-B cell. Large pre-B cells proliferate, producing small pre-B cells in which rearrangement of the light-chain (L chain) gene occurs. Successful light-chain gene rearrangement and expression of IgM on the cell surface define the immature B cell.

immunoglobulin chain made by a developing B cell. Once a B cell expresses a μ chain it is known as a **pre-B cell**. Pre-B cells represent two stages in B-cell development: the less mature **large pre-B cells** and the more mature **small pre-B cells** (Figure 6.4). Large pre-B cells have successfully rearranged a heavy-chain gene and make a μ heavy chain; they have stopped rearrangements of the heavy-chain genes but have yet to commence rearrangement of the light-chain genes.

Rearrangement of the light-chain genes occurs in the small pre-B cells. The κ light-chain genes are the first to rearrange, and only if those rearrangements fail to make a viable κ chain are the λ light-chain genes rearranged. When successful joining of light-chain V and J segments is achieved, a light-chain protein is synthesized and assembled in the endoplasmic reticulum with the μ chains to form membrane-bound IgM. The IgM further associates with Igα and Igβ to form a functional B-cell receptor complex, which is then transported to the cell surface (see Section 4-12, p. 111). Rearrangement of the light-chain genes stops and the small pre-B cell becomes an **immature B cell**.

6-2   B-cell development is stimulated by bone marrow stromal cells

B-cell development in the bone marrow is dependent on a network of non-lymphoid **stromal cells**, which provide specialized microenvironments for B cells at various stages of maturation (Figure 6.5). Stroma is the name given to the supportive cells and connective tissue in any organ. The stromal cells perform two distinct functions. First, they make specific contacts with the developing B cells through the interaction of adhesion molecules and their ligands. Second, they produce growth factors that act on the attached B cells, for example the membrane-bound stem-cell factor (SCF), which is recognized by a receptor called Kit on maturing B cells. Another important growth factor for B-cell development is interleukin-7 (IL-7), a cytokine secreted by stromal cells that acts on late pro-B and pre-B cells.

The most immature stem cells lie in a region of the bone marrow called the subendosteum, which is adjacent to the interior surface of the bone. As the B cells mature, they move, while maintaining contact with stromal cells, to the central axis of the marrow cavity. Later stages of maturation are less dependent on contact with stromal cells, allowing the B cells eventually to leave the bone marrow. The next stages of development occur after an immature B cell leaves the bone marrow, and they take place in a secondary lymphoid organ such as a lymph node, the spleen or a Peyer's patch. It is here that the immature B cell becomes a mature B cell that can respond to its specific antigen.

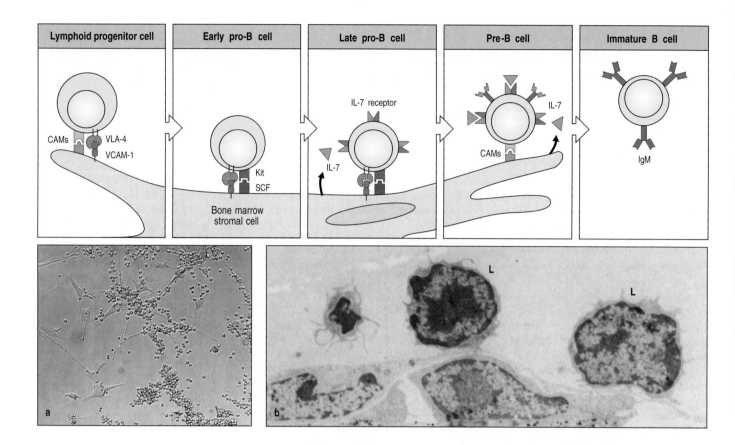

Figure 6.5 **The early stages of B-cell development are dependent on bone marrow stromal cells.** The top panels show the interactions of developing B cells with bone marrow stromal cells. Stem cells and early pro-B cells use the integrin VLA-4 to bind to the adhesion molecule VCAM-1 on stromal cells. This and interactions between other cell-adhesion molecules (CAMs) promote the binding of the receptor Kit on the B cell to stem-cell factor (SCF) on the stromal cell. Activation of Kit causes the B cell to proliferate. B cells at a later stage of maturation require interleukin-7 (IL-7) to stimulate their growth and proliferation. Panel a is a light micrograph of a tissue culture showing small round B-cell progenitors in intimate contact with stromal cells, which have extended processes fastening them to the plastic dish on which they are grown. Panel b is a high-magnification electron micrograph showing two lymphoid cells (L) adhering to a flattened stromal cell. Photographs courtesy of A. Rolink (a); Paul Kincade and P.L. Witte (b).

6-3 Pro-B-cell rearrangement of the heavy-chain locus is an inefficient process

Quality control is absolutely necessary in B-cell development because the process of gene rearrangement is inherently imprecise and inefficient. These complications come mainly from the random addition of N and P nucleotides, which occurs on forming the joints between V, D, and J gene segments (see Section 4-9, p. 108). Such additions can change the reading frame of the DNA sequence so that it no longer encodes an immunoglobulin heavy chain. Gene rearrangements that do not translate into a useful protein are called **nonproductive rearrangements**. Those rearrangements that preserve a correct reading frame and give rise to a complete and functional immunoglobulin chain are known as **productive rearrangements**. At each rearrangement event, there is only a one in three chance of the correct reading frame being maintained.

Every B cell has two copies of the immunoglobulin heavy-chain locus, which helps to increase the likelihood that a pro-B cell will make a productive heavy-chain gene rearrangement. The two copies of each locus are on homologous chromosomes; one is inherited from the mother and the other from the father. In the developing B cell, gene rearrangements can be made on both homologous chromosomes. Therefore, a B cell that has made a nonproductive rearrangement on one chromosome still has a chance of producing a heavy chain if it makes a productive rearrangement at the locus on the other homologous chromosome. If developing B cells make nonproductive rearrangements at both copies of the heavy-chain gene they lose their potential to make immunoglobulin; they do not develop further and they die in the bone marrow.

For an early pro-B cell to rearrange the immunoglobulin heavy-chain genes it must express the recombination-activating genes *RAG-1* and *RAG-2*, as well as other DNA-modifying enzymes needed to cut, paste, and add to the DNA (see

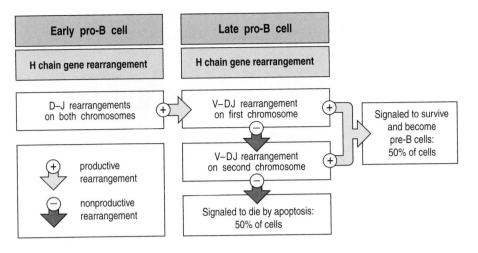

Figure 6.6 Immunoglobulin heavy-chain gene rearrangement in pro-B cells gives rise to both productive and nonproductive rearrangements. A productive rearrangement enables the B cell to proceed to the next stage of development. Rearrangements occur at the H-chain genes on both chromosomes, and if neither is successful the cell dies.

Section 4-9). Two transcription factors called E2A and EBF are responsible for these changes in gene expression. They also cause the expression of Pax-5, another transcription factor that is responsible for switching on the genes for many proteins that are expressed only in B cells, including Igα and a cell-surface protein called CD19, which in mature B cells forms part of the B-cell co-receptor that helps the cell respond to antigen.

The first rearrangement event is the joining of a D_H gene segment to a J_H gene segment, which occurs at the same time on the two copies of the heavy-chain locus (Figure 6.6). This rearrangement is relatively efficient because the human D segments give a functional protein sequence when read in any of the three reading frames. The second event in heavy-chain gene rearrangement is the joining of a V_H gene segment to the rearranged DJ_H sequence. This rearrangement is first targeted to only one of the heavy-chain loci. Only when V_H to DJ_H rearrangement on the first chromosome is nonproductive does rearrangement proceed on the second chromosome. The two-thirds failure rate in maintaining the reading frame, and the independent chances for success offered by having two chromosomes, mean that just over half the total number of pro-B cells make a functional heavy-chain gene. These cells advance along the developmental pathway to become large, dividing pre-B cells producing μ chains. Those pro-B cells that fail to make a μ chain die by apoptosis in the bone marrow. The cells consigned to die include pro-B cells that made two nonproductive V_H to DJ_H rearrangements and also ones that made two nonproductive D_H to J_H rearrangements. Despite their failure, the latter cells are allowed to proceed to V_H to DJ_H rearrangement even though it can have no useful effect. A general feature of lymphocyte development is for apoptosis to be the 'default' pathway that is followed unless a positive signal for survival and further differentiation is received. Such signals are called **survival signals**.

6-4   The pre-B-cell receptor monitors the quality of immunoglobulin heavy chains

One criterion that a pro-B cell must satisfy in order to survive is that it makes a μ heavy chain. The second criterion is that the μ chain must demonstrate the ability to combine with an immunoglobulin light chain. At this stage of B-cell development there are no genuine light chains with which to make such a test, but pro-B cells synthesize two proteins, called **VpreB** and **λ5**, which bind to μ heavy chains in a manner that mimics the immunoglobulin light chain. VpreB is structurally similar to the variable region, and λ5 is like a constant region; together they form what is known as the **surrogate light**

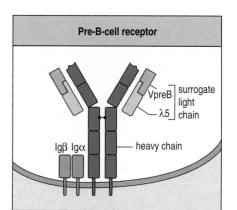

Figure 6.7 The pre-B-cell receptor resembles the B-cell receptor. The pre-B-cell receptor is distinguished from the B-cell receptor in that it lacks an immunoglobulin light chain, which is replaced by the surrogate light chain made up from the VpreB and λ5 polypeptides. It is thought that the pre-B-cell receptor does not appear on the cell surface but remains inside the cell in the cytoplasm, as part of membrane-enclosed vesicles.

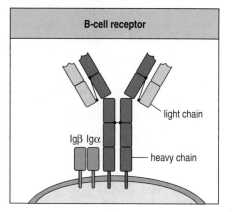

chain. VpreB and λ5 are encoded by conventional genes that do not rearrange and are separate from the immunoglobulin loci. Transcription of the VpreB and λ5 genes is controlled by the E2A and EBF transcription factors.

In the endoplasmic reticulum of the pro-B cell the μ chain forms disulfide-bonded homodimers and has the opportunity to assemble with VpreB, λ5, Igα, and Igβ to form a complex resembling the B-cell receptor (Figure 6.7). This receptor is called the **pre-B-cell receptor** because its presence sends the signal that allows a pro-B cell to become a pre-B cell. Gene rearrangement at the immunoglobulin heavy-chain locus stops and the cell undergoes several rounds of cell division that mark pro-B cells' transition to large pre-B cells. If the μ chain in a pro-B cell assembles poorly with the surrogate light chain a functional pre-B-cell receptor is not made and the cell dies by apoptosis. The complex of VpreB, λ5, and the μ chain does not have an antigen-binding site like that of IgM; neither is the pre-B-cell receptor well represented at the cell surface. These features suggest, but do not prove, that signaling from the pre-B-cell receptor does not require binding of a ligand or the appearance of the receptor at the cell surface, merely the assembly of the complex itself.

The importance of the pre-B-cell receptor for B-cell development is highlighted by the case of a young patient who inherited defective alleles of the λ5 gene from both parents. None of the developing B cells in this child could assemble a pre-B-cell receptor, even those that were making fully functional μ chains, and so all the B cells were consigned to die by apoptosis at the pro-B-cell stage. The consequence for the patient was a profound B-cell immunodeficiency. This resulted in persistent bacterial infections, which were cured with antibiotics and prevented by injections of antibodies obtained from the blood of healthy donors.

6-5   The pre-B-cell receptor causes allelic exclusion at the immunoglobulin heavy-chain locus

As well as eliminating B cells that cannot make functional μ chains, assembly of the pre-B-cell receptor also prevents B cells from making more than one functional μ chain. In a pro-B cell where rearrangement of the first immunoglobulin locus is successful, the synthesis of μ chain and assembly of the pre-B-cell receptor quickly signals for transcription of the *RAG* genes to stop. It also signals for RAG proteins to be degraded and for the structure of the chromatin of the heavy-chain locus to be reorganized into a state that resists gene rearrangement. These three effects synergize to prevent rearrangement of the second immunoglobulin heavy-chain locus and production of a second μ chain. This phenomenon, whereby a cell expresses only one of its two copies of a gene, is called **allelic exclusion**. Although any individual B cell expresses only one allele, the two alleles are expressed equally within the B-cell population.

The advantage of allelic exclusion is best appreciated by considering the alternative. A mature B cell making two μ chains with different antigen-binding specificities in equal quantities would express three types of B-cell receptor, of which the most abundant would have two different μ chains and thus two binding sites of different antigenic specificity. These receptors would be functionally substandard because they could not make strong bivalent attachments to multivalent antigens (Figure 6.8). The problems caused by this heterogeneity would be exponentially compounded when such a B cell produced plasma cells secreting pentameric IgM (see Figure 4.29, p. 115). The antibodies would be very heterogeneous, with less than 0.1% of them containing 10 identical μ chains. By this simple account, allelic antibodies made by B cells under allelic exclusion are estimated to be more than 1000-fold more effective than those made by hypothetical B cells without allelic exclusion.

These considerations were of practical importance in the early development of monoclonal antibodies for diagnostic and therapeutic applications (see Sections 4-5 and 4-6). The very first monoclonal antibodies were made by hybridomas that expressed two different heavy chains and two different light chains. One heavy chain and one light chain were derived from a splenic B cell from the mouse immunized with the desired target antigen, and together they formed an IgG specific for the immunizing antigen. The other heavy and light chains were made by the myeloma tumor cell to which the spleen cells had been fused (see Section 4-5) and which produced an immunoglobulin of different and unknown specificity. These hybridoma cells assembled 16 different forms of IgG, of which only one form had two identical binding sites specific for the desired antigen, and three forms had one specific binding site. The forms were not easily separated, and so most of the antibody produced (12 of the 16 forms) was inactive. A major advance in the production of monoclonal antibodies came with the use of myeloma fusion partners that were deliberately selected for the loss of functional immunoglobulin genes. Hybridomas derived with these fusion partners make homogeneous immunoglobulin that is specific for the immunizing antigen.

6-6   Rearrangement of the light-chain loci by pre-B cells is relatively efficient

Every large pre-B cell goes through several rounds of cell division, yielding a clone of around 100 small, resting pre-B cells that express identical μ chains but no longer make the surrogate light chain and the pre-B-cell receptor. At this stage the B cells turn their attention to making a proper light chain. The *RAG* genes, which were turned off in the dividing pre-B cells, are now reactivated; RAG proteins are produced and the immunoglobulin light-chain genes begin to rearrange.

Rearrangements take place at one light-chain locus at a time, with the κ locus usually being rearranged before the λ locus. Whereas two recombination events are required to bring together the V, D, and J segments of the heavy-chain locus, only a single event joining V with J is needed to rearrange a light-chain gene. Although this reduces the diversity achieved by combining different gene segments it has the advantage that several attempts to rearrange the same light-chain gene can be made by using V and J gene segments not involved in previous rearrangements (Figure 6.9). This is not possible at the heavy-chain locus, because the first two rearrangements that join a V to D and J excise all the other D segments. A number of attempts to rearrange the κ light-chain locus on one or other chromosome can therefore be made before commencing the rearrangement of a λ-chain locus on a different chromosome. The possession of four light-chain loci, and the opportunity for making several attempts to rearrange each locus, means that about 85% of the pre-B cell population makes a successful rearrangement of a light-chain gene (Figure 6.10).

Overall, less than half of the B-lineage cells produced in the bone marrow end up producing functional immunoglobulin heavy and light chains. Although there are no functional differences between κ and λ light-chain isotypes, the benefit of having the two isotypes is that it increases the rate at which small pre-B cells succeed in making immunoglobulin. That two-thirds of human

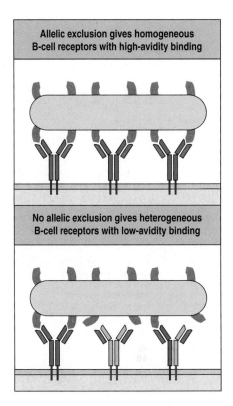

Figure 6.8 Allelic exclusion at the immunoglobulin loci results in B cells with antigen receptors of a single specificity. The top panel shows the binding to antigen of B-cell receptors produced in a B cell expressing immunoglobulin from one immunoglobulin heavy-chain locus and one immunoglobulin light-chain locus only. All the receptors have identical antigen-binding sites and bind their antigen with high avidity. The bottom panel shows the B-cell receptors formed in a hypothetical B cell expressing immunoglobulin from both the immunoglobulin heavy-chain loci and one light-chain locus. Hybrid immunoglobulins are formed with disparate antigen-binding sites and bind the antigen poorly and with low avidity. The disparities would be even greater in a B cell expressing two heavy-chain and two light-chain genes.

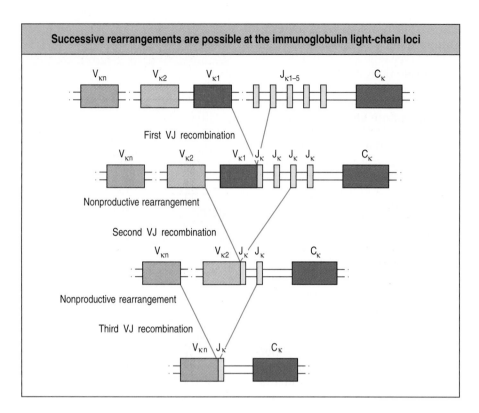

Successive rearrangements are possible at the immunoglobulin light-chain loci

First VJ recombination

Nonproductive rearrangement

Second VJ recombination

Nonproductive rearrangement

Third VJ recombination

Figure 6.9 Nonproductive light-chain gene rearrangements can be superseded by further gene rearrangement. The organization of the light-chain loci allows initial nonproductive rearrangements at one locus to be followed by further rearrangements of that same locus that can lead to production of a functional light chain. This type of rescue is shown for the κ light-chain gene. After a nonproductive rearrangement of V_κ to a J_κ, in which the translational reading frame has been lost, a second rearrangement can be made by $V_{\kappa 2}$, or any other V_κ that is on the 5' side of the first joint, with a J_κ that is on the 3' side of the first joint. When the second joint is made, the intervening DNA containing the first joint will be excised. There are five J_κ gene segments and many more V_κ gene segments, so that as many as five successive attempts at productive rearrangement can be made at each of the two κ light-chain loci.

immunoglobulins have κ chains and one-third have λ chains indicates that having the second isotype has improved the success rate by 50%.

When a functional light-chain gene has been formed, light chains are made and assemble with μ chains in the endoplasmic reticulum to form IgM. Association with the Igα and Igβ signaling components forms the mature B-cell receptor, which then moves to the cell surface (see Figure 6.7). The presence of a functional B-cell receptor at the surface sends signals to the cell's interior that quickly shut down light-chain gene rearrangement. This control, which is analogous to that exerted by the pre-B-cell receptor on the heavy-chain locus, ensures that a B cell expresses only one form of immunoglobulin light chain. Because there are two light-chain isotypes, the regulation consists of both allelic exclusion and isotype exclusion. With a functional B-cell receptor on its surface and recombination stopped in the nucleus, the pre-B cell becomes an immature B cell. If a cell makes a light chain that fails to associate with the heavy chain, it will usually be due to a fault in the light chain, because the heavy chain's competence has already been checked by the surrogate light chain. In these circumstances, recombination continues until either a functional light chain is made or no further rearrangements are possible because all V or J gene segments have been used up.

Light-chain rearrangement occurs independently in each of the 100 or so small pre-B cells that are the descendants of a single large pre-B cell and so express the same μ chain. Different light-chain rearrangements will be made in each of these cells. A success rate of 85% in light-chain manufacture means that around 85 of these small pre-B cells will become immature B cells making IgM. Two benefits arise from the clonal expansion of the large pre-B cells. First, it guarantees that the investment made to obtain each functional heavy chain is never lost through failure to make a light chain. Second, it creates a diverse population of immature B cells that have the same μ chain but a different κ or λ light chain. Clonal expansion has thereby allowed one successful heavy-chain gene rearrangement to give up to 85 clones of immature B cells with different antigenic specificities.

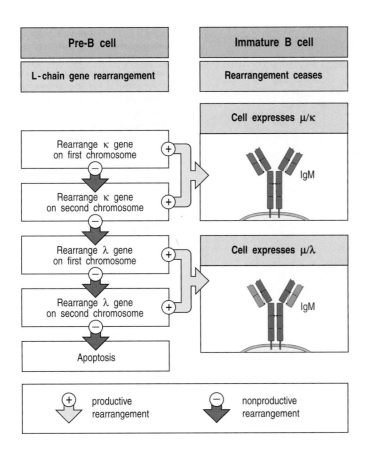

Figure 6.10 Rearrangement of the immunoglobulin light-chain genes in pre-B cells leads to the expression of cell-surface IgM.

6-7   B cells have to pass two main checkpoints in their development in the bone marrow

Because of the inherent inefficiency of the recombination process, B cells have evolved a conservative strategy for their development in the bone marrow. They first concentrate on making a heavy chain, and only when that is successful does a B cell turn to making a light chain. After each step the cells go through a **checkpoint**, when the quality of the immunoglobulin chains is assessed. As we saw in Section 6-4, the first of these occurs at the late pro-B-cell stage, with the formation, or not, of a functional pre-B-cell receptor. During the pro-B-cell stage, a mixture of productive and nonproductive gene rearrangements is randomly introduced into the immunoglobulin heavy-chain genes. The genetic variation this imparts to the B-cell population is then subjected to stringent selection for its capacity to make μ chains that can assemble a functional pre-B-cell receptor. If they can, they survive and multiply their kind; if not, they die. Selection by the pre-B-cell receptor is the first checkpoint in B-cell development (Figure 6.11).

The second checkpoint in B-cell development is at the stage of the small pre-B cell, when productive and nonproductive rearrangements have been randomly made in pre-B cells' light-chain genes. Cells are selected for survival by their capacity to make light chains that bind the existing μ chains and assemble a functional B-cell receptor. If they do not do so they die in the bone marrow; if they do, they survive to become an immature B cell (see Figure 6.11).

The only B cells that pass both checkpoints, therefore, are those that can make an immunoglobulin, an essential property for a B cell because this cell's only function in an immune response is to make immunoglobulin. Checkpoints at which selection determines cellular fate are features of both the development and the immune responses of B cells and T cells. One example we encountered in Chapter 4 is affinity maturation, in which antigen-activated B cells

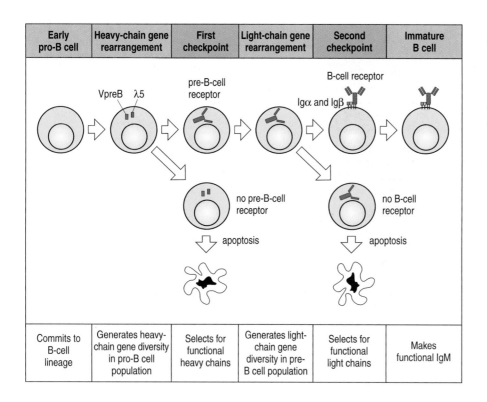

Early pro-B cell	Heavy-chain gene rearrangement	First checkpoint	Light-chain gene rearrangement	Second checkpoint	Immature B cell
Commits to B-cell lineage	Generates heavy-chain gene diversity in pro-B cell population	Selects for functional heavy chains	Generates light-chain gene diversity in pre-B cell population	Selects for functional light chains	Makes functional IgM

Figure 6.11 There are two fate-determining checkpoints during B-cell development in the bone marrow. Both checkpoints are marked by whether a functional receptor is made, which is a test of whether a functional heavy chain (at the first checkpoint) or a functional light chain (at the second checkpoint) has been produced. Cells that fail either of these checkpoints die by apoptosis.

undergo random somatic hypermutation and those cells that produce an immunoglobulin that binds antigen more tightly are selected to become plasma cells (see Section 4-14, p. 114).

6-8   A program of protein expression underlies the stages of B-cell development

The successive steps in B-cell development are accompanied by changes in the expression of proteins involved in rearranging the immunoglobulin genes, testing the quality of the immunoglobulin chains, and driving cellular proliferation (Figure 6.12). The RAG-1 and RAG-2 proteins are essential components of the recombination machinery (see Section 4-9, p. 108), and their genes are specifically turned on at the stages in B-cell development when heavy-chain or light-chain gene rearrangement occurs. Conversely, at the large pre-B-cell stage, when the quality of the heavy-chain gene rearrangements is examined and successful cells divide before rearrangement of the light-chain genes, RAG proteins are absent. Terminal deoxynucleotidyl transferase (TdT), the enzyme that adds N nucleotides at the junctions between rearranging gene segments, is expressed in pro-B cells when the heavy-chain genes begin to rearrange and is turned off in small pre-B cells when the light-chain genes begin to rearrange; this explains why N nucleotides are found in all VD and DJ joints of rearranged human heavy-chain genes but in only about half of the VJ joints of rearranged human light-chain genes.

Several proteins contribute to B-cell development by transducing signals from cell-surface receptors. Synthesis of the Igα and Igβ polypeptides is turned on at the pro-B-cell stage and continues throughout the life of a B cell, which means that they are always available to associate with immunoglobulin chains to form pre-B-cell and B-cell receptors that can transmit signals to the cell. This is important for allelic exclusion, because it ensures that gene rearrangement can be halted once a functional heavy chain, and then a light chain, has been made. Only when antigen-stimulated B cells differentiate into antibody-secreting plasma cells, and their antigen receptors are no longer required, are the Igα and Igβ genes turned off and immunoglobulin ceases to be present on the cell surface. For the same reason, the λ5 and VpreB components of the surrogate light chain are ready and waiting to form pre-B-cell receptors throughout the pro-B-cell stage (see Figure 6.12).

A variety of transcription factors control the genes involved in immunoglobulin gene rearrangement and B-cell development, only some of which are shown in Figure 6.12. Their coordinated activity ensures that the enzymatic machinery for gene rearrangement common to B and T cells is used in B cells to rearrange and express the immunoglobulin genes and not the T-cell receptor genes. Of particular importance in defining the B-cell lineage is the B-cell-specific transcription factor Pax-5, which is first synthesized in the early pro-B cell and continues to be synthesized throughout B-cell development and the lifetime of a mature B cell. It binds to the regulatory elements of the immunoglobulin genes and also to regulatory elements in the genes for many B-cell-specific proteins, such as λ5, VpreB, and CD19.

In all human cells except B cells the chromatin containing the immunoglobulin loci is kept in a 'closed' form that is never transcribed. In the early pro-B cell, Pax-5 binds to enhancer sequences located 3′ of the heavy-chain C-region

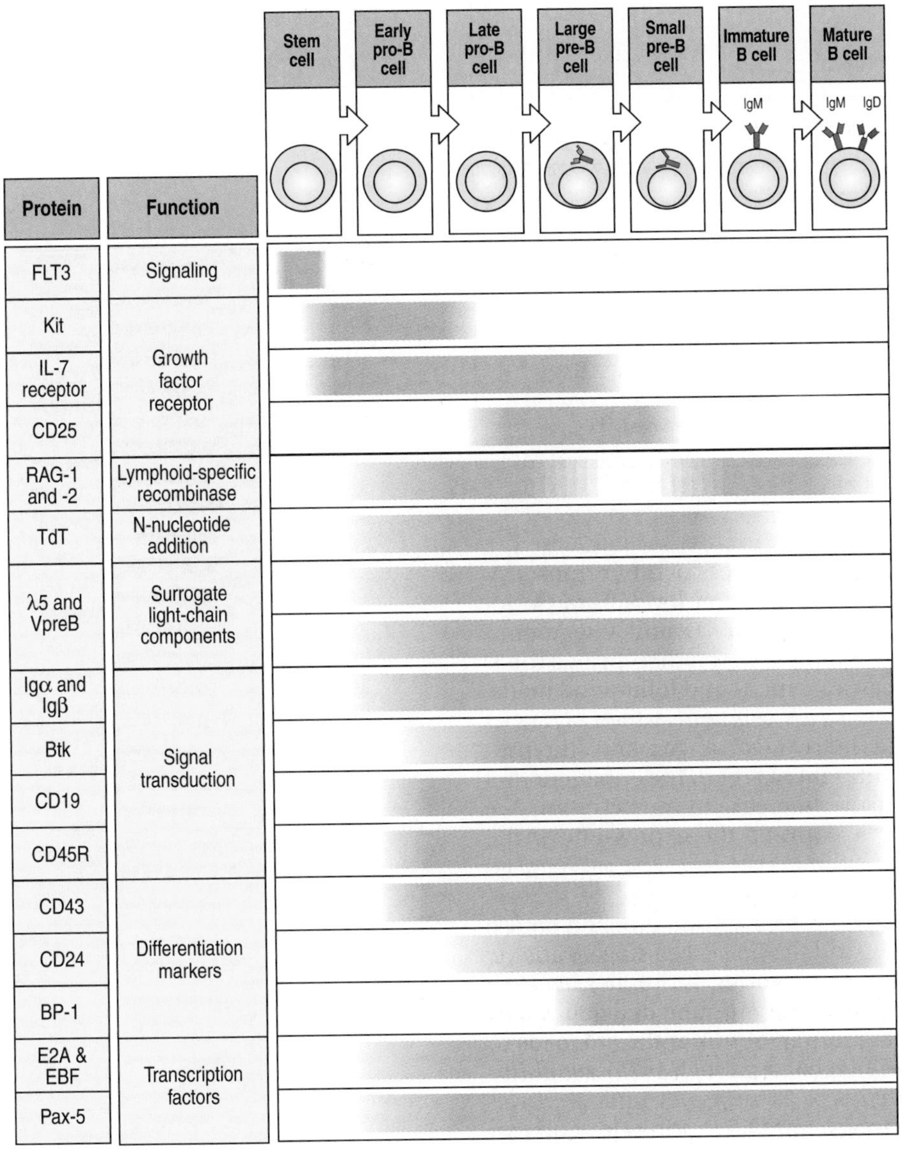

Figure 6.12 The expression of proteins involved in the rearrangement and expression of immunoglobulin genes changes during B-cell development. The rearrangement of immunoglobulin genes and the expression of the pre-B-cell receptor and cell-surface IgM requires several categories of specialized proteins at different times during B-cell development. Examples of such proteins are listed here, with their expression during B-cell development shown with red shading. FLT3 is a cell-surface receptor protein kinase that receives signals from stromal cells for the differentiation of common lymphoid progenitors. CD19 is a subunit of the B-cell co-receptor, which will be needed to cooperate with the antigen receptor to produce a strong signal that antigen has bound. CD45 is a cell-surface protein phosphatase that is also involved in modulating signals from an antigen-bound B-cell receptor. CD43, CD24, and BP-1 are cell-surface markers that are useful for distinguishing different stages in B-cell development. The roles of the other proteins listed here are explained in the text.

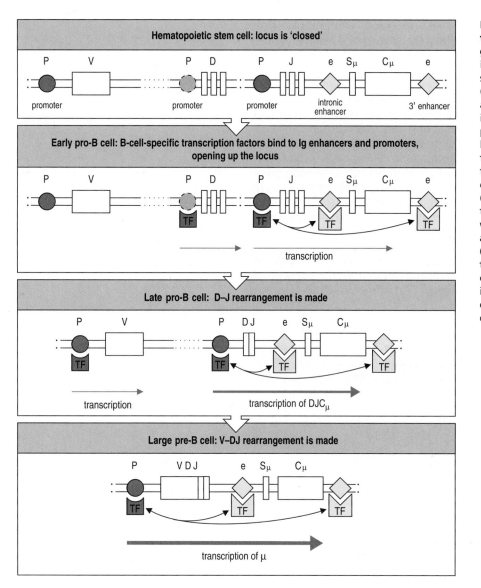

Figure 6.13 The rearrangement and transcription of immunoglobulin genes are linked. The heavy-chain locus is shown here. The blocks of V, D, and J segments each have their own promoter (P), and in addition two enhancers (e) are located flanking the C_μ gene and its isotype switch signal sequence S_μ (top panel). With commitment to the B-cell lineage the locus is opened up, allowing transcription factors (TF) such as Pax-5 to bind to promoter and enhancer elements and initiate transcription (second panel). This in turn facilitates the gene rearrangement reactions, which juxtapose regulatory elements and increase the levels of transcription (third and fourth panels). The levels of transcription are indicated by the width of the horizontal red arrows. Synergistic interactions between promoters and enhancers are shown by the curved double-headed black arrows.

genes. This opens up the chromatin and allows the access of other transcription factors that are not B-cell specific (Figure 6.13). A low level of transcription now occurs from the promoters upstream of the D and J segments. Transcription facilitates access of the RAG complex, which then brings the D and J segments together and allows the rearrangement and joint to be made. In turn, this initiates transcription upstream of a V segment, which favors its rearrangement to DJ. As a consequence of the two rearrangements, the promoter that is 5′ of the V-region segment is brought nearer to the enhancer that is 3′ of the C-region gene, resulting in a further increase in transcription. We can therefore see that gene rearrangement controls the expression of the immunoglobulin genes as well as generating their diversity.

One of the signaling proteins shown in Figure 6.12, Bruton's tyrosine kinase (Btk), is essential for B-cell development beyond the pre-B-cell stage. Patients who lack a functional *Btk* gene have almost no circulating antibodies because their B cells are blocked at the pre-B-cell stage. The immune deficiency suffered by these patients is called **X-linked agammaglobulinemia** and leads to recurrent infections from common extracellular bacteria such as *Haemophilus influenzae, Streptococcus pneumoniae, Streptococcus pyogenes,* and *Staphylococcus aureus.* The infections respond to treatment with antibiotics and can be prevented by regular intravenous infusions of immunoglobulin from pooled blood from healthy donors. The *Btk* gene is located on the X chromosome and the defective gene is recessive, so X-linked agammaglobulinemia is seen mostly in boys.

6-9 Many B-cell tumors carry chromosomal translocations that join immunoglobulin genes to genes that regulate cell growth

As B cells cut, splice, and mutate their immunoglobulin genes in the normal course of events, it is hardly surprising that this process sometimes goes awry to produce a mutation that helps convert a B cell into a tumor cell. Transformation of a normal cell into a tumor cell involves a series of mutations that release the cell from the normal restraints on its growth. In B-cell tumors, the disruption of regulated growth is often associated with an aberrant immunoglobulin gene rearrangement that has joined an immunoglobulin gene to a gene on a different chromosome. Events that fuse part of a chromosome with another are called **translocations** and, in B-cell tumors, the immunoglobulin gene has often become joined to a gene involved in the control of cellular growth. Genes that cause cancer when their function or expression is perturbed are collectively called **proto-oncogenes**. Many were discovered in the study of RNA tumor viruses that can transform cells directly. The viral genes responsible for transformation were named **oncogenes**, and it was only later realized that they had evolved from cellular genes that control cell growth, division, and differentiation.

The translocations between immunoglobulin genes and proto-oncogenes in B-cell tumors can be seen in metaphase chromosomes examined in the light microscope. Certain translocations define particular types of tumor and are valuable in diagnosis. In Burkitt's lymphoma, for example, the *MYC* proto-oncogene on chromosome 8 is joined by translocation to either an immunoglobulin heavy-chain gene on chromosome 14, a κ light-chain gene on chromosome 2, or a λ light-chain gene on chromosome 22 (Figure 6.14). The Myc protein is normally involved in regulating the cell cycle, but in B cells that carry these translocations its expression is abnormal, which removes some of the restraints on cell division. Because single genetic changes are rarely sufficient to transform cells malignantly, the onset of Burkitt's lymphoma requires mutations elsewhere in the genome in addition to the translocation between *MYC* and an immunoglobulin gene.

Another translocation found in B-cell tumors is the fusion of an immunoglobulin gene to the proto-oncogene *BCL2*. Normally, the function of the Bcl-2 protein is to prevent premature apoptosis in B-lineage cells. Overproduction of Bcl-2 in the mutant B cells having this translocation enables them to live longer than normal, during which time they accumulate additional mutations that can lead to malignant transformation. In addition to their value in immunological research, B-cell tumors have provided fundamental advances in knowledge of the proteins involved in the mechanism and regulation of cell division.

6-10 B cells expressing the glycoprotein CD5 express a distinctive repertoire of receptors

Not all B cells conform exactly to the developmental pathway described above. A subset of human B cells that arise early in embryonic development are distinguished from other B cells by the expression of CD5, a cell-surface glycoprotein that is otherwise considered a marker for the human T-cell lineage. B cells in this minority subset are termed **B-1 cells** because their development precedes that of the majority subset, whose development we have traced in

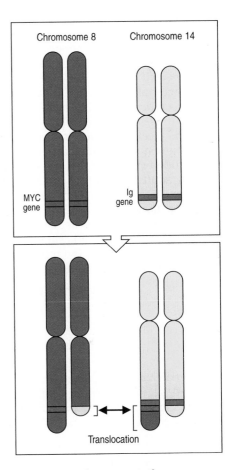

Figure 6.14 Chromosomal rearrangements in Burkitt's lymphoma. In this example from a Burkitt's lymphoma, parts of chromosome 8 and chromosome 14 have been exchanged. The sites of breakage and rejoining are in the proto-oncogene *MYC* on chromosome 8 and the immunoglobulin heavy-chain gene on chromosome 14. In these tumors it is usual for the second immunoglobulin gene to be productively rearranged and for the tumor to express cell-surface immunoglobulin. The translocation probably occurred during the first attempt to rearrange a heavy-chain gene. This counted as a nonproductive rearrangement and so the other gene was rearranged. In cases where the second rearrangement is also nonproductive, the cell dies and thus cannot give rise to a tumor.

previous sections and which are sometimes called **B-2 cells**. B-1 cells are also characterized by little or no IgD on the surface and a distinctive repertoire of antigen receptors. The B-1 cells are also known as **CD5 B cells**. Although comprising only around 5% of the B cells in humans and mice, B-1 cells are the main type of B cell in rabbits.

B-1 cells arise from a stem cell that is most active in the prenatal period. Their immunoglobulin heavy-chain gene rearrangements are dominated by the use of the V_H gene segments that lie closest to the D gene segments in the germline. As TdT is not expressed early in the prenatal period, the rearranged heavy-chain genes of B-1 cells are characterized by a lack of N nucleotides, and their VDJ junctions are less diverse than those in rearranged heavy-chain genes of B-2 cells. Consequently, the antibodies secreted by B-1 cells tend to be of low affinity and each binds to many different antigens, a property known as **polyspecificity**. B-1 cells contribute to the antibodies made against common bacterial polysaccharides and other carbohydrate antigens but are of little importance in making antibodies against protein antigens.

Those B-1 cells that develop postnatally use a more diverse repertoire of V gene segments and their rearranged immunoglobulin genes have abundant N nucleotides. With time, B-1 cells are no longer produced by the bone marrow and, in adults, the population of B-1 cells is maintained by the division of existing B-1 cells at sites in the peripheral circulation. This self-renewal is dependent on the cytokine IL-10. Most cases of chronic lymphocytic leukemia (CLL) are caused by B-1 cells, a propensity that might stem directly from their capacity for self-renewal. The properties of B-1 and B-2 cells are compared in Figure 6.15. In the remainder of this chapter, and in the rest of this book, B cell will refer to the population of B-2 cells, unless otherwise specified.

Property	B-1 cells	Conventional B-2 cells
When first produced	Fetus	After birth
N-regions in VDJ junctions	Few	Extensive
V-region repertoire	Restricted	Diverse
Primary location	Peritoneal and pleural cavities	Secondary lymphoid organs
Mode of renewal	Self-renewing	Replaced from bone marrow
Spontaneous production of immunoglobulin	High	Low
Isotypes secreted	IgM >> IgG	IgG > IgM
Requirement for T-cell help	No	Yes
Somatic hypermutation	Low–none	High
Memory development	Little or none	Yes

Figure 6.15 Comparison of the properties of B-1 cells and B-2 cells. B-1 cells develop in the omentum, a part of the peritoneum, as well as in the liver in the fetus, and are produced by the bone marrow for only a short period around the time of birth. A pool of self-renewing B-1 cells is then established, which does not require the microenvironment of the bone marrow for its survival. The limited diversity of the antibodies made by B-1 cells and their tendency to be polyspecific and of low affinity suggests that B-1 cells produce a simpler, less adaptive immune response than that involving B-2 cells.

Summary

B cells are generated throughout life from stem cells in the bone marrow. Different stages in B-cell maturation are correlated with molecular changes that accompany the gene rearrangements required to make functional immunoglobulin heavy and light chains. These changes are summarized in Figure 6.16. Many rearrangements are nonproductive, an inefficiency compensated for, in part, by the presence of two heavy-chain loci and four light-chain loci in each B cell, all of which can be rearranged in the attempt to obtain a functional heavy chain and light chain. The heavy-chain genes are rearranged first, and only when this produces a functional protein is a B cell allowed to rearrange its light-chain genes. After a successful heavy-chain gene rearrangement, the μ chain and surrogate light chain form the pre-B-cell receptor, which signals the cell to halt heavy-chain gene rearrangement and to proceed toward light-chain gene rearrangement. Similarly, after a successful light-chain gene rearrangement, the assembly of IgM signals the cell to halt light-chain gene rearrangement. These feedback mechanisms, whereby protein product regulates the progression of gene rearrangement, ensure that each B cell expresses only one heavy chain and one light chain and thus produces immunoglobulin of a single defined antigen specificity. In this manner, the program of gene rearrangement produces the monospecificity essential for the efficient operation of clonal selection. The success rate in obtaining a functional light-chain gene rearrangement is much higher than that for the heavy-chain gene because only two gene segments are involved (compared with three for the heavy-chain gene) and the B cell has four light-chain genes (two κ and two λ) with which to work. Errors in immunoglobulin gene rearrangement give rise to chromosomal translocations that predispose the B cells carrying them to malignant transformation. A minority class of B cells, termed B-1 cells, develops very early in embryonic life. They produce antibodies that tend to bind to bacterial polysaccharides, but are polyspecific,

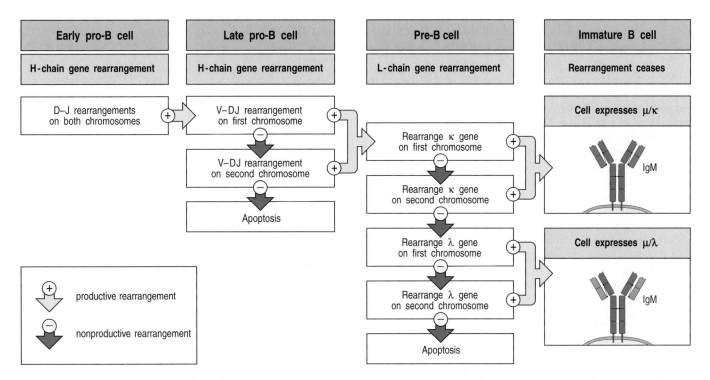

Figure 6.16 Summary of the order of gene rearrangements leading to the expression of cell-surface immunoglobulin. The heavy-chain genes are rearranged before the light-chain genes. Developing B cells are allowed to proceed to the next stage only when a productive rearrangement has been made. If a nonproductive rearrangement is made on one chromosome of a homologous pair, then rearrangement is attempted on the second chromosome.

generally of low affinity, and mainly of the IgM isotype. In aggregate, the properties of B-1 cells suggest that they represent a simpler and evolutionarily older lineage of B cell than the majority of the B-cell population.

Selection and further development of the B-cell repertoire

In the previous part of this chapter we saw how the population of immature B cells acquires a vast repertoire of receptors with different antigen specificities. This repertoire includes some immunoglobulins that bind to normal constituents of the human body and have the potential to initiate a damaging immune response. Receptors and cells that react against a person's own tissues are described as being **self-reactive** or **autoreactive**, and the components to which they bind are called **self antigens**. To prevent such responses, self-reactive immature B cells that encounter their self antigens either in the bone marrow or in the peripheral circulation are prevented from advancing from the immature B-cell stage to the mature B-cell stage.

In an immature B cell that does not react with a self antigen, alternative mRNA splicing of heavy-chain gene transcripts then produces the δ chain as well as the μ chain. This leads to the presence on the B-cell surface of both IgD and IgM (see Section 4-10). At this stage the immature B cells travel to secondary lymphoid organs, where they compete with each other to enter the primary follicles and complete their maturation. Competition is stiff, and only a minority of immature B cells become **mature B cells** that are competent to respond to their specific antigens. In this process of maturation, the B-cell receptor becomes capable of generating positive signals upon binding specific antigen. Only a tiny fraction of the mature B-cell population will ever meet their corresponding antigen and carry out their effector function of making antibodies against a pathogenic microorganism.

6-11   The population of immature B cells is purged of cells bearing self-reactive B-cell receptors

The mechanisms of gene rearrangement and somatic mutation described in the first part of this chapter produce a population of immature B cells that express surface IgM embracing a wide range of antigenic specificities. This repertoire of immature B-cell receptors includes many with affinity for self antigens that are components of healthy human tissue. Activation of mature B cells carrying such receptors by these antigens would produce self-reactive and potentially harmful antibodies that could disrupt the body's normal functions and cause disease. To prevent this from happening, the B-cell receptors of immature B cells are wired to generate negative signals within the cell when they bind to antigen. These signals cause the B cell either to die by apoptosis or to become inactivated. In this way, the immature B-cell population is subject to a negative selection process that prevents the maturation of self-reactive B cells. As a consequence, the resulting population of mature B cells in a healthy individual does not respond to self antigens and is said to be **self-tolerant**. As many as 75% of immature B cells have some affinity for self antigens, and so once they have been eliminated, the receptor repertoire of mature B cells is substantially different from that of immature B cells.

The lives of B cells are spent largely in the blood, the lymph, the secondary lymphoid organs, and the bone marrow. In these tissues, B-cell receptors can interact with molecules exposed on cell surfaces and in connective tissue and with soluble components of blood plasma. Because of their accessibility, these self antigens are the ones most likely to provoke an autoimmune B-cell response. Multivalent antigens (see Figure 4.10, p. 100) are most effective at

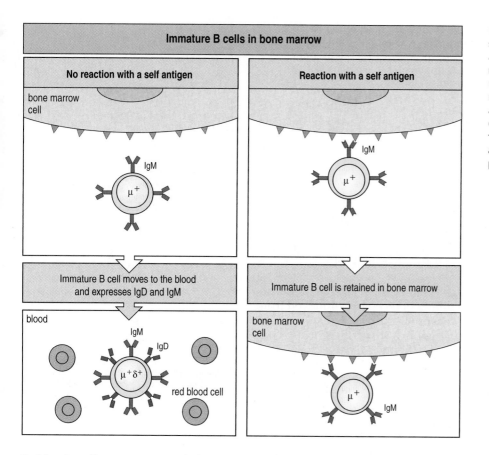

Figure 6.17 Immature B cells with specificity for multivalent self antigens are retained in the bone marrow. Immature B cells that are not specific for a self antigen in bone marrow mature further to express IgM and IgD and leave the bone marrow (left panels). Immature B cells specific for a self antigen on bone marrow cells are retained in the bone marrow (right panels).

linking B-cell receptors together and activating B cells, and many of the abundant glycoproteins, proteoglycans, and glycolipids present on human cells can act as multivalent self antigens.

To prevent the emergence of mature B cells specific for these common multivalent self antigens, the immature B-cell population is subject to a selection process that identifies such B cells and prevents their maturation. This selection starts in the bone marrow, where immature B cells are exposed to the wide variety of self antigens expressed by stromal cells, hematopoietic cells and macromolecules circulating in the blood plasma. Only immature B cells bearing receptors that do not interact with any of these self antigens are allowed to leave the bone marrow and continue their maturation in the peripheral circulation (Figure 6.17). Maturation first involves the alternative splicing of the heavy-chain mRNA so that the cell makes IgD as well as IgM (see Section 4-10, p. 110). This is followed by a shift from an excess of IgM over IgD on the cell surface to an excess of IgD over IgM in the mature B cell.

In contrast, immature B cells with receptors that bind a multivalent self antigen are signaled to arrest their developmental progression. They are retained in the bone marrow and given the opportunity to lose their self-reactivity for self antigen by altering their B-cell receptor.

6-12 The antigen receptors of autoreactive immature B cells can be modified by receptor editing

If the cell-surface IgM on an immature B cell binds a multivalent self antigen in the bone marrow, this signals the B cell to reduce the amount of IgM on its surface and to maintain expression of the RAG proteins, which permits the B cell to continue rearranging its light-chain loci. This further rearrangement leads to excision of the existing rearrangement, and so the original light chain

Figure 6.18 Many self-reactive B cells are rescued by receptor editing, which changes their antigen specificity. Binding to a self antigen in the bone marrow causes the immature B cell to continue to rearrange its light-chain locus. This deletes the existing self-reactive rearrangement and may also produce a new light-chain rearrangement giving immunoglobulin that is not specific for self antigen. Successive rearrangements occur until either a new non-self-reactive rearrangement is made, and the B-cell continues its development (right panels), or no more rearrangements are possible and the B cell dies (left panels).

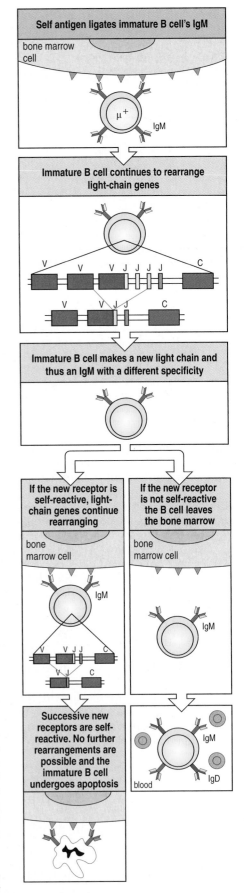

is no longer made. Continuing rearrangement may also produce a new functional light chain that assembles with the old heavy chain to make a new B-cell receptor. If this new receptor does not react with a self antigen, then the B cell can continue its developmental progression into a mature B cell.

If self-reactivity is retained, then further light-chain gene rearrangements are tried. This process of assessing the compatibility of receptors produced from successive gene rearrangements is called **receptor editing** (Figure 6.18). The multiplicity of V and J segments in the κ and λ light-chain genes provides considerable capacity for receptor editing. Inevitably, some B cells exhaust all the possibilities for light-chain gene rearrangement without ever gaining a receptor that is not self-reactive. These cells are signaled to die by apoptosis in the bone marrow and are phagocytosed by macrophages. The selective death of developing lymphocytes, and the consequent removal of their self-reactive receptor specificities from the B-cell repertoire, is called **clonal deletion**. Some 55 billion B cells die each day in the bone marrow because they fail to make a functional immunoglobulin or because they are autoreactive and subject to clonal deletion.

By enabling immature B cells to exchange a self-reactive light chain for one that is not self-reactive, receptor editing increases the overall efficiency of B-cell development and the probability that successful heavy-chain gene rearrangements will not be lost from the mature B-cell receptor repertoire. Numerous B cells are rescued by receptor editing, which constitutes the dominant mechanism for establishing the self-tolerance of the developing B-cell population.

6-13 Immature B cells specific for monovalent self antigens are made nonresponsive to antigen

In contrast with the multivalent self antigens, many other common self antigens are soluble proteins that carry only one copy of an epitope. Self-reactive B cells that bind to monovalent self antigens are not signaled to continue light-chain gene rearrangement or to die by apoptosis. Instead they become inactivated and unresponsive to their specific antigen. This state of developmental arrest is called **anergy** (Figure 6.19). Anergic cells make both IgD and IgM, but unlike in mature B cells, the IgM is prevented from assembling a functional B-cell receptor and is largely retained within the cell. Although normal amounts of IgD are present on the cell surface, they do not activate the B cell on binding to antigen. Anergic B cells are allowed to enter the peripheral circulation but their life-span of 1–5 days is short compared with the approximately 40-day half-life of mature B cells (cellular life-span is often expressed as the **half-life**, which is the period of time over which a population of cells reduces to half its original size).

B cells reactive to self antigens in the bone marrow therefore experience one of three fates. They either survive through receptor editing, which gets rid of their self-reactivity, die by apoptosis, or become anergic. These three mechanisms ensure that the B cells leaving the bone marrow are tolerant to all self antigens present in the bone marrow, many of which are also present in other tissues. This type of immunological tolerance is called **central tolerance** because it is developed in a primary lymphoid organ—the bone marrow in the case of B cells.

Not all self antigens are encountered in the bone marrow. Most organs and tissues of the body express tissue-specific cell-surface proteins, secreted proteins and other antigens that are accessible to circulating B cells. An immature B cell tolerant toward bone-marrow self antigens but reactive to a self antigen in the blood will encounter its antigens soon after leaving the bone marrow. Receptor editing is not an option for these cells, because by this stage the machinery for rearranging their immunoglobulin genes has been shut down permanently. Instead, self-reactive B cells outside the bone marrow either die by apoptosis or are rendered anergic when they encounter their self antigen. Tolerance induced to antigens outside the bone marow is called **peripheral tolerance**, and it removes circulating B cells reactive against the self antigens of tissues other than bone marrow.

Central and peripheral tolerance do not remove B cells that are reactive to self antigens present in places inaccessible to B cells, such as the insides of cells, or are present in amounts insufficient to trigger the B-cell to undergo apoptosis or become anergic. In situations of stress, disease, or trauma, self antigens to which the B-cell population has not become tolerant can become accessible and provoke an autoimmune response. One example is DNA, antibodies against which are associated with the autoimmune disease systemic lupus erythematosis.

6-14   Maturation and survival of B cells requires access to lymphoid follicles

After developing B cells leave the bone marrow, they begin to recirculate between the blood, the secondary lymphoid tissues and the lymph (see Figure 1.18, p. 17). At this stage of development the B cells are still not fully mature; they express high levels of surface IgM and low levels of surface IgD, whereas mature B cells have low surface IgM and high IgD. The final stage in B-cell maturation occurs when the immature B cells enter a secondary lymphoid tissue. Because the lymph node, spleen, Peyer's patches and other secondary lymphoid organs have a similar microanatomy and function, we focus here on the maturation of B cells as they circulate through a lymph node (Figure 6.20). B cells enter lymph nodes from the blood through the walls of high endothelial venules; stromal cells in the lymph node cortex secrete a chemokine called **CCL21**, for which B cells express a receptor, called **CCR7**. The mechanism underlying this homing of naive B cells to secondary lymphoid tissues is essentially the same as that used by neutrophils to enter infected tissue (see Figure 2.31) but uses different cell adhesion molecules and chemokines. Dendritic cells within the lymph node also secrete CCL21 and another chemokine, **CCL19**, to which B cells are responsive. The gradient of chemokines attracts blood-borne B cells to the high endothelial venules and they enter the lymph node by squeezing between the high endothelial cells (Figure 6.21).

Once inside the lymph node, B cells are further guided by chemokines to congregate in organized structures called **primary lymphoid follicles**. These areas consist principally of B cells enmeshed in a network of specialized stromal cells called **follicular dendritic cells** (**FDCs**). Despite their name, which comes from the long cell processes with which they touch B cells and each other, FDCs are not of hematopoietic origin and are unrelated to either the conventional dendritic cells that present antigens to T cells or the plasmacytoid dendritic cells that synthesize type I interferons. FDCs attract B cells into

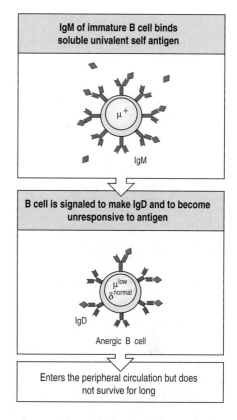

Figure 6.19 Immature B cells specific for monovalent self antigens develop a state of anergy.

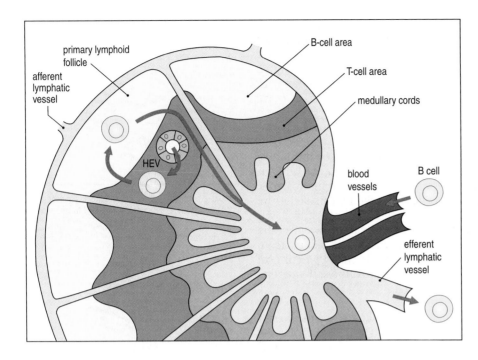

primary lymphoid follicle

afferent lymphatic vessel

B-cell area

T-cell area

medullary cords

HEV

blood vessels

B cell

efferent lymphatic vessel

stromal cell

Figure 6.20 The general route of B-cell circulation through a lymphoid tissue. B cells circulating in the blood enter the lymph node cortex via a high endothelial venule (HEV). From there they pass into a primary lymphoid follicle. If they do not encounter their specific antigen, they leave the follicle and exit from the lymph node in the efferent lymph. The circulation route is the same for immature and mature B cells, which all compete with each other to enter primary follicles.

the follicle by secreting the chemokine **CXCL13** (see Figure 6.21). The interaction between B cells and FDCs is mutually beneficial, providing signals that allow B cells to mature and survive while also preserving the integrity of the FDC network. The latter is mediated by a surface protein on B cells called lymphotoxin (LT), which is structurally related to the inflammatory cytokine tumor necrosis factor-α (TNF-α) and binds to a receptor on FDCs. Another member of the TNF family called **BAFF** (B-cell activating factor in the TNF family) also promotes B-cell survival. BAFF is made by several cell types in secondary lymphoid organs and is bound by the BAFF receptor expressed by B cells.

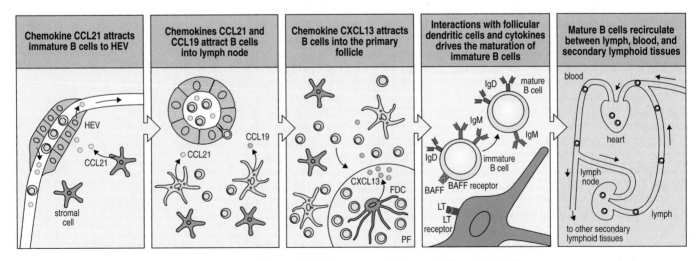

Chemokine CCL21 attracts immature B cells to HEV

Chemokines CCL21 and CCL19 attract B cells into lymph node

Chemokine CXCL13 attracts B cells into the primary follicle

Interactions with follicular dendritic cells and cytokines drives the maturation of immature B cells

Mature B cells recirculate between lymph, blood, and secondary lymphoid tissues

Figure 6.21 Immature B cells must pass through a primary follicle in a secondary lymphoid tissue to become mature B cells. Immature B cells enter the secondary lymphoid tissue through the walls of HEVs, attracted by the chemokines CCL21 and CCL19, and compete with other immature and mature B cells to enter a primary follicle (PF). Follicular dendritic cells (FDCs, turquoise) secrete the chemokine CXCL13, which attracts B cells

into the follicle. Those immature B cells that enter a follicle interact with proteins on FDCs that signal their final maturation into mature B cells. If they do not encounter their specific antigen in the follicle, mature B cells leave the lymph node and continue recirculating through the secondary lymphoid tissues via lymph and blood. Immature B cells that fail to enter a follicle also continue recirculating but soon die.

In the absence of its specific antigen, the B cell detaches from the FDC and leaves the lymph node in the efferent lymph to continue its circulation (see Figure 6.21). It is now a mature B cell, and if it meets its specific antigen it will respond by proliferating and undergoing differentiation into antibody-producing plasma cells, rather than becoming anergic or undergoing apoptosis. Mature B cells that have not yet encountered their antigen are called **naive B cells**. The survival of circulating mature naive B cells is dependent upon regular passage through the primary follicles of secondary lymphoid tissues.

Throughout life the production of B cells is enormous. In the bone marrow of a healthy young adult, about 2.5 billion (2.5×10^9) cells per day embark on the program of B-cell development. From the progeny of these progenitors, some 30 billion B cells leave the bone marrow to join the circulation. These immature B cells have to compete with the existing population of mature, circulating B cells for access to a limited number of follicular sites; those B cells that fail to gain regular access to a follicle die. This competition is weighted in favor of mature B cells and is so intense that the majority of the immature B cells fail to enter a follicle and die by apoptosis after only a few days in the peripheral circulation. Immature B cells that gain access to primary follicles and become mature B cells live longer, but they too disappear from the system with a half-life of around 100 days, unless they are activated by encounter with their specific antigen.

To reach the primary lymphoid follicles, B cells entering from the blood must pass through areas where T cells congregate (see Figure 6.20). Although anergic B cells can enter lymph nodes, they fail to reach a primary follicle and instead concentrate at the boundary between the follicle and the T-cell zone. This exclusion from the follicle means they do not receive the necessary survival signals and they eventually die by apoptosis.

6-15 Encounter with antigen leads to the differentiation of activated B cells into plasma cells and memory B cells

Secondary lymphoid tissues provide the sites where mature, naive B cells encounter specific antigen. When that happens, the antigen-specific B cells are detained in the T-cell areas and become activated by antigen-specific CD4 helper T cells, as briefly touched on in Chapter 1. These T cells provide signals that activate the B cells to proliferate and differentiate further. In lymph nodes and spleen, some of the activated B cells proliferate and differentiate immediately into **plasma cells**, which secrete IgM antibody (see Figure 6.21). Antibody secretion is effected by a change in the processing of the heavy-chain mRNA, which leads to the synthesis of the secreted form of immunoglobulin rather than the membrane-bound form (see Section 4-13, p. 113).

Plasma cells are completely specialized for the continuous synthesis and secretion of antibody; the cellular organelles involved in protein synthesis and secretion become highly developed, and 10–20% of the total cell protein made is antibody. As part of this program of terminal differentiation, plasma cells cease to divide and no longer express cell-surface immunoglobulin and MHC class II molecules. They therefore become unresponsive to antigen and to interaction with T cells.

Other activated B cells migrate to a nearby primary follicle, which changes its morphology to become a **secondary lymphoid follicle** containing a **germinal center** (Figure 6.22). Here, the activated B cells become large proliferating lymphoblasts called **centroblasts**; these mature into more slowly dividing B cells called **centrocytes**, which have undergone isotype switching and somatic hypermutation. Those B cells that make surface immunoglobulins with the highest affinity for the antigen are selected by the process of affinity

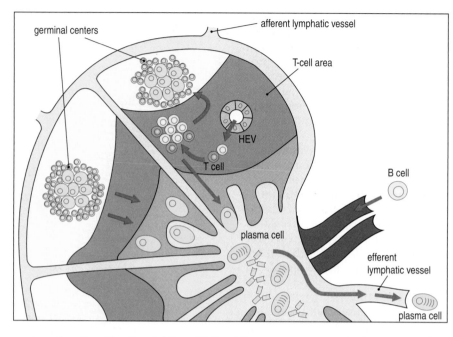

Figure 6.22 Mature B cells encountering antigen in secondary lymphoid tissues form germinal centers and undergo differentiation to plasma cells. A lymph node is illustrated here. A mature naive B cell entering the lymph node through an HEV and circulating through the lymph node encounters antigen within the cortex. Antigen is delivered in the afferent lymph that drains from infected tissue. The B cell is activated by CD4 helper T cells (blue) at the border between the follicle and the T-cell areas to form a primary focus of dividing cells. From this, some B cells migrate directly to the medullary cords and differentiate into antibody-secreting plasma cells. Other B cells migrate into a primary follicle to form a germinal center. B cells continue to divide and differentiate within the germinal center. Activated B cells migrate from the germinal center to the medulla of the lymph node or to the bone marrow to complete their differentiation into plasma cells.

maturation (see Section 4-14, p. 114), which occurs in germinal centers. A few weeks after a germinal center is formed, the intense cellular activity, called the germinal center reaction, dies down and the germinal center shrinks in size.

Cells that survive the selection process of affinity maturation undergo further proliferation as lymphoblasts and most migrate from the germinal center to other sites in the secondary lymphoid tissue or to bone marrow, where they complete their differentiation into plasma cells secreting high-affinity, isotype-switched antibodies. As the primary immune response subsides, germinal center B cells also develop into quiescent resting **memory B cells** possessing high-affinity, isotype-switched antigen receptors. The production of memory cells after a successful encounter with antigen establishes antigen specificities of proven usefulness permanently in the B-cell repertoire.

Memory B cells persist for long periods, and in their recirculation through the body they require only intermittent stimulation in the follicular environment. They are much more easily activated on encountering antigen than are naive B cells. Their rapid activation and differentiation into plasma cells on a subsequent encounter with antigen enable a secondary antibody response to an antigen to develop more quickly and become stronger than the primary immune response. It also explains why IgG and antibodies of isotype other than IgM predominate in secondary responses.

When B cells become committed to differentiation into plasma cells, they migrate to particular sites in the lymphoid tissues. In the lymph nodes these are the medullary cords and in the spleen these are the red pulp. In the gut-associated lymphoid tissues, prospective plasma cells migrate to the lamina propria, which lies immediately under the gut epithelium. Prospective plasma cells also migrate from lymph nodes and spleen to the bone marrow, which becomes a major site of antibody production. So, in one sense, the life of a B cell both starts and ends in the bone marrow.

6-16 Different types of B-cell tumor reflect B cells at different stages of development

The study of B-cell tumors has provided fundamental insights into both B-cell development and the control of cell growth generally. B-cell tumors arise from both the B-1 and B-2 lineages and from B cells at different stages of maturation and differentiation. The general principle that a tumor represents the uncontrolled growth of a single transformed cell is illustrated vividly by tumors derived from B-lineage cells. In a B-cell tumor, every cell has an identical immunoglobulin-gene rearrangement, which is proof of their derivation from the same ancestral cell. Although the cells of an individual B-cell tumor are homogeneous, the B-cell tumors from different patients have different rearrangements, which reflects the multiplicity of rearrangements found in the normal B cells of a healthy person (Figure 6.23).

Tumors retain characteristics of the cell type from which they arose, especially when the tumor is relatively differentiated and slow growing. This principle is exceptionally well illustrated by the B-cell tumors. Human tumors corresponding to all the stages of B-cell development have been described, from the most immature progenitor to the highly differentiated plasma cell (Figure 6.24). Among the characteristics retained by the tumors is their location at defined sites in the lymphoid tissues. Tumors derived from mature, naive B cells grow in the follicles of lymph nodes and form **follicular center cell lymphoma**, whereas plasma-cell tumors, called **myelomas**, propagate in the bone marrow.

Although **Hodgkin's disease** was one of the first tumors to be successfully treated by radiotherapy, the tumor's origin as a germinal center B cell was

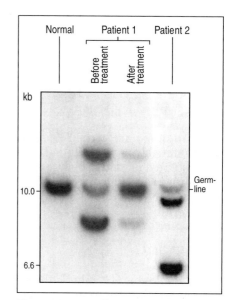

Figure 6.23 B-cell tumors are clonal in origin. The photograph shows agarose gel electrophoresis of a restriction enzyme digest of DNA samples from blood cells, which has then been probed to reveal the fragments containing the immunoglobulin heavy-chain J region. For a normal healthy person (left lane) only the band corresponding to the germline configuration is seen and it is due mainly to cell types other than B cells. The diversity of gene rearrangements in the B cells in the sample is such that the bands due to any individual rearrangement are too faint to be seen. By contrast, patients with a B-cell tumor (patient 1 and patient 2) give three bands, one corresponding to the unrearranged heavy-chain genes in non-B cells and two that correspond to rearrangement of the two heavy-chain gene alleles in the abundant clonal population of tumor cells. For each patient the two alleles are rearranged differently, thus giving bands in different positions. Anti-tumor treatment given to patient 1 is seen to reduce the tumor load, as judged by the relative intensities of the tumor-derived bands and the band representing the germline. Image courtesy of T. Vulliamy and L. Luzzatto.

Name of tumor	Normal cell equivalent and stage in development		Location	Status of Ig V genes
Acute lymphoblastic leukemia (ALL)	Lymphoid progenitor		Bone marrow and blood	Unmutated
Pre-B-cell leukemia	Pre-B cell	pre-B receptor		Unmutated
Mantle cell lymphoma	Resting, naive B cell			Unmutated
Chronic lymphocytic leukemia	Activated or memory B cell			Usually unmutated
Follicular center cell lymphoma / Burkitt's lymphoma	Mature, memory B cell / Resembles germinal center B cell		Periphery	Mutated, intraclonal variability
Hodgkin's lymphoma	Germinal center B cell			Mutated +/− intraclonal variability
Waldenström's macroglobulinemia	IgM-secreting B cell			Mutated, no variability within clone
Multiple myeloma	Plasma cell. Various isotypes		Bone marrow	Mutated, no variability within clone

Figure 6.24 The different B-cell tumors reflect the heterogeneity of developmental and differentiation states in the normal B-cell population. Each type of tumor corresponds to a normal state of B-cell development or differentiation. Tumor cells have similar properties to their normal cell equivalent, they migrate to the same sites in the lymphoid tissues, and they have similar patterns of expression of cell-surface glycoproteins.

discovered only recently. As a result of a somatic mutation, the tumor cells no longer have an antigen receptor. They often have a strange dendritic morphology and have been known as Reed–Sternberg cells. The disease presents clinically in several forms. In some patients it is dominated by nonmalignant T cells that are stimulated by the tumor cells, whereas in others there are both Reed–Sternberg cells and more normal-looking malignant B cells that have identical immunoglobulin-gene rearrangements.

B-cell tumors have proved invaluable in the study of the immune system. They uniquely provide cells that largely represent what happens normally, but they can be obtained in a large quantity, which is highly abnormal. The very first amino acid sequences of antibody molecules were obtained from patients with plasma-cell tumors, whose bodily fluids are dominated by a single species of antibody.

Summary

B-cell development is inherently wasteful because the vast majority of B cells die without ever contributing to an immune response. The first part of this chapter described how only around one-half of developing B cells succeed in making immunoglobulin and reaching the stage of an immature B cell bearing IgM at its surface. In this second part we have seen that the immunoglobulins made by a majority of immature B cells are autoreactive and rather than being a benefit for human health and survival are a potential cause of autoimmune disease. In the bone marrow, receptor editing enables many of the autoreactive B cells specific for multivalent self antigens to change their light chains, thereby modifying their immunoglobulins so that they no longer react with self antigens; those that fail to find a suitable light chain are eliminated by apoptosis. Autoreactive B cells specific for univalent self antigens are rendered anergic, a state in which their signaling is impaired and they die by neglect. As the immature B cells move out of the bone marrow and into other tissues they encounter additional self antigens not present in the bone marrow. B cells specific for these self antigens die quickly by apoptosis or slowly by anergy, but cannot be saved by receptor editing. The surviving B cells, which now express a little surface IgD as well as much IgM, remain immature and to complete their maturation they must enter a primary lymphoid follicle and interact with follicular dendritic cells. The competition between B cells for access to the follicles is such that only around 20% of the immature B cells mature and survive beyond a few days.

The mature B cells, which now have a high level of IgD and a low level of IgM on their surfaces, travel between the secondary lymphoid tissues in search of infection and pathogen-derived antigens that will bind their B-cell receptors. If in a period of 50–100 days they fail to find specific antigen, the B cells die. Only if a B cell encounters specific antigen will its clone be expanded and maintained for the rest of a person's life. No sooner has the clone expanded than individual B cells in the clone begin to differentiate and diverge. Some become plasma cells, which service the immediate need for antigen-specific antibody, while others become memory cells providing long-term immunity for the future. Over and above these changes, B cells within the clone hypermutate their immunoglobulins (see Section 4-14, p. 114). Most of the B cells expressing mutant immunoglobulins are abandoned, and only the few making antibodies of highest affinity are selected to survive and provide the B-cell response to the pathogen.

Summary to Chapter 6

B lymphocytes are highly specialized cells whose sole function is to recognize foreign antigens by means of cell-surface immunoglobulins and then to differentiate into plasma cells that secrete antibodies of the same antigen specificity. Each B cell expresses immunoglobulin of a single antigen specificity, but as a population B cells express a diverse repertoire of immunoglobulins. This enables the B-cell response to any antigen to be highly specific. Immunoglobulin diversity is a result of the unusual arrangement and mode of expression of the immunoglobulin genes. In B-cell progenitors the immunoglobulin genes are in the form of arrays of different gene segments that can be rearranged in many different combinations. Gene rearrangement occurs in bone marrow and is independent of a B-cell encounter with specific antigen. The gene rearrangements needed for the expression of a functional surface immunoglobulin follow a program; the successive steps define the antigen-independent stages of B-cell development as shown in Figure 6.25. The success rate for individual rearrangements is far from optimal, but the use of a stepwise series of reactions allows the quality of the products to be tested at critical checkpoints after heavy-chain and light-chain gene rearrangement. Failure at any step leads to the withdrawal of positive signals and death of the B cell by apoptosis. Gene rearrangement is controlled so that only one functional heavy-chain gene and one functional light-chain gene are produced in each cell. Thus, individual B cells produce immunoglobulin of a single antigen specificity. For a short period after successful immunoglobulin-gene rearrangement, any interaction with specific antigen leads to the elimination or inactivation of the immature B cell, thus rendering the mature B-cell population tolerant of the normal constituents of the body.

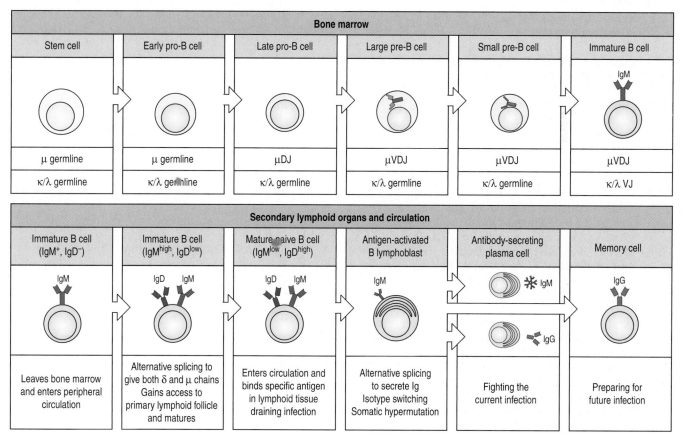

Figure 6.25 Summary of the main stages in B-cell development. The top panels summarize the early stages of development in the bone marrow. The status of the immunoglobulin heavy-chain (μ) and light-chain (κ/λ) genes is shown below each panel. The bottom panels summarize the development of B cells after they leave the bone marrow, enter secondary lymphoid tissues and are activated by pathogen-derived specific antigen. The diagram refers only to the development of B-2 cells.

Every day the bone marrow pushes out billions of new mature, naive B cells into the peripheral circulation. There is, however, limited space for B cells in the secondary lymphoid tissues and unless naive B cells encounter specific antigen they are likely to be dead within weeks. Thus, the B-cell repertoire is never static, and newly generated specificities are continually being tested against the antigens of microorganisms causing infection. Binding of antigen to the cell-surface immunoglobulin of a B cell in secondary lymphoid tissue initiates an antigen-dependent program of cellular proliferation and development and subsequent terminal differentiation, as shown in Figure 6.25. After encounter with antigen in secondary lymphoid tissues, B cells either differentiate directly into IgM-secreting plasma cells or undergo somatic hypermutation, isotype switching, and affinity maturation in germinal centers before differentiation into plasma cells or long-lived memory B cells. The end product of B-cell development is the plasma cell, in which surface immunoglobulin is no longer expressed and all the cell's resources are devoted to antibody secretion.

Questions

6–1   Place the following phases of a B cell's life history in the correct chronological order.
   a. Negative selection
   b. Attacking infection
   c. Finding infection
   d. Searching for infection
   e. Repertoire assembly
   f. Positive selection.

6–2   Place the following stages of B-cell development in the correct chronological order.
   a. Early pro-B cell
   b. Large pre-B cell
   c. Immature B cell
   d. Stem cell
   e. Late pre-B cell
   f. Small pre-B cell.

6–3
   A. Discuss the importance of the bone marrow stroma for B-cell development.
   B. What would be the effect of anti-IL-7 antibodies on the development of B cells in the bone marrow, and at which stage would development be impaired? Explain your answer.

6–4
   A. What are the two main checkpoints of B-cell development in the bone marrow?
   B. What is the fate of developing B cells that produce (i) functional or (ii) nonfunctional heavy and light chains?
   C. Explain how these two checkpoints correlate with the process of allelic exclusion that ensures that only one heavy-chain locus and one light-chain locus produce functional gene products.

6–5   What would be the consequence if terminal deoxynucleotidyl transferase (TdT) were expressed throughout the whole of small pre-B-cell development?

6–6   Which of the following would occur after the production of a functional μ chain as a pre-B-cell receptor?
   a. RAG proteins are degraded.
   b. The chromatin structure of the heavy-chain locus is reorganized to prevent gene rearrangement.
   c. Transcription of the *RAG1* and *RAG2* genes ceases.
   d. There is allelic exclusion of a second μ chain.
   e. All of the above are true.

6–7
   A. Give three properties that distinguish B-1 cells from B-2 cells.
   B. Do you think that B-1 cells should be categorized as participants in innate immune responses or as acquired immune responses? Explain your rationale.

6–8   Which of the following is true of centrocytes? (Select all that apply.)
   a. Somatic hypermutation has occurred.
   b. They are large proliferating cells.
   c. Isotype switching is complete.
   d. They produce secreted forms of immunoglobulins.
   e. They lack MHC class II molecules on the cell surface.

6–9   Which of the following statements regarding negative selection of B cells is correct?
   a. Negative selection is a process that occurs in secondary lymphoid organs.
   b. Negative selection is a process that occurs in the bone marrow but not in secondary lymphoid organs.
   c. Negative selection ensures that B cells bearing receptors for pathogens that will not be encountered in a person's lifetime are eliminated to make room for B cells bearing useful receptors.
   d. Negative selection eliminates B cells at the end of an infection as a means of terminating an immune

response once the pathogen has been removed from the body.
e. Negative selection ensures that autoreactive B cells are prohibited from emerging in the body.

6–10 Immunological tolerance in the B-cell repertoire is called _____ tolerance when it develops in primary lymphoid organs, and _____ tolerance when it is induced outside the bone marrow.
a. primary; secondary
b. apoptotic; anergic
c. stromal; follicular
d. receptor-mediated; systemic
e. central; peripheral.

6–11 What is the role of primary lymphoid follicles in eliminating B cells that have antigen receptors specific for soluble self antigen?

6–12 A plasma cell is characterized by which of the following features? (Select all that apply.)
a. It differentiates in the medulla of lymph nodes and the bone marrow.
b. It dedicates 10–20% of total protein synthesis to antibody production.
c. Levels of MHC class II molecules are elevated.
d. It undergoes extensive proliferation in germinal centers.
e. It produces secreted immunoglobulin instead of the membrane-bound form.

6–13
A. Explain why immunological memory is important in acquired immunity.
B. Describe how immunoglobulin expressed during a primary immune response differs qualitatively and quantitatively from the immunoglobulin expressed during a secondary immune response.

6–14 B-cell tumors originate during different developmental stages of B cells during their maturation in the bone marrow or after maturation and export to the periphery.
A. Explain why B cells isolated from a particular B-cell tumor all express the same immunoglobulin.
B. How might the immunoglobulin expressed on pre-B-cell leukemia cells be different from that expressed on immature B cells?

6–15 Yasuo Yamagata, a 63-year-old, experienced severe back pain for several weeks before visiting his family physician. He also complained of fatigue and looked pale. Blood analysis revealed a red blood cell count of 3.2 × 10^6/μl (normal 4.2–5.0 × 10^6/μl), a white blood cell count of 2800/μl (normal 5000/μl), a sedimentation rate of 30mm/h (normal <20mm/h) and a serum IgG of 4500 mg/dl (normal 600–1500 mg/dl). IgA and IgM levels were well below normal. Skeletal survey showed lytic lesions in vertebrae, ribs and skull. A bone marrow sample revealed 75% infiltration with plasma cells. Elevated protein in urine was confirmed to be Bence-Jones proteins (immunoglobulin light chains). The patient was diagnosed with IgG λ multiple myeloma and began an immediate chemotherapy regime. Which of the following would be consistent with this type of malignant tumor of plasma cells?
a. Serum IgG is polyclonal.
b. Anemia and neutropenia are present as the result of plasma-cell infiltration in the bone marrow and consequent limitation of space.
c. Susceptibility to pyogenic infections is unaffected.
d. Serum IgG consists of IgG1, IgG2, IgG3, and IgG4 in approximately equal proportions.
e. κ and λ light chains are found in excessive quantities in the urine.

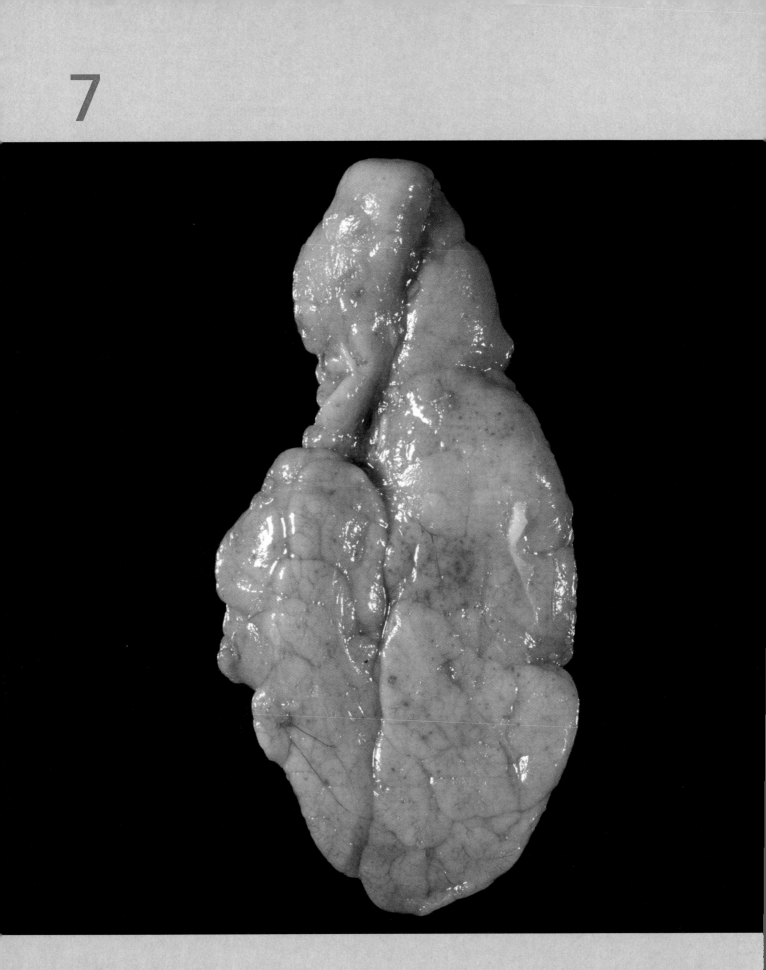

The thymus gland where T-cell development occurs.

Chapter 7

The Development of T Lymphocytes

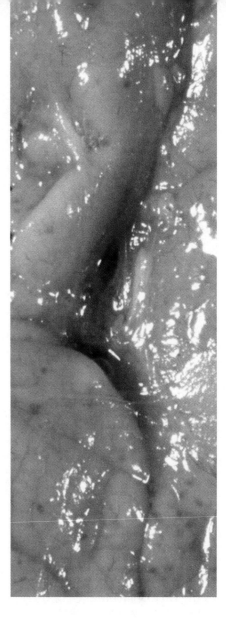

The paths of development of T and B lymphocytes have much in common: both types of cell derive from bone marrow stem cells and, during development, they must undergo gene rearrangement to produce their antigen receptors. However, whereas B cells rearrange their immunoglobulin genes while remaining in the bone marrow, the precursors of T cells have to leave the bone marrow and enter another primary lymphoid organ—the thymus—before they rearrange their T-cell receptor genes. Gene rearrangements in developing T cells proceed in a broadly similar fashion to those in B cells, but with some differences. The main difference is the formation of two distinct T-cell lineages: one expressing α:β receptors and the other γ:δ receptors (see Section 5-5, p. 130).

As we saw in Chapter 5, T-cell receptors do not recognize peptide antigens in isolation but in complexes with MHC molecules (see Section 5-14, p. 142). A major function of the thymus is to ensure that a person's mature T cells bear T-cell receptors that recognize peptides in the context of the particular MHC class I and class II isoforms expressed by that person, their self MHC. This is achieved by a process of positive selection in the thymus, which gives a survival signal to those immature T cells with receptors that interact with a self-MHC molecule and causes immature T cells that lack such receptors to die by neglect. The immature T cells chosen by positive selection then undergo an additional selective process, called negative selection, which induces the death of those T cells whose receptors bind too strongly to a self-MHC molecule and thus are autoreactive. Because of positive and negative selection, the mature T cells that leave the thymus to circulate through the secondary lymphoid organs are tolerant of self antigens, responsive to foreign antigens presented by self-MHC molecules, and ready to fight infection.

The first part of this chapter traces the stages in gene rearrangement that produce the primary repertoire of T-cell receptors. The second part of the chapter describes the processes of positive and negative selection that act on this repertoire in the thymus to produce the circulating population of mature naive T cells.

The development of T cells in the thymus

T cells are lymphocytes that originate from bone marrow stem cells but emigrate to mature in the thymus (Figure 7.1). With the discovery of this developmental pathway, these lymphocytes were called **thymus-dependent lymphocytes**, which soon became shortened to **T lymphocytes** or **T cells**. Two lineages of T cells develop in the thymus: the majority are α:β T cells and the

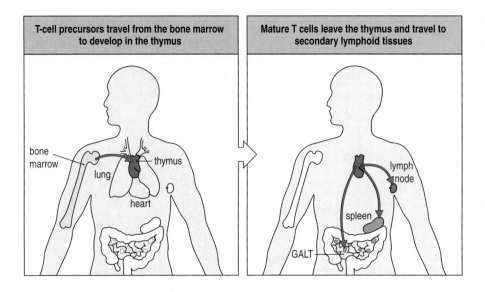

T-cell precursors travel from the bone marrow to develop in the thymus	Mature T cells leave the thymus and travel to secondary lymphoid tissues

Figure 7.1 **T-cell precursors migrate from the bone marrow to the thymus to mature.** T cells derive from bone marrow stem cells whose progeny migrate in the blood from the bone marrow to the thymus (left panel), where the development of T cells occurs. Mature T cells leave the thymus in the blood, from where they enter secondary lymphoid tissues (right panel) and then return to the blood in the lymph. In the absence of activation by specific antigen, mature T cells continue to recirculate between the blood, secondary lymphoid tissues, and lymph. GALT, gut-associated lymphoid tissue.

minority are γ:δ T cells. These lineages develop in parallel from a common **T-cell precursor**. During their development in the thymus, immature T cells also start to express other cell-surface glycoproteins related to their eventual functions in the immune response. Prominent among these are CD4 and CD8, the co-receptors that distinguish the two sublineages of α:β T cells that recognize peptide antigens presented by MHC class II and class I, respectively.

7-1   T cells develop in the thymus

The **thymus** is a lymphoid organ in the upper anterior thorax just above the heart. It is dedicated to T-cell development and contains immature T cells, called **thymocytes**, which are embedded in a network of epithelial cells known as the **thymic stroma** (Figure 7.2). Together with other cell types, these elements form an outer close-packed cortex and an inner, less dense, medulla (Figure 7.3). The thymus is designated a primary lymphoid organ because it is concerned with the production of useful lymphocytes, not with their application to the problems of infection. Unlike the secondary lymphoid organs, which perform the latter function, the thymus is not involved in lymphocyte recirculation; neither does it receive lymph from other tissues. The blood is the only route by which progenitor cells enter the thymus and by which mature T cells leave.

In the embryonic development of the thymus, the epithelial cells of the cortex arise from ectodermal cells, whereas those of the medulla derive from endodermal cells. Together, these two types of epithelial cell form a rudimentary thymus, called the **thymic anlage**, which subsequently becomes colonized by progenitor cells from the bone marrow. The progenitor cells give rise to thymocytes and to dendritic cells, the latter populating the medulla of the thymus (see Figure 7.3). Independently of these progenitors, the thymus is also colonized by bone marrow derived macrophages, which, although concentrated in the medulla, are also scattered throughout the cortex (see Figure 7.3). The T-cell progenitors enter the thymus at the junction between the cortex and the medulla. As they differentiate, the thymocytes first move out through the cortex to the subcapsular region and then move progressively back from the outer cortex to the inner cortex and the medulla.

The importance of the thymus in establishing a functional T-cell repertoire is clearly demonstrated by patients who have complete **DiGeorge's syndrome**. In this genetic disease, the thymus fails to develop, T cells are absent although

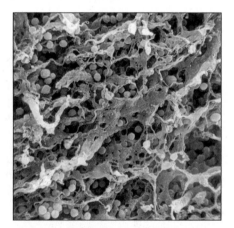

Figure 7.2 **The epithelial cells of the thymus form a network surrounding developing thymocytes.** In this scanning electron micrograph of the thymus, the developing thymocytes (the spherical cells) occupy the interstices of an extensive network of epithelial cells. Micrograph courtesy of W. van Ewijk.

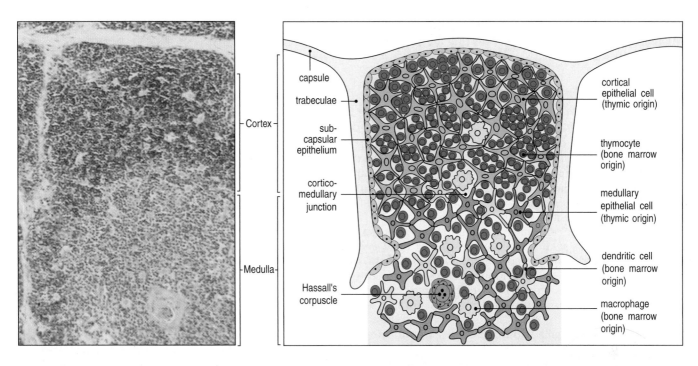

Figure 7.3 The cellular organization of the thymus. The thymus is made up of several lobules. A section through a lobule stained with hematoxylin and eosin and viewed with the light microscope is shown in the left panel. The cells in this view are shown in diagrammatic form in the right panel. In the left panel, the darker staining of cortex compared with the medulla can be discerned. As shown in the right panel, the cortex consists of immature thymocytes (blue), branched cortical epithelial cells (light orange), and a few macrophages (yellow). The medulla consists of mature thymocytes (blue), medullary epithelial cells (orange), dendritic cells (yellow), and macrophages (yellow). One of the functions of the macrophages in both cortex and medulla is to remove the many thymocytes that fail to mature properly. A characteristic feature of the medulla is Hassall's corpuscles, which are believed to be sites of cell destruction. Photograph courtesy of C.J. Howe.

B cells are made, and the resulting susceptibility to a wide range of opportunistic infections resembles that experienced by patients with severe combined immunodeficiency disease (SCID).

The human thymus is fully developed before birth and by one year after birth it begins to degenerate, with fat gradually claiming the areas once packed with thymocytes. This process, which continues steadily throughout life, is called the involution of the thymus (Figure 7.4). The reduced production of new T cells by the thymus with age does not grossly impair T-cell immunity, and neither does **thymectomy** (removal of the thymus) in adults. Once established, the repertoire of mature peripheral T cells seems to be long lived or

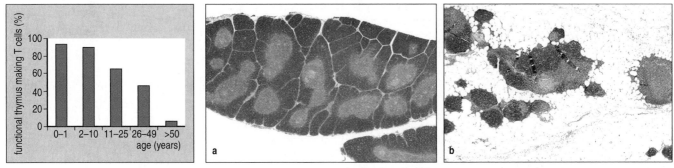

Figure 7.4 The proportion of the thymus that produces T cells decreases with age. Starting at birth, the T-cell-producing tissue of the thymus is gradually replaced by fatty tissue. This process is called the involution of the thymus. The graph shows the percentage of thymic tissue that is still producing T cells at different ages. The micrograph in panel a shows a section through the thymus of a 3-day-old infant; the micrograph in panel b shows a section through the thymus from a 70-year-old person for comparison. Tissue is stained with hematoxylin and eosin (red and blue). Magnification × 20. Micrographs courtesy of Yasodha Natkunam.

self-renewing, or both. In this it is different from the mature B-cell repertoire, which is composed of short-lived cells that are continually being replenished from the bone marrow.

7-2   Thymocytes commit to the T-cell lineage before rearranging their T-cell receptor genes

The progenitor cells that will eventually give rise to mature T cells are not committed to the T-cell lineage when they enter the thymus. At that time they express CD34 and other cell-surface glycoproteins that are characteristic of stem cells, and they lack all the characteristic cell-surface glycoproteins of mature T cells. On interaction with thymic stromal cells, the progenitor cells are signaled to divide and differentiate. After around a week, the cells have lost stem-cell markers and have become thymocytes that are committed to the T-cell lineage, as seen by their expression of the T cell-specific adhesion molecule CD2 and other glycoproteins expressed by T cells, such as CD5 (Figure 7.5). At this stage of development, these precursor thymocytes still do not express any component of the T-cell receptor complex (see Section 5-4, p. 129) or the T-cell co-receptors CD4 and CD8, but they are beginning to rearrange the T-cell receptor genes. Because these thymocytes express neither CD4 nor CD8 they are called **double-negative thymocytes** or **DN thymocytes**.

A critical cytokine in T-cell development is interleukin-7 (IL-7), which is secreted by thymic stromal cells and binds to the interleukin-7 receptor on the CD34-expressing progenitor cells. The importance of this interaction is demonstrated by the absence of T cells in immunodeficient patients who have inherited two defective alleles of the interleukin-7 receptor. Another major regulator of T-cell development is Notch 1, a cell-surface receptor on thymocytes that interacts with transmembrane ligands on thymic epithelial cells. At all stages in the early development of T cells, signals generated through

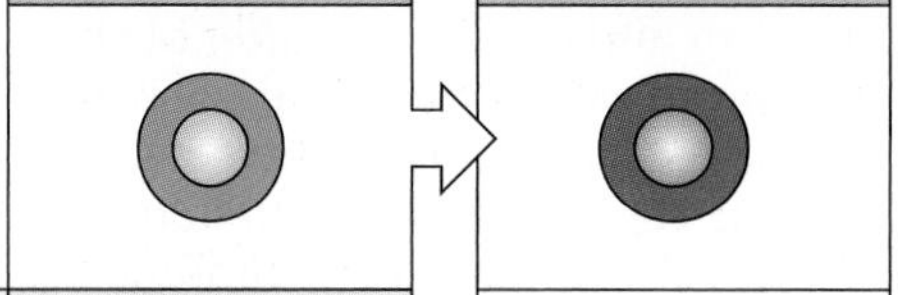

		Uncommitted progenitor cell	Double-negative thymocyte committed to the T-cell lineage
CD34	stem-cell surface marker	+	−
CD44	adhesion	+	−
CD2	adhesion and signaling	−	+
CD5	adhesion and signaling	−	+
IL-7 receptor (CD127)	cytokine receptor	−	+
CD1A	MHC class-I-like molecule	−	+
CD4	co-receptor	−	−
CD8	co-receptor	−	−
TCR genes	antigen receptor	germline	beginning to rearrange

Figure 7.5 Commitment to the T-cell lineage involves changes in gene expression and in cell-surface markers.

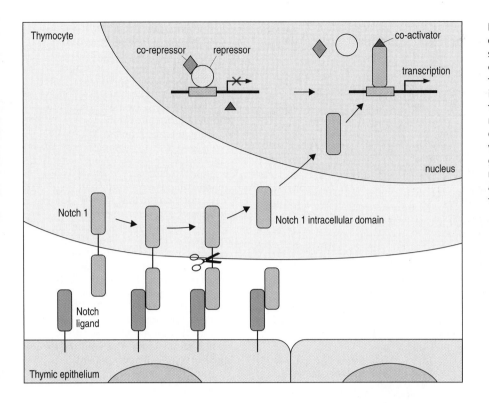

Figure 7.6 T-cell development is driven by the receptor Notch 1. The surface membrane-associated receptor called Notch 1 on the thymocyte binds to its ligand on thymic epithelium. This interaction induces a protease to cleave the intracellular domain from the plasma membrane. The soluble intracellular domain is translocated to the nucleus, where it turns on the expression of genes essential for T-cell development by removing repressive transcription factors and recruiting co-activating transcription factors.

Notch 1 are necessary to drive the cells along the pathway of T-cell differentiation. Notch 1 is one of a family of four Notch receptors in humans that control the development of many different cell types by deciding between one of two fates. In the unique environment of the thymus, Notch 1 keeps thymocytes on the pathway of T-cell differentiation and away from the pathway of B-cell differentiation. In this regard, Notch 1 has a role in T-cell development that is analogous to that of Pax-5 in B-cell development (see Section 6-8, p. 169).

Notch proteins are transmembrane receptors in which the extracellular and intracellular domains have distinct and complementary functions. When the extracellular domain of Notch 1 binds to its ligand on thymic epithelium, it initiates a proteolytic cleavage that releases the intracellular domain of Notch 1 from the membrane. The intracellular domain translocates to the thymocyte nucleus, where it becomes part of a transcription factor complex that initiates the transcription of genes needed for T-cell development. To do this it displaces repressive transcription factors from gene promoters and recruits activating factors (Figure 7.6).

7-3   The two lineages of T cells arise from a common thymocyte progenitor

T-cell development gives rise to two functionally different lineages of T cells that are distinguished by the expression of an α:β or a γ:δ T-cell receptor. Although their later stages of development in the thymus are very different, α:β and γ:δ T cells both derive from the common double-negative thymocyte precursor in which rearrangement of the genes encoding both of the T-cell receptors is initiated (Figure 7.7).

How T cells commit to either the α:β or γ:δ lineage is a complicated business, because individual cells are not restricted to rearranging either the α and β genes or the γ and δ genes. Instead, the double-negative thymocytes are programmed to begin rearrangements at the γ, δ, and β loci at roughly the same time. In effect, the γ and δ loci compete with the β locus in a race to make

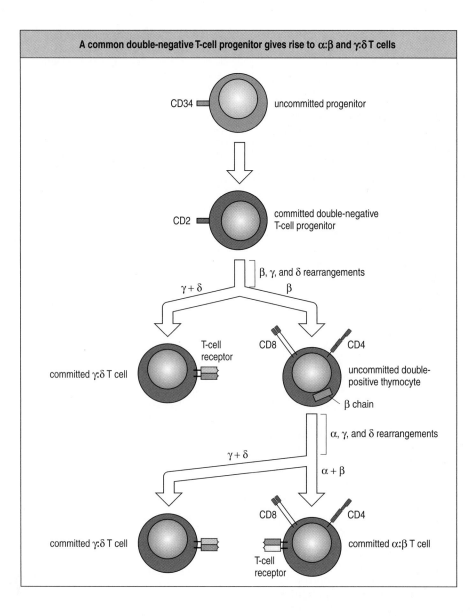

A common double-negative T-cell progenitor gives rise to α:β and γ:δ T cells

CD34 — uncommitted progenitor

CD2 — committed double-negative T-cell progenitor

β, γ, and δ rearrangements

γ + δ   β

committed γ:δ T cell   T-cell receptor   CD8   CD4   uncommitted double-positive thymocyte

β chain

α, γ, and δ rearrangements

γ + δ   α + β

committed γ:δ T cell   CD8   CD4   committed α:β T cell   T-cell receptor

Figure 7.7 α:β and γ:δ T cells develop from a common double-negative T-cell progenitor. T-cell precursors that enter the thymus express the hematopoietic stem-cell marker CD34 but none of the characteristic markers of mature T cells. Proliferation of these common progenitors followed by rearrangement of the δ-, γ-, and β-chain genes leads to early commitment of some cells to the γ:δ T-cell lineage, whereas others rearrange the β-chain gene first and temporarily halt gene rearrangement at this point. As soon as they produce a complete receptor, γ:δ cells can leave the thymus and travel to other tissues via the blood. In the β-chain-positive cells in the thymus, rearrangement of the α-, γ-, and δ-chain genes resumes and productive α-chain gene rearrangements in these cells produce double-positive CD4 CD8 α:β cells. A minority of the double-positive thymocytes give rise to additional γ:δ T cells. This ends the early stage of α:β T-cell development.

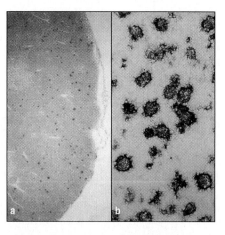

Figure 7.8 Immature T cells that undergo apoptosis are ingested by macrophages in the thymic cortex. Panel a shows a section through the thymic cortex (to the right side of the panel) and part of the medulla, in which cells have been stained for apoptosis with a red dye. Apoptotic cells are scattered throughout the cortex but are rare in the medulla. Panel b shows a section of thymic cortex at higher magnification that has been stained red for apoptotic cells and blue for macrophages. Apoptotic cells can be seen within the macrophages. Magnifications: panel a, × 45; panel b, × 164. Photographs courtesy of J. Sprent and C. Surh.

productive gene rearrangements and functional T-cell receptor chains (see Figure 7.7). If the thymocyte makes a functional γ:δ receptor before a functional β chain, then it commits to becoming a γ:δ T-cell. If a functional β chain is made before a γ:δ receptor it is incorporated into a protein called the pre-T-cell receptor. Although this outcome favors the α:β lineage it does not commit the cell to that lineage. Gene rearrangement stops at this point, while the cell proliferates and expresses the CD4 and CD8 co-receptors. Because they express both co-receptors, cells at this stage of development are called **double-positive thymocytes** or **DP thymocytes**. Rearrangement of the α-chain genes is now allowed and rearrangement at the γ- and δ-chain genes can also continue. Further competition ensues. If a double-positive thymocyte makes an α:β receptor before a γ:δ receptor it commits to the α:β lineage. Conversely, if a γ:δ receptor is made first, then the cell commits to the γ:δ lineage (see Figure 7.7).

Cells that fail to make productive T-cell receptor gene rearrangements die by apoptosis and are phagocytosed by macrophages in the thymic cortex (Figure 7.8). Apoptosis is the fate of all except a very few thymocytes (around 2%), and the macrophages of the thymus are continually removing dead and dying cells while not interfering with ongoing thymocyte development.

7-4   Gene rearrangement in double-negative thymocytes leads to assembly of either a γ:δ receptor or a pre-T-cell receptor

Like the immunoglobulin genes, T-cell receptor genes can make rearrangements that are either productive or nonproductive, and rearrangements can be attempted on both copies of each locus. The type of rearrangement is also analogous to those made by the immunoglobulin genes. For the β- and δ-chain loci, which contain V, D, and J segments, the first rearrangement joins D to J and the second joins V to DJ; for the α- and γ-chain loci, which contain only V and J segments, a single rearrangement joins V to J (see Figures 5.3 and 5.8, pp. 127 and 131).

The developmental pathway taken by a T cell is determined by the sequence in which the T-cell receptor genes make productive rearrangements. If a thymocyte makes productive γ- and δ-chain gene rearrangements before a productive β-chain gene rearrangement, then a γ:δ heterodimer is made. This assembles with the CD3 signaling complex (see Section 5-4, p. 130), moves to the cell surface and signals the cell to stop β-chain rearrangement. The cell is then committed to being a γ:δ T cell. As γ:δ T-cell receptors are not subject to the stringent positive and negative selection imposed upon the α:β receptor repertoire, γ:δ T cells soon leave the thymus in the blood and enter the circulation (Figure 7.9).

The more frequent outcome of the competition between the β-, γ- and δ-chain genes is for a productive β-chain gene rearrangement to be made before both productive γ- and δ-chain rearrangements occur. In this situation the β chain is made, translocated to the endoplasmic reticulum and tested there for its

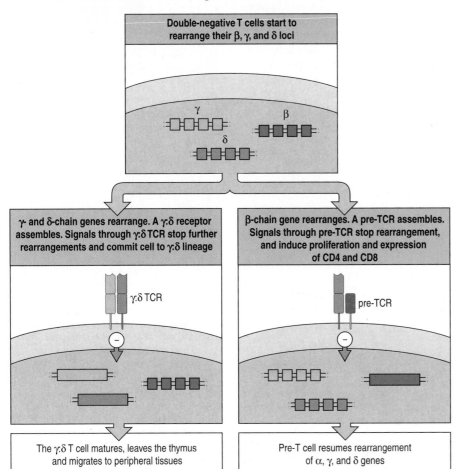

Figure 7.9 T-cell receptor gene rearrangements in double-negative thymocytes can lead to the expression of either a γ:δ receptor or a pre-T-cell receptor. In double-negative thymocytes, the β, γ, and δ genes rearrange (top panel). If successful γ- and δ-chain gene rearrangements occur first, then a γ:δ receptor is expressed and the cell is signaled to differentiate into a mature γ:δ cell (bottom left panel). If a successful β-chain gene rearrangement is made before both γ and δ gene rearrangements, a pre-T-cell receptor (pre-TCR) is assembled, which signals the cell to proliferate, express CD4 and CD8, and become a pre-T cell. At this stage the pre-T cell turns the recombination machinery back on to rearrange the α, γ, and δ genes.

Figure 7.10 Comparison of the structures of the pre-T-cell receptor and the T-cell receptor. The only difference is that the α chain of the T-cell receptor is replaced by the pTα chain in the pre-T-cell receptor.

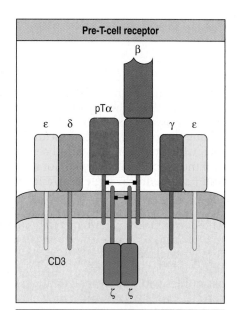

capacity to bind to an invariant polypeptide called **pTα**, which acts as a surrogate α chain analogous to the surrogate light chain of the pre-B-cell receptor (see Section 6-4, p. 163). If the β chain binds to pTα, this heterodimer assembles with the CD3 complex and ζ chain to form a protein complex called the **pre-T-cell receptor** (Figure 7.10), corresponding to the pre-B-cell receptor. Assembly of a pre-T-cell receptor signals the cell to halt rearrangement of the β-, γ-, and δ-chain genes (see Figure 7.9). Like the pre-B-cell receptor, assembly of the pre-T-cell receptor is sufficient for signaling and there is no requirement for binding a ligand. Assembly of the pre-T-cell receptor is a checkpoint in T-cell development that determines whether the β chain made by a developing T cell has the potential to bind to α chains, and if not, further development of the cell is stopped. Thymocytes that pass this test are called **pre-T cells** and proceed to the next stage of development.

7-5   Thymocytes can make four attempts to rearrange a β-chain gene

The competition to recruit thymocytes to the two T-cell lineages is biased to favor the α:β lineage, because commitment to that lineage requires only one productive gene rearrangement, whereas commitment to the γ:δ lineage requires a minimum of two productive rearrangements. Further increasing the probability of commitment to the α:β lineage is the potential for nonproductive β-chain gene rearrangements to be rescued by the further rearrangement made possible by the two C_β genes and their associated D_β and J_β segments (Figure 7.11). If a rearrangement at one β-chain locus is nonproductive, a thymocyte can attempt a rearrangement at the β-chain locus on the other, homologous, chromosome. A nonproductively rearranged β-chain gene can also be rescued by a second rearrangement at the same locus. This possibility is not available to the immunoglobulin heavy-chain genes, because a nonproductive rearrangement deletes all the nonrearranged D segments. The potential for trying out up to four β-chain gene rearrangements means that 80% of T cells make a productive rearrangement of the β-chain gene, compared with a 55% success rate for heavy-chain gene rearrangement by B cells.

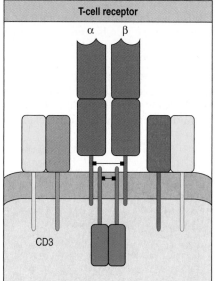

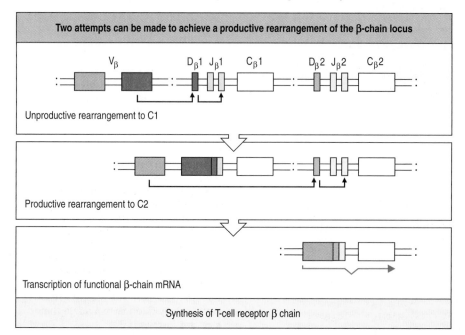

Figure 7.11 Rescue of unproductive rearrangements of the β-chain locus. Successive rearrangements can rescue an initial unproductive β-chain gene rearrangement, but only if that rearrangement involved D and J gene segments associated with the C_β1 gene segment. A second rearrangement is then possible, in which a second V_β gene segment rearranges to a DJ segment associated with the C_β2 gene segment, deleting C_β1 and the unproductively rearranged gene segments.

7-6   Rearrangement of the α-chain gene occurs only in pre-T cells

Successful rearrangement of a β-chain gene followed by signaling through the pre-T-cell receptor induces the pre-T cell to stop gene rearrangements by suppressing the expression of the *RAG-1* and *RAG-2* recombination-activating genes; the same phenomenon occurs during B-cell development (see Section 6-5, p. 164). This ensures that only one β chain gene has a productive rearrangement and is expressed, so that there is allelic exclusion at the β-chain locus. The pre-T cell is also induced to proliferate, which creates a clone of cells all expressing the same β chain. Proliferation is accompanied by the expression of first CD4 and then CD8 to give double-positive thymocytes. These cells, which constitute the majority of thymocytes, are found predominantly in the inner cortex of the thymus, where they interact intimately with the branching network of epithelial cells. On ceasing to proliferate, the large double-positive thymocytes become small double-positive thymocytes in which the recombination machinery is reactivated and targeted to the α-chain locus, as well as to the γ and δ loci, but not to the β-chain locus.

The T-cell receptor α-chain genes are comparable to the immunoglobulin κ and λ light-chain genes in that they do not have D segments and are rearranged only after their partner receptor chain has been expressed. As with the immunoglobulin light-chain gene, repeated attempts at α-chain gene rearrangement are possible. The presence of multiple V_α gene segments and about 60 J_α gene segments spread over some 80 kb of DNA allows many successive V_α to J_α rearrangements to take place at both α-chain alleles. This means that T cells with an initial nonproductive α-gene rearrangement are highly likely to be rescued by a subsequent rearrangement (Figure 7.12).

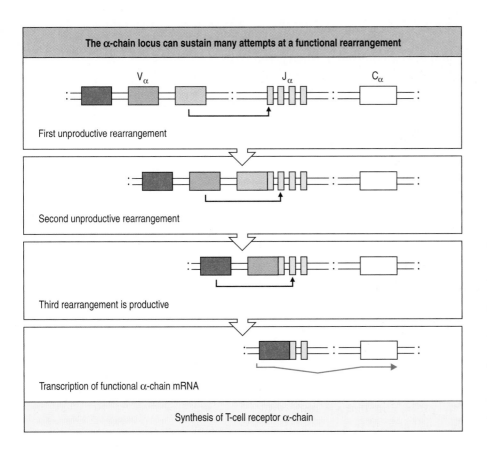

The α-chain locus can sustain many attempts at a functional rearrangement

V_α J_α C_α

First unproductive rearrangement

Second unproductive rearrangement

Third rearrangement is productive

Transcription of functional α-chain mRNA

Synthesis of T-cell receptor α-chain

Figure 7.12 Successive gene rearrangements allow the replacement of one T-cell receptor α chain by another. For the T-cell receptor α-chain genes, the multiplicity of V and J gene segments allows successive rearrangement events to jump over unproductively rearranged VJ segments, deleting the intervening gene segments. This process continues until either a productive rearrangement occurs or the supply of V and J gene segments is exhausted, whereupon the cell dies.

When an α-chain gene is rearranged, the δ locus situated within the α locus will be automatically deleted, irrespective of whether the rearrangement is productive or not (Figure 7.13); see also Figure 5.8 (p. 131) for a detailed diagram of the arrangement of the α and δ loci. This arrangement greatly reduces the probability that a T cell committed to the α:β lineage will end up expressing a γ:δ receptor as well as an α:β receptor.

When a double-positive T cell makes a productive α-chain gene rearrangement, the gene is transcribed and an α chain is made. After translocation to the endoplasmic reticulum the α chain is tested for its capacity to bind the β chain and assemble a T-cell receptor. This represents a second checkpoint during T-cell development. If successful, the double-positive cell is signaled to survive and to proceed to positive selection, the next step in the developmental pathway. If the α chain does not properly assemble with the β chain, further α-chain gene rearrangements are made until either a functional α chain is produced or the possibilities for gene rearrangement are exhausted. In the latter case, the pre-T cell will die by apoptosis.

7-7   Stages in T-cell development are marked by changes in gene expression

The outcome of this first part of α:β T-cell development is rearranged T-cell receptor genes and a diverse population of immature cells that each carry CD4, CD8 and a correctly folded and potentially useful α:β T-cell receptor at the cell surface. The stages in this development, which are usually defined by the readily detectable cell-surface phenotype, are also marked by changes in the expression of genes encoding intracellular molecules that contribute to the rearrangement and transcription of the T-cell receptor genes, to the expression of the pre-T-cell receptor, to signal transduction from the pre-T-cell receptor and the T-cell receptor, and to the expression of the CD4 and CD8 co-receptors (Figure 7.14).

The RAG-1 and RAG-2 proteins are essential for gene rearrangement and are selectively expressed at the two stages where the β and the α gene rearrangements, respectively, are made. With RAG-1 and RAG-2 controlling when gene rearrangement occurs, other enzymes involved in somatic recombination, such as the TdT that inserts the N-nucleotides, are expressed throughout this phase of development. The unique component of the pre-T-cell receptor, pTα, is also expressed throughout the period when rearrangements occur, so that a newly made β chain can be quickly assembled into a pre-T-cell receptor that signals the interior of the cell to stop recombination and to initiate cell division. Signals from the pre-T-cell receptor depend on the expression of the co-receptors CD4 and CD8, the signaling complex CD3, the tyrosine kinase ZAP-70, which is involved in relaying signals from the receptor, and the tyrosine kinase Lck, which is involved in signaling from the co-receptors. CD2 is an adhesion molecule on T cells that interacts with the cell-surface protein CD58 on other cells and generates signals that work in conjunction with those coming from the T-cell receptor. The transcription factors called Ikaros and GATA-3 are expressed in early T-cell progenitors and are essential for T-cell development. The transcription factor Th-POK is expressed late in development and is required for the development of single-positive CD4 T cells from double-positive thymocytes.

The early development of α:β T cells ends with the production of double-positive thymocytes that express CD4, CD8, and a functional T-cell receptor. The advantage of expressing both CD4 and CD8 is that it gives each immature T cell the option to use either co-receptor depending on whether its T-cell receptor is more fitted to recognizing peptide antigens presented by self-MHC class I or self-MHC class II. This increases the probability that a double-positive cell will complete its maturation by a factor of around two.

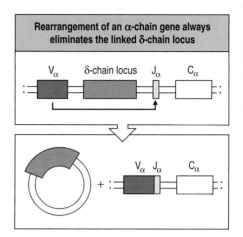

Figure 7.13 The δ-chain gene is sequestered within the α-chain gene and is deleted upon α-chain gene rearrangement. Upon rearrangement of an α-chain gene, the δ-chain gene it contains is eliminated as part of an extrachromosomal circle.

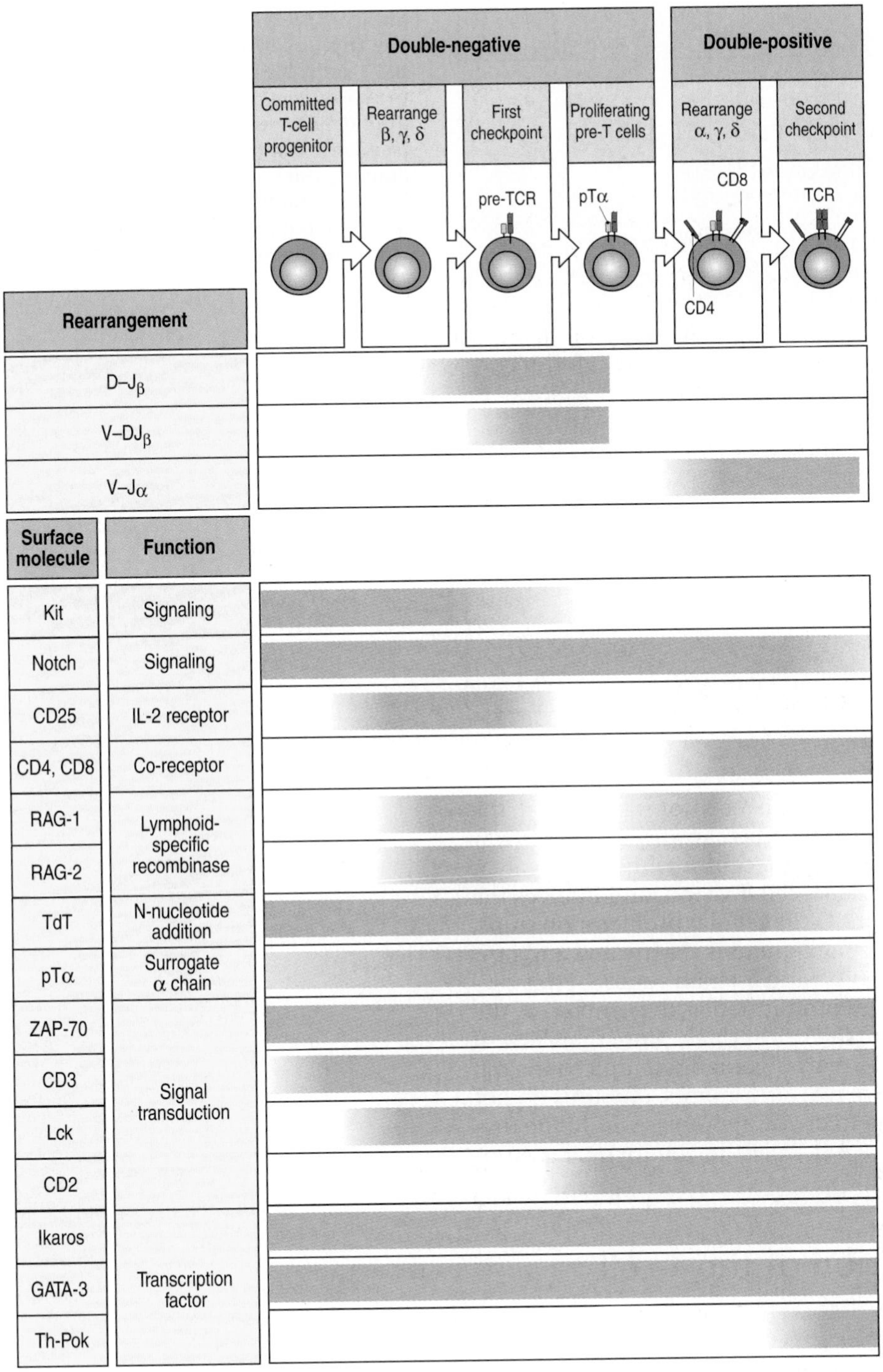

Figure 7.14 Stages of α:β T-cell development in the thymus correlate with T-cell receptor gene rearrangement and the expression of particular proteins by the developing T cell. All these proteins are described in the text.

	Double-negative				Double-positive	
	Committed T-cell progenitor	Rearrange β, γ, δ	First checkpoint (pre-TCR)	Proliferating pre-T cells (pTα)	Rearrange α, γ, δ (CD8 / CD4)	Second checkpoint (TCR)

Rearrangement	
D–J$_\beta$	
V–DJ$_\beta$	
V–J$_\alpha$	

Surface molecule	Function
Kit	Signaling
Notch	Signaling
CD25	IL-2 receptor
CD4, CD8	Co-receptor
RAG-1	Lymphoid-specific recombinase
RAG-2	
TdT	N-nucleotide addition
pTα	Surrogate α chain
ZAP-70	Signal transduction
CD3	
Lck	
CD2	
Ikaros	Transcription factor
GATA-3	
Th-Pok	

Summary

The thymus provides a sequestered and organized environment dedicated to T-cell development. Progenitor cells from the bone marrow migrate to the thymus, where they go through phases of division and differentiation associated with the rearrangement of T-cell receptor genes and the expression of other cell-surface glycoproteins involved in T-cell recognition and effector function. Commitment of thymocytes to either the γ:δ or the α:β lineage occurs as a consequence of the gene rearrangements made. The β-, γ-, and

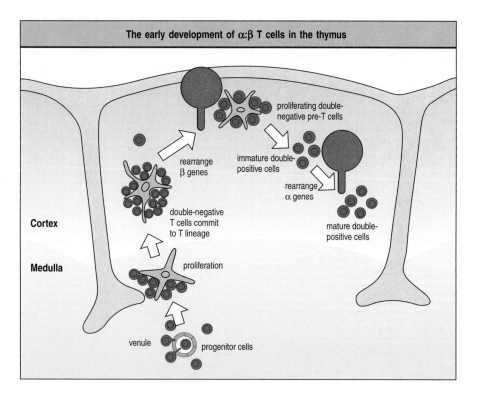

Figure 7.15 The development in the thymus of double-positive α:β T cells. This figure summarizes the early development of α:β T cells (round dark blue cells) in the thymus, from uncommitted progenitor cell to a double-positive cell bearing an α:β T-cell receptor. The two checkpoints of B-cell development are indicated. The paler blue cells are resident thymic cells.

δ-chain genes rearrange simultaneously, and if a γ:δ receptor is made first, the cell is committed to the γ:δ lineage. If a β-chain gene rearranges productively before this occurs, the β chain forms a pre-T-cell receptor with a surrogate α chain, pTα. This stops recombination, initiates cell division, and induces the expression of the CD4 and CD8 co-receptors. Once cellular proliferation stops, the machinery for T-cell receptor gene rearrangement is reactivated and now works on the α-chain gene in addition to the γ- and δ-chain genes. In this second phase of gene rearrangements, lineage commitment is determined by the expression of either γ:δ or α:β receptors on the cell surface. More than 90% of thymocytes that complete successful gene rearrangement commit to the α:β lineage. If cells fail to make successful gene rearrangements, and thus do not express a T-cell receptor, they are signaled to die by apoptosis within the thymus. The development of double-positive α:β T cells in the thymus is summarized in Figure 7.15.

Positive and negative selection of the T-cell repertoire

In the first phase of T-cell development just described, the role of the thymus is to produce T-cell receptors, irrespective of their antigenic specificity. The second phase of T-cell development involves a critical examination of the receptors produced and the selection of those that can work effectively with the individual's own MHC molecules in the recognition of pathogen-derived peptides. These selection processes involve only α:β T cells; γ:δ T cells seem indifferent to peptides presented by MHC molecules and recognize different types of antigen, which remain largely unknown. Once the γ- and δ-chain genes are rearranged productively, the development of γ:δ T cells within the thymus seems to be complete.

In this part of the chapter we see how the population of α:β double-positive thymocytes undergoes two types of screening. In the first screen, positive selection favors T cells that can recognize peptides presented by a self-MHC molecule; in the second, negative selection eliminates potentially autoreactive cells that could be activated by the peptides normally presented by MHC molecules on the surface of healthy cells.

7-8   T cells that recognize self-MHC molecules are positively selected in the thymus

T-cell receptors and MHC molecules have been co-evolving under natural selection for several hundreds of millions of years. Consequently, the primary T-cell receptor repertoire has a bias toward interaction with MHC molecules that is due to specificities built into the V gene segments. Thus, gene rearrangement provides an extensive repertoire of T-cell receptors that could be used with the hundreds of MHC class I and MHC class II isoforms present in the human population (see Chapter 5). However, the T-cell receptor genes of a given individual are not tailored specifically toward making receptors that interact with the particular forms of MHC molecule expressed by the same individual. Only a small subpopulation of the double-positive thymocytes, at most 2% of the total, have receptors that can interact with one of the MHC class I or II isoforms expressed by the individual, and will therefore be able to respond to antigens presented by these MHC molecules. **Positive selection** is the name given to the process whereby that small subpopulation is selected and signaled to mature further, leaving the vast majority of double-positive cells to die by apoptosis in the thymic cortex.

Positive selection takes place in the cortex of the thymus (see Figure 7.3). It is mediated by the complexes of self peptides and self-MHC molecules present on the surface of the cortical epithelial cells. As noted in Chapter 5, in the absence of infection MHC molecules assemble with self peptides derived from the normal breakdown of the body's own proteins. Thymic cortical epithelium expresses both MHC class I and class II molecules and thus presents self peptides in both MHC contexts. The cortical epithelial cells form a web of cell processes that envelop and make contact with the double-positive CD4 CD8 thymocytes. At regions of contact, potential interactions of the α:β receptor of a thymocyte with the self-peptide:self-MHC complexes on the epithelial cell are tested. If a peptide:MHC complex is bound within 3–4 days of the thymocyte's expressing a functional receptor, then a positive signal is delivered to the thymocyte, which continues its maturation. Cells that do not receive such a signal within this period die by apoptosis (Figure 7.16) and are removed by macrophages. The MHC isoform that positively selects the T cell is the one to which it becomes MHC restricted—that is, the T cell and all its progeny will recognize their specific peptide antigen only in the context of that particular MHC molecule (see Figure 5.31, p. 151).

The peptides presented on the surface of thymic epithelial cells are derived from those self proteins that are present in the thymus. The number of different self peptides that can be presented by an individual MHC molecule is estimated to be about 10,000, so for someone who is heterozygous for the six major polymorphic HLA genes, around 120,000 self peptides could be presented by the 12 different MHC molecules that person would possess. The mature T-cell receptor repertoire is estimated to be of the order of tens of millions or more, so most of these self-peptide:self-MHC complexes are likely to bind a T-cell receptor and contribute to positive selection.

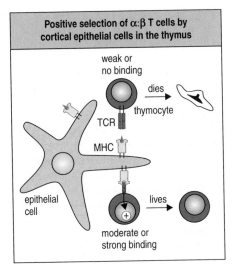

Figure 7.16 Positive selection of T cells in the thymus. T cells with a T-cell receptor (TCR) that binds to a self-MHC class I molecule on thymic cortical epithelial cells, macrophages, and other cells in the thymus are signaled to survive and proceed to negative selection. T cells with a T-cell receptor that binds to no self-MHC class I molecules are signaled to die.

7-9   Continuing α-chain gene rearrangement increases the chance for positive selection

If the first α-chain gene rearrangement that a pre-T cell makes is productive and leads to the assembly of an α:β T-cell receptor that interacts with a self-MHC molecule, positive selection will occur within a few hours. The cell is then signaled to turn off the recombination machinery, by degrading the RAG proteins and stopping transcription of the *RAG* genes, and to enter a phase of proliferation. In contrast, if the T-cell receptor does not bind a self-MHC molecule within this timeframe, rearrangement of the α-chain genes continues. This allows the T cell to try out a second α chain in its search for a functional T-cell receptor. A second productive rearrangement produces a different α chain that assembles with the β chain to give a T-cell receptor with a different binding-site specificity. The second T-cell receptor can then be assessed for its capacity to bind self MHC. Further rearrangements at the α-chain locus can continue throughout the 3–4 days of positive selection, during which time a double-positive thymocyte can explore the usefulness of several receptors in succession. This process improves the chance that the T cell will be positively selected.

After successfully rearranging a β-chain gene, a developing T cell immediately switches off the recombination machinery, with the result that there is allelic exclusion at the β-chain locus (see Section 7-6). Because the α-chain gene is not subject to allelic exclusion in the same way as the β-chain gene, rearrangements can occur at both copies of the α-chain locus, and double-positive thymocytes can express two α chains and, thus, two types of T-cell receptor. Such T cells can be positively selected through the engagement of either of these receptors, and around one-third of all mature α:β T cells carry two T-cell receptors. But the proportion of T-cell receptors that succeed in positive selection is so small that the proportion of T cells with two T-cell receptors that can both be activated by peptides presented by self-MHC molecules will be at most 1% or 2%. Thus, in almost all mature T cells with two receptors, one receptor will be nonfunctional, and for most practical purposes T cells can be said to have a single working antigen receptor. For immunoglobulins, which contain two or more heavy and light chains, allelic exclusion of both the heavy- and light-chain genes is essential to preserve their functional integrity (see Section 6-5, p. 164). Lack of allelic exclusion at the T-cell receptor α chain locus is not so disruptive because each T-cell receptor only contains one α chain.

The process by which a T-cell receptor tries out different α chains in order to become reactive with a self-MHC molecule is one of receptor editing, analogous to that used by B cells to change the light chain of their B-cell receptors (see Section 6-12, p. 175). The difference is that the B cell uses receptor editing to get rid of reactivity against self antigens, including self MHC, whereas the T cell uses receptor editing to acquire reactivity with self MHC.

7-10   Positive selection determines expression of either the CD4 or the CD8 co-receptor

Positive selection not only selects a repertoire of cells that can interact with an individual's own MHC allotypes, but it is also instrumental in determining whether a double-positive T cell will become a CD4 T cell or a CD8 T cell. As a result of positive selection, double-positive thymocytes mature into cells that express just one or other of the two co-receptors. They are then known as **single-positive thymocytes**.

As we saw in Chapter 5, CD4 interacts only with MHC class II molecules, and CD8 interacts only with MHC class I molecules. During positive selection, the

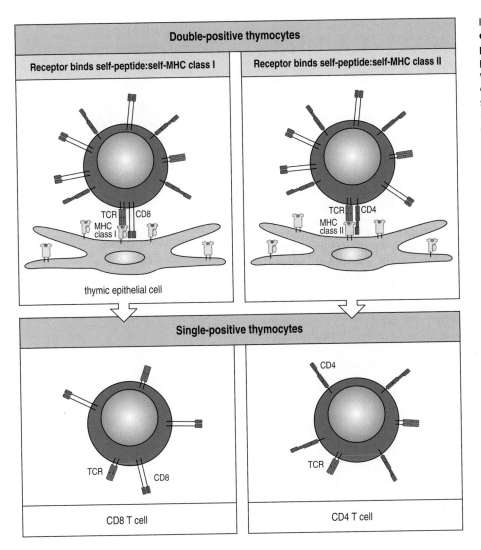

Double-positive thymocytes

Receptor binds self-peptide:self-MHC class I	Receptor binds self-peptide:self-MHC class II

TCR CD8
MHC class I

TCR CD4
MHC class II

thymic epithelial cell

Single-positive thymocytes

TCR CD8

CD8 T cell

CD4

TCR

CD4 T cell

Figure 7.17 Interaction of a double-positive T cell with a self-peptide:self-MHC complex during positive selection determines whether the T cell will become a CD4 or a CD8 T cell. The left panels show the selection of a T cell whose T-cell receptor (TCR) interacts with peptide:MHC class I complexes on a thymic epithelial cell. The right panels show the outcome for a cell bearing a receptor that interacts with peptide:MHC class II complexes.

double-positive CD4 CD8 T cell interacts through its α:β receptor with a particular peptide:MHC complex. When the interacting MHC molecule is class I, CD8 molecules are recruited into the interaction, whereas CD4 molecules are excluded. Conversely, when the selecting MHC molecule is class II, CD4 is recruited and CD8 excluded (Figure 7.17). The specific interactions between T-cell receptor, co-receptor, and MHC molecule and the differential signaling that is generated by the two co-receptors, particularly through the protein kinase Lck (see Section 7-7), commit the cell to either the CD4 or CD8 lineage and halt the synthesis of the other co-receptor. The single-positive cells initiate a program of gene expression that provides CD4 T cells with the capacity for helper function and CD8 T cells with the capacity for cytotoxic function. For example, the transcription factor Th-POK is essential for the development of CD4 T cells from double-positive thymocytes (see Figure 7.14).

The importance of MHC molecules in co-receptor selection and subsequent T-cell development is demonstrated by human immunodeficiency diseases called **bare lymphocyte syndromes**, which are characterized by a lack of expression of either MHC class I or MHC class II molecules by lymphocytes and thymic epithelial cells. Patients who lack MHC class I expression have CD4 T cells but no CD8 T cells, whereas patients lacking MHC class II have plenty of CD8 T cells but only a few abnormal CD4 T cells.

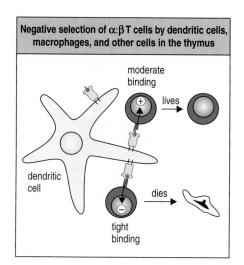

Figure 7.18 Negative selection of T cells in the thymus. T cells with a T-cell receptor (TCR) that binds too tightly to a self-MHC class I molecule on dendritic cells, macrophages, and other cells in the thymus are signaled to die. T cells with a receptor that binds moderately to a self-MHC class I molecule on dendritic cells, macrophages, and other cells in the thymus are signaled to survive, mature, and enter the peripheral circulation.

7-11   T cells specific for self antigens are removed in the thymus by negative selection

In Chapter 4 we saw how immature B cells whose immunoglobulin receptors bind to self antigens on the surface of bone marrow cells are eliminated from the repertoire by clonal deletion. A similar mechanism, called **negative selection**, deletes T cells whose antigen receptors bind too strongly to the complexes of self peptides and self-MHC molecules presented by cells in the thymus. Such T cells are potentially autoreactive, and if allowed to enter the peripheral circulation they could cause tissue damage and autoimmune disease. Whereas positive selection is mediated exclusively by epithelial cells in the cortex of the thymus, negative selection can be mediated by several cell types, including thymocytes themselves. The most important cells responsible for negative selection are, however, the bone marrow derived dendritic cells and macrophages (Figure 7.18). Engagement of the MHC molecules of one of these specialized thymic antigen-presenting cells by the receptors of an autoreactive T cell induces the T cell to undergo apoptosis. The dead cells are then phagocytosed by macrophages.

The mechanisms of positive and negative selection used by the thymus both involve the screening of interactions between T-cell receptors and self peptides bound to MHC molecules. How these interactions lead to the widely differing endpoints of cell death or cell growth has yet to be worked out. The processing and presentation of self antigens by thymic epithelium differs from that in other cells in several ways, including the use of the protease cathepsin L, whereas other cells use cathepsin S. Differences in the affinity of ligand–receptor interactions and the type of signaling pathways activated are also possible contributors.

7-12   Tissue-specific proteins are expressed in the thymus and participate in negative selection

The MHC molecules expressed by dendritic cells and macrophages in the thymus present peptides derived from all the proteins made by these cells and other cells they phagocytose, and also soluble proteins taken up from extracellular fluids. Negative selection will eliminate T cells specific for these self antigens, which include the ubiquitously expressed proteins. To extend negative selection to proteins that are specific to one or a few cell types, such as the insulin made only by the β cells of the pancreas, a transcription factor called **autoimmune regulator** (**AIRE**) causes several hundred of these tissue-specific genes to be transcribed by a subpopulation of epithelial cells in the medulla of the thymus. The presence of small amounts of these tissue-specific proteins in the thymus means that peptides derived from these proteins can be bound by MHC class I molecules to form complexes that participate in negative selection of the T-cell repertoire. AIRE was discovered as a result of the symptoms displayed by children who lack a functional *AIRE* gene. In these patients, T cells specific for tissue-specific antigens are not eliminated by negative selection and are permitted to mature and enter the peripheral circulation. Here they attack cells in a variety of tissues, causing a broad-spectrum autoimmune disease known as autoimmune polyglandular syndrome type 1 or autoimmune polyendocrinopathy–candidiasis–ectodermal dystrophy (APECED).

Negative selection in the thymus produces so-called central tolerance in the T-cell repertoire (see Section 6-13, p. 177). There are also mechanisms of peripheral tolerance that prevent the activation of self-reactive T cells that elude negative selection in the thymus and enter the peripheral circulation. As with B cells, one such mechanism is to render the self-reactive T cell anergic (see Section 6-13). In general, a T cell outside the thymus that encounters a self antigen in the absence of infection will receive signals that cause it either to be inactivated or to be briefly activated and then die; this latter mechanism is called activation-induced cell death. In patients lacking AIRE, the mechanisms of peripheral tolerance are overwhelmed by the unusually large number of self-reactive T cells that enter the peripheral circulation.

7-13 Regulatory CD4 T cells comprise a distinct lineage of CD4 T cells

So far in this book we have considered only the helper functions of CD4 T cells, which drive the immune response to infection by activating macrophages and B cells (see Section 5-6, p. 133). Another function carried by a distinct lineage of CD4 T cells is to suppress the response of self-reactive CD4 T cells to their specific self antigens: despite negative selection and the activity of AIRE, self-reactive CD4 T cells are present in the circulation of all people, even the most healthy. This subset of CD4 T cells is called **regulatory CD4 T cells** or just **regulatory T cells** (T_{reg}). Regulatory T cells have T-cell receptors specific for self antigens and are distinguished from other CD4 T cells by the expression of CD25 on the cell surface and the unique use of a transcriptional repressor protein called FoxP3. On contacting self antigens presented by MHC class II molecules, the regulatory T cells do not proliferate but respond by suppressing the proliferation of naive T cells responding to self antigens presented on the same antigen-presenting cell (Figure 7.19). These suppressive effects require contact between the two T cells and the secretion of cytokines that inhibit the activation and differentiation of effector T cells. Although once controversial, this active form of tolerance mediated by regulatory T cells is now considered one of the main mechanisms for protecting the integrity of the body's tissues and organs. The basis for the selection and development of this lineage of CD4 T cells in the thymus has yet to be defined.

Central to the development and function of regulatory T cells is FoxP3, which is encoded by a gene on the X chromosome and is expressed by these cells and by no others. A rare deficiency in FoxP3 principally affects boys, producing a fatal disorder that is characterized by autoimmunity directed toward a variety of tissues. The gut is almost always affected, and other common targets are the thyroid, pancreatic β cells and the skin. Affected children fail to thrive and suffer recurrent infections. This syndrome, which is caused by lack of regulatory T cells, is called the immune dysregulation, polyendocrinopathy, enteropathy, X-linked syndrome (**IPEX**). The only effective treatment for IPEX is a bone marrow transplant from a healthy HLA-identical sibling.

7-14 T cells undergo further differentiation in secondary lymphoid tissues after encounter with antigen

Only a small fraction of α:β T cells survive the obstacle course of positive and negative selection and leave the thymus as mature naive T cells. Like mature B cells, these mature T cells recirculate through the tissues of the body, passing from blood to secondary lymphoid tissues to lymph and then back to the blood. Mature T cells are longer lived than mature B cells and, in the absence of their specific antigen, continue to circulate for many years.

The T-cell rich areas of secondary lymphoid tissues provide specialized sites where naive T cells are activated by their specific antigens. Encounter with

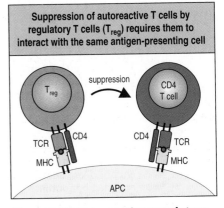

Figure 7.19 Autoreactive regulatory CD4 T cells prevent the proliferation of autoreactive helper CD4 T cells. Suppression of an autoreactive CD4 T cell by a regulatory T cell (T_{reg}) is dependent on the interaction of both T cells with the same antigen-presenting cell (APC).

antigen provokes the final phases of T-cell development and differentiation: the mature T cells divide and differentiate into effector T cells, some of which stay in the lymphoid tissues while others migrate to sites of infection.

Unlike B cells, which have just one terminally differentiated state—the antibody-secreting plasma cell—there are several different types of effector T cell. On activation by antigen, CD8 T cells become activated cytotoxic T cells, whereas CD4 T cells can be regulatory T cells and various types of helper cell that have differentiated under the influence of different combinations of cytokines. Which type of effector CD4 T cell predominates depends on the nature of the pathogen and the type of immune response required to clear it (see Chapter 5).

In healthy individuals, approximately twice as many CD4 T cells than CD8 T cells are present in the peripheral circulation. In patients with acquired immunodeficiency syndrome (AIDS) this proportion changes because the virus that causes the disease selectively infects and kills CD4 cells by exploiting the CD4 molecule as its receptor. The number of CD4 T cells declines, and this parameter is used by physicians to measure disease progression and to assess the effectiveness of therapy.

7-15   Most T-cell tumors represent early or late stages of T-cell development

Because of the DNA rearrangements that occur during T-cell development, T-cell tumors are not uncommon. Like B-cell tumors (see Section 4-12, p. 112), they correspond to defined stages in development. But unlike B-cell tumors, which cover all stages in B-cell development, T-cell tumors mostly correspond to either early or late stages in T-cell development. The absence of tumors corresponding to intermediate stages in T-cell development might be because immature T cells are programmed to die unless rescued within a short period by the next positive signal for maturation. Under these circumstances, there might not be time for thymocytes to accumulate the number of mutations necessary for malignant transformation.

Like B-cell tumors, T-cell lymphomas have characteristic rearrangements of their T-cell receptor genes, showing that they derive from single transformed cells (Figure 7.20). A patient's tissues can be analyzed for T-cell receptor gene rearrangements that define the tumor, which allows tumor growth and dissemination to be monitored over time and the results of therapy to be followed.

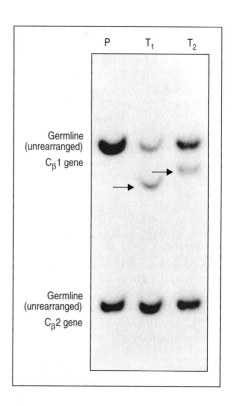

Figure 7.20 **The T-cell receptor gene rearrangements that characterize individual clones of T cells can be used to identify tumors of T cells.** Because a tumor is an outgrowth of a single transformed cell, the cells of a T-cell tumor all have the same rearranged T-cell receptor genes, which can therefore be detected and characterized. This figure shows a comparison of genomic DNA isolated from placenta (lane P) and two T-cell tumors (lanes T_1 and T_2). The DNAs were digested into fragments with a restriction endonuclease and separated by gel electrophoresis. They were then analyzed by Southern blotting with a radioactive T-cell receptor β-chain C-region cDNA probe. The fragments of genomic DNA that derive from the T-cell receptor β-chain genes show up on the film as dark bands. In placental cells the T-cell receptor genes are in the germline configuration, and bands corresponding to the $C_\beta 1$ and $C_\beta 2$ exons are seen. For the two tumors, bands corresponding to genes in the germline configuration are seen, but there are also bands corresponding to distinct rearrangements of a β-chain gene. Courtesy of T. Diss.

Summary

Once a developing thymocyte expresses an $\alpha{:}\beta$ receptor and CD4 and CD8 on its surface, it undergoes two types of selection—positive and negative—both of which involve testing the receptor's interactions with the complexes of self peptides bound by self-MHC molecules on the surface of thymic cells. Positive selection is the responsibility of epithelial cells in the cortex of the thymus. Double-positive thymocytes whose receptors engage self-peptide:self-MHC complexes on these cells continue their maturation. During positive selection the α-chain genes continue to rearrange, allowing the T-cell to try different receptors in order to find one that engages self MHC. Despite this receptor editing, the vast majority of double-positive thymocytes fail positive selection and die by apoptosis. The class of MHC molecule that drives positive selection determines which co-receptor—CD4 or CD8—is maintained on the single-positive T cell.

Negative selection is effected by other cells of the thymus, most importantly the dendritic cells and macrophages, which derive from bone marrow progenitors. Negative selection eliminates autoreactive cells whose receptors bind too strongly to a self-peptide:self-MHC complex, thereby helping to create a mature T-cell repertoire that does not react to the peptide:MHC complexes of normal healthy cells. Many tissue-specific proteins are also expressed by a subpopulation of thymic epithelial cells, which improves the efficiency of negative selection. Having survived both positive and negative selection the now mature T cells leave the thymus in the blood and circulate through the secondary lymphoid organs, where they encounter foreign antigen. On activation by antigen they differentiate further into effector T cells of various types.

Autoreactive T cells that avoid negative selection and enter the peripheral circulation are either anergized on contact with self antigen or actively suppressed by regulatory T cells. T-cell tumors correspond to early and late stages of T-cell development, but not to intermediate stages.

Summary to Chapter 7

In the thymus, three functionally distinct types of T cell develop from a common progenitor that comes from the bone marrow. One type of T cell expresses $\gamma{:}\delta$ receptors and is not restricted to the recognition of peptide antigens presented by MHC molecules. The other two types of T cell express $\alpha{:}\beta$ receptors and are distinguished by the co-receptors that they express—CD4 or CD8—and the class of MHC molecule to which their receptors are restricted. Commitment to either the $\gamma{:}\delta$ or the $\alpha{:}\beta$ T-cell lineage is determined by which pair of T-cell receptor loci successfully rearranges first. Subsequent phases of development in the thymus concern only the $\alpha{:}\beta$ T cells. The primary repertoire of T-cell receptors produced by gene rearrangement is acted upon by both positive and negative selection to produce a functional repertoire of mature naive T cells whose receptors can be activated by pathogen-derived peptides presented by self-MHC molecules but cannot be activated by the self peptides derived from the body's normal constituents. Positive selection tests the ability of $\alpha{:}\beta$ T-cell receptors to interact with self-MHC molecules expressed on the surfaces of cells in the thymus. Thymocytes with receptors that interact with self-peptide:self-MHC complexes are signaled to continue their maturation. Negative selection then eliminates those cells whose receptors interact too strongly with self-peptide:self-MHC complexes. The small fraction of thymocytes that survive both positive and negative selection leave the thymus to become mature circulating $\alpha{:}\beta$ T cells. These phases in T-cell development are summarized in Figure 7.21.

T-cell development produces several different types of T cell and is therefore more complicated than B-cell development. It also seems to be considerably

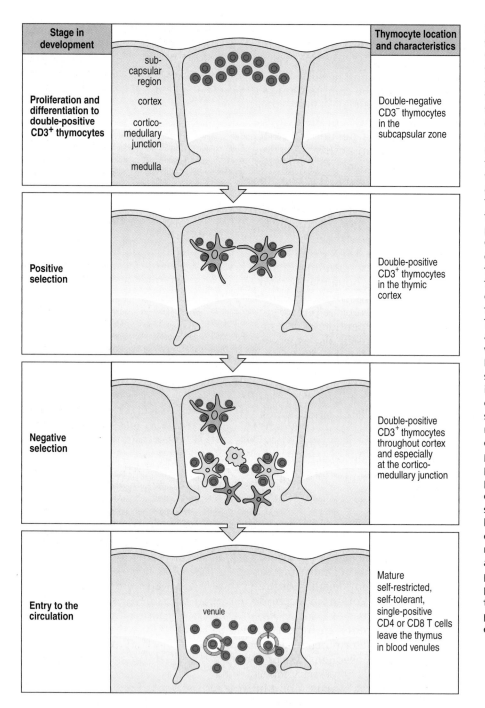

Stage in development	Thymocyte location and characteristics
Proliferation and differentiation to double-positive CD3+ thymocytes	Double-negative CD3− thymocytes in the subcapsular zone
Positive selection	Double-positive CD3+ thymocytes in the thymic cortex
Negative selection	Double-positive CD3+ thymocytes throughout cortex and especially at the cortico-medullary junction
Entry to the circulation	Mature self-restricted, self-tolerant, single-positive CD4 or CD8 T cells leave the thymus in blood venules

(Labels within figure: sub-capsular region, cortex, cortico-medullary junction, medulla, venule)

Figure 7.21 T cells develop in the thymus in a series of stages. In the first phase, thymocyte progenitors enter the thymus from the blood and migrate to the subcapsular region. At this stage, they do not express the antigen receptor, the CD3 complex, or the CD4 and CD8 co-receptors, and are known as double-negative thymocytes (top panel). These cells proliferate and begin to rearrange their β, γ, and δ T-cell receptor genes, which leads to the production of γ:δ cells and to cells expressing the pre-T-cell receptor. Stimulation through the pre-T-cell receptor leads to cell proliferation and the expression of both CD4 and CD8 co-receptors, producing double-positive cells. As the cells mature, they move deeper into the thymus. In the second phase of development, the double-positive thymocytes rearrange their α-chain genes, express an α:β T-cell receptor and the CD3 complex, and become sensitive to interaction with self-peptide:self-MHC complexes. Double-positive cells undergo positive selection in the thymic cortex through intimate contact with cortical epithelial cells (second panel). During positive selection, matching between the receptor specificity for MHC and the co-receptor molecules starts to take place, which eventually leads to single-positive CD4 or CD8 T cells. In the third phase of development, double-positive cells undergo negative selection for self-reactivity. Negative selection is believed to be most stringent at the cortico-medullary junction, where the nearly mature thymocytes encounter a high density of dendritic cells (third panel). Thymocytes that survive both positive and negative selection leave the thymus in the blood as mature single-positive CD4 or CD8 T cells and enter the circulation (bottom panel).

more wasteful of cells. The vast majority of developing thymocytes die without ever performing a useful task, and only a few percent of thymocytes fulfill the stringent requirements of selection and leave the thymus to enter the circulation. Whereas the bone marrow is continually turning over the B-cell repertoire during the whole of a person's lifetime, the thymus works principally during youth, when it serves to accumulate a repertoire of T cells that can then be used throughout life. This difference might reflect the magnitude of the body's investment in the development of each useful T cell, and the savings to be made by gradually shutting down the thymus with age.

Questions

7–1   The surrogate light chain operating during pre-B-cell development is made up of VpreB:λ5. Its expression with μ on the pre-B-cell surface is an important checkpoint in B-cell maturation. Name the T-cell analog of VpreB:λ5 and discuss how it is functionally similar.

7–2   Double-negative thymocytes initiate rearrangement at the _____ locus (loci) before all other T-cell receptor genes.
   a. γ and δ
   b. β
   c. α and β
   d. α, γ, and δ
   e. β, γ, and δ.

7–3   The function of negative selection of thymocytes in the thymus is to eliminate
   a. single-positive thymocytes
   b. double-positive thymocytes
   c. alloreactive thymocytes
   d. autoreactive thymocytes
   e. apoptotic thymocytes.

7–4   During the early developmental stages of α:β T cells in the thymus, there are two key checkpoints that must be satisfied to permit the progression of T-cell development. Explain what occurs at each checkpoint.

7–5   If a T-cell receptor on a double-positive thymocyte binds to a self-peptide:self-MHC class I complex with low affinity the result is
   a. negative selection and apoptosis
   b. cell proliferation
   c. rearrangement of the second β-chain locus
   d. positive selection of a CD4$^+$ T cell
   e. positive selection of a CD8$^+$ T cell.

7–6   The expression of MHC class II molecules is restricted to a small number of cell types.
   A. What are these cell types?
   B. Which of these cell types populate the thymus or circulate through it, and what role do they play in mediating positive and/or negative selection?
   C. Can you explain why it would be detrimental for non-circulating cells that populate tissues and glands to express MHC class II molecules?

7–7   In T cells, allelic exclusion of the α-chain locus is relatively ineffective, resulting in the production of some T cells with two T-cell receptors of differing antigen specificity on their cell surface.
   A. Will both these receptors have to pass positive selection for the cell to survive? Explain your answer.
   B. Will both receptors have to pass negative selection for the cell to survive? Explain your answer.
   C. Is there a potential problem having T cells with dual specificity surviving these selection processes and being exported to the periphery?

7–8   Mature B cells undergo somatic hypermutation after activation, which, after affinity maturation, results in the production of antibody with a higher affinity for antigen than in the primary antibody response. Suggest some reasons why T cells have not evolved the same capacity.

7–9   MHC class II deficiency is inherited as an autosomal recessive trait and involves a defect in the coordination of transcription factors involved in regulating the expression of all MHC class II genes (HLA-DP, HLA-DQ, and HLA-DR).
   A. What is the effect of MHC class II deficiency?
   B. Explain why hypogammaglobulinemia is associated with this deficiency.

7–10   As we age, our thymus shrinks, or atrophies, by a process called involution, yet T-cell immunity is still functional in old age.
   A. Explain how T-cell numbers in the periphery remain constant in the absence of continual replenishment from the thymus.
   B. How does this differ from the maintenance of the B-cell repertoire?

7–11
   A. Explain two ways in which the expression and processing of self antigens in thymic epithelium differs from the expression and processing of self antigens outside the thymus.
   B. In what way is the thymic situation advantageous for the purposes of negative selection?

7–12
   A. What is the role of regulatory CD4 T cells (T$_{reg}$)?
   B. How can T$_{reg}$ be distinguished from other non-regulatory CD4 T cells?

7–13   Which of the following statements is correct?
   a. In adults the mature T-cell repertoire is self-renewing and does not require a thymus for provision of new T cells.
   b. T cells and B cells are both short-lived cells and require continual replenishment from primary lymphoid organs.
   c. The human thymus is not fully functional until age 30, at which time it begins to shrink and atrophy.
   d. In DiGeorge syndrome the bone marrow takes over the function of the thymus and produces mature peripheral T cells.
   e. None of the above statements is correct.

7–14   Individuals with a defective autoimmune regulator gene (*AIRE*) exhibit
   a. DiGeorge syndrome
   b. autoimmune polyendocrinopathy–candidiasis–ectodermal dystrophy (APECED)
   c. severe combined immunodeficiency (SCID)
   d. MHC class I deficiency
   e. MHC class II deficiency.

7–15   Giulia McGettigan was born full term with a malformed jaw, cleft palate, a ventricular septal defect, and hypocalcemia. Within 48 hours of birth she developed muscle tetany, convulsions, tachypnea, and a systolic murmur. A chest X-ray showed an enlarged heart and the absence of a thymic shadow. Blood tests showed severely depleted levels of CD4 and CD8 T cells; B-cell numbers were low but within normal range. Parathyroid hormone was undetectable. Fluorescence *in situ* hybridization of the buccal mucosa revealed a small deletion in the long arm of chromosome 22. Giulia failed to thrive and battled chronic diarrhea and opportunistic infections, including oral candidiasis and *Pneumocystis carinii,* the latter infection causing her death. Giulia most likely had which of the following immunodeficiency diseases?

a. AIDS
b. DiGeorge syndrome
c. bare lymphocyte syndrome
d. chronic granulomatous disease
e. hyper IgM syndrome.

The lymph node, an example of a secondary lymphoid tissue where adaptive immune responses are produced.

Chapter 8

T Cell-Mediated Immunity

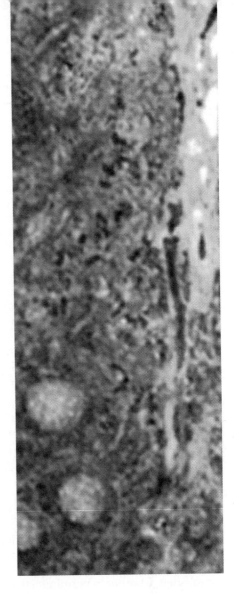

In Chapter 7 we saw how T cells develop in the thymus into a population of mature naive T cells. These naive T cells now circulate in the blood and the lymphatic system, passing through secondary lymphoid tissues where they can meet and be activated by their specific antigen. Pathogens and their antigens are brought from infected sites to the T-cell areas of the secondary lymphoid tissues in the draining lymph and by dendritic cells, which are uniquely proficient in the uptake and processing of antigens for presentation to naive T cells. Interaction of antigen on the dendritic cell surface with the T-cell receptor activates the T cell, which then undergoes clonal expansion and differentiation into effector T cells. Effector T cells either remain in the lymphoid tissue or migrate to sites of infection. Either way, a subsequent encounter with antigen triggers them to perform their effector function.

In the first part of this chapter we consider what happens when a naive T cell first encounters its specific antigen and is stimulated to differentiate into an effector T cell. This process of **T-cell activation**—also referred to as T-cell **priming**—is the first stage of a primary adaptive immune response. Effector T cell functions are examined in the second part of the chapter, where we also consider the role of regulatory T cells. Effector CD8 T cells are uniformly cytotoxic T cells that kill pathogen-infected cells. Effector CD4 T cells comprise several different functional subtypes, but they all secrete cytokines that activate other cells of the immune system. Because the function of effector CD4 T cells is principally to help other cells achieve their effector functions, they are often called helper T cells.

Cytotoxic CD8 T cells travel to infected tissues and kill any type of cell whose MHC class I molecules are presenting antigens to which the T cells are specific. Some effector CD4 T cells also travel to sites of infection, where they recognize their specific antigens presented by MHC class II molecules on macrophages. Through cell–cell contact and the delivery of cytokines, the CD4 T cells make the macrophages more proficient in capturing and killing pathogens. Other effector CD4 T cells stay in the lymphoid tissue and recognize antigens presented by MHC class II molecules on naive B cells. Again through cell–cell contact and cytokine production, these CD4 T cells activate the B cells to make pathogen-specific antibodies. These effector actions, which eventually lead to the removal and destruction of the pathogen, constitute the second stage of the primary immune response.

Activation of naive T cells on encounter with antigen

Once an infection begins, the immune system faces the challenge of quickly bringing the minute fraction of naive T cells that are specific for the pathogen into contact with pathogen antigens. This is accomplished in the secondary lymphoid tissues, in which antigens are brought from outlying tissues by dendritic cells and the lymph to meet naive T cells brought in by the blood. In this part of the chapter we shall examine the activation of naive T cells to become effector T cells by dendritic cells within secondary lymphoid tissues. Once activated, some antigen-specific effector T cells are sent out to the infected sites to work on the front line of defense, whereas others stay in the lymphoid tissues to help develop more weaponry. We shall also see how the interaction of a naive T cell with antigen presented by cells other than dendritic cells and other professional antigen-presenting cells leads to inactivation of the T cell rather than its activation. This response ensures that mature T cells reactive to self antigens are eliminated before they can become effector cells. Overall, the first phase of the primary immune response produces an expanded effector T-cell population that is ready to fight the infecting pathogen but is tolerant of self antigens.

8-1   Dendritic cells carry antigens from sites of infection to secondary lymphoid tissues

The immune system does not attempt to develop adaptive immune responses at the innumerable different sites in the body where pathogens set up an infection. Instead, its strategy is to capture some of the pathogen and take it to the nearest secondary lymphoid tissue, whose function is the generation of an adaptive immune response. Antigen capture and initiation of a primary immune response are carried out mainly by dendritic cells and to a lesser extent by macrophages—two related types of professional antigen-presenting cell. The course of events is essentially the same wherever an infection is established, whether in peripheral tissues, at mucosal surfaces, or in the blood (Figure 8.1).

Dendritic cells and macrophages are present as sentinels in all the body's tissues. During the innate immune response to infection, both these cell types become active in the uptake and degradation of pathogens and in the processing and presentation of their antigens on MHC class I and class II molecules. Whereas the macrophage has a range of functions in the defense and repair of damaged tissues, the major function of the dendritic cell is to trigger T-cell responses, for which it is highly specialized and highly effective. In particular, the dendritic cell is far superior to the macrophage at stimulating naive T cells and initiating a primary immune response. One reason for this superiority is that dendritic cells are migratory cells that carry their load of antigen from the site of infection to the nearest secondary lymphoid tissue, where they meet naive T cells. In contrast, macrophages are sessile cells that cannot move about. Consequently, the macrophages resident in infected tissue, and the cells most heavily exposed to the pathogen, have no opportunity to interact with naive T cells. Macrophages resident in the secondary lymphoid tissues, however, can contribute to T-cell activation if they ingest pathogens and their antigens that have been carried to the lymphoid tissue in the lymph, blood or extracellular fluid.

For infections in the skin and other peripheral tissues, the T-cell response is produced against antigens carried to the draining lymph nodes; for blood infections, antigens enter the spleen. The response to infections of the respiratory mucosa is in the tonsils or other bronchial-associated lymphoid tissues, whereas the response to gastrointestinal infections is in the Peyer's

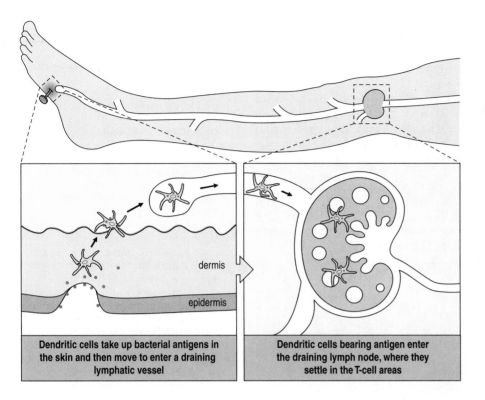

Dendritic cells take up bacterial antigens in the skin and then move to enter a draining lymphatic vessel

Dendritic cells bearing antigen enter the draining lymph node, where they settle in the T-cell areas

patches, appendix, or other gut-associated lymphoid tissues. In each of the secondary lymphoid organs a similar sequence of events occurs, which we shall illustrate by using the lymph node.

The movement of a dendritic cell from a peripheral site of infection to a secondary lymphoid organ is accompanied by changes in the dendritic cell's surface molecules, functions, and morphology (Figure 8.2). Whereas dendritic cells in tissues are active in the capture, uptake, and processing of antigens, they lose these properties on moving to a secondary lymphoid organ but gain the capacity to interact well with naive T cells. The dendritic cells in tissues are called **immature dendritic cells**, whereas those in lymph nodes are called **mature dendritic cells** or **activated dendritic cells**. On maturation, the finger-like processes called dendrites, after which the dendritic cell is named, become highly elaborated, which facilitates extensive interaction with T cells in the cortex of the lymph node. Dendritic cells are confined to the T-cell regions of the cortex, whereas macrophages are present in both the cortex and medulla. In addition to capturing free pathogens and other particulate matter being carried into the lymph node by the afferent lymph, macrophages are also responsible for eliminating the apoptotic lymphocytes produced as the inevitable by-products of the adaptive immune response.

8-2   Dendritic cells are adept and versatile at processing antigens from pathogens

To cope with the range of pathogenic microorganisms they encounter, dendritic cells have a variety of cellular mechanisms by which they obtain and process pathogen proteins (Figure 8.3). Like macrophages, dendritic cells bear numerous receptors for pathogen components on their surface, including phagocytic receptors such as the mannose receptor, and signaling receptors such as the Toll-like receptors. Receptor-mediated endocytosis is used to capture bacteria and virus particles from the extracellular fluid and target them for processing in the lysosomes. Antigens processed by this route are

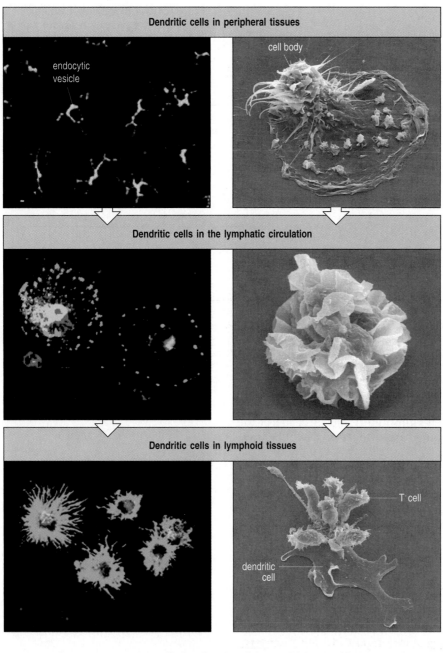

Figure 8.2 Dendritic cells change their functions on taking antigen from infected sites to secondary lymphoid tissues. In these images of dendritic cells in peripheral tissue (top panels), lymph (middle panels) and lymph node (bottom panels), the left panels are fluorescent micrographs of several cells with MHC class II molecules stained green and a lysosomal protein stained red, and the right panels are scanning electron micrographs of single cells. The dendritic cell body is clearly seen in the top right panel, but in the top left panel the cell bodies are difficult to discern. The observed yellow staining shows endocytic vesicles within the dendrites that contain both MHC class II molecules and lysosomal proteins: the combination of the red and green stains gives the yellow color. On activation and passage to the lymph the morphology of the dendritic cell changes, as shown in the middle panels. Phagocytosis has now stopped, a change indicated in the middle left panel by a partial separation of the green MHC class II staining from the red lysosomal protein staining. This shows that peptide-loaded MHC class II molecules are moving out of the endocytic vesicles and onto the cell surface. On reaching the T-cell area of a lymph node (bottom panels), the now mature dendritic cells concentrate on antigen presentation and T-cell stimulation instead of the uptake and processing of antigens. Here the MHC class II molecules, which are at high density on the surface of the numerous dendrites, are completely separated from the lysosomal protein in the intracellular vesicles. Photographs courtesy of I. Mellman, P. Pierre, and S. Turley.

presented to naive CD4 T cells by the dendritic cell's MHC class II molecules (see Section 5-12, p. 140). For pathogens that are not recognized by the phagocytic receptors, the nonspecific ingestion by the dendritic cell of large amounts of extracellular fluid by the process called **macropinocytosis** serves the same end.

Dendritic cells can also be infected by certain types of virus, which then make viral proteins that are degraded in the cytosol by the proteasome and delivered to the endoplasmic reticulum for presentation to naive CD8 T cells by MHC class I molecules. For viruses that do not infect dendritic cells, uptake of viral particles by receptor-mediated endocytosis or macropinocytosis can still lead to the presentation of viral peptides by MHC class I through cross-presentation from the endocytic pathway (normally leading to presentation on MHC class II) to the MHC class I pathway (see Section 5-16, p. 144). In viral infections such as influenza, in which infection can kill dendritic cells, the dendritic cell that carries the infection to the secondary lymphoid tissue may not be capable of presenting antigens to T cells. In such instances, antigens

Routes of antigen processing and presentation by dendritic cells				
Receptor-mediated endocytosis	Macropinocytosis	Viral infection	Cross-presentation after phagocytic or macropinocytic uptake	Transfer from incoming dendritic cell to resident dendritic cell
Type of pathogen presented Extracellular bacteria	Extracellular bacteria, soluble antigens, virus particles	Viruses	Viruses	Viruses
MHC molecules loaded MHC class II	MHC class II	MHC class I	MHC class I	MHC class I
Type of naive T cell activated CD4 T cells	CD4 T cells	CD8 T cells	CD8 T cells	CD8 T cells

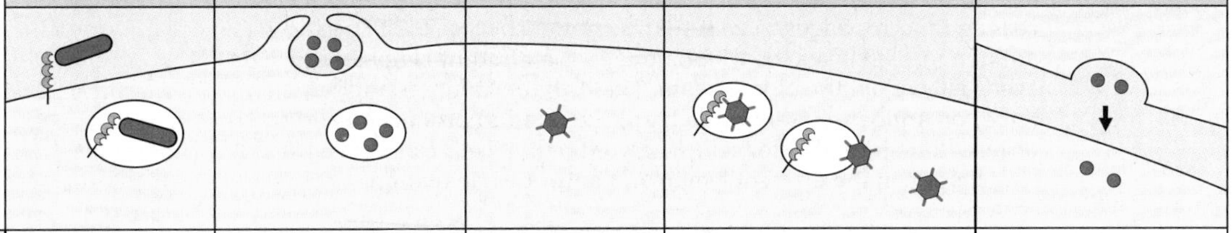

Figure 8.3 Dendritic cells use several pathways to process and present protein antigens. Uptake of antigens by phagocytosis or macropinocytosis delivers antigens to endocytic vesicles for presentation by MHC class II molecules to CD4 T cells (first two panels). Viral infection of the dendritic cell delivers peptides processed in the cytosol to the endoplasmic reticulum for presentation by MHC class I molecules to CD8 T cells (third panel). Viral particles taken up by the 'class II' pathways of phagocytosis and macropinocytosis can be delivered to the cytosol for processing and presentation to CD8 T cells by MHC class I (fourth panel). The mechanism of this cross-presentation is poorly understood. Lastly, antigens taken up by one dendritic cell can be delivered to a second dendritic cell for presentation by MHC class I molecules to CD8 T cells (fifth panel).

can be transferred from infected dendritic cells to healthy dendritic cells resident in the lymphoid tissue, and these present the antigens to CD8 T cells by cross-presentation.

Dendritic cells carry all of the Toll-like receptors except TLR9, making them highly sensitive to the presence of all manner of pathogens (see Chapter 2). Signals from the Toll-like receptors change the pattern of gene expression in the dendritic cell, leading to the cell's activation. One effect of activation is to increase the efficiency with which antigens are taken up and processed for presentation by MHC class II. Activation also causes expression on the dendritic cell surface of CCR7, the receptor for the chemokine CCL21, which is made in secondary lymphoid tissue. Signals induced by the interaction of CCL21 with CCR7 cause the pathogen-loaded and migrating dendritic cells to leave the draining lymph and enter the draining lymph node. These signals also induce the maturation of dendritic cells, so that when they reach the lymphoid tissue they no longer take up and process antigens. Instead, they now concentrate on the presentation of antigens to naive T cells. During maturation their expression of MHC class I and II molecules increases, leading to an abundance of stable, long-lived peptide:MHC complexes on the surface of the mature dendritic cell.

8-3 Naive T cells first encounter antigen presented by dendritic cells in secondary lymphoid tissues

Whereas the antigen-presenting dendritic cells enter the lymph node in the afferent lymph that drains from the site of infection, naive T cells can enter a lymph node in one of two ways. As we discussed in Chapter 1, one way is via the blood capillaries that provide oxygen and nutrients to the tissue. T cells in the blood bind to the endothelial cells of the thin-walled high endothelial venules (HEV), squeeze through the vessel wall, and enter the cortical region of the node. The naive T cell then passes through the crowded tissue, where it encounters dendritic cells and where its antigen receptor can examine the

Figure 8.4 Naive T cells encounter antigen during their recirculation through secondary lymphoid organs. Naive T cells (blue and green) recirculate through secondary lymphoid organs, such as the lymph node shown here. They leave the blood at high endothelial venules and enter the lymph-node cortex, where they mingle with professional antigen-presenting cells (mainly dendritic cells and macrophages). T cells that do not encounter their specific antigen (green) leave the lymph node in the efferent lymph and eventually rejoin the bloodstream. T cells that encounter antigen (blue) on antigen-presenting cells are activated to proliferate and to differentiate into effector cells. These effector T cells can also leave the lymph node in the efferent lymph and enter the circulation.

peptide:MHC complexes on their surfaces. When a T cell encounters a peptide:MHC complex to which its T-cell receptor binds, the T cell is retained in the lymph node and activated. It then proliferates and differentiates into a clone of effector T cells (Figure 8.4).

A second way in which a naive T cell can enter a lymph node is via the lymph. Lymph nodes are junctions in which several afferent lymphatic vessels unite to form a single efferent lymphatic vessel. The lymph draining from peripheral tissues will often pass through several lymph nodes on its way to the blood. Naive T cells that entered an 'upstream' lymph node from the blood, but did not encounter their specific antigen there, will leave in the efferent lymph and be carried to a 'downstream' lymph node which they will enter via one of the afferent lymphatic vessels. In this case they do not have to cross an epithelial barrier and enter the T-cell zone directly. If this lymph node is draining a site of infection, such naive T cells have the possibility of being stimulated by pathogen-laden dendritic cells that have entered the lymph node via a different afferent lymphatic, one that is draining the infected site (Figure 8.5).

In any given infection, naive T cells specific for the pathogen will represent only 1 in 10^4 to 1 in 10^6 of the total pool of circulating T cells. During most passages through a lymph node, a T cell does not find its specific antigen and leaves the medulla in the efferent lymph to continue recirculation. In the absence of specific antigen, circulating naive T cells live for many years as small non-dividing cells with condensed chromatin, scanty cytoplasm, and little synthesis of RNA or proteins. With time, the entire population of circulating T cells will pass through any given lymph node. The trapping of pathogens and their antigens in the lymphoid tissue nearest to the site of infection creates a concentrated depot of processed and presented antigens. This enables the small subpopulation of T cells specific for those antigens to be efficiently pulled out of the circulating T-cell pool and activated.

Once an antigen-specific T cell has been trapped in a lymph node by an antigen-presenting cell and activated, it takes several days for the activated T cell to proliferate and for its progeny to differentiate into effector T cells. This accounts for much of the delay between the onset of an infection and the appearance of a primary adaptive immune response. Most effector T cells leave the lymph node in the efferent lymph and on reaching the blood are rapidly carried to the site of infection, where they perform their effector functions.

8-4   Homing of naive T cells to secondary lymphoid tissues is determined by chemokines and cell-adhesion molecules

The specific entry of a naive T cell from the bloodstream into the cortex of a lymph node or other secondary lymphoid tissue—the process called homing—is analogous to that used by naive B cells to enter lymph nodes (see

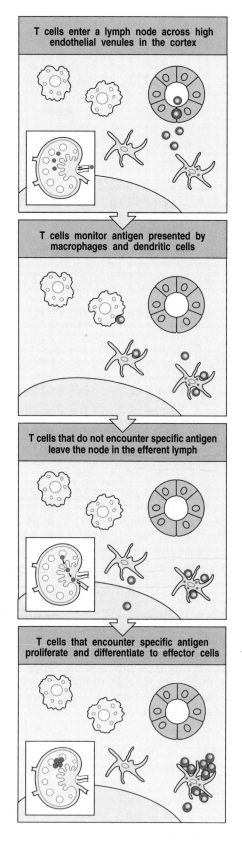

T cells enter a lymph node across high endothelial venules in the cortex

T cells monitor antigen presented by macrophages and dendritic cells

T cells that do not encounter specific antigen leave the node in the efferent lymph

T cells that encounter specific antigen proliferate and differentiate to effector cells

Figure 6.21, p. 178) or by neutrophils to enter infected tissues (see Figure 2.31, p. 55). The chemokines CCL21 and CCL19 are secreted by stromal cells and dendritic cells in the lymph-node cortex and are bound to the surface of the high endothelial cells of the venules, where they establish a concentration gradient along the endothelial surface. Like naive B cells, naive T cells express the receptor CCR7, which binds the chemokines CCL21 and CCL19 and by which the cell is guided along the chemokine gradient. The initial contact between the T cell and the endothelium allows interactions to be established between complementary adhesion molecules on the surfaces of the two cell types.

These interactions involve the four types of cell-adhesion molecule (see Figure 2.30, p. 54). **L-selectin** on the T-cell surface binds to the sulfated sialyl-Lewis^x carbohydrate of the vascular addressins **CD34** and **GlyCAM-1** on the surface of the high endothelial venules (Figure 8.6). The cooperative effect of many such interactions causes the naive T cell to slow down and attach to the high endothelial surface (Figure 8.7). After the initial attachment, the contact between the naive T cell and the endothelium is strengthened by additional interactions between the integrin LFA-1 on the T-cell surface and the adhesion molecules ICAM-1 and ICAM-2 on the vascular endothelium. The interaction between the chemokines secreted by the lymph node and the chemokine receptors on the T cell sends a signal that changes the conformation of LFA-1 so that it strengthens its hold on an ICAM. Through this array of cell-surface interactions the naive T cell is held at the surface of the endothelium and gradually moves toward higher and higher concentrations of chemokines. Eventually the T cell squeezes between the endothelial cells to enter the lymph-node cortex, the source of CCL21 and CCL19 (see Figure 8.7).

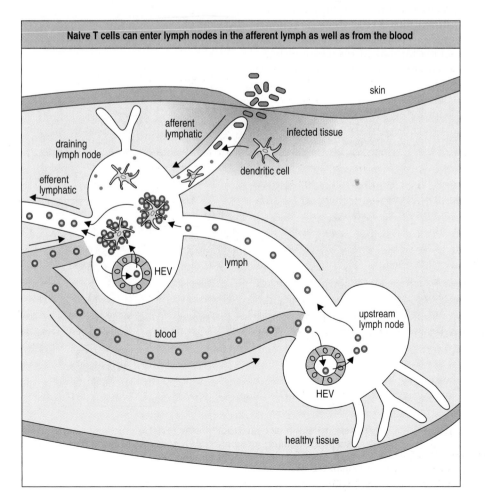

Figure 8.5 Naive T cells can enter lymph nodes from the blood or from the lymph. Recirculating naive T cells can enter a lymph node either directly from the blood or by moving from one lymph node to another via the lymphatics that connect them. In the case illustrated here, pathogen-specific T cells (blue) in the blood enter a lymph node that is draining an infected tissue. They encounter pathogen antigens, are activated, and leave as effector cells in the afferent lymphatic. At the same time, other pathogen-specific T cells (green) in the blood enter an 'upstream' lymph node that is draining healthy tissue. They do not encounter their antigen there, but they can be carried to the infected lymph node via a connecting lymphatic vessel. There they, too, will become activated by pathogen antigens.

Figure 8.6 Binding of L-selectin to mucin-like vascular addressins directs naive lymphocyte homing to lymphoid tissues. L-selectin on naive T cells and naive B cells binds to sulfated carbohydrate sialyl-Lewis^x moieties of vascular addressins CD34 and GlyCAM-1 on the high endothelial cells of lymph venules.

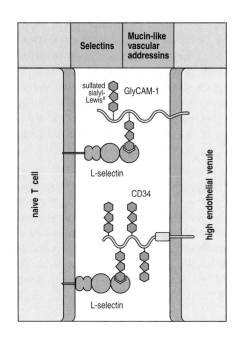

As naive T cells negotiate their way through the packed cells of the cortex, they bind transiently to the dendritic cells they meet. These interactions involve LFA-1 on the T cells binding to ICAM-1 and ICAM-2 on the dendritic cell, and LFA-1 on the dendritic cell binding to a third kind of ICAM—**ICAM-3**—on the T-cell surface. ICAM-3 also binds to the lectin **DC-SIGN**, an adhesion molecule unique to activated dendritic cells. Adhesion is strengthened by interaction between **CD2** on the T cell and **LFA-3** on the dendritic cell (Figure 8.8). These transitory cell–cell interactions enable the T-cell receptor to screen the peptide:MHC complexes on the surface of dendritic cells for ones that engage the receptor and activate the T cell.

When a naive T cell encounters a specific peptide:MHC complex, a signal is delivered through the T-cell receptor. This induces a change in the conformation of the T cell's LFA-1 molecules that increases their affinity for ICAMs (Figure 8.9). The interaction of the T cell with the antigen-presenting cell is stabilized and can last for several days, during which time the T cell proliferates, and its progeny, while also remaining in contact with the antigen-presenting cell, differentiate into effector cells.

If a naive T cell does not meet its antigen it will exit from a lymph node via the cortical sinuses, which lead into the medullary sinus and thence into the efferent lymphatic vessel. This is also the exit route for effector T cells after they have differentiated. The exit of T cells from a secondary lymphoid organ involves the lipid molecule **sphingosine 1-phosphate** (**S1P**), which has chemotactic activity. There seems to be a concentration gradient of S1P between the lymphoid tissues and lymph or blood, such that naive T cells expressing a receptor for S1P are drawn away from the lymphoid tissues and back into circulation.

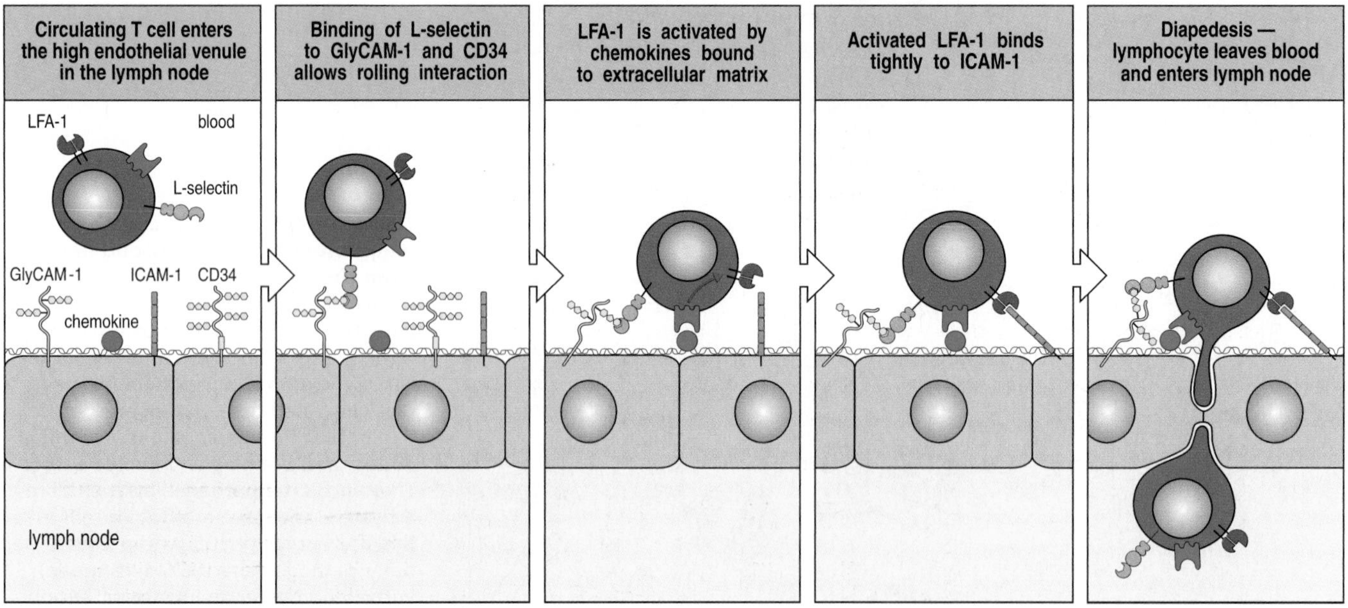

Circulating T cell enters the high endothelial venule in the lymph node	Binding of L-selectin to GlyCAM-1 and CD34 allows rolling interaction	LFA-1 is activated by chemokines bound to extracellular matrix	Activated LFA-1 binds tightly to ICAM-1	Diapedesis — lymphocyte leaves blood and enters lymph node

Figure 8.7 Naive T and B lymphocytes circulate in the blood and enter lymph nodes by crossing high endothelial venules. Lymphocytes bind to high endothelium in the lymph node through the interaction of L-selectin with vascular addressins. Chemokines, which are also bound to the endothelium, activate the integrin LFA-1 on the lymphocyte surface, enabling it to bind tightly to ICAM-1 on the endothelial cell. Establishment of tight binding allows the lymphocyte to squeeze between two endothelial cells, leaving the lumen of the blood vessel and entering the lymph node proper.

Figure 8.8 **Cell-surface molecules of the immunoglobulin superfamily initiate lymphocyte adhesion to professional antigen-presenting cells.** In the initial encounter of T cells with antigen-presenting dendritic cells, CD2, binding to LFA-3 on the antigen-presenting cell, synergizes with LFA-1 binding to ICAM-1 and ICAM-2. An interaction that seems to be exclusive to the interaction of naive T cells with dendritic cells is that between ICAM-3 on the naive T cell and DC-SIGN, a C-type lectin specific to dendritic cells, which binds ICAM-3 with high affinity.

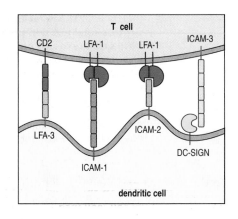

T cells activated by antigen suppress the expression of S1P receptors for several days. This means they cannot respond to the S1P gradient and so stay in the lymph node while they proliferate and complete their differentiation into effector T cells. After several days the effector T cells re-express S1P receptors and are then able to leave the lymph node, guided by the S1P gradient.

8-5   Activation of naive T cells requires a co-stimulatory signal delivered by a professional antigen-presenting cell

The intracellular signal generated by ligation of the T-cell receptor with a specific peptide:MHC complex is necessary to activate a naive T cell but is not sufficient. A second, **co-stimulatory**, signal is required. Co-stimulatory signals are delivered only by the professional antigen-presenting cells—dendritic cells, macrophages, and B cells—hence the obligatory role of these cells in activating naive T cells. A further requirement is that antigen-specific stimulation and co-stimulation must both be delivered by ligands on the same antigen-presenting cell.

The cell-surface protein on naive T cells that receives the co-stimulatory signal is called **CD28**. Its ligands are the structurally related **B7.1 (CD80)** and **B7.2 (CD86)** proteins, which are expressed only on professional antigen-presenting cells. B7.1 and B7.2 are collectively called **B7 molecules** and are also known as **co-stimulator molecules** or **co-stimulatory molecules**. CD28 and the B7 molecules are all members of the immunoglobulin superfamily. Activation of a naive T cell requires B7 molecules on a professional antigen-presenting cell to engage CD28 on the T-cell surface at the same time that peptide:MHC complexes on the same antigen-presenting cell engage T-cell receptors and co-receptors (either CD4 or CD8) (Figure 8.10). The combination of intracellular signals generated by these receptor–ligand interactions is essential to induce proliferation and further differentiation of the T cell.

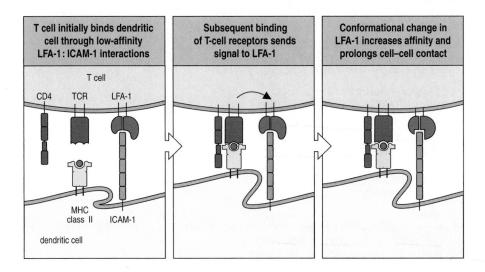

| T cell initially binds dendritic cell through low-affinity LFA-1 : ICAM-1 interactions | Subsequent binding of T-cell receptors sends signal to LFA-1 | Conformational change in LFA-1 increases affinity and prolongs cell–cell contact |

Figure 8.9 **Transient adhesive interactions between T cells and dendritic cells are stabilized by specific antigen recognition.** When a T cell binds to its specific ligand on an antigen-presenting dendritic cell, intracellular signaling through the T-cell receptor (TCR) induces a conformational change in LFA-1 that causes it to bind with higher affinity to ICAMs on the antigen-presenting cell. The T cell shown here is a CD4 T cell.

Figure 8.10 The principal co-stimulatory molecules on professional antigen-presenting cells are B7 molecules, which bind CD28 proteins on the T-cell surface. Binding of the T-cell receptor and its co-receptor CD4 to the peptide:MHC class II complex on the dendritic cell delivers a signal (arrow 1). This signal induces clonal expansion of T cells only when the co-stimulatory signal (arrow 2) is also given by the binding of CD28 to B7. Both CD28 and B7 are members of the immunoglobulin superfamily. There are two forms of B7, called B7.1 (CD80) and B7.2 (CD86), but their functional differences have yet to be understood.

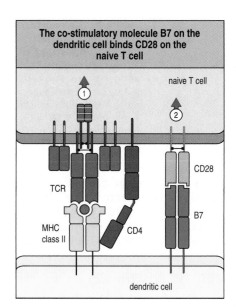

In the absence of infection, dendritic cells, macrophages and B cells do not express co-stimulatory molecules. Thus, the capacity of professional antigen-presenting cells to activate naive T cells is acquired only during infection. B7 expression is a direct consequence of infection, being induced by signaling from Toll-like receptors and other receptors of innate immunity that sense the presence of microbial products.

Although CD28 is the only B7 receptor on naive T cells, an additional receptor is expressed once T cells have been activated. This receptor, called **CTLA4**, is structurally similar to CD28 but binds B7 twentyfold more strongly than does CD28 and functions as an antagonist. Whereas B7 binding to CD28 activates a T cell, the engagement of CTLA4 dampens down activation and limits cell proliferation.

The role of CD28 in T-cell activation was tragically illustrated in 2006 by a phase 1 clinical trial in which six healthy volunteers were given a humanized anti-CD28 monoclonal antibody. In animal experiments this antibody had selectively activated regulatory T cells, which then nonspecifically suppressed other T cells. By binding to CD28 the antibody mimicked the natural B7 ligand to induce signals from CD28 that activated the regulatory T cells. The antibody was being developed as a possible therapy to reduce the T cell–mediated inflammation that characterizes arthritis and other autoimmune diseases. The phase 1 trial in healthy humans was a necessary preliminary test to determine how it behaved in humans before giving it to patients in further trials. The antibody was never given to patients, however, because of the catastrophic effects it had on the healthy volunteers—the opposite of what had been expected. Instead of inducing a suppressive response, the antibody activated vast numbers of effector T cells. These secreted massive quantities of cytokines and chemokines, causing systemic and life-threatening inflammation and autoimmunity. All six volunteers were hospitalized and became patients themselves; one remained in a coma for three weeks after experiencing failure of the heart, liver, and kidney, as well as suffering from septicemia, pneumonia, and gangrene.

8-6   Secondary lymphoid tissues contain three kinds of professional antigen-presenting cell

All three professional antigen-presenting cell types—dendritic cells, macrophages, and B cells—are present in secondary lymphoid tissues, but at different locations, as illustrated for a lymph node in Figure 8.11. Dendritic cells are present only in the cortical T-cell areas, macrophages are found throughout the cortex and medulla, and B cells are confined to the lymphoid follicles. These distributions reflect the different functions of the three cell types and their relative importance and potency in presenting antigens to naive T cells; dendritic cells are more effective than macrophages, which are more effective than B cells.

The **Langerhans cell** of the skin is a typical immature dendritic cell containing large granules, called Birbeck granules, which may be a form of phago-

	Professional antigen-presenting cells		
	Dendritic cell	**Macrophage**	**B cell**
Cell type	viral antigen / virus infecting the dendritic cell	bacterium	microbial toxin
Location in lymph node	T-cell areas		follicle
Antigen uptake	+++ Macropinocytosis and phagocytosis by tissue dendritic cells Viral infection	Phagocytosis +++	Antigen-specific receptor (Ig) ++++
MHC expression	Low on tissue dendritic cells High on dendritic cells in lymphoid tissues	Inducible by bacteria and cytokines − to +++	Constitutive Increases on activation +++ to ++++
Co-stimulator delivery	Constitutive by mature, nonphagocytic lymphoid dendritic cells ++++	Inducible − to +++	Inducible − to +++
Antigen presented	Peptides Viral antigens Allergens	Particulate antigens Intracellular and extracellular pathogens	Soluble antigens Toxins Viruses
Location	Ubiquitous throughout the body	Lymphoid tissue Connective tissue Body cavities	Lymphoid tissue Peripheral blood

Figure 8.11 Three types of professional antigen-presenting cell populate different parts of the lymph node. Dendritic cells are situated in the T-cell areas of the lymph-node cortex, whereas macrophages are distributed throughout the lymph node. B cells populate mainly the follicles. These distributions reflect differences in the functions of the three types of professional antigen-presenting cell.

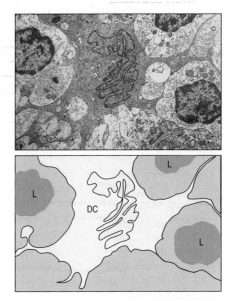

Figure 8.12 Mature dendritic cells in lymph nodes take the form of interdigitating reticular cells. Electron micrograph (top) of an interdigitating reticular cell in the T-cell area of a human lymph node, and an interpretative drawing (bottom). The interdigitating cell (DC; yellow) has a complex folded nucleus. Its cytoplasm makes a complex mesh around the surrounding T lymphocytes (L; blue). Magnification × 10,000. Photograph courtesy of N. Rooney.

some. On infection of an area of skin, the local Langerhans cells will take up and process microbial antigens before traveling to the T-cell areas in the cortex of the draining lymph node and maturing to become professional antigen-presenting cells. In lymph nodes, the mature dendritic cells have a characteristic morphology, which led to their being called **interdigitating reticular cells** (Figure 8.12). In addition to starting to express B7 molecules and producing more MHC molecules, activated dendritic cells also increase the expression of adhesion molecules. Mature dendritic cells in lymph nodes also secrete the chemokine **CCL18**, which specifically attracts naive T cells toward them. The adhesion molecule DC-SIGN is made uniquely by activated dendritic cells and, by binding tightly to ICAM-3 on the T cell (see Section 8-4), their interactions with naive T cells are strengthened.

Dendritic cells are found only in the T-cell areas of secondary lymphoid tissues, where their sole function is to activate T cells. Macrophages, in contrast, are found throughout the tissue of a lymph node because they have several different functions to perform. Macrophages are phagocytic cells that take up microorganisms and other particulate material from the extracellular environment and degrade them in phagolysosomes loaded with hydrolytic enzymes. One major function of macrophages in secondary lymphoid organs is to trap and degrade pathogens that arrive in the lymph from sites of infection. This prevents infection from reaching the blood and becoming systemic; it also enables macrophages to process and present pathogen-derived antigens to naive T cells. The removal of noninfectious particulates from the lymph by macrophages also prevents this material from entering the blood and blocking small blood vessels. A second major function of macrophages in secondary lymphoid tissues is to remove and degrade the numerous lymphocytes that are not selected to survive, and which die in these tissues.

In the absence of infection, macrophages express no B7 molecules and few MHC molecules. On their surface, however, are several receptors involved in innate immune responses. These recognize carbohydrates and other components of microbial surfaces that are not present on human cells. The receptors include the mannose receptor, scavenger receptor, complement receptors, and several Toll-like receptors (see Chapter 2). As with dendritic cells, when these receptors are engaged by their ligands, signals are transmitted to the macrophage to induce the expression of B7 and to increase the expression of MHC molecules. In this way, the presence of an infection results in macrophages becoming activated to professional antigen-presenting cell status (Figure 8.13). Because naive T cells circulate through secondary lymphoid tissues and not through peripheral sites of infection, macrophages activated in the site of infection do not serve to activate naive T cells. However, macrophages that become activated in the T-cell areas of the draining lymph node can present antigens and activate naive T cells.

The third type of professional antigen-presenting cell, the B cell, localizes to the lymphoid follicles, the B-cell areas of the secondary lymphoid tissues. The B cell specifically binds both particulate and soluble protein antigens from the extracellular environment by means of its surface immunoglobulin. The antigen is then internalized by receptor-mediated endocytosis; this is followed by its processing into peptides that bind to MHC class II molecules and can be presented to CD4 T cells. B cells therefore selectively present peptides of those antigens for which they are specific. In the primary immune response, B cells do not usually participate in the activation of naive T cells. Instead, naive B cells are themselves activated by helper CD4 T cells that recognize the pathogen antigens they present. These effector T cells have differentiated from naive antigen-specific T cells activated in the same secondary lymphoid organ.

8-7 When T cells are activated by antigen, signals from T-cell receptors and co-receptors alter the pattern of gene transcription

When a T cell binds the peptide:MHC complexes on a dendritic cell, the receptor–ligand interactions occur at localized and apposed areas of the two cell membranes. This region of contact and communication between the two cells is referred to as the **immunological synapse** or **T-cell synapse**. Within the synapse, the specific peptide:MHC complexes on the antigen-presenting cell and the T-cell receptors and co-receptors on the T cell cluster together, with cell-adhesion molecules forming a tight seal around the area. The signal that antigen has bound to the T-cell receptor is transmitted to the interior of the T cell by the cytoplasmic tails of the CD3 proteins, which are associated with the antigen-binding α and β chains of the receptor (see Figure 5.6, p. 130). The cytoplasmic tails of all the CD3 proteins contain sequences called **immunoreceptor tyrosine-based activation motifs** (**ITAMs**), which associate with cytoplasmic **protein tyrosine kinases**. These kinases are activated by receptor clustering and phosphorylate tyrosine residues in the ITAMs. Enzymes and other signaling molecules bind to the phosphorylated tyrosine

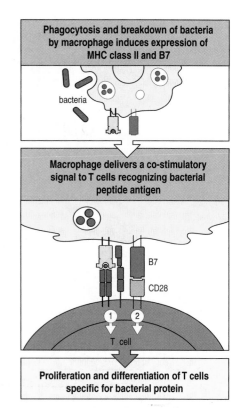

Figure 8.13 Microbial substances induce co-stimulatory activity in macrophages. Phagocytosis of bacteria by macrophages and their breakdown in the phagolysosomes lead to the release of substances such as bacterial lipopolysaccharide, which induce the expression of co-stimulatory B7 molecules on the surface of the macrophage. Peptides derived from the degradation of bacterial proteins in the macrophage vesicular system are bound by MHC class II molecules and presented on the macrophage surface. Activation of naive T cells is accomplished by the combination of B7 binding to CD28 and peptide:MHC complexes binding to the T-cell receptor.

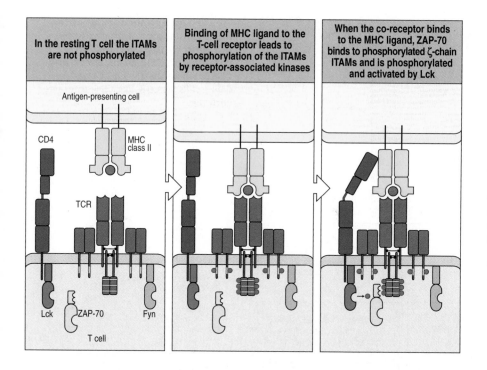

| In the resting T cell the ITAMs are not phosphorylated | Binding of MHC ligand to the T-cell receptor leads to phosphorylation of the ITAMs by receptor-associated kinases | When the co-receptor binds to the MHC ligand, ZAP-70 binds to phosphorylated ζ-chain ITAMs and is phosphorylated and activated by Lck |

Figure 8.14 Clustering of the T-cell receptor and a co-receptor initiates signaling within the T cell. When T-cell receptors become clustered on binding peptide:MHC complexes on the surface of an antigen-presenting cell, activation of receptor-associated kinases, such as Fyn, leads to phosphorylation of the CD3γ, δ, and ε ITAMs (yellow, with phosphorylated tyrosines shown as small red circles) as well as those on the ζ chain. The tyrosine kinase ZAP-70 binds to the phosphorylated ITAMs of the ζ chain but is not activated until the co-receptor binds to the MHC molecule on the antigen-presenting cell (here shown as CD4 binding to an MHC class II molecule), which brings the kinase Lck into the complex. This phosphorylates and activates ZAP-70.

residues and become activated in their turn. In this way, pathways of intracellular signaling are initiated that end with alterations in gene expression.

Signals from both the T-cell receptor and the CD4 or CD8 co-receptors combine to stimulate the T cell (Figure 8.14). The cytoplasmic tails of both CD4 and CD8 are associated with a protein tyrosine kinase called **Lck**. On formation of the T-cell receptor:MHC:co-receptor complex this kinase activates a cytoplasmic protein tyrosine kinase called **ZAP-70** (ζ chain-associated protein of 70 kDa molecular mass), which binds to the phosphorylated tyrosines on the ζ chain of the T-cell receptor complex. ZAP-70 is instrumental in initiating the intracellular signaling pathway. The importance of ZAP-70 for all subsequent signaling events was revealed by a patient with immunodeficiency due to the absence of functional ZAP-70. Although the patient had normal numbers of T cells, these were unable to develop intracellular signals on engagement of their antigen receptors.

The participation of a CD4 or CD8 co-receptor is essential for effective T-cell stimulation. A target cell minimally requires about 100 specific peptide:MHC complexes to trigger a naive T cell. Human cells express between 10,000 and 100,000 MHC molecules per cell, which means that 0.1–1% of the MHC molecules on a target cell would have to bind the same peptide for the cell to activate a T cell. In the absence of the correct co-receptor—CD8 for peptides presented by MHC class I and CD4 for peptides presented by MHC class II—stimulation of the T cell becomes highly inefficient, requiring about 10,000 specific peptide:MHC complexes on the target cell, a number almost never reached *in vivo*.

Within the immunological synapse an inner zone called the central supramolecular activation complex (c-SMAC) is where the T-cell receptors, co-receptors, co-stimulatory receptors (CD28), the CD2 adhesion molecules, and the signaling molecules concentrate. An outer zone called the peripheral supramolecular activation complex (p-SMAC) is where the integrin LFA-1, the cell-adhesion molecule ICAM-1, and talin, a cytoskeletal protein, are segregated (Figure 8.15).

Once activated, the T cell-specific kinase ZAP-70 triggers three signaling pathways that are common to many types of cell (Figure 8.16). In naive T cells, they

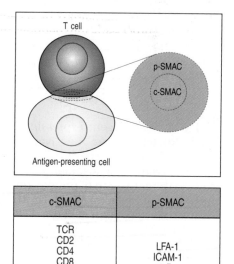

c-SMAC	p-SMAC
TCR CD2 CD4 CD8 CD28 PKC-θ	LFA-1 ICAM-1 talin

Figure 8.15 Protein–protein interactions in the area of contact between a T cell and an antigen-presenting cell form an ordered structure called the immunological synapse. The area of contact is divided into the central-supramolecular activation complex (c-SMAC) and the peripheral-supramolecular activation complex (p-SMAC), in which different sets of T-cell proteins are segregated.

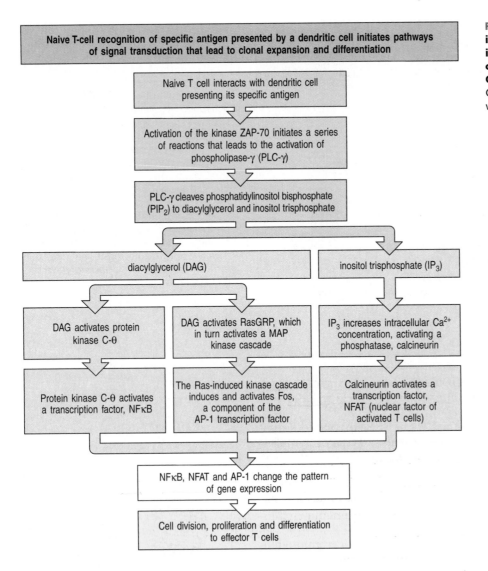

Figure 8.16 Simple outline of the intracellular signaling pathways initiated by the T-cell receptor complex, its CD4 co-receptor, and CD28. Similar pathways operate in CD8 T cells, as CD8, like CD4, interacts with Lck.

lead to changes in gene expression produced by the transcriptional activator **NFAT (Nuclear Factor of Activated T cells)** in combination with other transcription factors. One pathway initiated by ZAP-70 leads via the second messenger inositol trisphosphate to the activation of NFAT. A second pathway leads to the activation of protein kinase C-θ, which results in the induction of the transcription factor **NFκB**. The third signaling pathway initiated by ZAP-70 involves the activation of Ras, a GTP-binding protein, and leads to the activation of a nuclear protein called Fos, which is a component of the transcription factor **AP-1**. The co-stimulatory signals delivered through CD28 lead, among other things, to activation of the Jun protein, which, with Fos, forms the AP-1 transcription factor.

The combined actions of NFAT, AP-1, and NFκB turn on the transcription of genes that direct T-cell proliferation and the development of effector function. One of the most important of these genes is that for the cytokine interleukin-2.

8-8   Proliferation and differentiation of activated T cells are driven by the cytokine interleukin-2

Activation by a professional antigen-presenting cell initiates a program of differentiation in the T cell that starts with a burst of cell division and then leads to the acquisition of effector function. This program is under the control of a cytokine called **interleukin-2** (**IL-2**), which is synthesized and secreted by the activated T cell itself. IL-2 binds to IL-2 receptors at the T-cell surface to drive clonal expansion of the activated cell. IL-2 is one of a number of cytokines produced by activated and effector T cells that control the development and differentiation of cells in the immune response.

The production of IL-2 requires both the signal delivered through the T-cell receptor:co-receptor complex and the co-stimulatory signal delivered through CD28 (see Section 8-5). Signals delivered through the T-cell receptor complex activate the transcription factor NFAT, which activates transcription of the IL-2 gene. IL-2 and other cytokines have powerful effects on cells of the immune system, and their production is precisely controlled in both time and space. To avoid overproduction, cytokine mRNA is inherently unstable, and sustained production of IL-2 requires the stabilization of the mRNA. Stabilization is one of the functions of the co-stimulatory signal, and it causes a 20- to 30-fold increase in IL-2 production by the T cell. A second effect of co-stimulation is the activation of additional transcription factors that increase the rate of transcription of the IL-2 gene threefold. The principal effect of co-stimulation is therefore to increase the synthesis of IL-2 by the T cell by some 100-fold.

The IL-2 receptor on naive T cells is a heterodimer of a β and a γ chain that bind IL-2 with low affinity. On activation, a naive T cell begins to synthesize a third component of the receptor, the α chain, which on binding to the β:γ heterodimer forms a high-affinity IL-2 receptor, which is more responsive to IL-2 (Figure 8.17). On binding IL-2, the high-affinity receptor triggers the T cell to progress through cell division. T cells activated in this way can divide two to three times a day for about a week, enabling a single activated T cell to produce thousands of identical daughter cells. This proliferative phase is of crucial importance to the immune response because it produces large numbers of antigen-specific effector cells from rare antigen-specific naive T cells. In the response to certain viruses, nearly 50% of the CD8 T cells present at the peak of the response are specific for a single viral peptide:MHC complex.

The importance of IL-2 in the activation of the adaptive immune response is reflected in the mode of action of the immunosuppressive drugs cyclosporin A (cyclosporine), tacrolimus (also called FK506), and rapamycin (also called sirolimus), which are used to prevent the rejection of organ transplants. Cyclosporin A and tacrolimus inhibit IL-2 production by disrupting signals from the T-cell receptor, whereas rapamycin inhibits signaling from the IL-2 receptor. These drugs, therefore, suppress the activation and differentiation of naive T cells and all immune responses that require activated T cells.

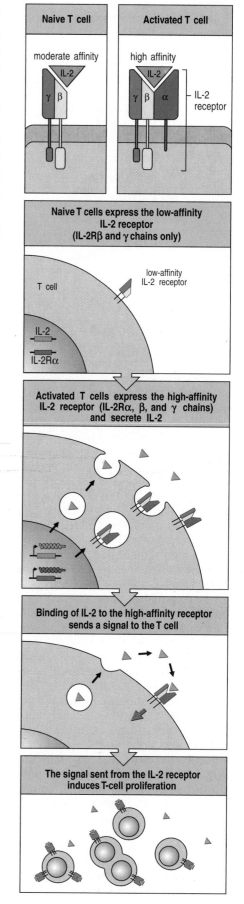

Figure 8.17 Activated T cells secrete and respond to interleukin-2 (IL-2). The receptor for IL-2 is made from three different chains, α, β, and γ. A receptor with low affinity for IL-2 is made from β and γ chains only, whereas the high-affinity receptor also requires the α chain (top panels). Naive T cells express the low-affinity receptor (second panel). Activation of a naive T cell by the recognition of a peptide:MHC complex accompanied by co-stimulation induces both the synthesis and secretion of IL-2 (orange triangles) and the synthesis of the IL-2 receptor α chain (blue). The α chain combines with the β and γ chains to make a high-affinity receptor for IL-2 (third panel). The cell also enters the first phase (G1) of the cell-division cycle. IL-2 binds to the IL-2 receptor (fourth panel), producing an intracellular signal that promotes T-cell proliferation (fifth panel).

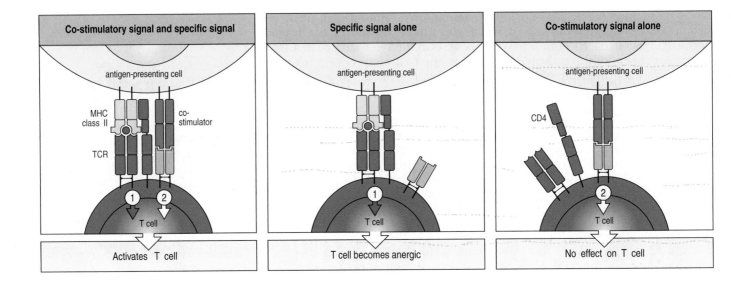

Figure 8.18 **T-cell tolerance to antigens expressed on non-professional antigen-presenting cells results from antigen recognition in the absence of co-stimulation.** A naive T cell can be activated only by an antigen-presenting cell carrying both a specific peptide:MHC complex and a co-stimulatory molecule on its surface. This combination results in the naive T cell's receipt of signal 1 from the T-cell receptor and signal 2 from the co-stimulator (left panel). When the antigen-presenting cell has the specific peptide:MHC complex to deliver signal 1, but no co-stimulator to deliver signal 2, the T cell enters a nonresponsive state called anergy (center panel). When the antigen-presenting cell has a co-stimulator to deliver signal 2, but no specific peptide:MHC complex to deliver signal 1, the naive T cell neither responds nor becomes anergic (right panel).

8-9   Antigen recognition by a naive T cell in the absence of co-stimulation leads to the T cell becoming nonresponsive

Among the mature naive T cells entering the peripheral circulation from the thymus are some that are specific for self proteins not encountered in the thymus. However, these T cells are unlikely to be activated because the cells presenting these self antigens will not express co-stimulatory molecules. When the T-cell receptor on a mature naive T cell binds to a peptide:MHC complex on a cell that does not express the co-stimulatory molecule B7, the T cell becomes anergic, that is, nonresponsive to antigen stimulation. In this state, the T cell cannot be activated even if it subsequently encounters its specific antigen presented by a professional antigen-presenting cell (Figure 8.18). The principal characteristic of anergic T cells is their inability to make IL-2; they are therefore unable to stimulate their own proliferation and differentiation. Thus, induction of tolerance in the mature T-cell repertoire seems to be based on the requirement that antigen-specific stimulation and co-stimulation of T cells are both delivered by the same antigen-presenting cell. The antigen-specific signals coming from the T-cell receptor, sometimes referred to as signal 1, are responsible for activating the naive T cells, whereas the antigen-nonspecific co-stimulatory signals, called signal 2, are needed for the proliferation and differentiation of the activated T cells.

Much research in immunology has involved studying the antibody and T-cell response to protein antigens in laboratory animals. Immunization with protein alone rarely led to an immune response. Reliable generation of a response required that protein antigens be mixed with certain bacteria or their breakdown products. It was discovered subsequently that the microbial components, known in this context as **adjuvants**, were inducing co-stimulatory activity in dendritic cells, macrophages, and B cells, a phenomenon that in an infection would occur naturally during the course of the innate immune response (see Section 8-5). Immunization in the absence of any microbial products to induce co-stimulation causes the antigen-specific T cells to become anergic, producing a temporary state of tolerance to the antigen. This explains why whole microorganisms are usually more effective vaccines than highly purified antigenic macromolecules. The induction of co-stimulatory activity by common microbial constituents is believed to be a major mechanism that allows the adaptive immune system to distinguish between antigens borne by infectious agents and antigens associated with innocuous proteins, including self proteins.

8-10   On activation, CD4 T cells acquire distinctive helper functions

Toward the end of their proliferation, activated T cells acquire the capacity to synthesize the proteins they need to perform the specialized functions of effector T cells (see Figure 8.17). For CD4 T cells these proteins comprise cell-surface molecules and soluble cytokines that activate and help other types of cell—principally macrophages and B cells—to participate in the immune response. Because of these facilitating functions, CD4 T cells are called helper cells. Mature helper CD4 T cells are heterogeneous in the cell-surface molecules they express, the cytokines they secrete and the functional effects they have on other immune-system cells. Defining the range of behavior are two types of helper cell called CD4 T_H1 cells or CD4 T_H2 cells. The main cytokines secreted by T_H1 cells—IL-2 and interferon (IFN)-γ—lead to macrophage activation, inflammation, and the production of opsonizing antibodies that enhance the phagocytosis of pathogens. The cytokines secreted by T_H2 cells—IL-4 and IL-5—lead mainly to B-cell differentiation and the production of neutralizing antibodies (Figure 8.19).

The differentiation pathway that an activated naive T cell will take is decided at an early stage of activation and determined by the local environment in the secondary lymphoid tissue. Particularly important in this decision are the cytokines present in the T-cell's immediate environment, which influence both the T cell and the dendritic cell that is guiding its activation. Dendritic cells are heterogeneous in the cell-surface molecules and cytokines they express, so the particular dendritic cell that activates the T cell will influence its subsequent differentiation. For both T_H1 and T_H2 cells, the pathway of differentiation is reinforced by the cytokines that they make themselves (Figure 8.20).

Stimulating T_H1 development are the cytokines IL-12 and IFN-γ, which are first produced in the course of the innate immune response: IL-12 by dendritic cells and macrophages, IFN-γ by NK cells (see Chapter 2). By binding to their receptors on the CD4 T cell, these cytokines induce changes in the gene expression and production of a transcription factor called T-bet, which turns on expression of the IFN-γ gene in the T cell. This commits the cell to being a CD4 T_H1 cell, and it then begins to secrete large quantities of IFN-γ. The increasing prevalence of IFN-γ in the environment in turn facilitates further differentiation of activated CD4 T cells along the T_H1 pathway.

In contrast, when IL-4 binds to the IL-4 receptor on an activated CD4 T cell this induces expression of the transcription factor GATA-3, which turns on expression of the genes for IL-4, IL-5, and other cytokines characteristic of CD4 T_H2 cells. This commits the cell to being a CD4 T_H2 cell. Once CD4 T_H2 cells begin to secrete IL-4, the local environment becomes one that facilitates the further differentiation of activated CD4 T cells along the T_H2 pathway. Which cell of innate immunity provides the IL-4 that starts differentiation along the T_H2 pathway has been uncertain. Currently, a strong candidate is the basophil, which is recruited into secondary lymphoid tissue responding to infection and makes IL-4. Such an initiating role for the basophil would also be consistent with its small numbers.

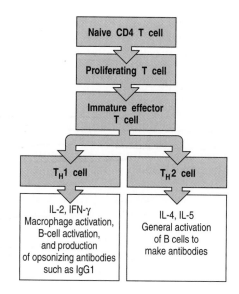

Figure 8.19 The stages of activation of CD4 T cells. Naive CD4 T cells first respond to peptide:MHC class II complexes by synthesis of IL-2 and proliferation. The progeny cells have the potential to become either T_H1 or T_H2 cells.

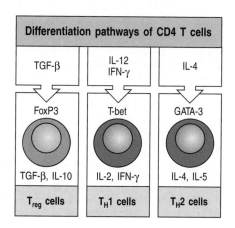

Figure 8.20 Different cytokine environments drive the differentiation of CD4 T cells that make different cytokines and have different functions. The principal cytokines that induce each type of effector T cell are shown in the top panels, the transcription factors that characterize these cell types are shown immediately above the cell, and the cytokines that the differentiated T cells produce are shown underneath. Differentiation of T_H1 and T_H2 is described in the text. The T_{reg} cells shown here are the regulatory CD4 T cells described in Section 7-13 (p. 203), whose function is to keep the activity of other effector T cells in check and prevent autoimmunity.

Although most adaptive immune responses involve contributions from both T_H1 and T_H2 cells, there are special circumstances in which the response becomes biased or **polarized** toward either a T_H1 or a T_H2 response. The fact that the cytokine product of each type of helper cell recruits further cells to the same pathway of differentiation gives positive feedback to such polarization. A response biased toward T_H1 cells corresponds to what has traditionally been described as **cell-mediated immunity**, a response dominated by the effector cells of the immune system. In contrast, a response biased toward T_H2 cells is dominated by antibodies, and corresponds to the traditional description of **humoral immunity**—'humors' being an alternative term for the body fluids in which antibodies are present.

Polarization of the CD4 T-cell response occurs in patients with leprosy, a disease caused by infection with *Mycobacterium leprae*, a bacterium that grows within the vesicular system of macrophages. In leprosy patients, the immune response becomes strongly biased toward either a T_H1 or T_H2 response, a choice that profoundly influences disease progression. A T_H1-biased response enables the infected macrophages to suppress bacterial growth and, although skin and peripheral nerves are damaged by the chronic inflammatory response, the disease progresses slowly and patients usually survive. For patients making a T_H2-biased response the situation is quite different. Inside macrophages the mycobacteria are inaccessible to specific antibody, and by growing unchecked they cause gross tissue destruction, which is eventually fatal (Figure 8.21). The visible symptoms of disease in patients making a T_H1 or T_H2 response to *M. leprae* are so dissimilar that the conditions are given different names: tuberculoid leprosy and lepromatous leprosy, respectively.

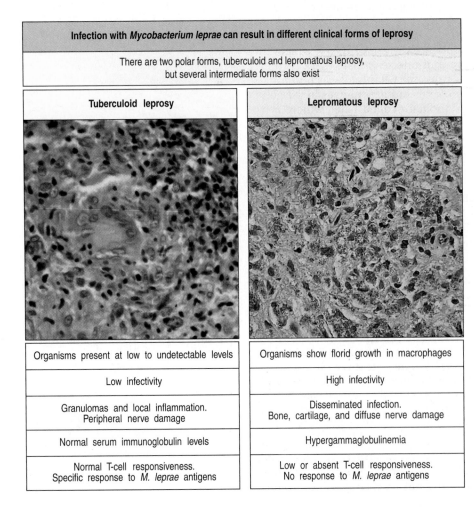

Infection with *Mycobacterium leprae* can result in different clinical forms of leprosy	
There are two polar forms, tuberculoid and lepromatous leprosy, but several intermediate forms also exist	
Tuberculoid leprosy	**Lepromatous leprosy**
Organisms present at low to undetectable levels	Organisms show florid growth in macrophages
Low infectivity	High infectivity
Granulomas and local inflammation. Peripheral nerve damage	Disseminated infection. Bone, cartilage, and diffuse nerve damage
Normal serum immunoglobulin levels	Hypergammaglobulinemia
Normal T-cell responsiveness. Specific response to *M. leprae* antigens	Low or absent T-cell responsiveness. No response to *M. leprae* antigens

Figure 8.21 Responses to *Mycobacterium leprae* are sharply differentiated in tuberculoid and lepromatous leprosy. The photographs show sections of lesion biopsies stained with hematoxylin and eosin. Infection with *M. leprae* bacilli, which can be seen in the right-hand photograph as numerous small dark red dots inside macrophages, can lead to two very different forms of the disease. In tuberculoid leprosy (left), growth of the microorganism is well controlled by T_H1-like cells that activate infected macrophages. The tuberculoid lesion contains granulomas and is inflamed, but the inflammation is localized and causes only local peripheral nerve damage. In lepromatous leprosy (right), infection is widely disseminated and the bacilli grow uncontrolled in macrophages. In the late stages there is severe damage to connective tissues and to the peripheral nervous system. There are several intermediate stages between these two polar forms. Photographs courtesy of G. Kaplan.

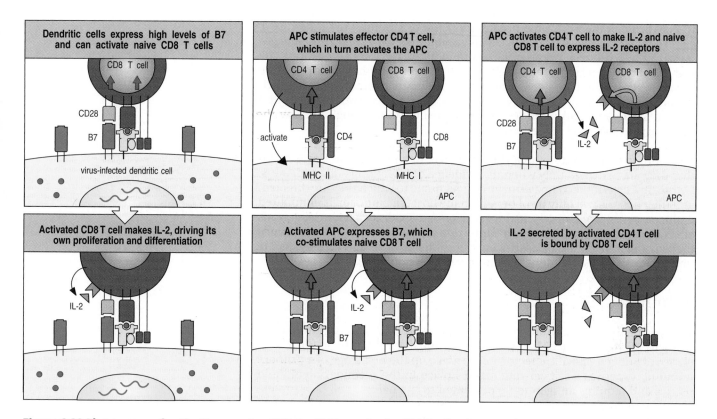

Figure 8.22 Three ways of activating a naive CD8 T cell. The left panels show how a naive CD8 T cell can be activated directly by a virus-infected dendritic cell. The center and right panels show two ways in which antigen-presenting cells (APC) that offer suboptimal co-stimulation can interact with a CD4 T cell to stimulate a naive CD8 T cell. One way is for cytokines secreted by the CD4 T cell to improve the co-stimulation of the antigen-presenting cell, for example by the induction of B7 expression (center panels). A second way is for cytokines secreted by the CD4 T cell, for example IL-2, to act directly on a neighboring CD8 T cell (right panels).

8-11 Naive CD8 T cells are activated to become cytotoxic effector cells in several different ways

The activation of naive CD8 T cells to become cytotoxic effector cells generally requires stronger co-stimulatory activity than is needed to activate CD4 T cells. Only dendritic cells, the most potent of the antigen-presenting cells, provide sufficient co-stimulation. When activated by antigen and co-stimulatory molecules on a dendritic cell, CD8 T cells are stimulated to synthesize both the cytokine IL-2 and its high-affinity receptor, which together induce the proliferation and differentiation of the CD8 T cells (Figure 8.22, left panels).

In circumstances in which the antigen-presenting cell offers suboptimal co-stimulation, CD4 T cells can help to activate naive CD8 T cells. To do this, the naive CD8 cell and the CD4 T cell must both recognize their specific antigens on the same antigen-presenting cell. When the CD4 T cell is already an effector cell, recognition of antigen causes it to secrete cytokines that then induce the antigen-presenting cell to increase its level of co-stimulatory molecules (see Figure 8.22, center panels). The activated antigen-presenting cell is then able to activate the naive CD8 T cell. In this mechanism, the CD4 T cell and the CD8 T cell can either interact simultaneously (as shown in Figure 8.22, center panels) or successively with the antigen-presenting cell. When the CD4 T cell is a naive cell, it is activated by the antigen-presenting cell to produce

IL-2, which can then drive the proliferation and differentiation of the CD8 T cell (see Figure 8.22, right panels). This mechanism works because engagement of the T-cell receptor is sufficient to induce CD8 T cells to express the high-affinity IL-2 receptor, although it is not enough to induce them to make their own IL-2. The two T cells must therefore interact simultaneously with the antigen-presenting cell; the close juxtaposition of the two T cells on the surface of the antigen-presenting cell then ensures that enough IL-2 is captured by the CD8 T cell to induce its activation.

The more stringent requirements for the activation of naive CD8 T cells mean that they are activated only when the evidence of infection is unambiguous. Cytotoxic T cells inflict damage on any target tissue to which they are directed, and their actions will only be of benefit to the host if a pathogen is eliminated in the process. Even then, the actions of cytotoxic T cells can have deleterious effects. For example, in fighting viral infections of the airways, cytotoxic T cells prevent viral replication by destroying the epithelial layer, but this makes the underlying tissue vulnerable to secondary bacterial infection.

Summary

All types of adaptive immune response are started by the activation of antigen-specific naive T cells. Their proliferation and differentiation to form large clones of antigen-specific effector T cells make up the first stage of a primary immune response. T-cell activation takes place in the secondary lymphoid tissues and requires antigen to be presented to the naive T cell by a professional antigen-presenting cell. In most instances this is a dendritic cell, because dendritic cells are uniquely specialized in transporting antigens from infected tissues to secondary lymphoid tissues and also in the processing and presentation of antigens from a wide range of pathogens to T cells. What distinguishes the professional antigen-presenting cells—dendritic cells, macrophages, and B cells—from other cells is that they express B7 co-stimulatory molecules that engage CD28 on the T-cell surface. Full T-cell activation and differentiation requires signals from the T-cell receptor, the co-receptor, and co-stimulatory molecules. The combination of these signals induces many changes in the T cells' pattern of gene expression, including the production of IL-2, a cytokine that is made by the T cells and is essential for their clonal expansion and differentiation into effector T cells. The whole process of T-cell activation and differentiation occurs in the immediate environment of the dendritic cell, during which time first the naive T cell and then its numerous progeny maintain contact with the dendrites of the dendritic cell.

B7 co-stimulatory molecules are not expressed ubiquitously by professional antigen-presenting cells but are induced by the presence of infection and the signals generated when innate immune receptors sense microbial products. This ensures that naive T cells do not respond to their specific antigens in the absence of infection and provides a mechanism of peripheral tolerance that prevents naive T cells with receptors that bind self antigens from being activated and differentiated into autoreactive effector T cells. When a naive T cell engages specific antigen on either a professional antigen-presenting cell or any other cell in the absence of infection, the lack of co-stimulation induces anergy, a state of tolerance.

Similar mechanisms are used to activate CD4 T cells and CD8 T cells, although CD8 T cells require stronger co-stimulation, often involving help from CD4 T cells. Whereas CD8 T cells are all destined to have cytotoxic effector function, CD4 T cells can differentiate along alternative pathways to produce effector cells that secrete different cytokines and drive the immune response in different directions. CD4 T_H1 cells secrete mainly cytokines that favor macrophage activation and cell-mediated immune responses, whereas CD4 T_H2 cells secrete mainly cytokines that stimulate B cells to produce antibodies.

The properties and functions of effector T cells

After differentiating in the secondary lymphoid tissues, effector T cells detach themselves from the antigen-presenting cell that nursed their differentiation. CD8 cytotoxic T cells and most CD4 T_H1 cells leave the lymphoid tissues and enter the blood to seek out the sites of infection, whereas CD4 T_H2 cells remain in the secondary lymphoid tissues. T-cell effector function is turned on when the T-cell receptors bind to peptide:MHC complexes on a target cell. This stimulates the T cell to release effector molecules that act on the target cell. As well as possessing specialized functions, effector T cells differ from naive T cells in ways that enable them to act more effectively as effector cells. We shall consider this first before discussing the specialized functions of CD8 T cells and CD4 T_H1 and T_H2 cells.

8-12 Effector T-cell responses to infection do not depend on co-stimulatory signals

When naive T cells differentiate into effector T cells they undergo changes in the types and abundance of the proteins present on the cell surface. A major functional change is that effector CD4 and CD8 T cells, unlike naive T cells, can respond to their specific antigen without the need for co-stimulatory signals. Because most human cell types never express the co-stimulatory molecule B7, this relaxation of the activation requirements means that cytotoxic CD8 T cells are able to kill any type of cell that becomes infected with a virus (Figure 8.23). Because macrophages and B cells, the cells with which CD4 effector T cells interact, express varying levels of co-stimulatory activity, relaxation of the requirement for co-stimulation also increases the numbers of these cells that can stimulate effector CD4 T cells and in this way strengthens the overall immune response. In the course of the innate and adaptive immune responses to infection, a variety of tissue cells are induced to express MHC class II molecules by IFN-γ. These cells are able to present antigens to pathogen-specific effector CD4 T cells and stimulate them to increase the inflammatory response in the infected tissue. Overall, the relaxed requirement for co-stimulation is an acknowledgement that activated T cells are evidence of a serious threat to the body and should be allowed to go about their business of removing it with less scrutiny than is given to naive T cells.

Effector T cells also express two to four times as much of the cell-adhesion molecules CD2 and LFA-1 as naive T cells (Figure 8.24), and so are able to interact with target cells expressing lower levels of ICAM-1 and LFA-3 than those found on the professional antigen-presenting cells. The interaction between an effector T cell and a target cell is short-lived unless the T-cell receptor is engaged by specific antigen. When this happens, a conformational change in LFA-1 strengthens the adhesion between the two cells.

Figure 8.23 Activation of effector T cells by antigen-presenting cells is less demanding than activation of naive T cells. Activation of naive T cells requires the T cell to recognize both specific antigen and a B7 co-stimulatory molecule on the antigen-presenting cell (left panel). Subsequent proliferation and differentiation produces effector T cells (middle panel) that can respond to cells having specific antigen on their surface but no B7. This is illustrated for a cytotoxic CD8 T cell (right panel).

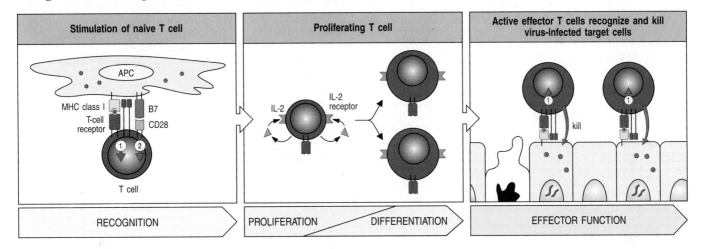

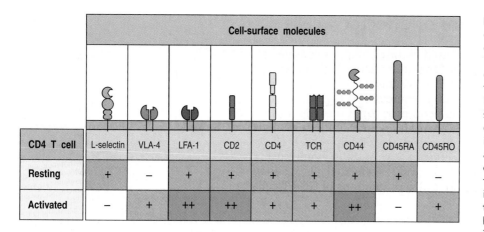

CD4 T cell	L-selectin	VLA-4	LFA-1	CD2	CD4	TCR	CD44	CD45RA	CD45RO
Resting	+	−	+	+	+	+	+	+	−
Activated	−	+	++	++	+	+	++	−	+

Figure 8.24 Activation of T cells changes the expression of several cell-surface molecules. A CD4 T cell is shown here. Resting naive T cells express L-selectin, by which they home to lymph nodes, and relatively low levels of other adhesion molecules, such as CD2 and LFA-1. Upon activation, expression of L-selectin ceases and increased amounts of the integrin LFA-1 are made. A newly expressed integrin called VLA-4 is a homing receptor for vascular endothelium at sites of inflammation; it guides activated T cells to infected tissues. Activated T cells have more CD2 on their surface, which increases adhesion to target cells, and also a higher density of the adhesion molecule CD44. Alternative splicing of RNA made from the CD45 gene causes activated T cells to express the CD45RO isoform, which associates with the T-cell receptor and CD4. This change makes the T cell more sensitive to stimulation by lower concentrations of peptide:MHC complexes.

The recirculation of naive T cells between lymph, blood, and secondary lymphoid tissues is determined by the adhesion molecules they express. With the activation of naive T cells by specific antigen and their maturation into effector T cells, changes in the expressed adhesion molecules allow the effector T cells to enter infected tissues and work there to reduce the infection. Effector T cells do not express L-selectin and cannot recirculate through lymph nodes by leaving the blood at high endothelial venules. Instead, effector T cells express the integrin **VLA-4** (Figure 8.25), which binds to the adhesion molecule VCAM-1 present on the activated endothelial cells of blood vessels in infected and inflamed tissues. By these changes in the adhesion molecules expressed, effector T cells are excluded from secondary lymphoid tissues and directed to enter the infected tissues where their effector functions are needed.

8-13 Effector T-cell functions are carried out by cytokines and cytotoxins

The molecules that carry out the effector functions of T cells fall into two broad classes: cytokines that alter the behavior of the target cells, and cytotoxic proteins or **cytotoxins**, which are used to kill infected cells. All effector T cells produce cytokines—of different types and in different combinations. Cytotoxins, in contrast, are the specialized products of cytotoxic CD8 T cells.

Cytokines are small secreted proteins and related membrane-bound proteins that act through cell-surface receptors, called cytokine receptors, and generally induce changes in gene expression within their target cell. In Chapter 2 we examined the effects of cytokines secreted by macrophages (IL-1, IL-6, and TNF-α) and NK cells (IFN-γ) in the innate immune response (see Section 2-13, p. 49). Secreted cytokines can act on the cell that made them (**autocrine** action), as with IL-2, or they can act locally on another type of cell (**paracrine** action). Many of the cytokines made by T cells are called **interleukins** and were assigned numbers according to the order of their discovery, for example IL-2. Cytokines made by lymphocytes are often called **lymphokines**, but in this book we shall use the general term cytokine for all these molecules.

Figure 8.25 Integrin VLA-4 enables effector T cells to home to inflamed tissue. The integrin LFA-1 (α$_L$:β$_2$) is present on all leukocytes, including T cells. It binds ICAMs and is important in the adhesive interactions that mediate cell migration and in the interactions of T cells with antigen-presenting cells (APC) or target cells; its expression is increased in effector T cells. The integrin VLA-4 (α$_4$:β$_1$) increases in abundance upon T-cell activation. It binds to the cell-adhesion molecule VCAM-1, which is selectively expressed on the endothelium of blood vessels in inflamed tissue and is important for recruiting effector T cells into sites of infection.

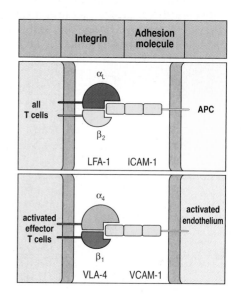

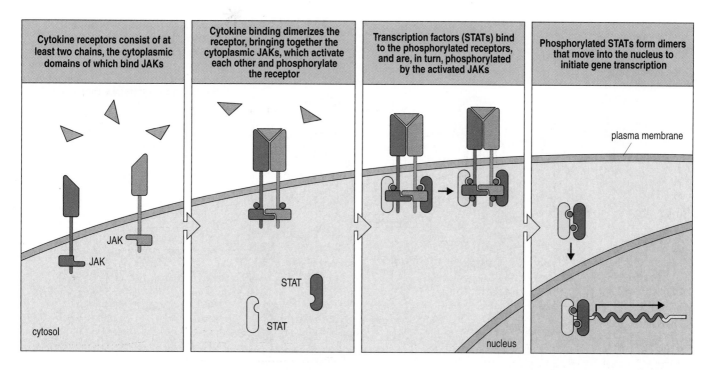

| Cytokine receptors consist of at least two chains, the cytoplasmic domains of which bind JAKs | Cytokine binding dimerizes the receptor, bringing together the cytoplasmic JAKs, which activate each other and phosphorylate the receptor | Transcription factors (STATs) bind to the phosphorylated receptors, and are, in turn, phosphorylated by the activated JAKs | Phosphorylated STATs form dimers that move into the nucleus to initiate gene transcription |

Figure 8.26 Many cytokine receptors signal through a pathway in which receptor-associated kinases activate transcription factors directly. These receptors consist of at least two chains, each associated with a specific Janus kinase (JAK) (first panel). Ligand binding and dimerization of the receptor chains brings together the JAKs, which transactivate each other, subsequently phosphorylating tyrosines in the receptor tails (second panel). Members of the STAT (signal transducer and activator of transcription) family of proteins bind to the phosphorylated receptors and are themselves phosphorylated by the JAKs (third panel). On phosphorylation, STAT proteins dimerize and go to the nucleus, where they activate transcription from a variety of genes important for adaptive immunity (fourth panel).

Cytokines work in the immediate vicinity of the effector T cell and for only short periods. Membrane-bound cytokines can have an effect only on the target cell in the localized area of the immunological synapse where the effector T-cell is bound. The secretion of soluble cytokines is similarly focused on the target cell by polarization of the T cell's intracellular secretory apparatus, which occurs on binding of the T-cell receptor. The advantages of a membrane-associated cytokine are that functional effects are obtained with smaller quantities of cytokine and with cytokines having lower affinities for the cytokine receptor. In addition, the interaction between the membrane-associated cytokine and its receptor can contribute to the adhesive interaction between an effector T cell and its target. Whereas membrane-associated cytokines can only exert their effect on the target cell to which the T cell is bound, a soluble cytokine has the advantage of being able to affect additional nearby cells by diffusing in the extracellular fluid. Some cytokines, such as TNF-α, are made in both membrane-associated and soluble forms.

The cytoplasmic tails of most cytokine receptors are associated with protein kinases known as **Janus kinases** (**JAKs**) (Figure 8.26). Cytokine binding causes dimerization of the cytokine receptors, which, in turn, activates the kinases to phosphorylate members of a protein family called **STATs** (**s**ignal **t**ransducers and **a**ctivators of **t**ranscription). On phosphorylation, two STATs dimerize and move from the cytoplasm to the nucleus. Here they activate specific genes, which differ according to the individual cytokine receptor–JAK–STAT pathway. These are short, direct, intracellular signaling pathways that enable cells to respond rapidly to cytokine stimulation.

T$_H$1 and T$_H$2 CD4 T cells are distinguished by the sets of cytokines they make and the effects that these have on the immune response. T$_H$1 cells work principally with macrophages in developing a cell-mediated immune response, whereas T$_H$2 cells work only with B cells in developing an antibody-mediated immune response. CD8 T cells act mainly through the cytotoxins they produce, but they also produce cytokines, which can have effects on other cells of the immune system (Figure 8.27).

Polarization of the human CD4 T-cell response to *M. leprae* is manifested in two quite different diseases: tuberculoid leprosy, a limited but chronic infection that is controlled by a T$_H$1 response, and lepromatous leprosy, a more

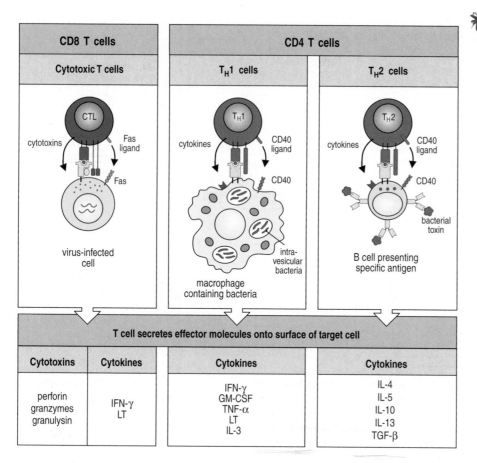

Figure 8.27 The three types of effector T cell produce distinct sets of effector molecules. The three main types of effector T cell are shown, as are the types of target cell with which they interact, and the effector molecules that they make.

severe condition in which the infection goes uncontrolled by a T$_H$2 response (see Section 8-10). In both conditions the skin lesions are infiltrated with T cells. Analysis of biopsies taken from such lesions reveals the underlying bias in cytokine production by the infiltrating T cells. In tuberculoid leprosy, the dominant cytokines are IL-2, IFN-γ, and lymphotoxin (LT), whereas in lepromatous leprosy they are IL-4, IL-5, and IL-10 (Figure 8.28).

8-14   Cytotoxic CD8 T cells are selective and serial killers of target cells at sites of infection

Once inside a cell, a pathogen becomes inaccessible to antibodies and other soluble proteins of the immune system. It can be eliminated either through the efforts of the infected cell itself or by direct attack on the infected cell by

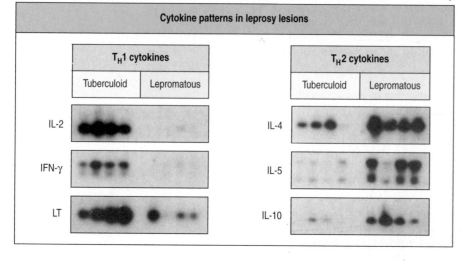

Figure 8.28 Cytokine patterns in leprosy lesions. The cytokine patterns in the two forms of leprosy differ, as shown by Northern blot analysis of the mRNA from lesions of four patients with lepromatous leprosy and four patients with tuberculoid leprosy. mRNAs for cytokines typically produced by T$_H$2 cells predominate in the lepromatous form, whereas mRNAs for cytokines produced by T$_H$1 cells predominate in the tuberculoid form. Cytokine blots courtesy of R.L. Modlin.

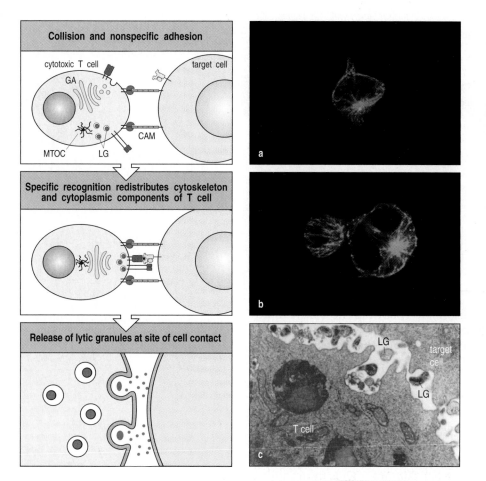

Figure 8.29 When cytotoxic T cells recognize specific antigen, the delivery of cytotoxins is aimed directly at the target cell. As shown in the panels on the left, initial adhesion to a target cell has no effect on the location of the lytic granules (LG) (top panel). Engagement of the T-cell receptor causes the T cell to become polarized: the cortical actin cytoskeleton at the site of contact reorganizes, enabling the microtubule-organizing center (MTOC), the Golgi apparatus (GA), and the lytic granules to align toward the target cell (center panel). Proteins stored in lytic granules are then directed onto the target cell (bottom panel). The photomicrograph in panel a shows an unbound, isolated cytotoxic T cell. The microtubules are stained green and the lytic granules red. Note how the lytic granules are dispersed throughout the T cell. Panel b depicts a cytotoxic T cell bound to a (larger) target cell. The lytic granules are now clustered at the site of cell–cell contact in the bound T cell. The electron micrograph in panel c shows the release of granules from a cytotoxic T cell. Panels a and b courtesy of G. Griffiths. Panel c courtesy of E.R. Podack.

the immune system. The function of cytotoxic CD8 T cells is to kill cells that have become overwhelmed by intracellular infection. Overall, the sacrifice of the infected cells serves to prevent the spread of infection to healthy cells. The importance of cytotoxic T cells in combating viral infection is seen in people who lack functional cytotoxic T cells and suffer from persistent viral infections.

Effector cytotoxic T cells contain stored **lytic granules**, which are modified lysosomes containing a mixture of specialized proteins called cytotoxins. CD8 T cells start to synthesize cytotoxins in inactive forms and to package them into lytic granules as soon as the T cells are activated by specific antigen in secondary lymphoid tissues. Effector CD8 T cells then migrate to sites of infection, where they recognize specific peptide:MHC class I complexes presented by infected cells. Binding via the T-cell receptor tells a T cell to secrete its cargo of lytic granules, which is delivered directly onto the surface of the infected target cell.

At sites of infection, cytotoxic CD8 T cells and the infected target cells are surrounded by healthy cells and cells of the immune system that have infiltrated the infected tissue. Because of their antigen specificity, cytotoxic T cells pick out only infected cells for attack and leave healthy cells alone. The T cell focuses granule secretion onto the small localized area of the target cell where it is attached to the T cell (Figure 8.29). In this way, cytotoxic granules neither attack healthy neighbors of an infected cell nor kill the T cell itself. As the target cell starts to die, the cytotoxic T cell is released from the target cell and starts to make new granules. Once new granules have been made, the cytotoxic T cell is able to kill another target cell. In this manner, one cytotoxic T cell can kill many infected cells in succession (Figure 8.30).

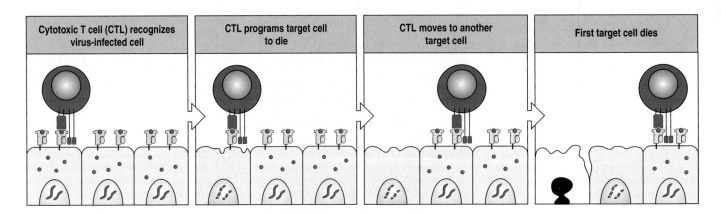

Cytotoxic T cell (CTL) recognizes virus-infected cell	CTL programs target cell to die	CTL moves to another target cell	First target cell dies

Besides their cytotoxic action, CD8 T cells also contribute to the immune response by secreting cytokines. One of these is IFN-γ, which inhibits the replication of viruses in the infected cells and increases the processing and presentation of viral antigens by MHC class I molecules. Another effect of IFN-γ is to activate macrophages in the vicinity of the cytotoxic T cells. These macrophages get rid of the dying infected cells, thereby allowing the T cells more room for maneuver and also helping the damaged tissue to heal and regenerate.

Figure 8.30 Cytotoxic CD8 T cells kill infected cells successively. Specific recognition of peptide:MHC complexes on an infected cell by a cytotoxic CD8 T cell (CTL) programs the infected cell to die. The T cell detaches from its target cell and synthesizes a new set of lytic granules. The cytotoxic T cell then seeks out and kills another target.

8-15 Cytotoxic T cells kill their target cells by inducing apoptosis

Cells killed by cytotoxic CD8 T cells do not lyse or disintegrate, like cells undergoing necrosis due to physical or chemical injury, but die by **apoptosis** (Figure 8.31). This form of cell death, which is widespread in the immune system (see, for example, Section 6-3, p. 163), induces a cell to commit suicide from within. By shriveling and shrinking without losing the cell's contents, it leaves a neat and tidy corpse. Apoptosis, also known as **programmed cell death**, prevents not only pathogen replication but also the release of infectious bacteria or virus particles from the infected cell.

Soon after contact with a cytotoxic T cell, the target cell's DNA starts to be fragmented by the cell's own nucleases. These enzymes cleave between the nucleosomes to give DNA fragments that are multiples of 200 base pairs in length and are characteristic of apoptosis. Eventually the nucleus becomes disrupted and there is a loss of membrane integrity and normal cell morphology. The cell destroys itself from within. It shrinks by the shedding of

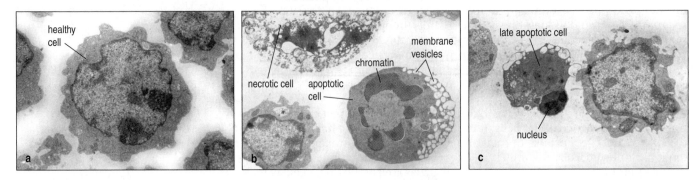

Figure 8.31 Apoptosis. Panel a shows an electron micrograph of a healthy cell with a normal nucleus. In the bottom right of panel b is an apoptotic cell at an early stage. The chromatin in the nucleus has become condensed (shown in red); the plasma membrane is well defined and is shedding vesicles. In contrast, the plasma membrane of the necrotic cell shown in the upper left part of panel b is poorly defined. The middle cell shown in panel c is at a late stage in apoptosis. It has a very condensed nucleus and no mitochondria, and the cytoplasm and cell membranes have largely been lost through vesicle shedding. Photographs courtesy of R. Windsor and E. Hirst.

Figure 8.32 Time course of programmed cell death. The four panels show time-lapse photographs of a cytotoxic T cell killing a target cell. The lytic granules of the T cell are labeled with a red fluorescent dye. In the top panel, the T cell has just made contact with the target cell and this event is designated as the Start. At this time, the T-cell granules are distant from the point of contact with the target cell. After 1 minute (second panel), the granules have begun to move toward the point of target-cell attachment, a move that is essentially completed after 4 minutes (third panel). After 40 minutes (bottom panel), the granules have been secreted and are seen between the T cell and the target cell. The target cell has now begun to undergo apoptosis, as shown by the fragmented nucleus. The T cell is now ready to detach from the apoptotic cell and seek out a further target cell. Photographs courtesy of G. Griffiths.

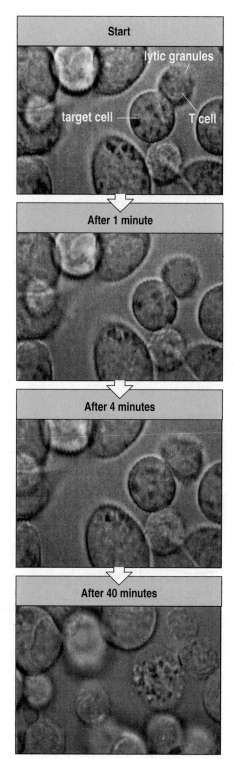

Start

lytic granules

target cell

T cell

After 1 minute

After 4 minutes

After 40 minutes

membrane-bound vesicles (see Figure 8.31, center panel) and the degradation of the cell contents until little is left. The changes that occur in the plasma membrane during apoptosis are recognized by phagocytes, which speed the dying cell on its way by ingesting and digesting it. The apoptotic processes that degrade the infected human cell also act on the infecting pathogen. In particular, the breakdown of viral nucleic acids prevents the assembly of infectious virus particles that might cause further infection were they to escape from the dying cell. A 5-minute contact between a cytotoxic T cell and its target cell is all it takes for the target cell to be programmed to die, even though visible evidence of death takes longer to become obvious (Figure 8.32).

Cytotoxic T cells induce apoptosis by two different pathways. The first is initiated by the release of cytotoxic granules that contain various cytotoxins. These include a family of serine proteases called **granzymes**, **perforin**—a protein that can make pores in membranes—and **granulysin**, a detergent-like protein that also associates with membranes. In a complex with the proteoglycan **serglycin**, perforin and granulysin make pores in the target-cell membrane and deliver the granzymes to the cell's interior. Once inside the target cell, the granzymes initiate a cascade of proteolytic cleavage reactions that leads to apoptosis.

A second way of inducing apoptosis is by interactions between cell-surface molecules on the cytotoxic T cell and the target cell. Activated cytotoxic T cells express the cell-surface cytokine **Fas ligand**, which binds to **Fas** molecules on the target-cell surface. This interaction sends signals to the target cell to undergo apoptosis. Although probably a minor pathway for killing infected cells, apoptosis induced by the interaction between Fas and Fas ligand is the main route by which unwanted lymphocytes are disposed of during lymphocyte development and in the course of an immune response. Individuals who lack functional Fas molecules can neither control the size of their lymphocyte population nor remove autoimmune cells. Consequently, they suffer from a disease in which the secondary lymphoid organs become swollen in the absence of infection (Figure 8.33) and in which autoimmune responses that attack healthy blood cells, platelets, and liver cells are typically made. This disease, which is usually caused by the inheritance of one nonfunctional copy of the Fas gene, is called autoimmune lymphoproliferative syndrome (ALPS).

8-16 T$_H$1 CD4 cells induce macrophages to become activated

A principal function of T$_H$1 cells is to help the resident macrophages at sites of infection become more proficient in the phagocytosis and killing of pathogens. To do this the T$_H$1 cells must leave the secondary lymphoid tissue where

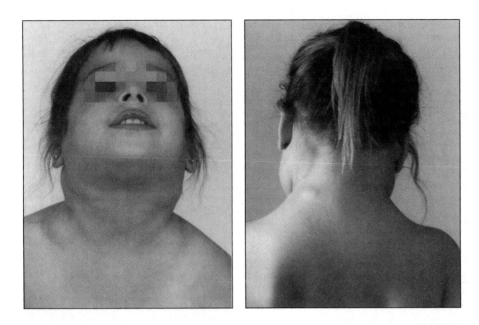

Figure 8.33 **Lymphadenopathy in autoimmune lymphoproliferative syndrome (ALPS).** Young girl with ALPS with very enlarged lymph nodes in her neck. Photograph courtesy of Jennifer Puck.

they differentiated and travel in the blood to the infected tissue. There the resident macrophages will have been phagocytosing and degrading pathogens from the beginning of the innate immune response and will be presenting pathogen-derived peptides on their MHC class II molecules. The antigen receptor of a T_H1 cell will recognize its antigen on the macrophage surface and bind to it. After this initial contact, the two cells form a strong adhesive interaction and an immunological synapse through which signals are sent from the T cell to the macrophage and vice versa. Two cells bound in this manner are called a **conjugate pair**. Cytokines secreted by the T_H1 cell induce changes in the macrophage that improves its ability to endocytose and destroy pathogens.

This enhancement of macrophage function is called **macrophage activation** and requires the interaction of peptide:MHC class II complexes on the macrophage with the T-cell receptor of the T_H1 cell. One effect of macrophage activation is to cause the phagosomes that contain captured microorganisms to be more efficiently fused with lysosomes, the source of hydrolytic degradative enzymes. Another effect is to increase the synthesis by macrophages of highly reactive and microbicidal molecules, such as oxygen radicals, nitric oxide (NO), and proteases, which together kill the engulfed pathogens. In patients with acquired immune deficiency syndrome (AIDS), the number of CD4 T cells decreases progressively, as does the activation of macrophages. In these circumstances, microorganisms such as *Pneumocystis carinii* and mycobacteria, which live in the vesicles of macrophages and are normally kept in check by macrophage activation, flourish as opportunistic, and sometimes fatal, infections.

Other changes that occur on macrophage activation help to amplify the immune response. Increased expression of MHC class II molecules and B7 on macrophages resident in secondary lymphoid tissues increases their ability to present antigen to and activate naive T cells, thus producing more effector T cells for the immune response. This, in turn, enhances macrophage activation and maintains increased numbers of macrophages in the fully activated state in the infected tissues.

Macrophages require two signals for activation, both of which can be delivered by effector T_H1 cells. The primary signal is provided by IFN-γ, the characteristic cytokine produced by T_H1 cells. The second signal makes macrophages responsive to IFN-γ and is delivered by **CD40 ligand**, a membrane-bound cytokine on the T-cell surface that binds to its receptor, **CD40**, on the macrophage (Figure 8.34). Macrophage activation increases the expression of both CD40 and receptors for the cytokine TNF-α, which raises the sensitivity of the macrophage to CD40 ligand and TNF-α. Activated macrophages also produce TNF-α (see Section 2-13, p. 49) in both membrane-bound and secreted forms; its effects synergize with those of IFN-γ to raise the level of macrophage activation.

CD4 T cells only make their effector molecules on demand, unlike CD8 T cells, which store them in granules. After encounter with antigen on a macrophage, an effector T_H1 cell takes several hours to synthesize the requisite effector cytokines and cell-surface molecules. During this time the T cell must maintain contact with its target cell. Newly synthesized cytokines are translocated into the endoplasmic reticulum of the T cell and delivered by secretory vesicles to the site of contact between the T cell and macrophage. Thus, they are focused on the target cell. Newly synthesized CD40 ligand is also expressed selectively at the region of contact with the macrophage. Together, this localized production of cytokines ensures the selective activation of those macrophages carrying the specific peptide:MHC complexes recognized by the T cell.

CD8 T cells are also an important source of IFN-γ, and because of this they can activate macrophages. Macrophage sensitization to IFN-γ need not require the action of CD40 ligand; small amounts of bacterial polysaccharide have a similar effect and can be of particular importance when CD8 T cells, which do not express CD40 ligand, are the principal source of IFN-γ.

The microbicidal substances produced by activated macrophages are also harmful to human tissues, which inevitably suffer damage from macrophage activity. For this reason, the activation of macrophages by CD4 T_H1 cells is under strict control. Cytokines secreted by CD4 T_H2 cells, which include **transforming growth factor-β (TGF-β)**, **IL-4**, **IL-10**, and **IL-13**, inhibit macrophage activation, which is an example of how cytokines secreted by T_H2 cells can control the T_H1 response. T_H1 cells stop producing IFN-γ if their antigen receptors lose contact with the peptide:MHC complexes on a macrophage, which is a further control on the T_H1 response.

8-17 T_H1 cells coordinate the host response to pathogens that live in macrophages

Certain microorganisms, including the mycobacteria that cause tuberculosis and leprosy, are intracellular pathogens that enjoy a protected life in the vesicular system of macrophages. The protozoan parasite *Leishmania* also survives for part of its life cycle inside vesicles in macrophages. Such microorganisms subvert the destructive mission of the macrophage for their own purposes. By sequestering themselves in this cellular compartment they cannot be reached by antibodies; neither are their peptides presented by MHC class I molecules, thus preventing the infected macrophage from being attacked by cytotoxic T cells. Mycobacteria avoid digestion by lysosomal enzymes by preventing the acidification of the phagolysosome that is required to activate the lysosomal hydrolases. Infections of this type are fought by T_H1 CD4 T cells, which help the macrophage to become activated to the point at which the intracellular pathogens are eliminated or killed.

The activation of macrophages by IFN-γ and CD40 ligand is central to the immune response against pathogens that proliferate in macrophage vesicles. In mice lacking functional IFN-γ or CD40 ligand, the ability of macrophages to kill intravesicular pathogens is impaired, and doses of mycobacteria or the parasite *Leishmania* that normal mice can withstand prove fatal. Although

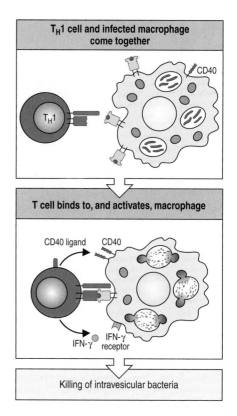

Figure 8.34 T_H1 CD4 cells activate macrophages to become highly microbicidal. When a T_H1 cell specific for a bacterial peptide contacts a macrophage that presents the peptide, the T_H1 cell is induced to secrete the macrophage-activating cytokine interferon-γ (IFN-γ) and also to express CD40 ligand at its surface. Together, these newly synthesized proteins activate the macrophage to kill the bacteria living inside its vesicles.

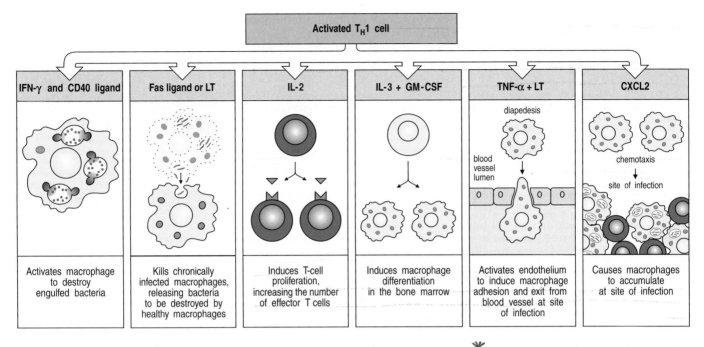

Figure 8.35 The immune response to intravesicular bacteria is coordinated by activated T$_H$1 cells. The activation of T$_H$1 cells by infected macrophages results in the synthesis of cytokines that activate the macrophage and coordinate the immune response to intravesicular pathogens. The six panels show the effects of different cytokines and chemokines secreted by T$_H$1 cells. CXCL2 is a chemokine. GM-CSF, granulocyte–macrophage colony-stimulating factor; LT, lymphotoxin.

IFN-γ and CD40 ligand are probably the most important effector molecules of T$_H$1 cells, other cytokines secreted by these cells help to coordinate responses to intravesicular bacteria (Figure 8.35). Macrophages chronically infected with intravesicular bacteria can lose the capacity to be activated. Such cells can be induced to undergo apoptosis by an effector T$_H$1 cell that uses Fas ligand to engage Fas on the macrophage.

The IL-2 produced by T$_H$1 cells induces T-cell proliferation and potentiates the production and release of other cytokines. Some of these recruit phagocytes—macrophages and neutrophils—to sites of infection. First, T$_H$1 cells secrete IL-3 and granulocyte–macrophage colony-stimulating factor (GM-CSF), which stimulate the increased production of macrophages and neutrophils in the bone marrow. Second, the TNF-α and LT made by T$_H$1 cells induce vascular endothelial cells at sites of infection to change the adhesion molecules they express so that phagocytes circulating in the blood can bind to them. At this point, the chemokine **CXCL2**, which is produced by T$_H$1 cells, guides the phagocytes between the endothelial cells and into the infected area. In total, the CD4 T$_H$1 cell orchestrates a multifaceted macrophage response that focuses on the destruction of pathogens taken up by macrophages.

When microbes resist the microbicidal effects of activated macrophages successfully, a chronic infection with inflammation can develop. Such areas of tissue often have a characteristic morphology, called a **granuloma**, in which a central area containing infected macrophages is surrounded by activated T cells. Giant cells resulting from the fusion of macrophages are present at the center of a granuloma; these contain the resistant pathogens. Large single macrophages, sometimes called epithelioid cells, form an epithelium-like layer around the center (Figure 8.36).

In tuberculosis, the centers of large granulomas can become cut off from the blood supply and the cells in the center die, probably from a combination of oxygen deprivation and the cytotoxic effects of macrophage products. The resemblance of the dead tissue to cheese led to this process being called **caseation necrosis**. It provides a vivid example of how CD4 T$_H$1 cells can produce a local pathology. Their absence, however, leads to death from disseminated infection, which is commonly seen in AIDS patients infected with opportunistic mycobacteria.

Figure 8.36 Granulomas form when an intracellular pathogen or its constituents resists elimination. In some circumstances mycobacteria (red) resist the killing effects of macrophage activation (top panel). A characteristic localized inflammatory response called a granuloma develops (second panel). The granuloma consists of a central core of infected macrophages, which can include multinucleated giant cells formed by macrophage fusions, surrounded by large single macrophages often called epithelioid cells. Mycobacteria can persist in the cells of the granuloma. The central core is surrounded by T cells, many of which are CD4 T cells. The photomicrograph (bottom panel) shows a granuloma from the lung. Photograph courtesy of J. Orrell.

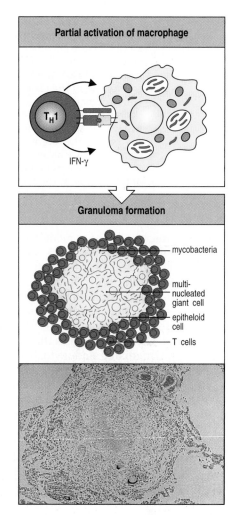

8-18 CD4 T$_H$2 cells activate only those B cells that recognize the same antigen as they do

During an infection, the T-cell zone of secondary lymphoid tissues contains pathogen-specific T$_H$2 effector cells that are the progeny of naive CD4 T cells activated by antigen-presenting dendritic cells. The main function of these T cells is to help B cells mount an antibody response against the infectious agent. Mature naive B cells passing through the lymphoid tissue pick up, process, and present their specific antigens. As the circulating B cells pass through the T-cell zones, they make transient interactions with the T$_H$2 cells, whose T-cell receptors screen the peptides presented by the MHC class II molecules on the surface of the B cell. When a B cell presents an antigen recognized by the T$_H$2 cell, the adhesive interactions are strengthened and the B cell becomes trapped by the T cell. This interaction gives rise to a primary focus of activated B cells and helper T cells (see Figure 6.22, p. 180).

When the T-cell receptor of a helper T cell recognizes peptide:MHC class II complexes on the surface of a naive B cell, the T cell responds by synthesizing CD40 ligand. This molecule is involved in all T-cell interactions with B cells, which express the corresponding receptor molecule CD40. The interaction of CD40 ligand with CD40 drives the resting B cell into the cycle of cell division. The characteristic cytokine secreted by T$_H$2 cells on stimulation by their target cell is IL-4, which works in concert with CD40 ligand to initiate the proliferation and clonal expansion of B cells that precede their differentiation into antibody-secreting plasma cells. T$_H$2 cells also produce the cytokines IL-5 and IL-6, which drive the further differentiation of B cells to plasma cells (Figure 8.37).

The principle governing T-cell help to B cells is that cooperation occurs only between B and T cells that are specific for the same antigen, although they usually recognize different epitopes. Such interactions are called **cognate**

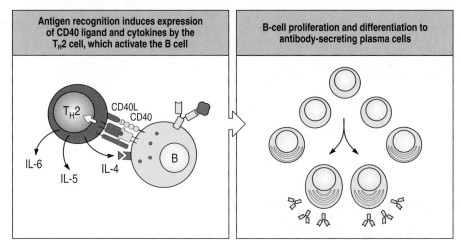

Figure 8.37 T$_H$2 cells stimulate the proliferation and differentiation of naive B cells. The specific interaction of an antigen-binding B cell with a helper T$_H$2 cell leads to the expression of CD40 ligand (CD40L) and the secretion of IL-4, IL-5, and IL-6. In concert, these T$_H$2 products drive the proliferation of B cells and their differentiation to form plasma cells dedicated to the secretion of antibody.

Figure 8.38 Molecular complexes recognized by both B and T cells make effective vaccines. The first panel shows a naive B cell's surface immunoglobulin binding a carbohydrate epitope on a vaccine composed of a *Haemophilus* polysaccharide (blue) conjugated to tetanus toxoid (red), a protein. This results in receptor-mediated endocytosis of the conjugate and its degradation in the endosomes and lyosomes, as shown in the second panel. Peptides derived from degradation of the tetanus toxoid part of the conjugate are bound by MHC class II molecules and presented on the B cell's surface. In the third panel, the receptor of a T$_H$2 cell recognizes the peptide:MHC complex. This induces the T cell to secrete cytokines that activate the B cell to differentiate into plasma cells, which produce protective antibody against the *Haemophilus* polysaccharide (fourth panel).

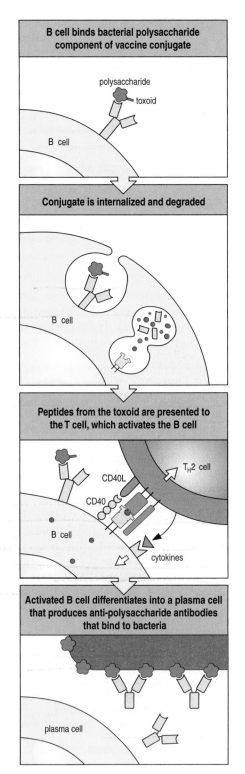

interactions. The peptide recognized by the T cell must be part of the same physical entity bound by the B cell's surface immunoglobulin. For example, the T cell might recognize a peptide derived from an internal protein of a virus, whereas the B cell recognizes an exposed carbohydrate epitope of a viral capsid glycoprotein. The specialized antigen-presenting function of a B cell makes it supremely efficient at presenting peptides that derive from any protein, virus, or microorganism that specifically binds to its surface immunoglobulin. Only those B cells that selectively internalize a pathogen antigen by receptor-mediated endocytosis will present enough of the pathogen-derived peptide to engage and stimulate an antigen-specific T$_H$2 cell. It is estimated that a B cell that can use receptor-mediated endocytosis to capture a particular antigen is 10,000-fold more efficient at presenting peptides derived from that antigen than a B cell that cannot use receptor-mediated endocytosis.

Knowledge of the mechanism by which B and T cells cooperate helps in the design of vaccines. An example illustrated in Figure 8.38 is the vaccine against *Haemophilus influenzae* B, a bacterial pathogen that is life-threatening to young children when it infects the lining of the brain—the meninges—producing a meningitis that, in severe cases, causes lasting neurological damage or death. Protective immunity against *H. influenzae* is provided by antibodies specific for the capsular polysaccharides. However, a child's antibody response is weakened by the lack of associated peptide epitopes that could engage T$_H$2 cells and provide help to polysaccharide-specific B cells. To enable the immune system to make antibodies against *H. influenzae*, a vaccine was made in which the immunizing antigen was the bacterial polysaccharide covalently coupled to tetanus toxoid, a protein containing good peptide epitopes that are bound by MHC class II molecules and presented to T$_H$2 cells.

8-19 Regulatory CD4 T cells limit the activities of effector CD4 and CD8 T cells

As we saw in Section 7-13 (p. 203), T-cell development in the thymus produces a population of regulatory autoreactive CD4 T cells whose function is to suppress the activation of naive autoreactive CD4 helper and CD8 cytotoxic T cells that have the potential to attack the body's healthy tissues. This is an example of a more general phenomenon in which regulatory T cells are used to control effector T cells in the response to infection, making sure that the collateral damage they cause is limited and that they are decommissioned once the pathogen has been defeated.

Although not uniquely defined by one effector molecule or cell-surface marker, these regulatory CD4 T cells express high levels of CD25, the α chain of the IL-2 receptor, Fox P3, and usually make cytokines that are immunosuppressive and anti-inflammatory, such as IL-4, IL-10, and TGF-β. Suppression depends on physical contact between the regulatory CD4 T cell and its target cells. One way in which regulatory T cells are thought to dampen the effector T-cell response is to enter the secondary lymphoid tissue that is generating the response. There the regulatory T cells interact with the dendritic cells, thus preventing them from interacting with and activating additional naive T cells.

A second way in which regulatory T cells function is through direct contact with effector T cells. On infection with hepatitis B virus only a minority of people eliminate the virus, whereas the rest develop a chronic liver disease associated with a chronic and ineffective adaptive immune response. The liver and blood of these people contain an unusually high level of regulatory T cells that are specific for hepatitic B virus antigens. Their suppressive activity involves the secretion of TGF-β, an immunosuppressive cytokine, and direct contact with the effector T cells. In these patients the regulatory T cells seem to have prematurely suppressed the effector T cells before the pathogen was defeated. Because a chronic hepatitis B virus infection is debilitating and can lead to liver cancer, the development of an effective vaccine against this virus was a major accomplishment.

Tuberculosis is another chronic disease; it results from an infection of the lungs with *Mycobacterium tuberculosis* that fails to be terminated by the adaptive immune response. The disease alternates between periods of quiescence and active disease. During active disease, the number of regulatory T cells at the infected sites increases, with a corresponding decrease in the numbers of effector T cells.

Summary

The three types of effector T cell—CD8 cytotoxic cells, CD4 T_H1 cells, and CD4 T_H2 cells—have complementary roles in the immune response. The common principle by which they function is to affect the behavior of other types of cell through intimate contact and the action of effector molecules. Naive T cells are activated to develop effector function in the secondary lymphoid tissues, whereupon cytotoxic CD8 T cells and CD4 T_H1 cells enter the blood and travel to sites of infection. In contrast, CD4 T_H2 cells remain in the secondary lymphoid tissue, where they activate naive B cells that have specificity for the same antigen as themselves. Linked recognition between T_H2 cells and B cells arises because a B cell efficiently internalizes and processes the antigen to which its surface immunoglobulin binds, and then presents peptide antigens to a T cell. When the T-cell receptor of the T_H2 cell binds to peptide:MHC complexes on the B-cell surface, the B cell becomes activated by interactions between CD40 ligand on the T cell and CD40 on the B cell, and by IL-4, the characteristic cytokine secreted by the T_H2 cell. Cytotoxic CD8 T cells induce cells overwhelmed by viral infection to die by apoptosis. This mode of death ensures that the infected cell's load of viruses is also destroyed rather than being released to infect healthy cells. Apoptosis is induced by the cytotoxic enzymes contained in secretory lytic granules that are stored by the cytotoxic T cell and are released onto the target cell membrane once contact has been established. After granule release, the T cell rapidly synthesizes new granules so it can kill several targets in succession. In contrast, the effector molecules of effector CD4 T cells are made to order once contact with an antigen-bearing target cell has been established. A major role of T_H1 CD4 cells is to activate macrophages, helping them to become more competent at destroying extracellular pathogens that have been taken into their vesicular system, including those that exploit the phagocytic pathway for their own survival. IFN-γ is the characteristic cytokine of T_H1 cells and is instrumental in activating macrophages. Once an infection has been brought under control, regulatory T cells prevent the production of new effector T cells and suppress the functions of existing effector T cells.

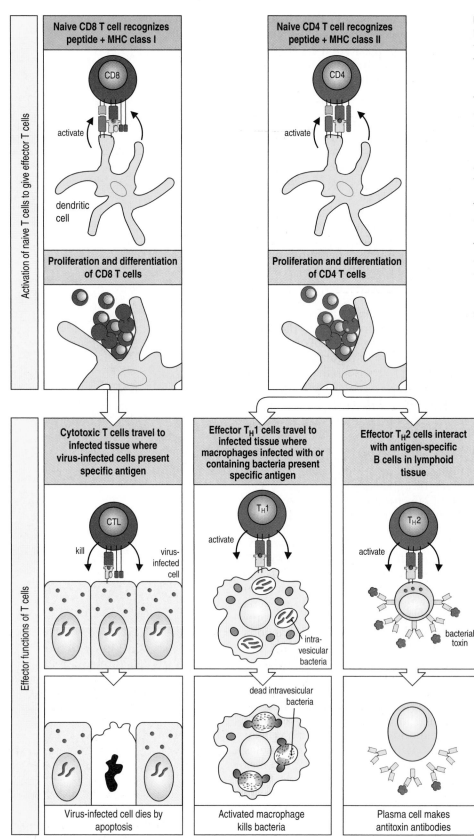

Figure 8.39 The T-cell response has two distinct stages. In the first stage, naive T cells encountering their cognate antigen on professional antigen-presenting cells are induced to proliferate and differentiate into effector T cells. In the second stage, these effector cells recognize target cells bearing specific antigen and interact with them. Naive CD8 T cells become CD8 cytotoxic effector cells (CTL), which kill target cells that present peptides derived from viruses and other cytosolic pathogens bound to MHC class I molecules. CD4 T cells differentiate into either T_H1 or T_H2 cells, which recognize antigen presented by MHC class II molecules. T_H1 cells activate macrophages, thus enhancing their general capacity to eliminate extracellular infection and, more specifically, to eliminate organisms colonizing the macrophages' vesicular system. T_H2 CD4 cells activate naive B cells and control many aspects of the antibody response.

Activation of naive T cells to give effector T cells

Naive CD8 T cell recognizes peptide + MHC class I

CD8

activate

dendritic cell

Proliferation and differentiation of CD8 T cells

Naive CD4 T cell recognizes peptide + MHC class II

CD4

activate

Proliferation and differentiation of CD4 T cells

Effector functions of T cells

Cytotoxic T cells travel to infected tissue where virus-infected cells present specific antigen

CTL

kill

virus-infected cell

Virus-infected cell dies by apoptosis

Effector T_H1 cells travel to infected tissue where macrophages infected with or containing bacteria present specific antigen

T_H1

activate

intra-vesicular bacteria

dead intravesicular bacteria

Activated macrophage kills bacteria

Effector T_H2 cells interact with antigen-specific B cells in lymphoid tissue

T_H2

activate

bacterial toxin

Plasma cell makes antitoxin antibodies

Summary to Chapter 8

All aspects of the adaptive immune response are initiated and controlled by effector T cells—CD4 T_H1 cells, CD4 T_H2 cells, and CD8 cytotoxic T cells. These cells differentiate from naive T cells that have been activated by specific antigen in the T-cell area of secondary lymphoid tissue. T-cell activation is accomplished by dendritic cells, the most professional antigen-presenting cells, which take up pathogens and antigens within the infected tissue and carry them in the draining lymph to the secondary lymphoid tissue. Activation leads to the production of IL-2, which drives cell proliferation and differentiation. All of this proceeds while the T cells remain in contact with the dendritic cell. Antigen-mediated signaling through the T-cell receptor is insufficient to activate a naive T cell. Additional co-stimulatory signals are required, which must come from interaction of CD28 on the naive T cell with B7 co-stimulatory proteins on the dendritic cell. That B7 proteins are expressed only in the presence of infection prevents primary T cell responses from being generated against self antigens.

Dendritic cells, macrophages, and B cells are the three types of professional antigen-presenting cell. They have distinctive roles in the primary immune response. Whereas dendritic cells present antigens to activate naive T cells and drive their differentiation to effector T cells, macrophages and B cells present antigens to effector CD4 T cells for them to become activated by those T cells. Effector CD4 T_H1 cells mostly migrate from the secondary lymphoid tissues to sites of infection. There they activate tissue macrophages. This increases the macrophages' capacity to phagocytose and kill pathogenic organisms in the extracellular space, and also their capacity to act as professional antigen-presenting cells. Activation also enhances the capacity of macrophages to eliminate microorganisms that deliberately parasitize macrophage vesicles. Within the secondary lymphoid tissues, effector CD4 T_H2 cells activate naive B cells that are specific for the same antigen as the T cell. Activated B cells divide and differentiate into antibody-secreting plasma cells, and also undergo isotype switching under the influence of effector CD4 T cells of both T_H2 and T_H1 types. The three types of effector T cell enable the human immune system to respond to different categories of infection and to different stages in the course of the same infection (Figure 8.39).

Questions

8–1
  A. At which anatomical sites do naive T cells encounter antigen?
  B. In which sites specifically would a pathogen or its antigens end up, and how, if they (i) entered the body through a small wound in the skin, (ii) entered the body from the gut, or (iii) got into the bloodstream?
  C. How do naive T cells arrive at these sites?
  D. Do all T cells leave these locations after priming, and if so, how?

8–2   Unlike innate immune responses, which can begin within hours of the onset of an infection, adaptive immune responses involving T cells usually take several days. What accounts for this delay between the initiation of an infection and the engagement of an adaptive immune response?

8–3
  A. Which selectins, mucin-like vascular addressins and integrins have a role in the circulation of T cells between the blood and lymphoid tissues?
  B. Describe in chronological order how T cells migrate across lymph node high endothelial venules (HEVs) from the blood by using these molecules.

8–4   Which of the following explains why dendritic cells are more efficient than macrophages at stimulating naive T cells?
  a. Macrophages do not express MHC class II molecules.
  b. Dendritic cells are migratory and transport antigen to neighboring secondary lymphoid tissue.
  c. Dendritic cells do not repair damaged tissues.
  d. Macrophages do not process antigens.
  e. Dendritic cells, but not macrophages, endocytose foreign antigen.

8–5

A. Which cell-surface glycoprotein distinguishes professional antigen-presenting cells from other cells and is involved in the co-stimulation of T cells?

B. What receptors can it bind on the T cell and what signal does it deliver in each case?

C. Explain the consequence of antigen recognition by T cells in the absence of this glycoprotein on the antigen-presenting cell.

8–6   An adjuvant enhances the effectiveness of vaccines by inducing the expression of _____ on _____.

a. co-stimulatory molecules; dendritic cells
b. CD28; macrophages
c. MHC class II molecules; T cells
d. T-cell receptor; T cells
e. immunoreceptor tyrosine-based activation motifs; dendritic cells.

8–7   The three main classes of effector T cells are specialized to deal with different classes of pathogens and produce different sets of cytokines.

A. Name these three classes.

B. For each class, describe how antigen is recognized and the corresponding effector functions.

C. Give a general example of an antigen for each class.

8–8   Virus-infected cells attacked and killed by effector cytotoxic T cells are often surrounded by healthy tissue, which is spared from destruction.

A. Explain the mechanism that ensures that cytotoxic T cells kill only the virus-infected cells (the target cells).

B. What cytotoxins do cytotoxic T cells produce?

8–9   Which of the following is a feature of regulatory T cells (T_{reg})? (Select all that apply.)

a. T_{reg} express CD8 and control effector cells by inducing apoptosis.

b. T_{reg} express high levels of CD25 (IL-2 receptor α chain) and secrete pro-inflammatory cytokines such as IFN-γ.

c. Physical association between T_{reg} and its target cells is mandatory for T_{reg} function.

d. By interacting with dendritic cells in secondary lymphoid tissue, T_{reg} prevent the interaction and activation of naive T cells.

e. T_{reg} secrete TGF-β and suppress effector T cell function.

8–10   What are the roles of the following molecules in the signal transduction pathway leading from the T-cell receptor: (i) the CD3 complex; (ii) protein tyrosine kinase Lck; (iii) CD45; (iv) ZAP-70; (v) the ζ chain; (vi) IP$_3$; (vii) calcineurin?

8–11   Antigen recognition by T cells in the absence of co-stimulation results in

a. upregulation of B7
b. expression of the high-affinity IL-2 receptor

c. T-cell anergy
d. T-cell apoptosis
e. phosphorylation of immunoreceptor tyrosine-based activation motifs.

8–12

A. Describe the morphology of a granuloma.

B. Which types of infection would lead to the formation of a granuloma?

C. Why is this type of pathology actually beneficial to the host?

8–13   Cyclosporin A is an immunosuppressive drug commonly used in transplant patients to prevent graft rejection by alloreactive T cells. It acts by interfering with the signaling pathway that leads from the T-cell receptor to transcription in the nucleus of the genes for the cytokine IL-2 and the α chain of the IL-2 receptor. Why does preventing the transcription of these genes lead to immunosuppression?

8–14   B cells are activated by CD4 T_H2 cells only if both cell types recognize the same antigen. The same epitope, however, does not need to be shared for recognition.

A. Discuss why this characteristic is important in vaccine design.

B. Provide an example of a conjugate vaccine used to stimulate the synthesis of IgG antibody against *Haemophilus influenzae* B polysaccharide.

8–15   Angelina Roebuck, a 24-month-old toddler, was experiencing frequent nose bleeds and skin rashes. When examined she had splenomegaly and hepatomegaly, and very enlarged lymph nodes around her neck and underarms. There was no evidence of any infection, but Angelina had higher levels than normal of circulating lymphocytes and an abnormally low blood platelet count. A lymph-node biopsy showed marked hyperplasia of the lymphoid follicles and numerous plasma cells in the lymph-node cortex. Which of the following genes would you analyze first for a mutation that is likely to be causing Angelina's condition?

a. CD40 ligand
b. granulysin
c. CD95 (Fas)
d. IL-2
e. CD25.

8–16   Vijay Kumar, a 19-year-old male who had emigrated to the United States from South India two years before, developed lesions in his nasal mucosa, and nodular skin lesions on his cheeks and buttocks. Examination of a stained biopsy of a skin lesion revealed numerous clumps of mycobacteria. T cells in the skin lesions were secreting IL-4, IL-5, and IL-10. What would be the most likely diagnosis?

a. tuberculosis
b. lepromatous leprosy
c. leishmaniasis
d. tuberculoid leprosy
e. allergic dermatitis.

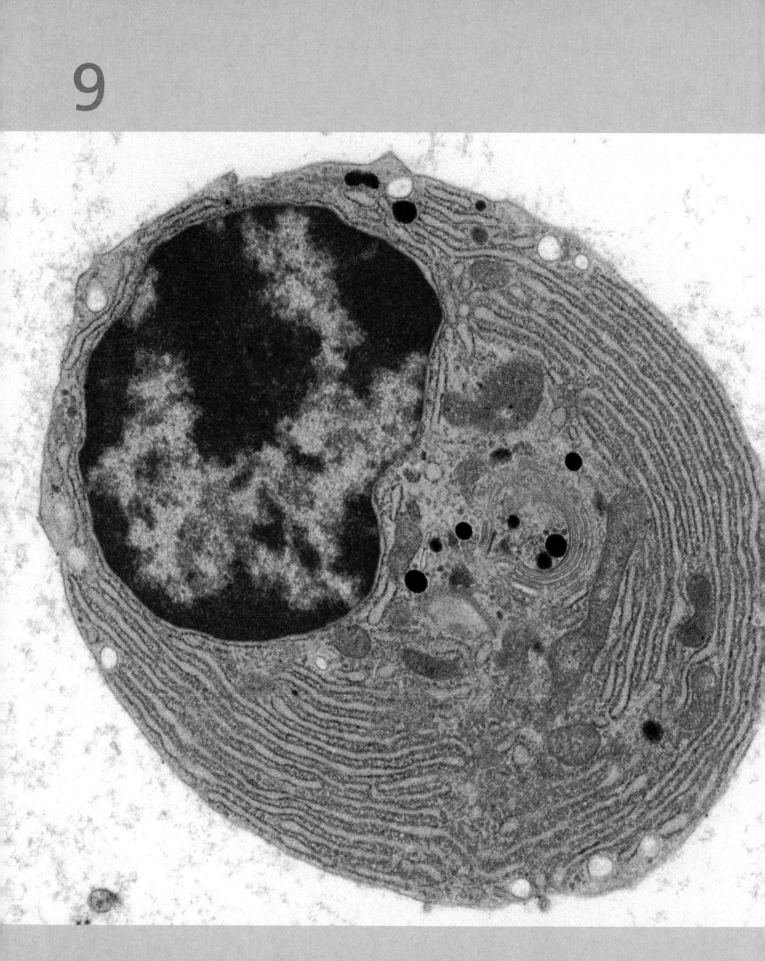

The plasma cell, the effector B cell that makes antibodies, the most potent weapon of adaptive immunity.

Chapter 9

Immunity Mediated by B Cells and Antibodies

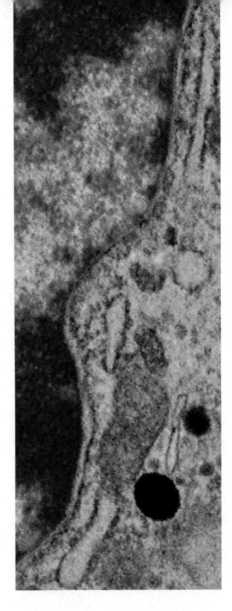

The production of antibodies is the sole function of the B-cell arm of the immune system. Antibodies are useful in the defense against any pathogen that is present in the extracellular spaces of the body's tissues. Some human pathogens, such as many species of bacteria, live and reproduce entirely within the extracellular spaces, whereas others, such as viruses, replicate inside cells but are carried through the extracellular spaces as they spread from one cell to the next. Antibodies secreted by plasma cells in secondary lymphoid tissues and bone marrow find their way into the fluids filling the extracellular spaces.

Antibodies are not in themselves toxic or destructive to pathogens; their role is simply to bind tightly to them. This can have several consequences. One way in which antibodies reduce infection is by covering up the sites on a pathogen's surface that are necessary for growth or replication, for example the viral glycoproteins that viruses use to bind to the surface of human cells and initiate infection. Such antibodies are said to **neutralize** the pathogen. In the development of a vaccine against an infectious agent or its toxic products, the gold standard that a company aims for is the induction of a **neutralizing antibody**. Antibodies also act as molecular adaptors that bind to pathogens with their antigen-binding arms and to receptors on phagocytic cells with their Fc regions. Thus, **opsonization** of a pathogen, or coating it with antibody, promotes its phagocytosis. Antibodies bound to the surface of pathogens also cause complement fixation through the classical pathway of complement activation. This further promotes the phagocytosis of the pathogen, because complement fragments deposited on the pathogen's surface bind to the complement receptors of phagocytes.

The structure, specificity, and other properties of antibodies were discussed in Chapter 4, and the development of B cells from their origin in bone marrow to differentiation into antibody-secreting plasma cells was the subject of Chapter 6. This chapter focuses on how antibodies clear infection by targeting destructive but nonspecific components of the immune system to an infecting pathogen. In the first part of the chapter we consider the antigens that provoke a B-cell response, how the response develops, and the generation of the different antibody isotypes. The structural differences between antibody isotypes provide a variety of adaptor functions that can target antibody-bound pathogen to different types of nonspecific effector cell; these aspects of the antibody-mediated immune response are discussed in the second part of this chapter.

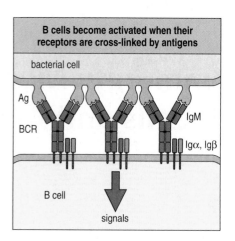

B cells become activated when their receptors are cross-linked by antigens

Figure 9.1 Cross-linking of antigen receptors is the first step in B-cell activation. The B-cell receptors (BCR) on B cells are physically cross-linked by the repetitive epitopes of antigens (Ag) on the surface of a bacterial cell. The B-cell receptor on a mature, naive B cell is composed of surface IgM, which binds antigen, and associated Igα and Igβ chains, which provide the signaling capacity.

Antibody production by B lymphocytes

The antibodies most effective at combating infection are those that are made early in an infection and bind strongly to the pathogen. On first exposure to an infectious agent, these two goals make competing demands on the immune system. As we saw in Chapters 6 and 8, B cells generally require help from activated T cells to mature into antibody-secreting plasma cells; this delays the onset of antibody production until around a week after infection. In addition, B cells take time to switch isotype and undergo affinity maturation, processes that are necessary for the production of the high-affinity antibodies that are most effective at dealing with pathogens. Thus, during the course of an infection, the effectiveness of the antibodies produced improves steadily. This experience is retained in the form of memory B cells and high-affinity antibodies, which provide long-term immunity to reinfection.

A faster primary response is made to certain bacterial antigens that are able to activate B cells without the need for T-cell help. However, the antibodies produced in such a response are predominantly of the IgM isotype and of generally low affinity. They do, however, provide an early defense, helping to keep the infection at a relatively low level until a better antibody response can be developed.

9-1   B-cell activation requires cross-linking of surface immunoglobulin

On binding to protein or carbohydrate epitopes on the surface of a microorganism, the surface IgM molecules of a naive, mature B cell become physically cross-linked to each other by the antigen and are drawn into the localized area of contact with the microbe. This clustering and aggregation of B-cell receptors sends signals from the receptor complex to the inside of the cell (Figure 9.1). Signal transduction from the B-cell receptor complex resembles, in many ways, the signaling from the T-cell receptor complex discussed in Section 8-7, p. 222. Both types of receptor are associated with cytoplasmic protein tyrosine kinases that are activated by receptor clustering, and both receptors activate similar intracellular signaling pathways.

Interaction of antigen with surface immunoglobulin is communicated to the interior of the B cell by the proteins Igα and Igβ, which are associated with IgM in the B-cell membrane to form the functional B-cell receptor. Like the CD3 polypeptides of the T-cell receptor complex, the cytoplasmic tails of Igα and Igβ each contain two immunoreceptor tyrosine-based activation motifs (ITAMs) with which the tyrosine kinases Blk, Fyn, and Lyn associate. The ITAMs become phosphorylated on tyrosine residues, which allows the tyrosine kinase Syk to bind to Igβ tails that are doubly phosphorylated. Interaction between bound Syk molecules initiates intracellular signaling pathways that lead to changes in gene expression in the nucleus (Figure 9.2).

9-2   B-cell activation requires signals from the B-cell co-receptor

Cross-linking of the B-cell receptor by antigen generates a signal that is necessary but not sufficient to activate a naive B cell. The additional signals required

Figure 9.2 Signals from the B-cell receptor initiate a cascade of intracellular signals. On clustering of the receptors, the receptor-associated tyrosine kinases Blk, Fyn, and Lyn phosphorylate the immunoreceptor tyrosine-based activation motifs (ITAMS, shaded yellow) on the cytoplasmic tails of Igα (blue) and Igβ (orange). Subsequently, Syk binds to the phosphorylated ITAMs of the Igβ chain. Because there are at least two receptor complexes in each cluster, Syk proteins become bound in close proximity and can activate each other by transphosphorylation, thus initiating further signaling. The signals produced are therefore ultimately relayed to the B-cell nucleus, where they induce changes in gene expression required for B-cell activation.

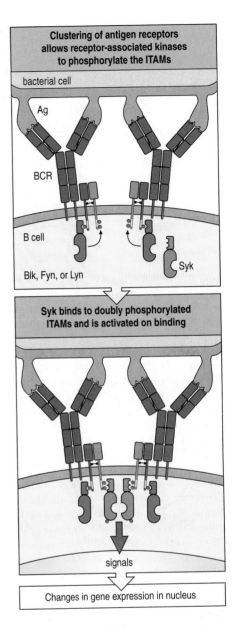

are delivered in several ways. One set of signals is delivered when the B-cell receptor becomes closely associated with another protein complex on the B-cell surface known as the **B-cell co-receptor** (Figure 9.3). The B-cell co-receptor is a complex of three proteins: the first is the complement receptor 2 (CR2 or **CD21**) that recognizes the iC3b and C3d breakdown products of the C3b fragments deposited on a pathogen; the second is the protein **CD19**, which acts as the signaling chain of the receptor; and the third is the protein **CD81**, whose function is not yet known. It does act, however, as a cell-surface receptor for the hepatitis C virus.

The generation of the iC3b and C3d ligands for the B-cell co-receptor involves the complement receptor CR1, which is also present on B cells. In the course of the innate immune response to infection, complement activation by the alternative, lectin, and classical pathways (see Chapter 2) leads to the deposition of C3b fragments on the pathogen's surface. C3b is the ligand for the complement CR1 receptor on B cells, which on binding to C3b facilitates its cleavage by factor I: first to the iC3b fragment and then to the more stable C3d fragment (Figure 9.4). By this cooperative process CR1 increases the abundance of B-cell co-receptor ligands on the pathogen. When the B-cell receptor binds to its specific antigen on the pathogen, the CR2 component of the B-cell co-receptor complex can bind to an adjacent C3d, which serves to bring the B-cell receptor and co-receptor into juxtaposition (Figure 9.5). The co-ligation of B-cell receptor and co-receptor brings the Igα-bound tyrosine kinase Lyn into close proximity with the cytoplasmic tail of CD19, which it phosphorylates. The phosphorylated CD19 then binds intracellular signaling molecules that generate signals that synergize with those generated by the B-cell receptor complex. Simultaneous ligation of the B-cell receptor and co-receptor increases the overall signal 1000–10,000-fold, thus greatly increasing the B cell's sensitivity to antigen.

9-3 The antibody response to certain antigens does not require T-cell help

For most primary immune responses, the activation of pathogen-specific B cells is dependent on helper CD4 T cells that recognize pathogen-derived peptides by the B cell's MHC class II molecules (see Section 8-18, p. 241). In this situation, activating signals emanate from the B-cell receptor and co-receptor, and also from cytokine receptors on the B-cell surface that bind to cytokines produced by the helper T cell.

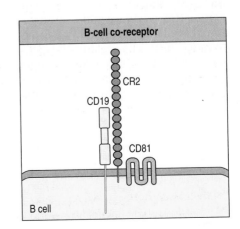

Figure 9.3 The B-cell co-receptor. The B-cell co-receptor is composed of three polypeptide chains, namely CD19, CD81, and CR2 (complement receptor 2). CR2 binds to complement on pathogen surfaces, CD19 is the signaling component, and the function of CD81 is unknown. Simultaneous signaling by B-cell receptor and B-cell co-receptor strengthens the signal to the B cell that an antigen has been bound.

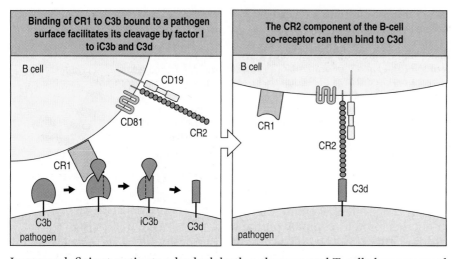

Figure 9.4 **Complement receptor CR1 facilitates the production of C3d, the ligand for complement receptor CR2, a component of the B-cell co-receptor.** B cells carry both CR1 and CR2. When CR1 binds to C3b fragments deposited on a pathogen's surface, it makes them susceptible to cleavage by complement factor I, a serine protease. The first cleavage yields the iC3b fragment, which is then further cleaved by factor I to give C3d. Both iC3b and C3d bind to the CR2 component of the B-cell co-receptor. It is through the binding of CR2 to either iC3b or C3d that the B-cell co-receptor detects the presence of pathogens. Because it facilitates the production of C3d from C3b, CR1 is said to be a co-factor or to have a co-factor role with regard to CR2.

Immunodeficient patients who lack both a thymus and T cells have normal numbers of B cells but fail to make antibody responses to most antigens. They do make antibodies against some microbial antigens, which were therefore termed **thymus-independent antigens** (**TI antigens**). TI antigens divide into two groups, **TI-1 antigens** and **TI-2 antigens**, according to the type of mechanism by which they activate B cells.

When naive B cells are activated in the absence of T-cell help, additional signals are required over and above those normally produced by the B-cell receptor and co-receptor. For TI-1 antigens these signals are provided by the signaling receptors of innate immunity such as the Toll-like receptors, which to varying degrees are expressed on B cells. The classic example of a TI-1 antigen is the lipopolysaccharide (LPS) of Gram-negative bacteria, which is recognized by the combination of TLR4 and CD14 (see Section 2-11, p. 45). The aggregate of signals generated when TLR4, the B-cell receptor and the B-cell co-receptor all recognize their ligands on a bacterial cell surface is sufficient to induce the B cells to divide and differentiate. When B cells are triggered by TI-1 antigens they produce only IgM antibodies, because cytokines produced by activated helper T cells are necessary for a B cell to switch its antibody isotype. A surface-associated TI-1 antigen such as LPS causes the T-cell-independent activation not only of B cells specific for epitopes of LPS (Figure 9.6, left panel) but also of B cells specific for other 'non-TI' antigens of the bacterial cell surface (Figure 9.6, center panels). In contrast, a soluble TI-1 antigen that binds to both a Toll-like receptor and the B-cell receptor would stimulate only the antibodies against its own epitopes.

Of the 10 Toll-like receptors, TLR9, which detects bacterial DNA, is the most strongly expressed by human B cells. If a TLR9-expressing B cell binds a bacterium with its B-cell receptor and co-receptor, it will internalize and degrade the bacterium, releasing the bacterial DNA into an endocytic vesicle, where it can be detected by TLR9 (see Section 2-11, p. 45). In the process, activating signals will be generated by all three receptors and the B cell will be activated to make antibodies against the cell-surface component of the bacterium recognized by the B-cell receptor (Figure 9.6, right panel). Because it stimulates TLR9, bacterial DNA also acts as a TI-1 antigen in its own right for those B cells that recognize it with their B-cell receptors, stimulating the production of

Figure 9.5 **Signals generated from the B-cell receptor and co-receptor synergize in B-cell activation.** Binding of CR2 to C3d fragments deposited on the surface of a pathogen brings together the B-cell co-receptor complex with the B-cell receptor. This causes them to cluster together on the B-cell surface. The cytoplasmic tail of CD19 is then phosphorylated by tyrosine kinases associated with the B-cell receptor. Phosphorylated CD19 binds intracellular signaling molecules whose signals synergize with those generated by the B-cell receptor.

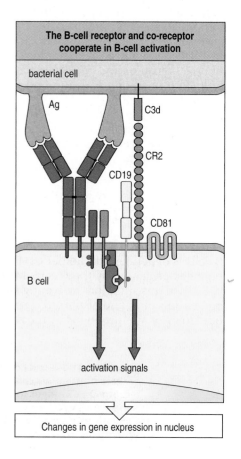

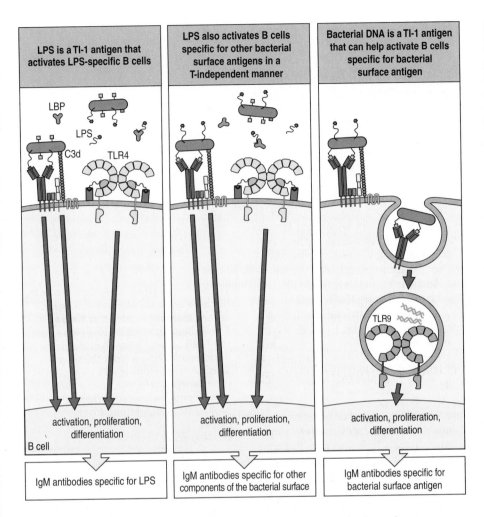

| LPS is a TI-1 antigen that activates LPS-specific B cells | LPS also activates B cells specific for other bacterial surface antigens in a T-independent manner | Bacterial DNA is a TI-1 antigen that can help activate B cells specific for bacterial surface antigen |

activation, proliferation, differentiation

IgM antibodies specific for LPS | IgM antibodies specific for other components of the bacterial surface | IgM antibodies specific for bacterial surface antigen

Figure 9.6 Thymus-independent (TI)-1 antigens activate B cells without T-cell help. In the activation of B cells by TI antigens, the signals produced by the B-cell receptor and the co-receptor are augmented by signals coming from another type of activating receptor expressed by the B cell, for example a Toll-like receptor. The combination of all these signals is sufficient to activate the B cell in the absence of antigen-specific helper T cells. Three examples of how this can happen are shown. In the left and center panels the other receptor is TLR4, which recognizes bacterial lipopolysaccharide (LPS), the classical TI antigen. Recognition of LPS by TLR4 also involves the soluble LPS-binding protein (LBP), and MD2 and CD14 on the B-cell surface (see Figure 2.23, p. 47). The left panel shows a B cell specific for LPS being activated by LPS to make antibody. Here, LPS molecules interact with both the B-cell receptor and TLR4. The center panel shows a B cell specific for another component of the bacterial surface. In this case, LPS interacts with TLR4, and the other bacterial antigen interacts with the B-cell receptor, activating the B cell to make antibody against the bacterial surface component. In the right panel, the B-cell receptor is again interacting with a bacterial cell-surface antigen, but the other activating receptor is TLR9, which recognizes the DNA of bacteria that have been internalized and degraded in endosomes.

anti-DNA antibodies. Some cell-surface bacterial antigens have been operationally defined as TI-1 antigens even though they would not act as such in isolation. This paradox arose because the concept of thymus-independent antigens was developed many decades before the mechanisms that allow a B-cell response to become independent of T cells were understood.

The second type of thymus-independent antigen, the TI-2 antigens, are typically composed of repetitive carbohydrate or protein epitopes present at high density on the surface of a microorganism. TI-2 antigens stimulate only those B cells that are specific for the antigen and are believed to act by cross-linking B-cell receptors and co-receptors so extensively that the need for additional signals is overridden (Figure 9.7). Typical antigens of this kind are the cell-wall polysaccharides of the pneumococcus (*Streptococcus pneumoniae*). The B cells responding to these antigens are often of the B-1 subpopulation (see Section 6-10, p. 171).

Although the TI-2 antigens on some bacteria induce an early antibody response that helps to contain the infection, this is limited in scope and duration. There is little isotype switching and so the antibodies are predominantly IgM. Neither is there somatic hypermutation, so there is no possibility of increasing the affinity for antigen of the antibodies produced.

Figure 9.7 TI-2 antigens activate B cells by extensive cross-linking of B-cell receptors and co-receptors. Unlike TI-1 antigens, TI-2 antigens do not activate additional signaling pathways but densely cross-link B-cell receptors and co-receptors to generate a signal above the level required to stimulate B-cell proliferation and differentiation.

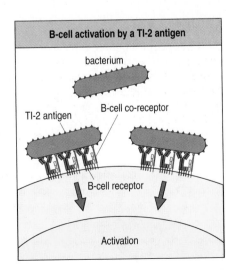

B-cell activation by a TI-2 antigen

9-4   Activation of naive B cells by most antigens requires help from CD4 T cells

Although the antibody response to a pathogen may be initiated by thymus-independent antigens, the bulk of the pathogen-specific antibody is eventually produced by B cells stimulated by thymus-dependent antigens. As introduced in Chapter 8, activation of these B cells occurs in the secondary lymphoid tissues where B cells, specific antigen, and helper CD4 T cells are all brought together (see Section 8-18, p. 241). We will continue to describe these processes using the lymph node as our example of a secondary lymphoid tissue.

Immature dendritic cells bring pathogens and their antigens from the infected tissue to the lymph node in the draining lymph. In effect, the dendritic cells spread the infection from its site of origin to the secondary lymphoid tissue. While doing this, the dendritic cells mature and take up residence in the T-cell area of the lymph node (see Section 8-3, p. 215) There they present antigens to the circulating T cells that enter the node either from the blood or in the afferent lymph coming from another lymph node (see Figure 8.5, p. 217). The receptors of an antigen-specific CD4 T lymphocyte engage the complexes of peptide antigens and MHC class II on the dendritic cell surface, initiating a prolonged interaction between the two cells during which the T cell is driven to divide and differentiate into a clone of effector helper T cells (Figure 9.8, left panel). When this differentiation occurs in the presence of the cytokine IL-4, the helper T cells become of the T_H2 type that stay in the lymph node, where their job is to activate antigen-specific B cells.

Circulating naive B cells home to lymph nodes from the blood or lymph, using the same mechanisms and the same cell adhesion molecules and chemokines (CCL21 and CCL19) as those used by naive T cells (see Sections 8-3 and 8-4, pp. 215–219). When B cells enter via the afferent lymph, they do not have to pass an epithelial barrier but are attracted directly into the T-cell zones by CCL21 and CCL19. Whether they come in from the lymph or the blood, the naive B cells first enter the T-cell zone, and in the absence of their specific antigens they are attracted by the chemokine CXCL13 into the B-cell zone. Here they compete to enter a primary follicle and receive survival signals from the follicular dendritic cells (FDCs) before leaving the node in the efferent lymph (see Section 6-14, p. 177).

If a B cell encounters its specific antigen, either in the lymph node (Figure 9.8, center panel) or during its earlier travels in the lymph and blood, cross-linking of the B-cell receptor and co-receptor induces signals that induce changes

Figure 9.8 Antigen-stimulated B cells become trapped in the T-cell zone, where they meet effector helper T cells. Recirculating naive T cells enter the T-cell zone of a lymph node draining a bacterial infection from the blood through high endothelial venules (HEV) (first panel). T cells that encounter specific antigen on a dendritic cell proliferate and differentiate to become helper T_H2 cells. Circulating naive B cells similarly enter the T-cell zone of the lymph node. B cells that encounter their specific antigen, often in the form of a complete bacterium, become trapped in the T-cell zone, where they process and present peptides from this antigen on MHC class II molecules (second panel). Helper T_H2 cells and B cells specific for antigens on the same bacterium form conjugate pairs in which the T cells begin to activate the B cells (third panel). This type of interaction between a T cell and a B cell that recognize different epitopes on the same antigen is called a cognate interaction.

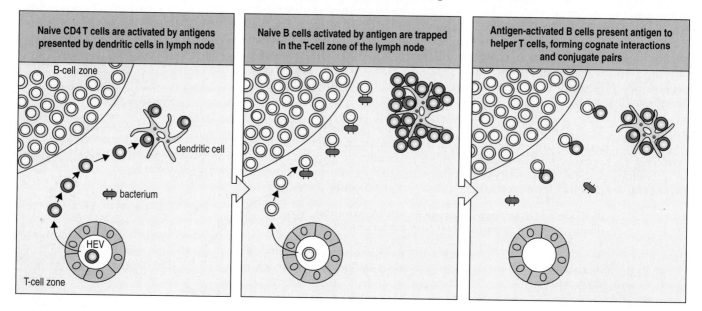

Figure 9.9 B-cell activation in response to thymus-dependent antigens requires cognate T-cell help. The first signal required for B-cell activation is delivered through the antigen receptor (top panel). With thymus-dependent antigens, the second signal is delivered by a cognate helper T cell that recognizes a peptide fragment of the antigen bound to MHC class II molecules on the B-cell surface (middle panel). The two signals together drive B-cell proliferation and differentiation into plasma cells.

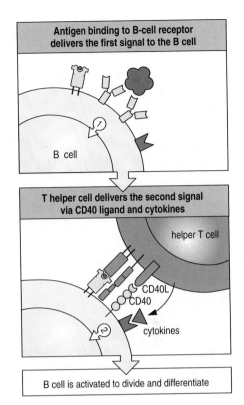

in the B-cell's expression of adhesion molecules and chemokine receptors. These cell-surface changes prevent the B cells from leaving the T-cell zone, where they remain trapped near the boundary with the B-cell zone. In this location, antigen-stimulated B cells are well placed to interact with newly differentiated antigen-specific helper CD4 T cells (Figure 9.8, right panel).

The B-cell receptor has two distinct roles in B-cell activation: binding antigen, which sends a signal to the B cell's nucleus to change gene expression; and internalizing antigen by receptor-mediated endocytosis, which facilitates its processing and presentation by MHC class II molecules. The receptors of the helper T cells sample the MHC class II molecules on the antigen-stimulated B cells trapped in the T-cell zone, and if they find their specific antigen the T cell and B cell form a conjugate pair. This interaction induces the T cell to express CD40 ligand, which then interacts with the B-cell's CD40 (Figure 9.9). This signals the B cell to activate the transcription factor NFκB and increase the surface expression of the adhesion molecule ICAM-1, which interacts with the integrin LFA-1 on the T cell (see Section 8-4, p. 217). These developments strengthen the cognate interaction between the B cell and the helper T cell. An immunological synapse is elaborated at the area of contact and the T cell's cytoskeleton and Golgi apparatus is reorganized, facilitating the efficient and focused delivery of cytokines onto the B cell (Figure 9.10). The most

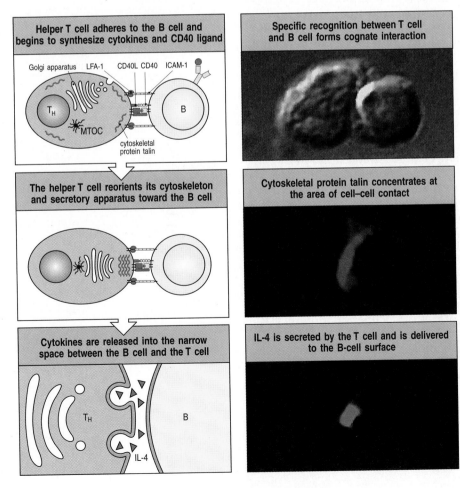

Figure 9.10 Helper T cells activate B cells through a focus of cell–cell contact where IL-4 and other cytokines are delivered to the B-cell surface. When the helper T-cell receptor binds its specific antigen on the B cell to form a cognate pair, the T cell is induced to express CD40-ligand (CD40L), a membrane-associated cytokine that is part of the tumor necrosis factor (TNF) family. The B cell already expresses CD40, a member of the TNF-receptor family, and the binding of CD40 ligand to CD40 helps activate the B cell. At the point of contact between the two cells a tight junction is formed by the interaction of T-cell LFA-1 with B-cell ICAM-1 (top panels). While the T-cell is synthesizing and accumulating IL-4 and other soluble cytokines, it reorients its cytoskeleton so that the secretory apparatus is directly facing the focus of contact with the B cell. As part of this process the cytoskeletal protein talin (shown in red in the middle and top left panels) becomes concentrated at the contact area (middle panels). The T cell then secretes IL-4 and other cytokines onto the B-cell surface at the focused region of contact (bottom panels). In the fluorescence micrographs (right panels), talin is stained red (middle right) and IL-4 is stained green (bottom right). MTOC, microtubule-organizing center. Photographs courtesy of A. Kupfer.

important of these cytokines is IL-4, which is the characteristic cytokine of the T_H2 response and is essential for B-cell proliferation and differentiation.

9-5 The primary focus of clonal expansion in the medullary cords produces plasma cells secreting IgM

Once an antigen-specific B cell and an antigen-specific T cell form a conjugate pair, they move together out of the T-cell zone in the cortex and into the medullary cords. Here both cells begin to divide, forming what is called the **primary focus** of clonal expansion. This period of cellular proliferation lasts for several days and produces dividing B lymphoblasts that start to secrete IgM. The antibody leaves the node in the efferent lymphatic vessel, which is contiguous with the medullary cords. The lymph delivers the IgM to the blood, from where it rapidly reaches the site of infection. For infections in which the pathogen carries no thymus-independent antigens, this will be the first antibody produced. It will start to appear several days after the onset of infection.

Some of the B lymphoblasts stay in the medullary cords, where they differentiate into plasma cells under the influence of the cytokines IL-5 and IL-6 secreted by T_H2 cells. They secrete predominantly IgM antibody, but some isotype switching can also occur in the primary focus.

The terminal differentiation of lymphoblasts to plasma cells is controlled by a transcription factor called B-lymphocyte-induced maturation protein 1 (BLIMP-1). This repressor stops transcription of the genes necessary for proliferation, somatic hypermutation and isotype switching. These cells then increase the expression of the immunoglobulin chains and of the factors involved in their synthesis and secretion. The morphological effects of activation are striking: the small resting B cell, which in appearance is all nucleus and no cytoplasm, gives rise to plasma cells, whose large active cytoplasm packed with rough endoplasmic reticulum is testimony to their function as antibody factories in which up to 20% of the protein made is immunoglobulin (Figure 9.11).

9-6 Follicular dendritic cells provide long-lasting depositories of B-cell antigens

In Chapter 6 we saw how interactions between B cells and FDCs in the primary follicles of secondary lymphoid tissues are critical for the maturation and survival of B cells. This reflects a more general dependence of B cells upon FDCs, which are dedicated to the development and immune response of B cells. As well as being functionally different from the myeloid dendritic cells that present antigens to naive T cells, FDCs are also developmentally distinct from them; they are not bone marrow derived hematopoietic cells but stromal cells that develop from fibroblast-like cells.

The characteristic microanatomy of the secondary lymphoid tissues, with their B-cell and T-cell zones and network of FDCs, is produced by dynamic interactions between these three types of cell. The differentiation of FDCs depends on cytokines of the TNF family made by lymphocytes, such as TNF-α and the lymphotoxins LT-α and LT-β, which interact with their corresponding receptors on the FDC precursors. Both T cells and FDCs are essential for producing B cells that have undergone affinity maturation and isotype switching.

One function of the FDCs is to provide a long-lasting depository of intact antigens that are available to interact with B-cell receptors. The extensive surface of the dendrites of FDCs is specialized in picking up intact antigens and

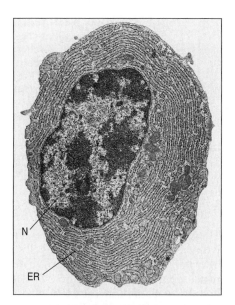

Figure 9.11 Plasma cell. Electron micrograph of a plasma cell. Note the characteristic 'clockface' pattern in the nucleus (N), which resembles the hands and face of a clock, as well as the extensive endoplasmic reticulum (ER), which is characteristic of cells synthesizing and secreting large amounts of protein, in this case antibodies. Photograph courtesy of C. Grossi.

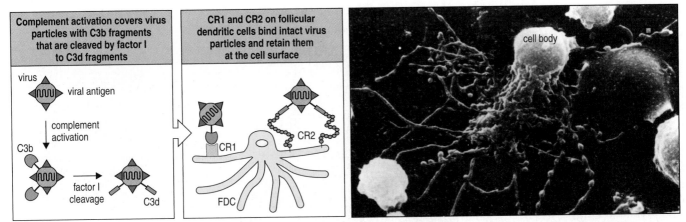

holding them in quantity for months, if not years. For example, in patients infected with the human immunodeficiency virus (HIV), viral particles are concentrated on FDCs. This is principally accomplished by the complement receptors, CR1 and CR2, which are expressed on FDCs as well as on B cells. Receptors CR1 and CR2 bind pathogens and antigens that have been tagged with complement C3b and C3d fragments, respectively (Figure 9.12, first two panels, and see also Section 9-2). Such complexes of a pathogen or an antigen with complement are called **immune complexes**. The immune complexes are not internalized or degraded by the FDC but are retained at its surface. The receptor CR2 is expressed only by B cells and FDCs and is thus dedicated to the B-cell response. The dendrites of the FDC can take on a form resembling bundles along a string. These bundles, called **iccosomes** (immune-complex coated bodies) can be bound and taken up by antigen-specific B cells, which then process and present the antigen (see Figure 9.12, right panel).

Figure 9.12 The dendrites of follicular dendritic cells use complement receptors to take up intact pathogens and antigens and preserve them for long periods. The first panel shows the deposition of C3b fragments on the surface of a virus particle and their subsequent cleavage to C3d. The second panel shows how this leads to binding of the virus by the complement receptors CR1 and CR2 present on the dendrites of a follicular dendritic cell (FDC) in a lymphoid follicle. The scanning electron micrograph on the right shows that FDCs have a prominent cell body and many dendritic processes. Pathogens and antigens are bound to complement receptors on the FDC surface and become clustered, forming prominent beads along the dendrites. The beads are shed from the cell as iccosomes and can be taken up by B cells. Photograph courtesy of A.K. Szakal.

9-7   Activated B cells undergo somatic hypermutation and isotype switching in the specialized microenvironment of the B-cell zone

Whereas some B lymphoblasts go directly from the primary focus to become IgM-secreting plasma cells in the medullary cords, other B lymphoblasts leave the primary focus and move into the primary follicles of the B-cell zone while still attached to their cognate helper T cells (Figure 9.13). Under the influence

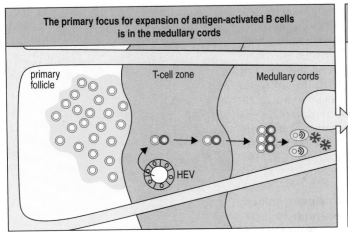
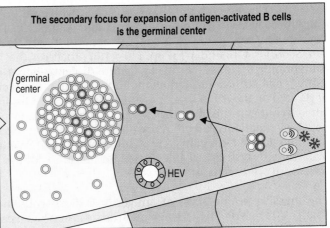

Figure 9.13 The primary and secondary foci for expanding antigen-activated B cells occur at different sites in the lymph node. Cognate pairs of antigen-activated B cells and helper T cells are formed in the T-cell zone and move to the medullary cords, where they divide and differentiate, forming the primary focus of B-cell expansion (left panel). Later, some of the cognate pairs of B and T cells move to a primary follicle (the B-cell zone), where they divide to form a germinal center, the secondary focus of B-cell expansion (right panel).

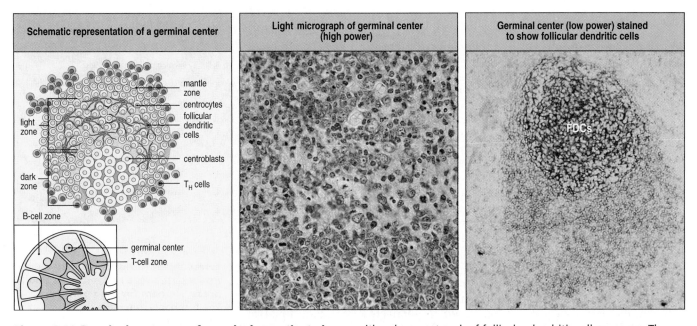

Schematic representation of a germinal center	Light micrograph of germinal center (high power)	Germinal center (low power) stained to show follicular dendritic cells

Figure 9.14 Germinal centers are formed when activated B cells enter lymphoid follicles. The germinal center, shown schematically in the left-hand panel, is a specialized microenvironment in which B-cell proliferation, somatic hypermutation, and selection for antigen binding all occur. Rapidly proliferating B cells in germinal centers are called centroblasts. Closely packed centroblasts form the so-called 'dark zone' of the germinal center, which can also be seen in the lower part of the center panel. As these cells mature, they stop dividing and become small centrocytes, moving out into an area of the germinal center called the 'light zone' (in the upper part of the center panel), where the centrocytes make contact with a dense network of follicular dendritic cell processes. The follicular dendritic cells are not stained in the center panel but can be seen clearly in the right panel, in which both follicular dendritic cells (FDCs, stained blue with antibody against Bu10, a marker of follicular dendritic cells) in the germinal center, as well as the mature B cells in the mantle zone (stained brown with an antibody against IgD) can be seen. The plane of this section reveals mostly the dense network of follicular dendritic cells in the light zone, although the less dense network in the dark zone can just be seen at the bottom of the figure. Photographs courtesy of I. MacLennan.

of the FDCs and the T cells, these B cells are subject to somatic hypermutation and isotype switching and will eventually produce plasma cells that make high-affinity antibodies of isotypes other than IgM. The FDCs make the cytokines IL-6, IL-15, 8D6, and BAFF, which force the B cells to divide rapidly—about once every 6 hours—and to become large, metabolically active centroblasts. The helper T cells also divide, making cytokines and interacting via their CD40 ligand with the CD40 on B cells, which induces the B cell to produce the enzyme activation-induced cytidine deaminase. This DNA-modifying enzyme is essential for both somatic hypermutation and isotype switching (see Sections 4-14 and 4-15, pp. 114–116), processes that now operate in the proliferating centroblasts. A key feature of centroblasts is that they do not express cell-surface immunoglobulin and cannot interact with the antigens deposited on the FDCs. They have already been specifically selected and activated by antigen, and this stage serves to generate a large population of B cells with switched isotype and many different combinations of mutation in the immunoglobulin V-region genes. This population of B cells will later form the basis for a further round of selection for immunoglobulin quality during the process of affinity maturation.

The massive proliferation of antigen-specific B cells in a primary follicle changes its morphology to become what is described as a secondary follicle. The dominant feature now becomes the **germinal center** containing all the rapidly dividing B and T cells (Figure 9.14, left and center panels). Surrounding the germinal center and pushed to the periphery of the follicle are the naive B cells that are passing through the lymph node in search of antigen and survival signals. These B cells now form a zone called the mantle zone. In a primary response, germinal centers appear in secondary lymphoid tissues about

1 week after the start of infection, and they are the cause of the characteristic swelling of lymph nodes draining an infected site (the so-called 'swollen glands'). The cellular and morphological events that form the germinal center and take place there are called the **germinal center reaction**. With time, the vast majority of lymphocytes present in a germinal center are clones derived from one or a few founder pairs of antigen-activated B and T cells.

As they divide, the centroblasts become increasingly closely packed and form a region that is darkly staining in histological sections and is called the **dark zone** of the germinal center. The centroblasts give rise to centrocytes; these divide more slowly and express surface immunoglobulin, which is now mutated and switched in isotype. The centrocytes leave the dark zone and move into the light zone, where there is a lower density of B cells and a higher density of FDCs and helper T cells. In germinal centers, about 80% of the cells are centroblasts, which have a CD38$^+$ CD44$^-$ CD77$^+$ phenotype, and 10–20% are centrocytes, which have a C38$^+$ CD44$^+$ CD77$^+$ phenotype.

Centrocytes are programmed to die by apoptosis within a short time unless their surface immunoglobulin is bound by antigen and their CD40 is bound by the CD40 ligand of a helper T cell. To survive, the centrocytes must compete with each other, first for access to antigen on FDCs and then for antigen-specific helper T cells.

9-8   Selection of centrocytes by antigen in the germinal center drives affinity maturation of the B-cell response

As we have seen in Chapters 6–8, a common theme in lymphocyte development is for a phase of activation and proliferation to be followed by one of selection. This is precisely what happens to the B cells maturing in a germinal center. Somatic hypermutation in a clone of expanding B cells produces centrocytes with a diversity of B-cell receptors. The surface immunoglobulin expressed by an individual centrocyte can have an affinity for its specific antigen that is higher, lower, or the same as that of the unmutated immunoglobulin. Thus, the population of centrocytes in a germinal center expresses immunoglobulins with a range of affinities for the specific antigen. The source of antigen available to the centrocytes is that deposited on the dendrites of the FDCs. Newly formed centrocytes move from the dark zone of the germinal center and enter the light zone, where they compete for access to the antigen displayed by the FDCs' iccosomes. If a centrocyte receptor binds antigen with sufficient strength it forms a synapse with the FDC that delivers antigen as well as survival signals to the B cell. As the B cell moves to the outer regions of the light zone, where helper T cells are concentrated, it processes and presents the antigen on MHC class II molecules. Engagement of peptide:MHC class II complexes by the T-cell receptor, and of CD40 on the centrocyte by CD40 ligand on the T cell, induces the centrocyte to express the protein Bcl-x$_L$, which prevents its death by apoptosis. If a centrocyte fails to obtain, internalize, and present antigen, it does not receive survival signals and dies by apoptosis (Figure 9.15).

The apoptotic centrocytes are then phagocytosed by macrophages in the germinal center. Macrophages that have recently engulfed apoptotic centrocytes are a characteristic feature of germinal centers and, because of their densely staining contents, are called **tingible body macrophages**—tingible simply meaning 'capable of being stained'. Somatic hypermutation can produce centrocytes bearing immunoglobulin that reacts with a self antigen on the surface of cells in the germinal center. When this happens, contact with helper T cells or other cells in the germinal center will render such centrocytes inactive or anergic—a mechanism similar to the one whereby self-reactive, immature B cells are inactivated in the bone marrow (see Section 6-13, p. 176).

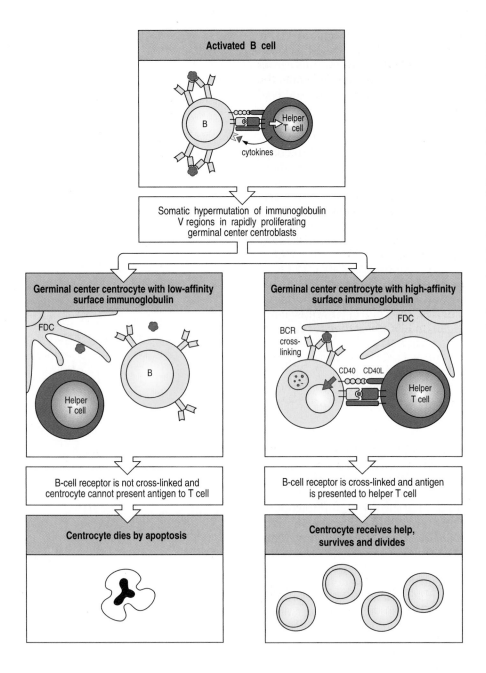

Figure 9.15 After somatic hypermutation, centrocytes with high-affinity receptors for antigen are rescued from apoptosis. In the germinal center, helper T cells induce dividing centroblasts to undergo somatic hypermutation (top panel). The resulting centrocytes, which have undergone somatic hypermutation, interact with follicular dendritic cells (FDCs) that display immune complexes on their surface (middle panel). Centrocytes whose receptors bind antigen poorly, or do not bind antigen at all because they have mutated beyond recognition, cannot compete for access to the FDCs and die by apoptosis (left panels). Centrocytes with receptors that bind antigen well receive signals from the FDCs and are induced to internalize, process, and present antigen to helper T cells. These cognate interactions enable the centrocyte to express the intracellular protein Bcl-x_L, which prevents apoptosis and ensures survival (right panels). BCR, B-cell receptor.

Only the centrocytes with the highest-affinity antigen receptors are selected for survival and further differentiation into antibody-producing plasma cells or into long-lived memory cells. In this way, the affinity of antibodies for the specific antigen increases during the course of an immune response and in subsequent exposures to the same antigen. This is the process known as **affinity maturation**. It ensures that the B cells concentrate their resources on making the best possible antibodies. In some situations, such as in tonsillitis, the secondary lymphoid organ becomes overwhelmed by a bacterial pathogen. With such abundance of antigen, there is less selection for the best antibodies. Inferior antibodies are made that prove ineffective in terminating the infection, which now follows a chronic course. Such infections can usually be overcome by a course of antibiotics, which sometimes needs to be combined with surgical removal of the infected lymphoid tissue.

The proliferating B cells of the germinal center are subject to mutational processes that change the sequence of the immunoglobulin genes and recombine their sequences with deletion of intervening DNA. Although these events are largely confined to the immunoglobulin genes, they do occur at a low frequency in other genes and in ways that favor malignant transformation. Such progression is encouraged by the nurturing environment provided by the FDCs. The consequence is that most B-cell lymphomas originate with germinal center cells.

9-9   The cytokines made by helper T cells determine how B cells switch their immunoglobulin isotype

In Chapter 4 we saw how the first immunoglobulins made by B cells are of the IgM and IgD classes, but that after activation by antigen B cells can switch their heavy-chain isotype to produce IgG, IgA, or IgE. Isotype switching takes place in activated B cells mainly within the germinal center, and the isotype to which an individual B cell switches is determined by cognate interactions with helper T cells. The particular isotype to which a switch is made depends on the cytokines secreted by the helper T cell. The roles of individual cytokines in switching the isotype of mouse immunoglobulin heavy chains are summarized in Figure 9.16. Cytokines secreted by T_H2 cells—IL-4, IL-5, and TGF-β—are the predominant players. They initiate the antibody response by activating naive B cells to differentiate into plasma cells secreting IgM, and also induce the production of other antibody isotypes including, in humans, the weakly opsonizing antibodies IgG2 and IgG4, as well as IgA and IgE. However, interferon (IFN)-γ, the characteristic cytokine produced by T_H1 cells, switches B cells to making the IgG2a and IgG3 subclasses of immunoglobulin (in mice) and the strongly opsonizing antibody IgG1 in humans.

T-cell cytokines induce isotype switching by stimulating transcription from the switch regions that lie 5′ to each heavy-chain C gene. For example, when activated B cells are exposed to IL-4, transcription from a site upstream of the switch regions of $C_\gamma1$ and C_ε can be detected a day or two before switching occurs. As with the low-level transcription that occurs in immunoglobulin loci before rearrangement (see Section 6-8, p. 170), this transcription could be opening up the chromatin and making the switch regions accessible to the somatic recombination machinery that will place a new C gene in juxtaposition to the V-region sequence.

The induction of isotype switching by cognate helper T cells also requires the ligation of CD40 on the B-cell surface by CD40 ligand on the T cells. The importance of helper T cells and the CD40–CD40 ligand interaction for isotype switching is apparent from the immunodeficiency of patients who lack CD40 ligand. These patients have abnormally high levels of IgM in their blood serum, which gives the name **hyper-IgM syndrome** to their condition, but

Influence of cytokines on antibody isotype switching							
Cytokine	IgM	IgG3	IgG1	IgG2b	IgG2a	IgA	IgE
IL-4	Inhibits	Inhibits	Induces		Inhibits		Induces
IL-5						Augments production	
IFN-γ	Inhibits	Induces	Inhibits		Induces		Inhibits
TGF-β	Inhibits	Inhibits		Induces		Induces	

Figure 9.16 Different cytokines induce B cells to switch to different immunoglobulin isotypes. Individual cytokines can either induce (green), augment (yellow), or inhibit (red) the switching of immunoglobulin synthesis to a particular isotype. The apparently inhibitory effects are due largely to the positive effect of the cytokine on switching to another isotype. This compilation is drawn from experiments on mouse B cells. There are differences in humans, but they are not yet as well worked out. For example, switching to IgA in humans involves TGF-β and IL-10, not IL-5.

almost no IgG and IgA because of the inability of their B cells to switch iso-type. They cannot make antibody responses to thymus-dependent antigens and their secondary lymphoid tissues contain no germinal centers (Figure 9.17), showing the general importance of the CD40–CD40 ligand interactions in T-cell help. Aspects of cell-mediated immunity are also impaired in these patients, who are mostly male because the gene for CD40 ligand is on the X chromosome.

9-10 Cytokines made by helper T cells determine the differentiation of antigen-activated B cells into plasma cells or memory cells

For mutated centrocytes that survive selection in the germinal center, the interaction with an antigen-specific helper T cell serves several purposes. The mutual engagement of ligands and receptors on the two cells generates an exchange of signals that induces the further proliferation of both B and T cells. This serves to expand the population of selected high-affinity, isotype-switched B cells. Individual B cells are also directed along pathways of differentiation leading to either plasma cells or memory B cells.

At the height of the adaptive immune response, when the main need is for large quantities of antibodies to fight infection, the centrocytes that win in this selection leave the germinal center and differentiate under the influence of IL-10 into antibody-producing plasma cells. These cells differ in many ways from the naive B cells that were selected by antigen (Figure 9.18). In the later stages of a successful immune response, as the infection subsides, centrocytes differentiate under the influence of IL-4 into long-lived, memory B cells, which now possess isotype-switched, high-affinity antigen receptors. Plasma cells are the effector B cells that provide antibody for dealing with today's infection, whereas the memory B cells represent an investment in preventing future infection by the same pathogen should the current infection be successfully resolved (Figure 9.19).

Summary

B cells respond to specific antigen by becoming activated, proliferating, and differentiating. They then become plasma cells that synthesize and secrete massive amounts of antibody. Activation of a mature but naive B cell requires signals delivered through its antigen receptor, and most B cells also need additional signals that are delivered only on cognate interaction with an antigen-specific, helper T cell. Activating signals are also delivered through the B-cell co-receptor when this is simultaneously ligated with the antigen

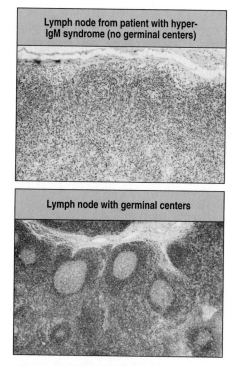

Figure 9.17 Comparison of normal and hyper-IgM syndrome lymph nodes. Bottom panel photograph courtesy of Antonio Perez-Atayde.

Property						
	Intrinsic			Inducible		
B-lineage cell	Surface Ig	Surface MHC class II	High-rate Ig secretion	Growth	Somatic hyper-mutation	Isotype switch
Resting B cell	Yes	Yes	No	Yes	Yes	Yes
Plasma cell	No	No	Yes	No	No	No

Figure 9.18 Comparison of resting B cells and plasma cells. The resting B cell expresses an antigen receptor in the form of surface immunoglobulin, and it can also take up protein antigen and present it as a peptide:MHC class II complex. Thus, it can activate helper T cells. Its immunoglobulin genes can also undergo somatic hypermutation, giving rise to progeny with altered immunoglobulin specificity. The plasma cell, in contrast, is a terminally differentiated B cell that is dedicated to the synthesis and secretion of soluble antibody. It no longer divides and its antibody specificity cannot be changed.

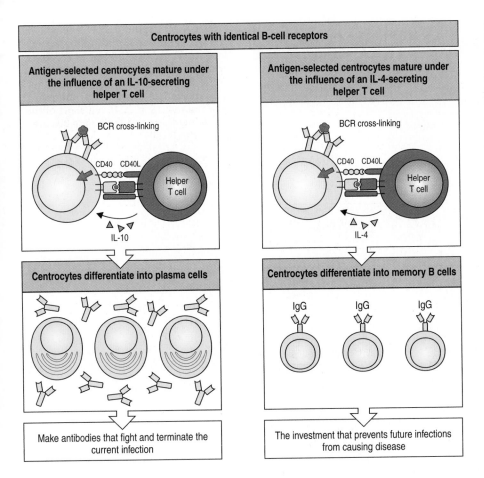

Figure 9.19 Cytokines made by helper T cells determine whether centrocytes differentiate into plasma cells or memory B cells. Centrocytes with identical high-affinity immunoglobulins can differentiate into either plasma cells (left panels) or memory B cells (right panels) depending on the cytokines secreted by their cognate helper T cells.

receptor. The signaling pathways used to activate B cells are summarized in Figure 9.20; they are similar to those used to activate T cells. The first antibodies made are always IgM; further contact with effector helper T cells is required for activated B cells to undergo isotype switching, somatic hypermutation, and affinity maturation within the germinal centers of secondary lymphoid organs. All this takes time, during which the pathogen can multiply, spread from the focus of infection, and cause disease. However, if the host survives, there will remain in the circulation expanded populations of high-affinity antibodies and memory B cells programmed to make them again should the need arise. Some antigens, notably certain components of bacterial cell walls and capsules, are capable of inducing a rapid antibody response that does not require T-cell help. These thymus-independent antigens are of two types. TI-1 antigens bind to a second receptor that contributes signals for mitosis and differentiation in addition to those generated through the B-cell antigen receptor. TI-2 antigens are generally microbial cell-surface macromolecules with repetitive epitopes that are present at high density on microbial surfaces and extensively cross-link the antigen receptors and co-receptors on the B-cell surface. Antibodies produced against TI antigens are predominantly IgM, and the cells producing them are often of the B-1 lineage. Responses to TI antigens induce neither immunological memory nor long-lasting immunity.

Antibody effector functions

As the B-cell response to infection gets under way, isotype switching diversifies antibody function by changing the heavy-chain constant region. The differences between the isotypes serve several purposes. By varying the number

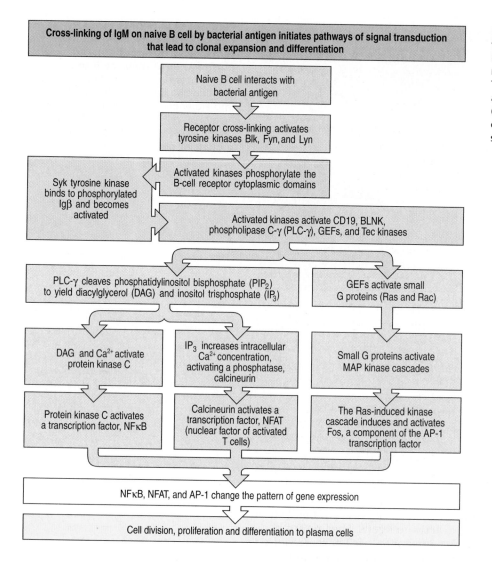

Figure 9.20 Simplified outline of the intracellular signaling pathways initiated by cross-linking of B-cell receptors by antigen in naive B cells. These pathways are similar to those that act in naive T cells exposed to antigen (see Figure 8.16, p. 224), but some of the components, such as the kinase Blk, are specific to B cells.

of antigen-binding sites and the flexibility of their movement, isotypic differences alter the strength with which antibodies bind to pathogens and antigens. They also vary the ability of antibodies to activate complement and deliver pathogens to phagocytes for destruction (see Figure 4.32, p. 118). Various cell types carry surface receptors that bind to the constant regions of antibodies of a particular class or subclass, irrespective of the antibody's antigen specificity. Some of these receptors are expressed by epithelium and deliver antibody to anatomical sites that would otherwise be inaccessible. Others are carried by effector cells, enabling the cells to capture antibody-bound pathogens and either kill them or eject them from the body. In this part of the chapter we consider how the complementary functions of antibodies of different isotypes provide effective immunity against the pathogens found in the many different environments in the human body.

9-11 IgM, IgG, and monomeric IgA protect the internal tissues of the body

In any antibody response, IgM is the first antibody to be produced. It is secreted as a pentamer by plasma cells in the bone marrow, the spleen, and the medullary cords of lymph nodes. IgM enters the blood and is carried to sites of infection, inflammation, and tissue damage throughout the body. The pentameric nature of IgM enables it to bind strongly to microorganisms and particulate antigens and quickly activate the complement cascade by the classical pathway. Complement activation coats the pathogen with C3b, which facilitates its uptake and destruction by a phagocyte. The main disadvantage of IgM is that its large size decreases the extent to which this antibody isotype can passively leave the blood and penetrate infected tissues.

As the immune response proceeds and B cells undergo somatic hypermutation and isotype switching, the need for the bulky IgM dwindles. Affinity maturation produces IgG and monomeric IgA molecules with two high-affinity binding sites for antigen that are just as effective as the 10 antigen-binding sites of IgM. These antibodies also have the advantage of being smaller and better able to get into infected tissue. IgG is the dominant blood-borne antibody, but monomeric IgA made by B cells activated in the lymph nodes or spleen also makes a contribution.

To improve the delivery of IgG to tissues, it is actively transported from the blood into the extracellular spaces within tissues. The endothelial cells of blood vessels carry out pinocytosis—the ingestion of small amounts of extracellular fluid—by which they take up plasma proteins for degradation in lysosomes. IgG is uniquely spared this fate, because in the acidic conditions of the endocytic vesicles the IgG associates with a membrane receptor that binds to the Fc portion of IgG. The receptor diverts the IgG away from the lysosomes and takes it to the basolateral surface of the cell, where the basic pH of the extracellular fluid induces the receptor to release the IgG (Figure 9.21). This transport receptor is called **FcRn**, or the Brambell receptor (FcRB) after the scientist who first described its function. FcRn is similar in structure to an MHC class I molecule, with the α_1 and α_2 domains forming a site that binds to the Fc region of the antibody. In the antibody:receptor complex, two molecules of FcRn bind to the Fc region of one IgG molecule. As well as maintaining a high level of IgG in the extracellular fluids of connective tissues, FcRn selectively protects IgG from the processes of degradation to which other plasma proteins are subject. As a consequence, IgG molecules have a longer half-life than most other plasma proteins. FcRn is just one of many different cellular receptors that bind to the Fc part of immunogloblins and are collectively called **Fc receptors**.

Besides providing a defense to all the tissues reached by the blood, an important function of circulating IgM, IgG, and IgA is to prevent blood-borne infection—septicemia—and the spread of microorganisms by neutralizing those that enter the blood. Because the blood circulation is so effective at distributing cells and molecules to all parts of the body, infections of the blood itself can have grave consequences.

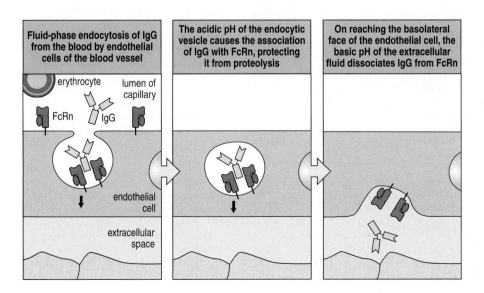

Figure 9.21 The receptor FcRn transports IgG from the bloodstream into the extracellular spaces. At the apical (luminal) side of the endothelial cell, IgG and other serum proteins are actively taken up by fluid-phase pinocytosis. In the endocytic vesicle the pH becomes acidic and each IgG molecule associates with two molecules of FcRn. The FcRn carries the IgG to the basolateral face of the cell and away from the degradative activity of the cell's lysosomes. At the basal side of the cell, the basic pH dissociates the complex of IgG and RcRn and the IgG is released into the extracellular space.

9-12   Dimeric IgA protects the mucosal surfaces of the body

Whereas IgM, IgG, and monomeric IgA provide antigen-binding functions within the fluids and tissues of the body, dimeric IgA protects the surfaces of the mucosal epithelia that communicate with the external environment and are particularly vulnerable to infection. These epithelia include the linings of the gastrointestinal tract, the eyes, nose, throat, the respiratory, urinary, and genital tracts, and the mammary glands. Dimeric IgA is made in the lamina propria at patches of mucosal-associated lymphoid tissues. The lamina propria is the connective tissue that underlies the basement membrane of the mucosal epithelium. In these tissues, antigen-specific B-cell and T-cell responses to local infections are developed. However, the IgA-secreting plasma cells are on one side of the epithelium and their target pathogens are on the other. To reach their targets, dimeric IgA molecules are transported individually across the epithelium by means of a receptor on the basolateral surface of the epithelial cells.

The dimeric form of IgA, but not the monomer, binds to a cell-surface receptor on the basolateral surface of epithelial cells, which is called the **poly-Ig receptor** because of its specificity for IgA dimers and IgM pentamers (Figure 9.22). The poly-Ig receptor itself is made up of a series of immunoglobulin-like domains and covalently binds to IgM and dimeric IgA via their J chains (see Figures 4.29 and 4.33, pp. 115 and 118), with which it makes a disulfide bond. On being bound, the IgA dimer is taken into the cell by receptor-mediated endocytosis and the antibody:receptor complex is carried across the cell to the apical surface in endocytic vesicles. Receptor-mediated transport of a macromolecule from one side of a cell to the other is known as **transcytosis**. Once receptor-bound IgA appears on the apical surface, a protease cleaves the poly-Ig receptor at sites between the membrane-anchoring region and the IgA-binding site. Dimeric IgA is released from the membrane still bound to a small fragment of the poly-Ig receptor, which is called the **secretory**

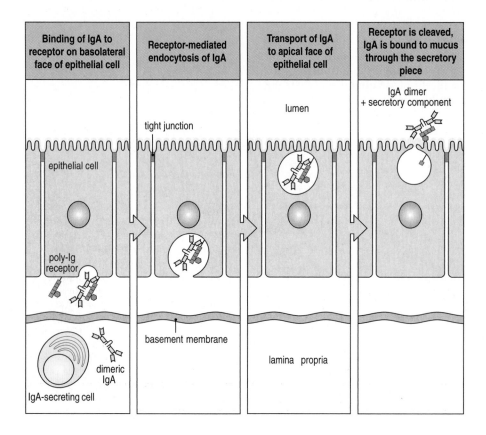

Figure 9.22 Transcytosis of dimeric IgA antibody across epithelia is mediated by the poly-Ig receptor. Dimeric IgA is made mostly by plasma cells lying just beneath the epithelial basement membranes of the gut, respiratory tract, tear glands, and salivary glands. The IgA dimer bound to the J chain diffuses across the basement membrane and is bound by the poly-Ig receptor on the basolateral surface of an epithelial cell. Binding to the receptor is via the C_H3 constant domains of the IgA heavy chains. The bound complex undergoes transcytosis across the cell in a membrane vesicle and is finally released onto the apical surface. There the poly-Ig receptor is cleaved, releasing the IgA from the epithelial cell membrane while still being bound to a fragment of the receptor called the secretory component or secretory piece. Carbohydrate (blue hexagon) of the secretory piece binds to mucus at the epithelial surface, thus preventing IgA from being washed away into the gut lumen. The residual membrane-bound fragment of the poly-Ig receptor is nonfunctional and is degraded.

Figure panels labels:
Binding of IgA to receptor on basolateral face of epithelial cell
Receptor-mediated endocytosis of IgA
Transport of IgA to apical face of epithelial cell
Receptor is cleaved, IgA is bound to mucus through the secretory piece

IgA dimer + secretory component
lumen
tight junction
epithelial cell
poly-Ig receptor
basement membrane
lamina propria
dimeric IgA
IgA-secreting cell

component, or **secretory piece**, of IgA. The IgA is then held at the mucosal surface, being bound to mucins—the glycoproteins in mucus—by the carbohydrate of the secretory piece. When IgA molecules bind to microorganisms at a mucosal surface they prevent their attachment to and colonization of the mucosal epithelium and thus facilitate the expulsion of pathogens in feces, sputum, tears, and other secretions.

9-13 IgE provides a mechanism for the rapid ejection of pathogens from the body

IgE antibodies are made by B cells that have been activated in lymph nodes and spleen; however, unlike IgM, IgG, and monomeric IgA, IgE spends very little time as soluble antibodies in the circulation. IgE is made in smaller quantities than the other isotypes and it is quickly bound by an Fc receptor, called FcεRI, that is carried by the mast cells resident in connective tissue and also by circulating basophils and the activated eosinophils present in mucosal tissues. **FcεRI** binds IgE with such strength and specificity that IgE cannot dissociate from the cell surface. Instead, IgE molecules coat mast cells, waiting to bind to pathogens and their antigens that come by. When a pathogen binds to the IgE on a mast cell and cross-links two or more FcεRI molecules in the mast-cell surface, it activates the cell to secrete active mediators that, among other effects, act on smooth muscle to cause violent reactions such as sneezing, coughing, vomiting, and diarrhea that forcibly eject pathogens from the respiratory and gastrointestinal tracts. IgE is the antibody isotype that works in the connective tissue, particularly that underlying the mucosal surfaces, and is specialized in causing the physical ejection of pathogens and toxic substances.

In many allergies and in asthma, the same violent reactions that eject dangerous pathogens are triggered by IgE responses made to innocuous substances such as grass pollens that pose no threat to the human body. These unnecessary and debilitating reactions are the side effects of having highly specialized and powerful antibodies that defend against pathogens attacking the different parts of the human body (Figure 9.23, left panel).

Figure 9.23 Immunoglobulin isotypes are selectively distributed in the body and passed to the young by their mothers. Left panel: in a healthy woman (or man), IgM, IgG, and monomeric IgA predominate in the blood (indicated schematically here by the colouring in the heart), whereas IgG and monomeric IgA are the major isotypes in the extracellular fluid. Dimeric IgA predominates in the secretions at mucosal epithelia. IgE is associated mainly with mast cells and is therefore found in the connective tissue beneath epithelial surfaces, particularly of the skin, the respiratory tract, and the gastrointestinal tract. The brain is devoid of immunoglobulin. Center panel: during its development in the mother, the fetus cannot make its own immunoglobulin. To provide the fetus with protective antibody, IgG is selectively transported from the maternal to the fetal circulation by FcRn in the placenta. Right panel: after birth, the gastrointestinal tract of the infant is supplied with protective maternal dimeric IgA, which is a major constituent of breast milk (right panel).

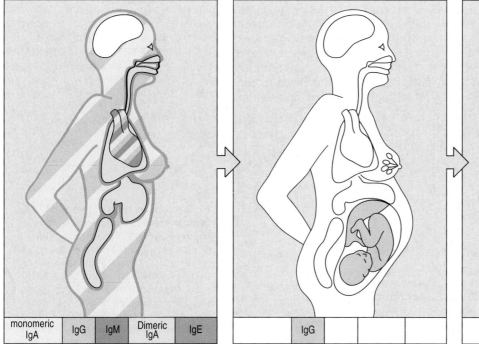
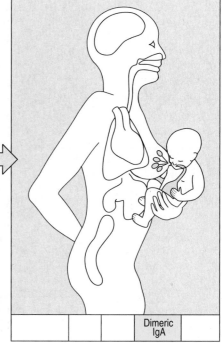

monomeric IgA	IgG	IgM	Dimeric IgA	IgE

	IgG		

		Dimeric IgA	

9-14   Mothers provide protective antibodies to their young, both before and after birth

During pregnancy, IgG from the maternal circulation is transported across the placenta and is delivered directly into the fetal bloodstream. This mechanism is so efficient that, at birth, human babies have as high a level of IgG in their plasma as their mothers, and as wide a range of antigen specificities. IgG is transported across the placenta by FcRn (see Figure 9.23, center panel).

As a result of this mechanism, at birth the baby will have protection against pathogens that provoke an IgG response but not against pathogens that infect mucosal surfaces and provoke a secreted, dimeric IgA response. To rectify this deficit, infants obtain dimeric IgA in their mother's milk, which contains antibodies against the microorganisms to which the mother has mounted an IgA response. On breast-feeding, the IgA is transferred to the baby's gut, where it binds to microorganisms, preventing their attachment to the gut epithelium and facilitating their expulsion in feces. The transfer of preformed IgA from mother to child in breast milk is an example of the **passive transfer of immunity** (Figure 9.23, right panel); another is the intravenous immunoglobulin given to patients with genetic defects in B-cell function (see Section 6-8, p. 171).

During the first year of life there is a window of time when all infants are relatively deficient in antibodies and especially vulnerable to infection. As the maternally derived IgG is catabolized and the consumption of breast milk diminishes, the antibody level gradually decreases until about 6 months of age, when the infant's own immune system starts to produce substantial antibody (Figure 9.24). Consequently, IgG levels are lowest in infants aged 3–12 months, and this is when they are most susceptible to infection. This problem is particularly acute in babies born prematurely, who begin life with lower levels of maternal IgG and take longer to attain immune competence after birth than babies born at term.

9-15   High-affinity neutralizing antibodies prevent viruses and bacteria from infecting cells

The first step in a microbial infection involves attachment of the organism to the outside surface of the human body, either some part of the skin or the mucosal surfaces. These attachments are established by specific interactions between a component on the microbial surface, which acts as a ligand, and a complementary component on the surface of human epithelial cells that the pathogen exploits as its receptor. High-affinity antibodies that bind to the microbial ligand and prevent the microbe's attachment to human epithelium

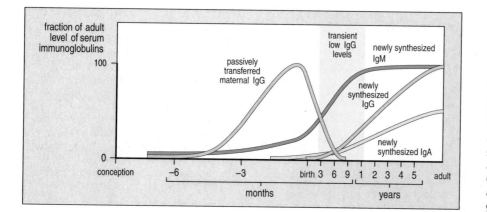

Figure 9.24 **In the first year of life, infants have a transient decrease in levels of IgG.** Before birth, high levels of IgG are provided by the mother, but after birth maternally derived IgG declines. Although infants produce IgM soon after birth, the production of IgG antibodies does not begin for about 6 months. The total level of IgG reaches a minimum within the first year and then gradually increases until adulthood.

stop the infection before it starts. For this reason, such antibodies are called neutralizing antibodies. Because many infections start at mucosal surfaces, neutralizing antibodies are often dimeric IgA.

The influenza virus, which usually enters the body by inhalation, infects epithelial cells of the respiratory tract by binding to the oligosaccharides on their surface glycoproteins. The virus binds to oligosaccharides on cell surfaces though a major protein component of its outer envelope. This is called the **influenza hemagglutinin**, because the protein can **agglutinate**, or clump together, red blood cells by binding to the oligosaccharides on the red cell surface. Neutralizing antibodies that have been developed during primary immune responses to influenza and other viruses are the most important aspect of subsequent immunity to these viruses. Such antibodies coat the virus, inhibit its attachment to human cells, and prevent infection (Figure 9.25).

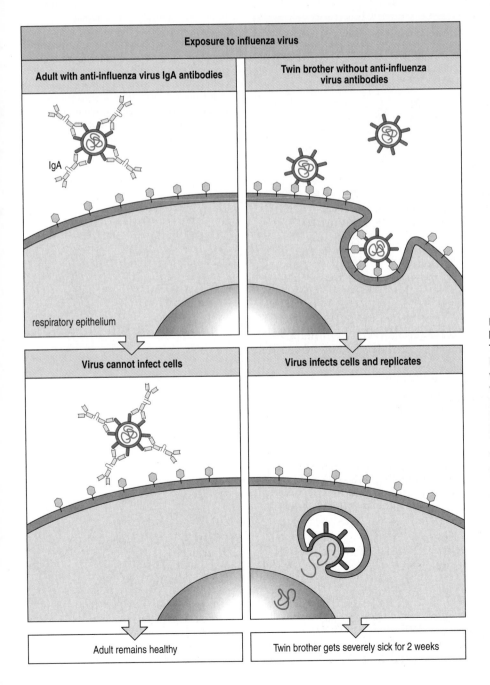

Figure 9.25 Viral infections can be blocked by neutralizing antibodies. The effect of exposure to this year's influenza virus is compared for an adult who has been vaccinated against the virus and has neutralizing anti-influenza IgA antibodies (left panels), and his identical twin brother who was too busy to get vaccinated and lacks anti-influenza antibodies (right panels). For the virus to replicate, it must enter a cell. It does this by binding to sialic acid on the cell surface via hemagglutinin on the viral surface. This is followed by internalization in an endosome and by fusion of the viral lipid envelope and the endosome membrane to release the viral RNA into the cytoplasm. Antibodies binding to the viral hemagglutinin prevent the virus from binding to the cell and halt the infection at the very first step. The antibodies shown here are IgA dimers, the form in which IgA is made and secreted at mucosal surfaces, such as the respiratory tract through which the influenza virus enters the body.

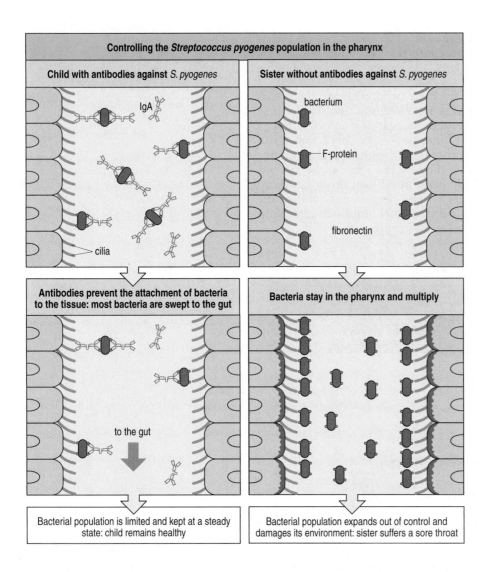

Figure 9.26 Disease-causing bacterial infections at mucosal surfaces can be prevented by neutralizing antibodies. The left panels depict the pharynx of a child who has previously suffered from a 'strep throat' and is making neutralizing IgA antibodies against the causal bacterium, *Streptococcus pyogenes*. The antibodies coat the bacteria and impair their ability to attach to the fibronectin in the extracellular matrix and stay in the pharynx. This keeps the bacterial population down to a size that does not cause disease. The right panels depict the pharynx of the child's younger sister, who has not previously experienced a strep throat and has not developed neutralizing antibodies against *S. pyogenes*. In the absence of antibodies, the size of the bacterial population is not tightly regulated and under favorable circumstances it can expand, causing damage to the mucosal surface and inducing inflammation. A sore throat ensues as well as an adaptive immune response that produces neutralizing antibodies against *S. pyogenes*.

The mucosal surfaces of the human body provide a variety of environments that are colonized by different bacteria. The bacteria attach to epithelial cells by using diverse surface components that are collectively called bacterial **adhesins**. For *Streptococcus pyogenes*, which inhabits the pharynx and is a common cause of a sore throat, the bacterial adhesin is a cell-surface protein called protein F that binds to fibronectin, a large glycoprotein component of the extracellular matrix. Secreted IgA antibodies specific for protein F limit the growth of the resident population of *S. pyogenes* and prevent them from causing disease. Only when this control is lost—by the emergence of a new bacterial strain that is not neutralized by the existing antibodies, or by the presence of an additional infection or other stress—will a sore throat develop. In general, IgA antibodies against adhesins limit bacterial populations within the gastrointestinal, respiratory, urinary, and reproductive tracts and prevent disease-causing infections in these tissues (Figure 9.26).

9-16 High-affinity IgG and IgA antibodies are used to neutralize microbial toxins and animal venoms

Many bacteria secrete protein toxins that cause disease by disrupting the normal function of human cells (Figure 9.27). To have this effect, a bacterial toxin must first bind to a specific receptor on the surface of the human cell. In some toxins, for example the diphtheria and tetanus toxins, the receptor-binding activity is carried by one polypeptide chain and the toxic function by another.

Disease	Organism	Toxin	Effects *in vivo*
Tetanus	*Clostridium tetani*	Tetanus toxin	Blocks inhibitory neuron action, leading to chronic muscle contraction
Diphtheria	*Corynebacterium diphtheriae*	Diphtheria toxin	Inhibits protein synthesis, leading to epithelial cell damage and myocarditis
Gas gangrene	*Clostridium perfringens*	Clostridial-α toxin	Phospholipase activation, leading to cell death
Cholera	*Vibrio cholerae*	Cholera toxin	Activates adenylate cyclase, elevates cAMP in cells, leading to changes in intestinal epithelial cells that cause loss of water and electrolytes
Anthrax	*Bacillus anthracis*	Anthrax toxic complex	Increases vascular permeability, leading to edema, hemorrhage, and circulatory collapse
Botulism	*Clostridium botulinum*	Botulinum toxin	Blocks release of acetylcholine, leading to paralysis
Whooping cough	*Bordetella pertussis*	Pertussis toxin	ADP-ribosylation of G proteins, leading to lymphocytosis
		Tracheal cytotoxin	Inhibits ciliar movement and causes epithelial cell loss
Scarlet fever	*Streptococcus pyogenes*	Erythrogenic toxin	Causes vasodilation, leading to scarlet fever rash
		Leukocidin Streptolysins	Kill phagocytes, enabling bacteria to survive
Food poisoning	*Staphylococcus aureus*	Staphylococcal enterotoxin	Acts on intestinal neurons to induce vomiting. Also a potent T-cell mitogen (SE superantigen)
Toxic shock syndrome	*Staphylococcus aureus*	Toxic-shock syndrome toxin	Causes hypotension and skin loss. Also a potent T-cell mitogen (TSST-1 superantigen)

Figure 9.27 Many common diseases are caused by bacterial toxins. Several examples of exotoxins, or secreted toxins, are shown here. Bacteria also make endotoxins, or nonsecreted toxins, which are usually only released when the bacterium dies. Endotoxins, such as bacterial lipopolysaccharide (LPS), are important in the pathogenesis of disease, but their interactions with the host are more complicated than those of the exotoxins and are less clearly understood.

Antibodies that bind to the receptor-binding polypeptide can be sufficient to neutralize a toxin (Figure 9.28), and the vaccines for diphtheria and tetanus work on this principle. They are modified toxin molecules, called **toxoids,** in which the toxic chain has been denatured to remove its toxicity. On immunization, protective neutralizing antibodies are made against the receptor-binding chain.

Bacterial toxins are potent at low concentrations: a single molecule of diphtheria toxin is sufficient to kill a cell. To neutralize a bacterial toxin, an antibody must be of high affinity and essentially irreversible in its binding to the toxin. It must also be able to penetrate tissues and reach the sites where toxins

Figure 9.28 Neutralization of toxins by IgG antibodies protects cells from toxin action. The protein toxins produced by many bacteria are usually of modular construction. One part of the toxin binds to a cellular receptor, which allows the toxin to be internalized, whereupon the second part poisons the cell. High-affinity neutralizing IgG antibodies bind to the receptor-binding part of the toxin and prevent its entry into the cell.

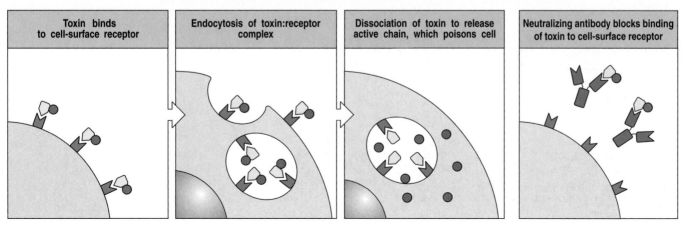

Toxin binds to cell-surface receptor	Endocytosis of toxin:receptor complex	Dissociation of toxin to release active chain, which poisons cell	Neutralizing antibody blocks binding of toxin to cell-surface receptor

are being released. High-affinity IgG is the main source of neutralizing antibodies for the tissues of the human body, whereas high-affinity IgA dimers serve a similar purpose at mucosal surfaces.

Poisonous snakes, scorpions, and other animals introduce venoms containing toxic polypeptides into humans through a bite or sting. For some venoms, a single exposure is sufficient to cause severe tissue damage or even death, and in such situations the primary response of the immune system is too slow to help. Because exposure to such venoms is rare, protective vaccines against them have not been developed. For patients who have been bitten by poisonous snakes or other venomous creatures, the preferred therapy is to infuse them with antibodies specific for the venom. These antibodies are produced by immunizing large domestic animals—such as horses—with the venom. Transfer of protective antibodies in this manner is known as **passive immunization** and is analogous to the way in which newborn babies acquire passive immunity from their mothers (see Section 9-14).

9-17   Binding of IgM to antigen on a pathogen's surface activates complement by the classical pathway

Only a fraction of antibodies have a direct inhibitory effect on a pathogen's capacity to live and replicate in the human body. The more common outcome is that antibodies bound to the pathogen recruit other molecules and cells of the immune system, which then kill the pathogen or eject it from the body. One route by which antibodies target pathogens for destruction is by activating complement through the classical pathway. In Chapter 2 we saw how the classical pathway is initiated when C-reactive protein binds to a bacterial surface. A similar activation of the classical pathway occurs when antibodies of some isotypes, but not all, bind to pathogen surfaces. The most effective antibodies at activating complement are IgM and IgG3 (Figure 9.29).

For IgM, the first antibody made in a primary immune response, activating complement is the major mechanism by which it recruits effector cells to the sites of infection. By itself, pentameric IgM does not activate complement, because it is in a planar conformation that cannot bind the C1q component of C1, the necessary first step in the classical pathway. On binding to the surface of a pathogen, the conformation of IgM changes to what is called the 'staple' form. In this form, the binding site for C1q on the Fc part of each IgM monomer becomes accessible to C1q. Multipoint attachment of C1q to IgM is necessary to obtain a stable interaction, but this is readily accomplished because IgM has five binding sites for C1q, and C1q has six binding sites for IgM (Figure 9.30). The interaction of C1q with IgM is similar to its interaction with

Antibody isotype	Relative capacity to fix complement
IgM	+++
IgD	–
IgG1	++
IgG2	+
IgG3	+++
IgG4	–
IgA1	+
IgA2	+
IgE	–

Figure 9.29 Classes and subclasses of antibodies differ in their capacity to activate and fix complement. The IgM and IgG3 isotypes are the most effective at activating the complement cascade.

Figure 9.30 Initiation of the classical pathway of complement activation by the binding of IgM to the surface of a pathogen. When soluble pentameric IgM in the 'planar' conformation establishes multipoint binding to antigens on a pathogen surface, it adopts the 'staple' conformation and exposes its binding sites for the C1q component of C1. Activated C1 then cleaves C2 and C4, and the C2a and C4b fragments form the classical C3 convertase on the pathogen surface. Conversion of C3 to C3b leads to the attachment of C3b to the pathogen surface and the recruitment of effector functions.

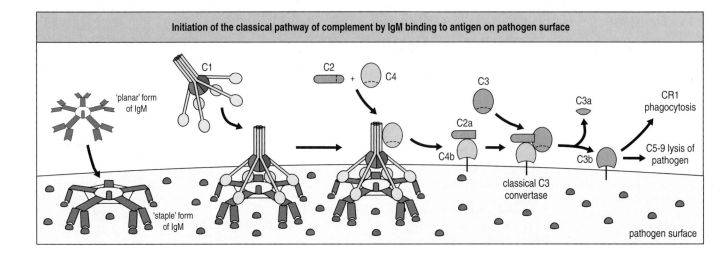

Initiation of the classical pathway of complement by IgM binding to antigen on pathogen surface

'planar' form of IgM

C1

C2 + C4

C3

C3a

CR1 phagocytosis

C2a

C4b

classical C3 convertase

C3b

C5-9 lysis of pathogen

'staple' form of IgM

pathogen surface

C-reactive protein, another pentameric protein with five binding sites for C1q (Figure 9.31).

Binding of IgM to C1q activates the C1r and C1s serine proteases that are the enzymatically active components of C1. First activated is C1r, which then cleaves and activates C1s. Activated C1s is the protease that binds, cleaves and activates both the C4 and C2 components of the classical pathway. C4b fragments that become covalently bonded to the pathogen surface bind C2a to assemble the classical C3 convertase, C4bC2a (see Figure 2.41, p. 61). The classical C3 convertase cleaves C3 to C3b and C3a, and once C3b has become covalently bonded to the pathogen it exponentially amplifies the response by assembling the C3 convertase of the alternative pathway (C3bBb). By the end of the complement-fixing reaction, most of the C3b fragments that cover the pathogen surface around the initiating antigen:antibody complex have been produced by the alternative convertase (Figure 9.32). This is another example of a situation in which adaptive immunity provides specificity to the reaction, and innate immunity provides the strength. Once the pathogen has been coated with C3b it can be efficiently phagocytosed by a neutrophil or macrophage using its CR1 complement receptors. Alternatively, for some pathogens, such as the bacterium *Neisseria*, the classical pathway of complement activation continues, leading to assembly of the membrane-attack complex and the death of the pathogen through perforation of its outer membrane (see Section 2-6, p. 39).

IgM fixes complement efficiently because its five binding sites for C1q allow each IgM molecule bound to a pathogen to fix complement independently. The drawbacks of IgM are its size, which restricts the extent to which it can penetrate infected tissues, and its limited capacity to recruit effector functions that are not dependent on complement. Once a primary immune response has progressed to the point at which B cells have undergone isotype switching and affinity maturation and have become plasma cells, the smaller and more versatile IgG molecules make up for the deficiencies of IgM.

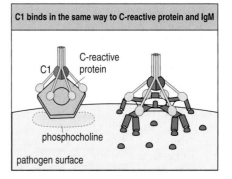

Figure 9.31 IgM and C-reactive proteins are both pentamers that bind to C1. In the innate immune response, the classical pathway of complement activation is initiated by the binding of the C1q component of C1 to the acute-phase protein C-reactive protein, which is a pentamer. In the adaptive immune response, the classical pathway is initiated by pentameric IgM. This is not a coincidence. In all likelihood the preexisting predilection of C1q for binding to pentameric C-reactive protein selected for the evolution of pentameric antibody molecules.

9-18   Two forms of C4 tend to be fixed at different sites on pathogen surfaces

The complement components that are uniquely used by the classical pathway are the binding protein C1, the protease C2, and the thioester-containing protein C4. Of these, C4 is usually present in two forms that are encoded by

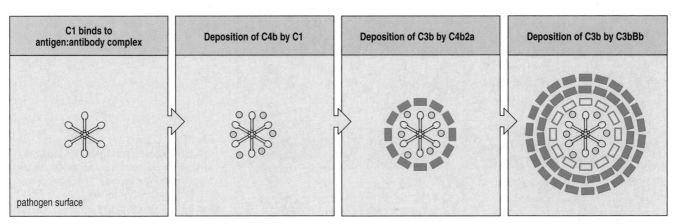

Figure 9.32 A bird's-eye view of the fixation of C4b and C3b fragments on a pathogen surface around an antigen:antibody complex. Antibody bound to an antigen on a microbial surface binds C1 (first panel), which leads to the deposition of C4b (pink circles) around the antigen:antibody complex (second panel). When C4b binds C2a to form the classical C3 convertase, a limited number of C3b molecules (green rectangles) are produced (third panel). These can bind Bb to form the alternative convertase, C3bBb (yellow rectangles), which leads to the deposition of many more C3b fragments on the microbial surface (green rectangles, fourth panel).

separate genes in the central region of the MHC. The two types of C4—C4A and C4B—have different properties. The thioester bond of C4A is preferentially attacked by the amino groups of macromolecules, whereas that of C4B is preferentially attacked by the hydroxyl groups. This complementarity increases the efficiency of C4 deposition and coverage of the whole pathogen surface. The two genes encoding C4A and C4B are closely linked and situated in the class III region of the MHC, where, through gene duplication and deletion, further diversification of the C4 genes has evolved (Figure 9.33). In humans, 13% of chromosomes lack a functional C4A gene and 18% of chromosomes lack a functional C4B gene; thus, more than 30% of the human population is deficient for one or other form of C4, and a partial lack of C4 is the most common human immunodeficiency. Reflecting the complementary functions of the two forms of C4, deficiency in C4A is associated with susceptibility to the autoimmune disease **systemic lupus erythematosus** (**SLE**), whereas deficiency in C4B is associated with lowered resistance to infection. As well as the simple presence or absence of the C4A and C4B genes, there are more than 40 different alleles of the C4 genes, which could be associated with further differences in C4 function.

9-19   Complement activation by IgG requires the participation of two or more IgG molecules

The binding of IgG to antigen on a pathogen's surface can also trigger the classical pathway of complement activation. Each IgG molecule has a single binding site for C1q in its Fc region. Unlike IgM, IgG does not need to sequester its C1q-binding site in the absence of antigen. The interaction of one IgG molecule with C1q is insufficient to activate C1 and it is therefore necessary for C1q to cross-link two or more IgG molecules bound to antigen on the surface of a pathogen, and the IgG molecules must be close enough together for C1q to span them (Figure 9.34, left panels). As a consequence, the activation of complement by IgG depends more on the amount and density of antibodies bound at a pathogen surface than does complement activation by IgM. After the complexes of antigen and IgG have activated C1, the classical pathway proceeds in exactly the same way as after activation by IgM.

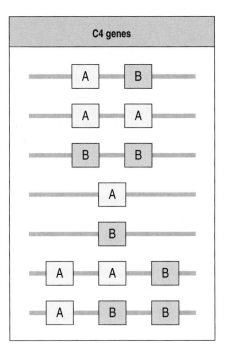

Figure 9.33 Humans differ in the number and type of genes for complement component C4. Complement proteins C4A and C4B have differences in the way they bond to pathogen surfaces. The genes for C4A and C4B are located in the central part of the MHC, between the class I region and the class II region. Although a majority of MHC haplotypes have one gene for C4A and one for C4B, a considerable minority have other arrangements involving the loss or duplication of one of the genes. These differences lead to variation in C4 function within the population and to immunodeficiency in some individuals.

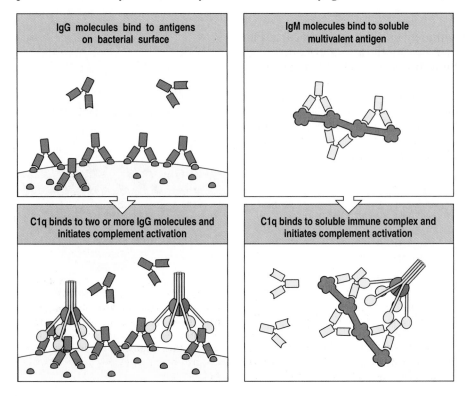

Figure 9.34 At least two molecules of IgG bound to pathogens or soluble antigens are required to activate the complement cascade. The left panels show the activation of complement by IgG bound to antigens on a pathogen surface. The C1q molecule needs to find pathogen-bound IgG molecules that are close enough to each other for the C1q molecule to span them. The right panels show the activation of complement by C1q binding to two IgG molecules in a soluble immune complex.

Because the antigen-binding sites on IgG are of higher affinity than those on IgM, they can form stable immune complexes with soluble multivalent antigens—for example, toxins secreted by pathogens or breakdown products from their death and degradation. These soluble immune complexes can also activate the classical pathway (Figure 9.34, right panels), leading to the deposition of C3b on the antigen and antibody molecules in the complex. Using their Fc receptors and complement receptors, phagocytic cells can readily take up these complexes of antigen, antibody, and complement from blood, lymph, and tissue fluids and clear them from the system.

9-20 Erythrocytes facilitate the removal of immune complexes from the circulation

Immune complexes that have become covered with fragments of C3b can now be bound by circulating cells that express the complement receptor CR1. Of these, the most numerous is the erythrocyte and the vast majority of immune complexes become bound to the surface of red blood cells. During their circulation in the blood, erythrocytes pass through areas of the liver and the spleen where tissue macrophages remove and degrade the complexes of complement, antibody, and antigen from the erythrocyte surface while leaving the erythrocyte unscathed (Figure 9.35). Although the vital function of erythrocytes is the transport of gases between the lungs and other tissues, they have also acquired functions in the defense and protection of tissues, of which the disposal of immune complexes is a crucial one.

If immune complexes are not removed, they have a tendency to enlarge by aggregation and to precipitate at the basement membrane of small blood vessels, most notably those of the kidney glomeruli, where blood is filtered to form urine and is under particularly high pressure. Immune complexes that pass through the basement membrane bind to CR1 receptors expressed by podocytes, specialized epithelial cells that cover the capillaries. Deposition of immune complexes within the kidney probably occurs at some level all the time, and mesangial cells within the glomerulus are specialized in the elimination of immune complexes and in stimulating the repair of the tissue damage they cause.

A feature of the autoimmune disease SLE is a level of immune complexes in the blood sufficient to cause massive deposition of antigen, antibody, and complement on the renal podocytes. These deposits damage the glomeruli, and kidney failure is the principal danger for patients with this disease. A similar deposition of immune complexes can also be a major problem for patients who have inherited deficiencies in the early components of the complement pathway and cannot tag their immune complexes with C4b or C3b. Such patients cannot clear immune complexes; these accumulate with successive antibody responses to infection and inflict increasing damage on the kidneys.

9-21 The four subclasses of IgG have different and complementary functions

A key feature of the IgG molecule that contributes to its potency and versatility is conformational flexibility. This allows the antigen-binding sites in the two Fabs and the effector-binding sites in the Fc to move in a partly independent manner and assume a wide range of different positions with respect to each other (Figure 9.36). This flexibility, due mainly to the hinge region, greatly improves the likelihood that an IgG molecule can simultaneously bind to two antigens on the surface of a pathogen and to effector molecules such as C1. The advantages of a flexible hinge are offset by the susceptibility of this loosely folded part of the heavy chain to proteolytic cleavage, which compromises

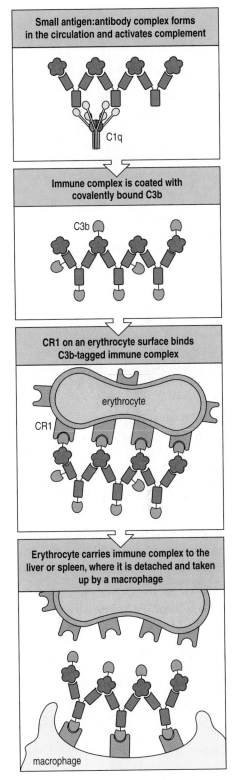

Figure 9.35 Erythrocyte CR1 helps to clear immune complexes from the circulation. Small soluble immune complexes bind to CR1 on erythrocytes, which transport them to the liver and spleen. Here they are transferred to the CR1 of macrophages and taken up for degradation.

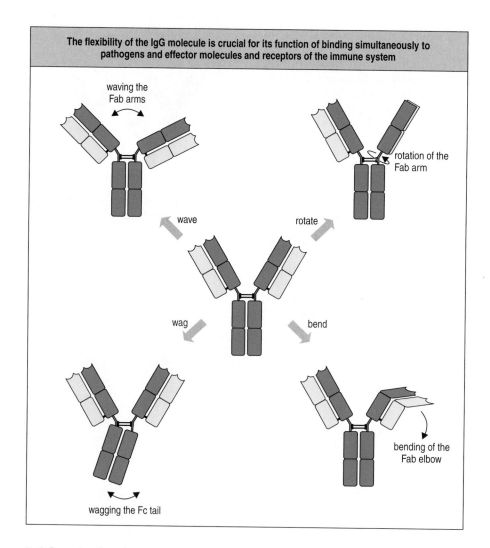

The flexibility of the IgG molecule is crucial for its function of binding simultaneously to pathogens and effector molecules and receptors of the immune system

Figure 9.36 IgG is a highly flexible molecule. The most flexible part of the IgG molecule is the hinge, which allows the Fab arms to wave and rotate and thus accommodate the antibody to the orientation of epitopes on pathogen surfaces. Adding to further flexibility in binding antigen is the 'elbow' within the Fab that allows the variable domains to bend with respect to the constant domains. Similarly, the wagging of the Fc tail allows IgG molecules that have bound to antigen to accommodate to the binding of C1q and other effector molecules.

IgG function by separating the Fc from the Fabs. In response to these conflicting pressures, four different IgG isotypes or subclasses have evolved. Called IgG1, IgG2, IgG3, and IgG4, they differ only in the constant region of the heavy chain and many of the differences are within the hinge (Figure 9.37).

IgG1 is the most abundant and versatile of the four subclasses and is intermediate in its flexibility, susceptibility to proteolysis, and capacity to activate complement (see Figure 4.32, p. 118). It is a good all-rounder that constitutes most of the antibody made against T-dependent protein antigens. In IgG2, the second most abundant subclass, the hinge is of similar length to that of IgG1 but contains additional disulfide bonds that reduce the flexibility, the

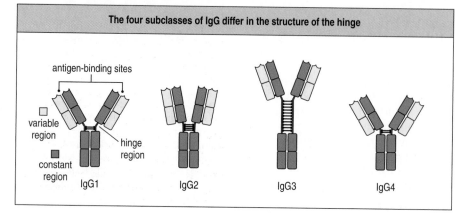

The four subclasses of IgG differ in the structure of the hinge

Figure 9.37 Different hinge structures distinguish the four subclasses of IgG. The relative lengths of the hinge and the number of disulfide bonds in the hinge that cross-link the two heavy chains are shown. Not shown are other differences in the amino acid sequence, particularly glycine and proline residues, that influence hinge flexibility.

	IgG subclass			
	IgG1	IgG2	IgG3	IgG4
Proportion of total IgG (%)	45–75	16–48	2–8	1–12
Length of heavy-chain hinge (amino acids)	15	12	62	12
Number of disulfide bonds in the hinge	2	4	11	2
Susceptibility of hinge to proteolytic cleavage	++	+	+++	+
Half-life in serum (days)	21	21	7	21
Capacity to bind C1q and activate complement	++	+	+++	+
Response to protein antigens	++	+	++	+
Response to carbohydrate antigens	+	++	–	–
Response to allergens	+	–	–	++
Number of allotypes	3	5	19	1

Figure 9.38 The four subclasses of IgG have different and complementary functions.

susceptibility to proteolysis, and the capacity to activate complement. IgG2 is preferentially made against the highly repetitive carbohydrate antigens of microbial surfaces, which put fewer demands on an antibody's flexibility and are often TI antigens (see Section 9-3). Consistent with this function for IgG2, infections with encapsulated bacteria are poorly controlled in individuals who are deficient in it (Figure 9.38).

Of the four IgG subclasses, IgG3 is the best at activating complement. It differs from the other subclasses by having a much longer hinge region: four times the length of the IgG1 hinge. The long hinge gives IgG3 greater flexibility in binding to antigens and also makes the Fc more accessible for binding to C1. Both these qualities contribute to its superior activation of complement. The disadvantage of the long hinge is that IgG3 is particularly susceptible to cleavage by proteases, as is reflected by its half-life in the circulation, which is one-third that of the other three subclasses (see Figure 4.31, p. 117). IgG3 deficiency is associated with recurrent infection leading to chronic lung disease.

By binding to pathogens and activating complement the IgG1, IgG2, and IgG3 subclasses of antibodies facilitate the activity of phagocytic cells and increase inflammation at a site of infection. In contrast, IgG4, the least abundant subclass of IgG, has an Fc region that does not activate complement. Further distinguishing the IgG4 molecule is a unique capacity to exchange a module composed of one heavy chain and one light chain with that of another IgG4 molecule. Because of the frequency with which this occurs, most IgG molecules have two different heavy chains, two different light chains and two antigen-binding sites of different specificity (Figure 9.39). This means that IgG4 is

Figure 9.39 IgG4 is present in the circulation in a functionally monovalent form. Like other IgG subclasses, IgG4 is synthesized in a form that has two heavy chains, two light chains, and two identical antigen-binding sites. Unlike other IgGs, however, molecules of IgG4 can interact in the circulation and exchange one heavy chain and its associated light chain. Because of this property, most IgG4 molecules in the circulation have two different binding sites for antigen. Thus they only interact with a pathogen or a protein antigen through one binding site.

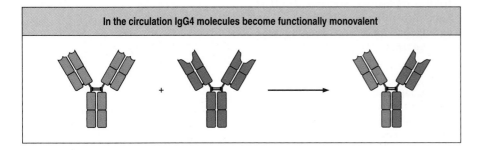

In the circulation IgG4 molecules become functionally monovalent

functionally monovalent and can only impede pathogens by the mechanism of neutralization. All these properties of IgG4 contribute to its being anti-in-flammatory in its effects. In allergic individuals, the levels of IgG4 are frequently elevated (see Figure 9.38), as well as those of IgE. When IgG4 binds to an allergen it can block the binding of IgE and thus reduce the severity of the allergic reaction. Thus, the function of IgG4 seems to be to damp down the immune response and prevent such over-reactions.

Before the unique exchange reaction of IgG4 was fully understood, IgG4 was considered the best isotype to use for a therapeutic monoclonal antibody because it would minimize the damaging effects of inflammation. The mono-valency of IgG4 is, however, a disadvantage in a therapeutic antibody, because its effective strength of binding is less than that of a bivalent IgG. The way round this problem is to mutate the C_H3 region of IgG4 so that it can no longer exchange binding sites, while preserving its poor activation of complement.

Allelic variation in the IgG constant regions gives rise to what are called the **Gm allotypes**. IgG3 is the most variable subclass, with 19 Gm3 allotypes, and IgG4 is the most conserved, with only a single Gm4 allotype. IgG1 and IgG2 have modest polymorphism, with three Gm1 and five Gm2 allotypes, respectively (see Figure 9.38).

9-22   Fc receptors enable hematopoietic cells to bind and be activated by IgG bound to pathogens

The power of the complement system is that it coats pathogens permanently with C3b fragments, delivering them to the complement receptors of phago-cytes for uptake and elimination. High-affinity antibodies hold onto patho-gens almost as tenaciously as C3b, and they too are ligands for receptors that are present on a range of cells of the immune system. These cells include neu-trophils, eosinophils, basophils, mast cells, macrophages, FDCs, and NK cells. The receptors are diverse and have specificity for different antibody isotypes, but they all bind to the Fc region and are known generally as Fc receptors. These Fc receptors of hematopoietic cells are functionally and structurally distinct from the FcRn of endothelial cells. Whereas FcRn has an MHC class I-like structure and transports antibodies across epithelium, the hematopoi-etic Fc receptors are made up of two or three immunoglobulin-like domains, and are signaling receptors that induce a response in the cells that express them.

A typical Fc receptor is **FcγRI**, which is specific for IgG and is constitutively expressed on monocytes, macrophages, and dendritic cells. At sites of infec-tion and inflammation, FcγRI expression is also induced in neutrophils and eosinophils. Expression of FcγRI is thus specific to myeloid cells. The α chain is anchored to the membrane and has three extracellular immunoglobulin-like domains that are all necessary for binding to the C_H2 domain and the lower part of the hinge of IgG (Figure 9.40). Associated with the α chain is a dimer composed of a second type of polypeptide, the γ chain, which trans-duces activating signals and is closely related to the ζ chain of the T-cell recep-tor complex, and like that chain it contains ITAM motifs. It is quite different from the γ chain associated with some cytokine receptors, such as the IL-2 receptor (see Section 8-8, p. 225). FcγRI binds with different affinities to the four IgG subclasses. The hierarchy of binding—IgG3 > IgG1 > IgG4 >>> IgG2—reflects the structural differences in the hinge and the C_H2 domain of the vari-ous subclasses.

The major function of FcγRI is to facilitate the uptake and degradation of pathogens by phagocytes and professional antigen-presenting cells. In Chapter 2 we saw how a variety of phagocytic and signaling receptors, includ-ing complement receptors, can enhance these processes during the innate

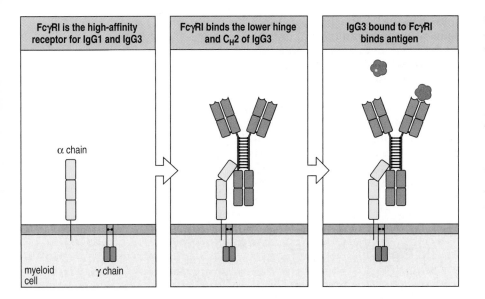

FcγRI is the high-affinity receptor for IgG1 and IgG3	FcγRI binds the lower hinge and C_H2 of IgG3	IgG3 bound to FcγRI binds antigen

α chain

myeloid cell   γ chain

Figure 9.40 FcγRI on myeloid cells is an Fc receptor that binds with high affinity to IgG1 and IgG3. The IgG-binding function of the receptor FcγRI is a property of the three extracellular, immunoglobulin-like domains of the receptor α chain. The signaling function of the receptor is the property of the γ chain, which forms a homodimer (left panel). Of the Fc receptors for IgG, only FcγRI can bind to IgG in the absence of antigen (shown here for IgG3, center panel). This enables IgG3 to trap pathogens at the surfaces of macrophages, dendritic cells, and neutrophils and target them for uptake and killing (right panel).

immune response. In the adaptive immune response, antibodies made against surface antigens will coat the pathogen with their Fc regions pointing outwards and free to bind to the FcγRI expressed on myeloid cell surfaces. On contact with a phagocyte, multiple ligand–receptor interactions are made, which produce a stable interaction and the clustering of receptors that is required to initiate intracellular signaling and phagocytosis (Figure 9.41). The necessity for antigen cross-linking of the bound IgG before a signal can be produced is illustrated by the behavior of IgG3 and IgG1. These subclasses have such a high affinity for FcγRI that individual molecules can bind to the receptor in the absence of antigen and are held transiently at the cell surface. Such interactions do not send a signal to the cell, however, because that requires the cross-linking of complexes of FcγRI and IgG by antigen.

After the pathogen has been bound to the phagocyte, interactions between antibody Fc regions and the FcγRI send signals to the phagocyte that facilitate the engulfment of the antibody-coated pathogen. The surface of the phagocyte gradually extends around the surface of the opsonized pathogen through cycles of binding and release between the Fc receptors of the phagocyte and the Fc regions projecting from the pathogen surface. Engulfment is an active process triggered by signals from the Fc receptors.

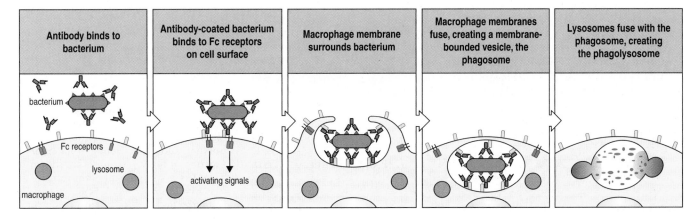

Antibody binds to bacterium	Antibody-coated bacterium binds to Fc receptors on cell surface	Macrophage membrane surrounds bacterium	Macrophage membranes fuse, creating a membrane-bounded vesicle, the phagosome	Lysosomes fuse with the phagosome, creating the phagolysosome

bacterium

Fc receptors

lysosome

macrophage

activating signals

Figure 9.41 Fc receptors on phagocytes trigger the uptake and breakdown of antibody-coated pathogens. Specific IgG molecules coat the pathogen surface, here a bacterium, and tether the bacterium to the surface of the phagocyte by binding to the Fc receptors. Signals from the Fc receptors enhance phagocytosis of the bacterium and the fusion of lysosomes containing degradative enzymes with the phagosome. The Fc receptors provide myeloid cells, such as phagocytes, with the means of efficiently capturing any infecting pathogen against which antibodies are made.

As an effector mechanism of adaptive immunity, antibody-mediated opsonization enhances the phagocytic mechanisms of innate immunity to greatly increase the speed at which pathogens are detected and devoured. In addition to reducing the pathogen load, antibody-mediated opsonization also increases the efficiency with which antigens are processed and presented to pathogen-specific T cells.

A coating of IgG antibodies makes different sorts of microorganisms appear similar to Fc-receptor-bearing phagocytes, and it enables these cells to deal with different pathogens by using just one ligand—the Fc region of IgG—and the same FcγRI receptor. Encapsulated bacteria such as *Streptococcus pneumoniae* have evolved cell-surface structures that are resistant to direct phagocytosis; for these species a coating with antibody that masks their surface is essential for their efficient phagocytosis.

9-23 A variety of low-affinity Fc receptors are specific for IgG

In addition to FcγRI there are two other types of Fc receptor that bind IgG, namely FcγRII and FcγRIII (Figure 9.42). FcγRII and FcγRIII bind IgG with much lower affinity than FcγRI, which is attributed to their having only two immunoglobulin-like domains (compared with the three of FcγRI). In the absence of cross-linking by antigen, IgG cannot bind to FcγRII or FcγRIII. This means that high concentrations of IgG of diverse antigenic specificities can circulate in the body's fluids without clogging up the Fc receptors of phagocytes in the absence of antigen.

FcγRII is a group of three receptors, each of which is encoded by a different gene. FcγRIIA is an activating receptor that promotes the uptake and destruction of pathogens by a wide range of myeloid cells. It does not require an associated signaling polypeptide because its cytoplasmic tail contains an ITAM signaling module like that of the γ chain. FcγRIIB2 is an inhibitory receptor expressed on macrophages, neutrophils, and eosinophils. It can antagonize the actions of activating Fc receptors in these cells and in this way it helps to control their inflammatory responses. FcγRIIB1 is an inhibitory receptor on mast cells and B cells, and similarly it behaves as a negative regulator of their responses. These inhibitory receptors bear **immunoreceptor tyrosine-based inhibitory motifs** (**ITIMs**) in their cytoplasmic tails, which associate with intracellular proteins that develop the inhibitory signals.

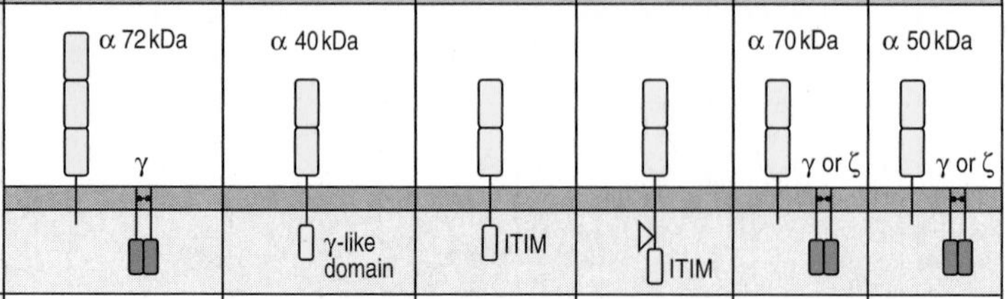

Receptor	FcγRI (CD64)	FcγRIIA (CD32)	FcγRIIB2 (CD32)	FcγRIIB1 (CD32)	FcγRIIIa (CD16)	FcγRIIIb (CD16)
Structure	α 72 kDa, γ	α 40 kDa, γ-like domain	ITIM	ITIM	α 70 kDa, γ or ζ	α 50 kDa, γ or ζ
IgG subclass specificity	3>1>4>>>2	R131: 3>1>>>2,4 H131: 3>1,2>>4	3>1>4>>2	3>1>4>>2	1,3>>>2,4	1,3>>>2,4
Relative binding strength to IgG1	200	4	4	4	1	1
Effect of ligation	Activation	Activation	Inhibition	Inhibition	Activation	Activation

Figure 9.42 **There are various receptors for the Fc regions of IgG.** The subunit structure, the relative binding strengths for different IgG isotypes, and the activating or inhibitory function of the Fcγ receptors are shown here. There are three different forms of FcγRII (CD32)—FcγRIIA, FcγRIIB1, and FcγRIIB2—and their different functions are described in detail in the text. R131 and H131 are two allotypes of FcγRIIA that differ in their specificity for IgG2. There are two different forms of FcγRIII—FcγRIIIa and FcγRIIIb—whose properties are described in detail in the text.

FcγRIIA is present in two main allotypes in the population. One of these allotypes is the only activating Fc receptor that effectively binds pathogens coated with IgG2, the second most abundant IgG isotype and the one that is enriched in response to bacterial polysaccharides. Because IgG2 cannot activate complement, it is entirely dependent on FcγRIIA for its effector function. Allotype H131 (with histidine at position 131) binds IgG2 as well as it binds IgG1, but allotype R131 (with arginine at position 131) is ineffective at binding IgG2 (see Figure 9.42). Neutrophils from individuals homozygous for R131, who comprise around 25% of African and European populations, are less effective at phagocytosing and killing IgG2-coated bacteria than are neutrophils from individuals heterozygous or homozygous for H131. Homozygosity for R131 is also associated with an increased risk of fulminant meningococcal disease or septic shock on infection with *Neisseria meningitidis*, indicating a role for IgG2 in protection against this bacterium. In Japan and other parts of East Asia, less than 4% of the population is homozygous for the R131 allotype.

FcγRIII has the lowest affinity for IgG and is made in two forms that are products of different genes: one form, FcγRIIIa, spans the membrane and is associated with a γ chain; the other form, FcγRIIIb, is attached to the outer face of the plasma membrane by a glycosylphosphatidylinositol (GPI) anchor and is not associated with a γ chain (see Figure 9.42). FcγRIII is the only Fc receptor expressed on NK cells, and its functions have mainly been studied in the context of killing by NK cells. NK cells can recognize and kill human cells coated with IgG1 or IgG3 antibodies specific for cell-surface components (Figure 9.43). This **antibody-dependent cell-mediated cytotoxicity (ADCC)** is the mechanism by which the therapeutic monoclonal antibody rituximab eliminates B cells and B-cell tumors (see Section 4-6, p. 104). Because ADCC requires the presence of isotype-switched IgG antibodies, this mechanism of NK-cell cytotoxicity can only occur at a late stage in the primary immune response, but in the secondary response—in which such antibodies are already present—it has the potential to contribute from the beginning of infections. Influenza-infected cells, for example, express viral glycoproteins at the surface that can bind influenza-specific IgG, which makes them targets for ADCC. Another situation in which preformed isotype-switched antibodies are present is in newborn infants, who have passively acquired IgG from their mother against many pathogens to which they have yet to be exposed.

Like FcγRIIA, each of the two FcγRIII genes is represented in human populations by two high-frequency alleles that encode proteins with different

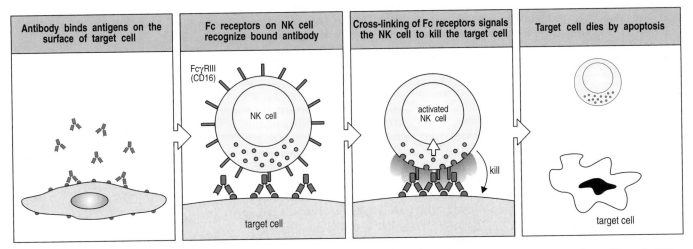

| Antibody binds antigens on the surface of target cell | Fc receptors on NK cell recognize bound antibody | Cross-linking of Fc receptors signals the NK cell to kill the target cell | Target cell dies by apoptosis |

Figure 9.43 Antibody-coated target cells can be killed by natural killer cells (NK cells) in antibody-dependent cell-mediated cytotoxicity (ADCC). NK cells are large granular lymphocytes that are distinct from B and T cells and have FcγRIII receptors (CD16) on their surface. When these cells encounter cells coated with IgG antibody, they rapidly kill the target cell.

functional properties. The two allotypes of FcγRIIIa differ in their affinity for IgG, with the higher-affinity allotype giving better antibody-dependent NK-cell cytotoxicity. The two allotypes of FcγRIIIb differ in their capacity to induce the phagocytosis of particles coated with IgG1 or IgG3. The occurrence of these dimorphisms suggests that there are benefits to having low-affinity Fc receptor allotypes with different affinities for IgG. In contrast, the high-affinity FcγRI exhibits no comparable dimorphism and is highly conserved.

9-24   IgE binds to high-affinity Fc receptors on mast cells, basophils, and activated eosinophils

IgE antibodies against a wide variety of different antigens are normally present in small amounts in all humans. They are produced in responses dominated by CD4 T_H2 cells, in which the cytokines produced favor switching to the IgE isotype. The FcεRI receptor for IgE on **mast cells**, **basophils**, and activated **eosinophils** (see Section 9-13) has an affinity (about 10^{10} M^{-1}) for the Fc region of IgE that is 100-fold higher than that of the high-affinity IgG receptor. As a consequence of this high affinity IgE molecules are tightly bound in the absence of antigen and the cells bearing FcεRI are almost always coated with antibody. In the absence of allergy or parasitic infection, a single mast cell carries IgE molecules specific for many different antigens.

Mast cells are sentinels posted throughout the body's tissues, particularly in the connective tissues underlying the mucosa of the gastrointestinal and respiratory tracts and in the connective tissues along blood vessels—especially those in the dermis of the skin. The cytoplasm of the resting mast cell is filled with large granules containing **histamine** and other molecules that contribute to inflammation, which are known generally as **inflammatory mediators**. Mast cells become activated to release their granules when antigen binds to the IgE molecules bound to FcεRI on the mast-cell surface (see Figure 9.44). To activate the cell, the antigen must cross-link at least two IgE molecules and their associated receptors, which means that the antigen must have at least two topographically separate epitopes recognized by the cell-bound IgE. Cross-linking of FcεRI generates the signal that initiates the release of the mast-cell granules. After degranulation, the mast cell synthesizes and packages a new set of granules.

Inflammatory mediators secreted into the tissues by activated mast cells, basophils, and eosinophils increase the permeability of the local blood vessels, enabling other cells and molecules of the immune system to move out of the bloodstream and into tissues. This causes a local accumulation of fluid, and the swelling, reddening, and pain that characterize inflammation. Inflammation in response to an infection is beneficial because it recruits cells and proteins required for host defense into the sites of infection.

The prepackaged granules and the high-affinity FcεRI receptor already armed with IgE make the mast cell's response to antigen impressively fast. The infections that are the 'natural' targets of IgE-activated mast cells and eosinophils are those caused by parasites. Parasites are a heterogeneous set of organisms that include the unicellular protozoa and multicellular invertebrates, notably the helminths—intestinal worms and the blood, liver, and lung flukes—and ectoparasitic arthropods such as ticks and mites. As a group, parasites establish long-lasting, persistent infections in human hosts and are well practised in the avoidance and subversion of the human immune system. Most parasites are much larger than any microbial pathogen. The largest human parasite is the tapeworm *Diphyllobothrium latum*, which can reach 9 meters in length and lives in the small intestine, causing vitamin B_{12} deficiency and, in some patients, megaloblastic anemia. Multicellular parasites cannot be controlled by the cellular and molecular mechanisms of destruction that work for microorganisms, so a different strategy based on IgE has evolved.

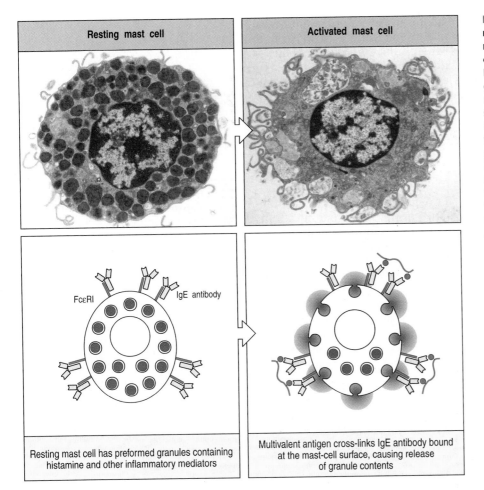

Figure 9.44 **IgE cross-linking on mast-cell surfaces leads to the rapid release of mast-cell granules containing inflammatory mediators.** Resting mast cells contain numerous granules containing inflammatory mediators such as histamine and serotonin. The cells have high-affinity Fc receptors (FcεRI) on their surface that are occupied by IgE molecules (left panels). Antigen cross-linking of bound IgE cross-links the FcεRI, triggering the degranulation of the mast cell and the release of inflammatory mediators into the surrounding tissue, as shown in the right panels. Photographs courtesy of A.M. Dvorak.

Inflammatory mediators released by mast cells, basophils, and eosinophils cause the contraction of smooth muscle surrounding the airways and the gut. In addition to violent muscular contractions that can expel parasites from the airways or gut, the increased permeability of local blood vessels supplies an outflow of fluid across the epithelium, which can help to flush out parasites. In summary, the combined actions of IgE, mast cells, basophils, and eosinophils serve to physically remove parasite pathogens and other material from the body.

Eosinophils can also use their Fcε receptors to act directly against multicellular parasites. Such organisms, even small ones such as the blood fluke *Schistosoma mansoni*, which causes schistosomiasis, cannot be ingested by phagocytes. However, if the parasite induces an antibody response and becomes coated with IgE, activated eosinophils will bind to it through FcεRI and then pour the toxic contents of their granules directly onto its surface (Figure 9.45).

For human populations in developed countries where parasite infections are rare, the mast cell's response is most frequently seen as a detriment, because its actions are the cause of allergy and asthma. People with these conditions make IgE in response to relatively innocuous substances, for example grass pollens or shellfish, that are often either airborne or eaten. Such substances are known as allergens. Having made specific IgE, any subsequent encounter with the allergen leads to massive mast-cell degranulation and a damaging response that is quite inappropriate to the threat posed by the antigen or its source. In extreme cases, the ingestion of an allergen can lead to a systemic life-threatening inflammatory response called anaphylaxis.

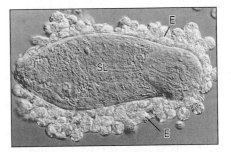

Figure 9.45 **Eosinophils attacking a schistosome larva in the presence of serum from an infected patient.** Large parasites, such worms or the schistosome larva (SL) shown here, cannot be ingested by phagocytes. However, when the worm is coated with antibody, especially IgE, eosinophils (E) can attack it by using their high-affinity Fc receptors (FcεRI). Similar attacks can be mounted by other Fc-receptor-bearing cells on large targets. Photograph courtesy of A. Butterworth.

Parasite infections that invoke a protective IgE response are not major health problems in the developed world, whereas allergy and asthma do not seem to be prevalent in the developing countries where infection with parasites is endemic. This is one of various pieces of circumstantial evidence suggesting that if elements of the immune system are left unstimulated by infection, they can respond in ways that are frankly unhelpful.

9-25 The Fc receptor for monomeric IgA belongs to a different family from the Fc receptors for IgG and IgE

Cells of the myeloid lineage also express an Fc receptor that binds to the monomeric form of IgA that is present in blood and lymph. This medium-affinity receptor, called FcαRI, has an α chain containing two immunoglobulin-like domains and it relies on the common γ chain for signal transduction. The main function of FcαRI is the same as that of FcγRI, with the difference that it facilitates the phagocytosis of pathogens coated with IgA, not IgG. Despite the functional similarities of these two receptors, they have different evolutionary histories. The Fc receptors for IgG and IgE are encoded by a family of genes on human chromosome 1, in which all the members derive from a common ancestral gene (Figure 9.46). In contrast, the gene for FcαRI is on chromosome 19, where it is part of a large, densely packed complex of gene families that encode receptors made up of immunoglobulin-like domains and expressed on myeloid cells or NK cells. This gene complex is called the leukocyte-receptor complex or LRC.

Figure 9.46 Comparison of structures and cellular distributions of the Fc receptors for IgG, IgE, and IgA. The binding specificities of the Fcγ receptors are shown in Figure 9.42. As well as the Fc-binding chain and the γ signaling chain, FcεRI has an additional transmembrane β chain, which is involved in signaling. Although it has a similar structure and belongs to the same large immunoglobulin superfamily, the IgA receptor FcαRI is not closely related to the rest of the Fc receptors and is encoded on a different chromosome. +, Darker pink, constitutive expression; (+), lighter pink, inducible expression.

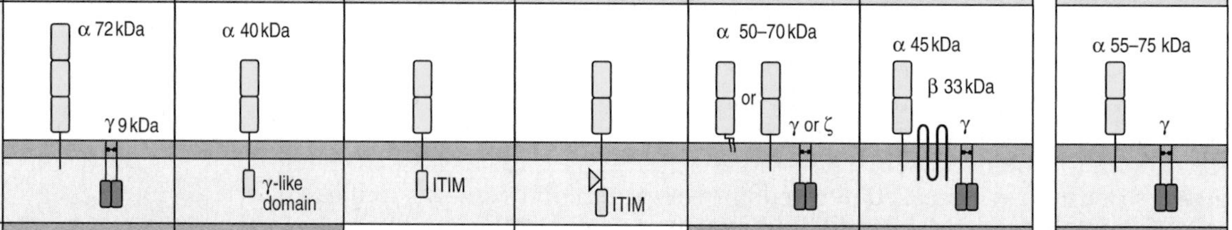

Ligand	IgG					IgE	IgA
Receptor name	FcγRI	FcγRIIA	FcγRIIB2	FcγRIIB1	FcγRIII	FcεRI	FcαRI
Receptor structure	α 72 kDa, γ 9 kDa	α 40 kDa, γ-like domain	ITIM	ITIM	α 50–70 kDa, or, γ or ζ	α 45 kDa, β 33 kDa, γ	α 55–75 kDa, γ
Activating	+	+			+	+	+
Inhibitory			+	+			
Macrophage	+	+	+		+		+
Neutrophil	(+)	+	+		+		+
Eosinophil	(+)	+	+		+	(+)	(+)
Mast cell				+		+	
Basophil						+	
B cell				+			
Dendritic cell	+						+
Langerhans' cell		+					
Platelet		+					
NK cells					+		
CD number	CD6		CD32		CD16	not assigned	CD89
Gene location	Chromosome 1						Chromosome 19

Fc receptors are not restricted to the adaptive immune response and the binding of immunoglobulin. We have seen in Chapter 2 how C-reactive protein, an acute-phase protein of the innate immune response, binds to phosphocholine on bacterial surfaces and activates complement by the classical pathway. C-reactive protein also binds to FcγRI and FcγRII, and by this mechanism it delivers *Streptococcus pneumoniae* for uptake and degradation by phagocytes. That these modern-day Fc receptors have the dual function of binding immunoglobulin and C-reactive protein indicates that the common ancestor of the IgG-specific and IgE-specific Fc receptors was a C-reactive protein receptor of innate immunity.

Summary

Secreted antibodies are the only effector molecules produced by B cells; their principal function is as adaptor molecules that neutralize pathogens and deliver them to effector cells for destruction. Antigen is bound by the variable Fab regions of the antibody, whereas the Fc region has binding sites that initiate complement activation and bind to Fc receptors on various types of effector cell. The isotype and oligomerization of an antibody determine where in the body an antibody will seek out antigens and the type of effector functions it can engage. The first immunoglobulin of the blood and tissues is pentameric IgM, which delivers pathogens to phagocytes by opsonizing them with complement. As the adaptive immune response matures, this function is taken over and improved on by high-affinity IgG and monomeric IgA, which on coating pathogens with antibody can deliver them directly to the IgG-specific and IgA-specific Fc receptors of phagocytes. Here the antibody acts as the opsonin. Four subclasses of IgG are further distinguished by their different capacities to activate complement, engage Fc receptors and respond to different types of antigen. All subclasses of IgG are actively transported from blood to the extracellular fluid in tissues by the transport receptor FcRn.

Dimeric IgA is made by lymphoid tissue lining mucosal surfaces and is actively transported across the mucosal epithelium by the poly-Ig receptor. In this way the mucosal surfaces of the respiratory tract, the gastrointestinal tract, and other anatomical sites have a continual supply of IgA that binds to the microorganisms that inhabit and infect those tissues. The Fc receptor specific for IgE is of such high affinity that the mast cells that bear it become loaded with IgE in the absence of specific antigen and are ready to respond to its introduction. When this occurs, the mast cells, which underlie the skin and mucosal surfaces, stimulate violent muscular contractions that expel pathogens from the body through coughing, sneezing, vomiting, or diarrhea.

During pregnancy, FcRn transports protective IgG from the mother's blood to the fetal circulation. After birth, the gastrointestinal tract of the child is supplied with protective dimeric IgA, an important constituent of breast milk. As the child's supply of maternal IgG and IgA declines during the first year of life, and before the child's immune system has fully developed, there is a period of enhanced susceptibility to infectious disease.

Summary to Chapter 9

The response of B lymphocytes to infection is the secretion of antibodies. These molecular adaptors bind and link the pathogen to effector molecules or cells that will destroy it. In developing an antibody response, the population of responding B cells can combine a quick but less than optimal response in the short term with a more effective response that takes time to develop. B-1 cells, which do not require cognate helper T cells, and IgM molecules, which are not dependent on high-affinity binding sites, both represent short-term strategies. In contrast, B-cell activation driven by cognate T-cell help, with resultant isotype switching and somatic hypermutation, is the long-term strategy that provides effective protection from subsequent infection by the pathogen. One function of the isotypic diversification of immunoglobulins is to provide antibody responses in different compartments of the human body. IgM, IgG, and monomeric IgA work in the blood, lymph, and connective tissues, providing antibody responses to infections within the body's tissues. IgE also travels to the connective tissues where it is bound tightly by the high-affinity IgE receptor on the surface of resident mast cells. In contrast, dimeric IgA is transported to the luminal side of the gut wall and other mucosal surfaces, where it provides antibody responses against the microorganisms that colonize these surfaces. A second function of antibody isotype is to recruit different effector functions into the immune response: complement activation combined with the complement and Fc receptors for IgG and IgA on neutrophils and macrophages deliver pathogens for phagocytosis, whereas the high-affinity receptor for IgE on mast cells, basophils, and eosinophils ensures that antigens binding IgE provoke an inflammatory response.

Questions

9–1   Cross-linking of immunoglobulin by antigen is essential but not always sufficient to initiate the signal cascade for B-cell activation. Stimulation of additional receptors is also necessary for the full activation and differentiation of naive B cells. Describe these receptors and their ligands, and outline how they help activate the B cell.

9–2
A. Explain the difference between thymus-dependent (TD) and thymus-independent (TI) antigens.
B. Give an example of a type 1 thymus-independent antigen (TI-1) and describe how such antigens bypass the requirement for T-cell help.
C. Repeat this for TI-2 antigens.
D. Some molecules seem to act as TI-1 antigens when part of a bacterial surface but need T-cell help if purified and administered on their own. Explain why this might be so.

9–3   Which of the following statements is true or false? If a statement is false, explain why.
A. Plasma cells produce secreted antibody, proliferate and undergo somatic hypermutation to produce antibodies with a higher affinity for antigen.
B. An immunodeficiency called hyper IgM syndrome is characterized by the lack of CD40 ligand expression on T cells.
C. Antibody-dependent cell-mediated cytotoxicity (ADCC) is mediated by NK cells, which use their Fc receptors to bind to antibody-coated target cells, which they kill through inducing apoptosis.
D. TI-2 polysaccharide antigens are commonly used in vaccines administered to infants because they stimulate strong antibody responses.

9–4   Explain why CD40 ligand expression on T cells is important in the T-cell zone of secondary lymphoid tissue and how this contributes to the formation of a primary focus.

9–5   Which of the following is a characteristic of follicular dendritic cells in the primary follicles of secondary lymphoid tissues? (Select all that apply.)
a. They are bone marrow derived hematopoietic cells.
b. They provide a stable depository of intact antigens able to bind to B-cell receptors.
c. They have a large surface area as a result of forming dendrites.
d. They internalize immune complexes through CR2 receptor cross-linking.
e. They bear bundles of immune complexes called iccosomes that are passed on to antigen-specific B cells.
f. They produce cytokines that induce B cells to proliferate and become centroblasts.

9–6
A. What is the main effector function of IgM antibody?
B. Why is IgM efficient at (i) preventing blood-borne infections and (ii) fixing complement, but (iii) less efficient than other antibody classes in inducing phagocytosis of immune complexes?

9–7
A. Explain how the poly-Ig receptor transports dimeric IgA antibodies across cellular barriers and specify the type of cell barrier involved.
B. What are the final locations of the transported material?

9–8
A. Explain how the receptor FcRn transports IgG antibodies across cellular barriers, and specify the type of cell barrier involved.
B. What is the final location of transported material?

9–9
A. What are the similarities between the activation of mast cells and NK cells via FcεRI and FcγRIII, respectively. Be specific.
B. What are the differences? Again, be specific.

9–10   Describe the course of events that results in the swollen lymph nodes characteristic of many infections. Use the following terms in your answer: B lymphoblasts, centroblasts, centrocytes, follicular dendritic cells, germinal center, primary focus, primary follicle, somatic hypermutation, T-cell area, and tingible body macrophages.

9–11
A. What is meant by the term "passive transfer of immunity," and how is it achieved? Give examples.
B. Give the isotype of the antibodies involved in (i) placental transfer and (ii) transfer into breast milk, and explain why these antibodies are important.
C. Do you think it is possible for a pregnant mother who has an autoimmune disease to transfer autoreactive antibodies to the developing fetus? Explain your answer.

9–12   Explain the origin of the secretory component and its significance after the release of dimeric IgA from the apical face of the gut epithelium.

9–13   How does IgE induce the forcible ejection of parasites and toxic substances from the respiratory and gastrointestinal tracts?

9–14   From an immunological viewpoint, why would it be inadvisable for a mother who has recently given birth to move with her newborn to a foreign country where there are endemic diseases not prevalent in her homeland?

9–15
A. Explain how bispecificity arises in IgG4.
B. What is the consequence of this process?

9–16   Amanda Chenoweth, 21 years of age, returned from a summer job as a pianist on a cruise ship where she was exposed daily to excessive sun; she developed a rash on her cheeks. She complained that her finger joints were stiff and painful, which made it difficult to play the piano, and that her hips became painful after sitting at the piano for long periods. Her blood sample tested positive for anti-nuclear antibodies and had decreased serum C3 levels. A urine albumin test showed elevated protein levels. A course of prednisone (an anti-inflammatory steroid) in combination with naprosyn (a nonsteroidal anti-inflammatory agent) was begun and her condition improved rapidly. What is the most likely cause and clinical name of her condition?
a. deterioration of the central nervous system; multiple sclerosis
b. cartilage destruction by bone-cell enzymes; rheumatoid arthritis
c. immune complexes fixing complement in kidney, joints and blood vessels; systemic lupus erythematosus
d. autoantibodies against acetylcholine receptor at the neuromuscular junction; myasthenia gravis
e. consumption of seafood to which she was allergic; acute systemic anaphylaxis.

Infection of the nose by Rhinovirus, a major cause of the common cold.

Chapter 10

The Body's Defenses Against Infection

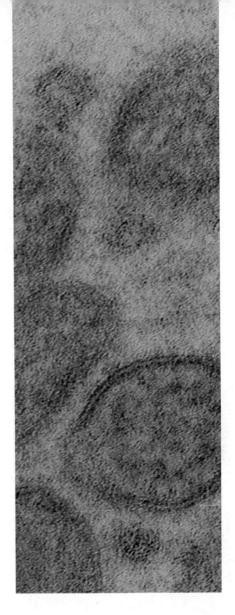

Most infectious diseases suffered by humans are caused by pathogens that are smaller than a human cell. For these agents, the human body constitutes a vast resource-rich environment in which to live and reproduce. In the face of such threats, the body deploys multifarious defense mechanisms that have accumulated over hundreds of millions of years of invertebrate and vertebrate evolution. In Chapter 1 we took a broad look at the nature of infections and of the body's defenses against them. We saw how these defenses fall into the fast-acting mechanisms of innate immunity, which attack infection from the beginning, and the slower processes of adaptive immunity, which respond with greater precision. Chapter 2 examined the cells and molecules of innate immunity, the order in which they are brought into play, and how they stop most infections at an early stage without calling upon adaptive immunity. As outlined in Chapter 3, the basis for the adaptive immune response is the clonal expression by lymphocytes of diverse antigen receptors of two sorts: the immunoglobulin receptors of B cells and the T-cell receptors of T cells. Underlying both the diversity and the clonal expression of these receptors is the unusual property that immunoglobulin and T-cell receptor genes must undergo somatic recombination and mutation to become functional. How these processes shape the development and function of adaptive immunity is described in Chapters 4–9.

Throughout these previous chapters we have examined the immune response in the context of pathogens that enter the body through skin wounds and stimulate an adaptive response in the draining lymph node. In the first part of this chapter we turn our attention to the immune response to pathogens, such as the common cold, that infect the human body by passing through the mucosal surface of tissues such as the bronchial and gastrointestinal tracts. A further focus of the previous chapters concentrated on the primary adaptive immune response that is made when a person first suffers the disease caused by a pathogen. In the second part of this chapter we see how a primary adaptive immune response produces immunological memory and protective immunity that lessens the impact of subsequent encounters with the same pathogen. In the last part of the chapter we examine minority populations of lymphocytes that form bridges between the innate immune response and the adaptive immune response by making use of adaptive immune mechanisms to contribute to innate immunity.

Preventing infection at mucosal surfaces

In previous chapters we concentrated our attention on infections in the connective tissue underlying the skin that stimulate adaptive immune responses in the draining lymph node. The value of this example is that it is simple and involves a tissue on which we have all observed the effects of wounds, infection, and inflammation. Until recently, these were the only responses studied by most immunologists, who commonly administered their experimental antigens by subcutaneous injection. But only a fraction of human pathogens enter the body's tissues by passage through the skin. Many more, including all viruses, gain access by passing through one of the mucosal surfaces. Although the immune response to infection of mucosal tissue has many elements and principles in common with infections of skin and connective tissues, there are also important differences, which provide the focus for this part of the chapter.

10-1   The communication functions of mucosal surfaces render them vulnerable to infection

Mucosal surfaces or the **mucosae** (singular **mucosa**) are found throughout much of the body, except the limbs, but they are predominantly out of sight. The mucosae are continually bathed in a layer of thick fluid that they secrete—the **mucus** that gives them their name. Mucus contains glycoproteins, proteoglycans, peptides, and enzymes that protect the epithelial cells from damage and help to limit infection. Mucosal epithelia line the gastrointestinal, respiratory, and urogenital tracts, and are also present in the exocrine glands associated with these organs: the pancreas, the conjunctivae and lachrymal glands of the eye, the salivary glands, and the mammary glands of the lactating breast (Figure 10.1). These tissues are all sites of communication, where material and information are passed between the body and its environment. Because of their physiological functions of gas exchange (lungs), food absorption (gut), sensory activity (eyes, nose, mouth, and throat), and reproduction (uterus,

Figure 10.1 Distribution of mucosal tissues. This diagram of a woman shows the mucosal tissues. The mammary glands are only a mucosal tissue after pregnancy, when the breast is lactating.

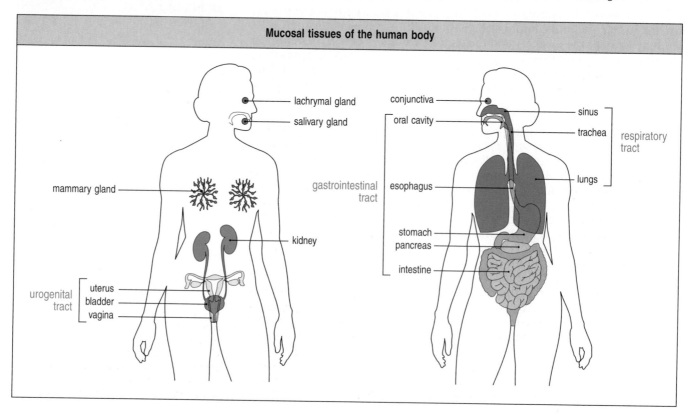

Mucosal tissues of the human body

lachrymal gland
salivary gland
conjunctiva
oral cavity
sinus
trachea
respiratory tract
lungs
mammary gland
gastrointestinal tract
esophagus
kidney
stomach
pancreas
intestine
urogenital tract
uterus
bladder
vagina

vagina, and breast), the mucosal surfaces are by necessity dynamic, thin, permeable barriers to the interior of the body. These properties make the mucosal tissues particularly vulnerable to subversion and breach by pathogens. This fragility, combined with the vital functions of mucosae, has driven the evolution of specialized mechanisms for their defense.

The combined area of the mucosal surfaces is much greater than that of the skin: the small intestine alone has a surface area 200 times that of the skin. Reflecting this difference, three-quarters of the body's lymphocytes are in secondary lymphoid tissues serving mucosal surfaces, and a similar proportion of all antibodies made by the body is secreted dimeric IgA, also known as **secretory IgA** (see Chapter 9). A distinctive feature of the gastrointestinal tract is its continuous contact with large populations of **commensal** microorganisms—the microbes that routinely inhabit the body and normally do not cause disease—as well as substantial quantities of proteins derived from the animals and plants that are our food. In this situation, the major challenge is to make immune responses that eliminate pathogenic microorganisms, limit the growth and location of commensal microorganisms, and do not attack our food. As most research on mucosal immunity has been on the gut, this will provide our main example of a mucosal tissue.

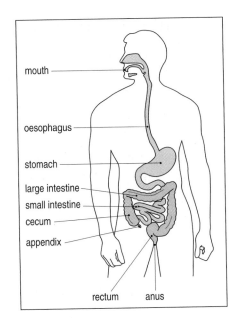

Figure 10.2 The human gastrointestinal tract.

10-2   The gastrointestinal tract is invested with distinctive secondary lymphoid tissues

The gastrointestinal tract extends from the mouth to the anus and is about 9 meters in length in an adult human being. Its physiological purpose is to take in food and process it into nutrients that are absorbed by the body and waste that is excreted; hence its alternative name, the alimentary canal. Different segments of the gastrointestinal tract have specialized functions (Figure 10.2). The mouth physically breaks food down, the stomach uses acid and enzymes for chemical degradation, in the small intestine (the duodenum, jejunum, and ileum) enzymatic degradation continues and nutrients are absorbed, and in the large intestine (the colon) waste is stored, compacted, and periodically eliminated. Essential contributions are made to these processes by the commensal microorganisms of the gut flora.

To provide prompt defense against infection, secondary lymphoid tissues and immune-system cells are spread throughout the gut and other mucosal tissues. They are present within the surface epithelium of the mucosa and also in the underlying connective tissue, called the **lamina propria**. In addition, the **mesenteric lymph nodes**, the largest nodes in the body, are dedicated to defending the gut. They are situated in a chain within the membrane of connective tissue (mesentery) that holds the gut in place. Although the gut-associated lymphoid tissues (GALT) come in a variety of sizes and forms, their microanatomy and organization into B-cell and T-cell zones are generally similar to those of other secondary lymphoid tissues. The presence of secondary lymphoid tissues within the mucosa means that adaptive immune responses to a mucosal infection can be initiated locally, as well as in the draining mesenteric lymph nodes. This is unlike the situation in the rest of the body, where adaptive immune responses are made in secondary lymphoid organs that are often distant from the site of infection.

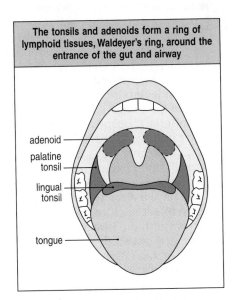

Figure 10.3 A ring of lymphoid organs guards the entrance to the gastrointestinal and respiratory tracts. Lymphoid tissues are shown in blue. The adenoids lie at either side of the base of the nose, and the palatine tonsils lie at either side of the palate at the back of the oral cavity. The lingual tonsils are on the base of the tongue.

At the back of the mouth and guarding the entrance to the gut and the airways are the **palatine tonsils**, **adenoids**, and **lingual tonsils**. These large aggregates of secondary lymphoid tissue are covered by a layer of squamous epithelium and form a ring known as Waldeyer's ring (Figure 10.3). In early childhood, when pathogens are being experienced for the first time and the mouth provides a conduit for all manner of extraneous material that is not food, the tonsils and adenoids can become greatly swollen because of recurrent infection. In the not so distant past, this condition was routinely treated by surgically

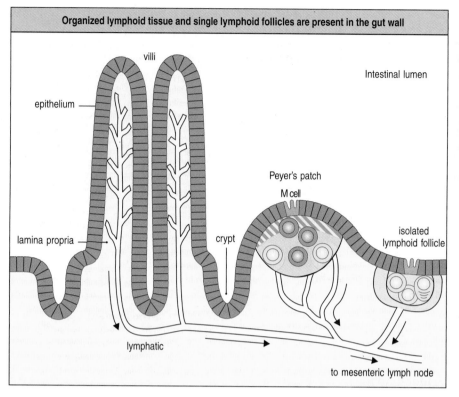

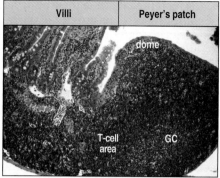

Figure 10.4 Gut-associated lymphoid tissues and lymphocytes. The diagram shows the structure of the mucosa of the small intestine. It consists of finger-like processes (villi) covered by a layer of thin epithelial cells (red) that are specialized for the uptake and further breakdown of already partly degraded food coming from the stomach. The tissue layer under the epithelium is the lamina propria, colored pale yellow in this and other figures in this chapter. Lymphatics arising in the lamina propria drain to the mesenteric lymph nodes, which are not shown on this diagram (the direction of lymph flow is indicated by arrows). Peyer's patches are secondary lymphoid organs that underlie the gut epithelium and consist of a T-cell area (blue), B-cell follicles (yellow), and a 'dome' area (striped blue and yellow) immediately under the epithelium that is populated by B cells, T cells, and dendritic cells. Antigen enters a Peyer's patch from the gut via the M cells. Peyer's patches have no afferent lymphatics, but they are a source of efferent lymphatics that connect with the lymphatics carrying lymph to the mesenteric lymph node. Also found in the gut wall are isolated lymphoid follicles consisting mainly of B cells. The light micrograph is of a section of gut epithelium and shows villi and a Peyer's patch. The T-cell area and a germinal center (GC) are indicated.

removing the lymphoid organs, a loss of immune capacity that resulted in a poorer secretory IgA response to oral polio vaccination.

The small intestine is the major site of food absorption and is the part of the gut most heavily invested with lymphoid tissue. Characteristic secondary lymphoid organs of the small intestine are the Peyer's patches, which are integrated into the intestinal wall and have a distinctive appearance, forming dome-like aggregates of lymphocytes that bulge into the intestinal lumen (Figure 10.4). The patches vary in size and contain between 5 and 200 B-cell follicles with germinal centers interspersed amid associated T-cell areas that also include dendritic cells. Overlying the lymphocytes and separating them from the gut lumen is a layer of follicle-associated epithelium. This contains conventional absorptive intestinal epithelial cells known as enterocytes and a smaller number of specialized epithelial cells called **microfold cells** (**M cells**). The name comes from their folded luminal surface with its absence of the microvilli that characterize enterocytes (Figure 10.5). Unlike enterocytes, M cells do not secrete digestive enzymes or mucus, lack a thick surface glycocalyx and have a weak system of lysosomes. These properties enable M cells to take up intact microorganisms and particulate antigens from the gut lumen and transfer them directly to a Peyer's patch to initiate an adaptive immune response. Between the epithelial cell layer and the lymphoid follicles of the Peyer's patch is a layer of tissue called the subepithelial dome, which is rich in dendritic cells, B cells and T cells. Lymphatics arising in the lamina propria and in the Peyer's patches drain to the mesenteric lymph nodes (see Figure 10.4).

In addition to the Peyer's patches, the small intestine contains numerous **isolated lymphoid follicles** (see Figure 10.4). These comprise a single follicle, consisting mostly of B cells, that is overlaid by epithelium containing M cells, as in a Peyer's patch. Isolated lymphoid follicles, but not Peyer's patches, are also present in the large intestine (see Figure 10.2). A distinctive secondary lymphoid organ of the large intestine is the appendix. It consists of a blind-ended tube about 10 cm in length and 0.5 cm in diameter that is attached to the cecum, the part of the large intestine that is contiguous with the small intestine. It is packed with lymphoid follicles, and when it is overrun by infection (appendicitis) the only treatment is surgical removal to prevent the appendix from bursting and causing a life-threatening infection of the peritoneum (peritonitis).

The immune responses that are generated against antigens in the gut-associated lymphoid tissues are qualitatively distinct from those stimulated in the spleen or in lymph nodes draining the skin or muscle. This is because gut lymphoid tissues have their own distinct content of lymphoid cells, hormones, and other immunomodulatory factors. During fetal development, the mesenteric lymph nodes and Peyer's patches differentiate independently of the spleen and the other lymph nodes (the so-called 'systemic immune system') and under the guidance of different chemokines and receptors for cytokines of the tumor necrosis factor (TNF) family. The differences between the gut-associated lymphoid tissues and the systemic lymphoid organs are thus imprinted early in life.

10-3 M cells and dendritic cells facilitate transport of microbes from the gut lumen to gut-associated lymphoid tissues

Whereas the healthy skin is impermeable to microorganisms, healthy gut epithelium actively monitors the contents of the gut lumen so that an adaptive immune response can be initiated against a pathogen before it has the chance to breach the epithelial barrier and colonize the lamina propria. The M cells of the Peyer's patches and isolated lymphoid follicles are specialized for the uptake of pathogens from the intestinal lumen and their transcytosis across the epithelium and into the lymphoid tissue, and are continuously sampling the gut flora in this manner. Transcytosis releases microorganisms at the M cell's basal membrane, which is extensively folded to form a pocket that encloses lymphocytes and dendritic cells. Here, dendritic cells can take up the microbes to process and present their antigens to naive T cells (Figure 10.6). Antigen-loaded dendritic cells either migrate from the dome region to the

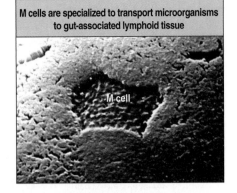

M cells are specialized to transport microorganisms to gut-associated lymphoid tissue

M cell

Figure 10.5 Microfold cells have characteristic membrane ruffles. This scanning electron micrograph of intestinal epithelium has a microfold or M cell in the center. It appears as a sunken area of the epithelium that has characteristic microfolds or ruffles on the surface. M cells capture microorganisms from the gut lumen and deliver them to Peyer's patches and the lymphoid follicles that underlie the M cells on the basolateral side of the epithelium. Magnification × 23,000.

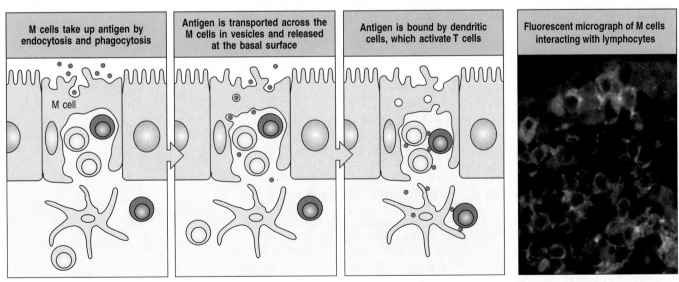

| M cells take up antigen by endocytosis and phagocytosis | Antigen is transported across the M cells in vesicles and released at the basal surface | Antigen is bound by dendritic cells, which activate T cells | Fluorescent micrograph of M cells interacting with lymphocytes |

M cell

Figure 10.6 Uptake and transport of antigens by M cells. Adaptive immune responses in the gut are initiated and maintained by M cells that sample the gut's contents (first panel) and deliver this material to 'pockets' on the basolateral side of the M-cell membrane (second panel). Here, dendritic cells and B cells take up antigen and stimulate the proliferation and differentiation of antigen-specific T cells (third panel).

The fourth panel is a micrograph of part of a Peyer's patch that shows epithelial cells (dark blue), some of which are M cells, with T cells (red) and B cells (green) in their pockets. The cells are distinguished by using fluorescently labeled antibodies specific for M cells, B cells, and T cells. Photograph courtesy of K.A. Hartwell and T.A. Ince.

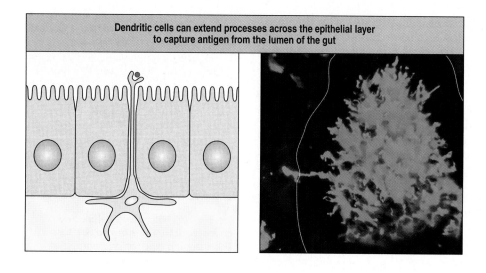

Dendritic cells can extend processes across the epithelial layer to capture antigen from the lumen of the gut

Figure 10.7 Capture of antigens from the intestine by dendritic cells. Dendritic cells in the lamina propria can sample the contents of the intestine by extending a process between the enterocytes without disturbing the barrier function of the epithelium (left panel). Such an event is captured in the micrograph (right panel) in which the mucosal surface is shown by the white line and the dendritic cell (stained green) in the lamina propria extends a single process over the white line and into the lumen of the gut. Magnification × 200. Micrograph courtesy of J.H. Niess.

T-cell areas of the Peyer's patch, or via the draining lymphatics to a mesenteric lymph node, where they meet and stimulate antigen-specific naive T cells.

Dendritic cells resident in the lamina propria outside the organized lymphoid tissues can also capture pathogens independently of M cells. In response to local infection, these dendritic cells become more mobile. They move into the epithelial wall or send processes through it that can capture microbes and antigens without disturbing the integrity of the epithelial barrier (Figure 10.7). Having obtained a cargo of antigens, the dendritic cells move into the T-cell area of the gut-associated lymphoid tissue, or travel in the draining lymph to the T-cell area of a mesenteric lymph node, to stimulate antigen-specific T cells.

Through this continual sampling of the contents of the gut lumen, T cells specific for pathogenic microorganisms, commensal microorganisms and food antigens are stimulated to become effector cells. Activated helper T cells then activate B cells to become plasma cells, as described in Chapter 9. These plasma cells secrete dimeric IgA specific for pathogens, commensals, and food antigens.

10-4 Effector lymphocytes populate healthy mucosal tissue in the absence of infection

Another characteristic of healthy mucosae that distinguishes them from healthy skin is that they are heavily populated with activated effector lymphocytes and other leukocytes (Figure 10.8). Most of the lymphocytes are effector T cells and plasma cells. Integrated into the epithelial layer of the small intestine is a distinctive type of CD8 T cell called the **intraepithelial lymphocyte**. On average there is about one intraepithelial lymphocyte for every 7–10 epithelial cells (Figure 10.9). Intraepithelial lymphocytes have been activated by antigen and contain intracellular granules like those of CD8 cytotoxic T cells. The intraepithelial lymphocytes can be either α:β CD8 T cells or γ:δ CD8 T cells. They express T-cell receptors with a limited range of antigen specificities, indicating that they have been activated by a relatively small number of antigens, and they have a distinctive combination of chemokine receptors and adhesion molecules that enables them to take up their unique position within the intestinal epithelium. In contrast to the epithelium, the lamina propria contains a variety of effector lymphocytes—CD4 T cells, CD8 T cells, and plasma cells—as well as dendritic cells and the occasional eosinophil or mast cell (see Figure 10.8). Neutrophils are rare in the healthy intestine, but they arrive in large numbers at sites of inflammation and infection.

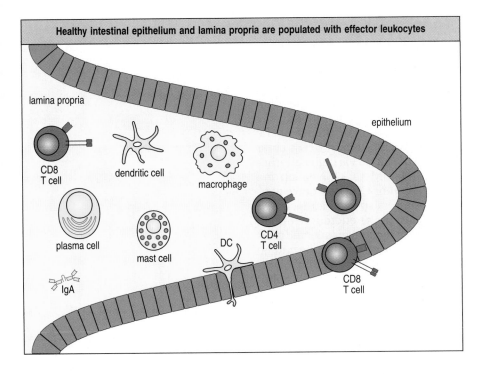

Figure 10.8 **Most immune-system cells in mucosal tisues are activated effector cells.** Outside the lymphoid tissues, the gut epithelium contains CD8 T cells, and the lamina propria is well populated with CD4 T cells, CD8 T cells, plasma cells, mast cells, dendritic cells (DC), and macrophages. These cells are always in an activated state because of the constant stimulation afforded by the gut's diverse and changing contents.

In contrast to the skin and other tissues, the healthy intestinal mucosa sustains a chronic adaptive immune response. This is achieved by the constant sampling of the gut's contents by M cells and dendritic cells, and the continual stimulation of B cells and T cells in a myriad of local immune responses in the gut-associated lymphoid tissue. Most of the foreign antigens are innocuous and the immune responses to them do not create a state of inflammation or symptoms of disease. This is in part due to the distinctive properties of the macrophages and dendritic cells that differentiate from blood-borne precursors in the specialized environment of the gut mucosa. Although the mucosa is well populated with macrophages that phagocytose and kill bacterial pathogens, they cannot make an inflammatory response because they lack TLRs and other signaling receptors necessary for the production of inflammatory cytokines (see Chapter 2). To similar effect, the dendritic cells resident in the mucosae are less likely than skin dendritic cells to activate an inflammatory T_H1-cell response. Overall, the collective purpose of these mucosal immune responses is not to remove microorganisms from the gut but to restrain them, keeping them in the gut lumen and out of the tissues. This applies just as much to commensal microorganisms as to pathogens, because when commensal microbes cross the epithelial barrier and propagate in the lamina propria they too can become disease-causing pathogens. Because of their chronic state of stimulation, the lymphocytes in the mucosal surface are able to respond quickly to any breach of the mucosal barrier and prevent inundation of the mucosal tissue with the gut contents.

During early childhood, the human body and its immune system grow and mature in the context of the body's microbial flora and the common pathogens in the environment. Like many other parts of the body, if the immune system is not used regularly to control infection it becomes impaired. This is well illustrated by laboratory animals that are deliberately born and raised under 'germ-free' conditions. In comparison to control animals with a normal gut flora, these 'gnotobiotic' animals have a stunted immune system—with smaller secondary lymphoid tissues, lower levels of serum immunoglobulin, and a generally reduced capacity to make immune responses.

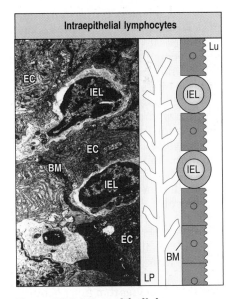

Figure 10.9 **Intraepithelial lymphocytes.** The electron micrograph on the left shows a section through mucosal epithelium. The orientation of the mucosal epithelium is shown diagrammatically on the right. In the micrograph, two prominent intraepithelial lymphocytes (IEL) are flanked by epithelial cells (EC). The epithelium is separated from the lamina propria (LP) by a basement membrane (BM). Lu, gut lumen. Magnification × 8000.

10-5   B cells and T cells commit to mucosal lymphoid tissues after they encounter their specific antigen

The naive B cells and T cells that emerge from the primary lymphoid organs into the bloodstream are not restricted in their recirculation to either the mucosa-associated lymphoid tissues or the spleen and lymph nodes that defend the non-mucosal tissues, but can recirculate via the lymph and the blood through both the mucosal and the systemic compartments of the immune system. As in lymph nodes and spleen, naive B cells and T cells enter Peyer's patches and mesenteric lymph nodes from the blood via high endothelial venules, attracted by the chemokines CCL21 and CCL19, which are released by the lymphoid organ and bind to the receptor CCR7 on naive lymphocytes. If a naive lymphocyte does not encounter its specific antigen in a Peyer's patch or mesenteric lymph node, it will leave in the efferent lymph to continue recirculation. In contrast, if the lymphocyte recognizes specific antigen it will be retained in the Peyer's patch or mesenteric lymph node and be activated to proliferate and differentiate into effector and memory lymphocytes. Activation is accompanied by the loss of CCR7 and the cell-adhesion molecule L-selectin, and thus of the ability of the lymphocyte to enter secondary lymphoid tissues from the blood. Instead, T cells and B cells activated in mucosal lymphoid tissues express chemokine receptors and cell-adhesion molecules that now commit them to working specifically in mucosal tissues.

B cells and T cells activated in a Peyer's patch leave in the lymph and travel via the mesenteric lymph nodes and the thoracic duct to the blood. Cells activated in a mesenteric lymph node leave in the efferent lymph and then similarly reach the blood. Once back in the bloodstream, these mucosally activated B cells and T cells home to the type of mucosa in which they were activated and enter it from the blood (Figure 10.10). Gut-homing effector lymphocytes, for example, express the integrin $\alpha_4{:}\beta_7$, which binds specifically to the mucosal vascular addressin **MAdCAM-1** present on endothelial cells of blood vessels in the gut wall. Guiding the lymphocytes into the gut tissue is the chemokine CCL25, which is secreted by the epithelium of the small intestine and binds to the receptor CCR9 expressed by effector lymphocytes (Figure 10.11, left panel). By this route the activated lymphocytes enter the lamina propria, where the B cells develop into plasma cells making secretory IgA and the effector T cells activate intestinal macrophages to improve their ability to phagocytose and destroy microorganisms in ways that minimize inflammation.

T cells destined to become intraepithelial lymphocytes express the chemokine receptor CCR9, but instead of the $\alpha_4{:}\beta_7$ integrin of other gut-homing T cells, they express the integrin $\alpha_E{:}\beta_7$, which binds to E-cadherin on the surface of epithelial cells (Figure 10.11, right panel). This adhesive interaction enables intraepithelial lymphocytes to intercalate within the layer of intestinal epithelial cells while maintaining the epithelium's barrier function.

Only lymphocytes that first encounter antigen in the gut-associated lymphoid tissue are induced to express gut-specific homing receptors and integrins. This induction is under the control of the intestinal dendritic cells and is mediated by retinoic acid, a derivative of vitamin A made by these cells. Analogous tissue specificities are also acquired by effector T cells activated in non-mucosal tissues. For example, skin-homing effector cells express the integrin $\alpha_4{:}\beta_1$ and the chemokine receptor CCR4. The tissue specificity of effector lymphocytes demands that vaccination against intestinal infections be made in a manner that delivers the antigen to those secondary lymphoid organs where natural responses to the pathogen are made. Thus polio vaccine is administered orally because natural polio infections are transmitted from one person to another by oral ingestion of the virus in fecal material.

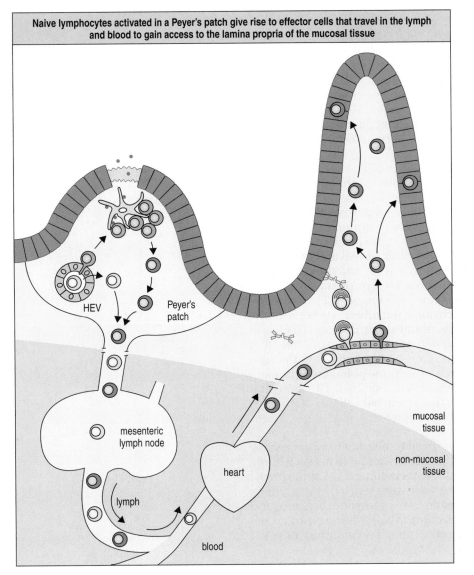

Naive lymphocytes activated in a Peyer's patch give rise to effector cells that travel in the lymph and blood to gain access to the lamina propria of the mucosal tissue

Figure 10.10 Lymphocytes activated in mucosal tissues return to those tissues as effector cells. Pathogens from the intestinal lumen enter a Peyer's patch through an M cell and are taken up and processed by dendritic cells. Naive T cells (green) and B cells (yellow) enter the Peyer's patch from the blood at a high endothelial venule. The naive lymphocytes are activated by antigen, whereupon they divide and differentiate into effector cells (blue). The effector cells leave the Peyer's patch in the lymph, and after passing through mesenteric lymph nodes, reach the blood, in which they travel back to the mucosal tissue where they were activated. The effector cells leave the blood and enter the lamina propria and the epithelium, where they perform their functions: killing and cytokine secretion for effector T cells, secretion of IgA for plasma cells.

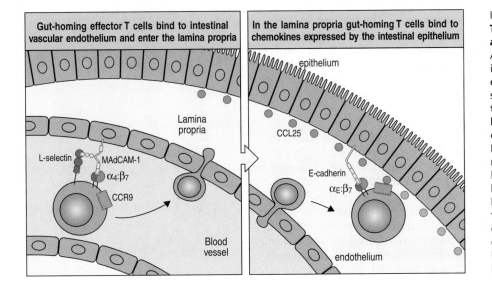

Gut-homing effector T cells bind to intestinal vascular endothelium and enter the lamina propria

In the lamina propria gut-homing T cells bind to chemokines expressed by the intestinal epithelium

Figure 10.11 Homing of effector T cells to the gut is controlled by adhesion molecules and chemokines. Antigen-mediated activation of T cells in mucosal lymphoid tissue produces effector cells that leave their activation site via the lymphatics and then return to mucosal tissue in the blood. This homing is mediated by integrin $\alpha_4{:}\beta_7$ on the effector cells, which binds to MAdCAM-1 on the blood vessels serving mucosal tissue (left panel). Once in the lamina propria, the T cells are guided by the chemokine CCL25, which is made by mucosal epithelium and binds to the chemokine receptor CCR9 on the effector T cells. Interaction with the gut epithelium is further established by integrin $\alpha_E{:}\beta_7$ on the effector T cells binding to epithelial cell E-cadherin.

10-6   Effector lymphocytes activated in any one mucosal tissue recirculate through all mucosal tissues

MAdCAM-1 is not restricted to intestinal blood vessels but is also present on the vasculature serving other mucosal tissues. As a result, B and T cells first activated in the gut-associated lymphoid tissue can, for example, recirculate as effector cells to the respiratory tract and vice versa. Thus the secondary lymphoid tissues of the mucosal surfaces form a unified and exclusive network of recirculation for effector cells that have been activated by microorganisms and antigens at mucosal surfaces.

Naive B cells activated in Peyer's patches and mesenteric lymph nodes preferentially undergo isotype switching to IgA, the dominant immunoglobulin in mucosal secretions. This process is directed by the cytokine transforming growth factor-β (TGF-β) and occurs in the organized lymphoid tissues of the gut, using the same molecular mechanisms as in the spleen and systemic lymph nodes (see Section 4-15, p. 115). After B-cell activation and differentiation the cells express the mucosal homing integrin $\alpha_4:\beta_7$ and the chemokine receptor CCR9. The effector B cells leave the secondary lymphoid tissue in the efferent lymph and travel to the blood. From the blood some will return to the lamina propria close to where they were activated, but others will home to the lamina propria at other sites in the same tissue and also in different tissues. In the lamina propria the activated B cells differentiate into plasma cells that synthesize intact J-chain-linked IgA dimers. These are secreted into the subepithelial space, from where they are transcytosed by the poly-Ig receptor and delivered to the mucosal surface (see Section 9-12, p. 266). The cells that express the poly-Ig receptor are the immature epithelial cells, or stem cells, located at the base of the intestinal crypts (see Figure 2.17, p. 43).

In nursing mothers, plasma cells originating from B cells that have been stimulated in the gut, the respiratory tract, and all other mucosal tissues will home to the lactating mammary gland and contribute their dimeric IgA to the breast milk. Thus the IgA in the breast milk will fully represent all the antibody responses that the mother has recently made to microorganisms, food, and infections in all the mucosal tissues. The suckling infant will receive maternal IgA that provides a comprehensive protective mucosal immunity directed toward the normal gut flora and endemic pathogens.

10-7   Dimeric IgA binds pathogens at various sites in mucosal tissues

We saw in Chapter 9 how monomeric IgA circulates in the blood and is produced in the bone marrow by plasma cells derived from B cells activated in systemic lymph nodes or the spleen. In contrast, dimeric IgA is secreted in mucosal tissues by plasma cells that derive from B cells activated in the mucosal immune system. Individual plasma cells make either monomeric or dimeric IgA, and in general the antigens that stimulate production of the two forms do not overlap.

Dimeric IgA can bind to antigens and pathogens at several different locations. If infection has penetrated the lamina propria, newly synthesized IgA can bind to microorganisms and carry them back to the gut lumen by transcytosis (Figure 10.12, first panel). The majority of IgA, which is transcytosed as free antibody, can bind to antigens in endosomal compartments, neutralize them and carry them to the lumen (Figure 10.12, second panel). After transcytosis, free IgA is retained at the epithelial surface through interaction with the mucus that covers the surface (see Section 9-12, p. 266). This creates a coating of IgA on the surface of the epithelium, binding microorganisms and preventing their attachment to epithelial cells, as well as neutralizing their toxins and destructive enzymes (Figure 10.12, third panel). IgA in the gut can also bind to

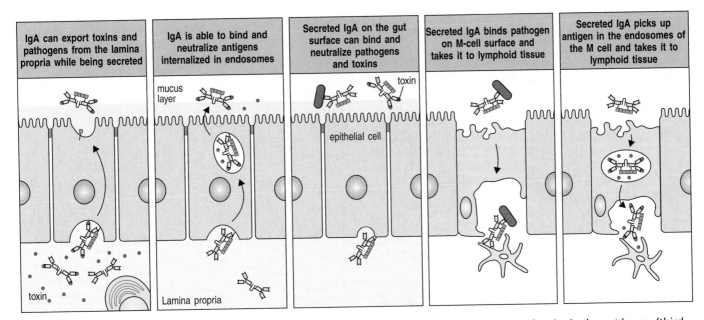

Figure 10.12 Secretory IgA can bind pathogens in several locations. In the lamina propria IgA can bind pathogens and antigens and carry them to the intestinal lumen on transcytosis (first panel). During transcytosis, free IgA can bind pathogens and antigens that been internalized into endosomes (second panel). IgA that has been secreted and is present at the mucosal surface can bind pathogens and toxins in the gut lumen (third panel). Secreted IgA can bind pathogens on an M-cell surface and by passing through the M cell can take them into the secondary lymphoid tissue (fourth panel). As free IgA passes through an M cell it can bind pathogens in the endosomes and take them into the secondary lymphoid tissue (fifth panel).

M cells and be transcytosed back into the mucosa. In this way, IgA bound to a microbe can target it for transport across the M cell and delivery to a follicle or Peyer's patch (Figure 10.12, fourth panel). Last, internalized free IgA can bind to microbes and antigens inside the M cells and neutralize them before delivering them to secondary lymphoid tissue (Figure 10.12, fifth panel).

Secretory IgA has little capacity or opportunity to activate complement or act as an opsin, and it cannot induce a state of inflammation. Instead it has evolved to be a non-inflammatory immunoglobulin that limits the access of pathogens, commensals, and food products to mucosal surfaces in a manner that avoids unnecessary damage to these delicate and vital tissues. Antibodies specific for commensal bacteria are well represented in the IgA secreted into the gut. By restricting commensal organisms to the lumen of the gut and limiting the size of their population, these antibodies have a crucial role in maintaining the symbiotic relationship with their human host.

10-8 Two subclasses of IgA have complementary properties for controlling microbial populations

Both monomeric and dimeric IgA can be made as either of two subclasses—or isotypes—of IgA. As we saw for the IgG subclasses (see Section 9-21, p. 275), the two IgA subclasses differ mainly in the hinge region, which is twice as long in IgA1 (26 amino acids) than in IgA2 (13 amino acids). The longer hinge in IgA1 makes it more flexible than IgA2 in binding to pathogens and thus more able to use multiple antigen-binding sites to bind to the same pathogen and deliver it to a phagocyte. The drawback to the longer IgA1 hinge is its greater susceptibility to proteolytic cleavage than the shorter IgA2 hinge. Major bacterial pathogens, including *S. pneumoniae*, *M. meningitidis* and *H. influenzae*, have evolved specific proteases that cleave the IgA1 hinge and thus separate the Fc and Fab regions. This prevents the antibody from targeting the bacteria to phagocyte-mediated destruction. Exactly the opposite effect can sometimes occur: bacteria coated with Fab fragments of IgA1 become more able to adhere to mucosal epithelium, penetrate the physical barrier, and gain access to the lamina propria to launch an infection.

In situations where IgA1 is ineffective because of the presence of specific proteases, the synthesis of IgA2 helps control bacterial infection. Although the IgA2 hinge is less flexible, it is highly protected by covalently linked carbohydrate, and bacteria have so far failed to evolve a protease that can cleave IgA2. In the blood, lymphatics, and extracellular fluid of the connective tissue, where bacterial populations are small and the IgA1-specific proteases pose less of a threat, most of the IgA made (93%) is of the IgA1 isotype. In contrast, in the colon, where bacteria are present at the highest density and IgA1-specific proteases are ubiquitous, the majority of IgA made (60%) is of the IgA2 isotype (Figure 10.13).

Normally the switch to IgA secretion goes from IgM to IgA1, but in the presence of the TNF-family cytokine APRIL, the isotype switches from IgM to IgA2. In the colon the epithelial cells make APRIL, which drives the switch to the IgA2 isotype in the resident B cells. In general, the synthesis of IgA2 is higher in mucosal lymphoid tissues than in the systemic lymphoid tissue, but the proportions of plasma cells making IgA1 and IgA2 also vary considerably between the different mucosal tissues (see Figure 10.13). The tissues most heavily populated with microorganisms—the large and small intestines, the mouth (supplied with IgA by the salivary glands), and the lactating breast (exposed to the heavily contaminated oral cavity of the suckling infant)—are those more focused on making IgA2. These differences show that the various mucosal tissues are not immunologically equivalent and face different challenges in balancing their burden of commensal and pathogenic microorganisms that make IgA1-specific proteases.

10-9   Humans with selective deficiency of IgA do not succumb to infection

In the caucasian population, about one person in 500 has little or no IgA but makes all the other immunoglobulin isotypes. This condition is called **selective IgA deficiency**. In African-Americans the frequency of this condition is only one-twentieth of that in caucasians, and in the Japanese it is even lower. Surprisingly, most people with IgA deficiency are healthy, and are often oblivious to their lack of an antibody that their friends and family members make in vast quantity. In IgA deficiency, the lack of IgA is compensated for by increased production of the other isotypes (Figure 10.14).

Most important of these is IgM, which can be secreted by mucosal tissues by the same mechanism as dimeric IgA because it contains the J chain necessary for binding to the poly-Ig receptor. Thus the secretion of pentameric IgM probably compensates for the absence of secretory IgA, at least in the relatively parasite-free environment of developed countries. The health and vigor of people with IgA deficiency might in part reflect the fact that in modern industrialized societies the eating of cooked and highly processed food means

The two IgA subclasses are differentially produced in tissues			
Tissue	IgA1 %	IgA2 %	IgA1/A2 ratio
Spleen, peripheral lymph nodes, tonsils	93	7	13.3
Nasal mucosa	93	7	13.3
Bronchial mucosa	75	25	3.0
Lachrymal glands (tear ducts)	80	20	4.0
Salivary glands	64	36	1.8
Mammary glands	60	40	1.5
Gastric mucosa (stomach)	83	17	4.9
Duodenum–jejunum (upper small intestine)	71	29	2.4
Ileum (lower small intestine)	60	40	1.5
Colon (large intestine)	36	64	0.6

Figure 10.13 IgA1 and IgA2 are differentially expressed in mucosal tissues. Shown here are the relative proportions of plasma cells making IgA1 and IgA2 in each tissue. The number of plasma cells correlates directly with the amount of antibody made and secreted. Data courtesy of Per Brandtzaeg.

	Percentage of B cells making antibody of four different isotypes							
	Normal individuals				IgA-deficient individuals			
	IgA	IgM	IgG	IgD	IgA	IgM	IgG	IgD
Nasal glands	69	6	17	8	0	20	46	34
Lachrymal and parotid glands	82	6	5	7	0	21	22	57
Gastric mucosa	76	11	13	0	0	64	35	1
Small intestine	79	18	3	0	0	75	24	1

Figure 10.14 Selective IgA deficiency. The table compares individuals who have normal production of IgA with individuals who have selective IgA deficiency. The percentages of B cells making IgA, IgM, IgG, and IgD in four different mucosal tissues are shown. The amount of antibody made is proportional to the number of B cells. Data courtesy of Per Brandtzaeg.

that the mucosal immune system is under far less pressure from the intestinal parasites, such as helminth worms, that were far more prevalent in those societies in the past and still affect 2 billion of the world's human population. In contrast, chronic lung disease is more common in people with IgA deficiency in industrialized countries, suggesting that the trend toward poorer air quality in the cities where most people live makes the role of IgA in the respiratory tract of increasing importance. The causes of selective IgA deficiency are poorly defined but are believed to be heterogeneous.

10-10 Intestinal epithelial cells contribute to innate defense of the gut

Intestinal epithelial cells are very active in the uptake of nutrients and other materials from the gut lumen. To detect the presence of bacteria in their cytoplasm, epithelial cells have intracellular sensors called the nucleotide-binding oligomerization domain proteins (**NOD proteins**), which are structurally related to the Toll-like receptors (see Section 2-11, p. 45). NOD1 recognizes a muramyl tripeptide that contains diaminopimelic acid, which is present only in the cell walls of Gram-negative bacteria. NOD2 recognizes a muramyl dipeptide present in the peptidoglycans of most bacterial cell walls. Binding to its ligand causes a NOD protein to form oligomers that activate the protein kinase RICK. Activated RICK turns on the NFκB signaling pathway in the epithelial cell (Figure 10.15), causing the cell to make and release cytokines, chemokines, and the antibacterial defensins (see Section 2-9, p. 43). The defensins kill the bacteria, whereas the chemokines attract neutrophils (via the chemokine CXCL8), monocytes (via CCL3), eosinophils (via CCL4), T cells (via CCL5), and immature dendritic cells (via CCL20). In this way, the onset of infection triggers an influx of inflammatory cells and lymphocytes into the mucosa from the blood, which forms the basis for an innate immune response and the potential for making an adaptive immune response should one be needed.

10-11 Intestinal helminth infections provoke strong T_H2-mediated immune responses

The intestines of virtually all humans, except those living in the developed world, are populated with parasitic helminth worms. These infections can cause debilitating and chronic disease. The parasites cause disease by competing with the host for nutrients or by causing local damage to epithelial cells or blood vessels. The host immune response against these parasites can also produce harmful effects. The nature of the immune response to a helminth depends on the parasite's life cycle. Some remain in the intestinal lumen, others enter and colonize the epithelial cells, and yet others invade beyond the intestine and spend part of their life cycle in another tissue, such as the liver, lungs, or muscle.

Characteristic of immunity to almost all helminth infections is that a response dominated by T_H2 CD4 T cells is protective, whereas a response dominated by T_H1 CD4 T cells fails to eliminate the pathogen and produces an inflammatory response that damages the mucosal tissues of the human host (Figure 10.16). The bias toward a T_H2 response is caused by worm products that influence the dendritic cells presenting worm antigens. In turn these dendritic cells influence the naive T cells activated by them to differentiate preferentially into T_H2 cells. The characteristic cytokines that T_H2 cells produce then lead to B cells' switching to making the IgE isotype (see Section 9-9, p. 261) and also recruit eosinophils and mast cells to the wall of the intestine. IL-5, for example, recruits eosinophils, which on activation by the binding of worm antigens to IgE release major basic protein and other cytototoxic molecules that damage and kill the worms. Mast cells activated by worm antigens bound

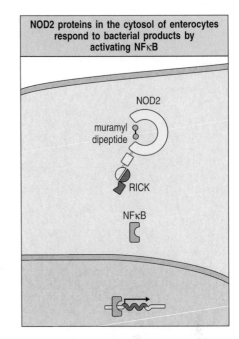

Figure 10.15 NOD proteins enable enterocytes to detect intracytoplasmic bacteria. NOD proteins in the cytoplasm of intestinal epithelial cells bind muramyl dipeptides from bacterial cell walls and cause the activation of NFκB. This induces the epithelial cells to produce chemokines, cytokines, and defensins. The functions that NOD proteins fulfill in the cytoplasm are analogous and complementary to those performed by Toll-like receptors at cell surfaces and in endosomes.

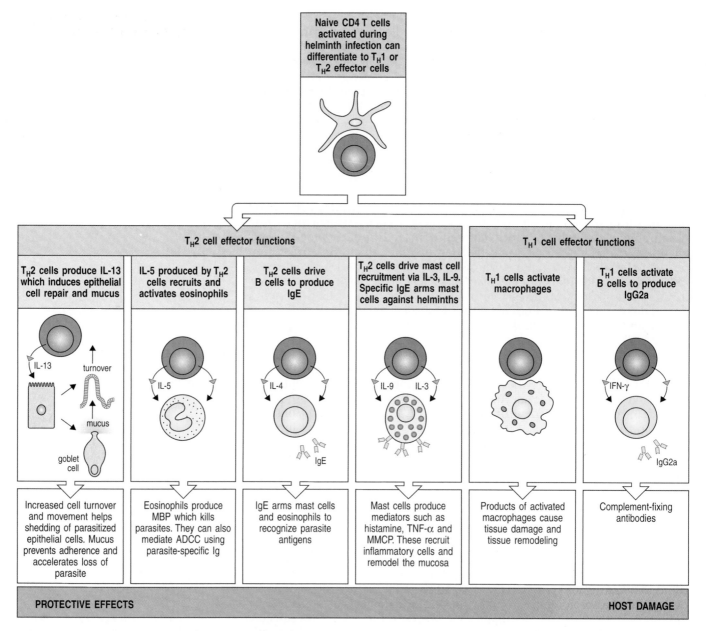

Figure 10.16 **Protective and pathological responses to intestinal helminths.** A feature of CD4 T-cell responses to intestinal helminths is that they can polarize, becoming either a protective T$_H$2 response (first four panels) or a T$_H$1 response (last two panels) that augments the disease and discomfort of the human host. T$_H$2 responses can lead to expulsion of the parasite, whereas T$_H$1 responses lead to persistent infection and chronic inflammation of the intestine. ADCC, antibody-dependent cell-mediated cytotoxicity; MBP, major basic protein; MMCP, matrix metalloproteinase.

to IgE release mediators such as histamine that cause the smooth muscle of the intestinal wall to contract, which helps expel the worms (see Section 9-24, p. 282). The T$_H$2 cytokines also increase the production of mucus by goblet cells, increase the contractility of the intestinal smooth muscle cells, and increase the migration and turnover of epithelial cells (see Figure 10.16).

Summary

The mucosal surfaces of the body cover vital organs that communicate material and information between the human body and its environment. In comparison to the skin, the mucosal surfaces are much larger and more vulnerable

Distinctive features of the mucosal immune system	
Anatomical features	Intimate interactions between mucosal epithelia and lymphoid tissues
	Discrete compartments of diffuse lymphoid tissue and more organized structures such as Peyer's patches, isolated lymphoid follicles, and tonsils
	Specialized antigen-uptake mechanisms provided by M cells in Peyer's patches, adenoids, and tonsils
Effector mechanisms	Activated effector T cells predominate even in the absence of infection
	Plasma cells are in the tissues where antibodies are needed
Immunoregulatory environment	Dominant and active downregulation of inflammatory immune responses to food and other innocuous environmental antigens
	Inhibitory macrophages and tolerance-inducing dendritic cells

Figure 10.17 Distinctive features of adaptive immunity in mucosal tissues.

to infection. Consequently some 75% of the immune system's cells resources are dedicated to defending the mucosae. The mechanisms and character of adaptive immunity in mucosal tissue, as exemplified by the gut, differ in several important respects from adaptive immunity in other tissues (Figure 10.17). Secondary lymphoid tissues, which are directly incorporated into the gut wall, continuously sample the gut's luminal contents and stimulate adaptive immune responses against pathogens, commensal organisms, and food. The effector T cells generated populate the epithelium and lamina propria of the gut, and the plasma cells produce dimeric IgA that is transcytosed to the lumen, where it coats the mucosal surface. In the healthy gut there is a chronic adaptive immune response, which is not inflammatory in nature. This response, in combination with the mechanisms of innate immunity, ensures that microorganisms are confined to the lumen of the gut and prevented from breaching the mucosal barrier.

Immunological memory and the secondary immune response

Having seen how immune responses are initiated in systemic and mucosal lymphoid tissues, we now consider how long-lived immunological memory to a pathogen develops. A successful primary immune response serves three purposes. It clears the infection, it temporarily strengthens defenses to prevent reinfection, and it establishes a state of long-term **immunological memory**. The last of these ensures that subsequent infections with the same pathogen will provoke a faster, stronger, **secondary immune response**. This anamnestic response or **memory response** is produced by circulating antibody and by expanded clones of long-lived B cells and T cells formed during the primary response. In a person with immunological memory, a second infection is usually cleared before it produces any symptoms. Even for pathogens that cause high mortality on first infection, for example the smallpox virus, there is little disease or mortality on secondary exposure. The purpose of vaccination is to produce a state of immunological memory against a particular pathogen before the pathogen itself is encountered. Immune responses made in all secondary lymphoid tissues—systemic and mucosal—can terminate infection and give immunological memory.

10-12 The antibodies formed during a primary immune response prevent reinfection for several months after disease

After successful termination of infection by the primary immune response, elevated levels of high-affinity pathogen-specific antibody will be present in the blood and lymph or at mucosal surfaces (Figure 10.18; see also Figure 1.26, p. 24). These levels are sustained for several months by plasma cells that replenish antibody that is used up or degraded. Many infectious diseases are seasonal, and during the winter a person can be exposed repeatedly to the same cold virus over a period of weeks or months, as it is passed around among family, friends, colleagues, and the community at large. During this period, the antibodies made against a cold caught at the start of the season prevent reinfection with the same virus. The antibody provides **protective immunity**. A second cold experienced during a winter will almost always be due to infection by a different type of cold virus.

After successful termination of an infection, high-affinity pathogen-specific antibodies will be present in the tissues or at the mucosal surfaces. If the pathogen reinvades during this period it will immediately be coated by IgA or IgG or captured by the IgE bound to mast cells and eosinophils. Viruses will be neutralized by antibody and will fail to infect cells, bacteria will be opsonized by antibody and complement and delivered for destruction to the Fc receptors and complement receptors of phagocytes, and parasites will be killed or ejected through the actions of mast cells and eosinophils activated by parasite-specific IgE. Such invading microorganisms are eliminated through the combined actions of specific antibody and innate immunity. In these circumstances, the pathogen gets little opportunity to grow and replicate, so the pathogen load never reaches a level at which lymphocytes are activated by antigen to give a secondary immune response (see Figure 10.18).

Almost all of the plasma cells made in the primary response live for only a few months. After they die, the amount of pathogen-specific antibody in the body

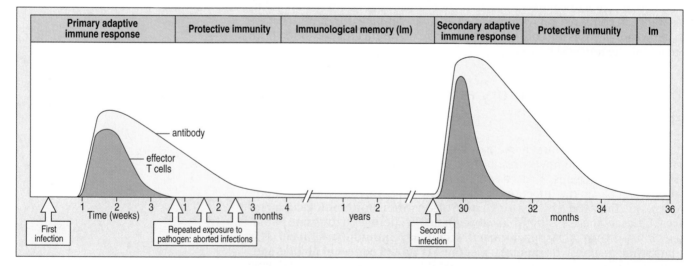

Figure 10.18 History of infection with a pathogen. Consider a student's history of infection with a pathogen. The student's first infection with the pathogen was not stopped by innate immunity and so a primary adaptive immune response developed. Production of effector T cells and antibodies terminated the infection. The effector T cells were soon inactivated, but antibody persisted, providing protective immunity that prevented reinfection despite frequent exposure to infected classmates. A year afterward, antibody levels had dropped and the pathogen would now be more likely to establish an infection. When a second infection did occur, a much faster and stronger secondary immune response was made; this eliminated the pathogen before it had a chance to disrupt tissue or cause signicant disease. This strong response was mediated by long-lived, pathogen-specific B cells and T cells that had been stockpiled during the primary immune response. The student's immune system had retained a 'memory' of that first infection.

gradually declines through degradation, and after one year it is mostly gone. In the winter after a first infection with a cold virus, this residual antibody may be insufficient to stop reinfection by that same virus. In that case the infection will provoke a secondary immune response that depends on immunological memory.

10-13 Immunological memory is sustained by clones of long-lived memory T cells and B cells

After the level of pathogen-specific antibody made during a primary immune response has declined, immune defenses are relaxed and the potential for that pathogen to reestablish an infection increases. In the case of a seasonal disease, reinfection might take place a year after the first infection. Even then, the residual level of antibody might, in combination with innate immunity, be sufficient to prevent an infection from becoming established. If that is not the case, however, the pathogen will grow and replicate and the infection will again spread to secondary lymphoid tissues. This time, however, it will stimulate a much stronger and quicker immune response—a secondary immune response.

During a primary immune response, the clonal expansion of pathogen-specific T cells and B cells gives rise both to short-lived effector cells that work to stop the infection and to long-lived **memory T cells** and **memory B cells** that will form the basis for future protection against that pathogen (Figure 10.19). In the secondary immune response, these memory cells are activated by antigen to proliferate and differentiate into effector cells. The molecular and cellular interactions that give rise to the secondary immune response are very similar to those occurring in the primary response. The differences lie in the increased speed and strength of the secondary response. Several factors contribute to this difference: memory cells are more sensitive to infection, more easily activated, and more abundant than naive lymphocytes specific for the same pathogen. Memory B cells have also undergone isotype switching and affinity maturation (see Chapter 9) and so will produce more effective antibodies than the IgM made at the beginning of the primary infection with the pathogen. As a consequence, the infection is cleared quickly by the secondary response, with few or no symptoms of disease.

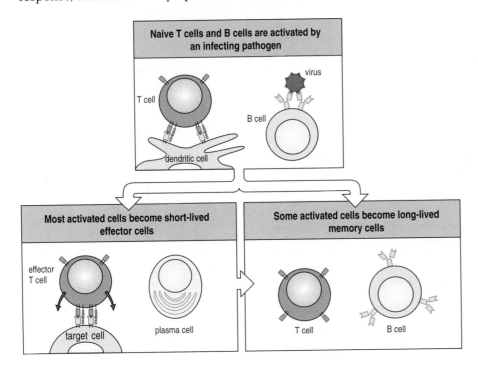

Figure 10.19 Both effector B and T cells and memory B and T cells are produced during a primary immune response.

In the course of the secondary immune response, the pathogen-activated memory B cells undergo further refinement through somatic hypermutation and affinity maturation of their immunoglobulins. Consequently, the populations of memory B cells generated during the secondary response, and available for a future, third, immune response to the pathogen, are different and improved over the memory B cells produced in the primary response. In this way, each successive infection with the pathogen further improves the defenses of adaptive immunity.

The phenomenon of **immunological memory** is classically illustrated by Peter Panum's nineteenth-century epidemiological study of the inhabitants of the Faroe Islands in the North Atlantic Ocean, who in 1781 were severely affected by a measles epidemic that infected the entire population. Over 60 years later, in 1846, when the measles virus was again introduced to the islands, almost all of the 5000 inhabitants who had been born after the original epidemic came down with the disease. Resistant to the virus were the 98 survivors of the 1781 epidemic, who retained immunological memory that prevented their second exposure to measles from becoming an established infection and causing disease.

10-14 Vaccination against a pathogen can generate immunological memory that persists for life

The goal of vaccination is to immunize people with a benign form of a pathogen and induce immunological memory, so that any infection with the real pathogen will meet with a secondary immune response that terminates the infection before it causes disease. Smallpox virus was once an effective killer of humankind: from 1850 to 1979 about 1 billion people died from smallpox. During this period, worldwide vaccination programs progressively reduced the spread of the virus to the point at which in 1972 mass vaccination was discontinued in the United States and in 1979 the smallpox virus was eradicated worldwide. Today, about half the population of the United States has been vaccinated against smallpox and half has not. Because neither group has been exposed to smallpox, comparison of the two groups reveals much about the immunological memory to vaccinia, the virus used for vaccination.

After vaccination, the level of vaccinia-specific antibody in the blood rapidly increases to a maximum level and over the next 12 months decreases to about 1% of the maximum. This steady-state level is maintained for up to 75 years and possibly for life (Figure 10.20, upper panel). As individual antibodies do not survive for such a long time, maintenance of this level means that some plasma cells or effector B lymphoblasts must remain active in the production of vaccinia-specific antibody throughout life. This is indeed what happens. After vaccination, the number of virus-specific B cells in the blood also increases rapidly to a maximum level and then declines over a 10-year period to reach a stable level of about 10% of the maximum, which is maintained for life. These cells represent a pool of memory B cells that would be used to respond to infecting smallpox virus or to a subsequent vaccination with vaccinia. Vaccination also causes the formation of pools of memory CD4 T cells

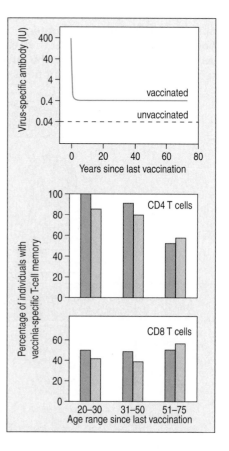

Figure 10.20 Retention of vaccina-specific antibodies and T cells after vaccination against smallpox virus. Specific anti-vaccinia antibodies continue to be made for as long as 75 years after the last exposure to vaccinia virus, the smallpox surrogate that is used for vaccination (top panel). The numbers represent international units (IU) of antibody, a standardized way of measuring an antibody response. Many vaccinated individuals retain populations of vaccinia-specific CD4 T cells and CD8 T cells (bottom panel). Only small differences are observed for individuals who received one (blue bars) or two (pink bars) vaccinations. Courtesy of Mark Slifka.

(Figure 10.20, middle panel) and CD8 T cells (Figure 10.20, lower panel) that can persist for up to 75 years. Like the memory B cells, it is these pools of memory T cells that would be used to respond to smallpox virus or a subsequent infection with vaccinia. Not all protective immunity is as persistent as that induced by the smallpox vaccine or measles virus. For example, after vaccination against diphtheria, a bacterial pathogen, the level of protective anti-diphtheria antibodies in the blood halves in about 19 years (as a comparison, the half-life of anti-measles protection is an estimated 200 years).

10-15 Pathogen-specific memory B cells are more abundant and make better antibodies than do naive B cells

During a primary infection, the proliferation and differentiation of antigen-specific naive B cells produces large numbers of antibody-secreting plasma cells to confront the ongoing infection, and a smaller number of memory B cells to deal with future infections. The first antibodies to be made in the primary response are low-affinity IgM, but as the response proceeds, somatic hypermutation, affinity maturation, and isotype switching give rise to high-affinity IgG, IgA, and IgE (see Chapter 9). Memory B cells are derived from the clones of B cells making the highest-affinity antibodies. A few weeks after a primary infection has been cleared, the number of memory B cells reaches a maximum, which is sustained for life. On a second infection, 10–100 times more pathogen-specific B cells respond than did naive B cells in the primary response. By these various mechanisms, the quantity and quality of the antibodies made in the secondary response are greatly improved compared with those in the primary response (Figure 10.21).

Memory B cells are also more sensitive than naive B cells to the presence of specific antigen, and they can also respond more quickly than naive B cells. The higher affinity of their antigen receptors makes memory B cells more efficient than naive B cells in binding and internalizing antigen for processing and presentation to helper T cells. Memory B cells also express higher levels of MHC class II and co-stimulatory molecules on their surface compared with naive B cells, which makes their cognate interactions with antigen-specific helper T cells more efficient. This has two effects. First, a smaller pathogen population is sufficient to trigger a B-cell response, which will therefore occur at an earlier stage in infection than in the primary response. Second, once activated, memory B cells take less time than activated naive B cells to differentiate into plasma cells. In the secondary response, new antibody is detectable in the blood after only 4 days, in comparison with 8 days in the primary response.

	Source of B cells	
	Unimmunized donor Primary response	Immunized donor Secondary response
Frequency of antigen-specific B cells	1 in 10^4 – 1 in 10^5	1 in 10^2 – 1 in 10^3
Isotype of antibody produced	IgM, IgG, IgA, IgE	IgG, IgA, IgE
Affinity of antibody	Low	High
Somatic hypermutation	Low	High

Figure 10.21 Comparison of the B-cell populations that participate in the primary and secondary adaptive immune responses. Key features that make the secondary response stronger than the primary response are the greater numbers of antigen-specific B cells present at the start of the secondary response and the preferential use of isotype-switched clones of B cells that express higher-affinity immunoglobulins as a result of somatic hypermutation and affinity maturation.

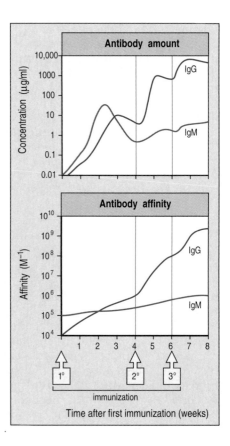

Figure 10.22 The amount and affinity of antibody increase after successive immunizations with the same antigen. This figure shows the results of an experiment on mice that mimics the development of specific antibodies when a person is given a course of three immunizations (1°, 2°, and 3°) with the same vaccine. The upper panel shows how the amounts of IgM (green) and IgG (blue) present in blood serum change over time. The lower panel shows the changes in average antibody affinity that occur. Note that the vertical axis of each graph has a logarithmic scale, because the observed changes in antibody concentration and affinity are so large.

10-16 Activation of a secondary response involves cell–cell interactions like those activating the primary response

A secondary immune response is only made if the reinfecting pathogen overcomes the defenses provided by innate immunity and the steady-state level of specific antibody. In these circumstances, the pathogen expands its numbers and is carried to the secondary lymphoid tissues by dendritic cells. Memory T cells differ from naive T cells in two ways that increase the speed of the secondary response. First, some recirculate to peripheral tissues, rather than through secondary lymphoid organs, and so can be activated directly at the site of infection by dendritic cells and macrophages presenting their specific antigen. Second, their activation requirements are less demanding than those of naive T cells because they do not require co-stimulation through CD28. Memory B cells recirculate similarly to naive B cells and, as in the primary response, the secondary B-cell response begins in a secondary lymphoid tissue at the interface between the B-cell zone and the T-cell zone. There the activation and proliferation of B cells is driven by cognate interactions with effector CD4 T cells produced by pathogen-activated memory T cells.

Memory B cells that have bound and endocytosed antigen present peptide:MHC class II complexes to their cognate helper T cells, which surround and infiltrate the germinal centers. Contact between the antigen-presenting B cells and helper T cells leads to an exchange of activating signals and the proliferation of both activated antigen-specific B cells and helper T cells. Competition for binding to antigen drives the selective activation of those B cells whose immunoglobulins have the highest affinities for antigen. As in a primary response, some of these cells develop immediately into plasma cells, whereas the others move to the follicles and participate in a germinal center reaction (see Section 9-7, p. 258). Here they enter a second round of proliferation during which they undergo somatic hypermutation and further isotype switching followed by affinity maturation. As a consequence, the average affinity of the antibodies made in the secondary response rises well above that of the antibodies made in the primary response (Figure 10.22).

10-17 Only memory B cells, and not naive B cells, participate in the secondary immune response

The secondary adaptive immune response to a pathogen involves the activation of clones of memory B cells that were products of the primary response. Activation of naive B cells specific for the pathogen is not permitted in the secondary immune response. This active suppression is mediated by immune complexes composed of the pathogen bound to antibodies made by B cells stimulated in the primary response. The complexes bind to the B-cell receptor of pathogen-specific naive B cells and also to the inhibitory Fc receptor, FcγRIIB1, which is expressed by naive B cells but not by memory B cells. This cross-linking of the B-cell receptor and the Fc receptor delivers a negative signal that inhibits activation of the pathogen-specific naive B cell (Figure 10.23).

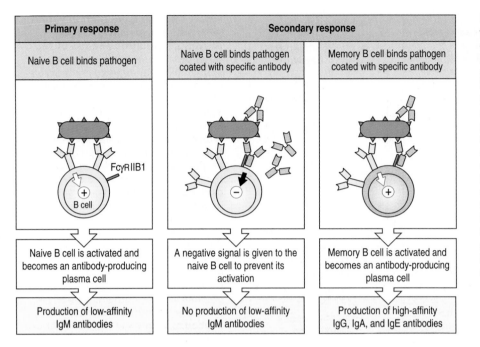

Figure 10.23 IgG antibody suppresses the activation of naive B cells by cross-linking the B-cell receptor and FcγRIIB1 on the B-cell surface. In a primary immune response, a pathogen binding to the antigen receptor of a naive B cell delivers a signal that activates the cell to become an antibody-producing plasma cell (left panel). In a secondary response, in which the antigen receptor and the Fc receptor FcγRIIB1 on a naive B cell can be cross-linked by a pathogen coated with IgG, this delivers a negative signal that prevents the activation of the cell (middle panel). In contrast, similar cross-linking of the antigen receptor and FcγRIIB1 on a memory B cell activates the cell to become an antibody-producing plasma cell (right panel).

The benefit of this negative regulation is that it prevents the production of low-affinity IgM antibodies in what would be a wasteful and interfering recapitulation of the primary response. Instead, resources are concentrated on improving the isotype-switched, affinity-matured IgG, IgA, and IgE antibodies developed during the primary response.

10-18 Immune-complex mediated inhibition of naive B cells is used to prevent hemolytic anemia of the newborn

The inhibition of naive B cells by immune complexes is put to practical use in preventing **hemolytic anemia of the newborn**, also called **hemolytic disease of the newborn**. This disease affects families in which the father is positive for the polymorphic erythrocyte antigen called **Rhesus (Rh)**, and the mother is negative. During a first pregnancy with a Rh⁺ baby, fetal erythrocytes cross the placenta and stimulate the mother's immune system to make anti-Rh antibodies. The antibodies made in this primary response cause little harm to the fetus, because they are mainly low-affinity IgM that cannot cross the placenta (Figure 10.24, left panel). During a second pregnancy with an Rh⁺ baby, however, fetal red cells again cross the placenta and induce a secondary response to Rh. This produces more abundant antibody, which is now high-affinity IgG that is transported across the placenta by FcRn (see Section 9-14, p. 267). These antibodies coat the fetal erythrocytes and cause them to be cleared from the circulation by the macrophages in the spleen. When born, such babies have severe anemia, which can have many other side-effects (Figure 10.24, center panel). Hemolytic anemia of the newborn is more common in caucasians, where 16% of mothers are Rh⁻ and 84% of babies are Rh⁺, compared with Africans or Asians, where less than 1% of the population is Rh⁻.

To prevent hemolytic anemia of the newborn, pregnant Rh⁻ women who have yet to make anti-Rh antibodies are infused with purified human anti-Rh IgG antibody, also called RhoGAM, during the 28th week of pregnancy. The amount of antibody infused is sufficient to coat all the fetal red cells that cross the placenta to enter the maternal circulation. Because all Rh antigen in the maternal circulation is in the form of complexes with human IgG, activation of the mother's Rh-specific naive B cells is prevented (Figure 10.24, right

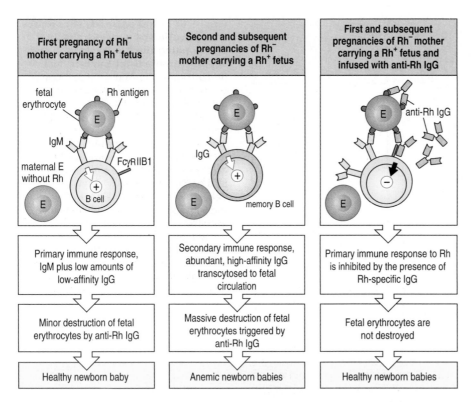

Figure 10.24 Passive immunization with anti-Rhesus antigen IgG prevents hemolytic anemia of the newborn. The Rhesus blood group antigen (Rh) expressed by erythrocytes is absent in a minority of individuals. Rh⁻ mothers carrying Rh⁺ fetuses become exposed to fetal erythrocytes and make Rh-specific antibodies that pass to the fetal circulation and cause fetal red cells to be destroyed. In a first pregnancy of this type the antibodies produced in the primary response cause only minor damage to the fetal red cells, and a healthy baby is born (left panel). In a second pregnancy, a secondary immune response ensues that produces antibodies causing massive destruction of fetal red cells, and at birth the baby is anemic (middle panel). This disease can be prevented if in the first and subsequent pregnancies the mother is passively infused with purified human anti-Rh antibodies before she has made her own response. The immune complexes of fetal erythrocytes coated with IgG prevent a primary B-cell response from being made to the Rh antigen (right panel).

panel). In effect, the mother's immune system is tricked into responding to this primary exposure to Rh antigen as though it were a secondary exposure. Within 3 days after the baby's birth, the mother is given a second infusion of anti-Rh IgG antibody, because during the trauma of birth she will have been further exposed to the baby's blood cells. This protects any future pregnancy from hemolytic anemia of the newborn.

Although the amount of infused anti-Rh antibody (300 micrograms) is in excess of that needed to coat the fetal red cells in the maternal circulation, almost none of it will be transported across the placenta and into the fetal circulation, where it could cause damage to fetal erythrocytes. This is because the anti-Rh is heavily diluted by the roughly 60 grams of IgG that is present in the mother's circulation and is not specific for the Rh antigen.

10-19 In the response to influenza virus, immunological memory is gradually eroded

Suppression of naive B-cell activation during the secondary response to a pathogen is a good strategy when dealing with invariant pathogens such as the measles virus, which does not change its antigens, but has drawbacks when confronting highly mutable pathogens such as influenza virus. Every year new influenza strains emerge that escape the protective immunity of some segment of the human population. In these variant strains, one or more of the epitopes targeted by the preexisting antibodies has been lost. Having successfully terminated a first infection with influenza, a person will have high-affinity antibodies against multiple epitopes of the viral capsid proteins. Together these antibodies neutralize the virus. During second and subsequent infections, the memory response limits the antibodies made to those directed at epitopes shared by the infecting strain and the original strain. With each passing year, the person will be exposed to influenza viruses that have fewer and fewer epitopes to which their immunological memory can respond. This allows the virus to gradually escape the host's protective immunity and to cause increasingly severe disease, while the host's immune system is

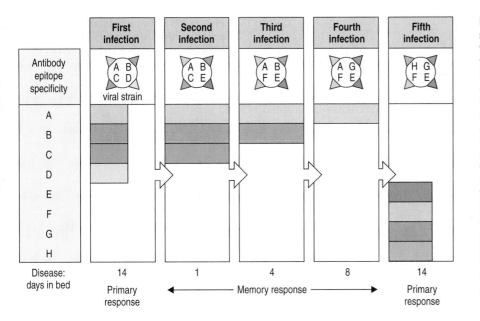

Figure 10.25 **Highly mutable viruses such as influenza gradually escape from immunological memory without stimulating a compensatory immune response.** A person's history of infection with influenza is shown here. The first infection is with a strain of influenza virus that elicits a primary antibody response to viral epitopes A, B, C, and D. The next four infections are with viruses that successively lose the epitopes of the first virus and gain in turn the epitopes E, F, G, and H. With each infection the strength of the person's memory response declines but cannot be compensated for by a new primary response until all the epitopes of the original strain are lost at the fifth infection. Then, no memory response is elicited and a primary immune response against all the new epitopes is made.

prevented from activating the many naive B cells that are specific for the changes in the virus. The imprint made by the original strain is broken only on infection with a strain of influenza that lacks all the B-cell epitopes of the original (Figure 10.25). Such a strain will produce a full-blown case of influenza and stimulate a primary B-cell response targeted against the full complement of new influenza epitopes. This phenomenon, whereby the first influenza strain to infect a person constrains the future response to other strains, has been described as **original antigenic sin**.

10-20 Several cell-surface markers distinguish memory T cells from naive T cells

Memory T cells have been harder to define and study than memory B cells, because T cells do not undergo isotype switching and somatic hypermutation, processes that indelibly mark memory B cells as different from the effector cells of the primary response. The challenge for immunologists has been to distinguish memory T cells from effector T cells.

During the immune response to a virus, for example cytomegalovirus, the number of virus-specific T cells increases dramatically with the production of effector cells. Subsequently, the number of virus-specific T cells declines to a plateau 100–1000-fold greater than the number of virus-specific T cells that were present at the very beginning of the response (Figure 10.26). These persisting cells are considered to be memory T cells. They are long-lived cells with characteristic cell-surface proteins, responsiveness to stimuli, and expression of genes that enable cell survival. Memory T cells have much in common with effector T cells, but there are differences, as shown in Figure 10.27.

Figure 10.26 **Generation of memory T cells during the response to a virus infection.** Cytomegalovirus (CMV) is a latent virus that usually remains quiescent but goes through periods of activation that are then quelled by the immune response. Such a period of reactivation is illustrated here in a patient who has undergone a hematopoietic stem-cell transplant. The increase in viral load that occurs when the virus is reactivated (lower panel) triggers a rapid increase in the numbers of virus-specific effector T cells present in the blood (upper panel). This falls back once the virus has been brought under control, leaving a sustained lower level of long-lived, virus-specific memory T cells. Data courtesy of G. Aubert.

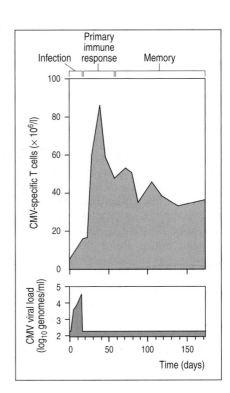

Protein	Naive	Effector	Memory	Comments
CD44	+	+++	+++	Cell-adhesion molecule
CD45RO	+	+++	+++	Modulates T-cell receptor signaling
CD45RA	+++	+	–	Modulates T-cell receptor signaling
CD62L	+++	–	Some +++	Receptor for homing to lymph node
CCR7	+++	+/–	Some +++	Chemokine receptor for homing to lymph node
CD69	–	+++	–	Early activation antigen
Bcl-2	++	+/–	+++	Promotes cell survival
Interferon-γ	–	+++	+++	Effector cytokine; mRNA present and protein made on activation
Granzyme B	–	+++	+/–	Effector molecule in cell killing
FasL	–	+++	+	Effector molecule in cell killing
CD122	+/–	++	++	Part of receptor for IL-15 and IL-2
CD25	–	++	–	Part of receptor for IL-2
CD127	++	–	+++	Part of receptor for IL-7
Ly6C	+	+++	+++	GPI-linked protein
CXCR4	+	+	++	Receptor for chemokine CXCL12; controls tissue migration
CCR5	+/–	++	Some +++	Receptor for chemokines CCL3 and CCL4; tissue migration

Figure 10.27 Comparison of proteins that are differentially expressed by naive T cells, effector T cells, and memory T cells. GPI, glycosylphosphatidylinositol.

Important changes in three cell-surface proteins in particular—L-selectin, CD44, and CD45—occur during the formation of memory T cells. Decreased expression of L-selectin and increased expression of CD44 change the pattern of T-cell homing so that some memory T cells can enter peripheral tissues rather than secondary lymphoid tissues. Isoforms of the tyrosine phosphatase CD45 with different extracellular domains are produced by alternative patterns of mRNA splicing. Naive T cells express predominantly the CD45RA isoform, which is associated with weaker signals in response to specific antigen; memory T cells express predominantly the CD45RO isoform, which has a smaller extracellular domain and is associated with stronger signals in response to antigen (Figure 10.28).

A healthy adult human has 10^{12} peripheral α:β T cells, of which half are naive T cells and half are memory T cells. From sequence analysis of T-cell receptors it is estimated that the naive T cells have 2.5×10^7 antigen specificities and the memory T cells have only 1.5×10^5 antigen specificities. The acquisition of T-cell memory for a pathogen means that, on average, 100-fold more T cells will respond to a secondary infection with a pathogen than responded to the primary infection.

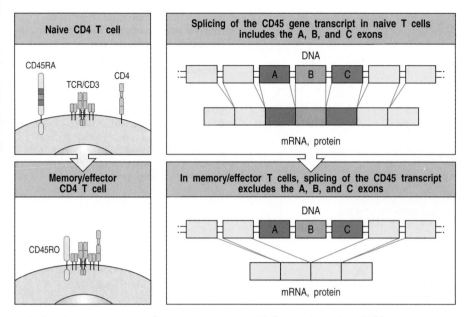

Figure 10.28 **Memory CD4 T cells express an altered CD45 isoform that works more effectively with the T-cell receptor and co-receptors.** CD45 is a transmembrane tyrosine phosphatase involved in T-cell activation and, by differential mRNA splicing, can be made in two isoforms, CD45RA and CD45RO, the former having a larger extracellular domain than the latter. Naive CD4 T cells express only CD45RA, effector T cells express predominantly CD45RO, and memory T cells express CD45RO (see Figure 10.27). The absence of the sequences encoded by exons A, B, and C in CD45RO enables it to associate with both the T-cell receptor and the CD4 co-receptor and improve the efficiency of signal transduction.

10-21 Two types of memory T cell function in different tissues

Two subsets of memory T cells with different activation requirements have been distinguished (Figure 10.29). One subset comprises the so-called **effector memory cells**. On encountering specific antigen, an effector memory cell quickly differentiates into a potent effector T cell secreting interferon (IFN)-γ, IL-4, and IL-5. With further differentiation, these effector T cells can show a bias toward either a T_H1 or T_H2 cytokine response. Effector memory cells are specialized for entering inflamed tissues and functioning within them, as shown by their expression of receptors for inflammatory chemokines and their high levels of integrins. In contrast, the second subset, the **central memory cells**, are more specialized toward entering the T-cell zones of secondary lymphoid tissues, as seen by their expression of the chemokine receptor CCR7. On encountering specific antigen, central memory cells respond by rapidly expressing CD40 ligand (with which they can interact with the CD40 on B cells; see Section 9-4, p. 255), but they take longer than effector memory cells to mature into functioning effector T cells. Within the population of CCR7-expressing central memory cells is a subset of cells expressing CXCR5, the receptor for the chemokine CXCL13 made in B-cell follicles. These cells, called **follicular helper cells**, produce IL-2 and provide help to B cells.

10-22 Maintenance of immunological memory is not dependent on antigen

A general property of lymphocytes is their requirement for regular stimulation if they are to survive. If such survival signals are not received, then lymphocytes die through apoptosis. During their development and recirculation, naive lymphocytes receive survival signals through their antigen receptors. This does not seem to be true of memory lymphocytes. The long-lived B-cell and T-cell memory to vaccinia, in the absence of any exposure to the virus, shows that antigen is not necessary for maintaining immunological memory (see Section 10-14). Memory is sustained by long-lived lymphocytes that were induced by the original exposure to antigen and then persist in its absence. At any one time most of the memory cells are in a quiescent state, but a small fraction of them are dividing and renewing the population. What provokes this regeneration is not fully known, but one proposal is that cytokines produced either constitutively or during immune responses against infections with unrelated pathogens provide the impetus to maintain memory. As with

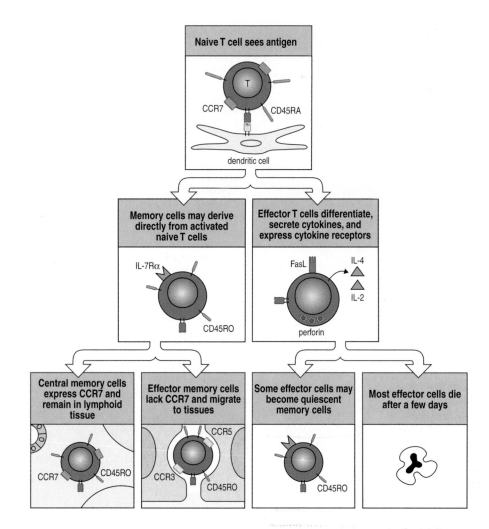

Figure 10.29 T cells differentiate into central memory and effector memory subsets distinguished by differential expression of the chemokine receptor CCR7. Quiescent memory T cells bearing the characteristic CD45RO surface protein can arise from effector T cells (right half of diagram) or directly from activated naive T cells (left half of the diagram). Two types of quiescent memory T cell can derive from the primary T-cell response: effector memory cells and central memory cells. After restimulation with antigen, effector memory cells mature rapidly into functional effector T cells that secrete large amounts of IFN-γ, IL-4, and IL-5. They do not express the receptor CCR7, but express receptors (CCR3 and CCR5) for inflammatory chemokines and home directly to sites of inflammation and infection. Central memory cells express CCR7 and remain in the secondary lymphoid tissues after restimulation, where they become effector T cells involved in helping B cells to produce antibody.

naive T cells, the survival and proliferation of memory CD4 and CD8 T cells depends on stimulation by the cytokines IL-7 and IL-15, but they divide more frequently than naive T cells. The number of memory cells for a given infectious agent is held constant, indicating that the maintenance of memory is a carefully regulated process for making the most of immunological experience.

Although the overall size of the memory CD8 T-cell population remains stable, the number of memory cells that represent memory to a particular pathogen changes with time and the history of infection. To accommodate the memory T cells acquired upon infection with a current virus, some of the memory T cells corresponding to previous infections with other viruses will be lost. Such losses can be recouped on subsequent reinfections. Although recent or recurrent infections will be better represented in the memory-cell population, this does not unduly compromise the memory response to other infections: elderly people can make effective secondary responses to pathogens that were encountered only in childhood.

Summary

In the course of the primary immune response, effector T cells and antibodies accumulate. Once an infection has been cleared, these cells and molecules will prevent reinfection by the same pathogen in the short term. In the long term, persistent clones of memory B and T cells expanded during the primary response afford protective immunity if the same antigen is encountered. When that happens they mount a secondary immune response that is both stronger and more rapid than the primary response. Antibodies produced in the secondary response are of higher affinity than in the primary response and are of isotypes other than IgM. Activation of naive lymphocytes is suppressed, and only memory lymphocytes are reactivated. In a secondary response, therefore, the immune system devotes all its resources to producing high-affinity antibodies and antigen-specific T cells that rapidly clear the invading pathogen before it can become established. The maintenance of memory cells does not seem to require the persistence of the original antigen; instead, survival signals for memory lymphocytes are provided by cytokines such as IL-7 and IL-15.

Bridging innate and adaptive immunity

In previous chapters of this book we first considered innate immunity and afterwards adaptive immunity. The logic to this arrangement is temporal: the innate immune response starts before the adaptive immune response. It also reflects natural history, because innate immune mechanisms such as complement and phagocytosis were used by invertebrates long before the evolution of vertebrates and adaptive immunity. It is, however, important to realize that the core components of adaptive immunity—antibodies, T-cell receptors, and MHC molecules—have existed for 400 million years, throughout which time the cellular and molecular mechanisms of innate and adaptive immunity have been coevolving and influencing each other. One striking example of this is that all the cells of innate immunity—macrophages, granulocytes, and NK cells—now carry receptors for the Fc regions of antibodies, the ultimate weapons of adaptive immunity (see Section 9-22, p. 278). As we shall see in this part of the chapter, there are other examples of a mix between innate and adaptive immunity that are more logically explained once the basics of both innate and adaptive immunity have been described.

10-23 γ:δ T cells contribute to the innate immune response

The γ:δ T cells of the immune system derive from the same precursor cells as α:β T cells and acquire their antigen receptors through a similar process of gene rearrangement (see Section 7-4, p. 193). Unlike α:β T cells, however, γ:δ T cells are not subjected to positive and negative selection in the thymus and their γ:δ receptors do not recognize complexes of peptides bound to MHC molecules. In human blood, fewer than 10% of the T cells are γ:δ T cells, but this population is dominated by cells having the same $V_\gamma 9$ and $V_\delta 2$ gene rearrangements (in an alternative nomenclature $V_\gamma 9$ is $V_\gamma 2$). These $V_\gamma 9$:$V_\delta 2$ T cells have some diversity in their receptors, which comes from the use of different J and D segments and from junctional diversity (see Chapter 5).

$V_\gamma 9$:$V_\delta 2$ T cells respond to a range of pathogens, including *Mycobacterium tuberculosis* and the malaria parasite *Plasmodium falciparum*. The antigens recognized by the $V_\gamma 9$:$V_\delta 2$ T cells are not peptides or proteins, however, but a variety of small phosphorylated intermediates made in the metabolic pathways of isoprenoid biosynthesis. In microorganisms, these antigenic intermediates, which are called **phosphoantigens** (Figure 10.30), are distinct from those made by the comparable metabolic pathways in human cells. Whereas α:β T cells have highly diverse receptors, each specific for a particular combination of peptide and MHC molecule, γ:δ T cells have receptors of limited diversity that each recognize a group of antigens sharing a common chemical feature. This mode of recognition is like that of the Toll-like receptors and other receptors of innate immunity.

More than 80% of the circulating $V_\gamma 9$:$V_\delta 2$ T cells lack the chemokine receptor CCR7 and so cannot enter secondary lymphoid tissues in the way that naive α:β T cells do. Instead, these γ:δ T cells express CCR5 and receptors for

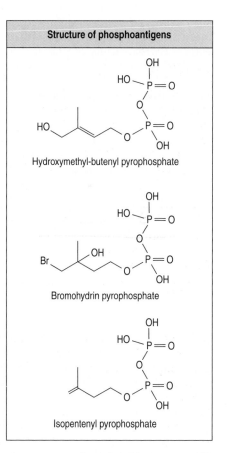

Figure 10.30 Phosphoantigens recognized by γ:δ T cells. The chemical structures of three phosphoantigens recognized by the $V_\gamma 9$:$V_\delta 2$ receptors of γ:δ T cells are shown here. All these antigens contain a pyrophosphate group. Two of these antigens are derived from pathogenic microbes, whereas bromohydrin pyrophosphate is a synthetic antigen being explored clinically as a stimulator of innate immunity.

inflammatory cytokines, which allows them to enter infected tissues and respond to infection. In this regard, γ:δ T cells are like the neutrophils, monocytes, immature dendritic cells, and NK cells of innate immunity. At the site of infection, $V_\gamma 9:V_\delta 2$ T cells activated directly by pathogen components or by pathogen-infected dendritic cells contribute to the inflammatory response by secreting the cytokines IFN-γ and TNF-α. The γ:δ T cells also kill infected cells and secrete granulysin, an antimicrobial peptide. As they mature further, γ:δ T cells acquire the characteristics of professional antigen-presenting cells: expression of MHC class II molecules, B7 co-stimulatory molecules and CD40 ligand is induced, and expression of MHC class I molecules increases. γ:δ T cells can also stimulate the maturation of dendritic cells through the secretion of TNF-α and the engagement of CD40 on the dendritic cell.

Activated γ:δ T cells can lose expression of the receptors for inflammatory chemokines and gain expression of CCR7, which then allows them to enter secondary lymphoid tissues and present antigens to α:β T cells. In the secondary response to a pathogen, γ:δ T cells at the site of infection can also stimulate pathogen-specific memory T cells that have entered the inflamed tissue.

A general feature of γ:δ T cells is that the populations present in different tissues express distinct and limited receptor repertoires. During fetal development and early life, γ:δ T cells with receptors composed of a $V_\delta 1$ chain in association with a variety of γ chains predominate in the blood. In adults, these cells are rare in the blood but common in the gut mucosa and in the spleen.

The ligands recognized by the $V_\gamma:V_\delta 1$ T-cell receptors are the stress-induced MIC proteins, which resemble MHC class I heavy chains and are ligands for the NKG2D lectin-like receptor. Injury or infection stress the enterocytes lining the gut and stimulate the expression of MIC-A and MIC-B (see Section 2-22, p. 66). NKG2D is expressed by many $V_\gamma:V_\delta 1$ T cells, and the signals received through both it and the $V_\gamma:V_\delta 1$ antigen receptor synergize in activating these T cells (Figure 10.31). NK cells and a subset of CD8-expressing α:β T cells also carry NKG2D receptors and recognize MICs, and all three cell types have the capacity to kill damaged or infected epithelial cells, thereby promoting repair of the injured mucosa.

In addition to their role in the inflammatory response to infection, γ:δ T cells are also implicated in the repair of tissue through the secretion of fibroblast growth factor, once the inflammatory response has subsided.

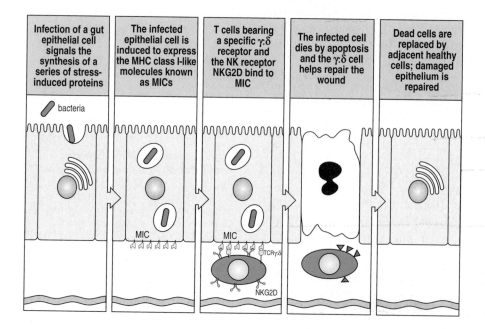

| Infection of a gut epithelial cell signals the synthesis of a series of stress-induced proteins | The infected epithelial cell is induced to express the MHC class I-like molecules known as MICs | T cells bearing a specific γ:δ receptor and the NK receptor NKG2D bind to MIC | The infected cell dies by apoptosis and the γ:δ cell helps repair the wound | Dead cells are replaced by adjacent healthy cells; damaged epithelium is repaired |

Figure 10.31 T cells with $V_\gamma:V_\delta 1$ and NKG2D receptors are resident in the intestinal mucosa, where they respond to infection. Infection or other injury to intestinal epithelial cells stimulates a stress response that induces cell-surface expression of MHC class I-like molecules called MICs (MIC-A and MIC-B). In the mucosa are γ:δ T cells (blue ovals) bearing a $V_\gamma:V_\delta 1$ T-cell receptor that is specific for MICs. These cells also express the receptor NKG2D, which also binds MICs. On NK cells, NKG2D is an activating receptor (see Section 2-22, p. 66), but on these γ:δ T cells it acts as a co-receptor for the γ:δ receptor. When an infected enterocyte expresses MICs the γ:δ T cells induce it to die by apoptosis. The dying enterocyte is removed from the epithelium, allowing the local tissue injury to be repaired. The γ:δ cells also contribute to repairing the wound through the secretion of fibroblast growth factor.

The properties and functions of γ:δ T cells parallel those of the B-1 subset of B cells (see Section 6-10, p. 171), and both cell types contribute to the innate immune response. They both have a distinctive cell-surface phenotype and express a restricted repertoire of antigen receptors, so that quite large numbers of cells will carry receptors of the same specificity; in other words, each lymphocyte clone is much larger than is usual for α:β T cells or conventional (B-2) B cells. This means that γ:δ T cells and B-1 cells can respond to a primary infection without first undergoing a lengthy period of clonal expansion and differentiation.

10-24 Individual NK cells express many different combinations of receptors belonging to one of two receptor families

NK cells are lymphocytes that function in the innate immune response, being particularly important in the defense against viral infections (see Section 2-21, p. 65). NK cells do not rearrange either their immunoglobulin genes or their T-cell receptor genes, which distinguishes them from the other types of lymphocyte. Apart from the T-cell receptor and its associated CD3 complex, almost all of the cell-surface proteins on T cells are also expressed by some NK cells, and almost all of the NK-cell receptors are expressed by some T cells, as exemplified by the presence of intestinal NKG2D-positive γ:δ T cells (see Figure 10.31). NK cells and cytotoxic T cells also have similar effector functions—killing infected cells and secreting cytokines—and the main difference is their participation in innate and adaptive immunity, respectively.

For practical purposes, the best definition of a human NK cell is a mononuclear cell that lacks CD3 and expresses the cell-surface protein CD56. CD56 is an adhesion molecule that is expressed principally in the brain, but among blood cells it is expressed only by NK cells. A further distinction is made between NK cells having a low or high level of CD56 on their surface. CD56^{low} NK cells predominate in the blood and are strongly cytotoxic. NK cells are also present in secondary lymphoid tissues; they are predominantly CD56^{high}, poorly cytotoxic, but strong secretors of cytokines. Current evidence suggests that the CD56^{high} NK cell is less mature than the CD56^{low} NK cell and that the development of NK cells, which starts in the bone marrow, is completed in secondary lymphoid tissues.

NK cells circulate in the blood as mature or partly activated cells that can rapidly enter inflamed tissue to perform their functions. To do this, an NK cell expresses a variety of activating and inhibitory receptors that in concert allow the cell to distinguish healthy cells from infected cells, and to spare the former while attacking the latter (see Figure 2.49). More than 30 different NK-cell receptors have been described. The receptor NKG2D is expressed by all NK cells, but most NK-cell receptors are expressed by subpopulations of NK cells. Consequently, individual NK cells express different combinations of receptors, imparting heterogeneity to a person's NK-cell population and providing a repertoire of responses to pathogens (Figure 10.32). Among other cell types involved in the innate immune response, such as macrophages, dendritic cells, and neutrophils, individual cells also express different combinations of receptors.

Many of the NK-cell receptors are either members of the immunoglobulin superfamily or are lectin-like receptors having extracelluar domains that resemble the carbohydrate-binding site of mannose-binding lectin (see Figure 2.48, p. 66). The families of genes encoding the lectin-like NK-cell receptors cluster together in a region of chromosome 12 called the natural killer complex (NKC; Figure 10.33). Similarly, genes for many of the NK-cell receptors in the immunoglobulin superfamily are found in a region of chromosome 14 called the leukocyte receptor complex (LRC). Both of these gene

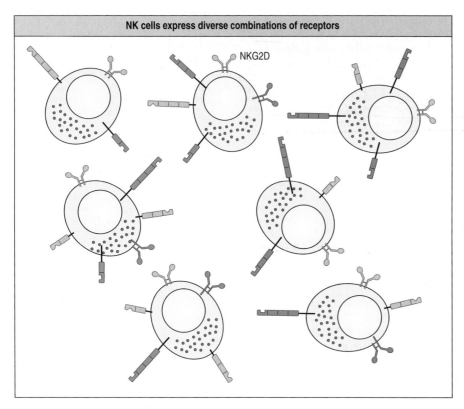

NK cells express diverse combinations of receptors

NKG2D

Figure 10.32 NK cells express different combinations of receptors. Receptors colored green are inhibitory; those colored gold are activating. Most NK-cell receptors are either structurally similar to C-type lectins, like NKG2D, or are made up of immunoglobulin-like domains (see Figure 2.48, p. 66).

complexes are filled with genes encoding receptors that contribute to innate immunity. Although we have so far considered MHC class I molecules primarily with regard to their presentation of peptide antigens to T cells and their association with adaptive immunity, they also provide ligands for various receptors encoded in the NKC and LRC regions, as we shall see next. The ability to recognize MHC class I molecules is yet another similarity between the NK cell and the cytotoxic T cell.

10-25 NK cells use receptors for MHC class I molecules to identify infected cells

We learned earlier that NKG2D binds the MIC proteins expressed by infected intestinal epithelium (see Figure 10.31). MIC proteins are structurally similar to MHC class I heavy chains, so it is not surprising that several other NK-cell receptors recognize the more conventional MHC class I molecules. Whereas NKG2D is an activating receptor expressed on all NK cells, many other MHC class I receptors have inhibitory functions and are expressed in different combinations on subpopulations of NK cells. Combinations of activating and inhibitory NK-cell receptors work together to detect and respond to infection.

Figure 10.33 Many NK-cell receptors are encoded by genes in one of two genetic complexes. The natural killer complex (NKC) on human chromosome 12 encodes lectin-like NK-cell receptors, whereas the leukocyte receptor complex (LRC) on chromosome 19 encodes NK-cell receptors that are members of the immunoglobulin superfamily. The LRC comprises the killer-cell immunoglobulin-like receptor (KIR), the leukocyte immunoglobulin-like receptor (LILR), and the leukocyte-associated immunoglobulin-like receptor (LAIR) gene families. Flanking the LRC are genes encoding other immunoglobulin-like molecules of the immune system, including the NKp46 receptor of NK cells, and the DAP10 and DAP12 signaling adaptor molecules that transduce signals from activating NK-cell receptors. Figure based on data courtesy of John Trowsdale.

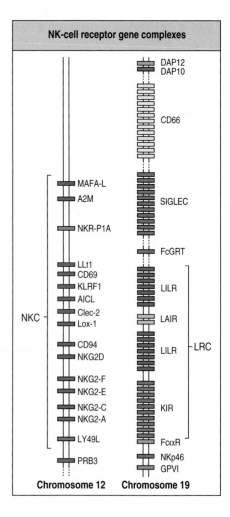

NK-cell receptor gene complexes

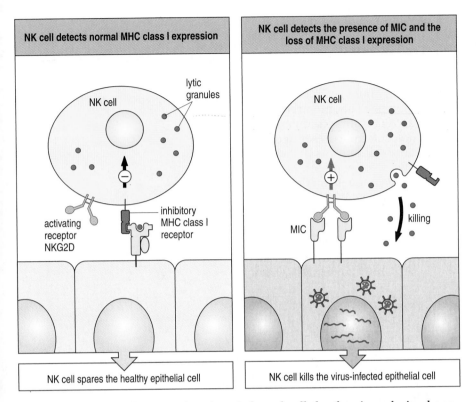

Figure 10.34 NK cells can detect the presence of infection by using the combination of an inhibitory receptor specific for MHC class I peptides and an activating receptor specific for MIC. NK cells survey the intestinal epithelium for infection. Healthy cells have a normal level of MHC class I molecules and do not express the MIC stress protein that binds NKG2D (left panel). On contacting a healthy epithelial cell the inhibitory MHC class I receptor delivers an inhibitory signal to the NK cell, which then leaves the epithelial cell untouched. In an epithelial cell infected with a virus the expression of MIC is induced, and expression of MHC class I molecules is lost. When the NK cell contacts the virus-infected cell, the binding of NKG2D to MICs on the infected cell's surface delivers an activating signal to the NK cell, which responds by killing the virus-infected cell (right panel).

Cytotoxic CD8 T cells recognize virus-infected cells by the virus-derived peptides bound to self MHC class I molecules on the cell surface (see Section 8-14, p. 235). Such peptide:MHC complexes engage the T-cell receptors and activate the T cells. NK cells also detect infection by monitoring MHC class I molecules, but they do it with less precision than CD8 T cells. NK cells respond to abnormally reduced expression of MHC class I molecules, a common characteristic of pathogen-infected cells. The NK cells examine the level of MHC class I molecules on cells in infected tissues by using their inhibitory receptors for these molecules. The inhibitory receptors work in conjunction with activating receptors such as NKG2D to determine whether the killing function of the NK cell is activated.

We shall illustrate this principle by considering the NK-cell response to a viral infection in intestinal epithelium. Healthy epithelial cells have normal levels of MHC class I molecules and very little MIC. When an NK cell interacts with a healthy cell, the inhibitory receptors are fully engaged by MHC class I molecules and deliver a strong negative signal, whereas NKG2D is hardly engaged at all by MIC and delivers only a weak activating signal. This combination of strong inhibitory and weak activating signals prevents the NK cell from killing the healthy cells. In infected epithelial cells, in contrast, levels of MHC class I molecules are reduced and high levels of MIC have been induced. In this case, the partial engagement of the inhibitory receptor by MHC class I molecules delivers a weak negative signal, whereas full engagement of NKG2D by MIC delivers a strong activating signal. The combination of weak inhibitory and strong activating signals drives the NK cell to kill the infected cells (Figure 10.34).

10-26 NK cells have inhibitory receptors with different specificities for MHC class I molecules

A common feature of infected cells is that the expression of MHC class I molecules is perturbed. This can involve changes in the abundance of MHC class I molecules at the cell surface or changes in the repertoire of peptides bound. NK cells use a variety of inhibitory receptors to detect alterations in the expression of MHC class I molecules. Some of these receptors monitor the overall level of MHC class I molecules, whereas others are specific for epitopes that are carried by particular subsets of MHC class I isoforms.

CD94:NKG2A is an inhibitory receptor that monitors the overall level of the MHC class I molecules HLA-A, HLA-B, and HLA-C on human cells and is not specific for any given HLA class I isotype or allotype. This monitoring is accomplished in an indirect fashion by CD94:NKG2A detecting complexes of peptide bound to the non-polymorphic class I molecule HLA-E (see Section 5-17, p. 146). HLA-E has the same ubiquitous tissue distribution as HLA-A, -B, and -C, but unlike them it has a restricted specificity for peptides, binding exclusively to peptides produced by the intracellular degradation of the leader sequences of the HLA-A, -B, and -C heavy chains. These nonamer peptides correspond to a relatively conserved sequence in the highly hydrophobic leader sequences. Because an HLA-E molecule needs to bind peptide to assemble properly and reach the cell surface, the amount of HLA-E on a cell's surface is a measure of the amount of HLA-A, -B, and -C being made by the cell (Figure 10.35). Interaction of CD94:NKG2A with peptide:HLA-E complexes

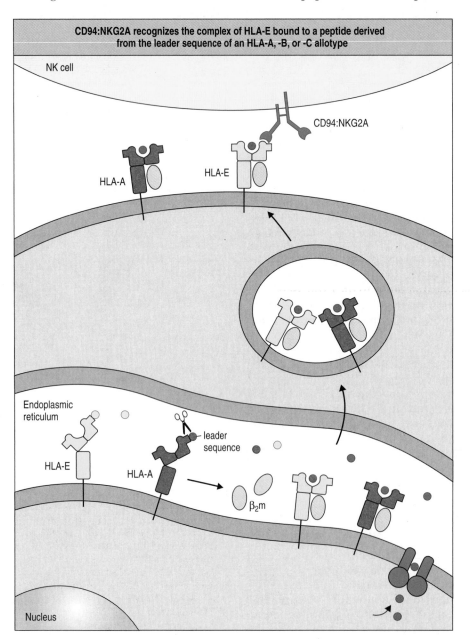

CD94:NKG2A recognizes the complex of HLA-E bound to a peptide derived from the leader sequence of an HLA-A, -B, or -C allotype

NK cell

CD94:NKG2A

HLA-A

HLA-E

Endoplasmic reticulum

leader sequence

HLA-E

HLA-A

β₂m

Nucleus

Figure 10.35 The lectin-like receptor CD94:NKG2A on NK cells binds complexes of HLA-E and peptides derived from the leader sequences of HLA-A, HLA-B, or HLA-C molecules. Leader peptides on polypeptides made on ribosomes in the cytosol allow the polypeptides to be translocated into the endoplasmic reticulum before they are folded into their proper conformation. During this process the leader peptides are cleaved off. The diagram shows how the leader peptide cleaved from an HLA-A heavy chain is bound by HLA-E and brought to the cell surface to interact with the inhibitory receptor CD94:NKG2A receptor on NK cells. β_2m, β_2-microglobulin.

on the cell surface thus enables the NK cell to assess the MHC class I status of a cell. In the extreme situation in which a cell makes no HLA-A, -B, or -C, the HLA-E heavy chain cannot assemble with peptide and β2-microglobulin and is retained within the endoplasmic reticulum.

Whereas CD94:NKG2A has a broad specificity for HLA class I and is sensitive to changes in expression of all HLA-A, -B, and -C allotypes, the family of **killer-cell immunoglobulin-like receptors** (**KIRs**) includes inhibitory receptors with narrower specificities. These receptors have either two or three immunoglobulin domains, and bind to polymorphic determinants on HLA-A, -B, or -C molecules. Individually, these KIRs are more able than CD94:NKG2A to detect the loss of expression of one HLA class I locus or allotype that has been targeted by a pathogen. Crystallographic structures show that KIRs, like the T-cell receptor, interact with the face of the MHC class I molecule formed by the two α helices and the bound peptide (Figure 10.36). The difference is that a KIR covers less of the face than the T-cell receptor, just that part containing the carboxy-terminal parts of the bound peptide and the α1 helix, and the amino-terminal part of the α2 helix. Polymorphisms at residues 77–83 in the HLA class I α1 helix, and particularly at position 80, determine whether a given HLA class I isoform can bind to a KIR and which ones it will bind to. Whereas all HLA-C allotypes are KIR ligands, only a minority of HLA-A and HLA-B allotypes are. This hierarchy indicates that HLA-C is more specialized toward the control of the NK-cell response, whereas HLA-A and HLA-B are more specialized at presenting peptide antigens to T cells. The specificities of the NK-cell receptors for MHC class I molecules are summarized in Figure 10.37.

10-27 Inhibitory receptors for self MHC class I make NK cells tolerant to self and responsive to loss of MHC class I

NK cells start to express inhibitory receptors for MHC class I molecules at late stages in their development. The CD94:NKG2A receptor is the first to be expressed, followed by the KIRs. Transcription of KIR genes is such that each NK cell expresses a random subset of the KIR genes, which creates enormous diversity of KIR expression in the NK-cell population (Figure 10.38). The expression pattern of KIRs is quite random and is independent of whether or not a person has a self MHC class I molecule that binds to the expressed KIRs. Consequently, most people have NK cells expressing KIRs that do not interact with their self MHC class I molecules. To perform their function, however,

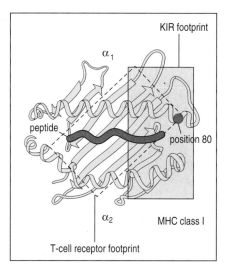

Figure 10.36 Killer-cell immunoglobulin-like receptors (KIRs) bind to the same face of the MHC class I molecule as the T-cell receptor. The ribbon diagram shows the structure of the α1 and α2 domains of an MHC class I molecule. The tops of the two α helices and of the bound peptide in the groove between the helices form the face that interacts with T-cell receptors and KIRs. The rectangle outlined with a dashed line shows the part of the MHC class I face that interacts with T-cell receptors; the rectangle outlined with a solid line shows the part that interacts with KIR.

Inhibitory NK-cell receptors for MHC class I	
Receptor	Ligand
CD94:NKG2A	Complexes of HLA-E with peptides derived from the leader peptides of HLA-A, -B, and -C
KIR2DL1	HLA-C allotypes having lysine at position 80
KIR2DL2/3	HLA-C allotypes having asparagine at position 80
KIR3DL1	HLA-A and HLA-B allotypes having the Bw4 serological epitope determined by sequence motifs at positions 77–83
KIR3DL2	Complexes of HLA-A3 and HLA-A11 with a specific peptide from Epstein–Barr virus
LILRB1	All HLA class I

Figure 10.37 HLA class I specificities of NK-cell receptors. The receptors vary in the breadth of their specificity. At one end of the spectrum is LILRB1, which binds a conserved epitope of HLA class I; at the other end is KIR3DL2, which is like an α:β T-cell receptor in its narrow specificity for a combination of a viral peptide and one of two structurally similar MHC class I allotypes.

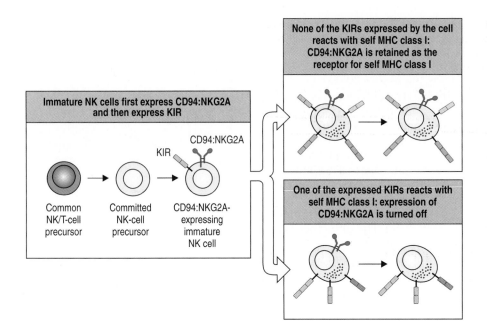

Figure 10.38 NK-cell acquisition of inhibitory receptors for self MHC class I. In NK-cell development, the inhibitory receptor CD94:NKG2A, with broad specificity for MHC class I, is the first receptor expressed. Subsequently, the family of KIR genes is expressed. NK cells express different numbers and combinations of KIRs. If none of the KIRs expressed by the NK cell reacts with self MHC class I, the NK cell retains expression of CD94:NKG2A as its only inhibitory receptor for self MHC class I (top right panel). If one or more of the KIRs expressed by the NK cell reacts with self MHC class I, then the cell usually, but not always, turns off the expression of CD94:NKG2A in favor of using KIR as the inhibitory receptor for self MHC class I.

each NK cell carries at least one type of inhibitory receptor that interacts with self MHC class I molecules. If none of the inhibitory KIRs expressed by the NK cell do so, then CD94:NKG2A can take that role. If an NK cell does have KIRs that interact with self MHC class I molecules, either they are coexpressed with CD94:NKG2A or, as happens more frequently, the expression of CD94:NKG2A is turned off.

The recognition of self MHC molecules by the inhibitory NK-cell receptors has two functions: first, it ensures that NK cells are restrained from attacking healthy cells; second, the magnitude of that restraint determines the force with which an NK cell will attack infected cells in which MHC class I expression is compromised. The interaction of CD94:NKG2A with self MHC class I molecules is similar in all individuals, whereas the interactions of KIRs with self MHC class I molecules are highly variable. For individuals with certain HLA types, the inhibitory interactions of KIRs with self MHC class I molecules are considerably greater than that achieved by CD94:NKG2A.

10-28 T cells recognizing lipid antigens protect against mycobacterial infection

Mycobacteria, including those that cause tuberculosis and leprosy, make many unusual lipids and glycolipids not made by human cells. In the response to mycobacterial infection, these unusual lipids serve as target antigens for effector CD4 T cells and CD8 T cells that express a diverse range of $\alpha{:}\beta$ receptors. What these receptors have in common is the recognition of lipid antigens presented by CD1a, Cd1b, or Cd1c, a group of β_2-microglobulin-associated MHC class I-like glycoproteins that are neither polymorphic nor encoded within the MHC. The CD1 gene family is on chromosome 1. The antigen-binding site formed by the α_1 and α_2 domains of CD1 is not a groove, as in HLA-A, -B, and -C molecules, but two or more hydrophobic channels running under the top surface of the protein. The long hydrophobic alkyl chains of glycolipid antigens, such as the glucose monomycolate presented by CD1b (Figure 10.39), fit into these channels, leaving the lipid's hydrophilic head-group to protrude and make contact with a T-cell receptor. CD1 isoforms differ in the number and length of the channels, which enables them each to present different groups of lipid antigens. Bound lipid is necessary for the integrity of the CD1 structure, and in the absence of pathogen-derived lipids the CD1 molecules assemble with self lipids.

Figure 10.39 The chemical structure of glucose monomycolate, a glycolipid antigen from the bacterium *Mycobacterium phlei*. This glycolipid antigen is part of the cell wall of *M. phlei*. It is presented to α:β T cells by CD1b. Its physical and chemical properties are very different from those of the peptide antigens presented by MHC class I molecules. Because much of the glycolipid antigen is not seen by the T-cell receptor but is hidden in hydrophobic channels in the MHC class I molecule, T cells stimulated by one glycolipid antigen will cross-react with other structurally related lipids that have similar hydrophilic headgroups.

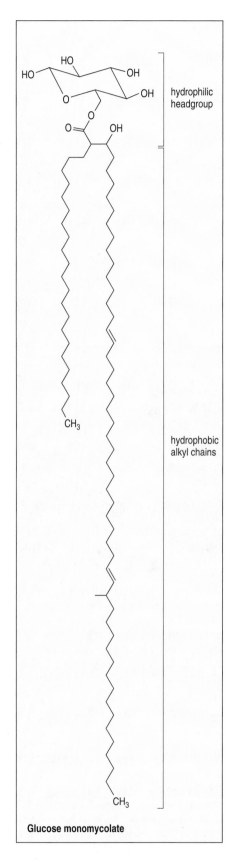

Glucose monomycolate

The expression of CD1a, CD1b, and CD1c is restricted to cell types that are infected by mycobacteria, such as dendritic cells and activated monocytes, and to cortical thymocytes. The latter expression allows T cells with receptors that bind CD1 too strongly to be eliminated in the thymus by negative selection. As ingested mycobacteria take up residence in intracellular vesicles, mycobacterial lipids will typically be present in endosomes. CD1 molecules can bind to lipids in both endoplasmic reticulum and endocytic vesicles. The CD1 molecules first assemble in the endoplasmic reticulum, usually in combination with a self lipid, and then travel to the cell surface. Cell-surface CD1 can be endocytosed in clathrin-coated vesicles and taken to endosomes and the MHC class II compartment, where the bound self lipid can be exchanged for a pathogen-derived lipid before the CD1 is returned to the cell surface (Figure 10.40). The α:β T-cell response to lipids presented by CD1 develops powerful effector T cells that secrete inflammatory cytokines and kill infected cells, and memory cells that provide long-lasting immunity.

10-29 NKT cells are cells of innate immunity that express α:β T-cell receptors

CD1d, a fourth isoform of CD1, is functionally different from the other isoforms. Like them, CD1d is specialized in the presentation of lipid antigens, but unlike them it is expressed by epithelial cells in many tissues, including the intestine, liver, pancreas, uterus, thymus, and tonsils. CD1d also presents lipid antigens from a wide range of pathogens. CD1d does not present antigens to conventional α:β T cells but to a subset of cells called **NKT cells** that express a variety of NK-cell receptors as well as an α:β T-cell receptor. Dominating the NKT-cell population are cells expressing receptors that carry α chains made from the same $V_\alpha 24$–$J_\alpha 8$ arrangement, associated with a variety of β chains. Like γ:δ cells, NKT cells can respond rapidly to infection, do not generate a memory response, are restricted in receptor repertoire, and participate in the innate immune response.

NKT cells can enter sites of infection, where their interactions with pathogen-loaded dendritic cells lead to their mutual activation. Increased secretion of IFN-γ and upregulation of CD40 ligand by the NKT cells induce an increased expression of IL-12 by the dendritic cells, which further activates the NKT cells. This helps initiate the inflammatory response and the recruitment of neutrophils, NK cells, and monocytes to the site of infection.

Summary

The γ:δ T cells have a complementary, and very different, role to the α:β T cells in the immune response. The TCRγ and TCRδ loci have far fewer V gene segments than the TCRα and TCRβ loci, and gene rearrangement is used not to create a vast diversity of receptors but to produce a relatively limited range of receptors against broad groups of microbial antigens. These receptors are expressed by large numbers of cells that can be activated quickly in the primary response to infection. The B-1 subclass of B cells has an analogous role

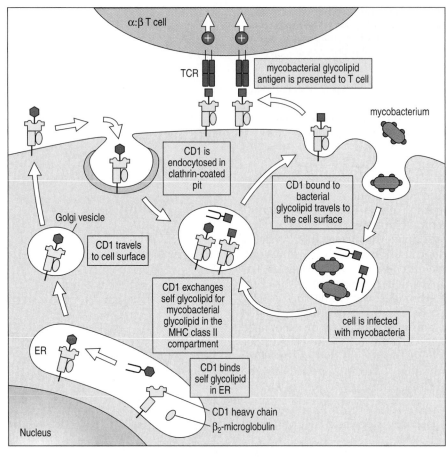

Figure 10.40 CD1 travels through both the endoplasmic reticulum and endocytic vesicles to bind glycolipid antigens. Sequence motifs in the cytoplasmic tails of CD1 molecules allow them to travel from the endoplamic reticulum (ER) through the Golgi to the plasma membrane and then to recycle from the plasma membrane in endosomal vesicles to the MHC class II compartment and back to the cell surface. Throughout this process, CD1 is bound to a lipid or glycolipid. In the example shown here of a cell infected with mycobacteria, CD1 first assembles with a self glycolipid (blue) in the endoplasmic reticulum and this is exchanged for a mycobacterial glycolipid antigen (red) in the MHC class II compartment. The complexes of mycobacterial glycolipid antigen and CD1 then move to the cell surface to engage the α:β receptors of a specific T cell, which is stimulated to kill the mycobacteria-infected cell. Within the MHC class II compartment are mechanisms for exchanging the glycolipids bound by CD1.

to that of the γ:δ T cells. NK cells are potentially cytotoxic lymphocytes that do not rearrange their immunoglobulin and T-cell receptor genes but have many similarities to CD8 T cells, including regulation of their development and effector function by MHC class I molecules. Whereas CD8 cytotoxic T cells are acutely sensitive to foreign peptides bound by MHC class I molecules, and thus recognize 'altered self,' NK cells are sensitive to the loss of MHC class I expression, and recognize 'missing self.' NK-cell functions are regulated by inhibitory receptors that recognize MHC class I molecules, which contrasts with the activating function of MHC class I recognition by T-cell receptor and co-receptor on CD8 T cells. Intracellular bacterial infections have selected for a third pathway of antigen presentation via the CD1 family of MHC class I-like molecules. These present lipid and glycolipid antigens to α:β T cells and to NKT cells. NKT cells express NK receptors and a restricted set of α:β T-cell receptors and function in the innate immune response. The lymphocytes of innate immunity—γ:δ T cells, NK cells, and NKT cells—all interact with dendritic cells at sites of inflammation and help determine if and when the innate immune response should recruit the adaptive immune response.

Summary to Chapter 10

The spleen and most lymph nodes provide the adaptive immune response to infections of the blood and the connective tissues that invest all the organs and tissues of the human body. Such infections can result from wounds, arising from accident or malevolent attack, that disrupt the protective barrier of the skin and provide the opportunity for microbes to invade. They can also result from pathogens that breach a mucosal surface but then travel to other tissues to establish their infection, such as many viruses. These infections are met with a short-lived and highly inflammatory immune response for which the goal is simply to eliminate the pathogen and repair the damaged tissues.

This form of aggressive and disruptive response is not the defense of choice for the mucosal surfaces, which constitute the human body's most extensive sites of interaction with the external environment and the microbial universe. These major interfaces, such as the linings of the gut and respiratory tract, are not tough physical barriers like the skin, but delicate tissues that communicate vital materials and information between the environment and the body.

Pathogens exploit these communication functions to gain entry to the body. Adding to the complexity of defending the gut is its vast population of commensal organisms that need to be controlled but should not be destroyed. Defense of mucosal surfaces is mounted by their own dedicated secondary lymphoid tissues, which provide mucosal immunity. Although the basic mechanisms of mucosal immunity resemble those used elsewhere in the body, the strategies of mucosal immunity are very different from those that work for systemic immunity. A chronic and subtle non-inflammatory immune response is maintained by secondary lymphoid tissues embedded within the mucosae. The overall aim is not to eliminate all microorganisms encountered but to keep them in their place and to deal with those that intrude through the mucosa without delay.

Most infections are cleared efficiently by the innate immune response and lead to neither disease nor incapacitation. In the minority of infections that escape innate immunity and spread from their point of entry, the pathogen then faces the combined forces of innate and adaptive immunity. The week or so required for the primary adaptive immune response to develop is the time when the body is most vulnerable and during which infections can progress to the point of causing disease, damage, and even death. This delay, while the pathogens have the upper hand and can be transmitted to other people, is largely responsible for the devastation caused by epidemic infections. Most infections, however, never reach that stage but are successfully terminated by the adaptable and specific recognition systems of adaptive immunity. A successful recovery from infectious disease and the consequent development of immunological memory provide the body with enhanced defenses that will repel future attack from the causal agent (Figure 10.41). Once a state of

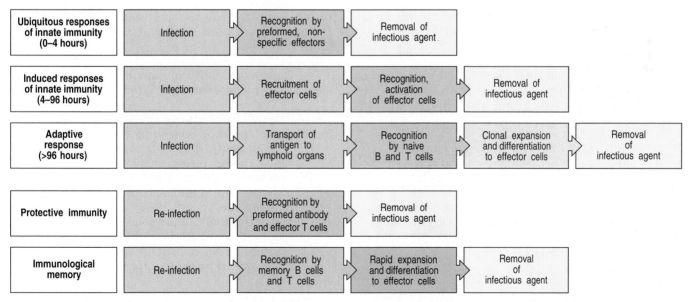

Figure 10.41 **The immune response to a pathogen.** The innate immune response can be divided into mechanisms that start immediately on infection, for example complement activation and macrophage phagocytosis, and those induced within a few hours by inflammatory cytokines. The effector functions of adaptive immunity begin to work only after 4 days

or more. In the period after the termination of an infection, the effector cells and molecules made in the primary response continue to provide protective immunity but gradually decline. In the long term, immunological memory enables the strength of the secondary response to far surpass the primary response.

immunity has been established, a subsequent encounter with the pathogen provokes a stronger and more rapid attack that clears the pathogen from the body before the infection can take hold.

Innate immunity evolved before adaptive immunity. So from their beginning the mechanisms of adaptive immunity have used and improved on those of innate immunity. As we have seen, many of the cells and molecules of innate immunity continue to work throughout the adaptive immune response. For the past 400 million years innate and adaptive immunity have been coevolving, and certain features that we associate with adaptive immunity have been incorporated into innate immunity. The γ:δ T cells, B-1 cells, and NKT cells are lymphocytes of the innate immune response that use their rearranging genes to make receptors that in their limited and cross-reactive specificities resemble the receptors of innate immunity. The activity of NK cells, the major lymphocytes of innate immunity, are controlled by invariant receptors for the highly polymorphic MHC class I molecules, components of the immune system that are usually considered part of the bedrock of adaptive immunity. The old distinctions between innate and adaptive immunity are becoming increasingly blurred.

Questions

10–1 Explain how secondary lymphoid tissues of the mucosa are (A) similar to and (B) different from secondary lymphoid tissues elsewhere in the body (the systemic immune system).

10–2 Why do children who have had their tonsils or adenoids removed respond less effectively to the oral polio vaccine than children who still have these tissues?

10–3 Describe two ways in which dendritic cells capture antigen from the intestine for presentation to T lymphocytes.

10–4 In which ways are macrophages in the intestinal mucosa (A) similar to and (B) different from macrophages in other anatomical locations?

10–5 Describe the route that a Peyer's patch-activated T lymphocyte follows, beginning with a naive T lymphocyte in a high endothelial venule and ending with an effector T lymphocyte in the lamina propria.

10–6 Identify four locations where secretory IgA can bind to antigens in mucosal tissue and for each give the fate of the antigen upon binding to secretory IgA.

10–7 What property of the mucosal immune system enables breast milk to contain antibodies against microorganisms encountered in the gut or other mucosal tissues? Explain your answer.

10–8 Explain why individuals who have the condition selective IgA deficiency do not succumb to repeated infection through mucosal surfaces.

10–9 Describe four actions of T_H2 effector cells that provide protection from infections by intestinal helminths and lead to expulsion of the parasites from the gastrointestinal tract.

10–10
A. Give at least two reasons why memory B cells respond more quickly than naive B cells to antigen.
B. Now do the same for memory T cells versus naive T cells.

10–11 Explain (A) why only memory B cells, and not naive B cells, participate in secondary immune responses to particular pathogens, and (B) why this is advantageous to the host.

10–12 Explain why suppression of naive B cells in secondary immune responses is advantageous for fighting the measles virus, but disadvantageous for fighting the influenza virus.

10–13 Explain briefly how immunological memory operates in (A) the short term and (B) the long term.

10–14 Natural killer cells (NK cells) carry activating and inhibitory receptors on their surface.
A. What property of NK cells do these receptors activate or inhibit, respectively? Explain your answer.
B. How are NK cells thought to use these receptors to recognize and eliminate virus-infected cells?
C. Why are the actions of NK cells categorized as innate immunity, and what do we know of their specificity for MHC class I molecules?
D. Why do the NK cells of the recipient of an organ transplant sometimes attack the transplanted tissue?

10–15
A. Describe the ligand for the NK-cell receptor CD94:NKG2A.
B. Why is the concentration of this ligand on the target cell an effective measure of the presence or absence of classical MHC class I molecules?

C. Why is this ligand considered a broad mechanism for the NK-cell detection of unhealthy cells that is relatively insensitive to MHC class I polymorphisms?

10–16 Fatima Ahmed, a 25-year-old recent immigrant from Sudan, is brought to an obstetrician by her husband Samir for a first-time visit at about 38 weeks of pregnancy. This is Fatima's first pregnancy, and she and the baby are in excellent health. She has had no previous antenatal care, fearing that if her condition had been revealed she would not have been permitted entry into the United States. She is seen weekly by her physician and delivered a healthy baby girl without complication 18 days later. The justification for the obstetrician administering RhoGAM postnatally is that:

a. Fatima is Rh$^+$ and the baby is Rh$^-$.
b. Fatima is Rh$^-$ and the baby is Rh$^+$.
c. Fatima is Rh$^+$ and Samir is Rh$^-$.
d. Samir is Rh$^-$ and the baby is Rh$^-$.
e. Fatima is Rh$^-$ and the baby is Rh$^-$.

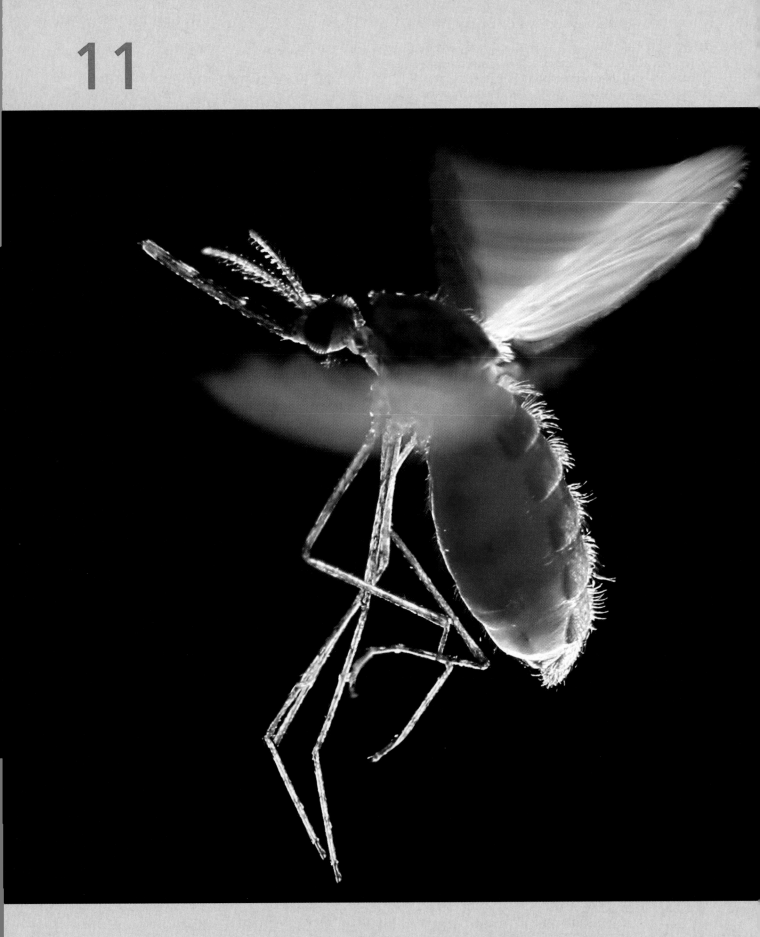

Mosquito, the insect vector that transmits the malarial parasite by feasting on human blood.

Chapter 11

Failures of the Body's Defenses

Most pathogens that threaten the human body are prevented from establishing infection, and those infections that do occur are usually terminated by the actions of innate and adaptive immunity. In this situation there is strong pressure on pathogens to evolve ways of escaping or subverting the immune response. Microorganisms with such advantages compete successfully against other potential pathogens to exploit the resources of the human body. The first part of this chapter describes examples of the different types of mechanism they use.

The body's defenses against infection can also fail because of inherited deficiencies of the immune system. Some of these are described in the second part of the chapter. Within the human population there are mutant alleles for many of the genes encoding components of the immune system. These mutant genes cause immunodeficiency diseases, which vary in severity depending on which gene is defective. Correlation of the molecular defects in immunodeficiency diseases with the types of infection to which patients become vulnerable reveals the effectiveness of the various arms of the immune response against different kinds of pathogen.

In the third part of the chapter we explore one particular host–pathogen relationship that combines themes from the first two parts of the chapter. This concerns the human immunodeficiency virus (HIV), which is extraordinarily effective at both escaping and subverting the immune response. During the course of an infection, which can last for decades, HIV gradually but inexorably wears down the immune system to the point at which it no longer works. The long-term consequence of HIV infection is that patients become severely immunodeficient and develop the fatal disease known as acquired immune deficiency syndrome (AIDS).

Evasion and subversion of the immune system by pathogens

The immune response to any pathogen involves complex molecular and cellular interactions between the pathogen and its host, and any stage in this interaction can be targeted by a pathogen and used for its own benefit. The systematic study of pathogen genomes reveals that most, if not all, pathogens have means of escaping or subverting immune defenses, and that some of them have many genes devoted to this purpose.

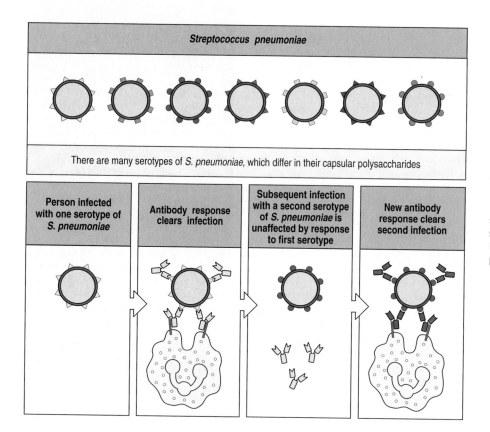

Figure 11.1 Protective immunity towards _Streptococcus pneumoniae_ is serotype-specific. Different strains, or serotypes, of _S. pneumoniae_ have antigenically different capsular polysaccharides, as shown in the upper panel. Antibodies against capsular polysaccharides opsonize the pathogen and enable it to be phagocytosed. A person infected with one serotype of _S. pneumoniae_ clears the infection with type-specific antibody, as shown in the lower panels. These antibodies, however, have no protective effect when the same person is infected with a different serotype of _S. pneumoniae._ The second infection can be cleared only by a new primary response.

11-1   Genetic variation within some species of pathogens prevents effective long-term immunity

Antibodies directed against macromolecules on the surfaces of pathogens are the most important source of long-term protective immunity to many infectious diseases. Some species of pathogen evade such protection by existing as numerous different strains, which differ in the antigenic macromolecules on their outer surfaces. One such is the bacterium _Streptococcus pneumoniae_, which causes pneumonia. Genetic strains of _S. pneumoniae_ differ in the structure of the capsular polysaccharides and compete with each other to infect humans. These strains, of which at least 90 are known, are called **serotypes** because antibody-based serological assays are used to define the differences between them. After resolution of infection with a particular serotype of _S. pneumoniae_, a person will have made antibodies that prevent reinfection with that type but will not prevent primary infection with another type (Figure 11.1). _S. pneumoniae_ is a common cause of bacterial pneumonia because its genetic variation prevents individuals from developing an effective immunological memory against all strains. Genetic variation in _S. pneumoniae_ has evolved as a result of selection by the immune response of its human hosts.

11-2   Mutation and recombination allow influenza virus to escape from immunity

Some viruses also display genetic variation, influenza virus being a well-studied example. This virus infects epithelia of the respiratory tract and passes easily from one person to another in the aerosols generated by coughs and sneezes. Protective immunity to influenza is provided principally by antibodies that bind to the hemagglutinin and neuraminidase glycoproteins of the viral envelope. These antibodies are made during the primary immune

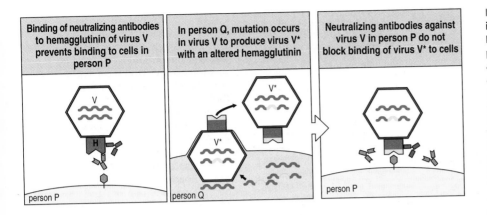

Binding of neutralizing antibodies to hemagglutinin of virus V prevents binding to cells in person P	In person Q, mutation occurs in virus V to produce virus V* with an altered hemagglutinin	Neutralizing antibodies against virus V in person P do not block binding of virus V* to cells

Figure 11.2 Evolution of new influenza variants by antigenic drift. Person P infected with influenza strain V produces neutralizing antibodies (green) against the viral hemagglutinin as well as antibodies against other epitopes on the protein (blue). When person P is further challenged with strain V, these antibodies bind to the hemagglutinin and prevent the virus from infecting cells (left panel). Influenza virus mutates frequently, and a point mutation (yellow) in the hemagglutinin gene of strain V during infection of person Q produces strain V* (center panel). This antigenic drift allows strain V* of influenza virus to infect cells of person P without obstruction from the antibodies previously made against strain V (right panel). For P to clear infection with strain V* requires a primary immune response, with the development of new specific neutralizing antibodies. Only the antigenic drift of the viral hemagglutinin is depicted and not that of the neuraminidase, which drifts similarly. Only two of the eight RNA molecules in each virus are drawn.

response to the virus. The course of a primary infection is short (1–2 weeks) and the virus is cleared from the system by a combination of cell-mediated immunity and antibodies. The pattern of infection of influenza virus characteristically causes **epidemics**, in which the virus spreads rapidly through a local population and then abruptly subsides. Long-term survival of the influenza virus is ensured by the generation of new viral strains that evade the protective immunity generated during past epidemics.

Influenza is an RNA virus with a genome consisting of eight RNA molecules. RNA replication is relatively error-prone and generates many point mutations on which selection can act. New viral strains that lack the hemagglutinin or neuraminidase epitopes that induced protective immunity in the previous epidemic emerge regularly and cause an influenza epidemic every other winter or so. An individual's protective immunity to influenza is determined by the strain of virus to which they were first exposed—the phenomenon of 'original antigenic sin' (see Section 10-9, p. 310). The history of exposure to particular strains of the virus differs within the population, largely according to age, and so there are subpopulations of people with differing degrees of immunity to the current strain of influenza. The people who suffer most at any particular time will be those whose protective immunity has been lost because of the new mutations present in the current strain. This type of evolution of influenza, which causes relatively mild and limited disease epidemics, is called **antigenic drift** (Figure 11.2).

In contrast, every 10–50 years an influenza virus emerges that is structurally quite different from its predecessors and is able to infect almost everyone. Besides spreading more widely to cause a **pandemic** (a worldwide epidemic), such viruses inflict more severe disease and a greater mortality than the viruses emerging from antigenic drift. The influenza strains that cause pandemics are recombinant viruses that derive some of their RNA genome from an avian influenza virus and the remainder from a human influenza virus. In these recombinant strains, the hemagglutinin and/or the neuraminidase are encoded by RNA molecules of avian origin and are antigenically very different from those against which people have protective immunity. New pandemic strains often arise in parts of south-east Asia where farmers live in close proximity to their livestock such as pigs, chickens, and ducks. One theory is that the recombinant viruses arise in pigs that have become simultaneously infected with both avian and human viruses. If such a recombinant jumps back into the human population it has a tremendous competitive advantage, and in sweeping through the human population will rapidly replace other influenza strains. Recombinant influenza viruses can similarly sweep through bird populations and are greatly feared by poultry farmers. This mode of evolution is called **antigenic shift** (Figure 11.3).

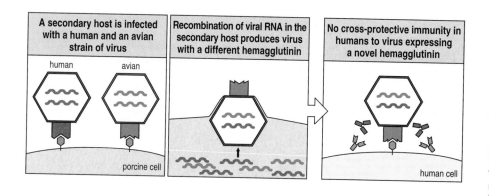

| A secondary host is infected with a human and an avian strain of virus | Recombination of viral RNA in the secondary host produces virus with a different hemagglutinin | No cross-protective immunity in humans to virus expressing a novel hemagglutinin |

Figure 11.3 Evolution of new influenza variants by antigenic shift. A human influenza virus (red) and an avian influenza virus (blue) simultaneously infect a cell within the secondary host, a pig (left panel), where their RNA segments become reassorted to produce a recombinant virus (center panel). The recombinant virus expresses hemagglutinin of avian origin that is antigenically very different from that in the original human virus. The recombinant virus readily infects humans because antibodies protective against the original hemagglutinin cannot bind to the new hemagglutinin (right panel).

11-3   Trypanosomes use gene rearrangement to change their surface antigens

Mutation and recombination are not the only methods by which pathogens can change the face they present to the immune system. Certain protozoans regularly change their surface antigens by a process of gene rearrangement, the most striking example being the African trypanosomes (for example, *Trypanosoma brucei*), which cause sleeping sickness. The life cycle of the trypanosome involves both mammalian and insect hosts. Insect bites transmit trypanosomes to humans, in whom the parasites replicate in the extracellular spaces. The trypanosome's surface is formed of a glycoprotein, of which there are numerous variants, each encoded by a different gene. The trypanosome genome contains more than 1000 genes encoding these **variable surface glycoproteins** (**VSGs**). At any time, an individual trypanosome produces only one form of VSG. This is because the rearrangement of a VSG gene into a unique site in the genome—the expression site—is required for its expression. Rearrangement occurs by a process of **gene conversion** in which the gene in the expression site is excised and replaced by a copy of a different but homologous gene (Figure 11.4). The vast majority of the rapidly replicating trypanosomes that emerge after initial infection will express the same dominant form of VSG. A very small minority will, however, have changed the expressed VSG gene and will now express other forms. The host makes an antibody response to the dominant form of VSG, but not to the minority forms. Antibody-mediated clearance of trypanosomes expressing the dominant VSG facilitates the growth of those expressing the minority forms, of which one will come to dominate the trypanosome population. In time, the numbers of trypanosomes expressing the new dominant form are sufficient to stimulate the production of antibodies, which clear the new dominant form. In turn this allows a further form to dominate and the cycle continues.

This mechanism of immune evasion causes trypanosome infections to produce a dramatic cycling in the number of parasites within an infected person (see Figure 11.4, bottom panel). The chronic cycle of antibody production and antigen clearance leads to a heavy deposition of immune complexes and

Figure 11.4 Antigenic variation by African trypanosomes allows them to escape from adaptive immunity. In the top panel, the VSGª gene (shown in red) is in the expression site and the VSGᵇ gene (yellow) and VSGᶜ gene (blue) are inactive. In the second panel, gene conversion has replaced VSGª with VSGᵇ at the expression site; in the third panel, VSGᶜ has replaced VSGᵇ at the expression site. The bottom panel shows how the patient's antibody response (thin black lines) to the VSG proteins selects for low-frequency variants and causes a cycle in which parasite populations (red, yellow, and blue lines) alternate between boom and bust over a number of weeks.

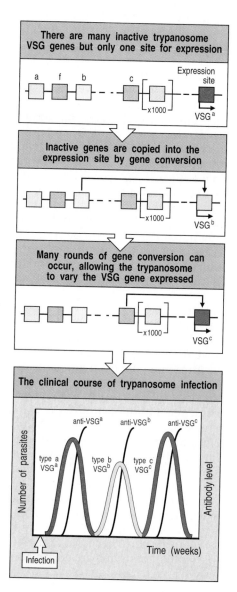

inflammation. Neurological damage occurs and eventually leads to coma, the so-called sleeping sickness. Trypanosome infections are a major health problem for humans and cattle in large parts of Africa. Indeed, it is largely because of trypanosomes that wild populations of big game animals still survive in Africa and have not been replaced by domestic cattle. Malaria, another disease caused by a protozoan parasite that escapes immunity by varying its surface antigens, is also a major cause of human mortality in equatorial Africa.

Gene conversion enables similar strategies of antigenic variation to be used by several species of bacteria whose ability to escape from the human immune response makes them successful pathogens and major public-health problems. *Salmonella typhimurium*, a common cause of food poisoning, can alternate the expression of two antigenically distinct flagellins, proteins of the bacterial flagella. This occurs by reversible inversion of part of the promoter of one of the flagellin genes, which inactivates that gene and allows the expression of the second gene. *Neisseria gonorrhoeae*, the cause of the widespread sexually transmitted disease gonorrhea, has several variable antigens, the most impressive being the pilin protein, a component of the adhesive pili on the bacterial surface. Like the VSGs of African trypanosomes, pilin is encoded by a family of variant genes, only one of which is expressed at a time. Different versions of the pilin gene introduced into the expression site provide a minority population of variant bacteria. When the host's immune response places pressure on the dominant type, another is ready to take its place.

11-4 Herpesviruses persist in human hosts by hiding from the immune response

To terminate an established viral infection, infected cells must be killed by cytotoxic CD8 T cells. For this to occur, some of the peptides presented by MHC class I molecules at the surface of infected cells must be of viral origin, a condition easily fulfilled by rapidly replicating viruses such as influenza. Consequently, influenza infections are efficiently cleared by the immune system by a combination of cytotoxic T cells and antibodies, the latter neutralizing extracellular virus particles. In contrast, some other viruses are difficult to clear because they enter a quiescent state within human cells, one in which they neither replicate nor generate enough virus-derived peptides to signal their presence to cytotoxic T cells. Development of this dormant state, which is called **latency** and does not cause disease, is a favored strategy of the herpesviruses. Later on, when the initial immune response has subsided, the virus will reactivate, causing an episode of disease.

Herpes simplex virus, the cause of cold sores, first infects epithelial cells and then spreads to sensory neurons serving the area of infection. The immune response clears virus from the epithelium, but the virus persists in a latent state in the sensory neurons. Various stresses can reactivate the virus, including sunlight, bacterial infection, or hormonal changes. After reactivation, the virus travels down the axons of the sensory neurons and reinfects the epithelial tissue (Figure 11.5). Viral replication in epithelial cells and the production of viral peptides restimulates CD8 T cells, which kill the infected cells, creating a new sore. This cycle can be repeated many times throughout life. Neurons are a favored site for latent viruses to lurk because they express very small numbers of MHC class I molecules, further reducing the potential for presentation of viral peptides to CD8 T cells.

The herpesvirus varicella-zoster (also called herpes zoster) remains latent in one or a few ganglia, chiefly dorsal root ganglia, after the acute infection of epithelium—chickenpox—is over. Stress or immunosuppression can reactivate the virus, which moves down the nerve and infects the skin. Reinfection causes the reappearance of the classic varicella rash of blisters, which cover the area of skin served by the infected ganglia. The disease caused by

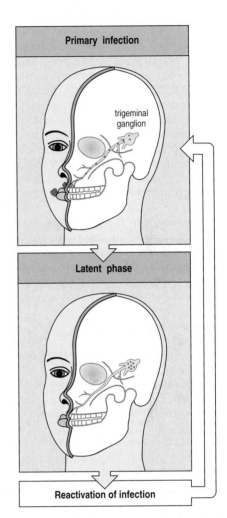

Figure 11.5 Persistence and reactivation of herpes simplex virus infection. The initial infection around the lips is cleared by the immune response and the resulting tissue damage is manifested as cold sores (upper panel). The virus (small red dots) has meanwhile entered sensory neurons, for example those in the trigeminal ganglion whose axons innervate the lips, where it persists in a latent state (lower panel). Various types of stress can cause the virus to leave the neurons and reinfect the epithelium, once again reactivating the immune response and causing cold sores. People infected with herpes simplex viruses periodically get cold sores as a result of this process. During its active phase the virus can be passed from one person to another.

reactivation of varicella-zoster is commonly known as shingles. In contrast to herpes simplex virus, reactivation of varicella-zoster usually occurs only once in a lifetime.

A third herpesvirus that causes a persistent infection is the Epstein–Barr virus (EBV), to which most humans are exposed. First exposure in childhood produces a mild cold-like disease, whereas adolescents or adults encountering EBV for the first time develop infectious mononucleosis (also known as glandular fever), an acute infection of B lymphocytes. EBV infects B cells by binding to the CR2 component of the B-cell co-receptor complex (see Figure 9.3, p. 251). Most of the infected B cells proliferate and produce virus, leading in turn to the stimulation and proliferation of EBV-specific T cells. The result is an unusually large number of mononuclear white blood cells (lymphocytes, mostly T cells), which gives the disease its name. After some time, the acute infection is brought under control by CD8 cytotoxic T cells, which kill the virus-infected B cells. The virus persists in the body, however, because a minority of B cells become latently infected. This involves shutting off the synthesis of most viral proteins except EBNA-1, which maintains the viral genome in these cells. Latently infected cells do not present a target for attack by CD8 cytotoxic cells because the proteasome is unable to degrade EBNA-1 into peptides that can be bound and presented by MHC class I molecules.

After recovery from the initial exposure to EBV it is unusual for reactivation of the virus to lead to disease. It seems likely that CD8 T cells quickly control episodes of viral reactivation. In immunosuppressed patients, however, reactivation of the virus can cause a disseminated EBV infection, and infected B cells can also undergo malignant transformation, causing B-cell lymphoproliferative disease.

11-5   Certain pathogens sabotage or subvert immune defense mechanisms

Pathogens also exploit the immune-system cells that are ranged against them. *Mycobacterium tuberculosis* commandeers the macrophage's pathway of phagocytosis for its own purposes. On being phagocytosed, *M. tuberculosis* prevents fusion of the phagosome with the lysosome, thus protecting itself from the bactericidal actions of the lysosomal contents. It then survives and flourishes within the cell's vesicular system. *Listeria monocytogenes*, in contrast, escapes from the phagosome into the macrophage's cytosol, where it grows and replicates. However, the intracytosolic way of life elicits cytotoxic CD8 T-cell responses against *L. monocytogenes*, which eventually terminate the infection.

The protozoan parasite *Toxoplasma gondii*, the cause of toxoplasmosis, creates its own specialized environment within the cells that it infects. It surrounds itself with a membrane-enclosed vesicle that does not fuse with other cellular vesicles or cell membranes. Such isolation prevents the binding of *T. gondii*-derived peptides to MHC molecules and their presentation to T cells. The spirochete *Treponema pallidum*, the cause of syphilis, evades specific antibody by coating itself with human proteins. This is also a strategy pursued by the schistosome, a parasitic helminth.

Of the four groups of pathogens (see Figure 1.4, p. 7), viruses have evolved the greatest variety of mechanisms for subverting or escaping immune defenses. This is because their replication and life cycle depend completely on the metabolic and biosynthetic processes of human cells. Viral self-defense strategies include the capture of cellular genes encoding cytokines or cytokine receptors, which when expressed by the virus can divert the immune response; the synthesis of proteins that inhibit complement fixation; and the synthesis of proteins that inhibit antigen processing and presentation by MHC class I

Viral strategy	Specific mechanism	Result	Viral examples
Inhibition of humoral immunity	Virally encoded Fc receptor	Blocks effector functions of antibodies bound to infected cells	Herpes simplex Cytomegalovirus
	Virally encoded complement receptor	Blocks complement-mediated effector pathways	Herpes simplex
	Virally encoded complement control protein	Inhibits complement activation of infected cell	Vaccinia
Inhibition of inflammatory response	Virally encoded chemokine receptor homolog	Sensitizes infected cells to effects of some chemokines; advantage to virus unknown	Cytomegalovirus
	Virally encoded soluble cytokine receptor, e.g., IL-1 receptor homolog, TNF receptor homolog, IFN-γ receptor homolog	Blocks effects of cytokines by inhibiting their interaction with host receptors	Vaccinia Rabbit myxoma virus
	Viral inhibition of adhesion molecule expression, e.g., LFA-3, ICAM-1	Blocks adhesion of lymphocytes to infected cells	Epstein–Barr virus
	Protection from NFκB activation by short sequences that mimic TLRs	Blocks inflammatory responses elicited by IL-1 or bacterial pathogens	Vaccinia
Blocking of antigen processing and presentation	Inhibition of MHC class I upregulation by IFN-γ	Impairs recognition of antigen-presenting cells by CD4 T cells	Herpes simplex Cytomegalovirus
	Inhibition of peptide transport by TAP	Blocks peptide association with MHC class I	Herpes simplex
Immunosuppression of host	Virally encoded cytokine homolog of IL-10	Inhibits T_H1 lymphocytes Reduces IFN-γ production	Epstein–Barr virus

Figure 11.6 Mechanisms by which herpesviruses and poxviruses subvert the immune response. Herpes simplex, cytomegalovirus, and Epstein–Barr viruses are herpesviruses; vaccinia and rabbit myxoma virus are poxviruses.

molecules. Examples of defensive mechanisms used by herpesviruses and poxviruses are shown in Figure 11.6.

Major players in the immune response to viral infections are NK cells and CD8 T cells, killer lymphocytes whose development and function are dependent upon MHC class I molecules. For this reason many viruses have evolved subversive mechanisms for interfering with the synthesis and expression of MHC class I. The herpesvirus **human cytomegalovirus** (**HCMV**) is particularly rich in such mechanisms: it has 10 proteins that interfere in diverse ways to diminish the capacity of MHC class I molecules to stimulate NK-cell and CD8 T-cell responses against HCMV-infected cells (Figure 11.7). One group of these saboteurs affect MHC class I by causing its degradation, by interfering with the proteasome, TAP or tapasin, or by retaining MHC class I in the endoplasmic reticulum, all mechanisms that prevent the presentation of viral antigens to CD8 T cells. Such mechanisms that reduce MHC class I expression should favor an NK-cell response against the infected cells that are now lacking self MHC class I (see Chapter 10). However, a second group of saboteurs interferes with the inhibitory CD94:NKG2A and LILRB1 NK-cell receptors that sense the missing self MHC class I, and with the activating NKG2D receptor, which recognizes the ligands MIC and ULBP.

HCMV protein	Subversive effect on the immune response
US2	Targets HLA class I molecules to the proteasome by transporting them to the cytosol
US3	Retains HLA class I in the endoplasmic reticulum by blocking tapasin function
US6	Inhibits TAP ATPase activity and function
US10	Binds HLA class I and delays its departure from the endoplasmic reticulum to the cell surface
US11	Targets newly synthesized HLA class I heavy chains for degradation in the cytoplasm
UL16	Inhibits NK-cell recognition of infected cells by binding to the ULBP ligands for NKG2D
UL18	MHC class I heavy-chain homolog that binds to the NK-cell receptor LILRB1
UL40	The leader peptide of UL40 binds to HLA-E and subverts the ability of CD94:NKG2A to monitor HLA-A, -B, -C expression
UL83	Blocks access to the proteasome and the generation of peptides for binding MHC class I
UL142	Downregulates the expression of the MIC-A and MIC-B ligands for NKG2D

Figure 11.7 Human cytomegalovirus interferes with the expression of MHC class I molecules in many different ways.

Human cytomegalovirus is an extremely well adapted human pathogen that infects more than half the population of the United States. Most of these 150 million people are unaware of their infection, because the virus causes few symptoms on initial infection and exists thereafter in a latent state, during which it is comfortably controlled by the combined activities of NK cells and CD8 T cells. In the healthy HCMV-infected person there is a well-tuned balance in which the virus survives and multiplies with little expense to the host. In contrast, HCMV causes life-threatening disease in immunocompromised individuals: the young, the elderly, the transplant recipient on immunosuppressive drugs and people infected with HIV. HCMV is the commonest infection affecting patients who have had a bone marrow transplant or hematopoietic stem-cell transplant, and if not treated with antiviral drugs it is fatal. HCMV infects a wide range of human cell types and is spread in bodily fluids by physical contact.

11-6 Bacterial superantigens stimulate a massive but ineffective T-cell response

Some pathogens induce a general suppression of a person's immune response. For example, staphylococci produce toxins such as the **staphylococcal enterotoxins** and **toxic shock syndrome toxin-1**, which can bind simultaneously to MHC class II molecules and T-cell receptors in the absence of a specific peptide antigen (Figure 11.8). By forming a bridge between a CD4 T cell's receptors and the MHC class II molecules of an antigen-presenting cell, these toxins mimic specific antigen and cause the T cell to divide and differentiate into effector T cells. Because the toxins bind to sites shared by many different T-cell receptors, they stimulate an excessive polyclonal response that can involve 2–20% of the total number of circulating CD4 T cells. Because of this property these toxins are called **superantigens**. The consequence of superantigen stimulation is a massive production and release of cytokines, particularly IL-1, IL-2, and TNF-α, which causes systemic shock (see Figure 2.29, p. 52). At the same time, a useful adaptive immune response is suppressed. After proliferation, T cells that have bound superantigens undergo apoptosis in the absence of any further specific stimulation, removing many antigen-specific T-cell clones from the peripheral circulation.

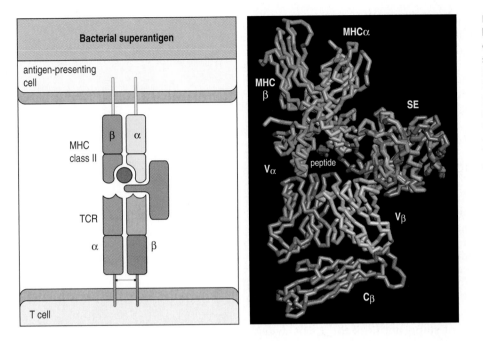

Figure 11.8 Bacterial superantigens bridge α:β T-cell receptors and MHC class II molecules in the absence of a specific peptide. In the diagram of the interaction (left panel) the superantigen is blue. The molecular model (right panel) shows the interactions of the superantigen staphyloccal enterotoxin (SE, blue) with an MHC class II molecule (yellow and green chains) and an α:β T-cell receptor (orange, gray and pink chains). Although the MHC class II molecule has a bound peptide (red) it is usually not one for which the T-cell receptor is specific.

11-7   Immune responses can contribute to disease

Because the immune response is a powerful and destructive force, some of the disease symptoms, or **pathology**, in most infections are due to the immune response. For some infectious diseases all pathology is due to the immune response, an example being the wheezy bronchiolitis caused by T_H2 cells responding to infection with **respiratory syncytial virus** (**RSV**). This virus accounts for a large proportion of hospital admissions for infants in the developed countries—some 90,000 admissions and 4500 deaths each year in the United States alone. A failed attempt at developing a vaccine against RSV revealed that infants vaccinated with a killed virus preparation suffered worse disease on subsequent infection with RSV than did unvaccinated infants. This unfortunate outcome was because the vaccine failed to induce neutralizing antibodies but successfully activated virus-specific T_H2 cells. On subsequent infection with RSV, the secondary response of the T_H2 cells produced IL-3, IL-4, and IL-5 in quantities that exacerbated the disease-causing aspects of the immune response—the induction of bronchospasm, the increased secretion of mucus and the recruitment of tissue-damaging eosinophils into the tissues of the respiratory tract.

Tissue damage and disease symptoms can also result from the immune response to parasites, as illustrated by *Schistosoma mansoni*. These blood flukes lay their eggs in the hepatic portal vein. Some eggs reach the intestine and are shed in the feces, thereby enabling the infection to spread to other people. Other eggs lodge in the portal circulation of the liver, where they elicit a powerful immune response that leads to chronic inflammation, hepatic fibrosis, and eventual liver failure. Underlying this progression is an excessive activation of T_H2 cells.

Summary

From the human perspective the ideal immune system would be one that terminates infection before the pathogen damages tissues or saps the body's resources. In contrast, an ideal situation for a pathogen is one in which the immune system does not interfere with growth and replication, while other parts of the body provide food and shelter. To further their cause, pathogens have evolved ways of reducing the effectiveness of the human immune response. Antigenic variation in the pathogen prevents the maturation of the adaptive response and the development of useful immunological memory. Latency, a means of avoiding the immune response, allows viruses to lie low within cells until immunity has waned. More active strategies are for pathogens to interfere with key elements of the immune response, either to inhibit normal immune function or to recruit the response to the pathogen's advantage. Immune responses to pathogens can themselves be a significant cause of pathology.

Inherited immunodeficiency diseases

Inherited defects in genes for components of the immune system cause **primary immunodeficiency diseases**, which reveal themselves by enhanced susceptibility to infection or autoimmunity. Primary immunodeficiency diseases are distinguished from **secondary immunodeficiency diseases**, which are due not to defective genes but to environmental factors, such as immunosuppressive drugs, that adversely impact the immune system. Before the advent of antibiotic therapy during the 1940s, most individuals with inherited immune defects died from infection during infancy or early childhood. Because so many normal infants also succumbed to infection in that earlier era, death from immunodeficiency did not stand out until the 1950s, when the first such disease was described. Since then, numerous inherited immunodeficiency diseases have been identified and correlated with susceptibility to particular classes of pathogen. Each disease is due to a defect in a particular protein or glycoprotein, and the precise symptoms depend on the role of that component in the immune response.

11-8   Rare primary immunodeficiency diseases reveal how the human immune system works

In dissecting the immune system of the laboratory mouse, immunologists 'knock out' selected genes and examine the immunodeficiency syndromes that they create. Equivalent human gene knockouts are provided by the more than 150 primary immunodeficiency syndromes that have been described so far. Treating and studying these patients have made invaluable contributions to knowledge of the human immune system. It is no coincidence that in almost all the previous chapters of this book, one or more primary immunodeficiency syndromes have been used to illustrate the functions of particular proteins and the effects of their absence (Figure 11.9). Most of these conditions are very rare and are caused by mutant genes that have no selective benefit for the people who carry them. They usually occur in small populations that are geographically or culturally isolated and where there are traditions of marriage within the group. Recent advances in genetics and genomics have made the precise identification of the genes responsible for immunodeficiency syndromes much easier; now the more challenging task is for doctors in the field to recognize a novel form of immunodeficiency syndrome when they see it. International collaborations help in the identification and treatment of these patients.

Whereas most of the primary immunodeficiencies listed in Figure 11.9 were discovered in patients with severe disease and are caused by exceedingly rare mutant alleles, defects in other immune-system genes are more frequent and have less dramatic effects. Examples of the latter are the lack of an A or a B isotype of complement component C4 and defective MHC class I alleles. Those immune-system genes that can be lost with little noticeable effect are usually members of multigene families in which another family member can, to some extent, compensate for the defective gene. In some cases this type of variability may represent a compromise, in that there are some benefits associated with having or lacking a particular gene.

11-9   Inherited immunodeficiency diseases are caused by dominant, recessive, or X-linked gene defects

All primary immunodeficiency diseases can be classified into three types: dominant, recessive, or X-linked. Syndromes due to a **dominant** defective allele show up in children who inherit a normal, functional allele from one parent and the defective allele from the other parent. Disease occurs because the abnormal properties of the defective allele interfere with and dominate over the functions provided by the normal allele. In contrast, a disease caused by a **recessive** allele is only manifested in patients who inherit the defective allele from both parents. Individuals who have one defective allele and one normal allele are healthy and are called **carriers** of the disease trait. In recessive diseases, the defective allele does not interfere with the function of the normal allele. A key difference between dominant and recessive disease traits lies in the fate of heterozygous individuals: for a dominant trait they get the disease, for a recessive trait they do not.

Figure 11.9 Mechanisms of human immunity are revealed by the study of inherited immunodeficiency syndromes. This figure shows those immunodeficiency syndromes mentioned previously in this book (with references to the relevant sections), their gene defects, and the effects they have on the immune system.

Name of deficiency	Affected genes	Immune defect	Susceptibility	Reference (section)
Asplenia	Not known	Absence of the spleen	Encapsulated extracellular bacteria	1-10, p. 21
C3 deficiency	C3	Lack of C3	Recurrent infection with Gram-negative bacteria	2-2, p. 33
Paroxysmal nocturnal hemoglobinuria	Somatic and germline mutations in genes involved in phosphatidylinositol glycan biosynthesis	Lack of DAF, HRF, and CD59	Lysis of erythrocytes by complement	2-6, p. 39
X-linked hypohydrotic ectodermal dysplasia and immunodeficiency	NEMO	Impaired activation of NFκB	Chronic bacterial and viral infections	2-12, p. 49
Chronic granulomatous disease	NADPH oxidase	Impaired neutrophil function	Chronic bacterial and fungal infections	2-16, p. 57
MBL deficiency	Mannose-binding lectin	Lack of mannose-binding lectin	Susceptibility to meningitis due to *Neisseria meningitidis*	2-18, p. 61
NK-cell deficiency	Not known	Absence of NK cells	Susceptibility to herpesvirus infections	2-21, p. 65
X-linked hyper IgM syndrome	AID or CD40 ligand or CD40 or NEMO	No isotype switching or somatic hypermutation in B cells	Extracellular bacterial and fungal infections	4-15, p. 116 9-9, p. 261
SCID	RAG1 or RAG2	No gene rearrangements in B cells and T cells	All types of infection	5-2, p. 128
Omenn syndrome	RAG1 or RAG2 or Artemis	Impaired RAG function	All types of infection	5-2, p. 128
Bare lymphocyte syndrome	TAP1 or TAP2	Low MHC class I expression	Respiratory viral functions	5-11, p. 139
Pre-B-cell receptor deficiency	λ5	Lack of B cells and antibodies	Persistent bacterial infections	6-4, p. 164
X-linked agammaglobulinemia	Bruton's kinase (BtK)	B cells blocked at pro-B-cell stage	Recurrent bacterial infections	6-8, p. 170
Complete DiGeorge's syndrome	Not known	Absence of the thymus and T cells	All types of infection	7-1, p. 188
Autoimmune polyendocrinopathy candidiasis ectodermal dystrophy	Autoimmune regulator (AIRE)	Reduced T-cell tolerance to self antigens	Autoimmune diseases	7-12, p. 202
IPEX	FOXP3	Lack of regulatory T cells and peripheral tolerance	Autoimmune diseases	7-13, p. 203
ZAP-70 deficiency	ZAP70	T cells that cannot signal through their receptors	All types of infection	8-7, p. 223
Autoimmune lymphoproliferative syndrome	Fas or Fas ligand	Enlarged spleen and lymph nodes	Lymphomas and autoimmunities	8-16, p. 238
IgG2 deficiency	Not known	Lack of IgG2	Encapsulated bacteria	9-21, p. 277
Selective IgA deficiency	Not known	Lack of IgA	No major susceptibility	10-9, p. 300

X-linked diseases are caused by recessive defects in genes on the X chromosome. Because males have only one X chromosome and females have two, the disease occurs in all males who inherit an X chromosome with a defective allele, but it will not show up in their sisters even if they inherit the same X chromosome. Disease occurs in females only when they inherit a defective X chromosome from both parents. X-linked diseases, of which we have already met three (see Figure 11.9), are therefore far more frequent in boys than in girls. For these traits only women serve as healthy carriers. Any disease caused by a dominant allele on an X chromosome would occur at equal frequency in boys and girls.

Dominance is most commonly seen when the defective gene encodes a protein that functions in a dimer or a larger protein complex. In such cases, the incorporation of one defective subunit can reduce or destroy the capacity of the complex to function. Before the 1950s, any dominant trait causing a severe immunodeficiency would probably have been eliminated from the population with the death of the child in whom the mutation first occurred. Thus, most of the severe inherited immunodeficiency syndromes that have been identified are due to recessive mutations in single genes. The known immunodeficiencies that are due to dominant mutations tend to be less severe and are caused by a reduction in a function rather than its loss.

11-10 Recessive and dominant mutations in the interferon-γ receptor cause diseases of differing severity

Interferon-γ (IFN-γ), the major cytokine that activates macrophages, is made by NK cells during the innate immune response and by T_H1 CD4 T cells and CD8 cytotoxic T cells during the adaptive immune response. When IFN-γ binds to the IFN-γ receptor on a macrophage surface, the macrophage is induced to change its pattern of gene expression and become better at taking up bacteria and killing them (see Section 8-16, p. 238). The IFN-γ receptor is a dimer composed of two polypeptides, IFNγR1 and IFNγR2, both of which associate with tyrosine kinases—Jak1 and Jak2, respectively. IFN-γ also functions as a dimer, and binding of the dimeric cytokine to sites on the IFNγR1 polypeptide cross-links two molecules of the receptor to initiate the signaling cascade (Figure 11.10, first panel).

The response of macrophages to IFN-γ is particularly important in the defense against intravesicular bacteria, such as mycobacteria, and both dominant and recessive mutations in IFNγR1 have been identified in patients suffering from persistent mycobacterial infections (Figure 11.10, second and third panels). Both types of mutation cause the condition of **IFN-γ receptor deficiency**. The recessive alleles contain mutations that prevent any expression of IFNγR1 at the cell surface. The macrophages and monocytes of patients with two recessive alleles carry only IFNγR2 at their surfaces and are unresponsive to IFN-γ (see Figure 11.10, second and fourth panels). For this group of patients the disease is generally more severe and appears at an earlier age. Heterozygotes are healthy because the protein made from the defective allele does not interfere with the product made from the normal allele, which assembles with IFNγR2 and moves to the cell surface as functional IFN-γ receptor (see Figure 11.10, first panel).

In the dominant mutants, IFNγR1 is truncated such that much of the cytoplasmic tail, which binds Jak1 and initiates signaling, is missing. The truncated IFNγR1 associates with IFNγ R2 protein and is taken to the surface as a receptor that binds IFN-γ but cannot transduce a signal. At the cell surface, these defective receptors compete for IFN-γ with the normal receptors that incorporate IFNγR1 made from the normal allele (see Figure 11.10, third panel). This competition is further weighted against the functional receptors

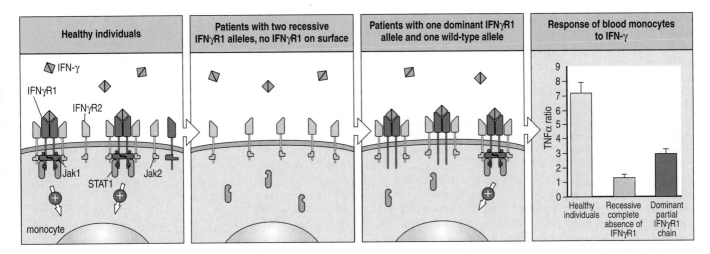

Figure 11.10 The impact of recessive and dominant mutations in the IFN-γ receptor on monocyte activation. Receptors for IFN-γ are composed of a dimer of IFNγR1 and IFNγR2. Two such dimers must be cross-linked by IFN-γ binding to the IFNγR1 chain for signaling to be triggered (first panel). Recessive mutant alleles of IFNγR1 produce a mutant chain that does not reach the surface. Thus, cells from patients homozygous for a recessive mutation have only IFNγR2 at the surface, lack IFNγR1 function and cannot respond to IFN-γ (second panel). Heterozygotes for such a mutation produce sufficient numbers of wild-type chains to assemble enough functional receptors for a normal response to IFN-γ, as in the first panel. Dominant mutant alleles of IFNγR1 produce a mutant chain lacking a signaling domain. This chain can assemble into a dimer and bind IFN-γ but cannot signal (third panel). Heterozygotes for a dominant mutation make a small number of functional receptors composed entirely of wild-type chains, but most receptors are non-functional. Thus their response to IFN-γ is defective (third panel). The fourth panel compares the results of IFN-γ stimulation of blood monocytes from normal, homozygous recessive and heterozygous dominant patients.

because the absence of the cytoplasmic domain from IFNγR1 prevents the mutant receptor from being recycled by endocytosis. It therefore accumulates at the cell surface, where it exceeds levels of the normal receptor fivefold. Because of the interference by the mutant receptors, the response of patients' macrophages and monocytes to IFN-γ is much reduced compared with healthy people, but it is greater than in patients carrying two recessive alleles (see Figure 11.10, fourth panel). Because of this difference, dominant mutants cause a less severe immunodeficiency, which tends to be detected at a later age.

11-11 Antibody deficiency leads to an inability to clear extracellular bacteria

The major threat to patients lacking antibodies is infection by pyogenic bacteria. These encapsulated bacteria, which include *Haemophilus influenzae*, *Streptococcus pneumoniae*, *Streptococcus pyogenes*, and *Staphylococcus aureus*, cannot be recognized by the phagocytic receptors of macrophages and neutrophils, so they frequently escape immediate elimination by the innate immune response. Such infections are normally terminated when the bacteria are opsonized by specific antibody and complement, whereupon the bacteria are readily taken up and killed by phagocytes. In patients lacking antibodies, infections with pyogenic bacteria tend to persist unless treated with antibiotics.

The first immunodeficiency disease to be described was characterized by antibody deficiency and X-linked inheritance and is named **X-linked agammaglobulinemia (XLA)**. The defect in XLA is in a protein tyrosine kinase, which is called Bruton's tyrosine kinase (Btk) to honor the discoverer of the syndrome. Btk contributes to intracellular signaling from the B-cell receptor and is necessary for the growth and differentiation of pre-B cells (see Figure

6.12, p. 169). Males inheriting an X chromosome with a mutant *Btk* gene that does not produce a protein therefore lack mature B cells. Btk is also expressed in monocytes and T cells, but these cells in patients with XLA are not obviously compromised by its absence.

Women with one functional and one nonfunctional copy of the *Btk* gene are themselves healthy, but they pass XLA on to half their male children. In all females, one X chromosome is randomly inactivated in every cell, and carriers of XLA can be identified by determining how the X chromosomes are inactivated in their B cells. In women who are not carriers, 50% of B cells inactivate one X chromosome and 50% the other at random. In carriers of XLA, only the B-cell precursors that inactivate the X chromosome containing the mutant *Btk* allele can develop. Thus, all mature B cells in heterozygous females have inactivated the X chromosome containing the nonfunctional *Btk* gene. By using genetic markers that distinguish between the two X chromosomes, the females in families with a history of XLA, or of any other X-linked syndrome, can be typed as carriers or noncarriers (Figure 11.11).

Patients who lack antibodies are vulnerable to infection with extracellular pyogenic bacteria that have polysaccharide capsules resistant to phagocytosis. In people who make normal antibody responses, such organisms are cleared by phagocytosis after opsonization by antibody and complement. Antibody-deficient patients are also more susceptible to viral infections, particularly those caused by enteroviruses, which enter the body through the gut and in normal individuals are neutralized by antibodies produced by the mucosal immune system.

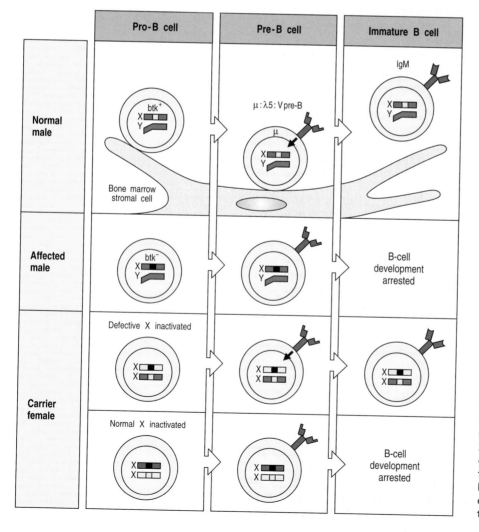

Figure 11.11 **In patients with X-linked agammaglobulinemia (XLA), B cells do not develop beyond the pre-B cell stage.** In XLA, Bruton's tyrosine kinase is defective. In patients with this disease, B cells become arrested at the pre-B-cell stage because intracellular signals cannot be generated by the pre-B-cell receptor. Most patients with XLA are males because males have only one copy of the X chromosome. Heterozygous females are carriers of the disease trait although healthy themselves. During development, female cells randomly inactivate one of their X chromosomes. Consequently, half of the developing B cells in a female carrier become arrested at the pre-B-cell stage because they have inactivated the X chromosome that carries the good copy of the *btk* gene; the other half develop to become functional B cells because they have inactivated the X chromosome that carries the bad copy of *btk* and thus use the X chromosome having the good copy.

Patients who have immunodeficiencies confined to B-cell functions are able to resist many pathogens successfully, and those to which they are susceptible can be treated with antibiotics. However, although pyogenic infections can be cured in this way, the successive rounds of infection and treatment sometimes lead to permanent tissue damage caused by the excessive release of proteases from both the infecting bacteria and the defending phagocytes. These effects are particularly pronounced in the airways, where the bronchi lose their elasticity and become sites of chronic inflammation. This condition, called **bronchiectasis**, can lead to chronic lung disease and eventual death. To prevent such developments, patients with XLA are given monthly injections of **gamma globulin**, an antibody-containing preparation made from the plasma of healthy blood donors. Such preparations contain antibodies against the range of common pathogens and provide what is called **passive immunity** against those pathogens.

11-12 Diminished production of antibodies also results from inherited defects in T-cell help

Diminished production of antibodies is also a symptom of defective genes for the membrane-associated cytokine CD40 ligand. As discussed in Chapter 9, interaction of CD40 ligand on activated T cells with B-cell CD40 is a crucial part of the T-cell help given to B cells. This stimulates B-cell activation, the development of germinal centers, and isotype switching. CD40 ligand is encoded on the X chromosome, so most patients with a hereditary deficiency in CD40 ligand are males. In the absence of CD40 ligand, virtually no specific antibody is made against T-cell dependent antigens. IgG, IgA, and IgE levels are extremely low, and IgM levels are abnormally high. Recognition of this latter characteristic led to the condition's being named **X-linked hyper IgM syndrome**. Patients with this immunodeficiency are inherently susceptible to infection with pyogenic bacteria, but these infections can usually be prevented by regular injections of gamma globulin and cleared by antibiotics when they do occur. A further consequence of the disease is the absence of germinal centers in the lymph nodes and other secondary lymphoid tissues (see Figure 9.17, p. 262).

Macrophage activation by T cells also depends on the interaction of CD40 ligand on the T cell with CD40 on the macrophage. The lack of this interaction in patients with X-linked hyper IgM syndrome impairs the inflammatory response and the mobilization of leukocytes by inflammatory cytokines. Whereas infection normally induces an increase in the number of white cells in the blood (**leukocytosis**), this cannot occur in patients lacking CD40 ligand. On the contrary, their blood can become profoundly deficient in neutrophils. This state, called **neutropenia**, leads to severe sores and blisters in the mouth and throat. These anatomical sites are always infested with bacteria and the integrity of these mucosal tissues depends on continual surveillance of their microbial populations by phagocytes. The symptoms of neutropenia can be cured by the intravenous administration of granulocyte–macrophage colony-stimulating factor (GM-CSF), a cytokine that stimulates the production and release of phagocytes by the bone marrow. In normal individuals, GM-CSF is secreted by macrophages in response to activation by T cells through interactions between CD40 and CD40 ligand.

11-13 Defects in complement components impair antibody responses and cause the accumulation of immune complexes

The effector functions recruited by antibodies to clear pathogens and antigens are all facilitated by complement activation. Consequently, the spectrum of infections associated with complement deficiencies overlaps substantially

344   Chapter 11: Failures of the Body's Defenses

Complement protein	Effects of deficiency
C1, C2, C4	Immune-complex disease
C3	Susceptibility to capsulated bacteria
C5–C9	Susceptibility to *Neisseria*
Factor D, properdin (factor P)	Susceptibility to capsulated bacteria and *Neisseria* but no immune-complex disease
Factor I	Similar effects to deficiency of C3
DAF, CD59	Autoimmune-like conditions including paroxysmal nocturnal hemoglobinuria
C1INH	Hereditary angioneurotic edema (HANE)

Figure 11.12 Diseases caused by deficiencies in the pathways of complement activation.

with that associated with defective antibody production. Defects in the activation of C3, and in C3 itself, are associated with susceptibility to a wide range of pyogenic infections, emphasizing the important role of C3 as an opsonin that promotes the phagocytosis of bacteria by macrophages and neutrophils. In contrast, defects in C5–C9, the terminal complement components that form the membrane-attack complex, have more limited effects, of which susceptibility to *Neisseria* is the best example. The most effective defense against *Neisseria* is complement-mediated lysis of extracellular bacteria, and this requires all the components of the complement pathway. Figure 11.12 lists the effects of the absence of complement components and complement inhibitory proteins.

The early components of the classical pathway are necessary for the elimination of immune complexes. As discussed in Section 9-20, p. 275, the attachment of complement components to soluble immune complexes allows them to be transported, or ingested and degraded, by cells bearing complement receptors. Immune complexes are mainly transported by erythrocytes, which capture the complexes with the CR1 complement receptor that binds to C4b or C3b. Deficiencies in complement components C1–C4 impair the formation of C4b and C3b and lead to the accumulation of immune complexes in the blood, lymph, and extracellular fluid and their deposition within tissues. In addition to directly damaging the tissues in which they deposit, immune complexes activate phagocytes, causing inflammation and further tissue damage.

Deficiencies in the proteins that control complement activation can also have major effects. People deficient for factor I in effect lack C3. Because factor I is absent, the conversion of C3 to C3b runs unchecked, and supplies of C3 are rapidly depleted (see Section 2-4, p. 36). Patients who lack properdin (factor P), a plasma protein that enhances the activity of the alternative pathway by stabilizing the C3 convertase, have a heightened susceptibility to *Neisseria*, because reduced deposition of C3 prevents the formation of the membrane-attack complex and bacterial lysis. In contrast, a deficiency in decay-accelerating factor (DAF) or CD59 causes an autoimmune-like condition. Lacking the protection conferred by DAF or CD59, the cells of these patients activate the alternative pathway of complement. The resultant complement-mediated lysis of erythrocytes causes the disease paroxysmal nocturnal hemoglobinuria (see Section 2-6, p. 40).

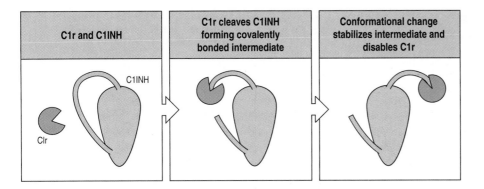

C1r and C1INH	C1r cleaves C1INH forming covalently bonded intermediate	Conformational change stabilizes intermediate and disables C1r

Figure 11.13 C1INH permanently inhibits C1r and C1s. The inactivation of C1r is shown here. C1s is inhibited in a similar way. C1INH is a member of the serpin family of protease inhibitors. They act as pseudosubstrates that bind to the active site of the protease and are cleaved by it. This forms a covalent bond between the protease and the inhibitor that stabilizes the protease and prevents it from releasing the pseudosubstrate and cleaving again. C1INH deficiency causes the syndrome hereditary angioneurotic edema.

Hereditary angioneurotic edema (**HANE**) is an autosomal dominant disease caused by deficiency in the complement regulator **C1 inhibitor** (**C1INH**). The disease is characterized by episodic bouts of subepithelial swelling of the face, larynx and abdomen. The swelling around the larynx can lead to death by suffocation. C1 inhibitor affects serine proteases such as C1r and C1s by binding to the active site and forming a covalent bond that irreversibly inactivates the protease. In patients deficient in C1 inhibitor the classical pathway is overactive, resulting in abnormally low levels of C2 and C4 in the blood and abnormally high production of the vasoactive C2a fragment. As well as participating in the regulation of complement activation, C1INH also controls serine proteases involved in blood clotting. In patients with HANE the blood-clotting pathways are also excessively active, resulting in abnormally high levels of the vasoactive peptide bradykinin. The combined actions of C2a and bradykinin cause fluid to leak out of the blood into the tissues, causing the edema that characterizes HANE.

C1 inhibitor is one example of a large family of serine and cysteine protease inhibitors called the **serpins**. They act as pseudosubstrates, each molecule of the inhibitor poisoning the active site of one protease molecule (Figure 11.13). The genetic dominance of C1 inhibitor deficiency does not arise from the protein's participation in a multisubunit complex, as is the case in IFN-γ receptor deficiency (see Section 11-10), but because one good copy of the *C1INH* gene cannot make sufficient inhibitor to control the complement and clotting cascades. HANE is treated with infusions of recombinant C1INH protein, some of which is purified from the milk of rabbits made transgenic for the human *C1INH* gene.

11-14 Defects in phagocytes result in enhanced susceptibility to bacterial infection

Phagocytosis by macrophages and neutrophils is the principal method by which the immune system removes and destroys bacteria and other microorganisms. Any defect that compromises phagocyte activity thus has a profound effect on the capacity to clear infections (Figure 11.14). One kind of deficiency arises from mutations in the gene encoding the protein CD18, which is the β₂ subunit of the leukocyte integrins LFA-1, CR3, and CR4. LFA-1 is needed for phagocytes to leave the blood and enter sites of infection (see Figure 2.31, p. 55). Phagocytes lacking functional integrins are unable to migrate to where they are needed. This syndrome is known as **leukocyte adhesion deficiency**.

Leukocyte adhesion deficiency is associated with persistent infection with extracellular bacteria, which cannot be cleared because of the defective phagocyte function. Children with this defect have recurrent pyogenic infections and problems with wound healing; if they survive long enough, they develop severe inflammation of the gums. Their neutrophils and macrophages cannot migrate into tissues and also, because CR3 and CR4 are complement receptors as well as adhesion molecules, the cells cannot take up and destroy

Syndrome	Cellular abnormality	Immune defect	Associated infections and other diseases
Leukocyte adhesion deficiency	Defective CD18 (cell adhesion molecule)	Defective migration of phagocytes into infected tissues	Widespread infections with capsulated bacteria
Chronic granulomatous disease (CGD)	Defective NADPH oxidase. Phagocytes cannot produce O_2^-	Impaired killing of phagocytosed bacteria	Chronic bacterial and fungal infections. Granulomas
Glucose-6-phosphate dehydrogenase (G6PD) deficiency	Deficiency of glucose-6-phosphate dehydrogenase. Defective respiratory burst	Impaired killing of phagocytosed bacteria	Chronic bacterial and fungal infections. Anemia is induced by certain agents
Myeloperoxidase deficiency	Deficiency of myeloperoxidase in neutrophil granules and macrophage lysosomes and impaired production of toxic oxygen species	Impaired killing of phagocytosed bacteria	Chronic bacterial and fungal infections
Chédiak–Higashi syndrome	Defect in vesicle fusion	Impaired phagocytosis due to inability of endosomes to fuse with lysosomes	Recurrent and persistent bacterial infections. Granulomas. Effects on many organs

Figure 11.14 Defects in phagocytic cells cause persistent bacterial infections.

bacteria opsonized with complement (see Section 2-5, p. 38). Patients deficient in CD18 suffer from infections that respond poorly to antibiotic treatment and persist despite the generation of normal B-cell and T-cell responses. Neutropenia caused by chemotherapy, malignancy, or aplastic anemia produces a similar susceptibility to severe pyogenic bacterial infections.

The capacity of phagocytes to kill ingested bacteria can also be blunted by a single defective gene. In **chronic granulomatous disease** (**CGD**), the antibacterial activity of phagocytes is compromised by their inability to produce the superoxide radical O_2^- (see Section 2-16, p. 57). Mutations affecting any of the four proteins of the NADPH oxidase system can produce this phenotype. The patients with this disease suffer from chronic bacterial infections, often leading to granuloma formation. Deficiencies in the enzymes glucose-6-phosphate dehydrogenase and myeloperoxidase also impair intracellular bacterial killing, leading to a similar but less severe phenotype. A different phenotype characterizes **Chédiak–Higashi syndrome**, in which phagocytosed materials are not delivered to lysosomes because of a defect in the vesicle fusion mechanism. This lack of phagocyte function has effects in many different organs as well as leading to persistent and recurrent bacterial infections. The mutations causing this disease are in the *CHS1* gene on chromosome 1, which is implicated in the generation of lysosomes.

11-15 Defects in T-cell function result in severe combined immune deficiencies

Whereas B cells contribute only to the antibody response, T cells function in all aspects of adaptive immunity. This means that inherited defects in the mechanisms of T-cell development and T-cell function have a general depressive effect on the immune system's capacity to respond to infection. Patients with T-cell deficiencies tend to be susceptible to persistent or recurrent infections with a broader range of pathogens than patients with B-cell deficiencies. Those patients who make neither T-cell-dependent antibody responses nor cell-mediated immune responses are said to have **severe combined immune deficiency** (**SCID**).

T-cell development and function depend on the action of many proteins, so the SCID phenotype can arise from defects in any one of a number of genes. Because of the unique inheritance pattern of the X chromosome, X-linked diseases are more easily discovered, and at least two forms of SCID are of this type. One is due to mutation in a gene on the X chromosome that encodes a protein subunit of several cytokine receptors, including those for IL-2, IL-4, IL-7, IL-9, and IL-15: it is called the **common gamma chain** (γ_c) and is quite different from the γ chain associated with the Fc receptors. The γ chain of the cytokine receptors interacts with the protein kinase Jak3 to induce signaling from the receptor when cytokine binds (a general view of cytokine receptor signaling via JAK protein kinases is shown in Figure 8.26, p. 233). Indeed, as would be predicted, patients defective in Jak3 kinase have an autosomally inherited immunodeficiency similar in phenotype to that of patients with X-linked SCID. The phenotype of SCID is so severe that affected infants survive only if they are kept isolated in a pathogen-free environment until their immune system has been replaced by bone marrow transplantation and the passive administration of antibodies (see Section 5-2, p. 128).

Another X-linked deficiency of T-cell function is **Wiskott–Aldrich syndrome** (**WAS**), which involves the impairment of platelets as well as lymphocytes. It also shows up in childhood as a history of recurrent infections but is less immunologically severe than SCID. Children have normal levels of T and B cells but do not make good antibody responses; they can be treated with regular injections of gamma globulin to replace the missing antibodies. The relevant gene on the X chromosome encodes a protein called Wiskott–Aldrich syndrome protein (WASP). This protein is involved in the cytoskeletal reorganization that is needed before T cells can deliver cytokines and signals to the B cells, macrophages, and other target cells with which they routinely interact during the immune response.

SCID due to an absence of T-cell function is also caused by defects in **adenosine deaminase** (**ADA**) or **purine nucleoside phosphorylase** (**PNP**), which are enzymes involved in purine degradation. Although the absence of these enzymes causes an accumulation of nucleotide metabolites in all cells, the effects of this storage are particularly toxic to developing T cells and, to a smaller extent, to developing B cells. Infants with these immunodeficiencies have an underdeveloped thymus that contains few lymphocytes. ADA and PNP deficiencies are autosomally inherited (Figure 11.15).

Lack of HLA class II molecules also causes SCID. The deficiency was named **bare lymphocyte syndrome** because the defect was first discovered on B lymphocytes, the major population of peripheral blood cells that expresses HLA class II. In these patients, CD4 T cells fail to develop (see Section 7-10, p. 201), which compromises all aspects of adaptive immunity. Bare lymphocyte syndrome arises from defects in transcriptional regulators essential for the expression of all HLA class II loci. A homozygous defect in any one of four proteins produces the condition. One protein is the class II transactivator (CIITA), the other three are components of RFX, a transcriptional complex that binds to a conserved sequence in the promoter of HLA class II genes called the X box.

A defect in either of the two genes encoding the TAP peptide transporter impedes the binding of peptides by HLA class I molecules, leading to an unusually low abundance of HLA class I molecules on cell surfaces. This form of immunodeficiency, called **bare lymphocyte syndrome (MHC class I)**, is less severe than the SCID caused by the absence of HLA class II, its principal effect being the selective loss of CD8 T cells (see Section 7-10, p. 201) and of cytotoxic T-cell responses to intracellular infections.

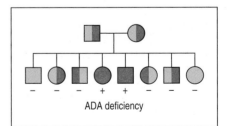

Figure 11.15 Adenosine deaminase (ADA) deficiency is inherited in an autosomal fashion. The inheritance of ADA deficiency in a family. Both parents are healthy carriers who have one functional copy of the ADA gene (red) and one defective copy (green). Two of the eight children inherited defective copies of the ADA gene from both parents and have ADA deficiency (+). Males are indicated by squares and females by circles.

Defects in various proteins and enzymes that contribute to the rearrangement of immunoglobulin and T-cell receptor genes also cause autosomally inherited forms of SCID or a related immunodeficiency called Omenn syndrome, depending on the particular defect (see Section 5-2, p. 128). These include the RAG-1 and RAG-2 proteins, the nuclease Artemis, and the DNA-dependent protein kinase (DNA-PK).

11-16 Some inherited immunodeficiencies lead to specific disease susceptibilities

Patients who lack the receptor for the cytokine IFN-γ suffer persistent and sometimes fatal infections of common intracellular bacteria, such as the ubiquitous nontuberculous strain of mycobacterium, *Mycobacterium avium* (see Section 11-10). **IL-12 receptor deficiency**, in which the receptor for the cytokine IL-12 is nonfunctional, produces a similar susceptibility to intracellular bacterial infections. In the innate immune response there is a mutual activation of NK cells and macrophages (Figure 11.16, left panel). This involves the IL-12 secreted by macrophages binding to the IL-12 receptor of NK cells and stimulating them to secrete IFN-γ. The IFN-γ then binds to the IFN-γ receptor on macrophages and activates phagocytosis and the secretion of proinflammatory cytokines. In the absence of a functioning IL-12 receptor, this cycle of mutual reinforcement cannot begin.

In the adaptive immune response, IL-12 secreted by macrophages binds to the IL-12 receptors of T cells, helping to induce the differentiation of T$_H$1 cells from activated antigen-specific naive CD4 T cells (see Section 8-10, p. 227). On interacting with antigen on the macrophage surface, T$_H$1 cells secrete IFN-γ, which acts on the macrophage to strengthen its activation, and thus leads to the destruction of the intracellular bacteria (Figure 11.16, right panel). IL-12 also acts on cytotoxic T cells to induce them to produce IFN-γ, which helps maintain macrophage activation (see Section 8-14, p. 236) and also provides an environment favoring the differentiation of T$_H$1 cells. So, again, in the absence of a functioning IL-12 receptor this mutual activation of macrophage and effector T cells cannot get started. Unable to make strong innate or adaptive immune responses to intracellular bacteria, people lacking an IL-12 receptor suffer persistent infections with strains of mycobacteria that are common in the environment. When individuals who had not yet been diagnosed with a deficiency of the IL-12 receptor or IFN-γ receptor were

Figure 11.16 Mutual activation of macrophages and effector lymphocytes in the innate and adaptive immune responses to intracellular bacterial infections. In the innate immune response, macrophages are activated by IFN-γ made by NK cells and in turn produce the cytokine IL-12. This binds to IL-12 receptors on the NK cells, inducing further secretion of IFN-γ and maintenance of macrophage activation (first panel). In an adaptive immune response, IL-12 secreted by activated macrophages acts on activated T$_H$1 cells, inducing their differentiation into IFN-γ secreting T$_H$1 cells, which interact with the macrophage to strengthen its activation. CD8 cytotoxic T cells (CTLs) are also responsive to the IL-12 produced by the macrophage and produce more IFN-γ, which also acts back on the macrophage to maintain and strengthen activation and destroy the intracellular bacteria. The IFN-γ receptor is shown in simplified form here (see Figure 11.10 for the complete functional receptor). In immunodeficient patients lacking either a functional IL-12 receptor or a functional IFN-γ receptor, this cycle of mutual activation cannot proceed and infection persists.

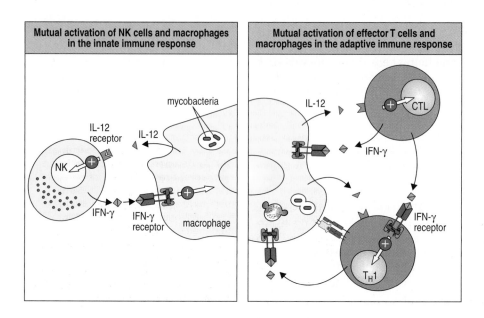

| Mutual activation of NK cells and macrophages in the innate immune response | Mutual activation of effector T cells and macrophages in the adaptive immune response |

vaccinated against tuberculosis with the Calmette–Guérin vaccine strain of live *Mycobacterium bovis*, the vaccine caused disseminated infection and disease, because these people could not control this normally nonpathogenic mycobacterium.

As we saw in Section 11-4, many healthy people maintain a persistent EBV infection of B cells, which is held in check by EBV-specific T cells. For patients, mostly boys, with a defect in a gene called *SH2D1A* on the X chromosome, a balance is never achieved and childhood EBV infections can become overwhelming and can even progress to lymphoma. This immunodeficiency is called **X-linked lymphoproliferative syndrome**. Although the SH2D1A protein is believed to be a regulator of lymphocyte-activating signals, its precise functions and contribution to the control of EBV infection are not yet clear.

11-17 Transplantation of hematopoietic stem cells is used to correct genetic defects of the immune system

Many immunodeficiencies are due to gene defects that principally affect hematopoietic cells. These conditions can be corrected by transplantation of hematopoietic stem cells in bone marrow—a **bone marrow transplant**. Such therapy is not undertaken lightly and the potential benefits of correcting the immunodeficiency must be weighed against the risks associated with the trauma of the procedure and the effects of the immunosuppressive drugs required for successful transplantation. In this procedure, a patient's own bone marrow is destroyed by a combination of radiation and cytotoxic chemotherapeutic drugs and is replaced by a graft from a healthy donor, involving the transfusion of bone marrow, which then reconstitutes the entire hematopoietic system (Figure 11.17).

The success of a bone marrow transplant is directly correlated with the degree of HLA matching between patient and donor. Matching serves two purposes. First, it reduces the extent of **graft-versus-host disease** (**GVHD**), a tissue-damaging reaction caused by mature T cells in the transplant that respond to the allogeneic MHC class I and II molecules of the recipient. Second, it ensures effective reconstitution of the adaptive immune system. After transplantation, T cells developing from stem cells in the bone marrow graft migrate to the thymus, where they mature into T cells under the influence of the thymic epithelial cells of the recipient and the HLA class I and II molecules they express. The developing T cells are positively selected in the recipient's thymus for interaction with the recipient's HLA allotypes. To reconstitute T-cell functions, the new T cells must be able to respond to antigens presented by the professional antigen-presenting cells (dendritic cells, B cells, and macrophages) derived from bone marrow, which after transplantation will all be of donor origin and donor HLA type. To satisfy this requirement, the donor and recipient must share at least one HLA class I allotype and one HLA class II allotype. In the hypothetical situation in which a transplant recipient shared

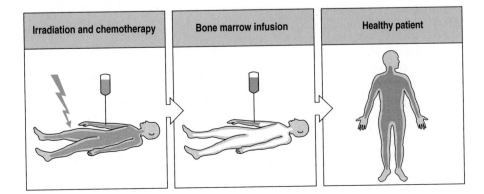

Irradiation and chemotherapy	Bone marrow infusion	Healthy patient

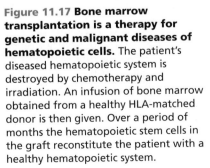

Figure 11.17 Bone marrow transplantation is a therapy for genetic and malignant diseases of hematopoietic cells. The patient's diseased hematopoietic system is destroyed by chemotherapy and irradiation. An infusion of bone marrow obtained from a healthy HLA-matched donor is then given. Over a period of months the hematopoietic stem cells in the graft reconstitute the patient with a healthy hematopoietic system.

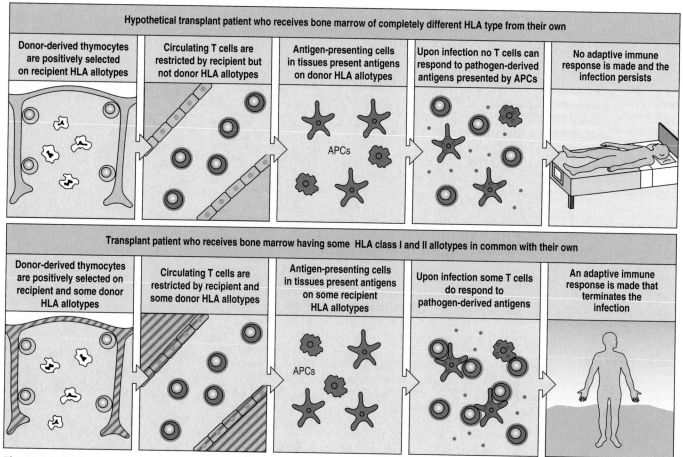

Figure 11.18 The donor and recipient of a bone marrow transplant must share HLA class I and II molecules if they are to reconstitute T-cell function in the recepient. After bone marrow transplantation, donor-derived thymocytes are positively selected on the HLA molecules carried by the recipient's thymic epithelium. The top panels show the hypothetical situation in which none of the recipient's HLA allotypes (red) are the same as the donor's HLA allotypes (blue). In this situation the recipient could not reconstitute a working T-cell system and would suffer from severe combined immunodeficiency. The lower panels show the situation when the recipient and donor share the HLA allotypes indicated by blue. In clinical practice, bone marrow transplant donors and recipients are chosen to share as many HLA class I and II allotypes as possible. APC, antigen-presenting cell.

no MHC allotypes with the donor, the recipient would end up with severe combined immunodeficiency, even if the transplant itself were successful. This is because the reconstituting T cells selected on the recipient's HLA molecules in the thymus would be unable to recognize peptide antigens presented by the donor's HLA molecules of the donor-derived dendritic cells, and other antigen-presenting cells, in the periphery (Figure 11.18). The extent to which the mature T cells respond to antigens presented by professional antigen-presenting cells of donor HLA type is directly correlated with the number of HLA class I and II allotypes that the donor and recipient share. For best immune reconstitution and minimal graft-versus-host disease, the donor of choice for any patient is a healthy sibling of identical HLA type.

Now that the genes responsible for immunodeficiency diseases are increasingly being identified, another type of therapeutic strategy is being explored. In **somatic gene therapy**, a functional copy of the defective gene is introduced into stem cells that have been isolated from the patient's bone marrow. The stem cells in which the defect has been corrected are then reinfused into the patient, where they provide a self-renewing source of immunocompetent lymphocytes and other hematopoietic cell types. Although attractive in principle, the practical development of gene therapy is at an early and experimental stage.

Summary

The best-characterized gene defects affecting the immune system are those that show up in early childhood and confer exceptional vulnerability to common infections. The characterization of immunodeficiency diseases and the gene defects that cause them is almost the only way in humans of determining the relative importance of different cells and molecules in immune defenses, and of testing current models of how the human immune system works. The most severe immunodeficiencies are due to gene defects that cause an absence of all T-cell function and thus, directly or indirectly, impair B-cell function as well. Such deficiencies are known as severe combined immune deficiencies (SCID). The absence of antibodies due to genetic defects in B-cell development or function leads to particular susceptibility to pyogenic bacteria. Deficiencies in the early components of complement pathways cause a failure to opsonize pathogens. This results in increased susceptibility to bacterial infection, as do defects in phagocytes.

Acquired immune deficiency syndrome

Acquired immune deficiency syndrome (**AIDS**) was first described by physicians early in the 1980s. The disease is characterized by a massive reduction in the number of CD4 T cells, accompanied by severe infections of pathogens that rarely trouble healthy people, or by aggressive forms of Kaposi's sarcoma or B-cell lymphoma. All patients diagnosed as having AIDS eventually die from the effects of the disease. In 1983 the virus now known to cause AIDS, the **human immunodeficiency virus** (**HIV**), was first isolated. Two types of HIV are now distinguished—HIV-1 and HIV-2. In most countries HIV-1 is the principal cause of AIDS. HIV-2 is less virulent, causing a slower progression to AIDS. It is endemic in West Africa and has spread widely through Asia.

AIDS is a disease new to the medical profession and also to the human species. The earliest evidence for HIV comes from samples from African patients obtained in the late 1950s. It is believed that the viruses first infected humans in Africa by jumping from other primate species—HIV-1 coming from the chimpanzee, HIV-2 from the sooty mangabey, a type of monkey. In neither of these species does the endogenous HIV-related virus cause disease.

As commonly occurs when a naive host population is hit with a new infectious agent, the effects of HIV on the human population have been immense and AIDS is now a disease of pandemic proportions. The World Health Organization currently estimates that 33 million people are infected with HIV (Figure 11.19). Although advances continue to be made in understanding the nature of the disease and its origins, the number of people infected with HIV continues to grow—2.5 million new infections in 2007—and tens of millions of people will die from AIDS in the years to come (see Figure 1.28, p. 25).

11-18 HIV is a retrovirus that causes slowly progressing disease

HIV is an RNA virus with an RNA nucleoprotein core (the nucleocapsid) surrounded by a lipid envelope derived from the host-cell membrane and containing virally encoded envelope proteins (Figure 11.20). HIV is an example of a **retrovirus**, so named because these viruses use an RNA genome to direct the synthesis of a DNA intermediate, a situation backwards or 'retro' from that used by most biological entities. One nucleocapsid protein is a protease used to cleave the gp41 and gp120 envelope glycoproteins from a larger precursor polyprotein; other nucleocapsid proteins are the enzymes reverse transcriptase and integrase, which are required for viral replication.

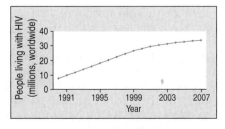
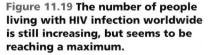

Figure 11.19 The number of people living with HIV infection worldwide is still increasing, but seems to be reaching a maximum.

Figure 11.20 The virion of human immunodeficiency virus (HIV). The upper panel is an electron micrograph showing three virions. The lower panel is a diagram of a single virion. gp120 and gp41 are virally encoded envelope glycoproteins of molecular masses 120 kDa and 41 kDa, respectively. Photograph courtesy of Hans Gelderblom.

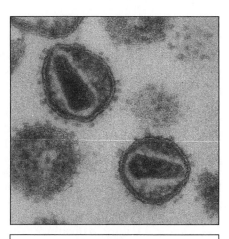

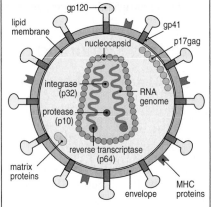

When HIV infects a cell, the RNA genome is first copied into a complementary DNA (cDNA) by reverse transcriptase. The viral integrase then integrates the cDNA into the genome of the host cell to form a **provirus**, a process facilitated by repetitive DNA sequences called long terminal repeats (LTRs) that flank all retroviral genomes. Proviruses use the transcriptional and translational machinery of the host cell to make viral proteins and RNA genomes, which assemble into new infectious virions. The genes and proteins of HIV are illustrated in Figure 11.21. HIV belongs to a group of retroviruses that cause slowly progressing diseases. They are collectively called the **lentiviruses**, a name derived from the Latin word *lentus*, meaning slow. In the course of infecting a human cell, HIV recruits 273 human proteins to serve its own purpose.

In almost all people, HIV produces an infection that cannot be successfully terminated by the immune system and continues for many years. Although the initial acute infection is controlled to the point at which disease is not apparent, the virus persists and replicates in a manner that gradually exhausts the immune system, leading to immunodeficiency and death.

11-19 HIV infects CD4 T cells, macrophages, and dendritic cells

Infection with HIV usually occurs after the transfer of bodily fluids from an infected person to an uninfected recipient. Provirus can be carried in infected CD4 T cells, dendritic cells, and macrophages, whereas virions can be transmitted via blood, semen, vaginal fluid, or mother's milk. Infection is commonly spread by sexual intercourse, intravenous administration of drugs with contaminated needles, breast-feeding, or transfusion of human blood or blood components from HIV-infected donors. Most infections take place across a mucosal surface.

Macrophages, dendritic cells, and CD4 T cells are vulnerable to HIV infection because they express CD4, which the virus uses as a receptor. Chimpanzees, our closest living relatives, are resistant to HIV infection because of a small difference in the structure of their CD4 glycoprotein in comparison with ours. The gp120 envelope glycoprotein of HIV binds tightly to human CD4, enabling virions to attach to CD4-expressing human cells. Before entry of the virus into the cell, gp120 must also bind to a co-receptor in the host-cell membrane. Once the co-receptor is bound, the gp41 envelope glycoprotein mediates fusion of the viral envelope with the plasma membrane of the host cell, allowing the viral genome and associated proteins to enter the cytoplasm.

The viral co-receptors are normal human chemokine receptors that HIV subverts to further its own propagation. There are different variants of HIV, and the cell types that they infect largely depend upon which co-receptor they bind. The HIV variants that spread infection from one person to another bind to the co-receptor CCR5 present on macrophages, dendritic cells, and CD4 T cells. Although they infect several types of human cell, these HIV variants are called 'macrophage-tropic' for want of a better term. The HIV variants that infect activated CD4 T cells bind to the co-receptor CXCR4 and are called 'lymphocyte-tropic.' Whereas infection by macrophage-tropic HIV variants requires only modest levels of cell-surface CD4, infection by the lymphocyte-

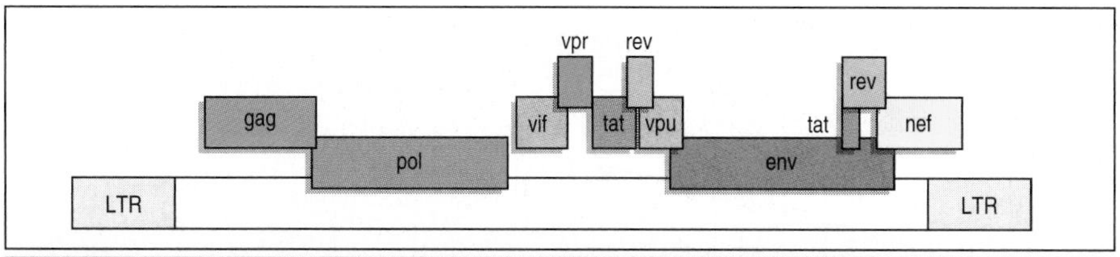

Gene		Protein
gag	Group-specific antigen	Core proteins and matrix proteins
pol	Polymerase	Reverse transcriptase, protease, and integrase enzymes
env	Envelope	Transmembrane glycoproteins. gp120 binds CD4 and CCR5; gp41 is required for virus fusion and internalization
tat	Transactivator	Positive regulator of transcription
rev	Regulator of viral expression	Allows export of unspliced and partially spliced transcripts from nucleus
vif	Viral infectivity	Affects particle infectivity
vpr	Viral protein R	Transport of DNA to nucleus. Augments virion production. Arrests cell cycle
vpu	Viral protein U	Promotes intracellular degradation of CD4 and enhances release of virus from cell membrane
nef	Negative-regulation factor	Augments viral replication *in vivo* and *in vitro*. Downregulates CD4 and MHC class II

Figure 11.21 The genes and proteins of HIV-1. HIV-1 has an RNA genome consisting of nine genes flanked by long terminal repeats (LTRs). The products of the nine genes and their known functions are tabulated. Several of the viral genes are overlapping and are read in different frames. Others encode large polyproteins that after translation are cleaved to produce several proteins having different activities. The gag, pol, and env genes are common to all retroviruses, and their protein products are all present in the virion.

tropic viruses requires the higher levels present on activated CD4 T cells. Macrophages and dendritic cells at the site of virus entry are the first cells to be infected. Subsequently, the virus produced by the macrophages starts to infect the CD4 T-cell population. In about 50% of cases, the viral phenotype switches to the lymphocyte-tropic type late in infection. This is followed by a rapid decline in CD4 T-cell count and progression to AIDS. In their mutual interaction with the human population, the two types of HIV variant have complementary roles: the lymphocyte-tropic viruses cause the disease, while the macrophage-tropic variants make it a pandemic.

The production of infectious virions from HIV provirus requires the infected CD4 T cell to be activated. Activation induces the synthesis of the transcription factor NFκB, which binds to promoters in the provirus. This directs the infected cell's RNA polymerase to transcribe viral RNAs. At least two of the proteins encoded by the virus serve to promote replication of the viral genome. Among other activities, the **Tat** protein binds to a sequence in the LTR of the viral mRNA, known as the transcriptional activation region (TAR), where it prevents transcription from shutting off and thus increases the transcription of viral RNA. The **Rev** protein controls the supply of viral RNA to the cytoplasm and the extent to which that RNA is spliced. At early times in infection Rev delivers RNA that encodes the proteins necessary for making virions. Later, complete viral genomes are supplied, which assemble with the viral proteins to form complexes that bud through the plasma membrane to give infectious virions (Figure 11.22).

11-20 Most people who become infected with HIV progress in time to develop AIDS

Immediately after infection with HIV a person can either be asymptomatic or experience a transient 'flu-like' illness. In either case, virus becomes abundant in the peripheral blood, while the number of circulating CD4 T cells

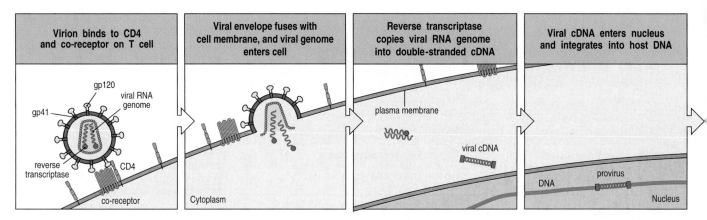

| Virion binds to CD4 and co-receptor on T cell | Viral envelope fuses with cell membrane, and viral genome enters cell | Reverse transcriptase copies viral RNA genome into double-stranded cDNA | Viral cDNA enters nucleus and integrates into host DNA |

Figure 11.22 The life cycle of HIV in human cells. The virus binds to CD4 on a cell surface by using the envelope protein gp120, which is altered by CD4 binding so that it now also binds a specific chemokine co-receptor on the cell. This binding releases gp41, which then causes fusion of the viral envelope with the plasma membrane and release of the viral core into the cytoplasm. The RNA genome is released and is reverse transcribed into double-stranded cDNA. This migrates to the nucleus in association with the viral integrase, and it is then integrated into the cell genome, becoming a provirus. Activation of a T cell causes low-level transcription of the provirus that directs the synthesis of the early proteins Tat and Rev. These then expand and change the pattern of provirus transcription to produce mRNA encoding the protein constituents of the virion and RNA molecules corresponding to the HIV genome. Envelope proteins travel to the plasma membrane, whereas other viral proteins and viral genomic RNA assemble into nucleocapsids. New virus particles bud from the cell, acquiring their lipid envelope and envelope glycoproteins in the process.

markedly declines (Figure 11.23). This acute viremia is almost always accompanied by activation of an HIV-specific immune response, in which anti-HIV antibodies are produced and cytotoxic T cells become activated to kill virus-infected cells. This response reduces the load of virus carried by the infected person and causes a corresponding increase in the number of circulating CD4 T cells. When an infected person first exhibits detectable levels of anti-HIV antibodies in their blood serum, they are said to have undergone **seroconversion**. The amount of virus persisting in the blood after the symptoms of acute viremia have passed is directly correlated with the subsequent course of disease.

The initial phase of infection is followed by an asymptomatic period, also called 'clinical latency.' During this phase, which can last for 2–15 years, there is persistent infection and replication of HIV in CD4 T cells, causing a gradual decrease in T-cell numbers. Eventually, the number of CD4 T cells drops below that required to mount effective immune responses against other infectious agents. That transition marks the end of clinical latency, the beginning of the period of increasing immunodeficiency, and the onset of AIDS. Patients with AIDS become susceptible to a range of opportunistic infections and some cancers, and it is from the effects of these that they die.

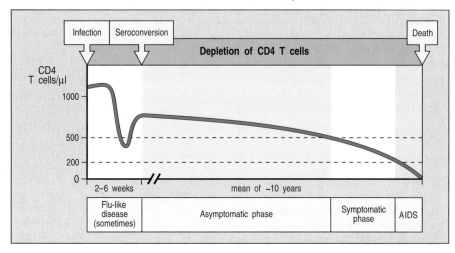

Figure 11.23 After infection with HIV there is a gradual extinction of CD4 T cells. The number of CD4 T cells (green line) refers to those present in peripheral blood. Opportunistic infections and other symptoms become more frequent as the CD4 T-cell count falls, starting at around 500 cells/μl. The disease then enters the symptomatic phase. When CD4 T-cell counts fall below 200 cells/μl the patient is said to have AIDS.

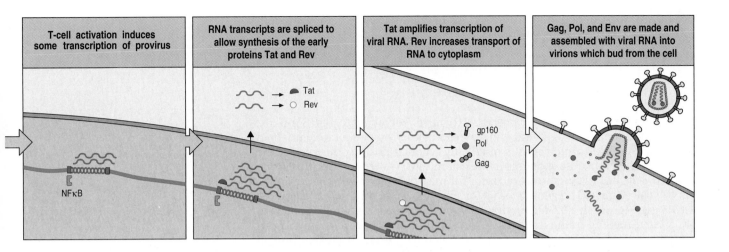

| T-cell activation induces some transcription of provirus | RNA transcripts are spliced to allow synthesis of the early proteins Tat and Rev | Tat amplifies transcription of viral RNA. Rev increases transport of RNA to cytoplasm | Gag, Pol, and Env are made and assembled with viral RNA into virions which bud from the cell |

At the beginning of the AIDS epidemic in North America and Europe, viral transmission through infected blood products caused hemophiliacs and other patients dependent on blood products to become infected with HIV. Because hemophiliacs are so dependent on the medical profession, whether they be HIV-infected or not, it was possible to study the progress of their HIV infections in a systematic and rigorous manner. The results, which were obtained during the time before effective treatments were available, demonstrate that most HIV-infected individuals are destined to progress to AIDS in the absence of effective medical intervention (Figure 11.24). Today, HIV infection through contaminated blood products has largely been eliminated in the richer countries by routine screening of individual units of blood for the presence of HIV.

Although the vast majority of infected people gradually progress to AIDS, a small minority do not; they are called 'long-term nonprogressors.' A few of them seroconvert, but their CD4 T-cell counts and other measures of immune competence are maintained. They have exceptionally low levels of circulating virus, and are being studied intensively to determine how they are able to control the infection. Another small group of people remain seronegative and disease-free despite extensive exposure to the virus, often as sex workers. Some of this group have specific cytotoxic lymphocytes and T_H1 cells directed against infected cells, which suggests that at some time they have either been infected with the virus or have been exposed to noninfectious HIV antigens. The mechanisms that allow these people to control HIV and resist AIDS are just beginning to be understood. As described in the next section, some peo-

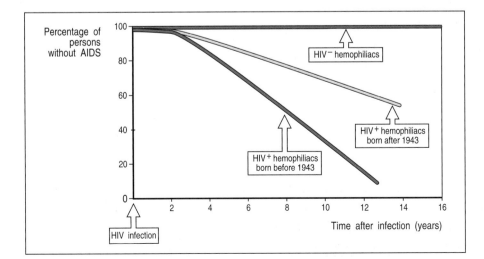

Figure 11.24 Once an HIV infection is established it usually leads to AIDS. Hemophilia is an inherited disease in which the blood clots poorly; any wound can cause excessive bleeding and be potentially fatal. Hemophilia is treated by regular intravenous infusions of clotting factors purified from the blood of healthy donors. In the early 1980s, when the AIDS epidemic was under way but its cause was still unknown, some seemingly healthy blood donors were infected with HIV. The virus from their donations contaminated several batches of clotting factor, and many hemophiliacs became infected with HIV. The graph shows the progression to AIDS of HIV-infected and uninfected hemophiliacs after infection. HIV infection almost inevitably leads to AIDS, as seen from the linear plots which do not taper off. The rate of progression to AIDS increases with the patients' age.

ple have a well-defined inherent resistance to infection by HIV: they are deficient in the viral co-receptor CCR5.

11-21 Genetic deficiency of the CCR5 co-receptor for HIV confers resistance to infection

Some people who are heavily exposed to HIV never become infected. Similarly, their isolated macrophages and lymphocytes cannot be deliberately infected with macrophage-tropic variants of HIV, the viral strains responsible for the spread of infection. These people are resistant to HIV infection because none of their cells has CCR5, the co-receptor for macrophage-tropic HIV variants, on the cell surface. The genetic basis for this defect is a mutated allele of the CCR5 gene in which a 32-nucleotide deletion from the coding region leads to an altered reading frame, premature termination of translation, and a non-functional protein. This deletion variant, called *CCR5-Δ32*, is present only in caucasian populations, in which 10% of the population is heterozygous for the variant and 1% is homozygous. Only people who are homozygous for *CCR5-Δ32* resist HIV infection. A few such homozygotes have become infected with HIV, apparently as a result of primary infection by lymphocyte-tropic strains of the virus that use the CXCR4 co-receptor.

CCR5 is a normal component of the human immune system, in which it is a receptor for the chemokines CCL3 (MIP-1α), CCL4 (MIP-1β) and CCL5 (RANTES). The fact that almost everyone has a functional gene for CCR5 strongly argues that this receptor has made useful contributions to human immunity, survival, and reproduction. However, for individuals facing exposure to HIV, the survival advantage of not having CCR5 clearly outweighs that of having it. In the absence of the HIV pandemic, homozygosity for the CCR5 deletion would be considered a mild form of immunodeficiency, like some of those described in the previous part of this chapter, but in today's world it becomes an important asset of disease resistance. This type of evolutionary process, in which an immune-system component that was useful in fighting past wars against pathogens becomes detrimental during a current conflict, has happened throughout human history. A pathogen subverts an immune-system component and consequently selects for the survival of those humans who have genetic variants that resist that subversion. What we experience is a never-ending arms race, forever selecting for polymorphism and change in the human immune system.

That the CCR5 deletion variant was already at a high frequency in the caucasian population before HIV began to colonize the human species is an indication that this mutation enhanced human survival in Europe during previous epidemics of infectious disease. Both plague and smallpox have been advanced as candidate selective agents. In the current age, HIV will probably have its strongest selective effect in sub-Saharan Africa, where some 30 million people are infected with HIV and constitute more than 30% of the population of several countries. As treatment is rarely available for African patients, mortality rates will be high and survival will be greatly enhanced by genetic variants in any immune-system component that prevents infection or reduces disease severity.

11-22 HLA and KIR polymorphisms influence the progression to AIDS

A feature of the early years of the AIDS epidemic was that a majority of those infected and diagnosed were young, well-educated people living in affluent countries. Thousands of these individuals were recruited into longitudinal studies that have been correlating the clinical progression of the infection with genetic polymorphisms of the immune system. In addition to showing that HLA homozygosity speeds the progression to AIDS (see Figure 5.35, p. 153), these studies showed that the HLA-B*27 and HLA-B*57 allotypes slow progression. Contributing to this effect, the HIV peptides presented by these HLA-B*27 and HLA-B*57 allotypes stimulate stronger CD8 T-cell responses to HIV-infected cells than do the peptides presented by other HLA-B allotypes.

Another property that the HLA-B*27 and HLA-B*57 allotypes have in common is the Bw4 motif, which is the ligand for the NK-cell receptor KIR3DL1/S1. Combinations of Bw4+ HLA-B allotypes and KIR3DL1/S1 allotypes are associated with different rates of progression to AIDS. The most favorable combinations are HLA-B*57 with high-expressing allotypes of KIR3DL1, and HLA-B*27 with low-expressing KIR3DL1 allotypes (Figure 11.25). These correlations suggest that the type of NK-cell response at the start of an HIV infection affects the relative success of the subsequent adaptive response.

11-23 HIV escapes the immune response and develops resistance to antiviral drugs by rapid mutation

People infected with HIV make adaptive immune responses that can prevent the overt symptoms of disease for many years. Included in these responses are T_H1 and T_H2 cells, B cells that make neutralizing antibodies, and CD8 cytotoxic T cells that kill virus-infected cells (Figure 11.26). However, the virus is rarely eliminated. One reason that the virus keeps ahead of the human immune response is its high rate of mutation during the course of an infection.

HIV and other retroviruses have high mutation rates because their reverse transcriptases lack proofreading mechanisms of the type possessed by cellular DNA polymerases. Consequently, reverse transcriptases are prone to making errors and these nucleotide substitutions soon accumulate to give new variant viral genomes. Even though the infection of a person might start from a single viral species, mutation throughout the infection produces many viral variants, called quasi-species, that coexist within the infected person.

The presence of variant viruses increases the difficulty of terminating the infection by immune mechanisms. Immune attack by neutralizing antibody will select for the survival of viral variants that have lost the epitope recognized by the antibody. Similarly, pressure from virus-specific cytotoxic T cells selects for viruses in which the peptide epitope recognized by the cytotoxic T cell has changed. In some instances the homologous peptide derived from the variant virus interferes with the presentation of the antigenic peptide from the original virus, thereby allowing both viral species to escape the cytotoxic T cell.

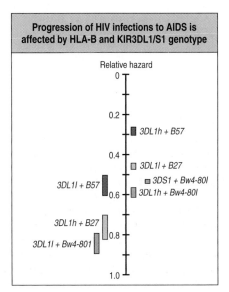

Figure 11.25 Combinations of KIR3DL1/S1 and HLA-B allotypes are associated with variable progression to AIDS for patients infected with HIV. KIR3DL1 and KIR3DS1 are alternative forms of the receptor, with the former being inhibitory and the latter activating. Relative hazard is a measure of the speed of progression to AIDS, with 1.0 indicating a faster rate than 0. Bw4, the ligand for KIR3DL1 and KIR3DS1, is an epitope carried by about one-third of HLA-B allotypes and is determined by polymorphisms at positions 77–83 of the HLA-B heavy chain. HLA-B27 and HLA-B57 both have a Bw4 epitope that has isoleucine at position 80. Bw4-80I denotes all HLA-B allotypes that have this Bw4 epitope. The inhibitory KIR3DL1 allotypes are divided into two groups—3DL1l and 3DL1h—that have low and high cell-surface expression, respectively. Patients who lack a Bw4 epitope have a relative hazard of 1.0. Data courtesy of Mary Carrington.

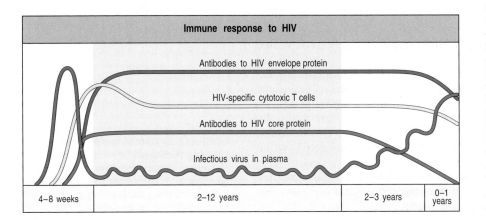

Figure 11.26 The adaptive immune response to HIV limits the effects of infection. During the course of infection, the levels of HIV virions and of components of adaptive immunity in the blood plasma change. Early in infection, while the adaptive immune response is being activated, the virus reaches high levels (red line). With the production of HIV-specific antibodies (blue lines) and HIV-specific cytotoxic T cells (yellow line), the virus is kept at a low level but is not eliminated. When the destruction of CD4 T cells outstrips their rate of renewal, adaptive immunity gradually collapses and virus levels increase again.

Figure 11.27 HIV can rapidly acquire resistance to protease inhibitor drugs. When a patient is treated with a single protease inhibitor drug the decrease in viral load (top panel) and increase in CD4 T-cell numbers (center panel) is only transient. The drug's benefit is short-lived because it selects for existing minority populations of drug-resistant HIV variants that rapidly expand and then continue the progression to AIDS (bottom panel).

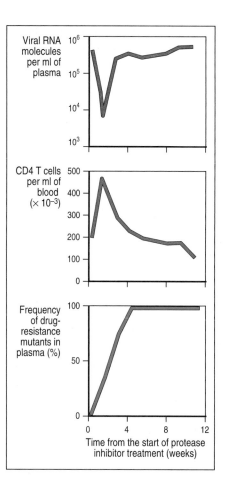

The high mutation rate of HIV greatly complicates the task of developing a vaccine against it. It also limits the effectiveness of antiviral drugs. Potential targets for drugs are the viral reverse transcriptase that is essential for the synthesis of provirus, and the viral protease that cleaves viral polyproteins to give viral enzymes and proteins. Inhibitors of reverse transcriptase and protease have been found, and these drugs prevent the further infection of healthy cells. Unfortunately, mutation inevitably produces HIV variants with proteins resistant to the action of the drugs. For protease inhibitors, resistance appears within days (Figure 11.27), whereas resistance to the reverse transcriptase inhibitor zidovudine (AZT) takes months to develop. The difference is because zidovudine resistance requires the accumulated effects of three or four independent mutations in the reverse transcriptase, whereas the protease can acquire resistance with just one mutation. One situation in which a relatively short course of drug treatment has long-term benefit is pregnancy. Treating HIV-infected pregnant women with zidovudine can prevent HIV transmission to their children *in utero* and at birth.

Because HIV can so easily escape from the effects of any single drug, several antiviral drugs are now used together in what is called **combination therapy**. The ideal is to destroy the entire population of viruses before any one of them has accumulated enough mutations to resist all the drugs. Such combination therapy, also called **highly active anti-retroviral therapy** (**HAART**) is effective at reducing the abundance of virus (the viral load) (Figure 11.28) and retarding disease progression. The drugs do not stop virus production by cells that are already infected, but they prevent new infections from forming a provirus and becoming productive. Two weeks after starting combination therapy the amount of virus in the blood has decreased to about 5% of the level before treatment. The rapidity of this decline is because both activated CD4 T cells and free virions have short lifetimes (Figure 11.29). At this point no naive or effector CD4 T cells are making virus and the level now present in the blood is due to production by longer-lived HIV-infected cells: macrophages, dendritic cells, and memory CD4 T cells. Continuing treatment further decreases the abundance of virus, but at a slower rate than before. Although the virus eventually becomes undetectable this does not reflect its eradication, because the virus re-emerges in patients who stop taking their medicine.

11-24 Clinical latency is a period of active infection and renewal of CD4 T cells

The administration of antiviral drugs to HIV-infected individuals has revealed the active nature of the infection during the period of clinical latency. Within 2 days of starting a course of drugs the amount of virus in the blood decreases markedly. At the same time, the number of CD4 T cells increases substantially (see Figure 11.27). This shows two things: first, that virus is being produced and cleared continuously within infected individuals, and second, that in the face of HIV infection the body continues to produce new CD4 T cells, which quickly become infected with HIV. Thus the period of clinical latency, when the overall numbers of CD4 T cells in the blood are gradually declining, is actually a time of immense immune activity. Vast numbers of T cells are produced and die, and virions are neutralized, the latter most probably through antibody-mediated opsonization and phagocytosis.

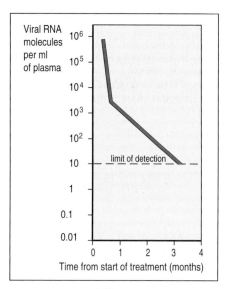

Figure 11.28 Combination drug therapy reduces HIV in the blood to below detectable levels. The abundance of HIV in patients' blood at different times after starting a course of combination drug therapy is shown.

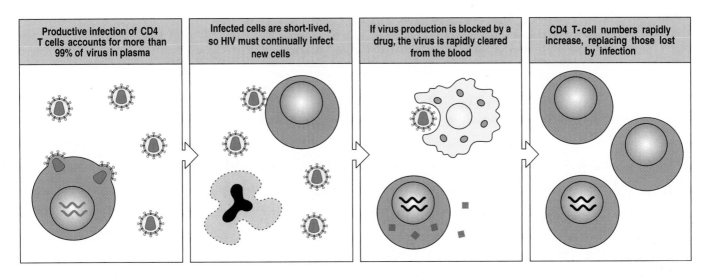

| Productive infection of CD4 T cells accounts for more than 99% of virus in plasma | Infected cells are short-lived, so HIV must continually infect new cells | If virus production is blocked by a drug, the virus is rapidly cleared from the blood | CD4 T-cell numbers rapidly increase, replacing those lost by infection |

Figure 11.29 Antiviral drugs rapidly clear virus from the blood and increase the number of circulating CD4 T cells. The first and second panels show that maintenance of HIV levels in the blood depends on the continual infection of newly produced CD4 T cells. This is because cells live for only a few days once infected. The third and fourth panels show the effects of administering a drug (red squares) that blocks the viral life cycle. The existing virions in the blood are rapidly cleared by the actions of neutralizing antibody, complement, and phagocytes. Newly produced CD4 T cells are not infected, whereupon they live longer and accumulate in the circulation.

Although measurement of lymphocyte and virion numbers in the blood is used clinically to monitor the progress of HIV infection, the secondary lymphoid tissues are the sites where most lymphocytes are found and where CD4 T cells are activated and produce virus. In HIV-infected people these tissues are loaded with virions, many of which are trapped on the surface of follicular dendritic cells.

11-25 HIV infection leads to immunodeficiency and death from opportunistic infections

Within a few days of being infected with HIV, a CD4 T cell dies. Three kinds of mechanism are thought to contribute to the death toll. One is direct killing as a result of the viral infection or virions binding to cell-surface receptors, the second is increased susceptibility of infected cells to apoptosis, and the third is killing by cytotoxic CD8 T cells specific for viral peptides presented by HLA class I molecules on the infected CD4 T cell.

Throughout clinical latency, the daily loss of CD4 T cells is for the most part compensated for by the supply of new CD4 T cells. With time, however, a steady decline in the number of CD4 T cells is evident, showing how the virus gradually wins this war of attrition. Eventually, the number of CD4 T cells becomes so low that immune responses to all foreign antigens are compromised and HIV-infected people become exceedingly susceptible to other infections. Because CD4 T cells are required for all aspects of adaptive immunity, the vulnerability of patients with AIDS due to late-stage HIV infection resembles that of children with inherited severe combined immune deficiency.

The infections that most frequently affect patients with AIDS are due to infectious agents that are already present in or on the body and that are actively kept under control in healthy people. Such agents are called opportunistic pathogens, and the infections they cause are known as opportunistic infections (Figure 11.30). There is a rough hierarchy in the times at which particular opportunistic infections occur in AIDS, which is correlated with the order

Figure 11.30 A variety of opportunistic infections kill patients with AIDS. The most common of the opportunistic infections that kill AIDS patients in developed countries are listed. The malignancies are listed separately but they are also the result of impaired responses to infectious agents.

Infections	
Parasites	*Toxoplasma* species *Cryptosporidium* species *Leishmania* species *Microsporidium* species
Bacteria	*Mycobacterium tuberculosis* *Mycobacterium avium* *intracellulare* *Salmonella* species
Fungi	*Pneumocystis carinii* *Cryptococcus neoformans* *Candida* species *Histoplasma capsulatum* *Coccidioides immitis*
Viruses	Herpes simplex Cytomegalovirus Varicella-zoster

Malignancies

Kaposi's sarcoma (associated with
  herpesvirus HHV8)
Non-Hodgkin's lymphoma, including
  EBV-positive Burkitt's lymphoma
Primary lymphoma of the brain

in which the different types of immunity collapse. Cellular immunity due to CD4 T$_H$1 cells tends to be lost before either antibody responses or cytotoxic CD8 T cells.

The oral and respiratory tracts are soft tissues loaded with organisms, and in many patients with AIDS they are the sites of the first opportunistic infections. For example, *Candida* causes oral thrush and *Mycobacterium tuberculosis* causes tuberculosis. Later, patients can suffer from diseases caused by the reactivation of latent herpesviruses that are no longer under control by CD8 T cells. Such diseases include shingles due to varicella-zoster, B-cell lymphoma due to EBV, and an endothelial tumor called Kaposi's sarcoma, which is caused by the herpesvirus HHV8. *Pneumocystis carinii*, a common environmental fungus that is hardly ever a problem to healthy people, frequently causes pneumonia and death in patients with AIDS. In the later stages of AIDS, reactivation of cytomegalovirus, a herpesvirus with similar effects to those of EBV, can cause B-cell lymphoproliferative disease. Infection with the opportunistic pathogen *Mycobacterium avium* also becomes prominent. The opportunistic infections suffered by individual patients with AIDS vary greatly. With the collapse of the immune system, only drugs and other interventions can be used to treat the opportunistic infections. These are rarely completely effective or without their own deleterious effects. Eventually, the accumulated tissue damage resulting from the direct effects of HIV infection, opportunistic infections, and medical intervention causes death.

Summary

A major cause of acquired immunodeficiency in the human population today is the human immunodeficiency virus (HIV). This is a slow-acting retrovirus that remains latent for years after the initial infection. HIV infects CD4 T cells, macrophages, and dendritic cells; all of these all carry on their surface the CD4 glycoprotein, which is the receptor for the virus. The effects of HIV are due chiefly to a gradual and sustained destruction of CD4 T cells, which leads eventually to a profound T-cell immunodeficiency known as acquired immune deficiency syndrome (AIDS). CD4 T cells proliferate as part of their normal function and are continually being renewed; these properties allow HIV to maintain a long-lasting infection in which individuals remain relatively healthy for years. Patients with AIDS succumb to a range of opportunistic infections and to some otherwise rare, virus-associated cancers. HIV has only begun to infect humans in the past 50 years, and the lack of any accommodation between host and pathogen is reflected in the strong correlation between infection and death. Genetic polymorphisms in the human population can also influence susceptibility to HIV infection. Because infection with HIV is usually dependent on the presence of the chemokine receptor CCR5, a viral co-receptor, people who lack this cell-surface protein have considerable resistance to HIV infection.

A feature of HIV infection that encourages its dissemination in the population is that infected individuals can live normal lives for many years without knowing they are infected. This poses problems for social approaches to reducing the frequency of sexual transmission. A second approach is to develop vaccines that provide protective immunity, terminating the initial acute HIV infection. Determination of the most effective aspects of the initial immune response will be crucial to this approach, which is also bedeviled by the mutability of HIV. The third approach is that of designing drugs that interfere with the growth and replication of HIV while not affecting normal functions of human cells. This pharmacological approach is the one that has seen most investment, and the results have been mixed. The mutability of the virus, which enables it to escape from individual drugs, is currently perceived as the major problem; its solution is the use of combination drug therapy.

Summary to Chapter 11

In the course of their long relationship with humans, successful pathogens have developed mechanisms that allow them to exploit the human body to the full. Indeed, a pathogen is, by definition, an organism that is habitually able to overcome the body's immune defenses to such an extent that it causes disease. One class of adaptations is those in which the pathogen changes itself or its behavior to evade the ongoing immune response. This prevents the immune system from adapting to the pathogen and improving the response. In a second type of adaptation, the pathogen is able to impair or prevent the immune response. Pathogens can have more than one such adaptation, and for some pathogens a considerable fraction of their genome is devoted to foiling the immune system. Highly successful pathogens are not necessarily the most virulent. Nonlethal ubiquitous host–pathogen relationships, such as that of humans and the Epstein–Barr virus, have generally evolved over a long period of association. However, even these pathogens can cause life-threatening disease when the immune system is compromised.

Inherited immunodeficiencies are caused by a defect in one of the genes necessary for the development or function of the immune system. Depending on the gene involved, immunodeficiencies range from manageable susceptibilities to particular pathogens to a general vulnerability created by the complete absence of adaptive immunity. The most severe inherited immunodeficiencies are very rare, which is evidence of the importance of the immune system to human survival.

Immunodeficiency can also be acquired as the result of infection. The human immunodeficiency virus (HIV) infects macrophages, dendritic cells, and CD4 T cells and eventually reduces the number of CD4 T cells to a level at which severe immunodeficiency results—the acquired immune deficiency syndrome (AIDS). This retroviral pathogen has only just begun to exploit the human species, but it already has a variety of effective adaptations acquired during its history in other primate hosts. Most people infected with HIV have no overt symptoms of disease for years, which facilitates the spread of HIV by sexual transmission, and HIV infection has now reached pandemic proportions within the human population. The discovery of natural genetic traits that endow individual humans with resistance to HIV indicates possibilities for accommodation in the longer term.

Questions

11–1
A. Explain what is meant by serotype-specific immunity.
B. How does *Streptococcus pneumoniae* exploit serotype-specific immunity to evade detection?

11–2
A. Which antigens are most important in the immune response to the influenza virus?
B. Explain the difference between antigenic drift and antigenic shift in the influenza virus.
C. Which is most likely to lead to a major worldwide pandemic?

D. What is the role of the phenomenon of 'original antigenic sin' in immunity to this virus?

11–3 Why does it benefit the African trypanosome (*T. brucei*) to maintain more than 1000 genes encoding surface glycoproteins, when only one of these glycoproteins is expressed on the surface of the parasite at any given time?

11–4   Using Table Q11–4, match the mechanism of evasion and subversion of the immune system in Column 1 with the pathogen in Column 2.

Table Q11–4

Column 1	Column 2
a. Variant pilin protein expression	1. *Staphylococcus aureus*
b. Induction of quiescent (latent) state in neurons	2. *Toxoplasma gondii*
c. Reactivation of infected ganglia after stress or immunosuppression	3. *Salmonella typhimurium*
d. Alternative expression of two antigenic forms of flagellin	4. Influenza virus
e. Recombination of RNA genomes of avian and human origins	5. *Mycobacterium tuberculosis*
f. Escape from phagosome and growth and replication in cytosol	6. Varicella-zoster
g. Survival in a membrane-bounded vesicle resistant to fusion with other cellular vesicles	7. *Neisseria gonorrhoeae*
h. Coating its surface with human proteins	8. *Treponema pallidum*
i. Inhibiting fusion of phagosome with lysosome and survival in the host cell's vesicular system	9. *Listeria monocytogenes*
j. Immunosuppression caused by nonspecific proliferation and apoptosis of T cells	10. Herpes simplex virus

11–5 Herpes simplex virus favors neurons for latency because of the low level of _____, which reduces the likelihood of killing by CD8 T cells.
  a. LFA-3
  b. Toll-like receptors (TLRs)
  c. transporter associated with antigen processing (TAP)
  d. MHC class I
  e. MHC class II.

11–6
  A. Deficiencies in antibody production can be due to a variety of underlying genetic defects. Name two immunodeficiency diseases, other than the severe combined immunodeficiencies, in which a defect in antibody production is the cause of the disease, and for which the underlying genetic defect is known. For each disease, say (i) how antibody production is affected, and (ii) what the underlying defect is and why it has this effect.
  B. What is the main clinical manifestation of immunodeficiency diseases in which antibody production is defective but cell-mediated immune responses are intact?

11–7 Explain why women who show no disease symptoms themselves can pass on some heritable diseases to their sons, whereas their daughters seem to be unaffected. Would a disease with this pattern of inheritance be caused by a recessive or a dominant allele?

11–8
  A. Explain why an inherited deficiency in complement components C3 or C4 can result in less efficient clearance of immune complexes from the blood and lymph than normal.
  B. What clinical symptoms can this lead to?
  C. What is the only known effect of deficiencies in complement components C5–C9? Explain this effect.

11–9
  A. Name three immunodeficiency diseases caused by defects in phagocytes.
  B. Which immunodeficiency disease is caused by a defect in the phagocyte NADPH oxidase system, and what is the cellular effect of this defect?
  C. What are the main clinical effects of defects in phagocyte function?

11–10
  A. What type of immune deficiency would you see in a child lacking the common γ chain of the receptor for cytokines IL-2, IL-4, and IL-7, among others? Explain your answer.
  B. Why would you see the same type of immunodeficiency in a child lacking Jak3 kinase function?
  C. What treatment might be possible to remedy this immunodeficiency?

11–11
  A. Explain the mechanism by which human immunodeficiency virus (HIV) enters a host cell.
  B. Explain the cellular tropism of HIV, discussing the difference between macrophage-tropic and lymphocyte-tropic HIV.
  C. Some people seem to be resistant to HIV infection because a primary infection cannot be established in macrophages. What is the reason for this?

11–12
  A. What does the term seroconversion mean in relation to an HIV infection?
  B. What relationship does seroconversion have to the time course of an HIV infection?

11–13 Which property of HIV renders the virus difficult to eradicate by the body's immune defenses, and also limits the efficacy of drug therapies?

11–14 Christiana Carter had no obvious problems until she was 18 months old, when she stopped gaining weight, her appetite became poor, and she had recurrent episodes of diarrhea. At 24 months, Christiana developed a cough with pulmonary infiltrates unresponsive to treatment with the antibiotics clarithromycin and trimethoprim/sulfamethoxazole. Within three months, she developed lymphadenopathy, hepatosplenomegaly, and fevers. A computed tomography scan revealed enlarged mesenteric and para-aortic lymph nodes. A biopsy of an enlarged axillary lymph node revealed acid-fast bacilli, and cultures from the lymph node and blood grew *Mycobacterium fortuitum*. HIV was ruled out after negative tests by ELISA and PCR. Serologic testing for tetanus antitoxoid antibody showed a normal post-vaccination level. Christiana's peripheral blood mononuclear cells (PBMCs) were cultured with interferon-γ plus LPS with no significant increase in TNF-α production. A variety of broad-spectrum and anti-mycobacterial antibiotics were administered, lowering the fever, and over the course of the next 2 months Christiana began to gain weight but continued to show signs of persistent infection. Which of the following is the most likely explanation for these clinical findings?

  a. leukocyte adhesion deficiency
  b. chronic granulomatous disease
  c. interferon-γ receptor deficiency
  d. X-linked agammaglobulinemia
  e. severe combined immune deficiency.

11–15 Doug Perkins, aged 33 years, received a third-degree burn on his forearm from a ruptured radiator hose in his car. Several days later he became confused and less mentally alert and his son brought him to the emergency room. He had a fever, rapid breathing, a generalized skin rash, hypotension, and a decreased white blood cell count. A blood culture tested positive for Group A β-hemolytic streptococci. Which of the following is the best explanation for his symptoms?

  a. Streptococcal superantigens caused the production of very high levels of T-cell cytokines, leading to systemic shock and apoptosis of superantigen-stimulated T-cell clones.
  b. Pathogen-specific T-cell activation resulted in massive proliferation and persistence of T cells secreting cytokines, leading to systemic shock.
  c. Superantigen cross-linking of MHC molecules on macrophages caused the death of macrophages and a corresponding decrease in phagocytosis and opsonization of the pathogen.
  d. Polyclonal activation of B cells caused hyperstimulation of B cells and subsequent immune-complex deposition in blood vessel endothelium, causing systemic shock.
  e. Massive levels of superantigens stimulated cross-linking and receptor-mediated endocytosis of MHC class I molecules resulting in ineffective CD8 T-cell activation and consequent septicemia.

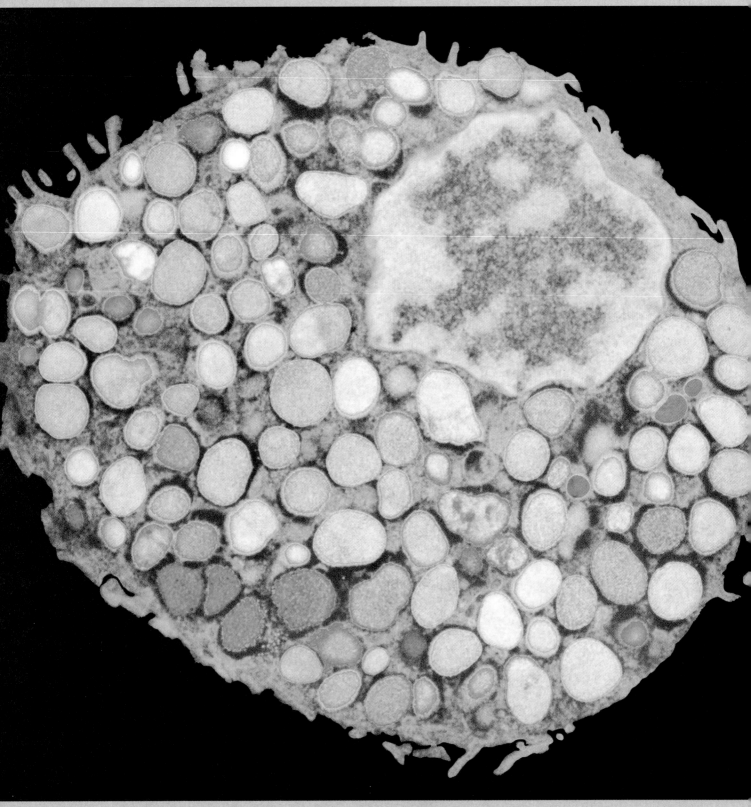

The mast cell, a major type of immune effector cell that causes the symptoms of allergies and asthma.

Chapter 12

Over-reactions of the Immune System

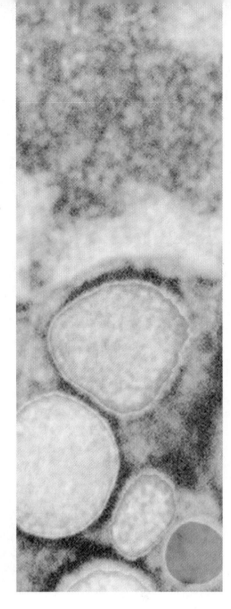

The protective functions of the immune system depend on recognition events that distinguish molecular components of infectious agents from those of the human body. In this context the immune system is said to recognize a pathogen's molecules as foreign. Besides infectious agents, humans come into daily contact with numerous other molecules that are equally foreign but do not threaten health. Many of these molecules derive from the plants and animals that we eat or are present in the environments where we live, work, and play. For most people for most of the time, contact with these molecules stimulates neither inflammation nor adaptive immunity.

In some circumstances, however, certain kinds of innocuous molecule stimulate an adaptive immune response and the development of immunological memory in predisposed members of the population. On subsequent exposures to the antigen the immune memory produces inflammation and tissue damage that is at best an irritation, and at worst a threat to life. The person feels ill, as though fighting off an infection, when no infection exists. The over-reactions of the immune system to harmless environmental antigens are called either **hypersensitivity reactions** or **allergic reactions**. The environmental antigens that cause these reactions are termed **allergens** and they induce a state of hypersensitivity or **allergy**, the latter word being of Greek derivation and meaning 'altered reactivity.'

Unfortunately, the antigens that provoke these over-reactions are often common in the human environment: in developed countries some 10–40% of the inhabitants are allergic to one or more environmental antigens, and the numbers continue to rise. Some of the antigens responsible for common hypersensitivities are listed in Figure 12.1.

12-1 Four types of hypersensitivity reaction are caused by different effector mechanisms of adaptive immunity

Hypersensitivity reactions are conventionally grouped into four types according to the effector mechanisms that produce the reaction (Figure 12.2). Some antigens (penicillin, for example) cause different types of hypersensitivity reaction, depending on the circumstances in which they are encountered. **Type I hypersensitivity reactions** result from the binding of antigen to antigen-specific IgE bound to its Fc receptor, principally on mast cells. This interaction causes the degranulation of mast cells and the release of inflammatory mediators. Type I reactions are commonly caused by inhaled particulate

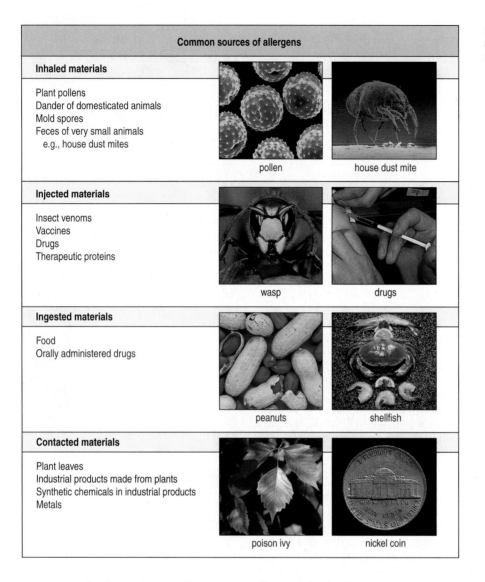

Figure 12.1 Some substances that are a common cause of hypersensitivity reactions.

antigens, of which plant pollens are good examples. Type I reactions have effects of varying severity, ranging from a runny nose to breathing difficulties and even death by asphyxiation. Type I hypersensitivity reactions are also described as **immediate hypersensitivity** because they occur immediately on exposure to antigen. Because of their prevalence and diversity, a large part of this chapter will be devoted to the IgE-mediated hypersensitivity reactions.

Type II hypersensitivity reactions are caused by small molecules that bond covalently to cell-surface components of human cells, producing modified structures that are perceived as foreign by the immune system. The B-cell response to these new epitopes produces IgG, which on binding to the modified cells causes their destruction through complement activation and phagocytosis. The antibiotic penicillin is an example of a small reactive molecule that can induce a type II hypersensitivity reaction.

Type III hypersensitivity reactions are due to small soluble immune complexes formed by soluble protein antigens binding to the IgG made against them. Some of these immune complexes become deposited in the walls of small blood vessels or the alveoli of the lungs. The immune complexes activate complement and initiate an inflammatory response that damages the tissue, impairing its physiological function. When antibodies or other proteins derived from nonhuman animal species are given therapeutically to patients, type III hypersensitivity reactions are a potential side-effect.

	Type I	Type II		Type III	Type IV		
Immune reactant	IgE	IgG		IgG	T_H1 cells	T_H2 cells	CTL
Antigen	Soluble antigen	Cell- or matrix-associated antigen	Cell-surface receptor	Soluble antigen	Soluble antigen	Soluble antigen	Cell-associated antigen
Effector mechanism	Mast-cell activation	Complement, FcR$^+$ cells (phagocytes, NK cells)	Antibody alters signaling	Complement Phagocytes	Macrophage activation	Eosinophil activation	Cytotoxicity
Example of hypersensitivity reaction	Allergic rhinitis, asthma, systemic anaphylaxis	Some drug allergies (e.g., penicillin)	Chronic urticaria (antibody to FcαRIα)	Serum sickness, Arthus reaction	Contact dermatitis, tuberculin reaction	Chronic asthma, chronic allergic rhinitis	Contact dermatitis

Figure 12.2 Hypersensitivity reactions fall into four classes on the basis of their mechanism. The types of preexisting immunity causing the reaction are listed (immune reactants), along with the types of antigen that provoke the response, the underlying effector mechanism, and the type of ailment produced.

The effector molecules initiating type I, II, and III hypersensitivity reactions are all antibodies. In contrast, **type IV hypersensitivity reactions** are caused by the products of antigen-specific effector T cells. Most reactions are caused by CD4 T_H1 cells. For example, the inflammatory reaction around the site of an insect bite or sting is caused by CD4 T_H1 cells that respond to peptide epitopes derived from venom and other insect proteins introduced by the bite. A minority of type IV hypersensitivity reactions are due to cytotoxic CD8 T cells. They arise when small, reactive, lipid-soluble molecules pass through cell membranes and bond covalently to intracellular human proteins. Degradation of these chemically modified proteins yields abnormal peptides that bind to HLA class I molecules and stimulate a cytotoxic T-cell response. For example, the allergic response to poison ivy involves cytotoxic T cells. These recognize peptides derived from intracellular proteins that are modified by chemical reaction with pentadecacatechol, a chemical acquired by touching the plant's leaves. Type IV hypersensitivity reactions are also described as delayed-type hypersensitivity or DTH because they are only manifested 1–3 days after exposure to antigen.

Type I hypersensitivity reactions

A prerequisite for a type I hypersensitivity reaction is that IgE antibody is made when a person first encounters the antigen. This is how the person becomes **sensitized** to the antigen. IgE differs from other antibody isotypes in being located predominantly in tissues, where it is bound to mast cells by high-affinity surface receptors known as FcεRI. We first look at the effector cells and molecules that produce a type I response, and then consider the characteristic properties of allergens and how sensitization occurs.

12-2 IgE binding to FcεR1 provides mast cells, basophils, and activated eosinophils with antigen receptors

IgE antibody causes allergic reactions because of its exceptionally high affinity for its Fc receptor and the cellular distribution of the receptor. Binding of the IgE constant region to its high-affinity receptor—**FcεRI**—is the tightest of the antibody–Fc receptor interactions ($K_d \approx 10^{10}$ M^{-1}, see Figure 9.46, p. 284) and can be considered for all practical purposes as irreversible. Also, unlike most other isotypes, IgE binds to its receptor in the absence of antigen. FcεRI is expressed constitutively by mast cells and basophils, and by eosinophils

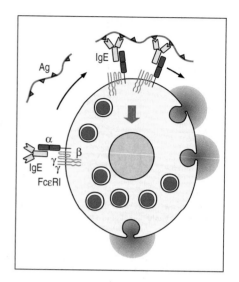

Figure 12.3 Cross-linking of FcεRI on the surface of mast cells by antigen and IgE causes mast-cell activation and degranulation. The high-affinity IgE receptor (FcεRI) on mast cells, basophils, and activated eosinophils is formed from one α chain, one β chain, and two γ chains. The binding site for IgE is formed by the two extracellular immunoglobulin-like domains of the α chain. The β chain and the two disulfide-bonded γ chains are largely intracellular and contribute to signaling. When receptors are cross-linked by antigen binding to IgE on the surface of a mast cell, a signal is transmitted that leads to the release of preformed granules (purple) containing histamine and other inflammatory mediators.

after they have been activated by cytokines. After a primary IgE response has subsided and the antigen has been cleared by the usual means, all antigen-specific IgE molecules that have not encountered antigen will be bound by their Fc regions to FcεRI on these cells. The stable complexes of IgE and FcεRI effectively provide mast cells, basophils, and eosinophils with antigen-specific receptors. All these cells have granules containing preformed inflammatory mediators. When the antigen is next encountered it will bind to the receptors and activate the cells to release their inflammatory mediators (Figure 12.3). Initial degranulation is followed by the induced synthesis of a wider range of mediators.

Two important features distinguish the 'antigen receptors' on mast cells, basophils, and activated eosinophils from those of B cells and T cells. The first is that the cell's effector function becomes operational immediately after antigen binds to the 'receptor' and does not require a phase of cell proliferation and differentiation. The second is that each cell is not restricted to carrying receptors of just one antigen specificity but can carry a range of IgE representing that present in the circulation. These features combine to quicken and strengthen the response to any antigen to which a person has been sensitized. When any one of those antigens enters a tissue, all the mast cells in the vicinity bearing a sufficient number of molecules of IgE specific for that antigen will be triggered to degranulate immediately.

IgE-mediated activation of mast cells, basophils, and activated eosinophils provides protection against parasites (see Section 9-24, p. 282), which are prevalent in tropical regions. This type of environment is where the human species originated and where it has lived for most of its existence. In the developing countries of the tropics, parasites are abundant and one-third of the world's population is infected by one or more species of parasite.

In contrast, in the developed countries where parasite infections are rare, IgE responses tend to be stimulated by contact with non-threatening substances in the environment. In North America and Europe, the impact of allergic disease is increasingly being felt in all aspects of human life. In the past 10–15 years the incidence of allergy has doubled and disproportionately affects children. Some physicians have called it an 'epidemic of allergy.' Because allergy has directed medical research in this area we know far more about the allergic effects of IgE than we do about its benefits in controlling parasite infections.

12-3 Mast cells defend and maintain the tissues where they live

Mast cells are resident in mucosal and epithelial tissues lining the body surfaces. Present in all vascularized tissues except the central nervous system and the retina, mast cells serve to maintain the integrity of the tissue where they reside, alert the immune system to local trauma and infection and facilitate the repair of damage caused by infection or wounds. Morphologically, the defining characteristic of the mast cell is the cytoplasm, which is filled with 50–200 large granules that contain inflammatory mediators. The name **mast cell** derives from the German word *Mastzellen*, meaning 'fattened cells' or 'well-fed cells,' which is how they appear under the microscope (Figure 12.4). Mast-cell granules contain principally histamine, heparin, tumor necrosis factor-α (TNF-α), chondroitin sulfate, neutral proteases, and other degradative enzymes and inflammatory mediators. Heparin, an acidic proteoglycan, is largely responsible for the characteristic staining of the mast-cell granules with basic dyes.

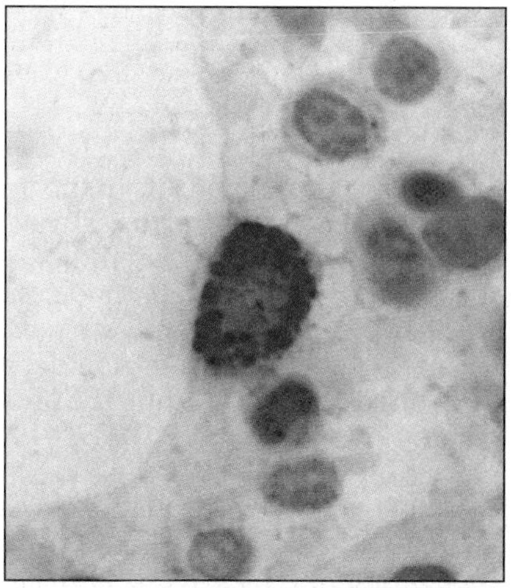

Although mast cells are known mainly for their destructive effects in IgE-mediated allergic reactions, they also respond in a constructive way to a much wider range of stimuli. As well as expressing FcεR1, mast cells express Toll-like receptors and Fc receptors for IgA and IgG. Mast cells can thus contribute to both the innate and the adaptive immune responses to infection. Whereas the IgE-mediated activation of mast cells via FcεR1 is limited to degranulation and to the synthesis of small inflammatory mediators known as eicosanoids, signaling through other receptors induces the manufacture and secretion of cytokines that recruit neutrophils, eosinophils, and effector T cells to the infected tissue, and also induces the secretion of growth factors that promote repair of the damage. Mast cells can thus tailor their cytokine response to the type of infecting pathogen (Figure 12.5).

Two kinds of human mast cell have been distinguished: a **mucosal mast cell**, which produces the protease tryptase, and a **connective tissue mast cell**, which produces chymotryptase. In patients with T-cell immunodeficiencies, only connective tissue mast cells are present, indicating that the development of mucosal mast cells is dependent on the effector T cells that populate the mucosal tissues (see Section 10-4, p. 294). Such tissue-specific differentiation of the mucosal mast cells in the gut probably specializes the mast cells for stimulating non-inflammatory responses against the gut flora and IgE-mediated responses against intestinal parasites.

Figure 12.4 Light micrograph of mast cell stained for the granule protease chymase to show the numerous granules that fill its cytoplasm. The mast cell is in the center of the picture: its nucleus is stained pink and the cytoplasmic granules are stained red. Magnification × 1000. Photograph courtesy of D. Friend.

Class of product	Product	Biological effects
Enzyme	Tryptase, chymase, cathepsin G, carboxypeptidase	Remodeling of connective tissue matrix
Toxic mediator	Histamine, heparin	Toxic to parasites Increase vascular permeability Cause smooth muscle contraction
Cytokine	TNF-α (some stored preformed in granules)	Promotes inflammation, stimulates cytokine production by many cell types, activates endothelium
	IL-4, IL-13	Stimulate and amplify T_H2-cell response
	IL-3, IL-5, GM-CSF	Promote eosinophil production and activation
Chemokine	CCL3	Chemotactic for monocytes, macrophages, and neutrophils
Lipid mediator	Leukotrienes C_4, D_4, and E_4	Cause smooth muscle contraction Increase vascular permeability Cause mucus secretion
	Platelet-activating factor	Chemotactic for leukocytes Amplifies production of lipid mediators Activates neutrophils, eosinophils, and platelets

Figure 12.5 Molecules released by mast cells on stimulation by antigen binding to IgE. The boxes shaded in red show the molecules (proteases, histamine, heparin, and TNF-α) that are prepackaged in granules and released immediately the mast cell is stimulated by antigen binding. TNF-α is also synthesized after mast-cell activation. The molecules listed in the white boxes are synthesized and released only as a consequence of mast-cell activation.

12-4   Tissue mast cells orchestrate IgE-mediated allergic reactions through the release of inflammatory mediators

Mast-cell activation occurs in the presence of any antigen that can cross-link the IgE molecules bound to FcεRI at the cell surface. This can be accomplished by antigens with repetitive epitopes that cross-link IgE molecules of the same specificity, or by antigens possessing two or more different epitopes that cross-link IgE molecules of different specificities. Once a mast cell's receptors have been cross-linked, degranulation occurs within a few seconds, releasing the stored mediators into the immediate extracellular environment (see Figure 12.5).

Prominent among the mediators is histamine, an amine derivative of the amino acid histidine (Figure 12.6). Histamine exerts a variety of physiological effects through three kinds of histamine receptor—H1, H2, and H3—which have been defined on different cell types. Acute allergic reactions involve the binding of histamine to the H1 receptor on nearby smooth muscle cells and on endothelial cells of blood vessels. Stimulation of the H1 receptor on endothelial cells induces vessel permeability and the entry of other cells and molecules into the allergen-containing tissue, causing inflammation. Smooth muscle cells are induced to contract on binding histamine, which constricts airways, for example, and histamine also acts on the epithelial linings of mucosa to induce the increased secretion of mucus. All these actions produce different effects depending on the tissue exposed to allergen. Sneezing, coughing, wheezing, vomiting, and diarrhea can all be induced in the course of an allergic reaction.

Other molecules released from mast-cell granules include mast-cell chymotryptase, tryptase, and other neutral proteases that activate metalloproteases in the extracellular matrix. Collectively, these enzymes break down extracellular matrix proteins. The action of histamine is complemented by that of TNF-α, which is also released from mast-cell granules. TNF-α activates endothelial cells, causing an increased expression of adhesion molecules, thus promoting leukocyte traffic from the blood into the increasingly inflamed tissue (see Figure 2.31, p. 55). Mast cells are unique in their capacity to store TNF-α and release it on demand. At the start of an inflammatory response, mast cells are usually the main source of this inflammatory cytokine.

Besides the preformed inflammatory mediators in granules, mast cells synthesize and secrete other mediators in response to activation (see Figure 12.5). These include chemokines, cytokines—IL-4 and more TNF-α—and eicosanoids. The latter, which include the prostaglandins and the leukotrienes, are lipids synthesized from fatty acids (Figure 12.7, top panel). All these mediators act locally within the area of tissue surrounding the activated mast cells. The leukotrienes have activities similar to those of histamine, but are more than 100 times more potent on a molecule-for-molecule basis. The two kinds of mediator are therefore complementary. Histamine provides a rapid response while the more potent leukotrienes are being made. In the later stages of allergic reactions, leukotrienes are principally responsible for inflammation, smooth muscle contraction, constriction of airways, and the secretion of mucus from mucosal epithelium. Mast cells also secrete prostaglandin D_2 (PGD$_2$), which promotes dilation and increased permeability of blood vessels and also acts as a chemoattractant for neutrophils. Aspirin reduces inflammation by inactivating prostaglandin synthase, the first enzyme in the cyclooxygenase pathway. The inactivation is irreversible because aspirin binds covalently to the active site of the enzyme (Figure 12.7, bottom panel).

The combined effect of the chemical mediators released by mast cells is to attract circulating leukocytes to the site of mast-cell activation, where they amplify the reaction initiated by antigen and IgE on the mast-cell surface.

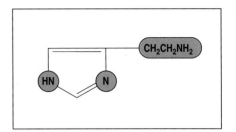

Figure 12.6 The chemical structure of histamine.

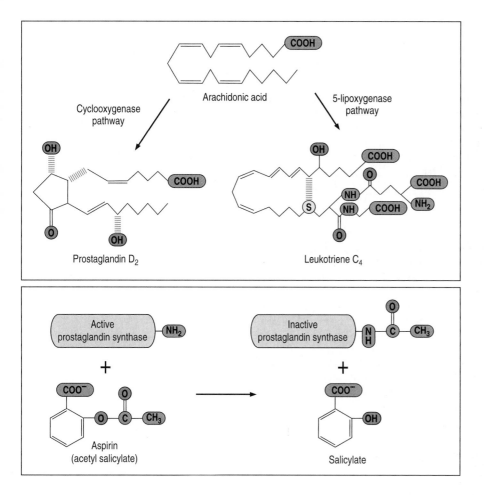

Figure 12.7 Mast cells synthesize prostaglandins and leukotrienes from arachidonic acid by different enzyme pathways. Arachidonic acid, an unsaturated fatty acid, is used as the substrate for the cyclooxygenase pathway to make prostaglandins, and as the substrate for the 5-lipoxygenase pathway to make leukotrienes, as shown in the top panel. Arachidonate is itself made from linolenic acid, one of the essential unsaturated fatty acids that must be provided in the diet because it cannot be made by human cells. Aspirin (acetyl salicylate) prevents the synthesis of prostaglandins by the cyclooxygenase pathway by irreversibly inhibiting the enzyme prostaglandin synthase, as shown in the bottom panel.

These effector leukocytes include eosinophils, basophils, neutrophils, and T_H2 lymphocytes. In parasite infections these cells work together to promote explosive reactions that either kill the parasite or expel it from the body (see Section 9-13, p. 267). In this situation, the benefit of eliminating the pathogen is worth suffering the damage caused by the immune response. In allergic reactions aimed at harmless antigens, the immune response provides no benefit at all, only detrimental side-effects—damage to the body's tissues and impairment of their function. For allergens that pervade the environment and provide a continual stimulation, the allergic response will gather in strength as increasing numbers of mast cells and eosinophils are recruited to the affected tissue and join in the frustrated attempts to eliminate the 'pathogen.'

12-5 Eosinophils are specialized granulocytes that release toxic mediators in IgE-mediated responses

Eosinophils are granulocytes whose granules contain arginine-rich basic proteins. In histological sections these proteins stain heavily with eosin, hence the name eosinophil, meaning 'liking eosin' (Figure 12.8). Normally, only a minority of eosinophils circulate in the blood. Most are resident in tissues, especially in the connective tissue immediately underlying epithelia of the respiratory, gastrointestinal, and urogenital tracts. As with mast cells, the activation of eosinophils by external stimuli leads to the staged release of toxic molecules and inflammatory mediators. Preformed chemical mediators and proteins present in the granules are released first (Figure 12.9, top panel). The normal function of these highly toxic molecules is to kill invading microorganisms and parasites directly. This is followed by a slower induced synthesis

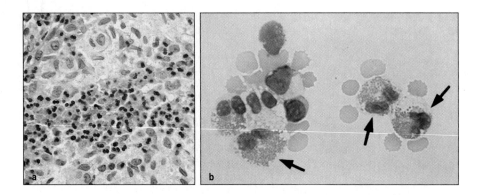

Figure 12.8 Eosinophils have a characteristic staining pattern in histological sections. Panel a shows a light micrograph of a section through skin affected by Langerhans cell histiocytosis—a condition in which Langerhans cells and eosinophils mutually stimulate each others' proliferation. The eosinophils have bilobed nuclei and stain pink with the stain eosin. Panel b shows a higher-power light micrograph of a blood smear in which partly degranulated eosinophils (arrows) are surrounded by erythrocytes. Panel a courtesy of T. Krausz; panel b courtesy of F. Rosen and R. Geha.

and secretion of prostaglandins, leukotrienes, and cytokines, which amplify the inflammatory response by the activation of epithelial cells and leukocytes, including more eosinophils (Figure 12.9, bottom panel).

The eosinophil response is highly toxic and is potentially damaging to the host as well as to parasites. Limiting the damage are various control mechanisms. When the body is healthy, the number of eosinophils is kept low by restricting their production in the bone marrow. When infection or antigenic stimulation activates T_H2 cells, the IL-5 and other cytokines released by the T_H2 cells stimulate the bone marrow to increase both the production of eosinophils and their release into the circulation. The migration of eosinophils into tissues is controlled by a group of chemokines (CCL5, CCL7, CCL11, and CCL13) that bind to the receptor CCR3 expressed by eosinophils. Of these chemokines, CCL11, also called **eotaxin**, is particularly important in the migration of eosinophils and is produced by activated endothelial cells, T cells, and monocytes.

Class of product	Product	Biological effects
Enzyme	Eosinophil peroxidase	Toxic to parasites and mammalian cells by catalyzing halogenation Triggers histamine release from mast cells
	Eosinophil collagenase	Remodeling of connective tissue matrix
Toxic protein	Major basic protein	Toxic to parasites and mammalian cells Triggers histamine release from mast cells
	Eosinophil cationic protein	Toxic to parasites Neurotoxin
	Eosinophil-derived neurotoxin	Neurotoxin
Cytokine	IL-3, IL-5, GM-CSF	Amplify eosinophil production by bone marrow Cause eosinophil activation
Chemokine	CXCL8	Promotes influx of leukocytes
Lipid mediator	Leukotrienes C_4, D_4, and E_4	Cause smooth muscle contraction Increase vascular permeability Cause mucus secretion
	Platelet-activating factor	Chemotactic to leukocytes Amplifies production of lipid mediators Activates neutrophils, eosinophils, and platelets

Figure 12.9 Activated eosinophils secrete toxic proteins contained in their granules and also produce cytokines and inflammatory mediators. The boxes shaded in red indicate the enzymes and toxic proteins that are contained preformed in the granules and are released immediately on eosinophil activation. The cytokines, chemokines, and lipid mediators (white boxes) are synthesized only after activation. Halogenation refers to the addition of halide atoms, such as chlorine and bromine, to molecules.

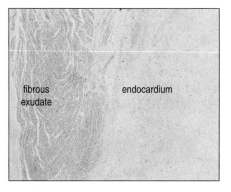

Figure 12.10 The presence of an abnormally large number of circulating eosinophils causes damage to the heart. The photograph shows a section of endocardium from a patient with hypereosinophilic syndrome. The damaged tissue is characterized by an organized fibrous exudate and an endocardium thickened by fibrous tissue. Although the patient has large numbers of circulating eosinophils, these cells are not seen in the injured endocardium, which is thought to be damaged by the contents of granules released by circulating eosinophils. Photograph courtesy of D. Swirsky and T. Krausz.

The activity of eosinophils is further controlled by modulation of their sensitivity to external stimuli. This is achieved through the regulated expression of FcεRI. In the resting state, eosinophils do not express FcεRI, do not bind IgE, and cannot therefore be induced to degranulate by antigen. Once an inflammatory response has been set in motion, cytokines and chemokines in the inflammatory site induce eosinophils to express FcεRI. The expression of Fcγ receptors and complement receptors on the eosinophil surface also increases, facilitating the binding of eosinophils to pathogen surfaces coated with IgG and complement.

The potential of eosinophils to cause tissue damage is clearly seen in patients who have abnormally high numbers of these cells. Certain T-cell lymphomas, for example, constitutively secrete IL-5, which continually expands the eosinophil population. This expansion is detected by a greatly increased number of eosinophils in the blood (**hypereosinophilia**). With so many circulating eosinophils, only a small proportion need to be activated for their effects to be seriously disruptive. Patients with hypereosinophilia can suffer damage to the endocardium of the heart (Figure 12.10) and to nerves, leading to heart failure and neuropathy. Both of these effects are thought to be due to cytotoxic and neurotoxic proteins released from eosinophil granules.

In localized allergic reactions, mast-cell degranulation and the activation of T$_H$2 cells cause the accumulation of activated eosinophils at the site. Their presence is characteristic of chronic allergic inflammation; for example, eosinophils are considered the principal cause of the airway damage that occurs in chronic asthma.

12-6 Basophils are rare granulocytes that initiate T$_H$2 responses and the production of IgE

Basophils are granulocytes whose granules stain with basic dyes such as hematoxylin (Figure 12.11). The granules of basophils package a similar, but not identical, set of mediators to those of mast cells. Although basophils resemble mast cells in some ways, they are developmentally more closely related to eosinophils. They have a common stem-cell precursor and require similar growth factors, including IL-3, IL-5, and GM-CSF. The production of eosinophils and basophils seems to be reciprocally regulated, so that a combination of TGF-β and IL-3 promotes the maturation of basophils while suppressing that of eosinophils. Basophils normally constitute less than 1% of the white blood cells, which for decades made them much more difficult to study than other leukocytes.

Knowledge of basophil biology is now advancing and points to the basophil as the key cell that initiates the T$_H$2 response through its unique capacity to secrete the T$_H$2 cytokines IL-4 and IL-13 at the beginning of an immune response. During the innate immune response, basophils are recruited to infected tissue, where they are activated through their Toll-like receptors and other receptors of innate immunity. In the infected tissue, basophils contribute effector functions similar to those of eosinophils. At the beginning of the adaptive immune response, basophils are also recruited to the secondary

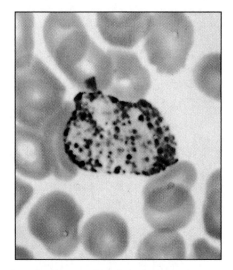

Figure 12.11 Light micrograph of basophil stained with Wright's Giemsa. The basophil in the center is surrounded by erythrocytes in this blood smear. Magnification × 1000. Photograph courtesy of D. Friend.

lymphoid tissues, where their secretion of IL-4 and IL-13 initiates the polarization of antigen-stimulated T cells towards a T_H2 response (see Section 8-10, p. 227). Basophils express CD40 ligand, which binds to the CD40 of antigen-stimulated B cells and, in combination with secreted IL-4 and IL-13, drives isotype switching to IgE and IgG4. Once specific IgE has been made, it binds to the receptor FcεRI on basophils and provides the main means for activating basophils in the secondary response to the pathogen.

Mast cells, eosinophils, and basophils often act in concert. Mast-cell degranulation initiates the inflammatory response, which then recruits eosinophils and basophils. Eosinophil degranulation releases **major basic protein**, which in turn causes the degranulation of mast cells and basophils. This latter effect is augmented by any of the cytokines—IL-3, IL-5, and GM-CSF—that affect the growth, differentiation, or activation of eosinophils and basophils.

12-7   Very few antigens that enter the human body are allergens that stimulate an IgE response

All type I hypersensitivity reactions are caused by allergen-specific IgE antibodies. Production of IgE is favored when the immune system is challenged with small quantities of antigen and when IL-4 is provided by activated basophils at the time when the naive CD4 T cells are presented with antigen. In these circumstances CD4 T cells tend to make a T_H2 response (see Section 8-10, p. 227), which then produces more IL-4 and additional cytokines that stimulate B cells to switch their immunoglobulin isotype to IgE (see Section 9-9, p. 261). Initial sensitization to an allergen is thus favored by circumstances that promote the production of antigen-specific T_H2 cells and the production of IgE, and is disfavored by conditions that produce T_H1 cells.

The allergens that provoke type I hypersensitivity responses are invariably proteins or chemicals, like some drugs, that chemically modify human proteins. Allergens enter the body by different routes, and the disease symptoms they cause vary with the site of entry (Figure 12.12). Most allergens are either

IgE-mediated allergic reactions			
Syndrome	**Common allergens**	**Route of entry**	**Response**
Systemic anaphylaxis	Drugs Serum Venoms Peanuts	Intravenous (either directly or after rapid absorption)	Edema Increased vascular permeability Tracheal occlusion Circulatory collapse Death
Wheal and flare	Insect bites Allergy testing	Subcutaneous	Local increase in blood flow and vascular permeability
Allergic rhinitis (hay fever)	Pollens (ragweed, timothy, birch) Dust-mite feces	Inhaled	Edema of nasal mucosa Irritation of nasal mucosa
Bronchial asthma	Pollens Dust-mite feces	Inhaled	Bronchial constriction Increased mucus production Airway inflammation
Food allergy	Shellfish Milk Eggs Fish Wheat	Oral	Vomiting Diarrhea Pruritis (itching) Urticaria (hives) Anaphylaxis

Figure 12.12 Allergic reactions mediated by IgE.

inhaled or ingested and cross a mucosal surface to activate mucosal mast cells. Others penetrate the body through wounds in the skin and activate connective tissue mast cells. Allergens represent only a minute proportion of all the proteins and chemicals that enter human bodies, and a central goal of research into allergy is to define what distinguishes proteins and chemicals that cause allergies from those that do not. Definitive answers have yet to be found, but there are certain properties that characterize the major inhaled allergens (Figure 12.13).

Most allergens are small, soluble proteins that are present in dried-up particles of material derived from plants and animals. Examples are pollen grains, the mixture of dried cat skin and saliva that forms dander, and the dried feces of the house dust mite *Dermatophagoides pteronyssimus*. The light dry particles become airborne and are inhaled by humans. Once inhaled, the particles become caught in the mucus bathing the epithelia of the airways and lungs. They then rehydrate, releasing the antigenic proteins. These antigens are carried to professional antigen-presenting cells within the mucosa. The antigens are then processed and presented by these cells to CD4 T cells, stimulating a T_H2 response that leads to the production of IgE and its binding to mast cells (Figure 12.14). Small soluble protein antigens are more efficiently leached out of particles and penetrate the mucosa. A substantial proportion of allergens are proteases, and it is likely that their enzymatic activities facilitate the breakdown of the particle, the release of allergen, and the generation of peptides that stimulate T_H2 cells.

The major allergen responsible for more than 20% of the allergies in the human population of North America is a cysteine protease derived from *D. pteronyssimus*. Advances in the heating and cooling of homes, offices, and other buildings are believed to be responsible for the prevalence of this allergy because they provide an environment that encourages both the growth of *D. pteronyssimus* and the desiccation of its feces. The air currents created by forced-air heating, air conditioners, and vacuum cleaners all help to move the particles into the air, where they will be breathed in by the human inhabitants of the buildings.

The cysteine protease of *D. pteronyssimus* is related to the protease papain, which comes from the papaya fruit and is used in cooking as a meat tenderizer. Workers involved in the commercial production of papain become allergic to the enzyme, an example of an occupational immunological disease. Similarly, the protease subtilisin, the 'biological' component of some laundry detergents, causes allergy in laundry workers. Chymopapain, a protease

Features of inhaled allergens that may promote the priming of T_H2 cells that drive IgE responses	
Molecular type	Proteins, because only they induce T-cell responses
Function	Allergens are often proteases
Low dose	Favors activation of IL-4-producing CD4 T cells
Low molecular weight	Allergen can diffuse out of particle into mucus
High solubility	Allergen is readily eluted from particle
High stability	Allergen can survive in desiccated particle
Contains peptides that bind host MHC class II	Required for T-cell priming

Figure 12.13 Properties of inhaled allergens.

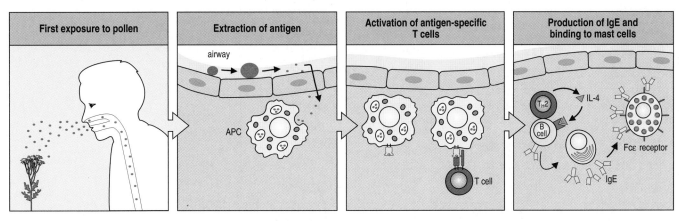

Figure 12.14 Sensitization to an inhaled allergen. Antigens that leach from inhaled pollen are taken up by antigen-presenting cells (APC) in the mucosa of the airways. These activate naive T cells to become T_H2 effector cells, which secrete IL-4. IL-4 binds to the B cell's IL-4 receptor, causing the B cell to switch its immunoglobulin isotype and secrete IgE. The IgE binds to FcεRI on mast cells.

related to papain, is used in medicine to degrade intervertebral disks in patients with sciatica. A rare complication of this procedure affects patients who are sensitized to chymopapain; they experience an acute systemic allergic response to the enzyme, an example of systemic anaphylaxis.

12-8   Predisposition to allergic disease has a genetic basis

In the caucasian populations of Europe and North America, up to 40% of people are more likely than the rest of the population to make IgE responses to common environmental antigens. Allergists call this predisposed state **atopy**. As a group, atopic people have higher levels of soluble IgE and circulating eosinophils than nonatopic people. Family studies and genomic analyses show that there is a genetic basis for atopy. It is estimated that half of the risk of getting an allergic diseases such as asthma is due to genetics and half is due to environmental factors.

Unlike the simplicity of the inherited immunodeficiencies, in which disease is due to defects in one or two copies of a gene, the genetics of susceptibility to asthma is complicated and involves polymorphisms of many different genetic loci that contribute to the adaptive immune response. Nine such loci are shown in Figure 12.15. Of these, six are directly involved in the stimulation of antigen-specific IgE and its function. They encode the MHC class II molecules, the T-cell receptor α chain, IL-4, the IL-4 receptor, the TIM proteins (implicated in the balance between T_H1 and T_H2), and FcϵRI. Of the remaining three loci, *ADAM33* encodes an enzyme involved in repairing damaged bronchial tissue, the β_2-adrenergic receptor (a receptor for epinephrine (adrenaline) and norepinephrine) influences bronchial activity, and 5-lipoxygenase is an enzyme on the pathway that synthesizes the leukotrienes released by mast cells, eosinophils, and basophils.

The IL-4 gene is part of a cluster of genes on chromosome 5 that contains the IL-3, IL-5, IL-9, IL-12, IL-13, and GM-CSF genes. These cytokines are all directly involved in isotype switching, eosinophil survival, and mast-cell

Gene	Nature of polymorphism	Possible mechanism of association
MHC class II genes	Structural variants	Enhanced presentation of particular allergen-derived peptides
T-cell receptor α locus	Non-coding variant	Enhanced T-cell recognition of certain allergen-derived peptides
TIM gene family	Promoter and structural variants	Regulation of the T_H1/T_H2 balance
IL-4	Promoter variant	Variation in expression of IL-4
IL-4 receptor α chain	Structural variant	Increased signaling in response to IL-4
High-affinity IgE receptor β chain	Structural variant	Variation in consequences of IgE ligation by antigen
5-Lipoxygenase	Promoter variant	Variation in leukotriene production
β_2-Adrenergic receptor	Structural variants	Increased bronchial hyper-reactivity
ADAM33	Structural variants	Variation in airway remodeling

Figure 12.15 Genes associated with susceptibility to asthma.

proliferation. Comparison of asthma and atopic dermatitis, another allergic disease, produces correlations with different groups of genes. Similarly, examination of the same disease but in different ethnic groups gives correlations with different groups of genes. This complexity is a reflection of the fact that allergic disease is caused by slight, but significant, perturbations of the balance between the T_H1 and T_H2 responses, which can be caused by many different combinations of gene polymorphisms in combination with environmental factors.

12-9 IgE-mediated allergic reactions consist of an immediate response followed by a late-phase response

One way in which clinical allergists detect a person's sensitivity to allergens is to observe the reaction when small quantities of common allergens are injected into the skin. Substances to which a person is sensitive produce a characteristic inflammatory reaction called a **wheal and flare** at the site of injection within a few minutes (Figure 12.16, arm on left). Substances to which the person is not allergic produce no such reaction. Because of its rapid appearance, the wheal and flare is called an **immediate reaction**. Such reactions are the direct consequence of IgE-mediated degranulation of mast cells in the skin. Released histamine and other mediators cause increased permeability of local blood vessels, whereupon fluid leaves the blood and produces local swelling (edema). The swelling produces the wheal at the injection site, and the increased blood flow into the surrounding area produces the redness that is the flare. Immediate reactions can last for up to 30 minutes, and the relative intensity of the wheal and flare varies. Some 6–8 hours after the immediate reaction has subsided, a second reaction—the **late-phase reaction**—occurs at the site of injection (Figure 12.16, arm on right). This consists of a more widespread swelling and is due to the leukotrienes, chemokines, and cytokines synthesized by mast cells after IgE-mediated activation.

Although skin tests are a good way of determining whether a patient has developed an allergen-specific IgE response, they cannot assess, for example, whether the response to a particular allergen is the cause of a patient's asthma—an allergic reaction in which the airways become inflamed, constricted, and blocked with mucus. To ascertain this, allergists make direct measurements of a person's breathing capacity in the absence and presence of inhaled allergen (Figure 12.17). On inhalation of an allergen to which a patient is sensitized, mucosal mast cells in the respiratory tract degranulate. The released

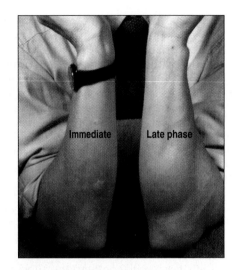

Figure 12.16 Allergic reactions consist of an immediate reaction followed later by a late-phase reaction. The photograph shows a person who has received an intradermal injection of grass-pollen extract in each arm. One injection (left) was made 15 minutes before the photograph was taken and shows the characteristic wheal-and-flare of the immediate reaction. The wheal is the raised area of skin with the injection site at the center, the flare is the redness (erythema) spreading out from the wheal. The other injection (right) was made 6 hours before the photograph was taken and shows the late-phase reaction. The swelling has spread from the site of injection to involve surrounding tissue. Photograph courtesy of S.R. Durham.

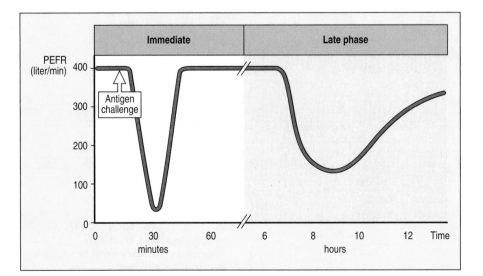

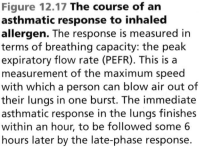

Figure 12.17 The course of an asthmatic response to inhaled allergen. The response is measured in terms of breathing capacity: the peak expiratory flow rate (PEFR). This is a measurement of the maximum speed with which a person can blow air out of their lungs in one burst. The immediate asthmatic response in the lungs finishes within an hour, to be followed some 6 hours later by the late-phase response.

mediators cause immediate constriction of the bronchial smooth muscle, which results in the expulsion of material from the lungs by coughing, and difficulty in breathing. As in the skin test, the immediate response in the lungs finishes within an hour, to be followed some 6 hours later by the late-phase response, which is due to leukotrienes and other mediators. In allergies to inhaled antigens, such as chronic asthma, the late-phase reaction is the more damaging. It induces the recruitment of leukocytes, particularly eosinophils and T_H2 lymphocytes, into the site and, if antigen persists, the late-phase response can easily develop into a chronic inflammatory response in which allergen-specific T_H2 cells promote eosinophilia and IgE production.

12-10 The effects of IgE-mediated allergic reactions vary with the site of mast-cell activation

When sensitized people are reexposed to an allergen, the effect of the IgE-mediated reaction varies depending on the allergen and the tissues with which it comes in contact. Only those mast cells at the site of exposure degranulate and, once released, the preformed mediators are short-lived. Their effects on blood vessels and smooth muscles are therefore confined to the immediate vicinity of the activated mast cells. The more sustained effects of the late-phase response are also restricted to the site of allergen exposure, because the leukotrienes and other induced mediators are also short-lived. The anatomy of the site of contact also determines how quickly the inflammatory reaction subsides. The tissues most commonly exposed to allergens are the mucosa of the respiratory and gastrointestinal tracts, the blood, and connective tissues. Airborne allergens irritate the respiratory tract and food-borne antigens the gastrointestinal tract; the blood and connective tissues receive allergens through insect bites and other wounds and also from absorption via the gut and respiratory mucosa (Figure 12.18).

For parasite infections, interactions between antigen, IgE, and mast cells trigger violent muscular contractions that can expel worms from the gastrointestinal tract and increase fluid flow to wash them out. In the lungs, the muscular spasms and increased secretion of mucus that result from mast-cell activation can be seen as a way to expel organisms attached to respiratory epithelium, such as lung flukes. In allergy, this crude defense is mistakenly ranged

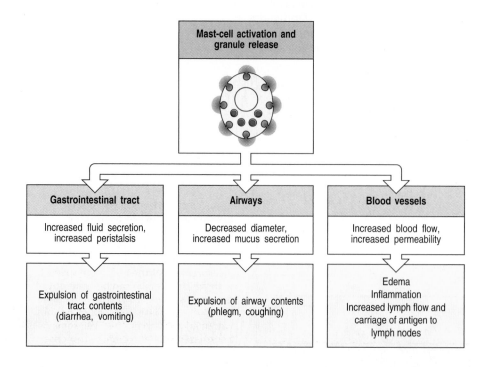

Figure 12.18 **The physical effects of IgE-mediated mast-cell degranulation vary with the tissue exposed to allergen.**

against non-threatening particles and proteins that are often smaller than any microorganism. The violence and power of the IgE-mediated response means that the clinical manifestations of an allergic reaction can vary markedly with the amount of IgE that a sensitized person has made, the amount of allergen triggering the reaction, and the route by which it entered the body. In the next four sections we examine how allergic reactions vary when they occur in different tissues.

12-11 Systemic anaphylaxis is caused by allergens in the blood

When an allergen enters the bloodstream it can cause widespread activation of the connective tissue mast cells associated with blood vessels. This causes a dangerous hypersensitivity reaction called **systemic anaphylaxis**. During systemic anaphylaxis, disseminated mast-cell activation causes both an increase in vascular permeability and a widespread constriction of smooth muscle. Fluid leaving the blood causes the blood pressure to drop drastically, a condition called **anaphylactic shock**, and the connective tissues to swell. Damage is sustained by many organ systems and their function is impaired. Death is usually caused by asphyxiation due to constriction of the airways and swelling of the epiglottis (Figure 12.19). In the United States, more than 160 deaths a year are the result of anaphylaxis, which is the most extreme over-reaction of the body's defenses. Because this form of immunity is fatal rather than protective, it was called '*ana*phylaxis,' meaning anti-protection, to contrast with the '*pro*phylaxis' afforded by protective immunity.

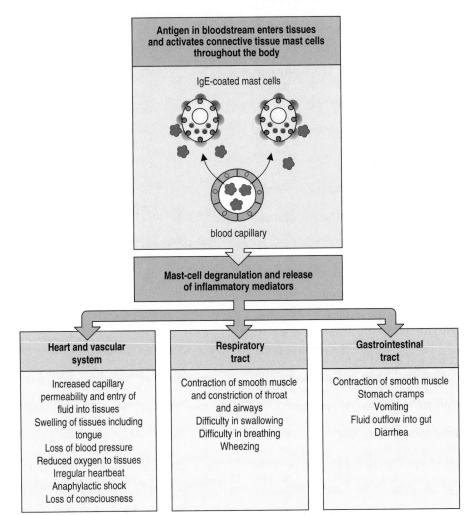

Figure 12.19 Systemic anaphylaxis is caused by allergens that reach the bloodstream and activate mast cells throughout the body.

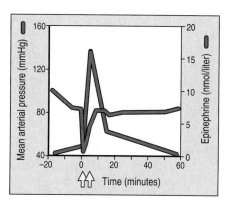

Figure 12.20 The change in blood pressure during systemic anaphylaxis and its treatment with epinephrine. Time 0 indicates the time at which the anaphylactic reaction was reported by the patient. The arrows indicate the times at which injections of epinephrine were given.

Potential allergens are introduced directly into the blood by stings from wasps, bees, and other venomous insects; these account for about a quarter of the fatalities from anaphylaxis in the United States. Drug injections can also lead to anaphylaxis. Systemic anaphylaxis can also be provoked by food or drugs taken orally, if the allergens they contain are absorbed rapidly from the gut into the blood. Foods that can cause anaphylaxis include peanuts and brazil nuts. Anaphylactic reactions can be fatal, but treatment with an injection of epinephrine will usually bring them under control. Epinephrine stimulates the reformation of tight junctions between endothelial cells. This reduces their permeability and prevents fluid loss from the blood, diminishing tissue swelling and raising blood pressure. Epinephrine also relaxes constricted bronchial smooth muscle and stimulates the heart (Figure 12.20). The danger of anaphylaxis is such that patients with known anaphylactic sensitivity to insect venoms or food are advised to carry a syringe full of epinephrine at all times.

The most common cause of systemic anaphylaxis is allergy to penicillin and related antibiotics, which accounts for about 100 fatalities a year in the United States. Penicillin is a small organic molecule with a reactive β-lactam ring. On ingestion or injection of the drug, the β-lactam ring can be opened up to produce covalent conjugates with proteins of the body, creating new 'foreign' epitopes. These modified proteins can then stimulate a variety of hypersensitivity reactions, as described in later sections. In some individuals, the response is dominated by T_H2 cells and thus by B cells producing IgE specific for the new epitopes. When penicillin is given to a person who has been sensitized in this manner, it causes anaphylaxis and even death. For this reason, clinicians take care to avoid prescribing any drug to patients who have a history of allergy to that drug. Unfortunately, penicillin cannot be modified to remove its allergic potential because the reactive β-lactam ring is essential for its antibiotic activity.

Reactions resembling anaphylaxis can occur in the absence of specific interaction between an allergen and IgE. Such **anaphylactoid reactions** are caused by other stimuli that induce mast-cell degranulation and can be occasioned by exercise or by certain drugs and chemicals. Anaphylactoid reactions are also treated with epinephrine.

12-12 Rhinitis and asthma are caused by inhaled allergens

Allergens most commonly enter the body by inhalation. Mild allergies to inhaled antigens are common, being manifested as violent bursts of sneezing and a runny nose, a condition called **allergic rhinitis** or hay fever. It is caused by allergens that diffuse across the mucous membrane of the nasal passages and activate mucosal mast cells beneath the nasal epithelium (Figure 12.21). Allergic rhinitis is characterized by local edema leading to obstruction of the nasal airways and a nasal discharge of mucus that is rich in eosinophils. There is also a generalized irritation of the nose due to histamine release. The reaction can extend to the ear and throat, and the accumulation of fluid in the blocked sinuses and eustachian tubes is conducive to bacterial infection. The same exposure to allergen that produces rhinitis can affect the conjunctiva of the eyes, where the reaction is called **allergic conjunctivitis**. It produces itchiness, tears, and inflammation. Although these reactions are uncomfortable and distressing, they are generally of short duration and cause no long-lasting tissue damage.

Figure 12.21 Allergic rhinitis is caused by allergens entering the respiratory tract. Histamine and other mediators released by activated mast cells increase the permeability of local capillaries and activate the nasal epithelium to produce mucus. Eosinophils attracted into the tissues from the blood become activated and release their inflammatory mediators. Activated eosinophils are shed into the nasal passages.

Much more serious is **allergic asthma**, a condition suffered by 130 million people worldwide, in which allergic reactions cause chronic difficulties in breathing, such as shortness of breath and wheezing. Asthma is triggered by allergens activating submucosal mast cells in the lower airways of the respiratory tract. Within seconds of mast-cell degranulation there is an increase in the fluid and mucus being secreted into the respiratory tract, and bronchial constriction due to contraction of the smooth muscle surrounding the airway. Chronic inflammation of the airways is a characteristic feature of asthma, involving a persistent infiltration of leukocytes, including T_H2 lymphocytes, eosinophils, and neutrophils (Figure 12.22). The overall effect of the asthmatic attack is to trap air in the lungs, making breathing more difficult. Patients with allergic asthma often need treatment, and asthmatic attacks can prove fatal.

Although allergic asthma is initially driven by a response to a specific allergen, the chronic inflammation that subsequently develops seems to be perpetuated in the absence of further exposure to the allergen. In **chronic asthma** the airways can become almost totally occluded by plugs of mucus (Figure 12.23). A generalized hyper-responsiveness or hyper-reactivity in the airways also develops, and environmental factors other than reexposure to specific allergen can trigger asthmatic attacks. Typically, the airways of chronic asthmatics are hyper-reactive to chemical irritants commonly present in air, such as cigarette smoke and sulfur dioxide. Disease can be exacerbated by immune responses to bacterial or viral infections of the respiratory tract, especially when they are dominated by T_H2 cells. For this reason, chronic asthma is classified as a type IV hypersensitivity reaction caused by T cells.

12-13 Urticaria, angioedema, and eczema are allergic reactions in the skin

Allergens that activate mast cells in the skin to release histamine cause raised itchy swellings called **urticaria** or **hives**. Urticaria means 'nettle-rash' and the word derives from *Urtica*, the Latin name for stinging nettles; the origin of the word hives is unknown. This reaction is essentially the same as the immediate wheal-and-flare reaction caused by the deliberate introduction of allergens into the skin in tests to determine allergy (see Section 12-9), which can also be produced by the injection of histamine alone (Figure 12.24). Activation of mast cells in deeper subcutaneous tissue leads to a similar but more diffuse swelling called **angioedema** (or angioneurotic edema). Urticaria and angioedema can arise as a result of any allergy to a food or drug if the allergen gets carried to the skin by the bloodstream, and they are among the many reactions that occur during systemic anaphylaxis. Insect bites are a common cause of urticaria, and such local reactions usually occur without inducing a more general anaphylaxis.

A more prolonged allergic response in the skin is observed in some atopic children. This condition is called **atopic dermatitis** or **eczema**. The word eczema is of Greek derivation and means to 'break out' or 'boil over.' The condition is characterized by an inflammatory response that causes a chronic and itching skin rash with associated skin eruptions and fluid discharge. This response has similarities to that occurring in the bronchial walls of asthmatics. Eczema frequently presents in families with a history of asthma and

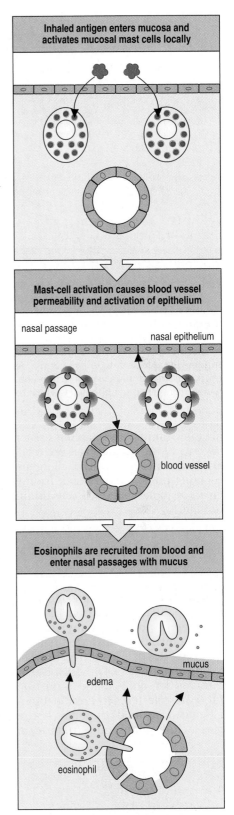

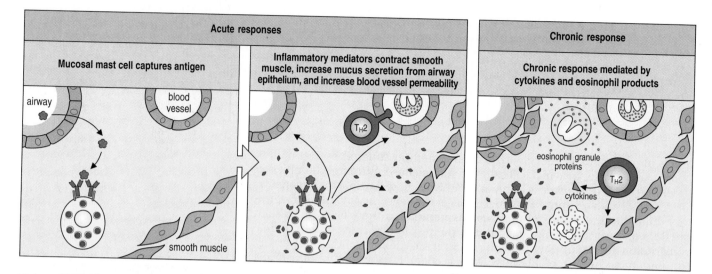

Figure 12.22 The acute response in allergic asthma leads to T$_H$2-mediated chronic inflammation of the airways. In sensitized individuals, mast cells carrying IgE specific for the allergen are present in the mucosa of the airways, as shown in the first panel. As shown in the second panel, cross-linking of specific IgE on the surface of mast cells by inhaled allergen triggers them to secrete inflammatory mediators, causing bronchial smooth muscle contraction and increased secretion of mucus from mucosal epithelium, which together lead to airway obstruction. Increased blood vessel permeability also caused by inflammatory mediators leads to edema and an influx of inflammatory cells, including eosinophils and T$_H$2 lymphocytes. Activated mast cells and T$_H$2 cells secrete cytokines that also augment eosinophil activation and degranulation, which causes further tissue injury and influx of inflammatory cells, as shown in the third panel. The end result is chronic inflammation, which can then cause irreversible damage to the airways.

allergic rhinitis, and is often associated with high IgE levels. However, the severity of the dermatitis is not readily correlated with exposure to particular allergens or to the levels of allergen-specific IgE. Thus the etiology of eczema remains poorly understood. Neither is it understood why eczema usually clears in adolescence, whereas rhinitis and asthma more often persist throughout life.

12-14 Food allergies cause systemic effects as well as gut reactions

Humans probably eat a greater variety of foodstuffs than any other animal. All human food is derived from plants and animals and contains a large number of different proteins, many of which are potentially immunogenic. As food passes down the gastrointestinal tract, the proteins are degraded by proteases into peptides of ever-decreasing size. These are a potential source of peptides for presentation to T$_H$2 cells. Despite the quantity and variety of food that humans eat, IgE is made against only an extremely small proportion of the proteins ingested. However, people sensitized to a particular protein will be allergic to any food containing that protein. Foods that commonly cause allergies include grains, nuts, fruits, legumes, fish, shellfish, eggs, and milk.

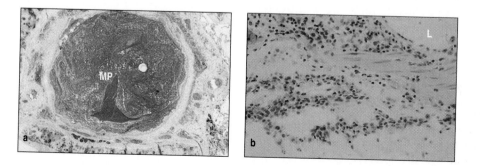

Figure 12.23 Inflammation of the airways in chronic asthma restricts breathing. Panel a shows a light micrograph of a section through the bronchus of a patient who died of asthma; there is almost total occlusion of the airway by a mucous plug (MP). The small white circle is all that is left of the lumen of the bronchus. In panel b, a light micrograph at higher magnification gives a closer view of the bronchial wall. It shows injury to the epithelium lining the bronchus, accompanied by a dense inflammatory infiltrate that includes eosinophils, neutrophils, and lymphocytes. L, lumen of the bronchus. Photographs courtesy of T. Krausz.

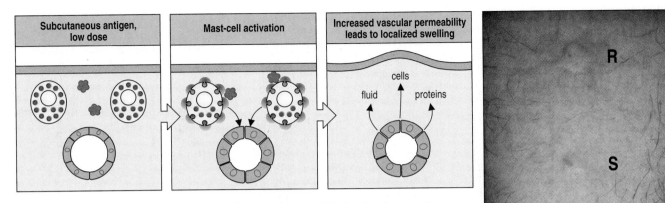

Figure 12.24 Allergen-induced release of histamine by mast cells in skin causes localized swelling. As shown in the panels on the left, allergen introduced into the skin of a sensitized individual causes mast cells in the connective tissue to degranulate. The histamine released dilated local blood vessels, causing rapid swelling due to leakage of fluid and proteins into tissues. The photograph shows the raised swellings (wheals) that appear 20 minutes after intradermal injections of ragweed pollen antigen (R) or histamine (H) into a person who is allergic to ragweed. The small wheal at the site of saline injection (S) is due to the volume of fluid injected into the dermis. Photograph courtesy of R. Geha.

Once sensitized to a food allergen, any subsequent intake causes a marked and immediate reaction that can drive the person from the dining room. The allergen passes across the epithelial wall of the gut and binds to IgE on the mucosal mast cells associated with the gastrointestinal tract. The mast cells degranulate, releasing their mediators, principally histamine. The local blood vessels become permeable, and fluid leaves the blood and passes across the gut epithelium into the lumen of the gut. Meanwhile, contraction of smooth muscle of the stomach wall produces cramps and vomiting; the same reaction in the intestine produces diarrhea (Figure 12.25). These reactions probably evolved originally to expel gut parasites; in the allergic reaction they serve to expel the allergen-containing food from the human body. This goal is indeed accomplished, but at the expense of dehydration, weakness, and waste of food.

In addition to allergic reactions localized in the gut, food allergens also produce reactions in other tissues, notably the skin. Depending on the timing and course of the gut reaction and of the uptake of allergen from the gut, allergen can enter the circulation and be transported elsewhere in the body. Mast cells in the connective tissue in the deeper layers of the skin tend to be activated by such blood-borne allergens, and their degranulation produces urticaria and angioedema. In this context, orally administered drugs behave similarly to food; they too can produce intestinal reactions, urticaria, and angioedema in sensitized individuals.

12-15 People with parasite infections and high levels of IgE rarely develop allergic disease

In tropical countries, parasitic helminth infections are endemic; more than 2 billion people worldwide are heavily and persistently infected. A universal feature of helminth infection is the stimulation of CD4 T_H2-like responses that produce raised levels of IgE, and increased numbers of eosinophils in the blood and mast cells in tissues. Only a small fraction of the IgE is parasite-specific; the remainder is highly heterogeneous and represents the product of nonspecific, polyclonal B-cell and T-cell activation by the parasite. Despite the abundance and variety of IgE in their system, people with helminth infections are rarely afflicted by allergic disease.

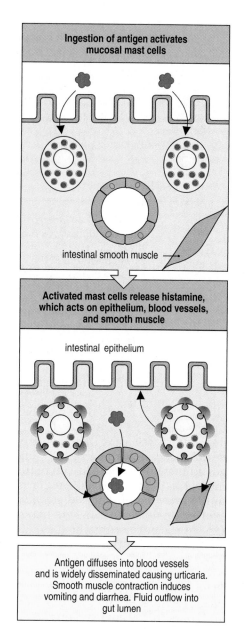

Figure 12.25 Ingested allergen can cause vomiting, diarrhea, and urticaria. Localized reactions are caused by the action of histamine on intestinal epithelium, blood vessels, and smooth muscles. Urticaria is caused by antigen that enters blood vessels and is carried to the skin.

Several facts may contribute to the resistance of helminth-infected individuals to allergy. One is that nonspecific IgE competes with parasite-specific IgE, or any allergen-specific IgE, for binding to the FcεRI on mast cells, basophils, and activated eosinophils. This limits the extent to which parasite antigen binding to specific IgE triggers IgE-mediated effector mechanisms, allowing the parasite to escape killing by activated eosinophils or expulsion from the body by the action of activated mast cells. Another contributing factor is that T-cell responses are generally suppressed in chronic parasitic infections, probably as a result of the participation of regulatory T cells and the inhibitory effects of IL-10, TGF-β, and nitric oxide.

For the populations of western Europe and North America, infections by helminths and many other parasites have largely been eradicated. In these populations the prevalence of IgE-mediated allergy and asthma has been steadily increasing. The 'hygiene hypothesis' proposes that this increase has been caused by better hygiene, vaccination to prevent infection, and the increased use of antibiotics and other drugs to stop infections. Children are exposed to fewer and less heavy infections than were their parents, with the result that their immune systems are poorly developed, insufficiently used and less successfully regulated. In other words, a lack of practice in dealing with real infections can reveal a propensity to perceive danger where it does not exist. Supporting the hygiene hypothesis are family studies showing that exposure of children to more infections, and at a younger age, reduces the likelihood that they will develop atopic allergic reactions.

12-16 Allergic reactions are prevented and treated by three complementary approaches

Three distinct strategies are used to reduce the effects of allergic disease. The first strategy is one of prevention—to modify a patient's behavior and environment so that contact with the allergen is avoided. Allergen-containing foods are avoided, houses are refurnished in ways that discourage mites, pets are kept outside, and desert vacations or sea cruises are taken during the pollen season.

The second strategy is pharmacological—to use drugs that reduce the impact of any contact with allergen. Such drugs block the effector pathways of the allergic response and limit the inflammation after IgE-induced activation of mast cells, eosinophils, and basophils. Antihistamines reduce rhinitis and urticaria by preventing histamine from binding to H1 histamine receptors on vascular endothelium and thus increasing vascular permeability. Corticosteroids, which suppress leukocyte function generally, are often administered topically or systemically to suppress the chronic inflammation of asthma, rhinitis, or eczema. Cromolyn sodium prevents the degranulation of activated mast cells and granulocytes and is inhaled by asthmatics as a prophylactic to prevent attacks. Epinephrine is used to treat anaphylactic reactions.

The third strategy in the treatment of allergy is immunological—to prevent the production of allergen-specific IgE. One way of achieving this is to modulate the antibody response so that it shifts from one dominated by IgE to one dominated by IgG. A procedure called **desensitization**, which was first described in 1911 and is used in a similar way today, can achieve this for some

Ingestion of antigen activates mucosal mast cells

intestinal smooth muscle

Activated mast cells release histamine, which acts on epithelium, blood vessels, and smooth muscle

intestinal epithelium

Antigen diffuses into blood vessels and is widely disseminated causing urticaria. Smooth muscle contraction induces vomiting and diarrhea. Fluid outflow into gut lumen patients and some allergies. Patients are given a series of allergen injections in which the dose is initially very small and is gradually increased. Successful treatments are associated with the production of allergen-specific antibodies of the IgG4 isotype—which, like IgE, is a product of a T_H2 response—and increased levels of IL-10. IgG4 is functionally monovalent and forms complexes with antigen that do not recruit effector cells (see Section 9-21, p. 277). An occasional consequence of the injections used for desensitization is anaphylaxis, because the patient is being exposed to the allergen to which they are sensitized. For this reason 'allergy shots' should always be given under carefully controlled conditions in which patients are monitored for the early symptoms of systemic anaphylaxis and, if need be, given epinephrine.

A more recent approach to desensitization is to vaccinate patients with allergen-derived peptides that are presented by HLA class II molecules to CD4 T_H2 cells. The aim is to induce anergy of allergen-specific T cells *in vivo* by decreasing the expression of the CD3:T-cell receptor complex at the T_H2 cell surface. In principle, the advantage of this method over the older approach is that anaphylaxis should never be triggered by the injections, because only the native allergen protein and not the peptide vaccine can interact with allergen-specific IgE. Factors complicating this approach are the HLA class II polymorphisms, which determine which allergen-derived peptides can be presented by any individual. Any vaccine with general applicability to a human population must contain sufficient different peptides so that an anergizing response can be produced irrespective of HLA class II type. Alternatively, vaccines could be custom-made from peptides selected on the basis of a patient's HLA class II type.

Because allergies are a prevalent and increasing problem for humans in the richer countries in the world, there is much interest within the biotechnology and pharmaceutical industries in developing new approaches to the relief or cure of allergy. One potential set of targets for drugs are the signaling pathways that cells of the immune system use to enhance the IgE response. For example, inhibitors of the cytokines IL-4, IL-5, or IL-13 could block such pathways. Conversely, the administration of cytokines that promote T_H1 responses could shift the antibody response away from IgE and towards IgG. In experiments on mice, IFN-γ and IFN-α have both been shown to reduce IL-4-stimulated IgE synthesis. The high-affinity IgE receptor is also a potential target for drugs that bind the receptor and prevent the arming of mast cells with allergen-specific IgE.

Summary

Type I hypersensitivity reactions are caused by protein allergens binding to IgE molecules and activating mast cells. The activated mast cells release a variety of chemical mediators that orchestrate a local state of inflammation. Smooth muscles constrict, blood vessels dilate, and eosinophils and basophils enter the affected area. The effects of IgE-mediated reactions vary with the route of allergen entry to the body and the tissue affected. Inhaled allergens, for example plant pollens and animal dander, activate mast cells in the respiratory tract. Rhinitis is caused by reactions in the upper airways; asthma by reactions in the lower airways. Insect stings deliver allergens into the skin, where mast-cell mediators cause hives and urticaria. Allergens in certain foods, such as peanuts or shrimp, trigger mast cells of the gastrointestinal tract, resulting in vomiting and diarrhea. When food allergens are absorbed into the blood they become disseminated throughout the body, causing systemic mast-cell activation and resulting in widespread urticaria or even systemic anaphylaxis. The violent reactions activated by IgE are believed to have evolved as a defense against parasites. In allergic reactions this defensive mechanism is misguidedly aimed at environmental proteins that pose no threat.

Type II, III, and IV hypersensitivity reactions

IgG antibodies and specific T cells can cause both acute and chronic adverse hypersensitivity reactions. A variety of different effector mechanisms are involved, which cause inflammatory reactions and tissue destruction.

12-17 Type II hypersensitivity reactions are caused by antibodies specific for altered components of human cells

Occasional side-effects seen after the administration of certain drugs are hemolytic anemia caused by the destruction of red blood cells, and thrombocytopenia caused by the destruction of platelets. These are examples of type II hypersensitivity reactions and have been associated with the antibiotic penicillin, quinidine (a drug used to treat cardiac arrhythmia), and methyldopa, which is used to reduce high blood pressure. In each case, chemically reactive drug molecules bind to surface components of red blood cells or platelets and create new epitopes to which the immune system is not tolerant (Figure 12.26). These epitopes stimulate the formation of IgM and IgG antibodies that are specific for the conjugate of drug and cell-surface component.

The penicillin-modified red cells acquire a coating of the complement component C3b as a side-effect of complement activation by the bacterial infection for which the drug has been given. This facilitates their phagocytosis by macrophages via complement receptors. These cells process the penicillin-modified protein and present peptides from it to specific CD4 T cells, which

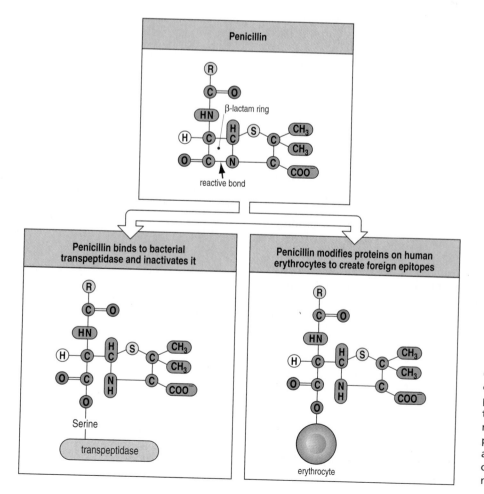

Figure 12.26 Penicillin and other small-molecule drugs can modify human cells so that they display foreign epitopes. Penicillin exerts its antibacterial action by mimicking the substrate for bacterial transpeptidase, which is required for bacterial cell wall synthesis. On binding to transpeptidase, a reactive bond in the β-lactam ring of penicillin opens and forms a covalent bond with an amino acid residue in the active site of the transpeptidase, thereby inactivating the enzyme permanently (lower left panel). The same mechanism occasionally results in molecules of penicillin becoming covalently bonded to surface proteins of human cells (lower right panel). This modification of human proteins creates new epitopes that can act like foreign antigens. Red blood cells (RBC) are the cells most commonly modified in this way.

Figure 12.27 Penicillin–protein conjugates stimulate the production of anti-penicillin antibodies. Red cells that have been covalently bonded to penicillin (P) are phagocytosed by macrophages, which process the penicillin-modified proteins and present peptide antigens to specific CD4 T cells. These are activated to become effector T_H2 cells, which stimulate antigen-specific B cells to produce antibodies against the penicillin-modified epitope.

are activated to become effector T_H2 cells. These then stimulate antigen-specific B cells to produce antibodies against the penicillin-modified epitope (Figure 12.27). Binding of antibodies to the drug-conjugated cells activates complement by the classical pathway, resulting in either cell lysis by the terminal complement components or receptor-mediated phagocytosis by macrophages in the spleen (Figure 12.28).

12-18 To avoid type II hypersensitivity reactions in blood transfusion, donors and recipients are matched for ABO antigens

Blood transfusion can give rise to life-threatening type II hypersensitivity reactions caused by antibodies in the recipient's circulation that bind to components on the surface of the erythrocytes in the transfused blood. The major immunogenetic barrier to transfusion with red blood cells arises from structural polymorphisms in the carbohydrates on glycolipids of the erythrocyte surface. The resulting antigenic differences in these carbohydrates are the basis for the **ABO system** of blood group antigens (Figure 12.29).

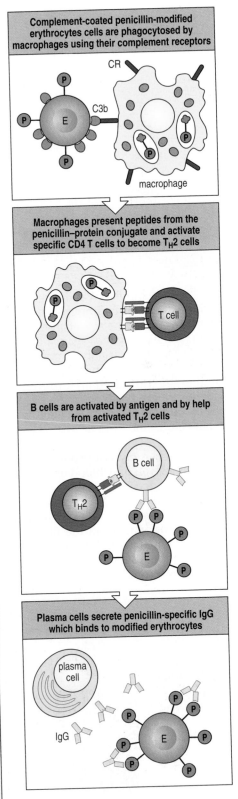

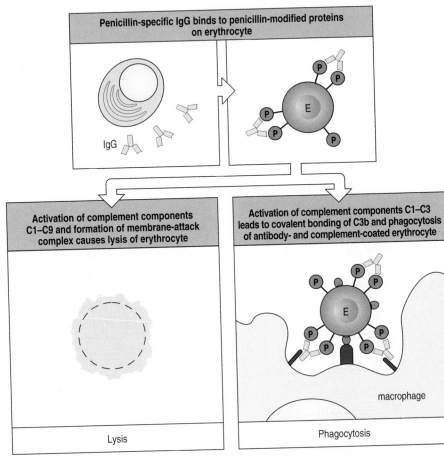

Figure 12.28 Binding of antibodies to penicillin-modified red cells makes them susceptible to complement- **mediated lysis or to phagocytosis via Fc receptors and complement receptors.**

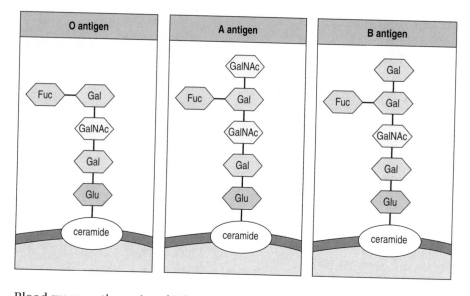

Blood group antigens A and B have structural similarities to some cell-surface carbohydrates of common bacteria. During infections with these bacteria, people who lack the A or B antigen, or both, and are therefore not tolerant of them, make antibodies against them. For example, the serum of people with blood group O invariably contains antibodies against blood-group antigens A and B. If such a person is transfused with group A or B blood, the antibodies bind to the transfused erythrocytes, causing complement fixation and rapid clearance of red cells from the circulation. Besides thwarting the purpose of the blood transfusion, these hemolytic reactions can cause fever, chills, shock, renal failure, and even death.

To prevent hemolysis and other transfusion reactions, patients needing transfusion are given only blood of compatible ABO type. Four blood types—O, A, B, and AB—account for the vast majority of people, which makes matching for this system of alloantigens much simpler than for HLA (Figure 12.30). Some 20 other polymorphic systems of blood group antigens have been defined, but apart from the Rhesus (Rh) blood group antigens, which also need to be matched, they are less important for matching in blood transfusion than the ABO system. To eliminate unexpected transfusion reactions, a recipient's serum is tested directly for its reactivity with the red cells selected

Figure 12.29 Structures of the ABO blood group antigens. The ABO antigens arise from a family of glycolipids at the erythrocyte surface. Their core structure consists of the lipid ceramide, attached to an oligosaccharide consisting of glucose (Glu), galactose (Gal), N-acetyl galactosamine (GalNAc), galactose, and fucose (Fuc). In people of blood group O this is the only glycolipid made. People of blood group A have an enzyme that can add an additional N-acetyl galactosamine, forming the A antigen. People of blood group B have an enzyme that can add an additional galactose to the core structure, forming the B antigen. The erythrocytes of people of blood groups A and B also express the core structure alone, which is why alloantibodies against O are not made.

Recipient	Potential donor			
	O	A	B	AB
O — Anti-A and anti-B antibodies				
A — Anti-B antibodies				
B — Anti-A antibodies				
AB — No antibodies against A or B				

Figure 12.30 Donors and recipients for blood transfusion must be matched for the ABO system of blood group antigens. Common gut bacteria bear antigens that are similar or identical to blood group antigens, and these stimulate the formation of antibodies against these antigens in individuals who do not bear the corresponding antigen on their own red blood cells (left column); thus, type O individuals, who lack A and B, have both anti-A and anti-B antibodies, whereas type AB individuals have neither. The combinations of donor and recipient blood groups that allow blood transfusion are indicated by green boxes; the combinations that would result in an immune reaction, and must be avoided, are indicated by the red boxes.

for transfusion on the basis of ABO typing. This direct assessment of compatibility is called a **cross-match test**. Cross-match tests are not performed between a recipient's cells and serum from the blood to be transfused, because when antibodies in transfused blood bind to recipient's erythrocytes they have no detrimental effect. The quantity of antibody transfused is insufficient to produce the density of antibody on the red-cell surface required to trigger hemolysis.

12-19 Type III hypersensitivity reactions are caused by immune complexes formed from IgG and soluble antigens

Complexes of soluble protein antigens and their high-affinity IgG antibodies are generated in almost all immune responses and in most situations they are cleared without causing tissue damage. Immune complexes vary greatly in size, from a simple complex of one antigen and one antibody molecule through to large aggregates containing millions of antigen and antibody molecules. The larger aggregates fix complement efficiently and are readily taken up by phagocytes and removed from the circulation. Smaller immune complexes are less efficient at fixing complement; they tend to circulate in the blood and become deposited in blood vessel walls. When these complexes accumulate at such sites, they become capable of fixing complement and initiating tissue-damaging inflammatory reactions through their interactions with the Fc receptors and complement receptors on circulating leukocytes and mast cells. This type of hypersensitivity is called type III hypersensitivity. Complement activation produces C3a and C5a. The former stimulates mast cells to release histamine, causing urticaria; the latter recruits inflammatory cells into the tissue. Platelets accumulate around the site of immune-complex deposition, and the clots that they form cause the blood vessels to burst, producing hemorrhage in the skin.

The size of the immune complexes formed at a particular time and place is strongly influenced by the relative concentrations of soluble antigen and antibody (Figure 12.31). Whether large immune complexes can be formed also depends on the size and complexity of the antigen; most antigens that will be

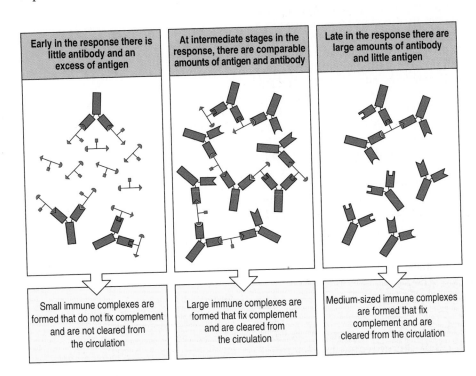

| Early in the response there is little antibody and an excess of antigen | At intermediate stages in the response, there are comparable amounts of antigen and antibody | Late in the response there are large amounts of antibody and little antigen |

| Small immune complexes are formed that do not fix complement and are not cleared from the circulation | Large immune complexes are formed that fix complement and are cleared from the circulation | Medium-sized immune complexes are formed that fix complement and are cleared from the circulation |

Figure 12.31 Immune complexes of different sizes and stoichiometries are formed during the course of an immune response.

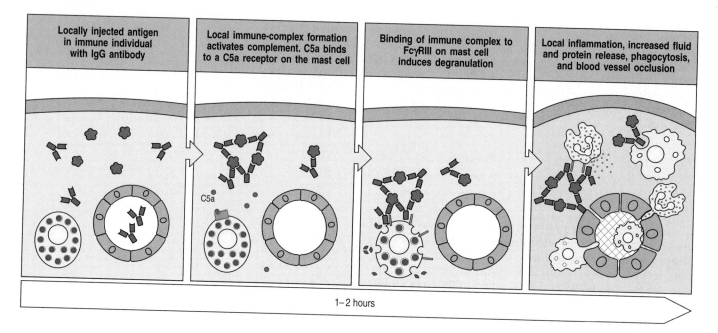

| Locally injected antigen in immune individual with IgG antibody | Local immune-complex formation activates complement. C5a binds to a C5a receptor on the mast cell | Binding of immune complex to FcγRIII on mast cell induces degranulation | Local inflammation, increased fluid and protein release, phagocytosis, and blood vessel occlusion |

1–2 hours

encountered in normal circumstances contain multiple epitopes and thus can, in principle, form extensive immune complexes by cross-linking antibodies. When antigen is in excess, as occurs early in the immune response, each antigen-binding site on an antibody binds an antigen molecule, producing small immune complexes that often contain a single antibody molecule and two antigen molecules. Late in the immune response, antibodies are in excess; in these circumstances each antigen molecule binds to several antibody molecules. At intermediate times, when the amounts of antigen and antibody are more evenly balanced, larger immune complexes are formed in which several antibody and antigen molecules are cross-linked. A minimum of two IgG molecules per complex is needed to fix complement, so it is at the beginning of the immune response, when soluble immune complexes fix complement poorly, that they are most likely to circulate in the blood and become deposited in blood vessel walls.

In people who have made IgG against a soluble protein, a type III hypersensitivity reaction can be experimentally induced in the skin by subcutaneous injection of the antigen. Specific IgG diffuses from the blood into the connective tissue at the site of injection and combines with antigen to form immune complexes. The complexes activate complement, creating an inflammatory reaction that draws leukocytes and antibodies into the site of injection. Here the Fc and complement receptors of leukocytes and mast cells engage the immune complexes, activating the cells and further propagating the local inflammatory reaction. This type of reaction was first described by Nicholas-Maurice Arthus and is called an **Arthus reaction** (Figure 12.32). In humans, Arthus reactions usually appear as localized areas of erythema and hard swelling (induration) that subside within a day. Such reactions can often be seen at the site of injections used to desensitize IgE-mediated allergies. The dependence of the Arthus reaction on immune-complex interactions with Fc receptors is demonstrated by the failure to produce Arthus reactions in mice lacking the γ chain common to all Fc receptors.

Figure 12.32 Localized deposition of immune complexes within a tissue causes a type III hypersensitivity reaction. In sensitized individuals, the introduction of allergen into a tissue leads to the formation of immune complexes with IgG in the extracellular fluid. The immune complexes activate complement and recruit inflammatory cells to the site, causing a hard swelling. Platelets accumulate in the capillary, leading to occlusion and rupture of the vessel, causing erythema. This inflammatory reaction is called the Arthus reaction.

12-20 Systemic disease caused by immune complexes can follow the administration of large quantities of soluble antigens

During the late nineteenth century and the first half of the twentieth century, diphtheria, scarlet fever, tetanus, and other life-threatening bacterial

Route	Resulting disease	Site of immune-complex deposition
Intravenous (high dose)	Vasculitis	Blood vessel walls
	Nephritis	Renal glomeruli
	Arthritis	Joint spaces
Subcutaneous	Arthus reaction	Perivascular area
Inhaled	Farmer's lung	Alveolar/capillary interface

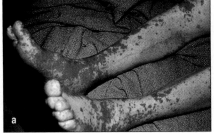

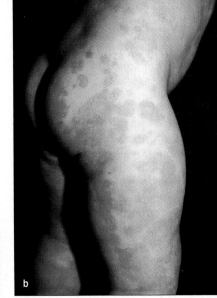

Figure 12.33 The pathology of type III hypersensitivity reactions is determined by the sites of immune-complex deposition. The table shows the types of reaction that result from different routes of antigen entry to the body. Serum sickness follows the intravenous administration of large amounts of foreign antigen. Photographs show hemorrhage in the skin (panel a) and urticarial rash (panel b) resulting from serum sickness. Photographs courtesy of R. Geha.

infections were treated by injecting patients with serum taken from horses that had been immunized with these bacteria or their toxins. The horse antibodies helped human patients to control and clear the infection but could also produce a systemic type III hypersensitivity reaction that became known as **serum sickness**. This condition occurred some 7–10 days after the administration of horse serum and was characterized by chills, fevers, rash, arthritis, vasculitis, and sometimes glomerulonephritis. The cause of serum sickness is the formation of antibodies against the foreign horse proteins and the deposition of small immune complexes in tissues; the symptoms depend on which tissues are affected (Figure 12.33).

Therapeutic administration of serum from immunized horses is rarely used today. One remaining application is the use of horses to prepare antivenom antisera for neutralizing the effects of snakebite. However, the symptoms of serum sickness are now seen in other circumstances in which patients have received an infusion of large amounts of a foreign protein. In the increasing number of patients receiving monoclonal antibodies as treatment for various diseases, serum sickness is a potential complication. It is caused by a patient's immune response to the monoclonal antibody and the formation of immune complexes containing human antibodies bound to the monoclonal antibody. Serum sickness also occurs occasionally in patients who have had a myocardial infarction (heart attack) and are treated with the bacterial enzyme streptokinase to degrade their blood clots. Serum sickness can also result from the intravenous administration of large amounts of a drug such as penicillin, which binds to host proteins on, for example, erythrocytes (see Section 12-17) and provokes an IgG response. This type of reaction can occur in people with no history of allergy to penicillin. Drug-induced serum sickness is now the most common example of this condition.

The onset of serum sickness coincides with the synthesis of antibodies, which form immune complexes with the antigenic proteins. Because the serum is loaded with antigen, large quantities of immune complexes are formed and dispersed throughout the body. The complexes fix complement and activate leukocytes bearing Fc receptors or complement receptors. These activated cells create an inflammatory response that causes widespread damage (Figure 12.34).

The formation of immune complexes induces the clearance of the antigenic proteins by the normal phagocytic pathways; consequently, serum sickness is of limited duration unless additional injections of the foreign antigen are given. If a second dose of antigen is given after the effects of the first dose have subsided, a secondary response will follow, with disease symptoms being manifested within a day or two of the second injection.

A disease similar to serum sickness can be seen in certain infections in which the immune system fails to clear the pathogen, and both the infection and the immune response persist. For example, in subacute bacterial endocarditis, or in chronic viral hepatitis, the multiplying pathogens continue to produce antigens, and plasma cells continue to make antibodies. Immune complexes are continually being generated, deposited, and cleared, processes that can cause injury to small blood vessels and nerves in many organs, including the skin and kidneys.

12-21 Inhaled antigens can cause type III hypersensitivity reactions

Some common inhaled antigens tend to provoke an IgG rather than an IgE response and cause type III hypersensitivity reactions. Continued exposure to the antigen leads to the formation of immune complexes and their deposition in the walls of the alveoli in the lungs. Such deposits stimulate an inflammatory response. The resulting accumulation of fluid, antigen, and cells impedes the lungs' normal function of gas exchange, and the patient experiences difficulty in breathing. Occupations in which workers are exposed daily to quantities of the same airborne antigens can lead to this condition. The immune systems of farm workers exposed to hay dust and mold spores are often provoked in this manner, giving rise to the occupational disease called **farmer's lung**. Without a change in work habits the continuing deposition of immune complexes in the alveolar membranes leads to irreversible lung damage.

12-22 Type IV hypersensitivity reactions are mediated by antigen-specific effector T cells

Hypersensitivity reactions caused by effector T cells specific for the sensitizing antigen are known as type IV hypersensitivity or **delayed-type hypersensitivity reactions** (**DTH**), because they occur 1–3 days after contact with antigen. This time course contrasts with that of antibody-mediated hypersensitivity reactions, which are generally apparent within a few minutes. The amount of antigen required to elicit a type IV hypersensitivity reaction is 100–1000 times greater than that required to produce antibody-mediated hypersensitivity reactions. This difference reflects the intrinsic inefficiency in generating peptide epitopes from protein antigens for presentation by HLA molecules. Antigens that cause common type IV hypersensitivity reactions are shown in Figure 12.35.

The best-studied example of a type IV hypersensitivity reaction is the **tuberculin test**, the clinical test used to determine whether a person has been infected with *Mycobacterium tuberculosis*. In this test a small amount of protein antigen extracted from *M. tuberculosis* is injected intradermally or intracutaneously. People with immunity to *M. tuberculosis*—those who have existing tuberculosis, or who have resolved an infection, or who have been vaccinated with the BCG strain—develop an inflammatory reaction around the site of injection 24–72 hours later. The response is mediated by T$_H$1 cells

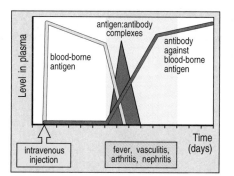

Figure 12.34 Serum sickness is a classic example of a transient immune-complex mediated syndrome. An injection of a large amount of a foreign antigen (yellow curve) into the circulation leads to an antibody response (red curve). These antibodies form immune complexes with the circulating foreign antigens (blue shaded area). The complexes are deposited in small blood vessels and activate complement and phagocytes, inducing fever and the symptoms of vasculitis, nephritis, and arthritis. All these effects are transient and resolve once the foreign antigen has been cleared.

Type IV hypersensitivity reactions are mediated by antigen-specific effector T cells		
Syndrome	**Antigen**	**Consequence**
Delayed-type hypersensitivity	Proteins: Insect venom Mycobacterial proteins (tuberculin, lepromin)	Local skin swelling: Erythema Induration Cellular infiltrate Dermatitis
Contact hypersensitivity	Haptens: Pentadecacatechol (poison ivy) DNFB Small metal ions: Nickel Chromate	Local epidermal reaction: Erythema Cellular infiltrate Vesicles Intraepidermal abscesses
Gluten-sensitive enteropathy (celiac disease)	Gliadin	Villous atrophy in small bowel Malabsorption

Figure 12.35 Examples of common type IV hypersensitivity reactions. Lepromin is an extract of human leprous tissue that is used in a skin test for leprosy infection. Hapten is the name given to any small molecule that when covalently bonded to a protein stimulates an immune response. DNFB, dinitrofluorobenzene.

that recognize peptides derived from the *M. tuberculosis* protein that are presented by HLA class II molecules. The peptides are presented by macrophages and dendritic cells in the vicinity of the injection and initially stimulate tuberculin-specific memory T cells that have left the blood and entered the tissue; these generate effector T_H1 cells locally. After activation, the T_H1 cells initiate further inflammatory reactions that recruit fluid, proteins, and other leukocytes to the site (Figure 12.36). Each of these phases takes several hours, accounting for the time taken before the response is seen or felt. The activated T_H1 cells produce cytokines that mediate these effects (Figure 12.37). In the United States the diagnostic value of the tuberculin skin test for existing infection currently outweighs the benefits of vaccination against tuberculosis, and hence vaccination is not routine. In the United Kingdom and some other European countries, BCG is given to children as one of their routine vaccinations. It is preceded by a tuberculin test, which serves to detect any preexisting infection or immunity.

Type IV hypersensitivity responses can be developed to a variety of antigens in the environment. For example, the dermatitis caused by contact with the North American plants poison ivy (*Toxicodendron radicans*; see Figure 12.1)

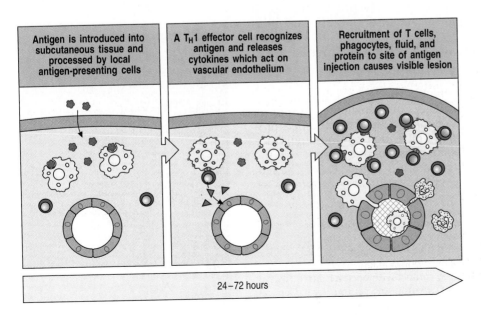

| Antigen is introduced into subcutaneous tissue and processed by local antigen-presenting cells | A T_H1 effector cell recognizes antigen and releases cytokines which act on vascular endothelium | Recruitment of T cells, phagocytes, fluid, and protein to site of antigen injection causes visible lesion |

24–72 hours

Figure 12.36 The stages and time course of a type IV hypersensitivity reaction. The first phase involves uptake, processing, and presentation of the antigen by local antigen-presenting cells. In the second phase, antigen-specific memory T cells produced during previous exposure to the antigen migrate into the site of injection and become activated. Because these antigen-specific cells are rare, and there is no inflammation to attract them into the site, it can take several hours for a T cell of the correct specificity to arrive. Activated T_H1 cells release mediators that activate local endothelial cells, recruiting an inflammatory cell infiltrate dominated by macrophages and causing the accumulation of fluid and protein. At this point, the lesion becomes apparent.

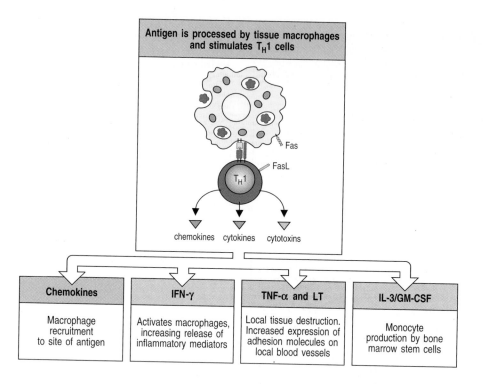

Antigen is processed by tissue macrophages and stimulates T_H1 cells

Fas

FasL

T_H1

chemokines   cytokines   cytotoxins

Chemokines	IFN-γ	TNF-α and LT	IL-3/GM-CSF
Macrophage recruitment to site of antigen	Activates macrophages, increasing release of inflammatory mediators	Local tissue destruction. Increased expression of adhesion molecules on local blood vessels	Monocyte production by bone marrow stem cells

Figure 12.37 Most type IV hypersensitivity reactions are orchestrated by the cytokines released by T_H1 CD4 cells in response to antigen. Macrophages or tissue dendritic cells recruited to the site of inflammation by chemokines present antigen and amplify the response. The release of TNF-α and the cytotoxin lymphotoxin (LT) affect local blood vessels through their effects on endothelial cells. IL-3 and GM-CSF stimulate the production of macrophages. IFN-γ and TNF-α activate macrophages. Macrophages are killed by LT and by the interaction of Fas on the macrophage with Fas ligand (FasL) on the T cell.

and poison oak (*T. diversilobium*, whose leaves are oval and untoothed, or shaped rather like oak leaves on young plants) is due to a reaction of this type and involves both CD4 and CD8 T cells. The reaction is caused by pentadeca-catechol, a small, highly reactive lipid-like molecule that is present in the leaves and roots of the plant and is easily transferred to human skin (Figure 12.38).

When a person contacts poison ivy, pentadecacatechol penetrates the outer layers of the skin and indiscriminately forms covalent bonds with extracellular proteins and skin cell-surface proteins. On degradation of the chemically modified proteins by skin macrophages and Langerhans cells, antigenic peptides that carry the pentadecacatechol adduct are generated and presented by HLA class II molecules to T_H1 cells. The cytokines secreted by the T_H1 cells activate macrophages and produce inflammation (see Figure 12.37). When pentadecacatechol penetrates the skin it also crosses the plasma membranes of cells and chemically modifies intracellular proteins. The processing of such modified proteins results in the presentation of chemically modified peptides by HLA class I molecules to CD8 T cells. On activation, the CD8 T cells have the potential to kill any cells that contacted the chemical and present modified peptides on their surfaces.

On first contact with poison ivy, a person might experience only a minor or undetectable reaction. During this primary or sensitizing response, Langerhans cells or dendritic cells carry pentadecacatechol-modified proteins to the draining lymph node, where the T-cell response is initiated. Once immunological memory has been developed, subsequent contacts with the plant yield unpleasant reactions of the type shown in Figure 12.38. The red, raised, weeping skin lesions are due to heavy infiltration of the contact sites

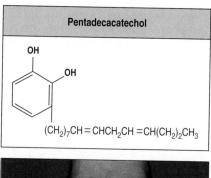

Pentadecacatechol

OH

OH

$(CH_2)_7CH=CHCH_2CH=CH(CH_2)_2CH_3$

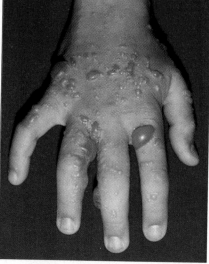

Figure 12.38 Physical contact with poison ivy transfers pentadecacatechol, which causes dermatitis. The chemical structure of pentadecacatechol is shown in the top panel. The photograph shows characteristic blistering skin lesions on the hand of a patient with dermatitis caused by contact with poison ivy (see Figure 12.1 for a photograph of the plant). Photograph courtesy of R. Geha.

with blood cells, combined with the localized destruction of skin cells and the extracellular matrix that holds the skin together. Because of the delayed nature of the reaction there is plenty of time for a person to transfer pentadecacatechol from the initial site of contact to other parts of the body, a process that often exacerbates the extent of the reaction.

The reaction to poison ivy is an example of **contact sensitivity**, so called because contact with the skin is necessary to initiate the allergic response. Contact sensitivity can also develop to coins, jewelry, and other metallic objects containing nickel. In this case the bivalent nickel ions are chelated by histidine in human proteins; processing of these proteins forms T-cell epitopes to which the immune system responds.

12-23 Celiac disease is caused by hypersensitivity to common food proteins

Celiac disease is an inflammatory autoimmune disease of the gut mucosa for which the interplay between genetic and environmental factors is understood in some detail. 'Celiac' means pertaining to the abdominal cavity. The disease is caused by an immune response to the gluten proteins of wheat flour, or the related proteins of barley and rye, all of which are major components of Western diets. CD4 T cells that respond to gluten-derived peptides in the gut-associated lymphoid tissues activate tissue macrophages, which secrete pro-inflammatory cytokines that produce inflammation in the small intestine. With persistent intake of gluten the inflammation becomes chronic and eventually causes atrophy of the intestinal villi, malabsorption of nutrients, and diarrhea (Figure 12.39). Children with the disease fail to thrive; adults can become anemic, depressed, and prone to other diseases, including intestinal cancer. Once diagnosed, the disease can be halted by strict adherence to a gluten-free diet, but it will return whenever gluten-containing food is eaten. It

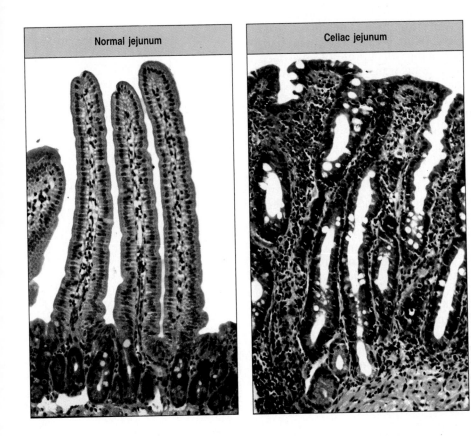

Figure 12.39 Comparison of healthy and celiac intestinal mucosa. Left: the surface of the normal small intestine is folded into finger-like villi, which provide an extensive surface for nutrient absorption. Right: in celiac disease the inflammation and immune response damages the villi. There is lengthening and increased cell division in the underlying crypts to produce new epithelial cells. There are greater numbers of lymphocytes in the epithelial layer and an increase in effector CD4 T cells, plasma cells, and macrophages in the lamina propria. The damage to the villi reduces the person's ability to utilize food and can cause life-threatening malabsorption and diarrhea. Right photograph courtesy of Allan Mowat.

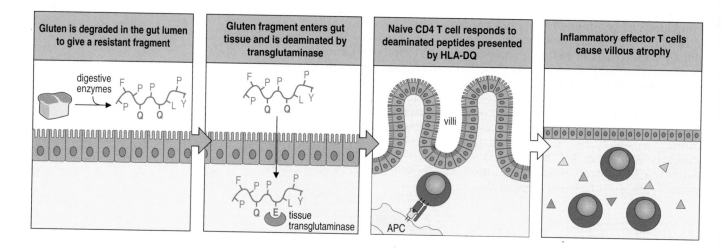

| Gluten is degraded in the gut lumen to give a resistant fragment | Gluten fragment enters gut tissue and is deaminated by transglutaminase | Naive CD4 T cell responds to deaminated peptides presented by HLA-DQ | Inflammatory effector T cells cause villous atrophy |

is quite clear that the major environmental factor that determines the onset of celiac disease is dietary gluten. For this reason the condition is also called gluten-sensitive enteropathy.

Genetic predisposition to celiac disease is strong, as shown by the 75% concordance between monozygotic twins. There is also a strong association with allotypes of the MHC class II locus HLA-DQ (see Chapter 5). A majority (about 80%) of patients with celiac disease (celiacs) have the HLA-DQ2 allotype, and most of the rest have DQ8. In the intestinal lesions of celiacs there are T_H1 cells that respond to gluten-derived peptides presented by either DQ2 or DQ8. These T cells are not detectable in the gut tissue of people who do not have celiac disease, nor are they found in the gut tissue of patients who have embraced a gluten-free diet and no longer have symptoms of disease. In celiac disease, oral tolerance to gluten has selectively become broken.

Gluten comprises two families of proteins rich in glutamine and proline—the glutenins and the gliadins. They give bread dough its elastic, gluey properties. In celiacs, T cells with specificity for a variety of glutenin- and gliadin-derived peptides presented by HLA-DQ can be present. What distinguishes these peptide epitopes is that some glutamine residues of the wheat proteins have been converted to glutamate by the human enzyme tissue transglutaminase. These changes, which give the peptides additional negative charges, are necessary for the peptides to bind to the positively charged binding pockets of the HLA-DQ2 or HLA-DQ8 allotypes. Important instigators of disease in some patients are T cells directed against epitopes contained in a particularly proline-rich 33-amino-acid fragment of gliadins. This fragment is resistant to proteolysis by digestive enzymes and the proteases involved in intracellular antigen processing, but binds strongly to tissue transglutaminase. After transglutaminase has converted certain glutamine residues (Q) to glutamate (E), the 33-residue peptide binds to HLA-DQ and provokes the activation of inflammatory effector T cells (Figure 12.40). The peptide can bind in different registers and present several different epitopes to T cells; these have a synergistic effect in initiating the disease-causing immune response. Once an inflammatory response has begun, it becomes exacerbated by increased production of transglutaminase and of the peptide antigens recognized by disease-causing T cells.

All celiacs make IgG or IgA autoantibodies specific for tissue transglutaminase. Most patients also make anti-gliadin antibodies. These antibodies are likely to be the products of B cells whose surface immunoglobulins bind and internalize complexes of transglutaminase and a gliadin fragment and then present modified gliadin fragments on HLA-DQ molecules to T cells. Patients suspected of having celiac disease are tested for the presence of autoantibodies; if the tests prove positive, a biopsy of the small intestine, the definitive diagnostic test, is authorized.

Figure 12.40 The mechanism of celiac disease. In celiac disease, inflammation of the small intestines is caused by CD4 T cells responding to peptides derived from gluten that are deaminated by tissue transaminase and presented by HLA-DQ8 or HLA-DQ2 molecules. Only part of the peptide epitope is shown. APC, antigen-presenting cell.

Although gluten-specific T cells are found specifically in the gut of celiacs but not in healthy controls, the peripheral blood T cells of both patients and controls contain cells that respond to gluten in culture. Unlike the disease-associated T cells of the gut, the gluten-specific T cells of the peripheral blood are specific for a wider range of glutenin and gliadin peptides. They are also presented by a wider range of HLA class II isoforms and are not necessarily dependent on transglutaminase-mediated modification. This truly emphasizes the importance of studying disease-causing immunity in the affected tissues.

As celiac-like diseases do not commonly arise from the consumption of rice or other dietary staples of non-caucasian peoples, the disease is practically specific to caucasians. Although almost all caucasians consume gluten and about 30% have HLA-DQ2, only 0.5–1% of them get celiac disease. These numbers show that other factors must contribute to the cause of disease. Although the concordance rate for celiac disease in monozygotic twins is impressively high, around 75%, the 25% of discordance also points to additional environmental factors that set the stage for gluten to trigger an inflammatory T-cell response.

12-24 Severe hypersensitivity reactions to certain drugs are strongly correlated with HLA class I allotypes

In a small minority of individuals a prescribed drug can cause a life-threatening disease called Stevens–Johnson Syndrome or SJS. During the first 2 weeks, this syndrome resembles an infection, with fever, aches, and a cough. Then a rash with blisters develops and the epidermis starts to peel away from the dermis and is easily rubbed off. Severe blisters also develop at the mucous membranes: eyes, mouth, throat, vagina, urethra, and anus. In SJS less than 10% of the skin is affected, but in more severe cases, called toxic epidermal necrolysis or (TEN), more than 30% of the skin is affected. The damage to the skin is equivalent to that caused by a severe burn.

Carbamazepine is an anticonvulsant drug that is used to treat epileptic patients and in south-east Asia it is the leading cause of SJS and TEN. All the patients receiving the drug who developed SJS or TEN had the *HLA-B*1502* allele, whereas this allele was present in only 4% of patients who had no adverse reactions to the drug (Figure 12.41). As well as being the strongest disease association ever made with an HLA polymorphism, this correlation suggests that the disease is caused by cytotoxic T cells responding to self peptides that are presented by HLA-B*1502 and have been chemically modified by carbamazepine. The *B*1502* allele is restricted to south-east Asian populations, explaining their higher incidence of carbamazepine-induced SJS and TEN. To prevent this disease, patients of south-east Asian origin should be typed for the presence of *HLA-B*1502* and carbamazepine not prescribed to those who are positive.

The drug allopurinol, used to treat gout and hyperuricemia (an excess of uric acid in the blood), can also induce SJS-TEN in populations worldwide. All south-east Asian patients who developed SJS-TEN after taking allopurinol

Drug	Disease treated	Disease caused	HLA association	Population	Statistical significance
Carbamazepine	epilepsy	SJS-TEN	HLA-B*1502	Han Chinese	$p = 3 \times 10^{-27}$
Allopurinol	gout, hyperuricemia	SJS-TEN	HLA-B*5801	Han Chinese	$p = 5 \times 10^{-24}$
Abacavir	HIV/AIDs	hypersensitivity syndrome	HLA-B*5702	Caucasian	$p = 5 \times 10^{-20}$

Figure 12.41 Hypersentitivity reactions to drugs affect individuals who have particular HLA-B allotypes. SJS-TEN, Stevens–Johnson syndrome with toxic epidermal necrolysis (see text).

had *B*5801*, whereas its frequency in the unaffected patients was 15%. This association has yet to be studied in other populations.

Abacavir is a nucleoside analog that inhibits the reverse transcriptase of HIV-1 and is commonly used to treat HIV-infected patients. Some 5–8% of patients who start therapy with abacavir become hypersensitive to it and develop an influenza-like illness and the potential for life-threatening renal failure or bronchoconstriction. In HIV-infected caucasian patients this hypersensitivity syndrome is strongly associated with *HLA-B*5701*, an allele that is found predominantly in caucasians (see Figure 12.41). Prospective screening of HIV-infected patients for the presence of *B*5701* to prevent them from being prescribed abacavir has been shown to reduce the incidence of drug-induced hypersensitivity.

Summary

Type II hypersensitivity reactions are mediated by IgG antibodies directed against cell-surface or matrix antigens and are commonly caused by drugs that are being given as treatment for other diseases. Because of their chemical reactivity, drugs bind to surface proteins of human cells, for example erythrocytes or platelets, creating new epitopes that stimulate an antibody response. On binding to the drug-modified cells, the antibodies activate complement, which leads to cell destruction. Type III hypersensitivity reactions are caused by soluble immune complexes formed by the binding of IgG to the soluble antigens against which they were made. They can be seen when non-human proteins, such as mouse monoclonal antibodies, are given therapeutically. Antibodies specific for the nonhuman proteins are made and form small immune complexes that are inefficiently cleared from the circulation. These tend to deposit in blood vessels, where they activate complement and inflammation. Depending on the sites of activation, these reactions can cause vasculitis, nephritis, arthritis, or lung disease. Type IV hypersensitivity reactions are caused by effector T cells, often responding to reactive chemicals transferred into the skin by physical contact to cause a contact dermatitis. In some conditions, tissue damage is caused by the activation of macrophages by T_H1 cells and the actions of cytotoxic T cells; in others, such as chronic asthma, cytokines produced by T_H2 cells specific for an inhaled allergen activate eosinophils and other inflammatory cells. Celiac disease is a chronic inflammatory disease of the intestine that is caused by an immune response directed against food proteins that derive from grain. Severe hypersensitivity reactions to several drugs are highly restricted to single HLA-B allotypes.

Summary to Chapter 12

The immune system provides the body with powerful defenses against infection. These include adaptive immune responses that have the potential to respond to any structure that is not a normal part of the body. Inevitably, all humans develop adaptive immunity to some foreign substances, such as animal and plant proteins in foods, that are not associated with infection. In general, such immune responses are harmless; however, this is not always so, and various allergies or hypersensitivities are caused by the immune system's overreactions to nonthreatening environmental antigens. A first exposure to an allergen is rarely noticeable, but hypersensitivity reactions are brought on by subsequent exposures in which the allergen interacts with previously formed antibodies or stimulates memory lymphocytes. Four types of hypersensitivity reaction are conventionally defined on the basis of the effector mechanisms that cause them. Hypersensitivity reactions of types I, II, and III are triggered by antibodies, whereas type IV hypersensitivity reactions are triggered by effector T cells. In all four types of reaction, recognition of the allergen triggers an unwanted inflammatory response of varying severity and duration.

Questions

12–1 Identify four different ways in which an individual may come into contact with an allergen, and provide two examples of allergens for each type of contact.

12–2 Match the types of hypersensitivity in list A with the immune system components involved (list B). There will be more than one item from list B for each answer.

List A	List B
a. Type IV	1. B cells
b. Type II	2. IgE
c. Type I	3. CD8 T cells
d. Type III	4. Soluble immune complexes
	5. CD4 T cells
	6. Mast cells
	7. IgG
	8. Complement binding to cell surfaces

12–3
  A. Describe in detail the mechanism responsible for mast-cell activation during a type I hypersensitivity reaction.
  B. What are the products of mast-cell activation?

12–4 Why are antihistamines used to treat allergic rhinitis and allergic asthma? What symptoms of each disease, respectively, do they alleviate?

12–5
  A. Describe three ways in which the immunoglobulins acting as antigen receptors on the surface of a mast cell differ from the immunoglobulins acting as the antigen receptors on the surface of a B cell.
  B. What is the essential difference in response of these two cell types when antigen binds to these surface immunoglobulins?

12–6 Some allergies can be treated by a procedure called desensitization. Explain (A) two current approaches to desensitization and (B) the main disadvantage associated with each.

12–7
  A. What type of hypersensitivity reaction typically occurs when penicillin is administered to an individual allergic to the drug?
  B. Describe in chronological order the events leading to this reaction.
  C. What type of hypersensitivity reaction can penicillin cause in a non-allergic individual?

12–8 Explain what causes farmer's lung and its pathology. State what type of hypersensitivity reaction is involved.

12–9 Describe four ways in which a type III hypersensitivity reaction differs from a type I reaction.

12–10
  A. What is the tuberculin test used for and how does it work?
  B. What type of hypersensitivity reaction is involved?
  C. Why does vaccination render this test useless?

12–11 Most extracellular antigens are taken up and processed by antigen-presenting cells via phagolysosomes and presented with MHC class II molecules to T_H1 and T_H2 cells.
  A. Explain how pentadecacatechol, the antigen responsible for the hypersensitivity reaction against poison ivy, is an exception to this route of antigen processing and gains entry into the cytosolic route of antigen processing.
  B. Which T cells are activated and what is the immunological consequence?

12–12 Give three ways in which a susceptible person can help to minimize the risk of having an allergic reaction.

12–13 Anita Garcia, 17 years old, and her roommate Rosa Rosario were celebrating a friend's birthday at a dessert buffet at a local restaurant when Anita developed acute dyspnea, and angioedema. She complained of an itchy rash and then had difficulty swallowing. Rosa drove Anita to the emergency room two blocks away rather than wait for an ambulance. As they approached the hospital, Anita lost consciousness. This medical emergency would most probably result in immediate _____ before any subsequent treatment.
  a. subcutaneous injection of epinephrine
  b. intravenous injection of corticosteroids
  c. intravenous injection of antihistamine
  d. intravenous injection of antibiotics
  e. intravenous injection of a nonsteroidal anti-inflammatory drug.

12–14 Look again at Question 12–13. What do you think Anita was suffering from? Given the circumstances in which the episode occurred, suggest a likely cause.

12–15 George Cunningham was diagnosed with Crohn's disease when 23 years old. He was experiencing acute abdominal pain, diarrhea, rectal bleeding, anemia and weight loss. He did not respond to conventional immunosuppressive therapies and was given a course of infliximab, an anti-TNF-α monoclonal antibody that suppresses inflammation by blocking TNF-α activity. On day 12 after receiving his first infusion, he developed a mild fever, generalized vasculitis, swollen lymph glands, swollen joints, and joint pain. Traces of blood and protein were detected in his urine. Which of the following is the most likely cause of these recent symptoms?

a. type I hypersensitivity involving anaphylaxis
b. type II hypersensitivity leading to hemolytic anemia
c. type III hypersensitivity caused by immune complex deposition in blood vessels
d. type IV hypersensitivity involving CD8 T-cell cytotoxicity
e. type II hypersensitivity leading to thrombocytopenia.

12–16 Anders Anderson, a 24-month-old, was seen by his pediatrician after a recent bout of diarrhea and vomiting. He had lost his appetite and complained that his stomach hurt. Anders was in the 5% centile for weight, had slender limbs, wasted buttocks, and a protuberant abdomen. Jejunal biopsy revealed abnormal surface epithelium, and villous atrophy with hyperplasia of the crypts. Which of the following would be a likely clinical finding in this patient?
a. glomerulonephritis
b. urticarial rash
c. anti-gliadin IgA antibodies
d. chronic wheezing
e. low blood pressure.

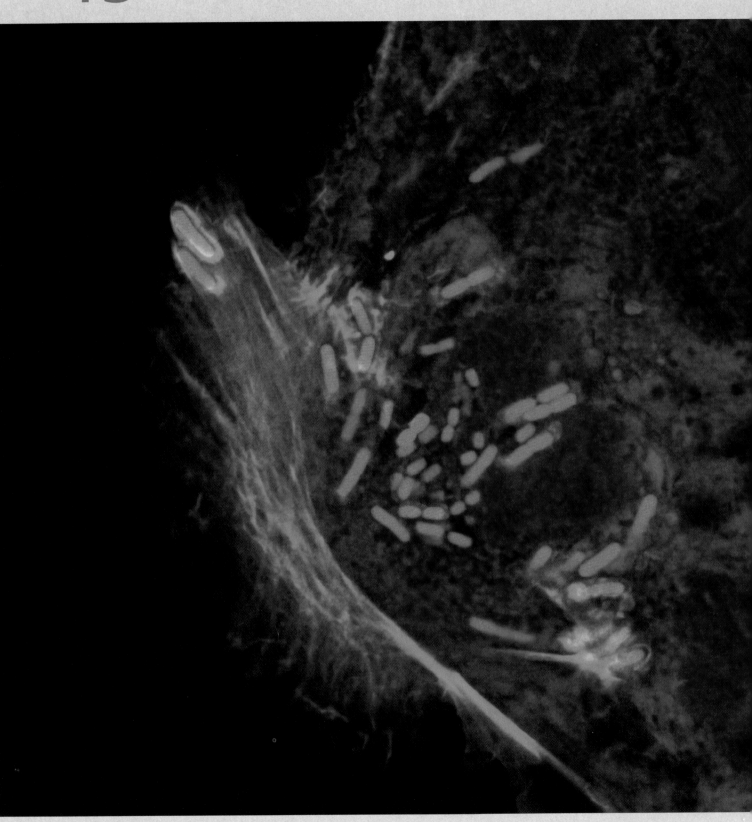

Intestinal infections with Shigella bacteria are correlated with the subsequent development of arthritis.

Chapter 13

Disruption of Healthy Tissue by the Immune Response

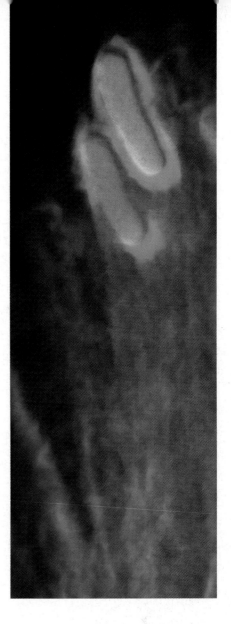

We have seen how hypersensitivity to harmless environmental antigens leads to acute or chronic disease, depending on the type of antigen and the frequency of exposure to it (see Chapter 12). In this chapter another set of chronic immunological diseases is considered, diseases caused by adaptive immunity that becomes misdirected at healthy cells and tissues of the body. Such diseases are known as **autoimmune diseases**, and many different types have been described. They are not uncommon; around 5% of the population of developed countries has one or more of these conditions, and their incidence is increasing. Autoimmune diseases can be caused by antibodies that perturb a normal physiological function or by inflammatory T cells that damage healthy cells or tissue at a rate that is beyond the capacity of the body to repair. When the targeted tissue is involved in essential routine functions of the body, the autoimmune disease can become a threat to life.

Autoimmune diseases are caused by unwanted adaptive immune responses; they represent failures of the mechanisms that maintain **self-tolerance**—the guard against attack on one's own cells and tissues—in the populations of circulating B and T cells. Although much is known of the effects of autoimmune disease, less is understood about the events that break **tolerance** and cause an autoimmune response. The first part of this chapter describes some of the more common autoimmune diseases, and the second discusses some of the predisposing factors for autoimmune disease and the mechanisms by which tolerance can be broken and lead to disease.

Autoimmune diseases

There are many chronic diseases in which the immune system is active, often causing the affected tissues to be inflamed and abnormally infiltrated by lymphocytes and other leukocytes, but in which there seems to be no associated active infection. These diseases are caused by the immune system itself, which attacks cells and tissues of the body as though they were infected, causing chronic impairment of tissue and organ function. Chronic diseases of this kind are collectively known as autoimmune diseases because they are caused by immune responses directed toward autologous (self) components of the body. An immune response that causes an autoimmune disease is called an **autoimmune response** and it produces a state of **autoimmunity**.

Autoimmune diseases vary widely in the tissues they attack and the symptoms they cause. Some focus on a particular organ or cell type, others act systemically. For most autoimmune diseases the incidence differs between females and males, with females being more commonly affected. A defining characteristic of autoimmune diseases is the presence of antibodies and T cells specific for antigens expressed by the targeted tissue. These antigens are called **autoantigens** and are a subset of the self antigens; the effectors of adaptive immunity that recognize them are known as **autoantibodies** and **autoimmune T cells**. The mechanisms of antigen recognition and effector function in autoimmunity are the same as those used in responding to pathogens and environmental antigens; the damaging responses are due to a variety of immune effector mechanisms.

13-1   In healthy individuals the immune system is tolerant of self antigens

Throughout the previous chapters of this book we have encountered mechanisms that prevent healthy cells and tissues of the body from either stimulating an immune response or being attacked by the effector mechanisms of the immune response. In aggregate these mechanisms make the immune system tolerant of self. Regulatory proteins in the blood and on cell surfaces prevent the fixation of complement on human cells, and the recognition molecules of innate immunity have been selected over hundreds of millions of years to distinguish microbial components from human components (see Chapter 2). For the B cells and T cells of adaptive immunity, which randomly evolve new receptors on a daily basis, there are a variety of mechanisms that eliminate or inhibit cells expressing receptors reactive to self antigens (Figure 13.1).

Negative selection of T cells in the thymus and of B cells in the bone marrow eliminates lymphocytes bearing self-reactive receptors before they can leave the primary lymphoid organs—the phenomenon of central tolerance (see Chapters 6 and 7). The efficiency of central tolerance is augmented by having tissue-specific proteins participate in negative selection in the thymus, through the activity of the transcription factor AIRE (see Section 7-12, p. 202). Those autoreactive cells that escape deletion by central tolerance and enter the peripheral circulation can be rendered non-responsive either through the induction of a state of anergy or through active suppression by regulatory T cells.

Another method that prevents an immune response being made to some self antigens is for them to be sequestered in immunologically privileged sites that are not accessible to circulating leukocytes. Examples of such sites are the brain, eye, and testis. During pregnancy the uterus is an immunologically privileged site in which the fetus is not exposed to the mother's leukocytes. In the evolution of placental reproduction in the mammals this was a necessity, because fetal cells express HLA class I and II molecules of paternal type, against which maternal B and T cells have the potential to make a strong alloreactive response.

13-2   Autoimmune diseases are caused by the loss of tolerance to self antigens

Whenever the effector cells and molecules of the immune system enter an infected tissue to battle a pathogen, there is inevitable destruction of uninfected cells and disruption of healthy tissue. To a considerable extent the symptoms and pathology of infectious disease are due to the activities of the immune system, not the pathogen, and many of the medicines we take to alleviate disease symptoms are ones that inhibit inflammation and other immune reactions. Normally, after successful elimination of a pathogen the inflammatory response is turned off and effector T cells either die or are shut off to become memory cells. At the site of infection, dead cells and debris will be removed and the tissue rebuilt, processes in which immune-system cells such as macrophages and γ:δ T cells are intimately involved.

Mechanisms that contribute to immunological self-tolerance
Negative selection in the bone marrow and thymus
Expression of tissue-specific proteins in the thymus
No lymphocyte access to some tissues
Suppression of autoimmune responses by regulatory T cells
Induction of anergy in autoreactive B and T cells

Figure 13.1 Mechanisms that contribute to immunological self-tolerance.

Autoimmune diseases occur when some aspect of self-tolerance is lost and an adaptive immune response is directed toward normal components of the healthy human body. In essence, the response strives to eliminate the target antigens from the body, and, until that is achieved or the patient dies, a chronic state of inflammation and lymphocyte infiltration exists in tissues where the target antigens are expressed. This severely interferes with tissue function. Although the regulatory mechanisms of the immune system respond to the situation and can provide temporary relief between episodes of disease, autoimmune diseases are rarely resolved or cured. They are chronic diseases in which the immune response remains suspended in its destructive phase and never reaches reconstruction.

13-3   The effector mechanisms of autoimmunity resemble those causing hypersensitivity reactions

Like the hypersensitivity reactions described in Chapter 12, autoimmune diseases can be classified according to the type of immunological effector mechanism causing the disease (Figure 13.2). Three kinds of autoimmune disease correspond to the type II, type III, and type IV hypersensitivities; no autoimmune diseases are caused by IgE, the source of type I hypersensitivity reactions. The autoimmune diseases corresponding to type II hypersensitivity (see Section 12-17, p. 386) are caused by antibodies directed against components of cell surfaces or the extracellular matrix; those corresponding to type III hypersensitivity (see Section 12-19, p. 389) are caused by soluble immune complexes deposited in tissues; those corresponding to type IV hypersensitivity (see Section 12-22, p. 392) are caused by effector T cells. Autoimmune diseases of all three kinds are described in Figure 13.2.

Autoimmunity corresponding to the type II hypersensitivity reaction is frequently directed at cells of the blood. In **autoimmune hemolytic anemia**, IgG and IgM antibodies bind to components of the erythrocyte surface, where they activate complement by the classical pathway, triggering the destruction of red cells. Complete activation of the classical pathway leads to formation of the membrane-attack complex and hemolysis—the lysis of red cells. Alternatively, erythrocytes coated with antibody and C3b are cleared from the circulation, principally by Fc receptors and complement receptors on phagocytes in the spleen (Figure 13.3). All mechanisms lead to a reduction in the red-cell count. This condition is called anemia, a term derived from Greek words meaning 'without blood.'

White blood cells can also be a target for autoantibodies and complement activation. Because nucleated cells are less susceptible than erythrocytes to complement-mediated lysis, the major effect of complement activation is to facilitate the phagocyte-mediated clearance of white blood cells in the spleen. Patients who develop antibodies against neutrophil surface antigens suffer from reduced numbers of circulating neutrophils, a state called **neutropenia**. Blood cells that have bound antibody and complement are, however, still able to function, so one treatment used for patients with chronic autoimmunity to blood cells is splenectomy, which reduces the rate at which the opsonized cells are cleared from the blood.

Antibody responses to components of the extracellular matrix are uncommon but can be very damaging when they occur. In **Goodpasture's syndrome**, IgG is formed against the α_3 chain of type IV collagen, the collagen found in basement membranes throughout the body. The antibodies are deposited along the basement membranes of renal glomeruli and renal tubules, eliciting an inflammatory response in the renal tissue (Figure 13.4). These basement membranes are an essential part of the blood-filtering mechanism of the kidney and, as IgG and inflammatory cells accumulate, kidney function becomes progressively impaired, leading to kidney failure and death if not treated.

Autoimmune disease	Autoantigen	Consequence
Antibody against cell-surface or matrix antigens (type II)		
Autoimmune hemolytic anemia	Rh blood group antigens, I antigen	Destruction of red blood cells by complement and phagocytes, anemia
Autoimmune thrombocytopenia purpura	Platelet integrin gpIIb:IIIa	Abnormal bleeding
Goodpasture's syndrome	Non-collagenous domain of basement membrane collagen type IV	Glomerulonephritis, pulmonary hemorrhage
Pemphigus vulgaris	Epidermal cadherin	Blistering of skin
Acute rheumatic fever	Streptococcal cell wall antigens. Antibodies cross-react with cardiac muscle	Arthritis, myocarditis, late scarring of heart valves
Graves' disease	Thyroid-stimulating hormone receptor	Hyperthyroidism
Myasthenia gravis	Acetylcholine receptor	Progressive weakness
Type 2 diabetes (insulin-resistant diabetes)	Insulin receptor (antagonist)	Hyperglycemia, ketoacidosis
Hypoglycemia	Insulin receptor (agonist)	Hypoglycemia
Immune-complex disease (type III)		
Subacute bacterial endocarditis	Bacterial antigen	Glomerulonephritis
Mixed essential cryoglobulinemia	Rheumatoid factor IgG complexes (with or without hepatitis C antigens)	Systemic vasculitis
Systemic lupus erythematosus	DNA, histones, ribosomes, snRNP, scRNP	Glomerulonephritis, vasculitis, arthritis
T cell-mediated disease (type IV)		
Type 1 diabetes (insulin-dependent diabetes mellitus)	Pancreatic β-cell antigen	β-cell destruction
Rheumatoid arthritis	Unknown synovial joint antigen	Joint inflammation and destruction
Multiple sclerosis	Myelin basic protein, proteolipid protein	Brain degeneration. Paralysis

Figure 13.2 A selection of autoimmune diseases, the symptoms they cause, and the autoantigens associated with the immune response. The autoimmune diseases are classified as types II, III, and IV because their tissue-damaging effects are similar to those of hypersensitivity reactions types II, III, and IV, respectively (see Chapter 12). snRNP, small nuclear ribonucleoprotein; scRNP, small cytoplasmic ribonucleoprotein.

Treatment involves plasma exchange to remove existing antibodies and immunosuppressive drugs to prevent the formation of new ones. Although autoantibodies are deposited in the basement membranes of other organs, the high-pressure filtering of blood by renal glomeruli is the vital function that is most sensitive to their effects. Indeed, kidney damage by the immune system is responsible for one-quarter of all cases of end-stage renal failure.

Glomerulonephritis can also be a consequence of autoimmune diseases that produce soluble immune complexes, as in type III hypersensitivity reactions.

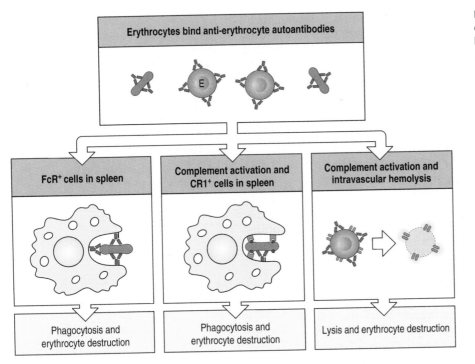

Figure 13.3 Three mechanisms destroy erythrocytes in autoimmune hemolytic anemia.

In the disease systemic lupus erythematosus (SLE), IgG made against common intracellular macromolecules forms soluble complexes with these autoantigens when they are released from damaged cells. Glomerulonephritis caused by deposition of the immune complexes in the blood capillaries of the renal glomeruli can lead to death. T cell-mediated immunity of the kind seen in type IV hypersensitivity reactions is responsible for the damage caused to the pancreas in type 1 diabetes (also known as insulin-dependent diabetes mellitus or IDDM), to joints in rheumatoid arthritis, and to the brain in multiple sclerosis. All three of these diseases, as well as SLE, will be described in more detail in later sections.

13-4 Endocrine glands contain specialized cells that are targets for organ-specific autoimmunity

The characteristic properties of endocrine glands make them particularly prone to involvement in autoimmune disease. First, the function of an endocrine gland—the synthesis and secretion of a particular hormone—involves tissue-specific proteins not expressed in other cells or tissues. Second, because they secrete their hormones into the blood, endocrine tissue is well vascularized, which facilitates interactions with cells and molecules of the immune

Figure 13.4 Autoantibodies specific for type IV collagen react with the basement membranes of kidney glomeruli, causing Goodpasture's syndrome. The panels show sections of a renal corpuscle in serial biopsies taken from a patient with Goodpasture's syndrome. Panel a, glomerulus stained for IgG deposition by immunofluorescence. Antibody against glomerular basement membrane is deposited in a linear fashion (green staining) along the glomerular basement membrane. The autoantibody causes the local activation of cells bearing Fc receptors, the activation of complement, and the influx of neutrophils. Panel b, hematoxylin and eosin staining of a section through a renal corpuscle shows that the glomerulus is compressed by the formation of a crescent (C) of proliferating mononuclear cells within the Bowman's capsule (B) and that there is an influx of neutrophils (N) into the glomerular tuft. Photographs courtesy of M. Thompson and D. Evans.

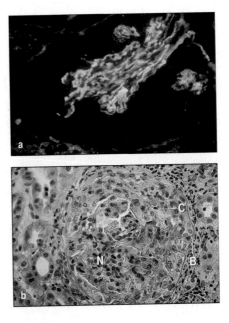

system. Loss of endocrine function has drastic systemic effects, causing pronounced disease and in some cases death. Each autoimmune disease of endocrine tissue is due to an impaired function of a single type of epithelial cell within an endocrine gland (Figure 13.5). For this reason such diseases are called **organ-specific autoimmune diseases**. To illustrate this principle we shall consider autoimmune diseases of the thyroid gland and of the islets of Langerhans in the pancreas in the following four sections.

13-5 Autoimmune diseases of the thyroid can cause either underproduction or overproduction of thyroid hormones

The thyroid gland regulates the basal metabolic rate of the body through the secretion of two related hormones, tri-iodothyronine and tetra-iodothyronine (thyroxine). These small iodinated derivatives of the amino acid tyrosine are produced in an elaborate way. The epithelial cells of the thyroid make a large glycoprotein called thyroglobulin, which is stored within follicles formed by the spherical arrangement of thyroid cells. Thyroid cells are uniquely specialized to take up iodide, which they use to iodinate and cross-link the tyrosine residues of stored thyroglobulin. When increased cellular metabolism is required, for example when the outside temperature drops, signals from the nervous system induce the pituitary, another endocrine gland, to secrete thyroid-stimulating hormone (TSH). Surface receptors on thyroid cells bind TSH, inducing the endocytosis of iodinated thyroglobulin and the release of thyroid hormones by proteolytic degradation of the protein. As blood levels of thyroid hormones rise they feed back on the pituitary, inhibiting the further release of TSH (Figure 13.6).

The proteins thyroglobulin, thyroid peroxidase, TSH receptor, and the thyroid iodide transporter are uniquely expressed in thyroid cells. Immune responses to all these autoantigens have been detected in autoimmune thyroid disease.

Autoimmune diseases of endocrine glands	
Thyroid gland	Hashimoto's thyroiditis Graves' disease Subacute thyroiditis Idiopathic hypothyroidism
Islets of Langerhans (pancreas)	Type 1 diabetes (insulin-dependent diabetes, juvenile-onset diabetes) Type 2 diabetes (insulin-resistant diabetes, adult-onset diabetes)
Adrenal gland	Addison's disease

Figure 13.5 Autoimmune diseases of endocrine glands.

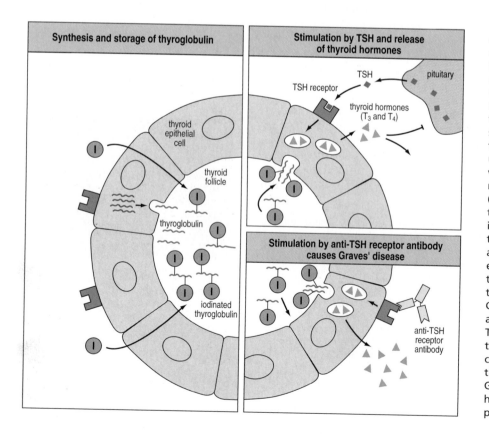

Figure 13.6 Autoantibodies against the TSH receptor cause the overproduction of thyroid hormones in Graves' disease. The diagram shows a thyroid follicle surrounded by epithelial cells. The left half of the figure shows uptake of iodide (green circles), the iodination of thyroglobulin and its storage in the follicle at times when thyroid hormones are not required. The upper right portion of the figure shows what occurs when thyroid hormones are needed. Thyroid-stimulating hormone (TSH) from the pituitary gland induces the endocytosis and breakdown of iodinated thyroglobulin, releasing the thyroid hormones tri-iodothyronine (T_3) and thyroxine (T_4). In addition to their effects on metabolism, T_3 and T_4 signal the pituitary to stop releasing TSH. At the lower right is shown the situation in Graves' disease. Autoantibodies, which are continually present, bind to the TSH receptor of thyroid cells, mimicking the action of TSH and inducing the continuous synthesis and release of thyroid hormones. In patients with Graves' disease, the production of thyroid hormones becomes independent of the presence of TSH.

However, the disease symptoms caused by different types of autoimmunity to thyroid antigens are very different. Some conditions cause a loss of thyroid hormones; others increase their production.

In **Graves' disease**, the autoimmune response is focused on antibody production, and the symptoms are caused by antibodies that bind to the TSH receptor. By mimicking the natural ligand, the bound antibodies cause a chronic overproduction of thyroid hormones that is independent of regulation by TSH and insensitive to the metabolic needs of the body (see Figure 13.6). This **hyperthyroid** condition causes heat intolerance, nervousness, irritability, warm moist skin, weight loss, and enlargement of the thyroid. Other aspects of Graves' disease are outwardly bulging eyes and a characteristic stare, symptoms caused by the binding of autoantibodies that react with the muscles of the eye. The autoimmune response in Graves' disease is biased toward a CD4 T_H2 response.

In **chronic thyroiditis**, also called **Hashimoto's disease**, the thyroid loses the capacity to make thyroid hormones. Hashimoto's disease seems to involve a CD4 T_H1 response, and both antibodies and effector T cells specific for thyroid antigens are produced. Lymphocytes infiltrate the thyroid, causing a progressive destruction of the normal thyroid tissue. By comparison, in Graves' disease there are fewer infiltrating lymphocytes and relatively little tissue destruction. Patients with Hashimoto's disease become **hypothyroid**, and eventually are unable to make thyroid hormone.

Treatment for Hashimoto's disease is replacement therapy with synthetic thyroid hormones taken orally on a daily basis. For Graves' disease the short-term treatment is drugs that inhibit thyroid function. The long-term treatment is removal of the thyroid by surgery, or its destruction by uptake of the radioisotope ^{131}I, followed by daily doses of thyroid hormones.

13-6 Ectopic lymphoid tissue can form at sites inflamed by autoimmune disease

A characteristic feature of Hashimoto's disease is that the immune-system cells that infiltrate the thyroid become organized into structures that anatomically resemble the common microstructure of secondary lymphoid organs (Figure 13.7). They contain T-cell and B-cell areas, dendritic cells, follicular dendritic cells, and macrophages. These structures are called **ectopic lymphoid tissues** or tertiary lymphoid organs; the process by which they form, termed lymphoid neogenesis, resembles the formation of secondary lymphoid tissues and is similarly driven by lymphotoxin (LT) (see Section 6-14, p. 178). Unlike secondary lymphoid tissues, the ectopic lymphoid tissue is not encapsulated, lacks lymphatics, and is exposed to the inflammatory environment caused by the autoimmune response. Ectopic lymphoid tissues also functionally resemble secondary lymphoid tissues. Within them, B cells and T cells can be stimulated by antigen to give effector cells, and B cells undergo somatic hypermutation and isotype switching.

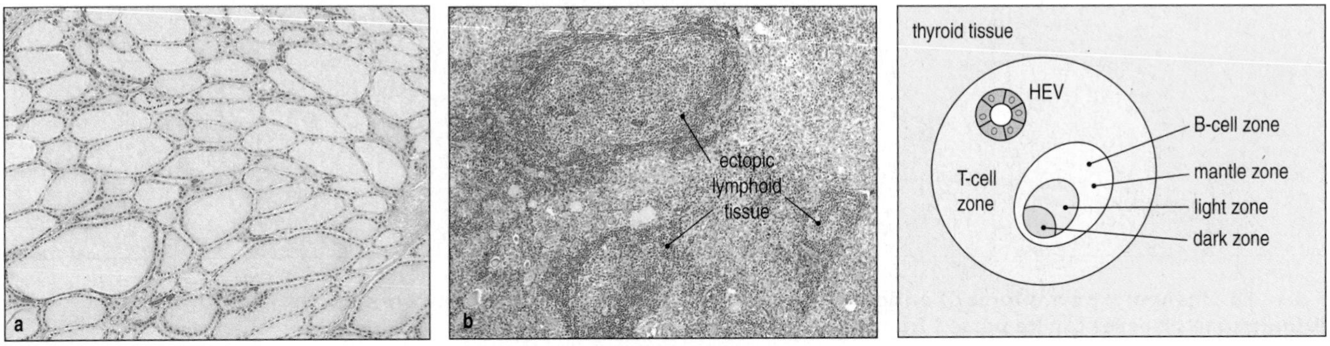

Figure 13.7 Hashimoto's thyroiditis. In a healthy thyroid gland the epithelial cells form spherical follicles containing thyroglobulin (panel a). In patients with Hashimoto's thyroiditis the thyroid gland becomes infiltrated with lymphocytes, which destroy the normal architecture of the thyroid gland and can become organized into structures resembling secondary lymphoid tissue (panel b), as shown in the schematic diagram at the right. Courtesy of Yasodha Natkunam.

Ectopic lymphoid tissue is formed in other autoimmune diseases, including rheumatoid arthritis, Graves' disease, and multiple sclerosis, but with less regularity than for Hashimoto's disease. Ectopic lymphoid tissue is also formed in tissues that are chronically infected with a pathogen, for example in the liver during chronic hepatitis C infection. In Chapter 10 we saw how the secondary lymphoid tissues in the gut are placed close to the sites of heavy infestation with microbes; the formation of ectopic lymphoid tissue may be a similar strategy that allows for amplification of the response to a persistent infection. For a tissue sustaining an autoimmune response this strategy can only exacerbate the disease.

13-7   The cause of an autoimmune disease can be revealed by the transfer of the disease by immune effectors

A central goal in the study of autoimmune diseases is the identification of the autoantigens and effector mechanisms responsible. This is no trivial task. One problem is that various types of autoimmunity are demonstrable in healthy people; a second is that once cell and tissue destruction has begun, this will often initiate further autoimmune responses that are consequences, not causes, of the disease.

Autoimmune diseases caused by antibody-mediated effector functions are more easily defined than those caused by effector T cells, because transfer of autoantibody from the patient to another human or to an experimental animal is sufficient to induce the disease. Pregnant women transport IgG molecules, but not lymphocytes, across the placenta to the fetal circulation (see Section 9-14, p. 268). If the mother has an antibody-mediated autoimmune disease such as Graves' disease, the baby will be born with the symptoms of disease. As the baby grows and the maternal IgG in its circulation degrades, the disease symptoms gradually go away. For diseases that could affect the baby's growth, treatment involves the removal of antibody from the circulation by total exchange of blood plasma (Figure 13.8).

When healthy rats are injected with serum from patients with Graves' disease they start to overproduce thyroid hormones and show symptoms of Graves' disease. In the past, such laboratory assays were used for the immunological diagnosis of Graves' disease. This whole-animal approach has now been superseded by assays that use a cultured line of rat thyroid cells. IgG fractions prepared from patients' plasma are added to thyroid-cell cultures and their

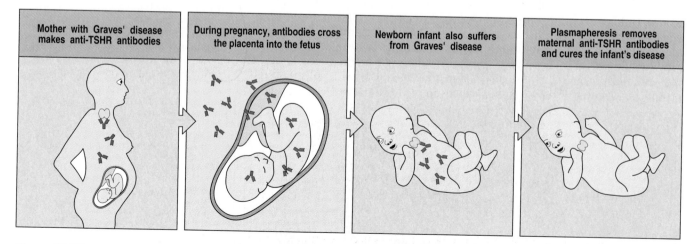

| Mother with Graves' disease makes anti-TSHR antibodies | During pregnancy, antibodies cross the placenta into the fetus | Newborn infant also suffers from Graves' disease | Plasmapheresis removes maternal anti-TSHR antibodies and cures the infant's disease |

Figure 13.8 Temporary symptoms of antibody-mediated autoimmune diseases can be passed from affected mothers to their newborn babies. TSHR, thyroid-stimulating hormone receptor.

capacity to cause cell activation and proliferation is assessed by measurement of the production of cyclic AMP and the synthesis of DNA, respectively. A complementary assay determines whether antibodies in a patient's plasma compete with TSH for binding to the TSH receptor of thyroid membranes isolated from pigs.

Because lymphocytes cannot pass from the maternal to the fetal circulation, babies born to mothers who have T cell-mediated autoimmune diseases do not show disease symptoms. Observations on human pregnancy therefore provide no information on the role of effector T cells in human autoimmune disease; neither do experiments in which T cells from patients are transferred into experimental animals. Such experiments fail to work because human T cells cannot recognize antigens presented by the MHC molecules of most other species. Only in models of autoimmune disease in inbred strains of rodents has it been possible to transfer disease with T cells and show they cause the disease.

13-8   Type 1 diabetes is caused by the selective destruction of insulin-producing cells in the pancreas

Insulin is secreted from the pancreas in response to the increased blood glucose level arising after a meal. By binding to surface receptors, insulin stimulates the body's cells to take up glucose and incorporate it into carbohydrates and fats. **Type 1 diabetes**, also called insulin-dependent diabetes mellitus (IDDM) or juvenile-onset diabetes, is caused by the selective autoimmune destruction of the insulin-producing cells of the pancreas. Because insulin is a major regulator of cellular metabolism it is essential for children's normal growth and development. Symptoms of disease are usually manifested in childhood or adolescence and they rapidly progress to coma and death in the absence of treatment. Type 1 diabetes principally affects populations of European origin, in which one person in 300 is a sufferer. This distribution, and the impact of the disease on young children, has made type 1 diabetes a major target for research in the countries of western Europe, North America, and Australia.

Scattered within the exocrine tissue of the pancreas are the **islets of Langerhans**, small clumps of endocrine cells that make the hormones insulin, glucagon, and somatostatin. The pancreas contains about half a million islets, each consisting of a few hundred cells. Each islet cell is programmed to make a single hormone: α cells make glucagon, β cells make insulin, and δ cells make somatostatin.

In patients with type 1 diabetes, antibody and T-cell responses are made against insulin, glutamic acid decarboxylase, and other specialized proteins of the pancreatic β cell. Which of these responses cause disease remains unclear. CD8 T cells specific for some peptide antigens unique to β cells are believed to mediate β-cell destruction, gradually decreasing the number of insulin-secreting cells. Individual islets become successively infiltrated with lymphocytes, a process called **insulitis**. β cells comprise about two-thirds of the islet cells; as they die, the architecture of the islet degenerates. A healthy person has about 10^8 β cells, providing an insulin-making capacity much greater than that needed by the body. This excess, and the slow rate of β-cell destruction, means that disease symptoms do not become manifest until years after the start of the autoimmune response. Disease commences when there are insufficient β cells to provide the insulin necessary to control the level of blood glucose (Figure 13.9).

The usual treatment for patients with type 1 diabetes is daily injection with insulin purified from the pancreata of pigs or cattle. Because of amino acid sequence differences between human and animal insulins, some patients

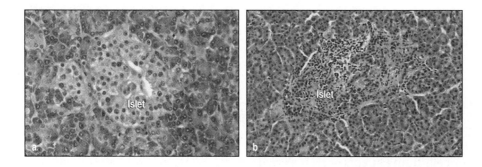

Figure 13.9 Comparison of histological sections of a pancreas from a healthy person and a patient with type 1 diabetes. Panel a shows a micrograph of a tissue section through a healthy human pancreas, showing a single islet. The islet is the discrete light-staining area in the center of the photograph. It is composed of hormone-producing cells, including the β cells that produce insulin. Panel b shows a micrograph of an islet from the pancreas of a patient with type 1 diabetes with acute onset of disease. The islet shows insulitis, an infiltration of lymphocytes from the islet periphery towards the center. The lymphocytes are the clusters of cells with darkly staining nuclei. Both tissue sections are stained with hematoxylin and eosin; magnification × 250. Photographs courtesy of G. Klöppel.

develop an immune response to animal insulin. The antibodies they make can have two bad effects when they bind to insulin—they reduce the activity of the insulin and form soluble immune complexes that can lead to further tissue damage and serum sickness (see Figure 12.33, p. 391). Recombinant human insulin produced in the laboratory from the cloned insulin gene is prescribed to patients who make antibodies against animal insulins.

13-9 Autoantibodies against common components of human cells can cause systemic autoimmune disease

So far, this chapter has concentrated on diseases in which a single type of cell is targeted for autoimmune attack. At the other end of the spectrum is **systemic lupus erythematosus** (**SLE**), a disease in which the autoimmune response is directed at autoantigens present in almost every cell of the body. SLE is an example of a **systemic autoimmune disease**.

Characteristic of SLE are circulating IgG antibodies specific for constituents of cell surfaces, cytoplasm, and nucleus, including nucleic acids and nucleoprotein particles. The binding of autoantibodies against cell-surface components initiates inflammatory reactions that cause cell and tissue destruction. In turn, these processes release soluble cellular antigens that form soluble immune complexes. On being deposited in blood vessels, kidneys, joints, and other tissues, such complexes can initiate further inflammatory reactions (Figure 13.10). All this provides a disrupted environment in which the immune system is increasingly stimulated to respond to common self components. SLE is thus a chronic inflammatory disease that can affect all tissues of the body. The disease commonly follows a course in which outbreaks of intense inflammation alternate with periods of relative calm. For individual patients

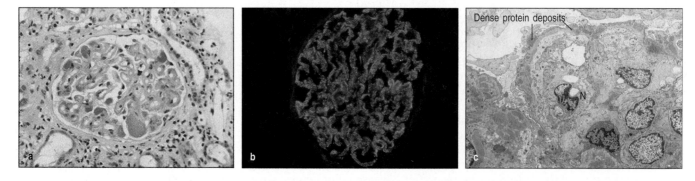

Figure 13.10 Deposition of immune complexes in the kidney glomeruli in systemic lupus erythematosus (SLE). Panel a shows a section through a glomerulus of a patient with SLE. Deposition of immune complexes causes thickening of the basement membrane (B). In panel b a similar kidney section is stained with fluorescent anti-immunoglobulin antibodies, revealing the presence of immunoglobulin in the basement membrane deposits. Panel c is an electron micrograph of part of a glomerulus. Dense protein deposits are seen between the glomerular basement membrane and the renal epithelial cells. Neutrophils (N) are also present, attracted by the deposited immune complexes. Photographs courtesy of H.T. Cook and M. Kashgarian.

the course of the disease is highly variable, both in its severity and in the organs and tissues involved. Many patients with SLE eventually die of the disease because of failure of vital organs such as the kidneys or brain.

Systemic lupus erythematosus was first described as 'lupus erythematosus' on the basis of the butterfly-shaped skin rash (erythema) that can occur on the face and gives the face an appearance like a wolf's head (*lupus* is Latin for wolf) (Figure 13.11). The rash is caused by the deposition of immune complexes in the skin and is similar to that caused by a type III hypersensitivity reaction (see Figure 12.33, p. 391). Further study of the disease revealed its systemic nature and so the word 'systemic' was added to the name. SLE is particularly common in women of African or Asian origin, of whom 1 in 500 has the disease. The factors that trigger the autoimmunity that leads to SLE remain unknown. However, they are likely to be simple in comparison with the complex autoimmunity and inflammation that develop during the course of disease. SLE is a disease in which the unwanted reactions of autoimmunity stimulate further autoimmunity that can send the immune system along a path of uncontrolled destruction.

13-10 Most rheumatological diseases are caused by autoimmunity

Common sites of disease in SLE are the joints, where the deposition of immune complexes causes inflammation and arthritis. Over 90% of patients with SLE suffer from arthritis and it is often the first symptom to be noticed. SLE is one of a number of rheumatic diseases, almost all of which are autoimmune in nature (Figure 13.12).

Rheumatoid arthritis is the most common of the rheumatic diseases, affecting 1–3% of the US population, with women outnumbering men by three to one. The disease involves chronic and episodic inflammation of the joints (Figure 13.13), usually starting between 20 and 40 years of age. A feature in 80% of patients with rheumatoid arthritis is the stimulation of B cells that make IgM, IgG, and IgA antibodies specific for the Fc region of human IgG. **Rheumatoid factor** is the name given to these anti-immunoglobulin autoantibodies.

In joints affected by rheumatoid arthritis there is a leukocyte infiltrate into the joint synovium consisting of CD4 and CD8 T cells, B cells, lymphoblasts, plasma cells, neutrophils, and macrophages. Among these are plasma cells making rheumatoid factor. Prostaglandins and leukotrienes produced by inflammatory cells are major mediators of the inflammation. In addition, neutrophils release lysosomal enzymes into the synovial space, causing damage to the tissue and proliferation of the synovium. Autoimmune CD4 T cells are activated by dendritic cells and these activate macrophages, which accumulate in the inflamed synovium and through secretion of pro-inflammatory cytokines such as TNF-α. IL-1, IL-6, and IL-7 recruit further effector cells into the joints, all adding to the tissue erosion. Proteinases and collagenases produced by inflammatory cells within the joint can also damage cartilage and supporting structures such as ligaments, tendons, and eventually the bones.

13-11 Rheumatoid arthritis can be treated with monoclonal antibodies that target either TNF-α or B cells

Rheumatoid arthritis is a chronic, painful, and debilitating disease, and patients often suffer it for many decades of their life. Treatment is usually a combination of physiotherapy with anti-inflammatory and immunosuppressive drugs. Two recently developed therapies targeting different aspects of the

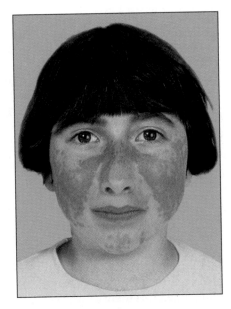

Figure 13.11 The characteristic facial rash of systemic lupus erythematosus. Although this butterfly-shaped rash was first used to recognize the disease, it is seen in only a proportion of patients who have the disease when defined immunologically. Photograph courtesy of M. Walport.

Rheumatic diseases caused by autoimmunity
Systemic lupus erythematosus (SLE)
Rheumatoid arthritis
Juvenile arthritis
Sjögren's syndrome
Scleroderma (progressive systemic sclerosis)
Polymyositis–dermatomyositis
Behçet's disease
Ankylosing spondylitis
Reiter's syndrome
Psoriatic arthritis

Figure 13.12 Rheumatic diseases are autoimmune in nature.

disease have proved successful in alleviating the symptoms of arthritis and in a few cases have terminated the disease. In the first approach the patient is infused with an anti-TNF-α monoclonal antibody. This chimeric antibody, called infliximab, eliminates the cytokine and reduces the inflammation as assessed by the level of C-reactive protein in the blood. It also reduces the joint swelling and the pain (Figure 13.14).

The second approach has used the anti-CD20 monoclonal antibody rituximab to deplete B cells in patients with rheumatoid arthritis. This treatment, which destroys B cells through the mechanism of antibody-dependent cytotoxicity (ADCC; see Section 9-23, p. 281) (Figure 13.15), reduces the number of circulating B cells by 98% and gives some benefit to around 80% of patients and major benefit to around 50%, a success rate comparable to that obtained with anti-TNF-α.

The discovery of rheumatoid factor led to the idea that rheumatoid arthritis was an antibody-mediated disease. Subsequent observations that 20% of patients with typical rheumatoid arthritis lack rheumatoid factor and that it is also a feature of other diseases (for example, it is found in 30% of SLE patients) challenged this theory. Later correlation of rheumatoid arthritis with particular HLA class II allotypes promoted the hypothesis that effector CD4 T cells were the cause of disease. The failure of therapies targeted to T cells to influence the course of rheumatoid arthritis and the success of B-cell depletion puts the ball back in the B-cell court. Both anti-TNF-α and anti-CD20 therapies are being explored in other autoimmune diseases and are showing promise. As these therapies are used more extensively, their side-effects, such as reduced resistance to infection, are also becoming apparent.

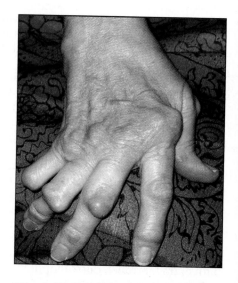

Figure 13.13 Inflamed joints in the hand of a patient with rheumatoid arthritis. Courtesy of J. Cush.

13-12 Multiple sclerosis and myasthenia gravis are autoimmune diseases of the nervous system

In **multiple sclerosis** an autoimmune response against the myelin sheath of nerve cells causes sclerotic plaques of demyelinated tissue in the white matter of the central nervous system. The plaques are readily revealed by magnetic resonance imaging. Symptoms of the disease include motor weakness, impaired vision, lack of coordination, and spasticity. The effects of multiple sclerosis are highly variable. It can take a slow progressive course, or it can involve acute attacks of exacerbating disease followed by periods of gradual recovery. The incidence of multiple sclerosis reaches 1 in 1000 in some populations, with onset commonly occurring at an age in the late twenties or early

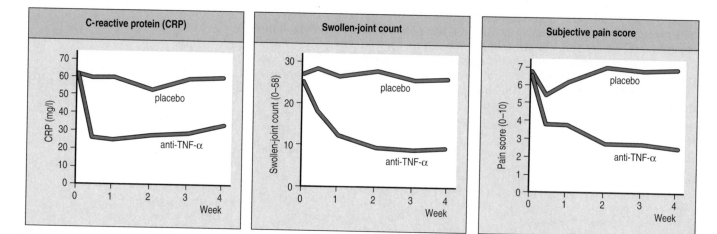

Figure 13.14 The effects of treatment of rheumatoid arthritis with anti-TNF-α. For each parameter measured (level of C-reactive protein, swollen joints, and pain), the values for patients given a placebo treatment are depicted by the blue curve and the values for patients treated with anti-TNF-α antibody are depicted by the red curve.

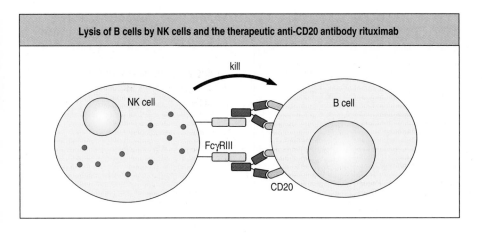

Figure 13.15 Killing of B cells by anti-CD20 antibody and NK cells. Patients with rheumatoid arthritis are treated with the anti-CD20 monoclonal antibody rituximab (dark purple). The antibody binds to the CD20 molecule (light purple) on the surface of B cells. The Fc region of the antibody engages the FcγRIII receptor on NK cells, activating the NK cell, which then kills the B cell. This is an example of antibody-dependent cell-mediated cytotoxicity (ADCC).

thirties. In extreme cases it leads to severe disability or death within a few years. In contrast, some patients with mild disease function well, with little neurological impediment.

Activated T_H1 CD4 cells and the interferon-γ (IFN-γ) they secrete are the effectors implicated in multiple sclerosis. These cells are enriched in the blood and cerebrospinal fluid. Activated macrophages present in sclerotic plaques release proteases and cytokines, which are the direct cause of demyelination. The effect of IFN-γ was most clearly demonstrated by a clinical trial of this cytokine as a possible treatment; tragically, the disease worsened in the patients treated with IFN-γ.

In 90% of patients with multiple sclerosis the sclerotic plaques contain plasma cells that secrete oligoclonal IgG into the cerebrospinal fluid. The autoantigens in multiple sclerosis are thought to be structural proteins of myelin, such as myelin basic protein, proteolipid protein, and myelin oligodendrocyte glycoprotein. Regular subcutaneous injection of IFN-β_1 reduces the incidence of disease attacks and the appearance of plaques. High doses of immunosuppressive drugs are used to reduce the severity of disease episodes.

Myasthenia gravis is an autoimmune disease in which signaling from nerve to muscle across the neuromuscular junction is impaired (Figure 13.16). Autoantibodies bind to the acetylcholine receptors on muscle cells, inducing their endocytosis and intracellular degradation in lysosomes. The loss of cell-surface acetylcholine receptors makes the muscle less sensitive to neuronal stimulation. Consequently, patients with myasthenia gravis suffer progressive muscle weakening as levels of autoantibody rise; the name of the disease means severe ('gravis') muscle ('myo') weakness ('asthenia'). Early symptoms of the disease are droopy eyelids and double vision. With time, other facial muscles weaken and similar effects on chest muscles impair breathing. This makes patients susceptible to respiratory infections and can even cause death. One treatment for myasthenia gravis is the drug pyridostigmine, an inhibitor of the enzyme cholinesterase, which degrades acetylcholine. By preventing

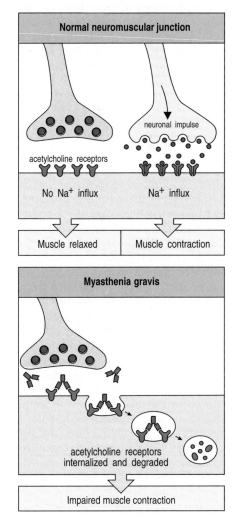

Figure 13.16 Autoantibodies against the acetylcholine receptor cause myasthenia gravis. In a healthy neuromuscular junction, signals generated in nerves cause the release of acetylcholine, which binds to the acetylcholine receptors of the muscle cells, causing an inflow of sodium ions that indirectly causes muscle contraction (upper panel). In patients with myasthenia gravis, autoantibodies specific for the acetylcholine receptor reduce the number of receptors on the muscle-cell surface by binding to the receptors and causing their endocytosis and degradation (lower panel). Consequently, the efficiency of the neuromuscular junction is reduced, which is manifested as muscle weakening.

Diseases mediated by antibodies against cell-surface receptors			
Syndrome	**Antigen**	**Antibody**	**Consequence**
Graves' disease	Thyroid-stimulating hormone receptor	Agonist	Hyperthyroidism
Myasthenia gravis	Acetylcholine receptor	Antagonist	Progressive muscle weakness
Insulin-resistant diabetes	Insulin receptor	Antagonist	Hyperglycemia, ketoacidosis
Hypoglycemia	Insulin receptor	Agonist	Hypoglycemia

Figure 13.17 Diseases mediated by antibodies against cell-surface receptors. Antibodies act as agonists when they stimulate a receptor on binding it, and as antagonists when they block a receptor's function on binding it.

acetylcholine degradation, pyridostigmine increases the capacity of acetylcholine to compete with the autoantibodies for the receptors. A second treatment for myasthenia gravis is the immunosuppressive drug azathioprine, which inhibits the production of autoantibodies.

Autoantibodies against cell-surface receptors can act either to stimulate or to inhibit the receptor (Figure 13.17). The autoantibodies in myasthenia gravis prevent the function of the acetylcholine receptor and are called receptor **antagonists**. In contrast, the autoantibodies in Graves' disease facilitate receptor function (see Section 13-5) and are called receptor **agonists**. Autoantibodies of both types can be made against the insulin receptor; they lead to different symptoms. The cells of patients with antagonistic autoantibodies are unable to take up glucose, which accumulates in the blood, causing hyperglycemia and a form of diabetes mellitus resistant to treatment with insulin. In contrast, in patients with agonistic antibodies, cells deplete blood glucose to abnormally low levels. This state of hypoglycemia deprives the brain of glucose, causing light-headedness.

Summary

In autoimmunity the mechanisms of antigen recognition and effector function are identical to those used in the response to pathogens. In contrast, the symptoms of autoimmune disease are highly variable, depending on the triggering autoantigen and the target tissue. Autoimmune diseases can be classified into three main types corresponding to the type II, type III, and type IV categories of hypersensitivity reactions, with which they share common effector mechanisms. Some autoimmune diseases, such as insulin-dependent diabetes or Graves' disease of the thyroid, are directed against antigens of one particular organ or tissue, and are known as organ-specific or tissue-specific autoimmune diseases. In contrast, systemic lupus erythematosus (SLE) is a systemic autoimmune disease, which is directed against components common to all cells. Autoimmune diseases caused by antibodies have been more readily studied than those caused by effector T cells, because antibodies can exert an effect when transferred from human patients to experimental animals. Understanding how effector T cells function in human autoimmune disease remains largely dependent on extrapolation from mouse models of autoimmune disease. Many autoimmune diseases tend to be chronic conditions in which episodes of acute disease are interspersed with periods of recovery. Once a tissue becomes inflamed and damaged by an autoimmune response, the resulting increase in the processing and presentation of self antigens frequently leads to expansion and diversification of the autoimmune response. In addition, infections can exacerbate autoimmunity in a nonspecific manner by inducing the production of inflammatory cytokines within the infected tissues.

Genetic and environmental factors predispose to autoimmune disease

At various stages in the development of the immune system and in the development and execution of an immune response, mechanisms are brought into play that prevent attack on healthy cells and tissues. Together, these result in the self-tolerance of the immune system. The tolerance-inducing mechanisms involved in adaptive immunity are more elaborate than those for innate immunity. This is because potentially self-reactive lymphocytes are continually being generated and a person's population of B cells and T cells changes according to their experience of infection, vaccination, and other antigenic challenge. Various mechanisms contribute to self-tolerance. During development, many clones of self-reactive B cells and T cells are deleted from the repertoire and die. Of the self-reactive cells that do enter the peripheral circulation, some become anergic, some remain physically separated from the self antigens to which they could respond, and others are suppressed by regulatory T cells. Naive T-cell activation requires co-stimulation, which itself depends on infection, and this limits the circumstances in which autoreactive T cells can be activated. In addition, in the absence of infection, T cells have very limited access to tissues other than the blood and lymphoid tissues, and thus they may never encounter tissues expressing potential autoantigens. All autoimmune diseases involve a breach of one of these mechanisms of self-tolerance. Both genetic and environmental factors contribute to loss of self-tolerance and the development of disease-causing autoimmunity. Particularly striking is the fact that many autoimmune diseases differentially affect males and females, with females having the greater propensity for these conditions overall (Figure 13.18). That a large majority of the human population never suffers autoimmune disease is testament to the protection provided by the immunological mechanisms of self-tolerance.

13-13 All autoimmune diseases involve breaking T-cell tolerance

Autoimmune diseases that are caused by autoreactive inflammatory T cells clearly involve a breach of T-cell tolerance. This is also true for autoimmune diseases caused by antibodies. The B cells involved have switched isotype and undergone affinity maturation by somatic hypermutation, and this means they will have received help from antigen-specific T cells.

During B-cell maturation in the bone marrow, clonal deletion and inactivation of self-reactive B cells prevent the emergence of cells with antigen receptors that bind common molecules of human cell surfaces or plasma. This process does not prevent the emergence of B cells with specificity for numerous other self antigens that are not present in the bone marrow or plasma. The activation of these autoreactive B cells is prevented by additional mechanisms, the most important of which is T-cell tolerance, which deprives them of T-cell help (Figure 13.19). When an autoreactive B cell is stimulated by its autoantigen it migrates to the T-cell area of a secondary lymphoid tissue. Because antigen-activated helper T cells are not available, the antigen-stimulated B cell fails to enter a primary lymphoid follicle and becomes trapped in the T-cell zone, where it dies by apoptosis.

13-14 Incomplete deletion of self-reactive T cells in the thymus causes autoimmune disease

Thymic selection of the T-cell repertoire provides the foundation for immunological self-tolerance. During T-cell development, negative selection removes T cells that respond to self peptides presented by the MHC molecules

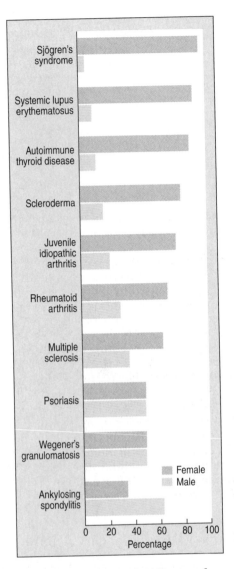

Figure 13.18 Relative incidences of autoimmune disease in females and males.

Figure 13.19 Arrest and death of autoreactive B cells in secondary lymphoid tissue in the absence of autoantigen-specific T cells. B cells enter the T-cell zone of a lymph node through high endothelial venules (HEVs). B cells with reactivity for foreign antigens (non-autoreactive) are shown in yellow, and autoreactive cells in gray. Non-autoreactive B cells enter the lymphoid follicle and receive signals to survive. If autoreactive B cells encounter their specific antigen they stop in the T-cell zone. Because there are no cognate T cells specific for the autoantigen, the autoreactive cells cannot leave the T-cell zone to enter the primary follicle. The autoreactive B cells thus fail to receive survival signals and undergo apoptosis *in situ* in the T-cell zone.

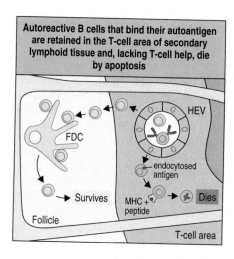

on thymic cells. In this way T-cell tolerance is produced to most of the peptides normally presented by cells of the body.

Emphasizing the importance of negative selection in the thymus for the prevention of autoimmunity is the nature of the inherited immunodeficiency disease caused by the absence of the transcription factor AIRE, which normally induces the expression of tissue-specific proteins in the thymus that contribute to negative selection of the developing T-cell repertoire (see Section 7-12, p. 202). Although defective alleles of the AIRE gene are very rare in most populations, they are sufficiently common in some groups—Finns, Sardinians, and Iranian Jews—for substantial numbers of children to be born who inherit two defective alleles. For these infants, the normal array of tissue-specific proteins is not made in the thymus and there is incomplete negative selection of the T-cell repertoire. Starting in childhood, autoimmune B-cell and T-cell responses develop against various peripheral tissues, including most endocrine glands. Patients differ in the tissues that are affected, and most patients suffer disorders of multiple organs. This syndrome is called inherited **autoimmune polyglandular disease** (APD) or **autoimmune polyendocrinopathy–candidiasis–ectodermal dystrophy** (APECED) (Figure 13.20). The ectodermal dystrophy manifests as abnormalities of teeth, hair, and fingernails (Figure 13.21).

Despite the range and severity of their symptoms, APECED patients have substantial life spans, the most life-threatening complications being squamous-cell carcinoma and fulminant autoimmune hepatitis. Thus, the lack of self-tolerance to major components of several organs does not cause an acute and catastrophic disease, but one that, although more complicated, has similarities to the more common autoimmune diseases. Like those diseases, the course of APECED is highly heterogeneous, indicating the involvement of other genetic and environmental factors. In Finnish, but not Iranian, patients, for example, the autoimmune symptoms are foreshadowed by chronic infection with the fungus *Candida albicans*, despite the presence of a strong antibody response to *Candida* antigens. Persistent *Candida* infection is the only symptom of disease that is common to all Finnish APECED patients (see Figure 13.20).

13-15 Insufficient control of T-cell co-stimulation favors autoimmunity

Even when negative selection is working normally, some autoreactive T cells escape deletion and enter the peripheral circulation. If these T cells encounter healthy tissue cells expressing complexes of MHC and autoantigenic peptides, they will usually not be activated, because tissue cells lack the B7 co-stimulators essential for the activation of naive T cells (see Section 8-5, p. 219). On the contrary, this manner of presentation can lead to tolerance by way of T-cell anergy (see Section 8-9, p. 226). This is one mechanism by which tolerance is induced in autoreactive T cells that reach the peripheral circulation.

APECED patients suffer a variety of autoimmune diseases and candidiasis	
Symptom	**Frequency in Finnish patients (%)**
Endocrine glands	
Hypoparathyroidism	85
Adrenal failure	72
Ovarian failure	60
Insulin-dependent diabetes mellitus	18
Testicular atrophy	14
Parietal cell atrophy	13
Hypothyroidism	6
Other tissues	
Candidiasis	100
Dental enamel hypoplasia	77
Nail dystrophy	52
Tympanic membrane calcification	33
Alopecia	27
Keratopathy	22
Vitiligo	13
Hepatitis	13
Intestinal malabsorption	10

Figure 13.20 Patients with deficiency of the autoimmune regulator protein AIRE suffer a wide range of autoimmune diseases. This condition is called autoimmune polyendocrinopathy–candidiasis–ectodermal dystrophy (APECED) or inherited autoimmune polyglandular disease (APD).

During T-cell activation the functions of the B7 co-stimulators are kept in check by CTLA-4 molecules, which compete with CD28 for binding B7 (see Section 8-5, p. 220). Alternative mRNA splicing of the genes encoding CTLA-4 produces soluble and membrane-bound forms of CTLA-4, both of which are functional in damping T-cell activation. Two alleles of the CTLA-4 gene encode proteins of identical amino acid sequence, but differ in the quantity of soluble CTLA-4 they make. The allele producing less of the soluble CTLA-4 is associated with susceptibility to Graves' disease, Hashimoto's disease, and type 1 diabetes, whereas the allele producing more of the soluble CTLA-4 is associated with resistance to these autoimmune diseases. Unlike in APECED, in which homozygosity for defective AIRE alleles guarantees that the disease will develop, neither CTLA-4 allele should be considered 'defective,' and the impact they have on disease is a subtle one. The 'susceptible' allele accounts for 63.4% of the alleles in patients with Graves' disease, compared with 53.2% in healthy controls. Nonetheless, the difference is statistically significant and points to the importance of the balance struck in T-cell activation through the actions of CD28 and CTLA-4.

Figure 13.21 Dystrophic fingernails in a patient with APECED. Photograph courtesy of Mark S. Anderson.

13-16 Regulatory T cells protect cells and tissues from autoimmunity

Potentially autoreactive circulating CD4 T cells against common autoantigens are present even in healthy people. These cells respond to autoantigens in culture but are kept in check in the body. Among these potentially autoreactive T cells are regulatory CD4 T cells that express CD25 and the α chain of the IL-2 receptor (see Section 7-13, p. 203 and 8-19, p. 242). On contacting self antigens presented by MHC class II molecules, the regulatory cells themselves do not proliferate but they can suppress the proliferation of naive T cells responding to autoantigens presented on the same antigen-presenting cell (Figure 13.22). This suppressive effect requires contact between the two T cells and also involves the secretion of non-inflammatory cytokines such as IL-4, IL-10, and transforming growth factor (TGF)-β. The suppressive function of these regulatory T (T_reg) cells is dependent on CTLA-4 but not on CD28, which is consistent with a mechanism in which co-stimulation of regulatory T cells involves the binding of B7 on the antigen-presenting cell to CTLA-4 on the

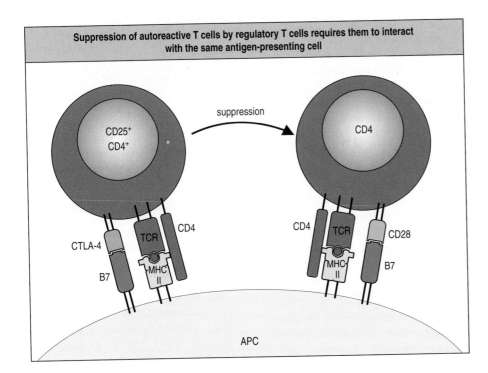

Figure 13.22 CTLA-4 is involved in the action of regulatory T cells. Suppression of an autoreactive T cell by a CD25+ CD4+ regulatory T cell is dependent on CTLA-4 on the regulatory T cell binding B7 on the antigen-presenting cell (APC), and on both T cells interacting with the same antigen-presenting cell.

regulatory T cell. The differential effects of CTLA-4 alleles on susceptibility to autoimmune disease may be manifested through this role of CTLA-4 in activating regulatory cells (see Figure 13.22).

What uniquely defines T_{reg} cells is their use of a transcriptional repressor called FoxP3 that is specified by a gene on the X chromosome. All T_{reg} cells express FoxP3 and no other cells do. A rare deficiency in FoxP3 principally affects boys, producing a fatal disorder that is characterized by autoimmunity directed toward a variety of tissues. The gut is almost always affected, and other common targets are the thyroid, pancreatic β cells and the skin. Affected children fail to thrive and suffer recurrent infections. This syndrome, which is caused by lack of T_{reg} cells, is called the **immune dysregulation, polyendocrinopathy, enteropathy, and X-linked syndrome** (**IPEX**). The only effective treatment for IPEX is a bone marrow transplant from a healthy HLA-identical sibling.

Current research has described an additional subpopulation of helper CD4 T cells that secrete large quantities of the cytokine IL-17, express the retinoid-related orphan receptor-γt (RORγt) and are designated **T_H17 cells**. Their development is dependent on IL-1 and IL-23, and they appear to share a common developmental precursor with FoxP3-expressing regulatory T cells, although T_H17 cells seem to have pro-inflammatory functions. IL-17 binds to the IL-17 receptor expressed by fibroblasts, epithelial cells and keratinocytes, inducing these cells to secrete cytokines that recruit inflammatory cells. In autoimmune diseases such as Crohn's disease, rheumatoid arthritis, and psoriasis, as well as in allergic asthma, T_H17 cells accumulate in the affected tissues, where they can represent up to 30% of the T cells present. T_H17 cells are absent from patients with Job's syndrome, an immunodeficiency due to defective STAT3, who are prone to recurrent infections including the staphylococcal abscesses called boils. Such a susceptibility indicates that T_H17 cells in healthy individuals have a pro-inflammatory role in the adaptive immune response to infection.

13-17 HLA is the dominant genetic factor affecting susceptibility to autoimmune disease

Susceptibility to autoimmune diseases runs in families and varies between populations. Such clinical observations first raised the possibility that genetic factors confer differential susceptibility to autoimmune diseases. Although natural variation in the *AIRE*, *CTLA4*, and *FOXP3* genes influences susceptibility to autoimmune disease in ways that shed light on the mechanisms of autoimmunity, their contribution to overall disease is minor. By far the most important genetic factors influencing susceptibility to autoimmune disease are located in the HLA complex.

A simple way of assessing the influence of HLA on susceptibility to an autoimmune disease is to compare the HLA types of pairs of siblings who have the same autoimmune disease with pairs of healthy siblings. Four HLA haplotypes segregate in nuclear families, two maternal haplotypes and two paternal haplotypes. Consequently, in the population at large, 50% of siblings will share one HLA haplotype, 25% will share two HLA haplotypes and 25% will be HLA disparate (Figure 13.23, upper panel). The distribution becomes distorted when pairs of siblings with type 1 diabetes are compared; the frequency of diabetic sibling pairs who share two HLA haplotypes, and are thus HLA identical, more than doubles, whereas the frequency of diabetic pairs sharing one HLA haplotype or no HLA haplotype is depressed (Figure 13.23, lower panel). Such an analysis demonstrates that HLA type is correlated with susceptibility to type 1 diabetes. Similar results are seen for many autoimmune and inflammatory diseases, although the strength of the effect varies substantially. Overall assessments of genetic make-up, including comparisons of

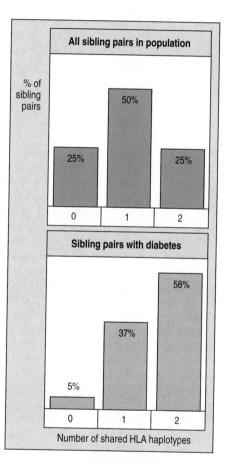

Figure 13.23 Family studies reveal that HLA type correlates with susceptibility to type 1 diabetes. The top panel shows the frequency with which two siblings share HLA haplotypes in the population as a whole. The percentages are those expected from a simple mendelian segregation of the two maternal and two paternal HLA haplotypes. In the bottom panel the analysis has been confined to pairs of siblings who both have type 1 diabetes. In these sibling pairs, the frequency distribution of HLA haplotypes differs greatly from that expected from simple mendelian segregation. Pairs of siblings with the disease are much more likely to have the same HLA type than are pairs of siblings who are healthy.

HLA-associated risk factors for autoimmune disease				
Disease	HLA allotype	Frequency (%)		Relative risk
		Patients	Control	
Ankylosing spondylitis	B27	> 95	9	> 150
Birdshot chorioretinopathy	A29	> 95	4	> 50
Narcolepsy	DQ6	> 95	33	> 40
Celiac disease	DQ2 and DQ8	95	28	30
Type 1 diabetes	DQ8 and DQ2	81	23	14
Subacute thyroiditis	B35	70	14	14
Multiple sclerosis	DQ6	86	33	12
Rheumatoid arthritis	DR4	81	33	9
Juvenile rheumatoid arthritis	DR8	38	7	8
Psoriasis vulgaris	Cw6	87	33	7
Addison's disease	DR3	69	27	5
Graves' disease	DR3	65	27	4
Myasthenia gravis	DR3	50	27	2
Type 1 diabetes	DQ6	< 0.1	33	0.02

Figure 13.24 Associations of HLA allotypes with autoimmune disease. The data are derived from the Norwegian population. In the column 'Relative risk,' numbers greater than 1 indicate that the HLA allotype confers increased susceptibility relative to the general population; numbers less than 1 indicate increased protection. Data courtesy of Erik Thorsby.

polymorphisms throughout the genome, estimate that HLA genes account for at least 50% of the genetic predisposition to autoimmune disease.

Although family analysis shows that genes within or linked to the HLA region are risk factors for autoimmune disease, it is not a good approach for identifying the individual genes. This is better done with large panels of unrelated patients and controls in which numerous HLA haplotypes with different combinations of linked alleles can be examined. Such studies show that the strongest associations are with alleles of the polymorphic HLA class I and class II genes (Figure 13.24), with weaker effects due to polymorphisms in other genes, for example the gene encoding TNF-α. More diseases are associated with HLA class II genes than with HLA class I genes, although the strongest association is of a class I molecule, HLA-B27, with ankylosing spondylitis, an arthritic disease.

HLA types were originally defined by using antibodies. For example, antibody typing identified several different HLA-DR alloantigens, which were called DR1, DR2, DR3, DR4, DR5, and so on. Subsequent analysis of the underlying genes by DNA sequencing showed that each alloantigen is heterogeneous, consisting of a group of related allotypes (the HLA-DR proteins) with β chains of different sequences. The standard nomenclature for HLA alleles is now based on DNA sequence. For example, the names of all alleles of the *DRB1* gene start with *DRB1*. Within the original HLA-DR4 grouping, 22 different allotypes of DRB1 have so far been identified. All these are known as DRB1*04. Individual allotypes are distinguished by two additional numerals: DRB1*0402, for example, is an allotype associated with a greater susceptibility to the autoimmune disease pemphigus vulgaris in some ethnic groups.

The assignment of HLA genes to disease susceptibilities is complicated by the fact that particular alleles of the different polymorphic genes are combined in HLA haplotypes at frequencies higher than expected by chance. This phenomenon of preferential allele associations is called **linkage disequilibrium (LD)**. One haplotype with unusually strong linkage disequilibrium includes the *HLA-A*01*, *HLA-B*08*, *HLA-DQ2*, and *HLA-DR3* alleles and is associated with several prominent autoimmune diseases, including type 1 diabetes, SLE, and myasthenia gravis. For geneticists, the problem with linkage disequilibrium is that one cannot separate out the individual contributions of the linked alleles to disease susceptibility. For type 1 diabetes, for example, distinguishing the contributions of *HLA-DQ* and *HLA-DR* has been tricky. One helpful approach was to compare populations of different ethnicities in which there are distinct sets of haplotypes with different patterns of linkage disequilibrium. The *HLA-A*01, HLA-B*08, HLA-DQ2, HLA-DR3* haplotype (the so-called *8.1* haplotype) is characteristic of northern European populations and increases in frequency the closer the population lives to the North Pole. In African and Asian populations many of the same alleles are present but are on different haplotypes. Thus, analysis of the disease associations in those populations provided an independent assessment of whether HLA-DQ2 or HLA-DR3 is associated with susceptibility to type 1 diabetes. The answer was HLA-DQ2.

Autoimmune diseases are more prevalent in caucasians than in other populations, which in part is due to the effects of the *8.1* haplotype. Despite its disadvantages, this haplotype is carried by more than 10 million Europeans and is well represented in the white populations of other continents. This illustrates a general point, which is that the large majority of people who have a disease-associated HLA type will never suffer from an autoimmune disease associated with that type. Conversely, not all patients who develop an HLA-associated disease have an HLA type that is correlated with disease susceptibility. Thus, the associations of common HLA alleles with autoimmune diseases do not derive from 'good' versus 'bad' alleles, as is true of a genetic disease such as APECED. Even for the strongest association of HLA type with disease, that of HLA-B27 with ankylosing spondylitis, only 2% of people who have HLA-B27 develop the disease and, conversely, 2% of patients with clinically defined disease do not have HLA-B27. Evidence for the contribution of other genetic factors is the fact that 75% of patients are male (see Figure 13.18), and that even in families with a history of ankylosing spondylitis, only a maximum of 20% of HLA-B27-expressing individuals develop disease.

13-18 Different combinations of HLA class II allotypes confer susceptibility and resistance to diabetes

To illustrate the difficulties of determining the genetic and molecular basis of susceptibility to autoimmune disease, we will look in detail at the association of HLA-DQ and HLA-DR protein isoforms with type 1 diabetes. Of the two, the effects due to HLA-DQ are stronger, but they are strongly influenced by the linked *HLA-DR* genes. Several HLA-DQ allotypes have associations with type 1 diabetes, whereas the influence of HLA-DR is limited to HLA-DR4. We saw in Chapter 5 that the HLA-DQ α and β chains are both polymorphic and encoded by neighboring genes. Thus, in a heterozygous individual, the HLA-DQ molecules can, in theory, be formed by the combination of α and β chains encoded by the same haplotype and/or by the α chain from one haplotype and the β chain from the other. In practice, however, there is a preferential association of the α and β chains encoded by the same haplotype, and only a very small amount of protein is produced by the combination of α and β chains from different haplotypes. Because of this, the naming of the HLA-DQ allotypes—HLA-DQ2 for example—refers specifically to the heterodimeric protein molecule formed by the combination of α and β chains commonly encoded on the same haplotype.

Common caucasian HLA haplotypes that encode either the HLA-DQ2 or HLA-DQ8 allotype confer susceptibility to type 1 diabetes. Heterozygous individuals who carry both the HLA-DQ2 and HLA-DQ8 haplotypes are, however, much more susceptible to the disease than those who have either haplotype alone. This augmented susceptibility is due to a novel HLA-DQ heterodimer, which is assembled only in heterozygous individuals and consists of the HLA-DQ8 α chain, DQA1*03, associated with the HLA-DQ2 β chain, DQB1*0201 (Figure 13.25). In northern Europeans this combination of α and β chains is never encoded in the same haplotype, and so can be produced only in heterozygotes. Haplotypes containing both susceptibility alleles can, how-

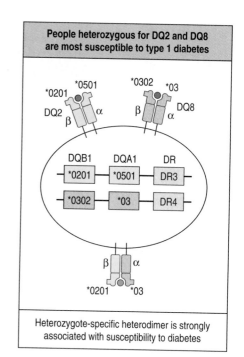

Figure 13.25 Certain HLA heterozygous individuals are more susceptible to diabetes than homozygous individuals. The person shown here has two HLA haplotypes that are independently associated with susceptibility to type 1 diabetes. The *DR3* haplotype contains *DQ* genes that encode the DQα*0501 chain and the DQβ*0201 chain, the *DR4* haplotype contains genes that encode the DQα*03 chain and the DQβ*0302 chain. The two α chains and two β chains made in this person's cells can associate in different combinations to form four different DQ isoforms, of which three (those shown in the figure) are associated with susceptibility to diabetes. The DQ heterodimer associated with the highest susceptibility is that comprising the DQα*03 chain made from the *DR4* haplotype and the DQβ*0201 chain made from the *DR3* haplotype. This heterodimer can be made only in *DR3/DR4* heterozygous individuals, whereas the two heterodimers with weaker disease association are also made in homozygous individuals: heterodimers of the DQα*0501 and DQβ*0201 chains (called the DQ2 molecule) in *DR3* haplotype homozygotes and heterodimers of the DQα*03 and DQβ*0302 chains (called the DQ8 molecule) in *DR4* haplotype homozygotes. The heterozygote is therefore more susceptible to disease than either homozygote. As a general rule, heterozygosity gives increased fitness over homozygosity, but in this situation the reverse is true.

	HLA locus			Amino acid position in DRβ chain			
Risk	DQB1	DQA1	DRB1	67	71	74	86
Protective	*0302 — *03 — *0403 *0201 — *0501 — *03			L	R	E	V
Moderate	*0302 — *03 — *0404 *0201 — *0501 — *03			L	R	**A**	V
High	*0302 — *03 — *0405 *0201 — *0501 — *03			L	R	**A**	**G**
High	*0302 — *03 — *0401 *0201 — *0501 — *03			L	**K**	**A**	**G**

Figure 13.26 Subtypes of _DRB1*04_ modify the susceptibility to type 1 diabetes conferred by the DQα*03:DQβ*0201 heterodimer. The leftmost panels show the risk of type 1 diabetes for individuals who have the haplotype combinations shown in the adjacent column. All these individuals are heterozygotes for _DR3_ and _DR4_ haplotypes and make the DQα*03:DQβ*0201 heterodimer that confers susceptibility to type 1 diabetes (see Figure 13.25). The only difference distinguishing these individuals is that they have different _DRB1*04_ alleles, as indicated by the differently colored boxes. The amino acid substitutions that distinguish the different DRβ*04 chains are shown on the right. The more common amino acid residue at each position is shown in gray; the less common residues are highlighted in red. The different HLA-DR heterodimers, formed by association of the four DRβ*04 chains with the invariant DRα chain, have profound effects on the disease susceptibility associated with the DQα*03:DQβ*0201 heterodimer.

ever, be present in Africans, and similarly confer susceptibility to type 1 diabetes. The HLA-DQ4 and HLA-DQ9 allotypes also confer susceptibility to type 1 diabetes, but their effects are weaker than those of HLA-DQ2 or HLA-DQ8.

The susceptibility associated with the DQA1*03:DQB1*0201 heterodimer is strongly influenced by the presence of the HLA-DR4 allotype, which is in linkage disequilibrium with DQ8. For HLA-DR, only the β chain is polymorphic, which simplifies its disease associations compared with HLA-DQ. The HLA-DR4 allotype consists of multiple subtypes that differ from each other by a few amino acid substitutions in the β chain. As seen in Figure 13.26, single amino acid substitutions can determine whether a DRB1*04 subtype confers susceptibility to type 1 diabetes or protection from it. In caucasian populations, DQ8 (_DQB1*0302, DQA1*03_) appears as a strong susceptibility factor because it is in linkage disequilibrium with DR4 subtypes that favor susceptibility, whereas in the Chinese population it is poorly associated with disease susceptibility because it is in linkage disequilibrium with DR4 subtypes favoring resistance.

The HLA-DQ6 allotype confers a strong resistance to type 1 diabetes. In heterozygous individuals who have HLA-DQ6 combined with a HLA-DQ type that confers susceptibility, the protection provided by HLA-DQ6 is dominant. Consequently, people who have HLA-DQ6 rarely get type 1 diabetes. This type of dominance has yet to be described for other autoimmune diseases. This may be a feature unique to type 1 diabetes autoimmunity or may reflect the fact that this is the most extensively studied autoimmune disease and is the only one for which this level of complexity has been recognized and addressed. HLA-DQ7 is also associated with a weak resistance, which in heterozygotes is not dominant over susceptibility allotypes.

Through the combination of maternal and paternal haplotypes at the HLA-DQ and HLA-DR loci, a diversity of human genotypes is produced that range from resistance to susceptibility in their contribution to genetic predisposition to type 1 diabetes.

13-19 Autoimmunity is initiated by disease-associated HLA allotypes presenting antigens to autoimmune T cells

Autoimmunity requires a breach of T-cell tolerance, which implies that the autoimmune response is started by autoreactive T cells being stimulated by specific peptide:MHC complexes. Because autoimmunity is exceptional and tolerance the rule, it is likely that when tolerance is broken it involves just one or a few autoreactive T-cell clones. This is consistent with the association of common autoimmune diseases with specific HLA allotypes, and it led to the commonly held view that autoreactive T cells activated by peptides presented by disease-associated HLA allotypes initiate the autoimmunity. The greater number of HLA class II associations is expected because they present antigens to CD4 T cells, which are both more abundant and more often the initiators of an immune response than CD8 T cells. This model has not been proved conclusively for any human disease, although the body of circumstantial and indirect evidence is persuasive. Moreover, the principle has been proved in animal models of autoimmune disease by demonstration that T-cell clones with well-defined autoantigen specificities can transfer disease.

Almost all of the disease-associated HLA allotypes comprise a series of structurally related subtypes differing in a few amino acid substitutions. Not uncommonly, one or more of the subtypes will not be associated with disease (Figure 13.27). The HLA-DRB1*0401, *0404, *0405, and *0408 allotypes confer genetic predisposition to rheumatoid arthritis, whereas the DRB1*0402 allele does not, and may even give protection. Although HLA-DRB1*0402 differs from HLA-DRB*0404 only by amino acid substitutions at positions 67, 70, and 71, these differences introduce negative charges that change the peptide-binding motif. Such correlations are consistent with peptide-presenting function being the basis for the association of HLA allotypes with autoimmune disease.

In patients with SLE, HLA class II polymorphism influences the specificity of autoantibodies made, presumably by its effect on the helper T-cell response. The greatest susceptibility to SLE is associated with HLA-DR3 haplotypes; patients with these haplotypes tend to make antibodies against proteins of a small cytoplasmic ribonucleoprotein complex. In contrast, patients with HLA-DR2 haplotypes make antibodies against double-stranded DNA, and patients with HLA-DR5 haplotypes make antibodies against the nuclear ribonucleoprotein complex known as the spliceosome.

Although the strongest association with autoimmune disease is of HLA-B27 with ankylosing spondylitis, the cause of this disease and the nature of the participation of HLA-B27 remain mysterious. Although T cells are implicated in causing this inflammatory disease, they are not thought to recognize peptide autoantigens presented by HLA-B27. In comparison with other HLA class I allotypes, the intracellular assembly of newly synthesized HLA-B27 molecules seems inefficient and leads to the accumulation of misfolded forms within the cell and on the cell surface. The possibility that these abnormal

DRB1*04 allele	Amino acid position in DRβ chain		
	67	70	71
*0401	L	Q	K
*0402	I	D	E
*0404	L	Q	R
*0405	L	Q	K
*0408	L	Q	R

Figure 13.27 Basic residues in the peptide-binding groove of the DRβ*04 chain are necessary to confer susceptibility to rheumatoid arthritis. With the exception of DRB1*0402, all the DRB1*04 alleles shown are associated with susceptibility to rheumatoid arthritis. What distinguishes the DRβ*0402 chain from those encoded by the other alleles is a cluster of amino acid substitutions at positions 67, 70, and 71. These substitutions change the localized charge environment within the peptide-binding groove by removing a basic (positively charged) residue (shown in blue) and inserting two acidic (negatively charged) residues (shown in red).

HLA-B27 molecules, which lack β₂-microglobulin and bound peptide, are the cause of disease is under investigation.

13-20 Noninfectious environmental factors influence the course of autoimmune diseases

The preceding sections show how the interaction of many different functional gene polymorphisms determines genetic predisposition to autoimmune disease. The fact that predisposed individuals develop disease with a maximum frequency of about 20% also emphasizes the importance of environmental factors in determining who actually gets ill. A simple illustration of the influence of an environmental factor is provided by Goodpasture's syndrome, the disease caused by autoantibodies against type IV collagen of basement membranes (see Section 13-3). Whereas all patients suffer glomerulonephritis, only 40% of them develop pulmonary hemorrhage as well. The patients who do are those who habitually smoke cigarettes. In non-smokers the basement membranes of lung alveoli are inaccessible to antibodies and there is neither antibody deposition nor disruption of the tissue. The alveoli in smokers' lungs are already damaged from their continual exposure to cigarette smoke. This lack of tissue integrity gives circulating antibodies access to the basement membranes, where deposition and complement activation burst blood vessels, causing hemorrhage.

Physical trauma can also give the immune system access to anatomical sites and autoantigens to which it is not normally exposed. One such immunologically privileged site is the anterior chamber of the eye, which contains specialized proteins involved in vision. If a blow ruptures an eye, proteins from the anterior chamber drain to the local lymph node and there induce autoimmunity. On occasion, this response will cause blindness in the damaged eye. Surprisingly, the immune response also gains access to the undamaged eye and can cause blindness there unless treatment is given. This condition is called **sympathetic ophthalmia** (Figure 13.28). The fact that both eyes are attacked shows that immunological privilege is due to mechanisms that prevent the induction of an immune response, rather than to a lack of access for effector cells and molecules once they have been activated. To preserve a patient's sight in the undamaged eye it is necessary to remove the damaged eye, which prevents the further induction of autoimmunity. Patients are also given immunosuppressive drugs to stop the ongoing autoimmune response.

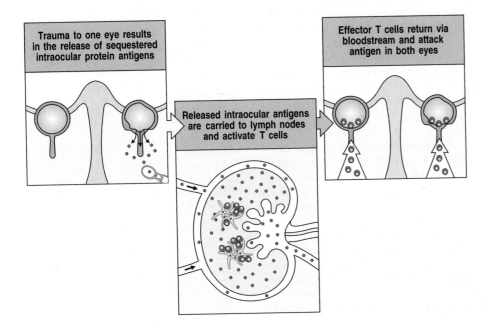

Figure 13.28 Physical trauma to one eye initiates autoimmunity that can destroy vision in both eyes.

Figure 13.29 The enzyme peptidyl arginine deiminase converts the arginine residues of tissue proteins to citrulline. In tissues stressed by wounds or infection, peptidyl arginine deaminase (PAD) activity is induced. By converting arginine residues to citrulline, PAD destabilizes proteins and makes them more susceptible to degradation. It also introduces novel B-cell and T-cell epitopes into tissue proteins that can stimulate an autoimmune response.

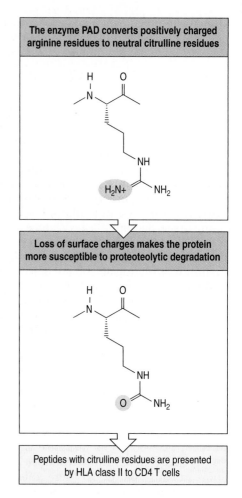

The enzyme PAD converts positively charged arginine residues to neutral citrulline residues

Loss of surface charges makes the protein more susceptible to proteoteolytic degradation

Peptides with citrulline residues are presented by HLA class II to CD4 T cells

Sympathetic ophthalmia shows how healthy tissues not involved in initiating autoimmunity can become vulnerable to attack when a response of the right specificity is started elsewhere. A similar effect seems important in the development of rheumatoid arthritis, as described in the next section.

13-21 Genetic and environmental effects combine to cause one form of rheumatoid arthritis

The recognition of rheumatoid factor as an immune complex originally led to the belief that rheumatoid arthritis is caused by autoantibodies (see Section 13-11). Subsequent discovery of the association with DRB1*04 (see Section 13-19) led to the competing theory that effector CD4 T cells cause the disease. More recent studies show that the autoimmune response in some rheumatoid arthritis patients is directed against self proteins and self peptides in which arginine residues have been converted into citrulline residues by inducible enzymes called peptidyl arginine deiminases (PADs) (Figure 13.29). This resembles the mechanism of celiac disease, in which the immune response is directed at epitopes of food proteins in which tissue transglutaminase has converted glutamine residues to glutamate (see Section 12-23, p. 396). In patients with rheumatoid arthritis who make antibodies against citrullinated epitopes, the association with DRB1*04 is strong, but in patients who lack such antibodies there is no association. This shows that two different immunological mechanisms are causing the disease diagnosed as rheumatoid arthritis. Such heterogeneity might explain the observation that only about 50% of rheumatoid arthritis patients get major benefit from therapy with anti-TNF-α or anti-CD20.

Citrullinated proteins provide a source of peptide antigens to which the T-cell repertoire is not tolerant. Presentation of citrullinated self peptides by MHC class II molecules could therefore activate T cells to help any B cells specific for the citrullinated self protein from which the T cell-stimulating peptide was derived. The activated B cells need not all be specific for citrullinated epitopes, because helper T cells specific for one epitope of a protein are able to provide help to B cells specific for all other epitopes of the protein. In the context of this mechanism, HLA-DRB1*04 is expected to be particularly good at presenting citrullinated peptides to T cells. Citrullination of proteins makes them more susceptible to proteolytic degradation, which also favors their ability to stimulate an autoimmune response.

Smoking is the major environmental factor associated with rheumatoid arthritis (Figure 13.30). However, this effect is seen only with the subset of patients who have antibodies against citrullinated self proteins. Thus smoking, HLA-DRB1*04, and an immune response to citrullinated proteins are all tied together in the same disease-causing mechanism. A working model is that the damage caused by smoking induces the expression of PAD in the respiratory tract and that this stimulates an autoimmune response to citrullinated self proteins. This immune response is not immediately destined to attack the joints, which is consistent with the observation that antibodies against citrullinated self proteins appear years before any symptoms of arthritis. The joints are attacked later, when some independent trauma such as a wound or an infection induces a state of inflammation in the joint and the activation of PAD. Under these circumstances effector and memory lymphocytes specific for citrullinated self proteins enter the inflamed joint tissue and respond to their specific antigens. Both the actions of effector T cells and the deposition of immune complexes will exacerbate the inflammation and develop the symptoms of rheumatoid arthritis.

13-22 Infections are environmental factors that can trigger autoimmune disease

In the previous section we saw how the tissue damage caused by habitually setting fire to cigarettes and inhaling their smoke and vapor sets the stage for an autoimmune response. In analogous fashion, pathogens are also environmental factors that cause inflammation and disrupted tissue when they set up an infection. For almost every autoimmune disease there are clinical observations suggesting that the autoimmunity is triggered by some sort of infection and the immune response that it provokes. For certain diseases the evidence is compelling; for others it is correlative, circumstantial, or anecdotal.

Experimental evidence that infection is necessary for inducing autoimmunity comes from the experience of immunologists at inducing autoimmune disease in animals by the injection of tissues, tissue extracts, or purified autoantigens. Injection of these preparations alone does not produce an autoimmune response. However, when such self components are mixed with microbial products that induce inflammation at the site of injection, autoimmunity is reliably produced. In models of spontaneous autoimmune disease, where no manipulation of the animals is required to induce disease, the incidence and severity of disease vary with the conditions under which the animals are housed and the microorganisms to which they are exposed, and for people there seem to be similar effects. The incidence of autoimmune disease in human populations increases with financial wealth, industrial development, and the changing ways of life that follow them.

A simple and well-defined example of an autoimmune disease that is a by-product of the specific response to infection is **rheumatic fever**. This involves inflammation of the heart, joints, and kidneys, which can follow 2–3 weeks after a throat infection with certain strains of *Streptococcus pyogenes*. In defeating the bacterial infection, antibodies specific for cell-wall components of *S. pyogenes* are made. Some of these antibodies happen to react with epitopes present on human heart, joint, and kidney tissue. On binding to the heart they activate complement and generate an acute and widespread inflammation—rheumatic fever—which sometimes causes heart failure (Figure 13.31). Such a chance resemblance of pathogen and host antigens is called **molecular mimicry**. This example shows how health-improving immune responses that successfully control infections can inadvertently become life-threatening autoimmunities. Rheumatic fever is a transient autoimmune disease; for lack of T-cell help the autoantigens cannot continue to stimulate antibody synthesis. This situation arises because the CD4 T cells that helped in the antibacterial response are not stimulated by the autoantigens. The incidence of rheumatic fever has greatly diminished since streptococcal infections began to be treated with antibiotics.

The transient nature of rheumatic fever emphasizes the necessity of T cells for the chronic autoimmunity that characterizes most autoimmune diseases. T-cell activation occurs only in the presence of inflammation, which is itself instigated by infection. It is therefore plausible that all destructive autoimmune reactions have their origins in infection. For instance, the physical blow that damages an eye and initiates sympathetic ophthalmia would certainly introduce microorganisms into the eye that could then initiate the necessary inflammation.

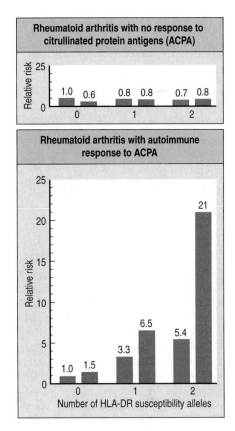

Figure 13.30 Patients with rheumatoid arthritis form two distinct groups. Upper panel: the relative risk of developing rheumatoid arthritis in which no autoimmune response to citrullinated protein antigens (ACPA) is made does not correlate with the presence of *HLA-DRB1*04* 'susceptibility' alleles or with smoking. The red columns indicate smokers, the blue columns nonsmokers. The numbers above each column are the relative risk of disease for that group. Lower panel: in contrast, the relative risk of developing rheumatoid arthritis in which an autoimmune response to ACPA has been made is increased by the presence of *HLA-DRB1*04* susceptibility alleles and by smoking. At highest risk are smokers who have any two HLA-DRB1*04 susceptibility alleles (see Figure 13.27). These data are from a cohort of Swedish rheumatoid arthritis patients. Data courtesy of Lars Klareskog.

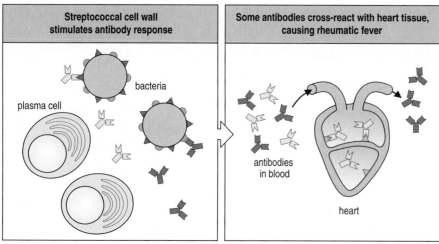

| Streptococcal cell wall stimulates antibody response | Some antibodies cross-react with heart tissue, causing rheumatic fever |

Figure 13.31 Antibodies against streptococcal cell-wall antigens cross-react with antigens on heart tissue. The immune response to the bacteria produces antibodies against various epitopes of the bacterial cell surface. Some of these antibodies (yellow) cross-react with the heart, whereas others (blue) do not. An epitope in the heart (orange) is structurally similar, but not identical, to a bacterial epitope (red).

Although less well established than for rheumatic fever, bacterial infections have been implicated in Reiter's syndrome and reactive arthritis, two of the arthritic diseases associated with HLA-B27. When cohorts of people were inadvertently infected with the same strain of food-poisoning bacteria, the frequency of autoimmune disease later developing among those expressing HLA-B27 in the infected group was higher than for the population as a whole. Such evidence for the involvement of gastrointestinal infection in autoimmunity has come from outbreaks of food poisoning among passengers on cruise ships and among police on crowd-control duty. Awaiting study is the large cohort of immunologists who were poisoned by a boxed lunch at a summer conference in the 1990s. Coxsackie B virus, which infects the β cells of the pancreas, has been implicated in triggering type 1 diabetes (Figure 13.32). It is also possible that candidiasis is an important trigger of autoimmunity in APECED (see Figure 13.20).

13-23 Autoimmune T cells can be activated in a pathogen-specific or nonspecific manner by infection

There are two distinct mechanisms by which infection could break down T-cell tolerance and activate autoreactive T cells. The first applies when circulating autoreactive T cells, which are anergized or suppressed through the

Associations of infection with autoimmunity		
Infection	**HLA association**	**Consequence**
Group A *Streptococcus*	Not known	Rheumatic fever (carditis, polyarthritis)
Chlamydia trachomatis	HLA-B27	Reiter's syndrome (arthritis)
Shigella flexneri, Salmonella typhimurium, Salmonella enteritidis, Yersinia enterocolitica, Campylobacter jejuni	HLA-B27	Reactive arthritis
Borrelia burgdorferi	HLA-DR2, DR4	Chronic arthritis in Lyme disease
Coxsackie A virus, Coxsackie B virus, echoviruses, rubella	HLA-DQ2, HLA-DQ8 DR4	Type 1 diabetes

Figure 13.32 Infections associated with the start of autoimmunity.

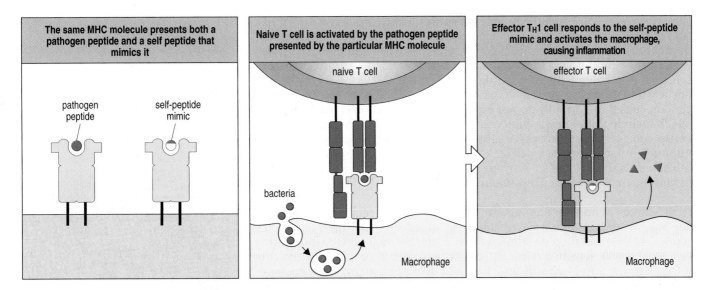

| The same MHC molecule presents both a pathogen peptide and a self peptide that mimics it | Naive T cell is activated by the pathogen peptide presented by the particular MHC molecule | Effector T$_H$1 cell responds to the self-peptide mimic and activates the macrophage, causing inflammation |

Figure 13.33 Autoimmunity may be caused by self peptides that mimic pathogen-derived peptides and stimulate a T-cell response. The first panel shows the same MHC molecule presenting two different peptides: one from the pathogen, and a self peptide that mimics it. The second panel shows a naive CD4 T cell being activated by the pathogen peptide. This activated CD4 T$_H$1 cell can then activate a macrophage presenting the self-peptide mimic, as shown in the third panel, thus initiating an inflammatory reaction.

activity of regulatory T cells, become activated in the context of the inflammatory state and cytokine environment produced by infection. In this mechanism the effect of the infection is not specific to the pathogen's antigens but is caused by nonspecific perturbation of the regulatory balance between T-cell populations that normally maintains peripheral tolerance to self antigens.

The second mechanism is antigen specific and is based on T-cell cross-reactivity. In rheumatic fever, antibodies made against a pathogen cross-react with a self antigen. T-cell receptors can also cross-react with different antigens. T-cell cross-reactivities are frequently found in experiments in which peptides with related sequences are bound to the same MHC allotype, or in which the same peptide is bound to very similar MHC allotypes. Cross-reactivities are also found for complexes in which neither the peptides nor the MHC allotypes are particularly closely related. Because all antigens recognized by T cells conform to the common structure of a peptide bound to an MHC molecule, T cells are intrinsically more cross-reactive than B cells, whose antigens are under no such structural limitation. Indeed, the developmental processes of positive selection, negative selection, and T-cell activation are all based on the inherent cross-reactivity of T-cell receptors.

The need for a T-cell response in order to produce persistent autoimmunity, and the inherent cross-reactivity of T-cell receptors, have reinforced the belief that chronic autoimmune diseases are caused by autoreactive T cells that arise in the course of combating infection. During an infection, each clone of T cells is activated by a pathogen-derived peptide bound to the MHC allotype that positively selected the T-cell clone. As naive cells, all these clones were tolerant of the complexes of self peptides and MHC allotypes to which they were exposed. Once activated, however, effector T cells can respond to other peptide:MHC complexes that bind with lower affinity than is required for activation. So T cells activated by a pathogen-derived peptide have the potential to attack cells presenting self-peptide:MHC complexes that are antigenically cross-reactive (Figure 13.33). This is the model of molecular mimicry applied to T-cell antigens. The sequences of human proteins and pathogen proteins reveal many potential T-cell epitopes that could provide such mimics.

Activated T cells gain access to most of the body's tissues, whereas naive T cells are restricted to the blood, secondary lymphoid tissues, and lymph. Within tissues, T cells activated by pathogens will encounter tissue-specific self-peptide:MHC complexes that did not participate in positive selection, negative selection, or the induction of peripheral tolerance. Some of these complexes might bind T-cell receptors with affinities even higher than that of the activating antigen.

When tissues become inflamed, some cells are induced to increase the expression of MHC class I and class II molecules. These changes, which increase the number of different peptide antigens presented by a cell and their density on the cell surface, can lead to interactions with clones of T cells that were not sensitive to the lower levels of antigen presentation. The greatest changes occur when the inflammatory cytokine IFN-γ induces MHC class II expression on cells that normally do not express these MHC molecules. Such cells, which include thyroid cells, pancreatic β cells, astrocytes, and microglia, are conspicuously present in tissues targeted by T cell-mediated autoimmunity. Although induced expression of MHC class II in the absence of co-stimulatory molecules is insufficient to activate naive T cells, it provides potential new targets for T cells activated by other means (Figure 13.34).

In the course of an infection, dendritic cells are able to pick and present tissue-specific proteins from dead, apoptotic, or infected cells. Naive autoimmune T cells could then be stimulated by a professional antigen-presenting cell, whereupon activated effector T cells would then be capable of entering infected tissue and of directly attacking healthy cells making the tissue-specific proteins.

13-24 In the course of autoimmune disease the specificity of the autoimmune response broadens

Pemphigus vulgaris and the milder variant **pemphigus foliaceus** are conditions characterized by blistering skin. They are caused by autoantibodies specific for desmogleins, which are adhesion molecules present in the cell junctions (the desmosomes) that bind skin keratinocytes tightly to each other. For certain native communities in rural Brazil, genetic predisposition and environmental factors cause pemphigus foliaceus to be endemic, with up to 1 in 30 people being affected. Long-term study of the autoantibody response in these people has shown that it evolves in a manner that correlates with the clinical progression of disease. The extracellular (EC) part of desmoglein comprises four structurally similar domains named EC1–EC4, and a fifth domain, EC5, which is structurally quite different. Autoantibodies present before the onset of disease are specific for epitopes in the EC5 domain, which is nearest to the cell membrane. These antibodies do not bind to cell-surface desmoglein, nor do they transfer disease when injected into mice. With disease onset, antibodies specific for the EC1 and EC2 domains are detectable (Figure 13.35). These IgGs bind to cell-surface desmoglein, and when injected into mice they induce disease. The process by which the immune response initially targets epitopes in one part of an antigenic molecule and then progresses to other, non-cross-reactive, epitopes of the same molecule is called **intramolecular epitope spreading**. The anti-desmoglein response in pemphigus foliaceus is an example of intramolecular spreading of B-cell epitopes. In this disease, it is only with epitope spreading that the disease-causing antibodies are made.

Intramolecular epitope spreading also occurs for the T-cell response in autoimmunity. The T-cell epitopes to which the response spreads are frequently ones to which the immune system is not tolerant, because those self peptides are not normally presented by MHC molecules at sufficient levels. These epitopes are called **cryptic epitopes** because they are normally hidden from the immune system and are revealed only under conditions of infection or inflammation.

Epitope spreading can also involve epitopes of different proteins, when they are part of a macromolecular complex or derive from the same cell type. This is called **intermolecular epitope spreading**. Macromolecular complexes facilitate epitope spreading in SLE, a disease in which many different autoantibodies are made against the components of intracellular nucleoprotein particles. The initiating event of an autoimmune response is loss of T-cell

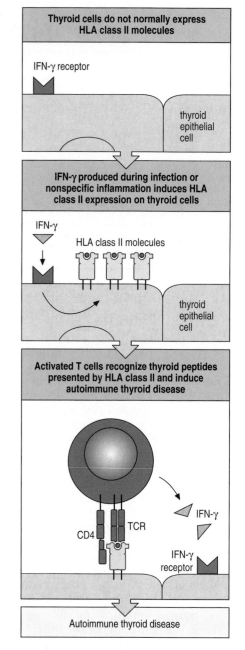

Figure 13.34 Induction of HLA class II expression on tissue cells facilitates autoimmunity. Thyroid epithelial cells do not normally express HLA class II molecules (top panel). They are induced to do so by IFN-γ (center panel). They are then able to present thyroid peptides to activated antigen-specific T cells, which induces autoimmune thyroid disease (bottom panel). IFN-γ produced by activated T cells feeds back to amplify HLA expression and antigen presentation.

Figure 13.35 The progressive involvement of desmoglein epitopes in pemphigus foliaceus. This skin-blistering disease is caused by antibodies that bind to desmoglein—an adhesion molecule in the cell junctions that hold keratinocytes together. When the autoimmune B-cell response begins, harmless antibodies are made against the extracellular domain of the desmoglein molecule nearest the keratinocyte surface (the EC5 domain). In time, the response spreads and antibodies are made against the extracellular domains EC1 and EC2. These antibodies cause disease and are of the IgG4 isotype.

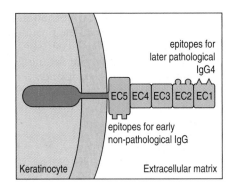

tolerance and the activation of a clone of autoreactive T cells that is specific for a peptide derived from the nucleoprotein complex. Cells from this one clone of T cells can activate B cells specific for many different epitopes on the surface of the complex, the only requirement being that the B-cell receptor can bind and internalize the complex and present that specific peptide to the T cell (Figure 13.36, left panel). This mechanism even allows T cells responding to peptide epitopes to activate B cells making high-affinity antibodies against nucleic acids, macromolecules that cannot be recognized by T cells.

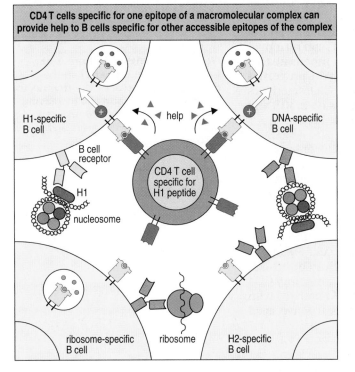

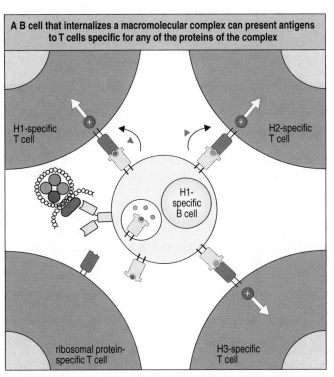

Figure 13.36 In systemic lupus erythematosus (SLE) the immune response is broadened in a antigen-specific manner. In patients with SLE, an ever-broadening immune response is made against nucleoprotein antigens such as nucleosomes, which consist of histones and DNA and are released from dying and disintegrating cells. The left panel shows how the emergence of a single clone of autoreactive CD4 T cells can lead to a diverse B-cell response to nucleosome components. The T cell in the center is specific for a specific peptide (red) from the linker histone H1, which is present on the surface of the nucleosome. The B cells at the top are specific for epitopes on the surface of a nucleosome, on H1 and DNA, respectively, and thus bind and endocytose intact nucleosomes, process the constituents, and present the H1 peptide to the helper T cell. Such B cells will be activated to make antibodies, which in the case of the DNA-specific B cell will be anti-DNA antibodies. The B cell at the bottom right is specific for an epitope on histone H2, which is hidden inside the intact nucleosome and is thus inaccessible to the B-cell receptor. This B cell does not bind the nucleosome and does not become activated by the H1-specific helper T cell. A B cell specific for another type of nucleoprotein particle, the ribosome (which is composed of RNA and specific ribosomal proteins), will not bind nucleosomes (bottom left) and will not be activated by the T cell. In reality a T cell interacts with one B cell at a time, but different members of the same T-cell clone will interact with the B cells of different specificity. The right panel shows the broadening of the T-cell response to the nucleosome. The H1-specific B cell in the center has processed an intact nucleosome and is presenting a variety of nucleosome-derived peptide antigens on its MHC class II molecules. This B cell can activate a T cell specific for any of these peptide antigens, which will include those from the internal histones H2, H3, and H4 as well as those from H1. This H1-specific B cell will not activate T cells specific for peptide antigens of ribosomes because ribosomes do not contain histones.

Once activated, B cells will broaden the T-cell response because they present peptides derived from all the proteins of the nucleoprotein complex and thus activate other clones of T cells specific for those peptides. This activation is achieved through cognate interactions between B cells and T cells (Figure 3.36, right panel) or by antibody binding to the nucleoprotein complex and facilitating its uptake and processing and the presentation of the resulting peptides by macrophages or dendritic cells.

13-25 Senescence of the T-cell population can contribute to autoimmunity

Maintenance of T-cell tolerance to self is essential for preventing autoimmune disease. An important and unanswered question is how this is affected by the gradual decay of the thymus as an organ for producing new naive T cells. Involution of the human thymus begins soon after birth. Although the size of the organ remains constant, there is a steady decay in the fraction of the tissue devoted to the production of naive T cells, which is first replaced by tissue containing mature T cells and subsequently by fat. By age 50 the capacity to produce new T cells is reduced to 20%, and by age 70 it is practically gone (see Figure 7.4, p. 189). By their nature, T-cell populations are dynamic: the survival of T cells requires that they divide periodically, and it is estimated that around 1% of the body's T cells are replaced each day. When the thymus can no longer satisfy the need for naive T cells, the immune system compensates by expanding the populations of existing T-cell clones and altering the properties of some T cells so that they become less susceptible to apoptosis. Characteristics of the latter type of T cells are the lack of CD28 and the expression of KIR and other receptors that are more usually associated with NK cells.

Rheumatoid arthritis is generally associated with people aged 50 or over. The incidence of the disease has a negative correlation with thymus function; the disease increases in frequency with age, whereas thymus function decreases (Figure 13.37). The T cells of patients with rheumatoid arthritis have antigen receptors that are around one-tenth as diverse as those of healthy people of similar age. In their blood and affected joints can be found large expanded clones of autoreactive CD4 T cells that lack CD28 and express NK-cell receptors, notably the activating receptor KIR2DS2. Although these cells are insensitive to co-stimulation, they are not anergic and can be activated through KIR2DS2 to produce large amounts of IFN-γ. These highly inflammatory CD4 T cells are implicated in contributing to rheumatoid arthritis. In patients with rheumatoid arthritis, therefore, the T-cell population seems to have aged prematurely.

13-26 Does the current increase in hypersensitivity and autoimmune disease have a common cause?

Like allergic disease, the incidence of autoimmune disease is increasing in the richer nations, to the point of causing widespread concern. In both types of disease the immune system works in ways that are counterproductive and the disease-causing mechanisms are the same as those used to control pathogens. The hygiene hypothesis was initially conceived to explain the increase in allergies, but it can also account for the increasing autoimmunity.

The essential proposition is that modern practices of hygiene, vaccination, and antibiotic therapy cause some children's immune systems to develop in ways that are different from those achieved in former times. Because children's immune systems are not tried out against such a range or variety of childhood infection, they do not become so skilled in attacking infection while maintaining tight T-cell tolerance. At the cellular level this involves balancing the functions of inflammatory T cells, non-inflammatory T cells, and

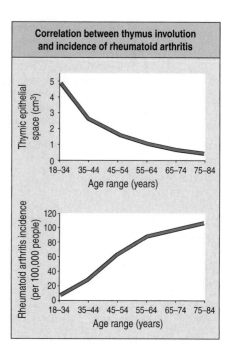

Figure 13.37 With age there is an inverse correlation between the decreasing capacity of the thymus to make new T cells and the increasing incidence of rheumatoid arthritis. Data courtesy of C.M. Weyand and J.J. Goronzy.

regulatory T cells. In this context, the kinetics and time-frame of thymic involution clearly advise that facility with one's immune system, like language, sport, and numerous other skills, is probably best acquired in childhood. Although today's children are much less likely than their forebears to die from a childhood infection, their reduced experience in making constructive inflammatory responses against real-life infections means they are more likely to over-react when infection comes their way. In turn, the increased over-reactions stretch T-cell tolerance and occasionally it breaks, starting a chronic and damaging disease.

Summary

Autoimmune diseases start with a minor breach of T-cell tolerance and end up with a persistent, powerful adaptive immunity that attacks one or more of the body's tissues. Normally T-cell tolerance is achieved by negative selection in the thymus and a combination of peripheral mechanisms that prevent the activation of potentially autoreactive T cells. The system usually works well, because most people worldwide never suffer from autoimmune disease. Both genetic polymorphisms and environmental factors are implicated in predisposing people to autoimmune disease. The most important polymorphic genes are those for the HLA class I and II molecules that present peptide antigens to T cells. Different HLA allotypes and isotypes are correlated with different disease susceptibilities and may be instrumental in presenting autoantigens to tolerance-breaking T cells. HLA accounts for about half of the genetic predisposition to autoimmune disease. Other genetic factors include proteins involved in thymic selection and in the control of T-cell activation. The environmental factors are ones that damage the integrity of human tissues and stimulate the adaptive immune system. These comprise trauma, chemicals, irradiation, and certain foods, all of which may also be accompanied by some form of infection. Rare T cells that are activated in response to these events continue to be activated once they are over and initiate a chronic inflammatory state. In these circumstances there is increased loss of tolerance, recruitment of additional autoreactive B and T cells, and expansion of the autoimmune response to an increasing range of autoantigens.

Even for HLA genes the correlations with disease susceptibility are weak in absolute terms: most people having a predisposing HLA type do not develop the associated disease, and not all people who develop the disease have the associated HLA type. The same is true for the environmental factors that affect autoimmune disease. No known infection inevitably leads to autoimmunity, and different pathogens can trigger clinically similar autoimmune diseases in people of different HLA types. Identifying the infections that trigger autoimmune diseases is inherently difficult, first because the infections are successfully resolved and rarely recorded, and second because the symptoms of autoimmune disease usually begin long after the event that triggered the autoimmune response.

Summary to Chapter 13

Central to the success of adaptive immunity is the provision of strong responses to foreign antigens while sustaining tolerance toward human macromolecules. Among the mechanisms that prevent autoreactivity are clonal deletion and inactivation of potentially autoreactive lymphocytes, anatomical barriers that prevent lymphocytes from reaching certain tissues, and lymphocyte activation requirements that depend on the presence of infection. Failure of any of these mechanisms can lead to autoimmune responses that attack the body's tissues and cause disease. Autoimmune diseases are complicated multifactorial diseases for which no single entity can be identified as the necessary and sufficient cause of a particular disease. They arise as a consequence of unlucky combinations of genetic and environmental factors, each of which by itself is weakly correlated with disease. With rare exceptions—APECED and IPEX—autoimmune diseases do not arise from obvious genetic or functional deficiencies in the immune system but are the occasional and perhaps inevitable side-effects of successfully fighting infections. A major challenge in treating autoimmune diseases is that patients usually come to medical attention only when the destructive effects of autoimmunity are at an advanced stage and when the initiating autoimmune response has become expanded and diversified. With better knowledge of the mechanisms operating in human autoimmune disease it should be possible to identify patients at earlier stages of disease, when tissue damage is limited, the autoimmune response is more narrowly defined, and it should be easier to control the response without resorting to nonspecific immunosuppression.

Questions

13–1
A. What is the rationale for categorizing the autoimmune diseases as types II, III, and IV? Be specific.
B. In this categorization, why is there no type I autoimmune disease?

13–2 Which of the following describes processes by which self-reactive lymphocytes are rendered incapable of mounting an autoimmune response? (Select all that apply.)
a. sequestration of autoantigens in immunologically privileged sites
b. induction of anergy in peripheral compartments
c. positive selection of autoimmune T lymphocytes in secondary lymphoid tissues
d. negative selection of T lymphocytes in the thymus
e. suppression by regulatory T cells
f. negative selection of B lymphocytes in the bone marrow
g. expression of AIRE in the bone marrow
h. induction of alloreactive responses it the thymus

i. somatic hypermutation to an alternative antigen specificity
j. apoptosis in primary lymphoid tissue
k. deprivation of T-cell help.

13–3 From Table Q13–3 below match the autoimmune disease in column 1 with the corresponding antigen in column 2 and the consequence in column 3. Use each answer only once. Then indicate whether the autoimmune disease is categorized as type II, III, or IV.

13–4 Describe the three immunological mechanisms responsible for the destruction of red blood cells in autoimmune hemolytic anemia.

13–5 Characterize two properties of endocrine glands that render them susceptible to autoimmune attack.

13–6 Hashimoto's and Graves' diseases both impair normal functioning of the thyroid gland but do so using different immunopathological mechanisms. Compare and contrast these mechanisms.

Table Q13–3

Column 1	Column 2	Column 3
a. Rheumatoid arthritis	1. Myelin basic protein, proteolipid protein	A. Destruction of red blood cells by complement and phagocytosis, anemia
b. Subacute bacterial endocarditis	2. DNA, histones, ribosomes, snRNP, scRNP	B. Joint inflammation and destruction
c. Autoimmune hemolytic anemia	3. Thyroid-stimulating hormone receptor	C. Pancreatic β cell destruction
d. Mixed essential cryoglobulinemia	4. Bacterial antigen	D. Glomerulonephritis
e. Multiple sclerosis	5. Rheumatoid factor IgG complexes	E. Hyperthyroidism
f. Systemic lupus erythematosus	6. Epidermal cadherin	F. Blistering of skin
g. Type 1 diabetes	7. Synovial joint antigen	G. Glomerulonephritis, vasculitis, arthritis
h. Graves' disease	8. Rh blood group antigens	H. Systemic vasculitis
i. Pemphigus vulgaris	9. Pancreatic β cell antigen	I. Brain degeneration, paralysis

13–7 Indicate whether each of the following statements is true (T) or false (F).

___a. During pregnancy IgG antibodies and activated lymphocytes can cross the placenta and enter the circulatory system of the fetus.

___b. Blood plasma exchange (plasmapheresis) can be used to remove maternal IgG from the newborn.

___c. All autoimmune diseases involve a breach of T-cell tolerance.

___d. Newborns of mothers with T cell-mediated autoimmune diseases exhibit the same symptoms as their mothers.

___e. Autoimmune diseases can be induced after an infection.

13–8

A. What mechanism of self-tolerance is broken in the autoimmune syndrome APECED?

B. What is the underlying genetic defect in APECED? Explain why it leads to a reduction in self-tolerance.

13–9 You have isolated a subset of CD25⁺ CD4⁺ T cells from the blood that have T-cell receptors specific for a self antigen but do not proliferate when challenged with the antigen *in vitro*. What is the name given to these T cells and what role are they thought to have in preventing autoimmunity?

13–10

A. In general, which genes have been found to correlate best with susceptibility or resistance to autoimmune diseases? (Do not specify individual genes or diseases.)

B. What general hypothesis has been proposed to account for this association?

13–11 People who are heterozygous for HLA-DQ2 and HLA-DQ8 allotypes are at greater risk of developing type 1 diabetes than those who are homozygous for HLA-DQ2 or HLA-DQ8.

A. Explain the reason for this increased susceptibility.

B. Why is the above statement true mainly for people of northern European origin but not for some other ethnic groups?

13–12

A. Which patients affected by Goodpasture's syndrome also succumb to pulmonary hemorrhage?

B. Explain the reason for this complication.

13–13 Explain the relationship between HLA-DRB1*04, smoking, the expression of peptidyl arginine deaminase, and rheumatoid arthritis.

13–14 In the context of autoimmunity ,(A) define molecular mimicry and (B) provide an example.

13–15 A recent therapy developed for the treatment of rheumatoid arthritis includes the use of _____ monoclonal antibodies that suppress the autoimmune response. (Select all that apply.)

a. anti-TNF-α

b. anti-C-reactive protein

c. anti-CD20

d. anti-rheumatoid factor

e. anti-CD3.

13–16 Seventeen-year-old Lisa Montague practiced piano 3–4 hours each day while preparing for music college auditions. Some of her pieces required sustained arm-muscle activity and she began to find them hard to play, even though she had previously played them easily. When she also started to have difficulty swallowing and chewing, she told her mother, who took her to the emergency room, where the physician noticed drooping eyelids and limitation of ocular motility. An electromyogram detected impaired nerve-to-muscle transmission. Administration of pyridostigmine rapidly improved Lisa's symptoms. Which of the following blood-test results would be most consistent with her condition?

a. elevated rheumatoid factor

b. elevated anti-myelin basic protein antibodies

c. elevated anti-acetylcholine receptor antibodies

d. elevated anti-nuclear antibodies

e. elevated anti-Rh antibodies.

A map showing the relatedness of pathogenic strains of influenza virus isolated in different places and at different times.

Chapter 14

Vaccination to Prevent Infectious Disease

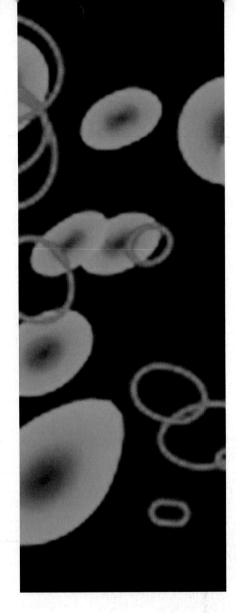

The preceding four chapters of this book have described the various ways in which the immune system responds helpfully or otherwise to environmental antigens. Given the spectrum of disease that results from inadequate or inappropriate immunity, clinical immunologists are driven to search for ways of manipulating the immune response to the benefit of their patients. The next three chapters each examine a different area of medicine where this quest is being pursued—vaccination, organ transplantation, and cancer. In this chapter we examine vaccination. This is a procedure in which the adaptive immune system is manipulated in an antigen-specific manner to mimic infection by a particular pathogen and stimulate protective immunity against it without causing the disease itself. Vaccination has protected billions of people against certain infectious diseases and is by far the most successful, frequent, and widely used manipulation of the immune response.

The modern era of vaccination began in the 1780s with Edward Jenner's use of cowpox to vaccinate against smallpox. The material used for vaccination is called a **vaccine**. Although Jenner's procedure was widely embraced, vaccines for other diseases did not emerge until well into the nineteenth century. That was the period when microorganisms were first isolated, grown in culture, and shown to cause disease. Largely by a process of trial and error, methods of inactivating microbes or impairing their ability to cause disease were developed and vaccines were produced. Vaccines were eventually developed against most of the epidemic diseases that had plagued the populations of western Europe and North America. Indeed, by the middle of the twentieth century it began to seem that the combination of vaccines and antibiotics would solve the problem of infection once and for all. Such optimism was soon tempered by the difficulty of developing effective vaccines for some diseases, and by the emergence of new diseases and antibiotic-resistant forms of old ones. This chapter surveys the vaccines in current use and then turns to the challenge of pathogens that have yet to be tamed by vaccination.

14-1 Viral vaccines are made from whole viruses or viral components

1st prescribed vaccine

The first medically prescribed vaccine was against smallpox, a disease characterized by a rash of spots that develop into scarring pustules. The very first smallpox vaccines were made from dried pustules taken from people who seemed to have less severe symptoms of the disease. This early vaccine thus contained the smallpox virus itself. Small amounts of this material were given

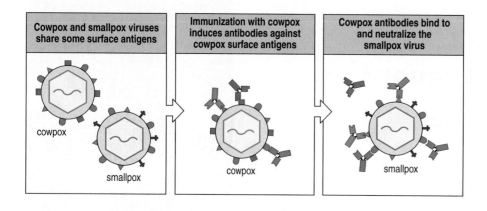

Figure 14.1 Vaccination with cowpox virus elicits neutralizing antibodies that react with antigenic determinants shared with smallpox virus. Shared antigenic determinants of cowpox also elicit protective T-cell immunity against smallpox (not shown here).

to healthy people, either intranasally or intradermally through a scratch on the arm—a procedure known as **variolation**, from the word **variola**, the Latin name given to both the pustule and to the disease itself. Although successful in many cases, the drawback of variolation was the frequency with which it produced full-blown smallpox, resulting in the death of around 1 in 100 of those vaccinated. Despite the risk, variolation was widely used in the eighteenth century because the threat from smallpox was so much greater. At that time smallpox killed one in four of those infected, and there were regular epidemics. In London, for example, more than one-tenth of all deaths were due to smallpox.

Jenner's innovation toward the end of the eighteenth century was to use the related cowpox virus as a vaccine for smallpox. The cowpox virus, called **vaccinia**, causes only very mild infections in humans, but the immunity produced gives effective protection against smallpox as well as cowpox because the two viruses have some antigens in common (Figure 14.1). In the nineteenth century, Jenner's vaccine replaced variolation and was eventually responsible for the eradication of smallpox in the twentieth century. The words vaccinia and vaccination derive from *vaccus*, the Latin word for cow. Although the term vaccination was once used specifically in the context of smallpox, it now refers to any deliberate immunization that induces protective immunity against a disease.

Jenner's strategy for vaccination against smallpox is not possible for most pathogenic viruses because very few have a natural 'safe' counterpart. Most vaccines in use today are composed of preparations of the disease-causing virus in which its ability to cause disease has been destroyed or weakened. One type of vaccine consists of virus particles that have been chemically treated with formalin or physically treated with heat or irradiation so that they are no longer able to replicate. These are called **killed** or **inactivated virus vaccines**. The vaccines for influenza and rabies, and the Salk polio vaccine, are of this type. Only viruses whose nucleic acid can be reliably inactivated make suitable killed virus vaccines. Such vaccines also have the drawback that large amounts of pathogenic virus must be produced during their manufacture.

most pathogenic viruses have NO safe counterpart

A second type of vaccine consists of live virus that has mutated so that it has a reduced ability to grow in human cells and is no longer pathogenic to humans. These vaccines, called **live-attenuated virus vaccines**, are usually more potent at eliciting protective immunity than killed virus vaccines because the attenuated virus can usually replicate to a limited extent and thus mimics a real infection. Most of the viral vaccines currently used to protect humans are live-attenuated vaccines. Attenuated virus is produced by growing the virus in cells of non-human animal species. Such conditions select for variant viruses that grow well in the non-human host and are consequently less fit for growth in humans (Figure 14.2). The measles, mumps, polio (Sabin vaccine), and

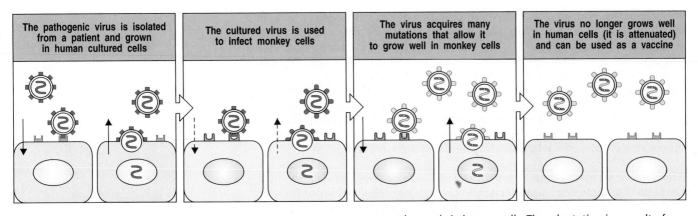

Figure 14.2 **Attenuated viruses are selected by growing human viruses in non-human cells.** To produce an attenuated virus, the virus must first be isolated by growing it in cultured human cells. This in itself can cause some attenuation; the rubella vaccine, for example, was made in this way. In general, however, the virus is then grown in cells of a different species, such as a monkey, until it has become fully adapted to those cells and grows only poorly in human cells. The adaptation is a result of mutation, usually a combination of several point mutations. It is usually hard to tell which of the mutations in the genome of an attenuated viral stock are critical to attenuation. An attenuated virus grows poorly in the human host, where it produces immunity but not disease.

yellow fever vaccines consist of live-attenuated viruses. Attenuated viral strains can also arise naturally. As a viral infection passes though the human population the virus generally diversifies by mutation, sometimes producing a strain with reduced pathogenicity. These strains represent natural forms of live-attenuated virus and also provide candidates for vaccines. One of the three live-attenuated poliovirus strains that make up the oral polio vaccine is of this type (the Sabin strain 2).

In the human immune response to viruses, protective neutralizing antibodies tend to be directed toward particular surface components of the virus. Vaccines against some viruses have been made by using just these antigenic viral components; they are known as **subunit vaccines**. For example, the vaccine for hepatitis B virus (HBV) consists of the hepatitis B surface antigen (HBsAg), with which most anti-HBV antibodies react. The hepatocytes of patients infected with HBV secrete the surface antigen into the blood as minute particles. The first anti-hepatitis B vaccine was made from HBsAg purified from the blood of infected people, a procedure requiring the careful removal of infective viral particles, also known as virions. The vaccine now used is made from HBsAg produced by recombinant DNA technology, which avoids this problem. This vaccine provides about 85% of those vaccinated with protective immunity against HBV infection. People who do not respond may lack an effective CD4 T-cell response because their HLA class II allotypes fail to present HBsAg-derived peptides.

There remain, however, viruses for which no vaccine has yet been developed, despite considerable effort. Some are those, such as HIV, that naturally persist in the body for long periods, evading the efforts of the immune system to dislodge them. Others, like the viruses that cause the common cold, naturally provoke rather weak immune responses and also exist in such a large number of different strains that it is virtually impossible to develop a vaccine that will protect against all of them.

14-2   Bacterial vaccines are made from whole bacteria, their secreted toxins, or capsular polysaccharides

Vaccines against bacterial diseases have been developed with the approaches outlined above for viruses. Despite much research and development, the number of live-attenuated bacterial vaccines remains small. The oldest and most commonly used is the Bacille Calmette–Guérin (BCG) vaccine, which was derived from a bovine strain of *Mycobacterium tuberculosis* and provides protection against tuberculosis. The efficacy of this vaccine varies in different populations and although routinely given to children in some European countries, BCG is not used in the United States. More recently, live-attenuated vaccines against species of *Salmonella* have been introduced for both medical and veterinary use. An attenuated strain of *Salmonella typhi*, the bacterium that causes typhoid fever, was made by mutagenesis and selection for loss of a lipopolysaccharide necessary for pathogenesis. In this vaccine strain, an enzyme necessary for lipopolysaccharide synthesis is defective.

Many bacterial diseases are due to the effects of toxic proteins secreted by the bacteria. Examples are diphtheria, caused by *Corynebacterium diphtheriae*, and tetanus, caused by *Clostridium tetani*. In particular, vaccines against these diseases need to stimulate the generation of toxin-specific neutralizing antibodies (see Section 9-16, p. 270), so that the toxin will be removed before it can have an effect. Such vaccines are made by purifying the respective toxin—**diphtheria toxin** or **tetanus toxin**—and treating it with formalin to destroy its toxic activity. The inactivated proteins, called **toxoids**, retain sufficient antigenic activity to provide protection against disease. Diphtheria and tetanus vaccines are thus comparable to the viral subunit vaccines. In medical practice, tetanus and diphtheria toxoids are combined in a single vaccine with a killed preparation of the bacterium *Bordetella pertussis*, the agent that causes whooping cough. The **combination vaccine** is called DTP, signifying *d*iphtheria, *t*etanus, *p*ertussis; it provides protection against all three diseases. The immune response to the diphtheria and tetanus toxoids is enhanced by the adjuvant effect of the whole pertussis bacteria, which produces a strong inflammatory reaction at the site of injection. In some countries—Japan was the first—the DTP vaccine has been superseded by DTaP vaccines in which the pertussis bacteria are replaced by an acellular pertussis component (aP). This component contains a modified inactivated form of pertussis toxin—pertussis toxoid—and one or more of the other pertussis antigens, such as filamentous hemagglutinin, pertactin, and fimbrial antigen.

Many pathogenic bacteria have an outer capsule composed of polysaccharides that define species-specific and strain-specific antigens. For these encapsulated bacteria, which include the pneumococcus (*Streptococcus pneumoniae*), salmonellae, the meningococcus (*Neisseria meningitidis*), *Haemophilus influenzae*, *Escherichia coli*, *Klebsiella pneumoniae*, and *Bacteroides fragilis*, the capsule determines both the pathogenicity and the antigenicity of the organism. In particular, the capsule prevents the fixation of complement by the alternative pathway. Only when antibodies bind to the capsule does complement fixation lead to bacterial clearance. Consequently, the aim of vaccination against such bacteria is to produce complement-fixing antibodies that bind to the capsule.

Subunit vaccines composed of purified capsular polysaccharides have been developed for some of these bacteria and they are effective in adults, who make T-independent responses against them (see Section 9-3, p. 251). In contrast, these vaccines are ineffective in children under 18 months old, probably because children do not develop good responses to polysaccharide antigens until a few years after birth, and in the elderly. Both these groups are particularly susceptible to infection by encapsulated bacteria. The solution has been to convert the bacterial polysaccharide into a T-dependent antigen. This is done by covalently coupling the polysaccharide to a carrier protein, for example tetanus or diphtheria toxoid, which provides antigenic peptides for stimulating a CD4 T-cell response. On vaccination, T cells responding to the protein carrier can then stimulate polysaccharide-specific B cells to make antibodies. Vaccines of this type are called **conjugate vaccines**. A conjugate vaccine against *H. influenzae* (see Figure 8.38, p. 242) has reduced the incidence and severity of childhood meningitis caused by this bacterium.

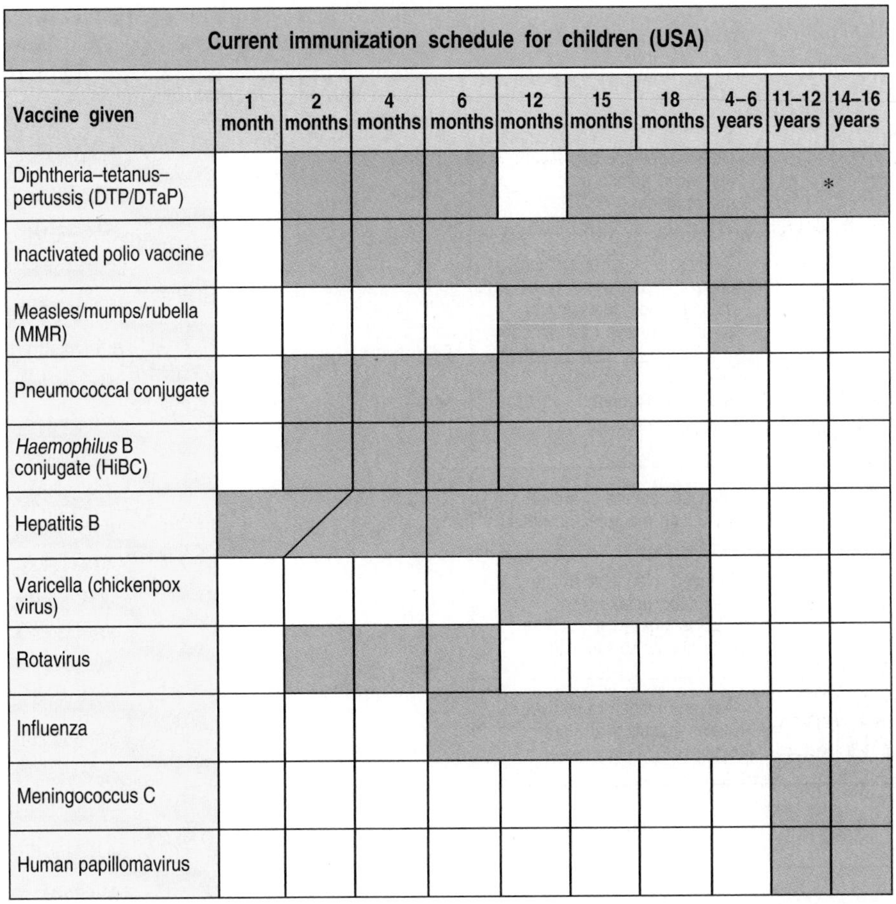

Current immunization schedule for children (USA)										
Vaccine given	1 month	2 months	4 months	6 months	12 months	15 months	18 months	4–6 years	11–12 years	14–16 years
Diphtheria–tetanus–pertussis (DTP/DTaP)		■	■	■		■	■	■		*
Inactivated polio vaccine		■	■	■	■	■	■	■		
Measles/mumps/rubella (MMR)					■	■		■		
Haemophilus B conjugate (HiBC)		■	■	■	■	■				
Pneumococcal conjugate		■	■	■	■	■				
Hepatitis B	■	■		■	■	■	■			
Varicella (chickenpox virus)					■	■	■			
Rotavirus		■	■	■						
Influenza				■	■	■	■	■		
Meningococcus C									■	
Human papillomavirus									■	

Figure 14.3 Childhood vaccination schedules in the United States. Each red bar denotes a time at which a vaccine dose should be given. Bars spanning multiple months indicate a range of times during which the vaccine may be given. DTaP, diphtheria, tetanus, and an acellular pertussis vaccine. *Tetanus and diphtheria toxoids only.

A current schedule for recommended childhood immunizations in the United States is given in Figure 14.3. Multiple immunizations with most vaccines are recommended because they are needed to develop the memory B cells and T cells necessary for long-term protection.

14-3 Adjuvants nonspecifically enhance the immune response

A prerequisite for a good immune response is a state of inflammation. During an infection this is initiated by microbial products that activate the innate immune response (see Chapter 2). To work effectively, vaccination must also create a state of inflammation at the site in the body where the antigens are injected. In general, immunization with purified protein antigens leads to a feeble immune response. Although these proteins are genuine antigens in that they can, in principle, be recognized by the immune system and provoke the formation of specific antibodies, they are not **immunogens**: that is, they cannot stimulate an effective immune response on their own. A response to such antigens can be enhanced by substances that induce inflammation by antigen-independent mechanisms. Such enhancing substances are called adjuvants, a word meaning 'helpers' (Figure 14.4).

In experimental immunology the most effective adjuvant is **Freund's complete adjuvant**, an emulsion of killed mycobacteria and mineral oil into which antigens are vigorously mixed. The physical characteristics of the emulsion cause soluble protein antigens to aggregate and precipitate, forming particles. This prevents the antigen and adjuvant from being rapidly dispersed and degraded throughout the body, and provides a localized and persistent

Adjuvants that enhance immune responses		
Adjuvant name	Composition	Mechanism of action
Freund's incomplete adjuvant	Oil-in-water emulsion	Delayed release of antigen; enhanced uptake by macrophages
Freund's complete adjuvant	Oil-in-water emulsion with dead mycobacteria	Delayed release of antigen; enhanced uptake by macrophages; induction of co-stimulators in macrophages
Freund's adjuvant with MDP	Oil-in-water emulsion with muramyl dipeptide (MDP), a constituent of mycobacteria	Similar to Freund's complete adjuvant
Alum (aluminum hydroxide)	Aluminum hydroxide gel	Delayed release of antigen; enhanced macrophage uptake
Alum plus *Bordetella pertussis*	Aluminum hydroxide gel with killed *B. pertussis*	Delayed release of antigen; enhanced uptake by macrophages; induction of co-stimulators
Immune stimulatory complexes (ISCOMs)	Matrix of lipid micelles containing viral proteins	Delivers antigen to cytosol; allows induction of cytotoxic T cells
MF59	Squalene–oil–water emulsion	Delayed release of antigen

Figure 14.4 Some adjuvants used in experimental immunology. Only two adjuvants—alum and MF59—are licensed for use in vaccines given to humans.

stimulation of the immune system. The particulate nature of the antigen also facilitates its uptake by macrophages and dendritic cells. The biologically active components contributed by the mycobacteria are a glycolipid, *N*-acetyl-muramyl-L-alanine-D-isoglutamine, known for short as muramyl dipeptide, and elements of the bacterial cell-wall skeleton. When antigen and adjuvant are taken up by macrophages and dendritic cells, the cells' NOD2 receptors will recognize the muramyl dipeptide and activate the cells by the NFκB pathway (see Section 10-10, p. 301). This causes the macrophages to create a state of inflammation and the dendritic cells to travel to the draining secondary lymphoid tissue and start an adaptive immune response to the antigen. In this way the adjuvant tricks the immune system into perceiving that the innocuous protein antigen is causing a dangerous infection.

Any microbial product that is detected by a NOD receptor, a Toll-like receptor, or other activating receptors of innate immunity has the potential to act as an adjuvant. For example, the *B. pertussis* component of the DTP vaccine contains bacterial products that serve as adjuvants and enhance the antibody response to the diphtheria and tetanus toxoids as well as to pertussis itself.

A promising refinement to the design of vaccines is to administer cytokines that will bias the immune response to give the helper T cells—T_H1 or T_H2—that best favor a protective immune response to the particular pathogen.

Although many substances and preparations are known to be adjuvants, the only adjuvants approved for use in human vaccines are alum—a form of aluminum hydroxide—and an emulsion of squalene, oil, and water called MF59. Because vaccines are given to large numbers of healthy people, the safety standards for vaccines are high and do not tolerate the side-effects of the most potent adjuvants. Although alum is a safe adjuvant, it is not particularly powerful, especially in stimulating immunity mediated by CD4 T_H1 cells and CD8

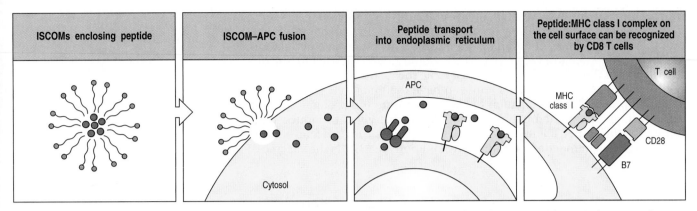

| ISCOMs enclosing peptide | ISCOM–APC fusion | Peptide transport into endoplasmic reticulum | Peptide:MHC class I complex on the cell surface can be recognized by CD8 T cells |

Figure 14.5 ISCOMs (immune stimulatory complexes) can be used to deliver peptides to the MHC class I processing pathway. ISCOMs are lipid micelles that will fuse with cell membranes. Peptides trapped in ISCOMs can be delivered to the cytosol of an antigen-presenting cell (APC), allowing the peptide to be transported into the endoplasmic reticulum, where it can be bound by newly synthesized MHC class I molecules and hence transported to the cell surface as peptide:MHC class I complexes. This is a possible method of delivering vaccine peptides to activate CD8 cytotoxic T cells. ISCOMs can also be used to deliver proteins to the cytosol of other types of cell, where they can be processed and presented as though they were a protein produced by the cell.

T cells. **ISCOMs (immune stimulatory complexes)** are lipid carriers that act as adjuvants but have minimal toxicity. They seem to load peptides and proteins into the cell cytoplasm, allowing class I-restricted T-cell responses to peptides (Figure 14.5). These carriers are used in veterinary medicine but have yet to be approved for use with any human vaccine.

Another important practical aspect of vaccination is the route by which a vaccine is introduced into the human body. Today, most vaccines are given by injection or scarification. These procedures are disliked because of the pain that they cause, and for most pathogens they do not mimic the normal route of infection, which is via mucosal surfaces. Vaccination by the oral or nasal routes would be less traumatic and potentially more effective in stimulating protective immunity. It would also be cheaper, because less skill and time would be required of the person administering the vaccine. That oral vaccines can be effective is shown by the live-attenuated polio vaccine. Moreover, just as the disease-causing poliovirus can be transmitted by the orofecal route, so can the vaccine, for example by fecal contamination of swimming pools.

14-4   Vaccination can inadvertently cause disease

Live-attenuated viruses have made the best vaccines because they challenge the immune system in ways that are most like the natural pathogen. Consequently, the immune system is best prepared for a real infection after vaccination with a live virus vaccine. However, because of their similarity to the pathogen, attenuated viruses can revert to becoming pathogenic. For example, the Sabin polio vaccine, which markedly reduced the incidence of polio in North America and Europe, induces polio and paralysis in three people per million vaccinated. The Sabin vaccine (trivalent oral polio vaccine, TVOP) consists of three different live-attenuated polio strains, all of which are needed to give protection against the natural variants of poliovirus. Of these, strain 3 is the one responsible for causing disease after vaccination. In 2007, the people of northern Nigeria experienced one of the largest recorded outbreaks of polio caused by a vaccine strain.

Strain 3 differs from a natural strain of polio by 10 nucleotide substitutions. Unfortunately, mutation back to the original nucleotide at just one of the substituted positions in strain 3 is sufficient to cause reversion to pathogenicity, and this happens at a low frequency either during preparation of the vaccine (which is carefully monitored) or after vaccination. In contrast, strain 1 differs from natural strains by 57 or more nucleotide substitutions, and reversion to pathogenicity is much less likely because it requires more than one back-mutation. As the incidence of natural polio infection decreases in a population, fear of the side-effects of vaccination can become greater than fear of the disease itself. For this reason there is considerable social pressure to improve the polio vaccine and some parents refuse to have their children vaccinated. A modified protocol for vaccination helps to overcome the problem; a killed polio vaccine (inactivated polio vaccine; IPV) is first given to induce some immunity; this is then followed by the live vaccine. In the United States, TVOP is no longer recommended for routine vaccination, and IPV is the vaccine of choice.

14-5   The need for a vaccine and the demands placed on it change with the prevalence of the disease

In eighteenth-century Europe, the high probability of death or permanent facial scarring from smallpox made the risk of variolation acceptable to those who could afford it. Later, the side-effects of the cowpox vaccine were tolerated while smallpox still posed a threat and terrified the population. This fear persisted into living memory, as is recounted in the eyewitness account on the opposite page by someone who had a brush with smallpox as a child. In the late twentieth century, smallpox was eradicated and vaccination was discontinued. Smallpox vaccination seemed to be an example of a preventive measure that was so successful it put itself out of business. In the early twenty-first century, however, a new fear that smallpox might be used as a biological weapon by terrorists and 'rogue' states has led to a program of renewed vaccine production and vaccina-tion.

Concern for the safety of a vaccine can sometimes lead to a resurgence of disease, as seen for whooping cough in the 1970s. At the beginning of the twentieth century, 1 in 20 children in the USA died from whooping cough. The DTP vaccine containing whole killed *B. pertussis* bacteria was introduced in the 1940s and was routinely given to infants at 3 months of age. The vaccination program produced a hundredfold decline in the annual incidence of whooping cough, from 2000 cases per million to 20 cases per million.

But as people's fear of the disease abated, their concern for the side-effects of the vaccine increased. All children vaccinated with pertussis vaccine develop inflammation at the injection site, and some develop a fever that induces persistent crying. Very rarely, the vaccinated child suffers fits and either a short-lived sleepiness or a transient floppy unresponsive state, all of which cause parental anxiety. In the 1970s, awareness of these established neurological side-effects was heightened by anecdotal reports of vaccination causing encephalitis and permanent brain damage, a connection that has never been conclusively proved. Distrust of pertussis vaccination grew, most notably in Japan.

In Japan, DTP vaccination was introduced in 1947. By 1974 the incidence of pertussis had been reduced by more than 99% and in that year no deaths were ascribed to the disease. In the following year, two children died soon after vaccination, raising fears that the vaccine was the direct cause of the deaths. During the next five years, the number of Japanese children being vaccinated fell from 85% to 15% and, as a consequence, the incidence of whooping cough increased about 20-fold, as did the number of deaths from the disease. Japanese companies then developed vaccines containing antigenic components of pertussis instead of whole bacteria; acellular pertussis vaccines replaced the whole-cell vaccines in 1981 in Japan. By 1989 the incidence of pertussis was again at the very low levels of 1974. Acellular vaccines are now being increasingly used in other countries. These vaccines have a reduced incidence of the common side-effects such as inflammation, pain, and fever.

An eventful voyage from Sydney, Australia, to London, UK, on the SS Mooltan in 1949.

In 1949 my mother was taking myself, then 6 years old, and my younger sister 'home' from Australia on the SS Mooltan to see our English relatives. During our voyage from Sydney to London there was an outbreak of smallpox on board the ship, although we only learned of this after dropping anchor some distance off the English coast.

An announcement was made over the PA system for all passengers to assemble in the lounge, where the captain told us "we need 'health clearance' and that staff from the Immigration and Quarantine Services will be coming aboard." Small boats came alongside, and a parade of equipment-carrying officials embarked and came into the lounge. A doctor then told us that three cases of smallpox had been diagnosed and a crew member had died the previous day of the illness.

Tables were set up in the lounge in a semicircle with doctors and nurses at each. We lined up to be vaccinated. My memory is of a 'cork' with a needle set in its center being heated on a Bunsen burner, cooled and then a rough circle on our upper arms was scratched and criss-crossed just piercing the skin. At the next table a small piece of gauze impregnated with a white powder was placed on the scratches and rubbed in. At the third table a small dressing was applied. In time I had an incredibly painful upper arm at the vaccination site. A huge lobulated blister developed, surrounded by an extensive erythema as far as my elbow and shoulder. My whole upper arm was so grossly swollen that I had to wear a scarf as a sling. Most other passengers had the same reaction. As time went on many had burst blisters with nasty suppurating infections and high temperatures. Moving about was difficult. If someone bumped against the painful arm it could be agony and even getting dressed was difficult. Parents with babies had an especially difficult time.

Over the next few days the story of what had happened became clearer and spread throughout the ship. A Singalese crew member had become sick, but he had been hidden away for several days by his shipmates and his work shifts covered. He died of his illness, but by the time his death was reported several others were also seriously ill with high fever, delirium and the terrible rash. A woman in a cabin along the corridor from us became ill and was taken to the sick bay. She later died. As the days passed more people succumbed. We stayed anchored far from the shore until no new cases had occurred for several days. My mother told me of a terrible fear throughout the ship and of people confined to their tiny cabins except at meal times. I am not sure how long we stayed off-shore, but before coming into port at Tilbury, a huge amount of the ship's stock was dumped into the sea: china, glassware, cutlery, bedding, lounges and curtains were thrown overboard. It seemed no risks were being taken. Altogether 11 people had died and were buried at sea.

On going ashore we were taken straight to my grandparents' house. A quarantine official was assigned to visit us there every day and we were not to leave the house and garden for 10 days. The man who came was a tiny, nervous fellow, who was not the least reassuring to our family. In fact he seemed quite terrified about visiting us. He would ring the front door bell and then stand 3 or 4 metres up the path away from the door to ask if we were all well. My grandparents said they found out who their friends were at this time. Some people were kind and brought food and groceries, leaving them at the front gate and asking if we needed anything else. Others would walk along the street and cross to the other side of the road until well past the house. Luckily no symptoms appeared, and after 2 weeks we emerged to start leading a normal life.

About two decades later as a newly graduated physiotherapist, I was a witness to the last poliomyelitis epidemic in Australia and was involved in a rehabilitation programme for the many, mostly young people who had this dreadful disease. When I hear of opponents of vaccination, I do my best to tell them of my experiences and to urge them to do more research before refusing to vaccinate their children.

Iris Loudon

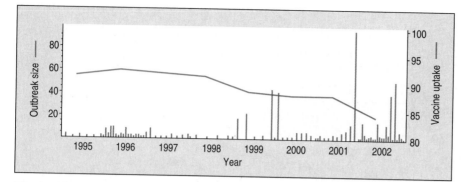

Figure 14.6 As fewer children were vaccinated against measles virus, the incidence of measles infections increased. Data are from the UK population. Since 1998, when the MMR vaccine was introduced, the number of children vaccinated (red line) decreased and outbreaks of measles (blue bars) increased in both frequency and size. The vaccine uptake is the percentage of children completing a primary course of the MMR vaccine at their second birthday. Data courtesy of V.A.A. Jansen and M.E. Ramsey.

The incidence of measles infections in the UK population is currently increasing owing to public distrust of the combined vaccine against the measles, mumps, and rubella viruses (MMR vaccine). Measles is a highly infectious and potentially dangerous disease, which caused around 100 deaths a year in the UK before mass vaccination against measles was started in 1968. The combined MMR vaccine was introduced in 1988 and in the early 1990s 91% of children were being vaccinated against measles. Ten years later it was claimed, on the basis of 12 cases in which children were diagnosed as autistic soon after being vaccinated, that there was a link between autism and the MMR vaccine. Although the link has not been substantiated, distrust of MMR has meant that the proportion of children being vaccinated against measles has steadily decreased, with the result that outbreaks of measles infections in the UK have begun to increase in frequency and size (Figure 14.6).

In 1998 the vast majority of the UK population had protective immunity against measles virus as a result of previous vaccination or infection. In this situation, a population has what is called **herd immunity**, which also indirectly protects the minority of people who have not been vaccinated. The pathogen cannot create a epidemic because of the low probability of finding susceptible individuals and creating a chain of infection. If the proportion of children being immunized against measles continues to decrease, however, the proportion of the UK population with no immunity to measles will become sufficiently large for herd immunity to be lost. An outbreak of measles could then more easily become an epidemic.

14-6 Vaccines have yet to be found for many chronic pathogens

The diseases for which we already have vaccines are ones in which the infection is acute and resolves in a matter of weeks, either by elimination of the pathogen or by the death of the patient. In the absence of vaccination, many of those infected would clear the infection, showing that the human immune system can defeat the invader. Even at its worst, smallpox killed only one-third of those it infected. For such diseases, a vaccine that mimics the pathogen and provokes an immune response similar to that raised against the pathogen itself is likely to provide protective immunity. The diseases for which vaccines are available are listed in Figure 14.7.

In contrast, most of the diseases for which vaccines have been difficult to find are due to chronic infections (Figure 14.8). Such pathogens are adept at evading and subverting the immune system, and so can live for years in a human host (see Chapter 11). Despite considerable research and investment, the approaches to vaccine design that have worked so well for acute bacterial and viral infections have failed to find vaccines for the chronic infections that plague humanity, especially infections with parasites. In most chronic infections there is little evidence that the immune system can clear the infection unaided, and, as a group, pathogens that cause chronic infections divert the immune system into making responses that do not clear the infection.

Available vaccines for infectious diseases in humans			
Bacterial diseases	**Types of vaccine**	**Viral diseases**	**Types of vaccine**
Diphtheria	Toxoid	Yellow fever	Attenuated virus
Tetanus	Toxoid	Measles	Attenuated virus
Pertussis (*Bordetella pertussis*)	Killed bacteria. Subunit vaccine composed of pertussis toxoid and other bacterial antigens	Mumps	Attenuated virus
Paratyphoid fever (*Salmonella paratyphi*)	Killed bacteria	Rubella	Attenuated virus
Typhus fever (*Rickettsia prowazekii*)	Killed bacteria	Polio	Attenuated virus (Sabin) or killed virus (Salk)
Cholera (*Vibrio cholerae*)	Killed bacteria or cell extract	Varicella (chickenpox)	Attenuated virus
Plague (*Yersinia pestis*)	Killed bacteria or cell extract	Influenza	Inactivated virus
Tuberculosis	Attenuated strain of bovine *Mycobacterium tuberculosis* (BCG)	Rabies	Inactivated virus (human). Attenuated virus (dogs and other animals). Recombinant live vaccinia-rabies (animals)
Typhoid fever (*Salmonella typhi*)	Vi polysaccharide subunit vaccines. Live-attenuated oral vaccine	Hepatitis A	Subunit vaccine (recombinant hepatitis antigen)
Meningitis (*Neisseria meningitidis*)	Purified capsular polysaccharide	Hepatitis B	Subunit vaccine (recombinant hepatitis antigen)
Bacterial pneumonia (*Streptococcus pneumoniae*)	Purified capsular polysaccharide	Human papillomavirus	Subunit vaccine (virus coat proteins)
Meningitis (*Haemophilus influenzae*)	*H. influenzae* polysaccharide conjugated to protein	Rotavirus	Attenuated virus Recombinant live virus

Figure 14.7 Diseases for which vaccines are available. Not all of these vaccines are equally effective, and not all are in routine use.

Consequently, successful vaccines against these diseases must stimulate different immune responses from those resulting from most natural exposures to the pathogen.

For many chronic infections, a minority of people who are exposed to the pathogen become resistant to long-term infection. A study of their resistance should reveal examples of successful immune responses that might be emulated by vaccination. For example, of people infected with the hepatitis C virus, less than 30% clear the infection quickly, whereas the majority develop a chronic infection in which the liver goes through cycles of destruction and regeneration (Figure 14.9). Vaccine design for hepatitis C virus could be helped by a comparison of the unsuccessful immunity in patients with chronic hepatitis with the successful immunity in people who clear hepatitis C infection. This, however, is more easily said than done. Studying people who make unsuccessful responses is relatively easy, because they become sick and seek medical help. In contrast, people who successfully repel the virus are healthy, do not show up in a doctor's office, and are consequently difficult to identify. As a result, knowledge of the immune response against pathogens causing chronic infections encompasses mainly the failures of the immune system and not its successes.

14-7 Genome sequences of human pathogens open up new avenues of vaccine design

Complete genome DNA sequences for an increasing number and variety of human pathogens have been determined. In the past 10 years, 140 bacterial

Some diseases for which effective vaccines are not yet available		
Disease	Estimated annual mortality	Estimated annual incidence
Malaria	1.1 million	300–500 million
Schistosomiasis	15,000	No numbers available
Worm infestation	12,000	No numbers available
Tuberculosis	1.6 million	8 million
Diarrheal disease	1.8 million	4–5 billion
Respiratory disease	3.9 million	~360 million
HIV/AIDS	2.7 million	5 million
Measles*	611,000	30–40 million
Hepatitis C	46,000	~170 million [†]

Figure 14.8 Diseases for which better vaccines are needed. *The measles vaccines currently used are effective but they are heat sensitive and require carefully controlled refrigeration, as well as reconstitution, before use. In some tropical countries this reduces their usefulness. †This figure is the number of people worldwide estimated by the World Health Organization to be chronically infected with the hepatitis C virus. Annual mortality is difficult to estimate because hepatitis C is associated with chronic conditions such as cirrhosis of the liver and liver cancer, from which death eventually results. Estimated mortality data for 2002 are from *World Health Report 2004* (World Health Organization).

genomes and 1600 viral genomes have been sequenced, many of them from human pathogens. Knowledge of the genetic blueprint for a pathogen provides a foundation for defining the biology of its pathogenesis and its interactions with the human immune system. With this knowledge, direct manipulation of a pathogen's genes can be designed to produce attenuated strains having the properties required of a vaccine (Figure 14.10). Such an approach should enable strain 3 of the Sabin polio vaccine to be mutated in ways that reduce its reversion frequency.

Cloned genes from pathogens can be safely expressed in bacterial or other cell cultures to produce subunit vaccines, thus avoiding the handling of large amounts of infectious agents or material purified from infected blood, as was used for the first hepatitis B vaccine. The ease with which genes can be moved from one organism to another opens up the possibility of using current bacterial and viral vaccine strains as vehicles for antigens that could stimulate protective immunity against other pathogens. Natural pathogens have acquired many mechanisms for evading and escaping the immune system; the removal of these functions could also generate improved strains for vaccination. In a complementary fashion, microbial genes encoding adjuvants could be incorporated into vaccine strains of other organisms.

The ideal starting point for vaccine design would be a comprehensive knowledge of how the human immune system responds to the particular infection and which mechanisms enable the pathogen to be quickly cleared from the body. With this understanding, candidate vaccines could be designed to stimulate the clones of B and T cells that make up a successful immune response. Most of the traditional vaccines have seemed to work by the induction of protective antibodies and, until recently, vaccines have been evaluated purely in those terms. An understanding of the importance of T-cell responses in protective immunity has stimulated the study of vaccines containing peptide epitopes that bind MHC molecules and thus can stimulate T cells specific for the pathogen. Although this approach has been shown to work, its application is complicated by the diversity of human HLA types and the peptides they are able to present.

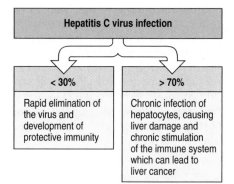

Figure 14.9 A minority of people exposed to hepatitis C virus resist the virus, whereas the majority develop chronic infection.

The success of vaccines against chronic infections will depend critically on their ability to steer the CD4 T-cell response in ways that are helpful. In natural infections the bias toward either a CD4 T_H1 or T_H2 response can determine whether an infection resolves promptly or degenerates into chronic disease (see Section 8-10, p. 227). The inclusion of cytokines in vaccines might help to drive immunity in the desired direction.

14-8   A useful vaccine against HIV has yet to be found

Since HIV was discovered in 1983 there have been intensive efforts in several countries to make a vaccine against it. The ideal vaccine would be one that induced neutralizing antibodies that bind tightly to the outside surface of HIV and prevent it from infecting human cells. Because good neutralizing antibodies against other viruses bind to glycoproteins of the viral capsid, initial HIV vaccines were targeted to make antibodies against gp120, the viral capsid protein that binds to CD4 and facilitates infection (see Section 11-19, p. 352). Such a vaccine was tested in 5000 volunteers whose daily lives put them at risk of exposure to and infection by HIV. The result of this trial, which was announced in 2003, showed that vaccination gave no protection. This modern example shows how unpredictable and difficult vaccine development can be. After this failure the goals have been lowered from preventing infection to enhancing the immune response once HIV infection has occurred. Various strategies seek to bolster the responses of NK cells, T cells, and B cells to HIV-infected cells. The target antigens need not be confined to virally encoded proteins, because HIV infection profoundly affects the expression of numerous human genes (see Section 11-18, p. 352), including those of endogenous retroviruses.

In the development of vaccines against polio, measles, and other viruses, it was known from the beginning that infected humans had the capacity to recover from acute infection with the acquisition of protective immunity. For HIV there is no documented case of a person who has recovered from acute infection and for whom the protective immunity can be analyzed and set as the goal to be attained through vaccination. Thus, the types of immune effectors that can terminate HIV infection and provide an immunological memory are unknown.

From the study of antibodies in the serum of HIV-infected people, several monoclonal antibodies have been made that can neutralize a range of HIV strains in experimental situations. The challenge is to find an immunogen that will stimulate the production of such antibodies when humans are vaccinated. A further complication is that although neutralizing antibodies are made in HIV-infected people, the virus readily mutates and escapes the antibodies.

Because HIV-infected individuals make more effective T-cell responses than antibody responses, much vaccination effort is now directed at the induction of T-cell responses. This approach is based on the assumption that effective CD8 and CD4 T cells can terminate HIV infection at an early stage, or keep it under control and at a low level. Evidence to support this idea comes from the approximately 5% of people who remain uninfected despite being frequently exposed to HIV. These people make a T-cell response to HIV but not an antibody response.

Although more than 50 different HIV vaccines have been tested in clinical trials, none of these has proved to be an effective approach to HIV vaccination. The most recent failure, announced in September 2007, was of a widely touted vaccine designed to stimulate cytotoxic T-cell responses against peptide antigens of the HIV proteins Gag, Nef and Pol (see Figure 11.21, p. 353). The vaccine was given to thousands of healthy uninfected people who were at risk of

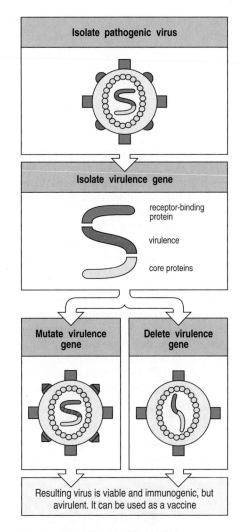

Figure 14.10 Production of live-attenuated viral strains by recombinant DNA techniques. If a viral gene that is necessary for virulence but not for growth or immunogenicity can be identified, this gene can be either mutated (left lower panel) or deleted (right lower panel) by using recombinant DNA techniques. This procedure creates an avirulent (nonpathogenic) virus that can be used as a vaccine. The mutations made in the virulence gene are usually many, so that the virus cannot easily revert to the wild type.

being infected with HIV, but it proved no better than a placebo in reducing either the incidence of HIV infection or the amount of virus in the blood. More worrying was the suggestion in the data that the vaccine made some individuals more susceptible to HIV infection.

14-9 An effective and acceptable rotavirus vaccine has been developed

Rotavirus, which has a wheel-like appearance in the electron microscope, was discovered in 1973 and shown to be a major cause of severe childhood diarrhea. Almost all children are exposed to rotavirus, and the illness it causes is responsible for more than 2 million visits to the hospital and 610,000 deaths per year worldwide (Figure 14.11). Efforts to make a vaccine commenced soon after the discovery of rotavirus, but it is only now, more than 30 years later, that effective and acceptable vaccines are finally available, and there are two of them. Along the way were promising vaccines that proved too weak, or too strong, or too likely to produce unacceptable side-effects.

The rotavirus genome consists of 11 molecules of double-stranded RNA. VP4 and VP7 are coat proteins that are the major targets for neutralizing antibodies. Rotaviruses undergo frequent mutation, and so VP4 and VP7 are variable proteins and generate different serotypes of the virus, which further diversify by reassortment of the different segments of the RNA genome, as in the influenza virus (see Section 11-2, p. 331). Although there are 42 naturally occurring rotavirus variants, only five of them account for 90% of disease and both vaccines combine antigens from common strains. One vaccine, called Rotarix, comprises an attenuated human virus with common VP4 and VP7 variants. The other, called RotaTeq, is in the Jennerian tradition and contains a mixture of five cattle rotavirus strains, which do not cause disease in humans, each of which has been engineered to express a different common human VP4 or VP7 glycoprotein (Figure 14.12). These vaccines give 85–98% protection against severe rotavirus diarrhea. Like the oral poliovirus vaccine they can be made without sophisticated biotechnology by simple growth in tissue culture, and they are readily transferable to the poorer countries, where most mortality from rotavirus infections occurs (see Figure 14.11).

14-10 Vaccine development faces greater public scrutiny than drug development

Two big issues make the development and success of a vaccine inherently more difficult than that of a drug. The first is that drugs are given to people who are already sick and grateful for any improvement in their condition, whereas vaccines are administered to healthy individuals, often the very young, for whom the adverse affects of the procedure are more likely to be uppermost in their parents' minds than any perceived benefit. The second is that drugs are administered on a case-by-case basis by the individual physician to the individual patient, whereas vaccination is usually part of a larger program that is mandated by governments and other higher authorities and applied to the population at large, as exemplified by the worldwide vaccination programs aimed at eradicating smallpox and polio. If an individual has had a painful strep throat for two weeks and is given a course of antibiotics, they will feel relief within days and give thanks for the benefit of such drugs. On the other hand, if that same individual has been immunized against polio and never catches the paralytic disease that made it notorious, they would never know if they had been spared because of protective immunity or because they were never exposed to poliovirus. The only situation in which a vaccine's benefit can truly be felt is when exposure is obvious, because other unvaccinated members of the local community are suffering and succumbing in numbers to the disease.

| Annual childhood mortality from rotavirus infection ||
Country	Annual deaths
India	146,000
Nigeria	47,500
China	41,000
Pakistan	36,000
Ethiopia	29,000
Congo	29,000
Bangladesh	19,000
Afghanistan	18,000
Indonesia	15,000
Tanzania	11,000

Figure 14.11 Childhood mortality from rotavirus infection in the 10 most affected countries. Data courtesy of the Centers for Disease Control and Prevention.

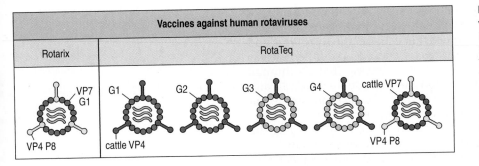

Figure 14.12 Two effective rotavirus vaccines. The Rotarix vaccine (left panel) is an attenuated human rotavirus that has common P8 and G1 variants of the VP4 and VP7 glycoproteins, respectively. The RotaTeq vaccine is based on a cattle rotavirus that is non-pathogenic to humans. The vaccine consists of five viral strains, four expressing different human VP7 variants (G1, G2, G3, and G4) and cattle VP4, and one expressing cattle VP7 and the human P8 variant of VP4. Cattle components are in purple.

All drugs and all vaccines have some unwanted side-effects, and the risk of these side-effects has to be weighed against the benefits of protective immunity. Because the side-effects of a vaccine cause disease in previously healthy children, the standards set for vaccines are always higher than those set for drugs. It has also meant that throughout the modern era of vaccination, there have been episodes of unsubstantiated attribution of side-effects to various vaccines, as recently occurred for the MMR vaccine and autism (see Section 14-5). Because of the nature of herd immunity, individuals can gain the benefit of mass vaccination programs without sharing the risk. This again is another difference from drugs, for which the cost and the benefit are inseparable. However, if enough individuals take this selfish strategy then herd immunity is lost, and in the ensuing epidemic they too will lose out.

Although vaccines are arguably the most successful of medical interventions, their acceptance and use has been greatly influenced by factors having nothing to do with science or medicine. For example, an early version of a rotavirus vaccine was abandoned in 1999 because of its side-effect, a bowel obstruction that if not treated with an enema or surgery could be fatal, which reportedly occurred in 1 out of 7000 children vaccinated. Although the decision to abandon the virus can be justified for the population of the United States, where only 1 in 100,000 deaths is due to rotavirus infection, the vaccine would still have greatly benefited the populations of other countries, such as India, where as many as 1 in 200 children were dying of rotavirus infection. Thus the decision to discontinue rotavirus vaccination in some countries led to many more childhood deaths from rotavirus infection than lives saved by avoiding the vaccine's side-effect.

Because of the costs and difficulties of vaccine development and the political and legal complications associated with their side-effects, both real and imaginary, commercial vaccine manufacture has been a declining industry for 40 years. For example, since 1967 the number of vaccine manufacturers in the United States declined from 26 to 5. With the emergence of new diseases such as HIV and SARS, and with governmental concerns that old foes such as smallpox will be resuscitated as biological weapons, the situation is changing. Vaccines are now busily being manufactured and stockpiled.

Summary to Chapter 14

Vaccination is preventive medicine. It involves the deliberate immunization of healthy people with some form of a pathogen or its component antigens and induces a protective immunity that prevents any infection with the pathogen from causing disease. Vaccines can consist of killed whole pathogens, live-attenuated strains, nonpathogenic species related to the pathogen, or secreted and surface macromolecules of the pathogen. Vaccination has saved millions of lives and reduced the incidence of many common infectious diseases, particularly in the industrialized countries. The prevention of infectious disease by vaccination illustrates superbly how manipulation of the immune response can benefit public health. The pathogens for which effective vaccines have been found are those that cause acute infections and are not highly mutable. When these pathogens caused epidemic disease, many people survived and developed immunity, showing that the human immune system could respond to the infection in a productive way. Death, when it occurred, was because the immune system was just too slow. What vaccination achieved was to start the immune response ahead of the infection, giving the immune system an edge on the pathogen. By reducing the incidence of disease, successful vaccination programs have inevitably led to decreasing public awareness of the effects of disease and increasing concern with the safety of vaccination.

Vaccine development has largely been a process of trial and error, one in which a knowledge of immunological mechanisms played little part and the guiding principle was for the vaccine to resemble the natural pathogen as closely as possible. Although this approach has worked for pathogens causing acute infections, it has failed to produce vaccines against pathogens that establish chronic infections and cause chronic disease. For such diseases we know little about what constitutes a successful immune response. Against some of these pathogens the human immune response is uniformly unsuccessful; for others a minority of people seem to be able to clear the infection soon after infection; and for yet others, some immunity does develop but what it consists of, and when and how it is produced, are still unclear. A greater understanding of the successful immune responses against such pathogens should help in the development of vaccines against them. These pathogens are adept at fooling the immune system, and effective vaccines might need to push the immune system to respond in ways that are different from those seen in most natural infections with the pathogen.

Questions

14–1 Differentiate between the following types of vaccine and give an example of each: (A) inactivated virus vaccines; (B) live-attenuated virus vaccines; (C) subunit vaccines; (D) toxoid vaccines; and (E) conjugate vaccines.

14–2
  A. "An antigen is not necessarily an immunogen." Explain this statement.
  B. Explain why adjuvants are used in experimental immunology.
  C. Which adjuvants are used for human vaccination?

14–3 What risks are associated with live-attenuated virus vaccines?

14–4 Bacterial vaccines differ from viral vaccines in that only in bacterial vaccines are _____ used. (Select all that apply.)
  a. subunit components
  b. toxoids
  c. whole infectious components
  d. capsular polysaccharides
  e. capsule:carrier protein conjugates.

14–5 Reasons complicating the development of vaccines to combat chronic diseases include _____. (Select all that apply.)
  a. evasion of the host's immune system by the pathogen

  b. the polymorphic diversity of MHC class I and class II molecules
  c. the generation of inappropriate immune responses that do not eradicate the pathogen
  d. survival of the infectious agent for long periods inside the host
  e. high mutation rates in the pathogen.

14–6
  A. What is the risk to a population that reduces its use of particular vaccines over a period of time?
  B. Identify two cases in which this has happened and the underlying reason for distrust in the benefit of the vaccine.

14–7
  A. Explain the difference between the Rotarix and the RotaTeq vaccines used to protect against rotavirus infections.
  B. Which vaccine provides broader protection?
  C. Why is this important?

14–8 Why is determining the genome sequences of human pathogens important in the development of new and more effective vaccines?

14–9 Madison Tavistock, a healthy 2-year-old living in Cincinnati, had attended the Wee Folks Daycare Facility for a year. Her parents joined an anti-vaccination group when she was 9 months old, 3 months before the recommended immunization schedule for the MMR vaccine, but after Madison had already been vaccinated with DPT. They strongly believe that the unsubstantiated risk of autism reported for MMR vaccination outweighs the benefits, and consequently have opted not to have their daughter immunized. The best explanation for why Madison has not contracted measles even though she has regular contact with other children in a large city is that _____.

a. Madison is tolerant to measles antigens.
b. The measles virus may have infected Madison, but is dormant.
c. The DPT vaccine provides cross-protective immunity against measles.
d. The other children in the day care center have been vaccinated and she has herd immunity.
e. The attenuated measles virus in the MMR vaccine received by the other children in her daycare facility was transmitted to Madison and she developed asymptomatic natural immunity.

14–10 Jenny O'Mara was five months pregnant when she stepped on a rusty piece of scrap metal while hauling rotted wood from a dilapidated shed in her garden. The sliver of metal cut through her sneaker and pierced her heel deeply. Her physician gave her a tetanus booster. When Jenny's baby was born she decided to breastfeed. If the baby's antibodies were tested for specificity to tetanus 2 months after birth, what would be the expected finding?

a. the presence of anti-tetanus toxoid IgA antibodies
b. the presence of anti-tetanus toxoid IgM antibodies
c. the presence of anti-tetanus toxoid IgG antibodies
d. the presence of IgM antibody specific for *Clostridium tetani* cell-wall components
e. the presence of IgG antibody specific for *Clostridium tetani* cell-wall components.

14–11 On an otherwise uneventful sunny Sunday afternoon, an extremist group enters your city in a large van and drives to the front entrance of the Convention Center where the annual flower show is taking place. The occupants unload large crates resembling flats of assorted flowers, and then drive off. Within minutes the crates explode, showering the visitors with an opaque powder. Medical teams are called to the scene to care for the injured, and CDC officials wearing level-4 containment suits arrive in a few hours to test the contents of the powder for human pathogens using multiplex PCR methodology (a rapid method for identifying pathogens by their DNA). Which of the following potential bioterrorism agents would pose the most serious threat to those exposed?

a. *Bacillus anthracis* (anthrax)
b. *Corynebacterium diphtheriae* toxin (diphtheria)
c. *Yersinia pestis* (plague)
d. variola major (smallpox)
e. *Clostridium botulinum* toxin (botulism).

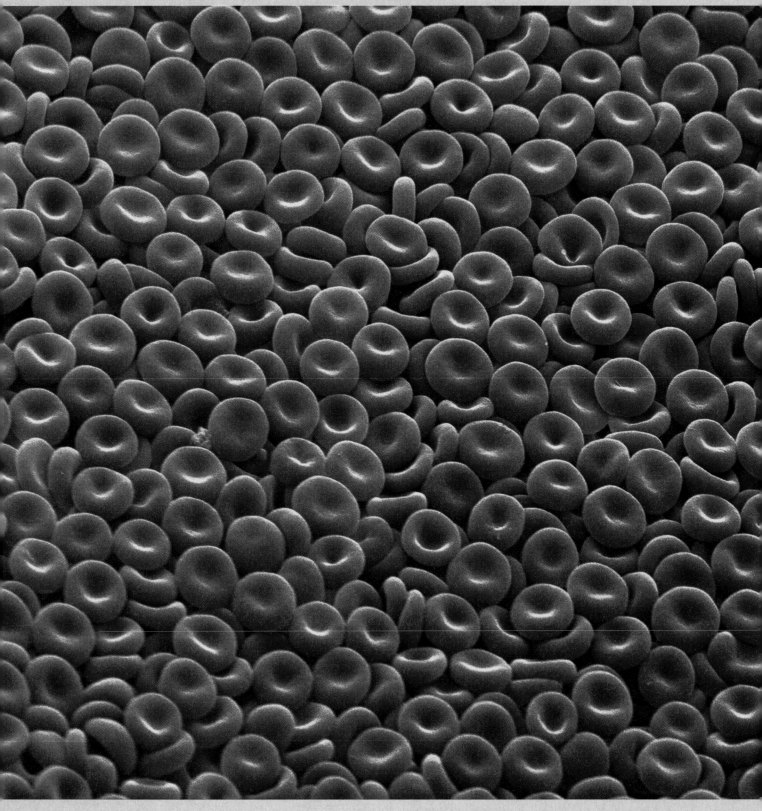

Red blood cells are the most commonly transplanted type of human cell or tissue.

Chapter 15

Transplantation of Tissues and Organs

The replacement of diseased, damaged or worn-out tissue was for centuries a dream of the medical profession. Achieving the reality required the solution of three basic problems. First, transplants must be introduced in ways that allow them to perform their normal functions. Second, the health of both the recipient and the transplant must be maintained during surgery and the other procedures used in transplantation. Third, the immune system of the patient must be prevented from developing detrimental adaptive immune responses that can result in rejection of the transplanted tissue, destruction of healthy tissues and other complications.

During the past 60 years, solutions to these problems have been found, and transplantation has progressed from being an experimental procedure to the treatment of choice for a variety of conditions. In clinical practice, selective suppression of the response to the transplanted tissue has yet to be achieved and so nonspecific suppression is accomplished by using a variety of drugs and antibodies. In contrast to vaccination, which selectively stimulates immunity to a particular pathogen, transplantation involves manipulations that cause widespread inactivation of the immune response.

Two introductory sections in this chapter describe how the tissues and organs divide into three groups according to the type and severity of the immunological barriers encountered in transplantation. The pioneering transplantation procedure was blood transfusion, followed by the transplantation of solid organs, notably the kidney, and lastly the transplantation of hematopoietic stem cells in bone marrow. After these introductory sections, the chapter is divided into two parts: the first examines solid organ transplantation and the second hematopoietic stem cell transplantation.

15-1 Transplant rejection and graft-versus-host reaction are immune responses caused by genetic differences between transplant donors and recipients

Immune responses against transplanted tissues or organs are caused by genetic differences between donor and recipient, the most important of which are antigenic differences in the highly polymorphic HLA class I and class II molecules. This is why these antigens are generally known as **transplantation antigens** or the major histocompatibility antigens (**histocompatibility** means tissue compatibility). The complex of genes that encodes the transplantation antigens has the general name of the major histocompatibility complex

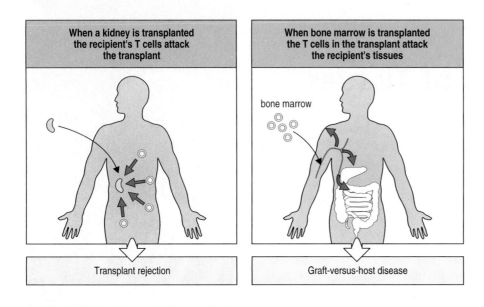

Figure 15.1 Alloreactions in transplant rejection and graft-versus-host reaction. As shown in the left panel, rejection of a transplanted organ occurs when the recipient makes an immune response against it. Graft-versus-host disease occurs when T cells in transplanted bone marrow attack the tissues of the recipient or host, principally the skin, liver, and intestines, as shown in the right panel.

(MHC), although it goes under different names in different species (for example, the HLA complex in humans). Antigens such as these, which vary between members of the same species, are known as **alloantigens**; the immune responses that they provoke are known as alloreactions (see Section 5-22, p. 153). A subfield of immunology, called **immunogenetics**, is devoted to the genetics of alloantigens.

In clinical transplantation, two distinct types of alloreaction can occur depending on the type of tissue transplanted. After transplantation of solid organs, such as kidney and heart, alloreactions developed by the recipient's immune system are directed at the cells of the graft and can kill them, a process called **transplant rejection**. A different situation pertains in bone marrow transplantation, in which the recipient's immune system is destroyed and eventually replaced with one reconstituted from the hematopoietic stem cells in the bone marrow graft (see Section 11-17, p. 349). Here the major type of alloreaction arises from mature T cells in the grafted bone marrow that attack and reject the recipient's healthy tissue. This type of alloreactive response by donor lymphocytes is called a **graft-versus-host reaction** (**GVHR**). It causes **graft-versus-host disease** (**GVHD**), which, with varying degrees of severity, affects almost all patients who have a bone marrow transplant (Figure 15.1).

Because people do not normally make an immune response to their own tissues, tissues transplanted from one site to another on the same person are not rejected. This type of transplant, called an **autograft**, is used to treat patients who have suffered burns. Skin from unaffected parts of the body is grafted into the burnt areas, where it facilitates wound healing. Immunogenetic differences are also avoided when tissue is transplanted between identical twins. The first successful kidney transplant, in 1954, involved the donation of a kidney by a healthy twin to his brother, who was suffering from kidney failure. A transplant between genetically identical individuals is called a **syngeneic** transplant or an **isograft**. A transplant made between two genetically different individuals is called an **allograft** or an **allogeneic** transplant.

15-2 Blood transfusion is the most widespread kind of transplantation in clinical medicine

The genetic barriers to transplanting blood are fewer than for most other tissues and, in 1812, blood transfusion was the first transplant to save a life. Blood transfusion is the most common clinical transplantation procedure, being given to one in four people at some time during their lives. It is used when trauma, surgery, childbirth, or disease causes severe blood loss and patients need the immediate replacement of fluid, proteins, and cells. Today, donated blood is commonly separated into erythrocytes, plasma, and platelets, and transfusions are made of only those components that a patient needs. Plasma replaces fluid and prevents bleeding, erythrocytes improve respiration and metabolism, and platelets facilitate clotting to prevent bleeding. Transfused blood components are usually needed only in the short term, because within a few weeks the patient's bone marrow will make up the loss. This demand is less stringent than that placed on transplanted organs such as hearts or kidneys, which need to function for years.

Human erythrocytes lack HLA class I and II molecules, which are therefore of little consequence in blood transfusion. However, blood transfusion can give rise to life-threatening type II hypersensitivity reactions, as described in previous chapters for the ABO (see Section 12-18, p. 387) and rhesus (see Section 10-18, p. 309) red-cell alloantigens. Rhesus D is the main alloantigen in the latter system. The problems occur when the transfused patient has preformed antibodies that bind to alloantigens on the transfused erythrocytes and cause their lysis. To avoid such reactions, patients and donors are serologically typed for the erythrocyte alloantigens and matched according to the compatibilities shown in Figure 15.2. And before transfusion of any unit of blood, a cross-match test determines whether the patient has antibodies that react with the donor's red cells. In this way the transfusion of blood that is compatible for ABO and rhesus but reactive with other antibodies in the patient's circulation is also avoided.

Although serological typing for ABO and rhesus antigens allows most transfusion hypersensitivity reactions to be avoided, there are numerous other polymorphic antigens on the red-cell surface. Antibodies against these antigens can be stimulated by blood transfusion and are more frequent in patients who have received multiple blood transfusions. Some of these patients have made so many different alloantibodies that it becomes difficult to find any donor to whom they are not reactive. Ongoing trials are examining the value of using DNA-based methods to type blood donors for the polymorphisms of all 31 genes that determine red-cell alloantigens. In principle, this could lead to more precise and selective matching of blood donors to recipients.

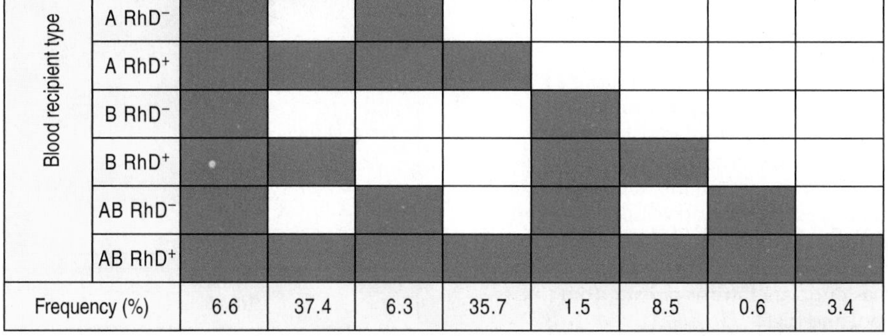

Figure 15.2 Matching the ABO and rhesus erythrocyte antigens in blood transfusion. Although there are many different red-cell antigens, the important ones in clinical practice are A, B, and O and the rhesus D antigen (RhD). O does not represent a specific antigen but the absence of both the A and B antigens. The green squares show the combinations of blood donor and recipient that permit successful transfusion. O RhD⁻ people lack all three antigens (A, B, and rhesus D) and are 'universal donors' who can provide a transfusion to any human being but can only receive blood from other O RhD⁻ donors. In contrast, AB Rh⁺ people have all three antigens and are 'universal recipients' who can receive blood from any donor but can only donate to other AB Rh⁺ donors. The figure's bottom line gives the frequency of the eight blood types in the US population.

Transplantation of solid organs

Throughout much of the world, organ transplantation is now a routine clinical procedure that saves and extends many thousands of human lives. This part of the chapter examines the immunological mechanisms that can cause the rejection of organ transplants and the genetic strategies, immunosuppressive drugs, and other therapies that are used to avoid, prevent, and treat rejection. Emphasis will be on the kidney because it is frequently transplanted and is relatively vulnerable to rejection.

15-3   Antibodies against ABO or HLA antigens cause hyperacute rejection of transplanted organs

ABO antigens are also expressed on the endothelial cells of blood vessels, and this is an important factor in the transplantation of solid organs such as kidneys. For example, were a type O recipient to receive a kidney graft from a type A donor, then anti-A antibodies in the recipient's circulation would quickly bind to blood vessels throughout the graft. By fixing complement throughout the vasculature of the graft, the antibodies would produce a very rapid rejection of the graft (Figure 15.3). This type of rejection, called **hyperacute rejection**, can occur even before a transplanted patient has left the operating room. Hyperacute rejection is the most devastating form of rejection for organ grafts. It is directly comparable to type III hypersensitivity reactions (see Section 12-19, p. 389), in which immune-complex deposition causes complement activation within blood vessel walls. To avoid hyperacute rejection, transplant donors and recipients are typed and cross-matched for the ABO blood group antigens.

Because HLA class I molecules are expressed constitutively on vascular endothelium, preexisting antibodies against HLA class I polymorphisms can also cause hyperacute rejection. It is therefore essential that transplant recipients do not have antibodies that bind to the HLA class I allotypes of the transplanted organ. Antibodies against HLA class II contribute to a lesser extent to hyperacute rejection. HLA class II molecules are not normally expressed on endothelium but are induced by infection, inflammation, or trauma, all of which occur during transplantation.

No reliable method of reversing hyperacute rejection has been found, so this type of rejection is avoided by choosing compatible transplant donors and recipients. Compatibility is assessed with a **cross-match test**, in which blood serum from the prospective recipient is assessed for the presence of antibodies that bind to the white blood cells of the prospective donor. The traditional cross-match test detects antibodies in the patient's serum that trigger complement-mediated lysis of the donor's lymphocytes. The assay is usually performed on separated B cells and T cells so that reactivities due to antibodies against HLA class I and II molecules can be distinguished: anti-HLA class I antibodies react with both B cells and T cells, whereas antibodies against HLA class II react only with B cells. A more sensitive cross-match assay uses flow

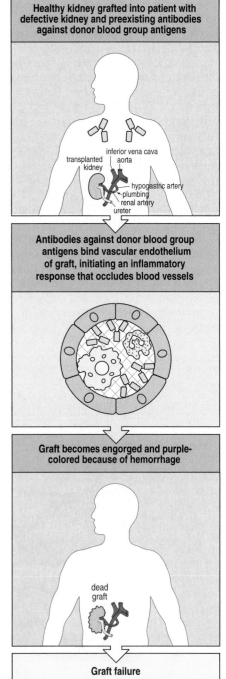

Figure 15.3 Hyperacute rejection is caused by preexisting antibodies binding to the graft. Before transplantation, some recipients have made antibodies that react with donor ABO or HLA class I antigens. When the donor organ is grafted into such a recipient, the antibodies immediately bind to vascular endothelium, initiating the complement and clotting cascades. Blood vessels in the graft become obstructed by clots and leak, causing hemorrhage of blood into the graft. The graft becomes engorged, turns purple from the presence of deoxygenated blood, and dies.

Figure 15.4 Pregnancy is the one natural situation that leads to the production of anti-HLA antibodies. With very few exceptions, the mother and father in human families have different HLA types (top panel). When the mother becomes pregnant she carries for 9 months a fetus that expresses one HLA haplotype of maternal origin (pink) and one HLA haplotype of paternal origin (blue) (center panel). Although the paternal HLA class I and II molecules expressed by the fetus are alloantigens against which the mother's immune system has the potential to respond, the fetus does not provoke such a response during pregnancy and is protected from preexisting alloreactive antibodies or T cells. The trauma associated with childbirth, which physically separates the mother, the child, and the placenta, allows fetal cells and antigens to enter the maternal circulation and stimulate an adaptive immune response to the paternally inherited HLA molecules (bottom panel).

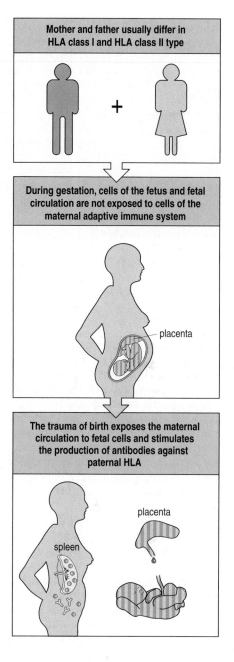

cytometry to examine the binding of the patient's antibodies to the prospective donor's lymphocytes (see Figure 4.14, p. 103). This allows all antibody isotypes to be detected, not just those that fix complement.

15-4 Anti-HLA antibodies can arise from pregnancy, blood transfusion, or previous transplants

In some cases, prospective transplant patients have already made anti-HLA class I and II antibodies. The natural situation in which this happens is pregnancy. In almost all pregnancies the fetus expresses paternal HLA isoforms that are not part of the mother's HLA type and have the potential to stimulate an alloreactive immune response. During gestation the anatomy of the placenta segregates the fetal and maternal circulations and prevents the mother's B and T cells from being stimulated by fetal alloantigens (Figure 15.4). During the trauma associated with birth, when the mother, child and placenta are forcibly separated, cells and material of fetal origin enter the maternal circulation and stimulate an immune response. Antibodies can be made against any paternal HLA allotype expressed by the baby and not carried by the mother. With successive pregnancies, increasing levels of anti-HLA antibodies can develop, and multiparous women are the main source of the anti-HLA sera used in serological HLA typing. The presence of anti-HLA antibodies in the maternal circulation has no detrimental effect on subsequent pregnancies, but will complicate any future search for a compatible organ transplant should it be needed.

Blood transfusions can also lead to the production of anti-HLA antibodies. In routine blood transfusion, HLA type is not assessed and so although the donor and recipient are matched for ABO type they are not matched for HLA type. The infusion of HLA-incompatible leukocytes and platelets in a blood transfusion can therefore generate antibodies specific for the donor HLA allotypes. Patients who have had multiple blood transfusions have been stimulated by many HLA allotypes and can develop antibodies that react with the cells of most other people in the population. The degree to which a patient seeking a transplant has been sensitized to potential donors is assessed by testing their sera against a representative panel of individuals from the population and expressing the number of positive reactions as a percentage **panel reactive antibody** (**PRA**). The higher the value of a patient's PRA, the more difficult it is to find a suitable transplant donor.

A third way in which patients develop anti-HLA antibodies is from a previous organ transplant. Now that transplantation has been routine practice for more than 35 years, many patients have had more than one transplant. As with blood transfusions, the more transplants a person has had, the greater their percentage PRA tends to be.

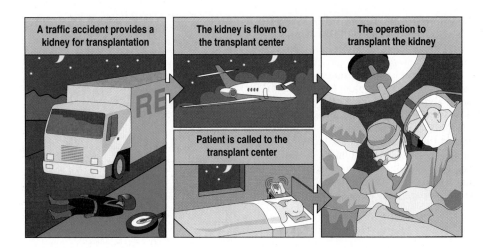

Figure 15.5 Clinical transplantation involves both a donated organ and a transplant recipient that are stressed and inflamed. This cartoon depicts a typical sequence of events that occurs before transplantation of a cadaveric organ transplant.

15-5   Organ transplantation involves procedures that inflame the donated organ and the transplant recipient

Patients receiving an organ transplant usually have a history of disease in which the organ to be replaced has gradually degenerated. Often there is an immune component to this decay, such as the deposition of immune complexes, which leads to kidney damage and failure. Before transplantation, patients with renal failure are maintained by dialysis, a procedure that induces inflammation through the interaction of dialysis membranes and serum proteins. Consequently, the transplant patient is already inflamed before transplantation, a state that is exacerbated by the damage and disruption caused by the surgery involved in transplantation. Thus, on receiving an organ transplant the recipient's body is both prepared and ready to direct innate and adaptive immunity toward the transplanted tissue.

Donated organs are also inflamed. In the case of cadaveric organs—that is, those taken from a deceased donor—the donor will usually have died in a violent or stressful manner, and the procedures used to collect organs and transport them to the transplant center add to the stress (Figure 15.5). During this time, the organs are deprived of blood, a state called **ischemia**. Ischemia causes damage to the blood vessels and tissue of the organs through activation of the endothelium and of the complement system, infiltration with inflammatory leukocytes, and the production of cytokines. The success of transplantation depends heavily on limiting the damage caused by ischemia. For kidney or liver transplantation it is possible to use a living healthy donor, and this confers great advantage. Donation and transplantation are performed at the same time and in the same place, with minimal time of ischemia. This approach has mainly been used for related donors and recipients, but it is being extended to unrelated family members, for example husbands and wives. Because of the reduced inflammation and tissue damage associated with organs from living healthy donors, the success of the transplantation is less sensitive to HLA mismatch than transplantation from cadaveric donors.

15-6   Acute rejection is caused by effector T cells responding to HLA differences between donor and recipient

Most organ transplants are performed across some HLA class I and/or II difference. In this situation the transplant recipient's T-cell population includes clones of alloreactive T cells that are specific for the HLA allotypes of the

Figure 15.6 Gross appearance of an acutely rejected kidney. The rejected graft is swollen and has deep-red areas of hemorrhage and gray areas of necrotic tissue. Courtesy of B.D. Kahan.

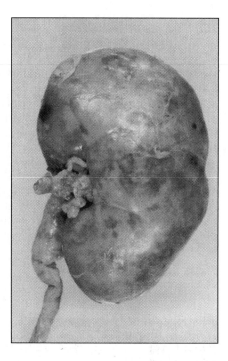

transplanted tissue that are not shared with the recipient. CD8 T cells respond to the HLA class I differences, and CD4 T cells respond to the HLA class II differences. The alloreactive T-cell response produces effector CD4 and CD8 T cells, both of which can attack the organ graft and destroy it (Figure 15.6). This is called **acute rejection**. Unlike hyperacute rejection, it takes days to develop and can be reduced or prevented by intervention. To prevent acute rejection, all transplant patients are conditioned with immunosuppressive drugs before transplantation and maintained on them after transplantation. Patients are carefully monitored for early signs of acute rejection and are treated with additional immunosuppressive drugs or anti-T-cell antibodies when it occurs. The effector mechanisms underlying acute rejection are just like those causing type IV hypersensitivity reactions (see Section 12-22, p. 392).

The inflamed state of the transplanted organ activates the organ's dendritic cells. These donor-derived dendritic cells migrate to the draining secondary lymphoid tissue, where they settle into the T-cell zone and present complexes of donor MHC and donor self peptides to the recipient's circulating T cells. Because different HLA allotypes bind different sets of self peptides, each allotype selects a different repertoire of T-cell receptors during thymic selection. Consequently, the T-cell repertoire selected by the recipient's HLA type contains numerous T-cell clones that can respond to the HLA:self-peptide complexes presented by donor cells of different HLA type. For this reason, alloreactive T-cell responses stimulated by HLA differences are stronger than T-cell responses to a vaccine or to a pathogen. Many clones of alloreactive T cells have a memory phenotype, revealing that they were originally stimulated and expanded in response to pathogens and are cross-reactive with allogeneic HLA. This type of alloreactive response, in which recipient T cells are stimulated by direct interaction of their receptors with the allogeneic HLA molecules expressed by donor dendritic cells, is called the **direct pathway of allorecognition** (Figure 15.7). It produces effector T cells that migrate to the transplanted tissue, where T_H1 cells activate the resident macrophages to inflame the tissue further and CD8 T cells systematically kill the cells of the transplanted tissue.

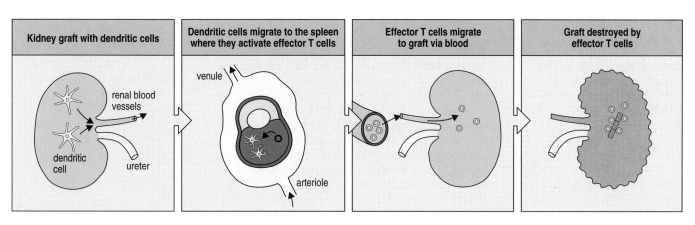

| Kidney graft with dendritic cells | Dendritic cells migrate to the spleen where they activate effector T cells | Effector T cells migrate to graft via blood | Graft destroyed by effector T cells |

Figure 15.7 Acute rejection of a kidney graft through the direct pathway of allorecognition. Donor dendritic cells in the graft (in this case a kidney) carry complexes of donor HLA molecules and donor peptides on their surfaces. The dendritic cells are carried to a draining secondary lymphoid organ (the spleen is illustrated here), where they move to the T-cell areas. Here, they activate the recipient's T lymphocytes whose receptors can bind specifically to the complexes of allogeneic donor HLA (both class I and class II) in combination with donor peptides. After activation, the effector T cells travel in the blood to the grafted organ, where they attack cells that display the peptide:HLA-molecule complexes for which the T cells are specific.

15-7   HLA differences between transplant donor and recipient activate numerous alloreactive T cells

We have seen how the cross-match test is used prospectively to examine how a patient's antibodies could react with transfused blood or transplanted tissue. The **mixed lymphocyte reaction** (**MLR**) assesses the extent to which a patient's T cells will respond to transplanted tissue. Peripheral blood cells from the recipient are mixed in tissue culture with lethally irradiated cells from a potential transplant donor. Both the proliferation of T cells and the effector function of donor-specific cytotoxic T cells can be measured (Figure 15.8). Around 5% of the T-cell population can be activated by the complexes of self peptides and allogeneic HLA class I and II molecules encoded by a disparate HLA haplotype. The strength of this response, comparable to that induced by a bacterial superantigen (see Section 11-6, p. 336), emphasizes the benefit of HLA matching and the need for potent immunosuppressive drugs.

15-8   Negative selection in the thymus limits the number of expressed MHC isoforms

The magnitude of the alloreactive response to nonself MHC provides a measure of the T cells that are negatively selected within the thymus of an individual. If the MHC molecules from another person activate a certain fraction of the mature T cells, then a similar stimulation by self MHC during T-cell development should lead to the deletion of a similar fraction of the positively selected T cells. The effects on positive and negative selection of changing the number of different MHC molecules are illustrated in Figure 15.9. In Chapter 5 we saw that MHC diversity is advantageous to an individual because it broadens the repertoire of pathogen-derived peptides presented (see Section 5-21, p. 151). Here we see how a similar expansion in the number of self peptides presented leads to an equivalent increase in the population of positively selected T cells in the thymus. However, with each additional type of MHC molecule, the proportion of T cells that are negatively selected goes up, not in an arithmetic manner as with positive selection, but in a geometric manner that increases with the square of the total number of MHC molecules. The result is that the increase in an individual's mature T-cell repertoire afforded by an additional MHC molecule rapidly declines as the number of MHC molecules increases. Beyond a certain number, the presence of additional MHC molecules will begin to affect the mature T-cell repertoire adversely. These counterbalancing effects of MHC diversity on the T-cell repertoire can explain why vertebrate species tend to limit the number of expressed MHC class I or II isotypes in the genome to three of each or less. Because of the expression of haplotypes from both chromosomes, this results in the expression of a maximum of around 12 different MHC isoforms in humans.

15-9   Chronic rejection of organ transplants is due to the indirect pathway of allorecognition

In addition to hyperacute and acute rejection, transplanted human organs can be rejected by a third mechanism called **chronic rejection**. This phenomenon, which occurs months or years after transplantation, is characterized by reactions in the vasculature of the graft that cause thickening of the vessel walls and a narrowing of their lumina (Figure 15.10). The blood supply to the graft gradually becomes inadequate, causing ischemia and loss of function of the graft, and the graft eventually dies. Chronic rejection causes the failure of more than half of all kidney and heart grafts within 10 years after transplantation. Chronic rejection is correlated with antibodies specific for the HLA class I molecules of the graft. Consistent with antibodies being the cause of chronic rejection, grafts undergoing chronic rejection are infiltrated with CD40-expressing B cells and T helper cells expressing CD40 ligand.

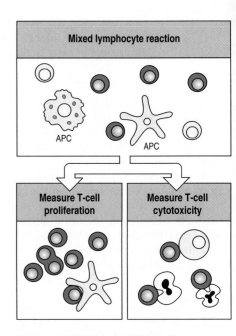

Figure 15.8 The mixed lymphocyte reaction (MLR) can be used to detect HLA differences. Peripheral blood mononuclear cells, which include lymphocytes and monocytes, are isolated from the two individuals to be tested. The cells from the person who serves as the stimulator (yellow) are first irradiated to prevent their proliferation. Then they are mixed with the cells from the other person who serves as the responder (blue) and cultured for five days (top panel). In the culture, responder lymphocytes are stimulated by allogeneic HLA class I and II molecules expressed by the stimulator's monocytes and the dendritic cells that differentiate from the monocytes. The stimulated lymphocytes proliferate and differentiate into effector cells. Five days after mixing, the culture is assessed for T-cell proliferation (bottom left panel), which is due to CD4 T cells recognizing HLA class II differences, and for cytotoxic T cells (bottom right panel) produced in response to HLA class I differences. The mixed lymphocyte reaction was instrumental in distinguishing MHC class II from MHC class I.

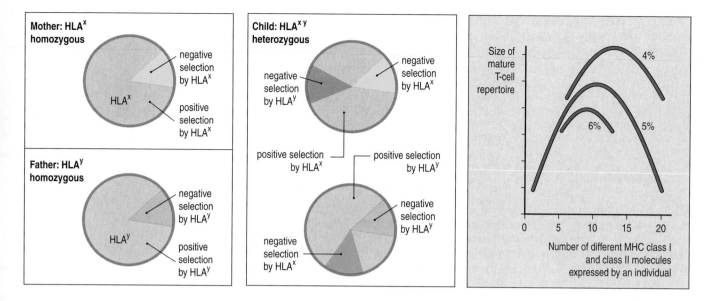

Figure 15.9 Analysis of alloreactivity shows how too many MHC isoforms will deplete the T-cell repertoire through excessive negative selection in the thymus. The left and center panels show how positive and negative selection have different cumulative effects as the number of MHC molecules (HLA in humans) expressed by an individual increases. In this family, the mother is homozygous for the HLAˣ haplotype, whereas the father is homozygous for the disparate HLAʸ haplotype. Their child expresses both the x and y haplotypes. In the parents, the x or the y set of HLA molecules positively selects a set of T-cell receptors, which are then reduced by negative selection on the same set of HLA molecules. Whole circles (pie charts) are the populations of thymocytes positively selected by the HLA type indicated; the colored wedges are the proportions of these cells that are subsequently negatively selected by each HLA haplotype. Populations of cells negatively selected by HLAˣ are indicated in yellow and those negatively selected by HLAʸ in orange. The child can be considered as positively selecting the sum of the T-cell receptors selected by the two parents. However, in the child, each of these two positively selected repertoires is then negatively selected on both the x and y sets of HLA molecules. Whereas the number of positively selected subsets of T cells in the child is the sum (2) of those in the two parents (1 + 1), the number of negatively selected subsets of T cells in the child is the square (4) of the number in each parent (2). Thus, the effect of negative selection increases disproportionately as the number of different HLA molecules expressed by an individual increases. This effect is shown in the graph on the right, in which the relationship between the size of the mature T-cell repertoire and the number of HLA isoforms expressed by an individual is plotted. The three plots vary according to the estimated value (4%, 5%, and 6%) for the percentage of T cells positively selected by one HLA isoform that are negatively selected by a second HLA isoform. A value of 4% gives an optimal number of HLA molecules of 12–13, which is comparable to the number of different HLA molecules expressed by most people.

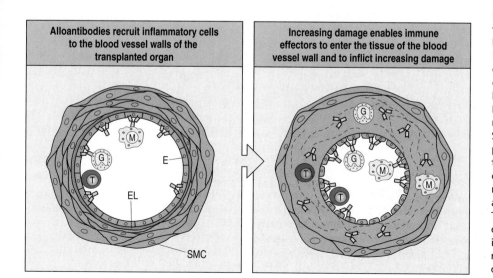

Figure 15.10 Chronic rejection in the blood vessels of a transplanted kidney. Left-hand panel: chronic rejection is initiated by the interaction of anti-HLA class I alloantibodies with blood vessels of the transplanted organ. Antibodies bound to endothelial cells (E) recruit Fc receptor-bearing monocytes and neutrophils. EL, internal elastic lamina; SMC, smooth muscle cells. Right-hand panel: accumulating damage leads to EL thickening and to infiltration of the underlying intimas with SMCs, macrophages (M), granulocytes (G), alloreactive T cells (T), and antibodies. The net effect is to narrow the lumen of the blood vessel and create a chronic inflammation that intensifies tissue remodeling. Eventually the vessel becomes obstructed, ischemic, and fibrotic.

The CD4 helper T cells that initiate the response leading to chronic rejection of organ transplants do not recognize their specific antigens by the direct pathway of allorecognition but by another pathway called the **indirect pathway of allorecognition**. The differences in mechanism underlying the two pathways of allorecognition are compared in Figure 15.11. In the indirect pathway, some of the donor-derived dendritic cells that migrate to the draining lymphoid tissue die there by apoptosis. Membrane fragments containing HLA molecules from these apoptotic cells are taken up by dendritic cells of the transplant recipient and processed so that peptides derived from donor HLA allotypes are presented by the recipient's HLA allotypes. Because of the endocytic mode of uptake, most of these peptides, which can be derived from HLA class I or II molecules, will be presented by the HLA class II allotypes of the recipient. If the peptides are different in amino acid sequence from those produced by degradation of the recipient's own HLA allotypes, these complexes will stimulate a CD4 T-cell alloreaction. The responding alloreactive CD4 T cells are specific for the complex of a peptide derived from a donor HLA allotype bound to a recipient HLA class II allotype. This way of stimulating alloreactive T cells is called the indirect pathway because the alloreactive

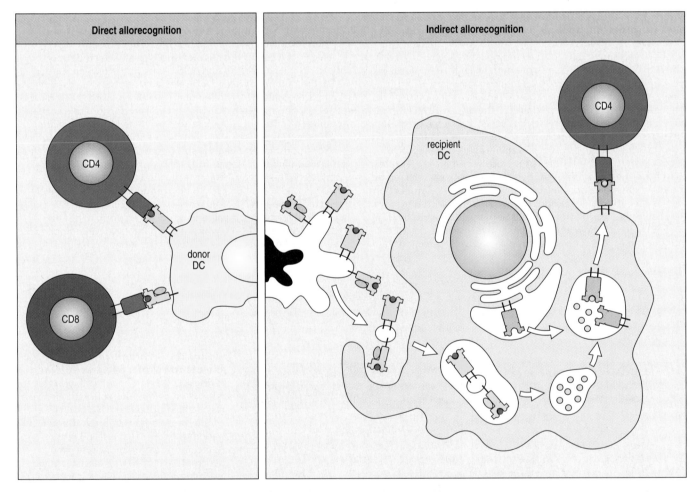

Figure 15.11 Direct and indirect pathways of allorecognition contribute to graft rejection. Dendritic cells from an organ graft stimulate both the direct and indirect pathways of allorecognition when they travel from the graft to the draining lymphoid tissue. The left-hand panel shows how the allogeneic HLA class I and II allotypes of donor type on a donor dendritic cell (donor DC) will interact directly with the T-cell receptors of alloreactive CD4 and CD8 T cells of the recipient (direct allorecognition). The right-hand panel shows how the death of the same antigen-presenting cell produces membrane vesicles containing the allogeneic HLA class I and II allotypes, which are then endocytosed by the recipient's dendritic cells (recipient DC). Peptides derived from the donor's HLA molecules (yellow) can then be presented by the recipient's HLA molecules (orange) to peptide-specific T cells (indirect allorecognition). Presentation by HLA class II molecules to CD4 T cells is shown here. Peptides derived from donor HLA can also be presented by recipient HLA class I molecules to CD8 T cells (not shown).

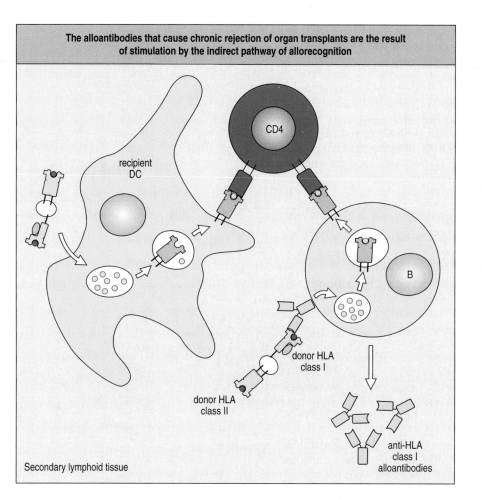

The alloantibodies that cause chronic rejection of organ transplants are the result of stimulation by the indirect pathway of allorecognition

CD4

recipient
DC

B

donor HLA
class I

donor HLA
class II

anti-HLA
class I
alloantibodies

Secondary lymphoid tissue

Figure 15.12 The indirect pathway of allorecognition is responsible for stimulating the production of the anti-HLA antibodies that cause chronic rejection. The processing and presentation of allogeneic HLA class I by a dendritic cell (DC) of the recipient is shown. The dendritic cell activates helper CD4 T cells, which in turn activate B cells that have bound and internalized allogeneic donor HLA molecules. Shown here is a cognate interaction that leads to the production of an anti-HLA class I antibody. Anti-HLA class II antibodies can be produced similarly. Because activated endothelium expresses both HLA class I and II molecules, antibodies against both classes of HLA molecule can contribute to chronic rejection.

T cells do not directly recognize the transplanted cells but recognize subcellular material that has been processed and presented by the recipient's own cells (see Figure 15.11).

The indirect pathway of allorecognition is a special case of the normal mechanism by which T cells recognize the protein antigens of pathogens; in transplantation, the foreign peptide antigens come from the proteins of another human body. T cells stimulated by the indirect pathway of allorecognition can also contribute to the acute rejection of transplanted organs, although they are usually less numerous than those stimulated by the direct pathway.

The alloreactive T-cell response stimulated by the direct pathway wanes with time after transplantation. This correlates with the elimination of dendritic cells of donor origin and the repopulation of the transplanted organ with immature dendritic cells of recipient origin. These latter cells can, however, still increase the stimulation of alloreactive T cells through the indirect pathway of allorecognition. Transplant recipients are selected for not having serum antibody that reacts with the transplanted organ; this means they will not have memory B cells that can respond to the allogeneic HLA allotypes. They may, however, have naive B cells reactive for the allogeneic HLA antigens. After transplantation, the stimulation of helper CD4 T cells through the indirect pathway can thus initiate an antibody response against the allogeneic HLA allotypes (Figure 15.12). The stimulated helper CD4 T cells will help naive B cells that are specific for HLA allotypes of the graft and thus provoke a B-cell response in which alloantibodies specific for donor HLA class I are formed. The HLA-specific CD4 T cells will also provide help to B cells specific for other alloantigens that are incorporated into the HLA-containing subcellular fragments. This can lead to expansion and epitope spreading of the allogeneic

B-cell and T-cell response and the gradual impairment of the function of the transplanted organ through the same mechanisms that operate in autoimmune disease (see Section 13-24, p. 430).

The indirect pathway of allorecognition can also give rise to regulatory CD4 T cells that suppress alloreactive CD4 and CD8 effector T cells and improve the clinical outcome after kidney transplantation. Such regulatory T cells seem more active in patients who previously received blood transfusions that, fortuitously, shared an HLA-DR allotype with the transplanted kidney. This phenomenon of previous blood transfusions improving the outcome of organ transplantation is known as the **transfusion effect**.

15-10 Matching donor and recipient for HLA class I and class II allotypes improves the outcome of organ transplantation

The first successful organ transplant was achieved with a kidney transplant between identical twins, in whom there was no risk of alloreactivity leading to graft rejection. As very few patients have an identical twin, additional approaches were needed to make transplantation more generally available to patients with kidney disease. The combination of two complementary approaches proved successful. The first is to select a transplant donor who is as similar as possible to the recipient in HLA class I and II types. This reduces the number of alloreactive T cells that can be triggered by the transplanted tissue. The second is the use of a battery of immunosuppressive drugs that prevent and interfere with the activation and proliferation of T cells.

Clinical transplantation was pioneered with the kidney for two principal reasons. First, patients whose kidneys had failed could be sustained by the well-established procedure of dialysis. This meant that graft failure or rejection did not inevitably lead to the patient's death. The second was the simple fact that everyone has two kidneys but can manage with only one, so healthy relatives could donate a kidney to a needy patient. Immunogenetic differences within a family are much smaller than within the population at large, and so the probability of finding an HLA-matched person within the family is higher. Analysis of the fate of kidney transplants performed between HLA-identical and nonidentical family members was instrumental in showing that the better the HLA class I and II match, the better is the clinical outcome. Matching for HLA-A, HLA-B, and HLA-DR is the most important. Most clinical HLA typing is currently performed by DNA analysis, although serological HLA typing is still used for some purposes and in some transplant centers.

After the success of transplantation between living relatives, methods were developed to transplant kidneys from unrelated donors who had been killed in accidents (cadaveric donors). Over 100,000 kidney transplants have been performed worldwide, and a statistical analysis of these data shows that both graft performance and long-term health of the recipient increase with the degree of HLA match (Figure 15.13).

15-11 Allogeneic transplantation is made possible by the use of three types of immunosuppressive drug

The limited supply of donated organs and the diversity of HLA types in the population mean that a majority of patients receive organs mismatched at one or more HLA loci. Drugs are used to suppress the alloreactions that would otherwise lead to transplant rejection. The **immunosuppressive** drugs used in clinical transplantation are of three kinds. The **corticosteroids** are steroids with anti-inflammatory properties like those of the natural glucocorticosteroid hormones made in the adrenal cortex. The second kind of drug consists of

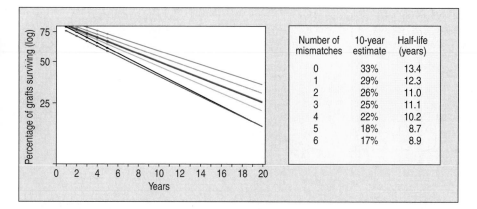

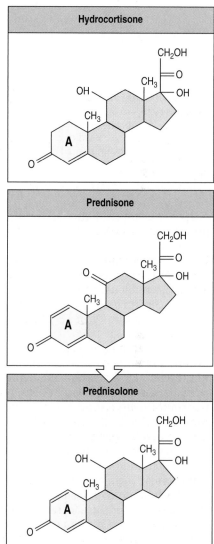

Figure 15.13 **HLA matching improves the survival of transplanted kidneys.** The colored lines in the left panel represent the actual (to 5 years) and projected survival rates of kidney grafts in patients with no (blue), 1 (orange), 2 (red), 3 (dark blue), 4 (green), 5 (black), and 6 (brown) HLA mismatches, plotted on a semi-logarithmic scale. Data courtesy of G. Opelz.

cytotoxic drugs that, by interfering with DNA replication, kill proliferating lymphocytes activated by graft alloantigens. The third type of immunosuppressive drug comprises microbial products that inhibit the signaling pathways of T-cell activation.

Any drug powerful enough to inhibit alloreactions also inhibits the normal immune responses to infecting pathogens. Consequently, administration of these drugs, which is greatest during the period immediately before and after transplantation, renders transplant patients highly susceptible to infection. Patients are initially cared for in conditions under which their exposure to pathogens is reduced. As their immune systems accommodate to the graft, the dose of immunosuppressive drugs is gradually reduced to 'maintenance levels' that prevent rejection while sustaining active defenses against infection. Aggressive early treatment has reduced the incidence of acute rejection in clinical transplantation. However, with the reduction of immunosuppression and the restoration of a patient's immunocompetence, the likelihood of chronic rejection increases.

All immunosuppressive drugs are also toxic to other tissues in varying degrees. Because these 'side-effects' vary for the different drugs, immunosuppressive drugs are generally used in combination so that their immunosuppressive effects are additive but their toxic effects are not. Certain side-effects emerge only after patients have taken immunosuppressive drugs for long periods. These include a higher incidence of certain types of malignant disease, particularly carcinomas of the skin and the genital tract, lymphoma, and Kaposi's sarcoma. The incidence of cancer in transplant recipients is on average three times that of similarly aged people who have not received a transplant.

15-12 Corticosteroids change patterns of gene expression

Hydrocortisone, also called cortisol, is the principal steroid made by the adrenal cortex; for more than 50 years it has been used clinically to reduce inflammation. The steroid commonly given to transplant patients is **prednisone**, a synthetic derivative of hydrocortisone that is about four times more potent in reducing inflammation. Prednisone has no biological activity until it is enzymatically converted *in vivo* to **prednisolone** (Figure 15.14). Prednisone is an example of a **pro-drug**, a name given to drugs that are given to patients in an inactive form and become chemically or enzymatically converted to the active form within the body. By itself, prednisone is insufficiently immunosuppressive to prevent graft rejection, but it works well in combination with a cytotoxic drug.

Corticosteroids have wide-ranging physiological effects and affect all white blood cells, not just lymphocytes, as well as other cells of the body. Unlike many other biologically active molecules, steroid hormones do not act on

Figure 15.14 **Chemical structures of hydrocortisone, prednisone, and prednisolone.** Prednisone is a synthetic analog of the natural adrenocorticosteroid hydrocortisone, or cortisol. It is converted *in vivo* to the biologically active form, prednisolone. Introduction of the 1,2 double bond into the A ring increases anti-inflammatory potency approximately fourfold compared with hydrocortisone, without modifying the sodium-retaining activity of the compound.

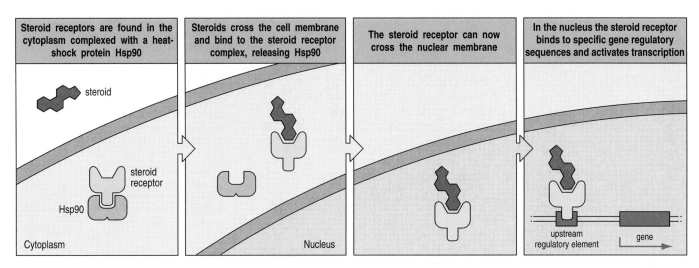

Figure 15.15 Steroids act at intracellular receptors. Corticosteroids are lipid-soluble compounds that diffuse across the plasma membrane and bind to their receptors in the cytosol. The binding of corticosteroid to the receptor displaces a dimer of a heat-shock protein named Hsp90, exposing the DNA-binding region of the receptor, which then enters the nucleus and binds to specific DNA sequences in the promoter regions of steroid-responsive genes. Corticosteroids exert their effects by modulating the transcription of a wide variety of genes.

receptors at the cell surface but diffuse across the plasma membrane and bind to specific receptors in the cytoplasm. Before steroid binding, the receptors are associated with another cytoplasmic polypeptide called Hsp90 (*heat-shock protein* of 90 kDa molecular weight). Steroid binding induces a conformational change in the receptor, which then dissociates from Hsp90 and enters the nucleus. There, the complex of receptor and steroid binds selectively to certain genes, activating their transcription (Figure 15.15). The transcription of about 1% of a cell's genes can be influenced by corticosteroids.

In the context of their anti-inflammatory action, an important effect of corticosteroids is inhibition of the function of NFκB, a transcription factor important for cellular activation and cytokine production in the immune response (see Chapter 2). In quiescent cells, NFκB is held in the cytoplasm through its association with a protein called IκBα. On cellular activation, IκBα becomes phosphorylated, allowing NFκB to dissociate and enter the nucleus, where it initiates cytokine gene transcription. Corticosteroids increase the production of IκBα, thereby preventing NFκB from gaining access to the nucleus. By this mechanism they suppress the production of cytokines, such as IL-1 by monocytes, which stimulate inflammation and immune responses (Figure 15.16). Another major effect of steroids is to alter the homing of lymphocytes so that they do not enter the secondary lymphoid tissues and sites of inflammation but congregate in the bone marrow. This prevents naive lymphocytes from being stimulated by alloantigens, and prevents effector T cells from entering and attacking the transplanted tissue.

Figure 15.16 Effects of corticosteroids on the immune system. Corticosteroids regulate the expression of many genes, with a net anti-inflammatory effect. First, they reduce the production of inflammatory mediators, including some cytokines, prostaglandins, and nitric oxide (NO). As well as downregulating the cytokines listed here, corticosteroids also indirectly cause a decrease in IL-2 synthesis by activated lymphocytes, by their effects on other cytokines. Second, they inhibit inflammatory cell migration to sites of inflammation by inhibiting the expression of adhesion molecules. Third, corticosteroids promote the death by apoptosis of leukocytes and lymphocytes. NOS, nitric oxide synthase.

Corticosteroid therapy	
Activity	**Effect**
↓ IL-1, TNF-α, GM-CSF ↓ IL-3, IL-4, IL-5, CXCL8	↓ Inflammation ↓ caused by cytokines
↓ NOS	↓ NO
↓ Phospholipase A₂ ↓ Cyclo-oxygenase type 2 ↑ Lipocortin-1	↓ Prostaglandins ↓ Leukotrienes
↓ Adhesion molecules	Reduced emigration of leukocytes from vessels
Induction of endonucleases	Induction of apoptosis in lymphocytes and eosinophils

Because of their multifarious effects on gene expression and cellular metabolism, corticosteroid drugs have many adverse side-effects, including fluid retention, weight gain, diabetes, loss of bone mineral, and thinning of the skin. Because of their mechanism of action, corticosteroids are most effective as immunosuppressive drugs when they are first administered before transplantation. With this approach, the patterns of cytokine gene expression are already changed in the recipient's cells at the time of alloantigenic challenge. They are used as an acute immunosuppressive agent during episodes of rejection, which are often caused by infection, but their continued use is to be avoided wherever possible.

15-13 Cytotoxic drugs kill proliferating cells

A cytotoxic drug commonly used in solid organ transplantation is **azathioprine**, a pro-drug that is first converted *in vivo* to 6-mercaptopurine and then to 6-thioinosinic acid (Figure 15.17). The latter compound inhibits the

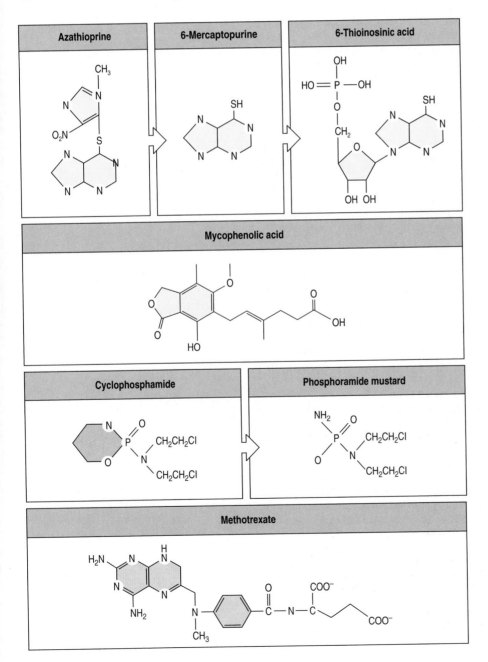

Figure 15.17 The chemical structures and metabolism of cytotoxic drugs. Azathioprine was developed as a modification of the anti-cancer drug 6-mercaptopurine; by blocking the reactive thiol group, the metabolism of this drug is slowed down. It is slowly converted *in vivo* to 6-mercaptopurine, which is then metabolized to 6-thioinosinic acid; this blocks the pathway of purine biosynthesis. Mycophenolate is a newer drug that also blocks purine biosynthesis after being metabolized to mycophenolic acid. Cyclophosphamide was developed as a stable pro-drug, which is activated enzymatically in the body to phosphoramide mustard, a powerful and unstable DNA-alkylating agent. Methotrexate blocks DNA synthesis by interfering with thymidine synthesis.

production of inosinic acid, an intermediate in the biosynthesis of adenine and guanine nucleotides, which are essential components of DNA. The principal effect of azathioprine is therefore to inhibit DNA replication. Azathioprine has no effect on cells until they attempt to replicate their DNA, whereupon they die. While helpfully inhibiting the proliferation of alloantigen-activated lymphocytes, azathioprine, like other cytotoxic drugs, damages all tissues of the body that are normally active in cell division. Principally affected are the bone marrow, the intestinal epithelium, and hair follicles, leading to anemia, leukopenia, thrombocytopenia, intestinal damage, and loss of hair. When pregnant women have to take cytotoxic drugs, fetal development can be adversely affected. Because azathioprine cannot act until a patient's immune system has been stimulated by alloantigen, it need only be administered after transplantation. Mycophenalate mofetil, a product of the fungus *Penicillium stoloniferum*, is a more recently developed drug with similar effects to azathioprine. It is metabolized in the liver to **mycophenolic acid** (see Figure 15.17), which prevents cell division by inhibiting inosine monophosphate dehydrogenase, an enzyme necessary for guanine synthesis.

Cyclophosphamide is one of the nitrogen mustard compounds that were developed as chemical weapons and saw heavy use during World War I. It is a pro-drug that is converted in the body to phosphoramide mustard; this alkylates and cross-links DNA molecules (see Figure 15.17). These covalent modifications render cells incapable of normal division and also affect transcription. Consequently, cyclophosphamide is equally immunosuppressive given before or after antigenic stimulation.

Cyclophosphamide has many toxic effects that limit its clinical application. In addition to side-effects shared with other cytotoxic drugs, cyclophosphamide specifically damages the bladder, sometimes causing cancer or a condition called hemorrhagic cystitis. Unlike azathioprine, cyclophosphamide is not particularly toxic to the liver, and for patients who have sustained liver damage or become otherwise sensitized to azathioprine, it is a useful alternative. Cyclophosphamide is most effective when used in short courses of treatment.

Methotrexate was one of the first cytotoxic drugs shown to be effective in treating cancer cells. It prevents DNA replication by inhibiting dihydrofolate reductase, an enzyme essential for the cellular synthesis of thymidine. Methotrexate is the drug of choice for inhibiting GVHD in bone marrow transplant recipients (see Figure 15.17).

15-14 Cyclosporin A, tacrolimus, and rapamycin selectively inhibit T-cell activation

During the 1960s and 1970s, transplant physicians depended on combinations of corticosteroids and cytotoxic drugs to prevent the rejection of transplanted organs. Toward the end of the 1970s, new kinds of immunosuppressive drug were introduced that selectively inhibited T-cell activation. These drugs had a marked impact on clinical transplantation in the 1980s and 1990s, leading to improved graft survival, a wider range of tissues and organs being transplanted, and transplantation being recommended for an increased range of diseases. This period became known as the 'cyclosporin era,' because cyclosporin A was the first of these drugs to be introduced.

Cyclosporin A (also known as **cyclosporine**) is a cyclic decapeptide derived from a soil fungus, *Tolypocladium inflatum*, that was originally isolated in Norway. It inhibits the activation of T cells by antigen by disrupting the transduction of signals from the T-cell receptor. Signals from the T-cell receptor normally lead to hydrolysis of membrane lipids to give inositol trisphosphate and the consequent release of Ca^{2+} from intracellular stores. The elevation in cytosolic Ca^{2+} concentration activates the cytoplasmic serine/threonine

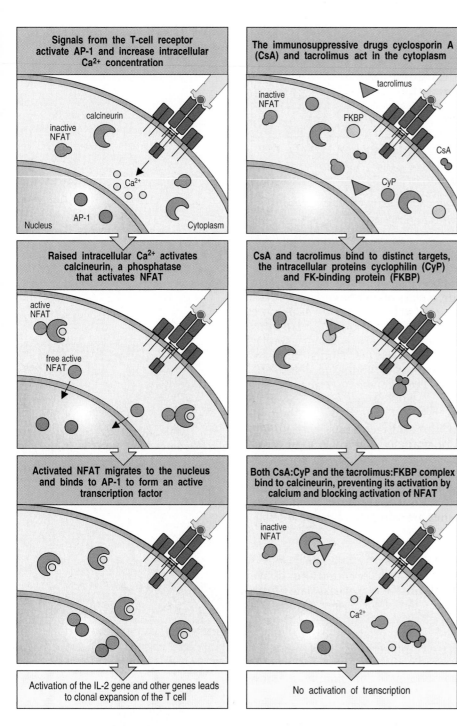

Signals from the T-cell receptor activate AP-1 and increase intracellular Ca²⁺ concentration

calcineurin

inactive NFAT

Ca^{2+}

Nucleus   AP-1   Cytoplasm

Raised intracellular Ca²⁺ activates calcineurin, a phosphatase that activates NFAT

active NFAT

free active NFAT

Activated NFAT migrates to the nucleus and binds to AP-1 to form an active transcription factor

Activation of the IL-2 gene and other genes leads to clonal expansion of the T cell

The immunosuppressive drugs cyclosporin A (CsA) and tacrolimus act in the cytoplasm

tacrolimus

inactive NFAT

FKBP

CsA

CyP

CsA and tacrolimus bind to distinct targets, the intracellular proteins cyclophilin (CyP) and FK-binding protein (FKBP)

Both CsA:CyP and the tacrolimus:FKBP complex bind to calcineurin, preventing its activation by calcium and blocking activation of NFAT

inactive NFAT

Ca^{2+}

No activation of transcription

Figure 15.18 Cyclosporin A and tacrolimus inhibit T-cell activation by interfering with the serine/threonine phosphatase calcineurin. Signaling via T-cell receptor-associated tyrosine kinases leads to the activation and increased synthesis of the transcription factor AP-1, as well as increasing the concentration of Ca²⁺ in the cytoplasm (left panels). The Ca²⁺ binds to calcineurin and thereby activates it to dephosphorylate the cytoplasmic form of the nuclear factor of activated T cells (NFAT). Once dephosphorylated, the active NFAT migrates to the nucleus to form a complex with AP-1; the NFAT:AP-1 complex can then induce the transcription of genes required for T-cell activation, including the gene encoding IL-2. When cyclosporin A (CsA) or tacrolimus is present, it forms complexes with its immunophilin target, cyclophilin (CyP) or FK-binding protein (FKBP), respectively (right panels). The complex of cyclophilin with cyclosporin A can bind to calcineurin, blocking its ability to activate NFAT. The complex of tacrolimus with FKBP binds to calcineurin at the same site, also blocking its activity.

phosphatase **calcineurin**, which in turn activates the transcription factor **NFAT** (see Section 8-7, p. 224). In resting T cells, NFAT is present in the cytoplasm in a phosphorylated form. Calcineurin removes the phosphate, permitting NFAT to enter the nucleus, where it binds to the transcription factor AP-1 to form a transcriptional regulatory complex that turns on transcription of the gene encoding IL-2.

Cyclosporin interferes with calcineurin activity. It diffuses across the plasma membrane into the cytosol, where it binds peptidyl-prolyl isomerase enzymes, which in this context are called **cyclophilins**. The complex of cyclosporin A and cyclophilin binds to calcineurin, inhibiting its phosphatase activity and preventing it from activating NFAT. In the presence of cyclosporin, therefore, IL-2 cannot be made and the program of T-cell activation, proliferation, and differentiation is shut down at a very early stage (Figure 15.18).

Cell type	Effects of cyclosporin A and tacrolimus
T lymphocyte	Reduced expression of IL-2, IL-3, IL-4, GM-CSF, TNF-α Reduced cell division because of decreased IL-2 Reduced Ca^{2+}-dependent exocytosis of cytotoxic granules Inhibition of antigen-driven apoptosis
B lymphocyte	Inhibition of cell division because T-cell cytokines are absent Inhibition of antigen-driven cell division Induction of apoptosis after B-cell activation
Granulocyte	Reduced Ca^{2+}-dependent exocytosis of granules

Figure 15.19 Immunological effects of cyclosporin A and tacrolimus.

Tacrolimus, also called **FK506**, was isolated from the soil actinomycete *Streptomyces tsukabaensis*. It is a macrolide, a class of compound with structures based on a many-membered lactone ring that is attached to one or more deoxy sugars. Although structurally distinct from cyclosporin A, tacrolimus suppresses T-cell activation by the inhibition of calcineurin through a similar mechanism. The peptidyl-prolyl isomerases to which tacrolimus binds are distinct from the cyclophilins and are known as FK-binding proteins. The cyclophilins and **FK-binding proteins** are known collectively as **immunophilins**.

Although the principal effect of cyclosporin A and tacrolimus is to inhibit T-cell activation, the activation of B cells and granulocytes is also suppressed (Figure 15.19). A major advantage of these drugs is that they do not target proliferating cells, so the reduced hematopoiesis and intestinal damage seen with cytotoxic drugs does not occur. A side-effect associated with the continued administration of cyclosporin A or tacrolimus is nephrotoxicity, and some patients can no longer tolerate the drug—they are said to have become sensitized to it.

The success of cyclosporin A and tacrolimus in transplantation encouraged the search for similarly selective drugs. One find is an immunosuppressive macrolide called **rapamycin** (also called **sirolimus**), which was isolated from *Streptomyces hygroscopicus*, a soil bacterium found on Easter Island. The island's Polynesian name, 'Rapa ui,' was used to name the drug. Although rapamycin binds to FK-binding proteins, it does not interfere with calcineurin but blocks T-cell activation at a later stage by preventing signal transduction from the IL-2 receptor. Rapamycin is more toxic than either cyclosporin A or tacrolimus but has become a useful component of combination therapy.

15-15 Antibodies specific for T cells are used to prevent and control acute rejection

After transplantation, patients are maintained on a combination of immunosuppressive drugs. Because of their toxicity and the immunodeficiency caused by these drugs, physicians seek to lower the dosage gradually to the minimum that will maintain tolerance of the transplant. Inevitably, there are times when the balance is upset and early symptoms of rejection appear (Figure 15.20). Such episodes can be treated with a 5–15-day course of daily injections of T cell-specific antibodies, as well as an increased dose of immunosuppressive drugs.

Polyclonal anti-T-cell antibodies are made in sheep and goats that have been immunized with human thymocytes or lymphocytes. Antibody-containing

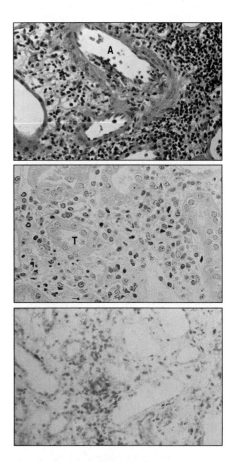

Figure 15.20 Acute rejection in a kidney graft. Top panel shows lymphocytes around an arteriole (A) in a kidney undergoing rejection. The middle panel shows lymphocytes surrounding the renal tubules (T) of the same kidney, and the bottom panel shows the staining of T lymphocytes with anti-CD3 (brown staining) in the same section. Photographs courtesy of F. Rosen.

fractions called **antithymocyte globulin** (**ATG**) or **antilymphocyte globulin** (**ALG**) are prepared from the animals' blood. Another source of anti-T-cell antibodies is hybridoma cell lines making mouse monoclonal antibodies specific for proteins present only on the T-cell surface, for example CD3.

Immunosuppressive antibodies work in one of two ways. ALG and ATG cause the destruction of the lymphocytes to which they bind, through complement fixation and phagocytosis. By contrast, monoclonal anti-CD3 interferes with the function of the T cells to which it binds, causing reduced expression of the CD3:T-cell receptor complex on the cell surface and depletion of these cells from the circulation. Because the immunosuppressive antibodies are **xenogeneic**—that is, they come from a nonhuman species—they tend to stimulate an antibody response against them. This reduces their immunosuppressive activity when used on subsequent occasions. In such situations the patient's antibodies form immune complexes with the immunosuppressive antibodies, clearing them from the circulation before they can bind to T cells. Such reactions can also lead to serum sickness (see Section 12-20, p. 391). For this reason, physicians generally use each xenogeneic immunosuppressive antibody to counter just one episode of acute rejection per patient.

In recent years, anti-CD3 and other immunosuppressive mouse monoclonal antibodies have been humanized by incorporating the minimal mouse antigen-binding site into the framework of a human antibody (see Section 4-6, p. 104). This permits them to be given on multiple occasions to the same patient and to be used prophylactically to prevent acute rejection as well as treating it after transplantation. Daclizumab, a humanized monoclonal antibody specific for the α chain of the high-affinity IL-2 receptor, reduces the incidence of acute rejection in kidney transplantation by 40%. One dose of daclizumab is given 1 hour before transplantation, and four subsequent doses are given at 2-week intervals. An advantage in using an antibody specific for this form of the IL-2 receptor is that it targets only activated T cells (see Section 8-8, p. 225), whereas anti-CD3 antibodies target all T cells.

15-16 Patients needing a transplant outnumber the available organs

Kidney transplantation has advanced to where it is now possible to transplant cadaveric kidneys across considerable HLA mismatches. This progress helped the development of heart transplantation, for which only cadaveric donors could be considered. Heart transplantation is inherently more difficult than kidney transplantation, principally because the failure of a grafted heart is fatal, whereas patients with failed kidney grafts can go back to dialysis. The use of cyclosporin A and tacrolimus has increased the success of heart transplantation, mainly by preventing death from acute rejection or infection during the first few months after transplantation. As a consequence, the number of heart transplants increased considerably after 1979. In the USA, some 2000 patients each year receive a heart transplant, and more than half of them are projected to be still alive 10 years after the operation.

Liver transplantation has similarly progressed in the cyclosporin era, from being a relatively risky procedure to one offering considerable benefit. In 1979 only 30–40% of patients survived a liver transplant for more than a year; today 70–90% of patients are surviving after 1 year and 60% are alive after 5 years. A similar improvement has been seen for lung transplantation.

The very success of solid organ transplantation has created its own problem, namely that the number of patients who could benefit from a kidney, heart, or liver transplant greatly exceeds the organs available from live and deceased donors (Figure 15.21). Patients are therefore placed on waiting lists and chosen for transplantation on the basis of various criteria, including the disease

Organ	Patients on waiting list		
	1999 August	2003 November	2008 March
Kidney	42,875	83,284	75,004
Liver	13,698	17,237	16,367
Heart	4287	3556	2654
Lung	3343	3907	2117
Pancreas	502	1404	1620

Figure 15.21 The need for tissue transplants outruns the supply of donated organs. Patients who are eligible for a transplant have to wait 2–3 years on average in the United States or United Kingdom before they receive the transplant. The numbers of waiting patients in the United States for three time points in the period 1999–2008 are shown. In 2007, a total of 28,353 transplants was performed. Each year more than 6000 patents on the waiting list die before they can receive a transplante. Data courtesy of the United Network for Organ Sharing.

severity and the HLA match with available organs. To increase the organ supply, some countries have instituted a policy whereby cadaveric organs from accident victims become automatically available for clinical transplantation unless the person has deliberately opted out. Other countries retain the policy that organs can be used for transplantation only if an accident victim has deliberately opted in by previously signing a consent form. Even then, relatives can overrule the victim's wishes. In providing organs for transplantation, the 'opt out' system works much better than 'opt in' (Figure 15.22).

The unsatisfied demand for kidneys has inevitably led to a flourishing and unregulated international trade in human kidneys, generally involving donors from the poorest countries selling one of their kidneys for transplantation to patients from the richest countries. These operations can involve so-called 'transplant tourism' to medical centers in countries of intermediate wealth, where the donor, the recipient, and the surgeon are brought together by a broker. Although many national and international organizations prohibit such trade, it now accounts for several thousand transplants each year.

The limiting supply of organs could be overcome by using organs from animals. This type of transplantation, in which donor and recipient are of different species, is called **xenotransplantation**, and the grafted tissue a **xenograft**. Pigs are considered the most suitable donor species for humans: first because pig and human organs are of similar sizes, and second because they are already farmed, slaughtered, and consumed by humans in large numbers. The immunological barriers facing xenotransplantation are formidable. As a start, most humans have circulating antibodies that bind to pig endothelial cells and would cause hyperacute rejection. In this context these antibodies are called **xenoantibodies** and the carbohydrate antigens on pig endothelium to which they bind are called **xenoantigens**. As with the alloantibodies that humans make against ABO blood group alloantigens, the xenoantibodies are probably induced by infections with common bacteria whose surface carbohydrates resemble those of pig cells. Exacerbating the potential problems of hyperacute rejection is the fact that the complement regulatory proteins on the surface of pig cells (the pig versions of CD59, DAF, and MCP; see Section 2-4, p. 36) do not inhibit human complement.

To improve the compatibility of their organs with the human immune system, pigs would need to be genetically modified with several human genes. Over and above the immunological barrier is the concern that immunosuppressed patients receiving pig xenotransplants could provide a route for endogenous pig retroviruses to infect the human population and have an effect like HIV. As on many previous occasions, the development of procedures to save lives by organ transplantation raises complicated ethical questions.

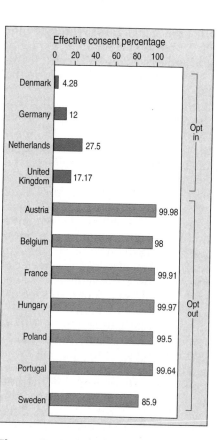

Figure 15.22 The supply of cadaveric donor organs is higher in countries where opting out of donation requires effort than in countries where opting in requires effort. Data courtesy of E.J. Johnson and D. Goldstein.

15-17 The need for HLA matching and immunosuppressive therapy varies with the organ transplanted

Each organ has a unique anatomy, function, and vasculature, properties that affect both the strength of antigenic stimulation induced by transplantation and the nature of the alloreactive response. At one end of the spectrum is the cornea of the eye, the first solid organ to be successfully transplanted (in 1905) and the one transplanted with greatest frequency: more than 30,000 corneal transplants from cadaveric donors are performed each year in the United States alone. Because of the eye's distinctive immunological environment, corneal transplants are transplanted with 90% success in the absence of HLA matching or immunosuupressive therapy (Figure 15.23).

For the eye to function properly, the cornea and the anterior chamber of the eye must remain clear and allow a precise passage of light onto the retina. As any inflammation can impair vision, the eye has evolved an immunological environment that suppresses inflammation while maintaining sufficient protection against pathogens. The cornea is transparent and lacks vasculature, while the aqueous humor of the anterior chamber contains immunomodulatory factors that inhibit the activation of T cells, macrophages, neutrophils, and complement. In particular, the cytokine TGF-β causes the resident dendritic cells to downregulate T-cell co-stimulatory factors such as CD40 and prevents the secretion of IL-12. When antigens are introduced into the eyes of experimental animals, they are carried to the spleen by these dendritic cells, where they generate a T-cell response that is skewed toward regulatory T cells and the production of IL-4 and TGF-β. This capacity to generate an active and systemic state of tolerance to foreign antigens in the eye is called **anterior chamber-associated immune deviation (ACAID)**. It allows the eye to tolerate corneal grafts of all HLA types and provides a natural example of the specific tolerance that transplant physicians aspire to induce against other transplanted tissues.

The liver is an organ that is also successfully transplanted across major differences in HLA class I and II, and it has even been claimed that transplant outcome is inversely correlated with the degree of HLA match. As a consequence, HLA type or cross-match are not assessed before liver transplantation; ABO type is the only genetic factor affecting donor selection. Clinical experience indicates that the liver is relatively refractory to either acute or hyperacute rejection, yet the use of cyclosporin A and tacrolimus has markedly improved the success of liver transplants. The liver has a specialized architecture and vasculature, and hepatocytes express very low levels of HLA class I proteins and no HLA class II. These properties, and the daily exposure of liver cells to the digestion products of a myriad of foreign proteins from the intestines, with their characteristic anti-inflammatory environment (see Chapter 10), could all contribute to the distinct immunobiology of the transplanted allogeneic liver.

At the other end of the spectrum from the eye and the liver is the bone marrow, the tissue for which transplantation is most sensitive to HLA disparity. The reasons for this are explained in the next part of this chapter.

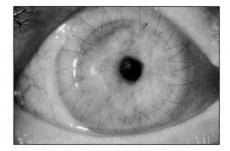

Figure 15.23 A successful corneal allograft. Successful transplantation of corneas from cadaveric donors requires no assessment of HLA type and no administration of immunosuppressive drugs. The lack of any rejection response is because of the naturally immunosuppressive environment in the anterior chamber of the eye and the lack of blood vessels in the cornea. Courtesy of Jerry Y. Niederkorn and James P. McCulley.

Summary

Of the many problems faced by transplant physicians, the most challenging has been the alloreactive immune response against ABO and HLA antigens that rejects transplanted organs. Hyperacute rejection caused by preformed antibodies is avoided by selecting donor organs that do not react with the recipient's antibodies. Acute rejection arises by the direct pathway of allorecognition, in which the recipient's T cells respond to allogeneic HLA class I and II molecules of the transplanted organ. The potential strength of the response is reduced by selecting a donor who is as closely as possible HLA-matched to the recipient. This tactic is complemented by a portfolio of immunosuppressive drugs and anti-T-cell monoclonal antibodies, which is used to prevent and swiftly treat any episode of acute rejection. After transplantation there is a gradual accommodation between the organ graft and the recipient that allows the doses of immunosuppressive drugs to be reduced. With time, most patients can be maintained on a level of immunosuppression that allows the immune system to recover and provide defense against infection. In the long term the major threat to health is the chronic rejection caused by the antibody response that the recipient makes to the allogeneic HLA molecules of the organ graft. This response is initiated by the indirect pathway of allorecognition, in which dendritic cells of the recipient process and present to T cells the allogeneic HLA molecules of the graft. The potential for rejection varies with the type of organ transplanted, as does the need to match donor and recipient for HLA type. The success of organ transplantation is such that an inadequate supply of donated organs is a major problem faced today.

Transplantation of hematopoietic stem cells

Whereas solid organ transplantation involves a partnership between transplant surgeons who carry out the procedures and transplant physicians who care for the recipients afterwards, bone marrow transplantation involves hematologists, oncologists, and radiologists. Like blood, bone marrow is a liquid transplant that is infused intravenously into patients whose own bone marrow has been weakened or destroyed by chemotherapy and irradiation. The key component of a **bone marrow transplant** is the hematopoietic stem cell (see Section 1-7, p. 12), which in time repopulates the haematopoietic system of the transplant recipient. In contrast to solid organ transplantation, the number of bone marrow donors far exceeds the number of patients needing a transplant. But a further difference, the greater sensitivity of bone marrow transplantation to HLA difference, means that many patients are unable to find a donor with an optimal HLA match. Bone marrow transplantation is the therapy of choice for an increasing variety of genetic and malignant diseases.

15-18 Bone marrow transplantation is a treatment for genetic diseases of blood cells

Bone marrow transplantation, which permanently replaces an individual's entire hematopoietic system, including the immune system, was initially developed as treatment for a variety of genetic diseases that impair the function of one or more types of hematopoietic cell. In Chapter 11 we saw how it is used to reconstitute healthy immune systems in children with genetic immunodeficiences (see Section 11-17, p. 349); it is also the therapy of choice for red-cell deficiencies, such as thalassemia major, sickle-cell anemia, and Fanconi's anemia, as well as a variety of conditions affecting leukocytes and causing immunodeficiency (Figure 15.24). In a bone marrow transplant, the important cells are the pluripotent stem cells that reconstitute the patient's immune system and also their red cells and platelets. In 2–3 weeks after a successful transplant, new circulating blood cells begin to be produced from the

Genetic diseases treatable by bone marrow transplatation	
Disease	Deficiency
Wiskott–Aldrich syndrome	Defective leukocytes and platelets
Fanconi anemia	Failure of bone marrow to make blood cells
Kostmann syndrome	Low neutrophil count (neutropenia)
Osteopetrosis	Defective bone modeling and remodeling by osteoclasts
Ataxia telangiectasia	Neurological impairment and immunodeficiency
Diamond–Blackfan syndrome	Low erythrocyte count (anemia)
Mucocutaneous candidiasis	Ineffective T-cell response to fungal infections
Cartilage–hair hypoplasia	Short limbs, fine sparse hair and immunodeficiency
Mucopolysaccharidosis	Various deficiencies of lysosomal enzymes
Gaucher's syndrome	Deficiency of the lysosomal enzyme glucocerebrosidase
Thalassemia major	Defective hemoglobin, impared erythrocyte function
Sickle-cell anemia	Defective hemoglobin, impaired erythrocyte function

Figure 15.24 Genetic diseases for which bone marrow transplantation is a therapy. As well as the diseases listed here, bone marrow transplantation is a therapy for many genetically determined immunodeficiencies, such as SCID (see Chapter 11).

transplanted marrow. This is a sign that the pluripotent stem cells have colonized the bones, the process known as **engraftment**. With time the transplant replaces the defective hematopoietic system with one that is normal.

The logistics of bone marrow transplantation are different from those of organ transplantation, more resembling a blood transfusion. The donors are alive and healthy, and the transplanted tissue is given by an intravenous infusion that involves no surgery. The immunology of bone marrow transplantation is also different from that of other types of transplantation. This is because it involves transplantation of the immune system, cells of which are present in almost every organ and tissue of the body. Whereas kidney transplantation involves suppression of the recipient's T-cell response to prevent rejection of the graft, in bone marrow transplantation the recipient's immune system is deliberately destroyed to the point where the patient would not survive without the transplant. This conditioning regimen is called **myeloablative therapy**, because it destroys the bone marrow, and is accomplished by a combination of cytotoxic drugs and irradiation. Myeloablative therapy serves two purposes: the first is to prevent rejection of the grafted cells by the recipient's T cells; the second is to kill all the hematopoietic cells within the recipient's bone marrow and provide room for the transplanted hematopoietic stem cells to interact with bone marrow stromal cells.

Within a few weeks after transplantation, the hematopoietic system begins to be reconstituted. The cells of innate immunity, for example granulocytes and NK cells, recover more quickly than the B and T cells of adaptive immunity. When fully reconstituted the patient is a chimera in which the hematopoietic cells are of donor genotype and the rest of the cells are of recipient genotype. The patient's new immune system can then become tolerant of both donor and recipient HLA allotypes. A critical feature of the T cells of the patient's new immune system is their positive selection by thymic epithelial cells expressing the recipient's HLA allotypes. For those T cells to be activated by infections, they also need to interact with donor-derived dendritic cells that present pathogen-derived antigens on donor HLA allotypes (see Section 11-17, p. 349). This can occur only if the recipient and donor have HLA allotypes in common; the more HLA allotypes they share, the better it works.

15-19 Allogeneic bone marrow transplantation is the preferred treatment for many cancers

As well as being used to replace defective hematopoietic systems with functional ones, bone marrow transplantation is also an important treatment for many cancer patients, particularly those with tumors of immune-system cells. In treating cancer patients with chemotherapy and irradiation, the advantages gained by killing off the malignant cells have to be balanced against the damage caused to normally proliferating vital tissues. Of these, the bone marrow is the most susceptible. Bone marrow transplantation offers the opportunity to increase anticancer treatment beyond the point where it becomes lethal, after which the patient is rescued with an allogeneic transplant from a healthy HLA-matched donor. Since the first bone marrow transplants were performed some 35 years ago, the procedure has been used against an increasing range of cancers (Figure 15.25).

15-20 The alloreactions in bone marrow transplantation attack the patient, not the transplant

Because of the severe conditioning regimen, rejection mediated by recipient T cells responding to HLA allotypes of a bone marrow graft is not the problem that it is in organ transplantation. Instead, the main cause of damaging alloreactions are mature T cells in the transplanted bone marrow that respond to

Malignant diseases treatable by bone marrow transplantation	
Allogeneic transplant	Autologous transplant
Aplastic anemia	Leukemia
Leukemia	Acute myelogenous
Acute myelogenous	leukemia (AML)
leukemia (AML)	Acute lymphocytic
Acute lymphocytic	leukemia (ALL)
leukemia (ALL)	Multiple myeloma
Chronic myelogenous	Non-Hodgkin's
leukemia (CML)	lymphoma
Myelodysplasia	Hodgkin's disease
Multiple myeloma	Solid tumors
Non-Hodgkin's	Ovarian
lymphoma	Testicular
Hodgkin's disease	Neuroblastoma

Figure 15.25 Cancers for which bone marrow transplantation is a therapy.

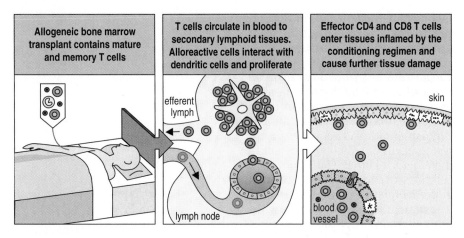

| Allogeneic bone marrow transplant contains mature and memory T cells | T cells circulate in blood to secondary lymphoid tissues. Alloreactive cells interact with dendritic cells and proliferate | Effector CD4 and CD8 T cells enter tissues inflamed by the conditioning regimen and cause further tissue damage |

Figure 15.26 Graft-versus-host disease is due to donor T cells in the graft that attack the recipient's tissues. After bone marrow transplantation, any mature donor CD4 and CD8 T cells present in the graft that are specific for the recipient's HLA allotypes become activated in secondary lymphoid tissues. Effector CD4 and CD8 T cells move into the circulation and preferentially enter and attack tissues that have been most damaged by the conditioning regimen of chemotherapy and irradiation: skin, intestines, and liver.

the recipient's HLA allotypes in a graft-versus-host reaction (Figure 15.26). This GVHR is the major cause of morbidity and mortality after bone marrow transplantation. The condition it causes, acute graft-versus-host disease (GVHD), can attack almost every tissue of the body but principally involves the skin, the intestines, and the liver. In essence, GVHD is an acute autoimmune disease that can prove fatal. The severity of acute GVHD varies, four grades being defined in clinical diagnosis (Figure 15.27). The characteristic skin rash of GVHD tends to develop with the kinetics of a primary immune response during the 10–28 days after transplantation. The fine, diffuse erythematous rash begins on the palms of the hands, the soles of the feet, and on the head, and then spreads to the trunk. The intestinal reaction causes cramps and diarrhea, and inflammation of the bile ducts in the liver causes hyperbilirubinemia and a rise in the levels of liver enzymes in the blood. Methotrexate in combination with cyclosporin A is used to reduce the incidence and severity of GVHD.

Mature T cells from the transplant will circulate in the recipient's blood and enter secondary lymphoid tissues, where they interact with the recipient's dendritic cells. Alloreactive T cells will be stimulated to divide and differentiate into effector cells, which will travel via the lymph and the blood to inflamed tissues. The conditioning regimen, which seeks to destroy the rapidly dividing bone marrow cells, also damages other tissues in which there is normally much cellular proliferation. Prominent among these are the skin, the intestinal epithelium, and the hepatocytes of the liver: the targets for GVHD. The conditioning regimen preferentially creates inflammation in these tissues, which has been described as a 'cytokine storm.' This environment causes dendritic cells to activate and migrate from the inflamed tissues to draining secondary lymphoid tissue, where they stimulate alloreactive T cells. More importantly, it makes these tissues more accessible to alloreactive effector T cells.

Tissue reactions in the four grades of graft-versus-host disease			
Grade	Skin	Liver	Gastrointestinal tract
I	Maculopapular rash on <25% of body surface	Serum bilirubin 2–3 mg/dl	>500 ml diarrhea/day
II	Maculopapular rash on <25–50% of body surface	Serum bilirubin 3–6 mg/dl	>1000 ml diarrhea/day
III	Generalized erythroderma	Serum bilirubin 6–15 mg/dl	>1500 ml diarrhea/day
IV	Generalized erythroderma with blistering and desquamation	Serum bilirubin 15 mg/dl	Severe abdominal pain with or without intestinal obstruction

Figure 15.27 Characteristics of the four grades of graft-versus-host disease.

Because the number of mature T cells present in the transplant is limited, the duration of acute GVHD is usually restricted to the first few months after transplantation. The chronic rejection reactions seen in kidney transplant patients have their analog in the chronic GVHD experienced by 25–45% of bone marrow transplant patients who survive longer than 6 months after the transplant. The course of this condition resembles that of an autoimmune disease, and the cumulative effect is to produce severe immunodeficiency, leading to recurrent life-threatening infections.

15-21 Matching donor and recipient for HLA class I and II is particularly important in bone marrow transplantation

Almost all bone marrow transplant patients suffer to some extent from GVHD. The severity of GVHD correlates strongly with the degree of HLA mismatch (Figure 15.28), and, because of the potentially fatal consequences of GVHD, the clinical outcome and success of bone marrow transplantation is more sensitive to HLA mismatching than is solid organ transplantation. Whereas donors are in short supply for organ transplantation, numerous bone marrow donors are available, because bone marrow, like blood, is donated by healthy individuals without compromising their immunological or hematological functions. Marrow is usually aspirated from the iliac crests of the pelvis under anesthesia. For patients without HLA-identical siblings to donate bone marrow, there are national and international registries of HLA-typed people that are searched to identify histocompatible donors. Worldwide, some 5 million potential donors have been HLA typed for this purpose.

Despite such numbers, about 30% of cancer patients who are clinically eligible for a transplant do not find a suitable HLA-matched donor. To help these patients, a procedure called **autologous bone marrow transplantion** was devised. Here, samples of the patient's own bone marrow are taken before the remainder is destroyed by the treatments given for the cancer. After separation of the bone marrow stem cells from any remaining tumor cells, the stem cells are reinfused into the patient. Autologous transplants avoid the problems of histoincompatibility and GVHD and thus patients do not require immunosuppression. However, their use is limited by the rate of relapse of malignant disease, which is greater than with allogeneic transplants.

Hematopoietic stem cells for transplantation can now be obtained by less invasive procedures than bone marrow aspiration. One method is to mobilize hematopoietic stem cells from the donor's bone marrow into the peripheral blood by treatment with granulocyte colony-stimulating factor (G-CSF) and granulocyte–macrophage colony-stimulating factor (GM-CSF), which are sometimes combined with cyclophosphamide chemotherapy in autologous

Figure 15.28 The clinical outcome of treatment with bone marrow transplantation is improved as the HLA match becomes closer. Two parameters of clinical outcome are shown: survival of the patient (left panel) and the probability of getting severe GVHD (right panel). Severe GVHD is defined as grades III and IV (see Figure 15.27). HLA class I and class II antigens were initially defined by serological HLA typing, and these antigens were subsequently subdivided into different alleles by DNA sequencing. For example, the HLA-A2 antigen comprises more than 130 alleles that differ from each other by one or a few nucleotide substitutions. An 'allele match' means that the donor and recipient have identical alleles at the HLA-A, HLA-B, HLA-C, and HLA-DR loci, the major factors in transplantation. A single class I mismatch means that the donor and recipient differ by one allele for one HLA class I locus, and similarly for HLA class II. Transplants with a class I and class II mismatch have one mismatched allele for HLA class I and one for class II. Overall, the fewer the number of mismatches, the better the survival and health of the transplanted patient. Data courtesy of E. Petersdorf.

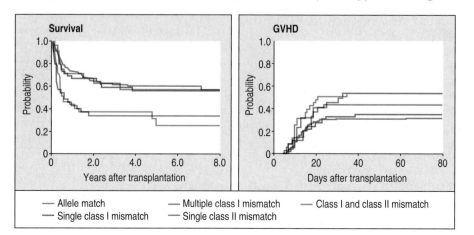

Survival — Years after transplantation — Probability

GVHD — Days after transplantation — Probability

— Allele match
— Single class I mismatch
— Multiple class I mismatch
— Single class II mismatch
— Class I and class II mismatch

transplants. Leukocytes are then selectively removed from the blood by a process called leukapheresis, which involves an apheresis machine to which the donor's circulation is connected for several hours. The CD34 surface glycoprotein is used as a marker for hematopoietic stem cells. A cell fraction enriched for the CD34-expressing stem cells is isolated from the removed leukocytes and used as the transplant. Between a quarter and half a billion CD34-positive cells are needed to ensure prompt engraftment after transplantation.

Another source of hematopoietic stem cells is umbilical cord blood, obtained from the placenta after birth. Umbilical cord blood is rich in stem cells and also contains fewer cells that contribute to the GVHR. Although engraftment is slower with a cord blood transplant, there is less GVHD and a greater HLA disparity can be tolerated than in transplantation with bone marrow or blood-derived stem cells. Since its first use in 1988 to treat a child with Fanconi anemia, more than 7000 cord blood transplants have been performed and cord blood banks have been established in several countries. Because stem cells for transplantation are now being obtained from sources other than bone marrow, the more general terms **hematopoietic stem cell transplantation** and **hematopoietic cell transplantation** are increasingly used instead of bone marrow transplantation.

15-22 HLA-identical bone marrow transplants cause GVHD through recognition of minor histocompatibility antigens

When seeking an allogeneic bone marrow donor, the 'gold standard' is a healthy HLA-identical sibling. Thousands of such transplants have been carried out. But despite the HLA compatibility, many of these patients were still affected by GVHD, particularly males receiving bone marrow transplants from their HLA-identical sisters. In this circumstance, GVHD is caused by T cells specific for self-peptide antigens that are presented by HLA class I or II on the brother's cells, but not on his HLA-identical sister's cells. In the sister, such T cells are not deleted during thymic selection and will be transferred to the brother in the bone marrow transplant. The peptide antigens involved are known as H-Y antigens because they derive from proteins encoded on the Y chromosome, which is carried only by males, and differ in sequence from the homologous protein encoded on the X chromosome. Several H-Y antigens are defined and include peptides presented by HLA-A, HLA-B, HLA-DQ and HLA-DR isoforms (Figure 15.29). Alloantigens such as the H-Y antigens, in which the allogeneic difference is due to the bound peptide and not to the MHC molecule, are called **minor histocompatibility antigens**, and the genes encoding them **minor histocompatibility loci**.

Minor histocompatibility antigens arise not only because of the differences between the sexes but also from the wide variety of polymorphic human proteins encoded by autosomal genes for which the allotypic differences affect the way in which peptides derived from the self protein are processed and presented by HLA class I or II (Figure 15.30). The majority of minor

Figure 15.29 The minor histocompatibility H-Y antigens are diverse and expressed only by males. For each peptide antigen listed here, its amino acid sequence, the HLA class I or class II molecule that presents it, and the gene encoding the protein from which it is derived are shown. Where known, the amino acid substitutions that distinguish the H-Y antigen (upper letters) from the homologous sequence encoded by the X chromosome (lower letters) are indicated. The H-Y antigens presented by A*01 and A*02 have an additional cysteine disulfide bonded to the cysteine indicated by C⁺ in the peptide sequence.

Minor histocompatibility H-Y antigens		
Peptide antigen	HLA restriction	Gene
IVD$\overset{C^+}{\underset{S}{}}$LTEMY	A*01	USP9Y
FIDSYIC⁺QV	A*02	SMCY
EVLLRPGLHFR	A*33	TMSB4Y
SP$\overset{S}{\underset{A}{}}$VDKA$\overset{R}{\underset{Q}{}}$AEL	B*07	SMCY
LP$\overset{H}{\underset{R}{}}NHT\overset{D}{\underset{N}{}}$L	B*08	UTY
$\overset{R}{\underset{G}{}}$ESEE$\overset{E}{\underset{A}{}}S\overset{V}{\underset{P}{}}$SL	B*4001	UTY
TIRYPDP$\overset{V}{\underset{L}{}}$I	B*52	RPS4Y1
HIE$\overset{N}{\underset{S}{}}FSD\overset{FD}{\underset{VE}{}}$MGE	DQB*05	DOX3Y
SKGRYIPPHLR	DRB1*1501	DOX3Y
VIKVNDTVQI	DRB3*0301	RPS4Y1

Autosomal minor histocompatibility antigens				
Name	Peptide antigen	HLA restriction	Chromosome	Gene
HA-3	V$\frac{T}{M}$EPGTAQY	A*01	15	LBC oncogene
HA-2	YIGEVLVS$\frac{V}{M}$	A*02	7	Myosin 1G
HA-8	$\frac{R}{P}$TLDKVLEV	A*02	9	KIAA020
HA-1	VL$\frac{H}{R}$DDLLEA	A*02	19	KIAA0223
ACC-1	DYLQ$\frac{Y}{C}$VLQI	A*24	15	BCL2A1
UGT2B17	AELCNPFLY	A*29	13	UGT2B17
LRH-1	TPNQRQNVC	B*07	17	P2X5
ACC-2	KEFED$\frac{D}{G}$IINW	B*44	15	BCL2A1
HB-1	EEKRGSL$\frac{H}{Y}$VW	B*44	5	Unknown

Figure 15.30 Minor histocompatibility antigens encoded by autosomal genes. For each antigen listed here, the amino acid sequence, the HLA class I or class II molecule that presents it (HLA restriction), the chromosomal location of the gene that encodes it, and the name of the gene are given. Where known, the amino acid substitutions that distinguish the H-Y antigen (upper letters) from the homologous sequence encoded by the X chromosome (lower letters) are indicated.

histocompatibility antigens are presented by HLA class I, which is consistent with these peptides being the breakdown products of intracellular proteins (Figure 15.31). For some, the allelic difference determines whether or not the peptide will bind to the HLA molecule, but for others it can alter the stability of the peptide, its post-translational modification and whether a protein is actually made.

15-23 Some GVHD helps engraftment and prevents the relapse of malignant disease

One way to reduce the severity of GVHD is to remove T cells from the bone marrow graft before it is given to the patient. This is done by using lectins or monoclonal antibodies that bind selectively to mature T cells. Although markedly reducing GVHD, this procedure leads to a higher incidence of graft failure and, for cancer patients, a higher incidence of disease recurrence (relapse). This indicates that T-cell alloreactions help engraftment by subduing residual

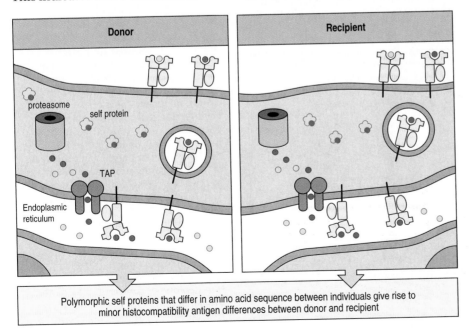

Polymorphic self proteins that differ in amino acid sequence between individuals give rise to minor histocompatibility antigen differences between donor and recipient

Figure 15.31 Minor histocompatibility antigens are peptides derived from polymorphic proteins other than HLA class I and class II molecules. Self proteins are routinely digested by proteasomes within the cell's cytosol, and peptides derived from them are delivered to the endoplasmic reticulum, where they can bind to MHC class I molecules and be delivered to the cell surface. If a polymorphic protein differs between the graft donor (shown in red on the left) and the recipient (shown in blue on the right), it can give rise to an antigenic peptide (red on the donor cell) that can be recognized by the recipient's T cells as nonself and elicit an immune response. Such antigens are the minor histocompatibility antigens.

activity of the recipient's immune system and that they also eliminate cancer cells that escaped the conditioning regimen. Consistent with this are observations that graft failure and disease relapse are higher for bone marrow transplants in which alloreactions cannot occur—autologous transplants and transplantation between identical twins. Also pointing to the same conclusion is the experience that the acute GVHD due to minor histocompatibility antigens improves the long-term clinical outcome for transplants between HLA-identical siblings.

When alloreactive T cells in the graft help to rid the patient of residual leukemia cells it is called a **graft-versus-leukemia** (**GVL**) **effect** or a **graft-versus-tumor** (**GVT**) **effect**. A current movement in hematopoietic stem cell transplantation is to use new protocols that promote the GVL reaction and place less emphasis on using chemotherapy and irradiation to eliminate the tumor. This allows less severe conditioning regimens to be used that do not completely disable the patient's hematopoietic system; after treatment the patient is less severely immunocompromised and recovers more quickly. Patients treated with the older protocols would often need to be isolated for several weeks in intensive care, whereas patients receiving these so-called 'mini-transplants' need not be isolated. They spend less time in hospital, and some can even be treated as outpatients. The mini-transplant has the potential to provide treatment to many cancer patients who for reasons of age or previous treatment history are unlikely to survive therapy involving the prior destruction of their bone marrow.

One approach to producing a GVL effect is to give patients transfusions of donor lymphocytes or T cells after they have received the hematopoietic stem cell graft. Such donor lymphocyte transfusions are given at a time when the inflammation caused by the conditioning regimen has subsided and the likelihood of severe GVHD is diminished.

15-24 NK cells also mediate GVL effects

Although the trend in HLA-matched transplants is to use gentler protocols with nondestructive conditioning regimens, a very different strategy is proving helpful to some of the 30% of patients with leukemia who cannot find an HLA-matched donor. Most of these patients have a willing donor in their family who shares one HLA haplotype with the patient but differs in the second. Potential donors are mothers, fathers, and 50% of siblings. This type of transplant is called a **haploidentical transplant** (Figure 15.32). Incompatibility for

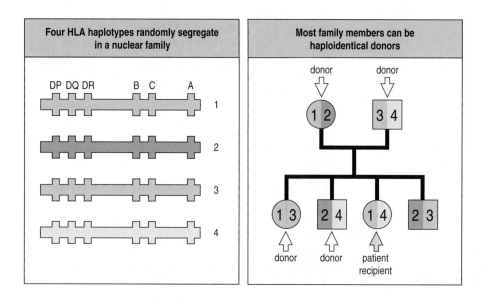

Figure 15.32 Almost all patients who need a hematopoietic cell transplant have an HLA-haploidentical family member who is willing to be the donor. In nuclear families, where the children derive from the same two parents, there are four different HLA haplotypes (left panel). For any child in the family, either parent can provide an HLA-haploidentical transplant as, on average, can 50% of the siblings (right panel). In contrast, neither parent and only 25% of siblings, on average, can provide an HLA-identical transplant. In determining the suitability of siblings as a donor for a brother or sister needing a hematopoietic stem-cell transplant, it is not uncommon to detect the presence of HLA haplotypes contributed by neither official parent. Such discoveries are treated with the utmost discretion.

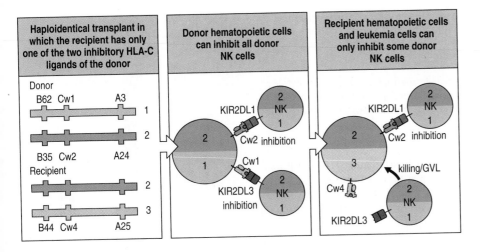

Figure 15.33 Alloreactive NK cells can provide a graft-versus-leukemia effect in patients receiving a haploidentical hematopoietic cell transplant. The HLA class I genotypes of the donor and of the recipient, a patient with acute myelogenous leukemia, are shown in the left panel. The key difference is at the HLA-C locus. The donor has Cw1 (with asparagine at position 80) and Cw2 (with lysine at position 80); the recipient has Cw2 and Cw4, both of which have Lys80. The NK cells of donor genotype that emerge soon after transplantation can be divided into two groups according to whether they are inhibited by Cw2 interacting with the NK-cell receptor KIR2DL1 or by Cw1 interacting with the receptor KIR2DL3 (center panel). Hematopoietic cells of the recipient, including residual leukemia cells, can inhibit only the first group of NK cells and not the second group, which kill the residual leukemia cells (right panel). This graft-versus-leukemia (GVL) effect stems from the fact that the recipient's HLA type lacks an inhibitory HLA-C ligand that is part of the donor's HLA type, and so donor NK cells can attack and kill the recipient's leukemia cells.

a complete HLA haplotype has the potential to generate strong, fatal alloreactions. To prevent GVHD, the graft is stringently depleted of T cells and the recipient is infused with anti-T-cell antibodies. Graft rejection is prevented through a combination of an intensive conditioning regimen and a larger than normal dose of hematopoietic stem cells.

Patients given a haploidentical transplant get little or no GVHD and require no further immunosuppressive treatment after transplantation. As their immune systems reconstitute, alloreactive NK cells can emerge. These provide a GVL effect that reduces the incidence of leukemic relapse. The occurrence and specificity of the NK-cell-mediated alloreactions are determined by the interactions of inhibitory KIR receptors with HLA-B and HLA-C ligands and are predictable from the HLA types of the donor and recipient (see Sections 10-26 and 10-27, pp. 319 and 321) (Figure 15.33). As a rule, NK-cell alloreactions occur when the recipient's HLA class I allotypes provide ligands for fewer types of inhibitory KIR than the donor's HLA class I allotypes. Patients with acute myelogenous leukemia (AML) benefit from such NK-cell alloreactions, whereas patients with acute lymphocytic leukemia (ALL) do not. The alloreactive NK-cell response wanes and is undetectable 4 months after transplantation. With full reconstitution of the immune system, the NK-cell population becomes tolerant of both recipient and donor cells.

15-25 Hematopoietic cell transplantation can induce tolerance to solid organ transplants

During gestation the blood circulation of dizygotic twins is joined in about 8% of cases. After birth, the hematopoietic system of each twin is a chimera that contains cells of both genotypes. In situations where the twins have different HLA types, they become tolerant of each other's cells, as shown by successful skin transplantation and lack of stimulation in a mixed-lymphocyte culture (Figure 15.34). Blood cell chimerism is also observed in the fraction of patients receiving solid organ transplants who have been successfully weaned off immunosuppressive drugs without adverse effect on either the graft or the recipient. For these patients, chimerism is due to the persistence of donor hematopoietic cells that were introduced with the organ transplant. The transfusion effect (see Section 15-9) could also be due to chimerism developed by hematopoietic stem cells in transfused blood. Such observations have suggested that combining a mini bone marrow transplant with a solid organ transplant could facilitate the development of active long-lasting tolerance that would not depend on immunosuppressive therapy. This is an active area of transplantation research in which some success has been seen both with HLA-matched donors and recipients and with family donors mismatched for one HLA haplotype. After such combined transplants there is an initial

Figure 15.34 Dizygotic twins who have had a common blood circulation during gestation are tolerant of each other's tissues. The stimulation index is derived from the results of mixed lymphocyte reactions (MLRs) in which the lymphocytes of each twin are stimulated by the other twin and an unrelated control. When the sister's lymphocytes are used as stimulators in the MLR against the brother's, and vice versa, a negligible response is obtained in comparison with the strong response that both make to lymphocytes from an unrelated control. Data courtesy of Frans Claas.

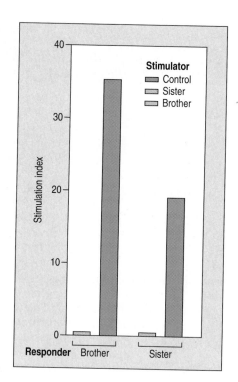

phase of alloreactivity and a requirement for immunosuppressive drugs, followed by an accommodation that then leads to a state of tolerance associated with FoxP3-expressing regulatory T cells. Immunosuppressive therapy can be discontinued after a year, and a stable state of tolerance and health has been maintained for up to 5 years.

Summary

Hematopoietic stem cell transplantation is used to correct genetic immunodeficiencies and to eliminate leukemias, lymphomas, and other tumors of hematopoietic cells. Before the transplant the patient undergoes a conditioning regimen that destroys the existing bone marrow, and tumor cells in the case of cancer patients. The patient is also given immunosuppressive drugs and antibodies that prevent rejection of the graft and allow the hematopoietic stem cells to seed the bone marrow, where they divide and differentiate. The immunosuppression also reduces the graft-versus-host disease caused by mature T cells in the graft that respond to any differences in the major and minor histocompatibility antigens of the patient and the donor. The success of bone marrow transplantation is well correlated with the extent of the HLA match, and in all circumstances the donor of choice is an HLA-identical sibling with some incompatibility for minor histocompatibility antigens. The resultant graft-versus-host reaction to the minor histocompatibility antigens helps in the establishment of the graft and, by eliminating residual tumor cells, prevents relapse.

Summary to Chapter 15

Many human diseases involve the malfunction of a single organ or tissue that causes incapacitation or death. With transplantation, diseased tissues are replaced by healthy ones, and this can lead to improved health and longer life. Although blood transfusions are short-lived, they demonstrated the clinical value of transplantation long before it was possible for other tissues and organs. Similarly, the observation of life-threatening transfusion reactions and the serological analysis of polymorphic red-cell antigens revealed the importance of immunogenetics. For transplantation of other tissues the proof of principle came from successful transplantations between genetically identical twins, but for such procedures to have any general application it was necessary to deal with the highly polymorphic white-cell antigens—HLA class I and class II—that by presenting peptides to T cells initiate adaptive alloreactive immune responses with the potential to destroy the transplant in solid organ transplantation or to destroy the recipient in hematopoietic stem cell transplantation.

In the practice of transplantation, a general principle is to avoid preformed antibodies that can bind to the transplanted tissues. Antibodies are potent weapons and can neutralize a transplant with the same effect with which they neutralize a pathogen. Avoidance is achieved empirically by cross-match tests. The potential for stimulating an alloreactive response in transplantation is reduced by finding a donor whose HLA type matches that of the recipient.

For patients with common HLA types this is easily achieved, but for patients with one of the many rare HLA types a compromise is often necessary. To prevent the stimulation of an alloreactive response transplant, patients are given a variety of immunosuppressive drugs and antibodies, both before and after transplantation, that prevent the activation and proliferation of T cells. The patients are carefully monitored and any evidence of graft rejection or graft-versus-host disease is treated with additional immunosuppression. The drug doses are gradually lowered to a maintenance level.

The impressive improvement in transplantation success over the past 30 years has come principally from more effective drugs and antibodies that nonspecifically suppress the T cells of the immune system. Throughout this period the goal of transplant immunologists has been to selectively suppress the immune response to the allogeneic transplant and release the transplant patient from a lifetime of immunosuppressive drugs. Various lines of evidence suggest this might be achieved by establishing a chimeric hematopoietic system in the transplant recipient that is populated by hematopoietic stem cells of both donor and recipient origin.

Questions

15–1 How do the clinical objectives of transplantation differ from those of vaccination?

15–2 Explain why, in principle, an organ transplanted from any donor other than an identical twin is almost certain to be rejected in the absence of any other treatment.

15–3 The term _____ is used to describe polymorphic antigens that vary between individuals of the same species.
   a. xenoantigens
   b. immunoantigens
   c. alloantigens
   d. histoantigens
   e. autoantigens.

15–4 Contrast acute rejection and chronic rejection.

15–5
   A. Explain why an organ transplant made between a donor of blood group AB and a recipient of blood group O will always be rejected, even if it is perfectly HLA matched and the recipient has been given immunosuppressant drugs. What is this type of rejection called?
   B. Give another example of ABO incompatibility between donor and recipient that would lead to this type of rejection.
   C. What other antigen incompatibilities, other than those of blood group, are most likely to provoke this type of rejection?
   D. Which pre-surgical laboratory test should be performed to prevent this type of rejection?

15–6 We learned in Chapter 5 that the benefit of having and expressing multiple MHC class I and class II genes is that it increases the number and variety of pathogen-derived peptide antigens that can potentially be presented to T cells. If more is better, then why has natural selection not favored the evolution of more than three genes each for MHC class I and MHC class II?

15–7
   A. Identify three general classes of drug that are used to suppress acute transplant rejection, and provide examples of each class.
   B. What side-effects and toxic effects are associated with each class of drug?

15–8 Explain how cyclosporin A acts as an immunosuppressant drug.

15–9 Graft-versus-host disease (GVHD) is a consequence of _____.
   a. mature T lymphocytes from the donor mounting an immune response against tissue of the recipient
   b. mature T lymphocytes from the recipient mounting an immune response against tissue of the donor
   c. mismatching A, B, and O antigens between donor and recipient
   d. mismatching rhesus antigen between donor and recipient
   e. antibodies of the donor stimulating NK cell antibody-dependent cell-mediated cytotoxicity (ADCC) of tissues of the recipient.

15–10
   A. Explain how mouse monoclonal antibodies (MoAbs) can be used to suppress acute graft rejection.
   B. What feature of these mouse antibodies compromises their effectiveness _in vivo_ and limits their use?

15–11 Which of the following best explains why a bone marrow donor needs to be HLA-matched to the recipient?
   a. The bone marrow transplant contains enough mature T cells to reconstitute the recipient and the recipient provides the antigen-presenting cells.
   b. The recipient's MHC molecules mediate positive selection of thymocytes in the thymus that interact with donor-derived MHC molecules in the periphery.
   c. Reconstituted T cells are restricted by donor, not recipient, HLA allotypes.
   d. Without an HLA match, the donor-derived thymocytes undergo negative selection.
   e. If the donor is not HLA matched, the reconstituted T cells will be autoreactive.

15–12
   A. Explain why a boy with leukemia who receives a bone marrow transplant from his sister that is perfectly matched for MHC class I and class II is still likely to get graft-versus-host disease.
   B. Which effector T cells are usually involved in this reaction, and why?

15–13
   A. Are the criteria for selecting suitable donors the same for liver and bone marrow transplants?
   B. Why or why not?

15–14 From a clinical perspective explain how the logistics of organ transplantation differ from those for a bone marrow transplant.

15–15 Indicate whether each of the following statements is true (T) or false (F).
   ___a. ABO or rhesus antigen mismatches stimulate cytotoxic T-cell responses.
   ___b. There are polymorphic antigens other than ABO and rhesus antigens that can cause type II hypersensitivity reactions.
   ___c. Cross-matching has now been replaced with routine use of DNA-based methods.
   ___d. Lymphocytes and erythrocytes express HLA class I and II molecules.
   ___e. Platelet transfusions are used to replace fluid and prevent bleeding.

15–16 Carter Petersen received an apparently successful bone marrow transplant from his HLA-identical sister. Twenty-five days later he developed watery diarrhea, a patchy rash on his face and neck that spread to his trunk, and jaundice. He showed improvement after treatment with cyclosporin A and methotrexate. What is the most likely explanation for his symptoms?
   a. acute rejection of the graft
   b. graft-versus-host disease
   c. cytomegalovirus infection
   d. type II hypersensitivity reaction
   e. host-versus-graft disease.

15–17 Richard French, 53 years old, was diagnosed with chronic myelogenous leukemia. His elder brother Don is HLA-haploidentical and will donate bone marrow. Richard's oncologist has recommended him to a medical center that favors using bone marrow depleted of mature T cells prior to infusion. The most likely rationale for employing the practice of T-cell depletion is that _____.
   a. T-cell depletion will remove alloreactive T cells from the donor and prevent the potential for graft-versus-host disease (GVHD)
   b. mature T-cell chimerism is required to establish long-term tolerance
   c. because Don is HLA-haploidentical and male, there is no risk of alloreactivity toward major or minor histocompatibility antigens
   d. because of Don's age, the expected bone marrow harvest is already marginal for successful engraftment, and depletion measures would compromise the yield of stem cells
   e. the benefit of using a cocktail of immunosuppressive drugs outweighs the risk of contaminating the bone marrow during T-cell depletion.

15–18 Forty-four-year old Danielle Bouvier is on the waiting list for a kidney transplant and is receiving weekly dialysis. Her HLA type is: HLA-A: 0101/0301; HLA-B: 0702/0801; HLA-DRB1: 0301/0701. Today, Danielle's physician informed her that several potential kidney donors are available. Which of the following would be the most suitable?
   a. A: 0301/0201; B: 4402/0801; DRB1: 0301/0403
   b. A: 0301/2902; B: 1801/0801; DRB1: 0301/0701
   c. A: 2902/0201; B: 0702/0801; DRB1: 0301/13011
   d. A: 0101/0101; B: 5701/0801; DRB1: 0701/0701
   e. A: 0101/0301; B: 0702/5701; DRBA: 0403/0301.

Melanoma, an aggressive form of skin cancer that is difficult to treat and for which an effective immunotherapy is sought.

Chapter 16

Cancer and Its Interactions with the Immune System

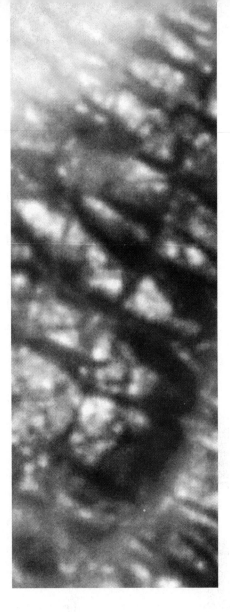

Cancer is a diverse collection of life-threatening diseases that is caused by abnormal and invasive cell proliferation. It accounts for about 20% of deaths in the industrialized countries; worldwide there are some 6 million new cases of cancer each year, and half of these people will die from the disease. Cancer is largely a disease of older people—the median age for cancer is 70 years—and in the industrialized societies where both life expectancy and the average age of the population are increasing, so is fear of cancer.

For a normal cell to evolve into a disease-causing cancer is a lengthy process. It entails the accumulation of several independent mutations, which individually have escaped the cell's intrinsic safety mechanisms that repair mutated DNA and cause abnormally behaving cells to die by apoptosis. Cancer cells resemble virus-infected cells in that their internal affairs have been perturbed and reordered from those of normal cells, and indeed some cancers are the consequence of viral infection. The immune system can identify, control, and eliminate cancer cells by using the same cells and molecules that are employed against virus-infected cells. Most emerging cancers are likely to be eliminated by the immune system before they are detectable or cause any symptoms. Only a minority of cancers defeat the immune system and progress to cause the diseases that are so feared.

In treating cancer, physicians resort to surgery, radiation, and cytotoxic drugs, sometimes referred to 'slash, burn, and poison.' Although these treatments give remission or cure to some patients, more often they are limited by the incomplete elimination of cancer cells and the deleterious side-effects of the treatment. Today, fewer than 10% of patients with metastatic colon cancer can expect to survive for more than 5 years after diagnosis, and for pancreatic cancer the percentage is 5%. For over a century, cancer immunologists have sought to harness a patient's immune system to augment conventional therapies. Although ideas for boosting cancer immunity have been regularly shown to work to some extent in animal models, the development of useful routine immunotherapies for human cancer is still at an early stage. Increased understanding of the mechanisms by which an immune response to a cancer is started and sustained has led to several promising new initiatives. In this chapter we first consider the processes by which cancers emerge in the body. Then we examine the immune response to cancer. Finally, we turn to the ways in which the immune system is being manipulated and stimulated to make stronger and more specific responses to cancer cells.

16-1 Cancer results from mutations that cause uncontrolled cell growth

Maintaining the human body involves continuing cell division to replace worn-out cells, repair damaged tissue, and mount immune responses against invading pathogens. Of the order of 10^{16} cell divisions are estimated to occur in a human body during a lifetime. An essential preparation for cell division is DNA replication, which makes two identical copies of the diploid genome. Although the enzymes that replicate DNA are extremely accurate and use proofreading mechanisms to correct mistakes, rare errors are still made. Additional changes in DNA arise from chemical damage that escapes the DNA repair machinery.

Changes in DNA are called **mutations**. They include the substitution, insertion, and deletion of nucleotides, recombination between different members of a gene family, and chromosomal rearrangements. Mutations in the germline—the eggs and sperm—provide the variation that allows the human species to diversify and evolve. In contrast, mutations in somatic cells affect only the individual in which they arise. Most somatic mutations are never noticed because their effects, if any, are manifested in a single cell. There are, however, certain mutations that abolish the normal controls on cell division and cell survival. When that happens, the mutant cell proliferates to form an expanding population of mutant cells that eventually disrupts the body's physiology and organ function, causing the diseases that are collectively called cancer.

Tumor, which means swelling, and **neoplasm**, which means new growth, are words used synonymously to describe tissue in which cells are multiplying abnormally. The branch of medicine that deals with tumors is called **oncology**, the prefix *onco* deriving from the Greek word for swelling, *ogkos*. Not all tumors are malignant: **benign tumors**, such as warts, are encapsulated, localized, and limited in size; the **malignant tumors**, in contrast, can continually increase their size by breaking through basal laminae and invading adjacent tissues (Figure 16.1).

The word **cancer**, meaning an evil spreading in the manner of a crab (Latin, *cancer*), is used to describe the diseases caused by malignant tumors. In addition to spreading out locally from their site of origin, cancer cells can be carried by lymph or blood to distant sites, where they initiate new foci of cancerous growth. This mode of spreading is called **metastasis**, the site of origin being called the primary tumor and the sites of spreading being called secondary tumors. Cancers arise most commonly in tissues that are actively

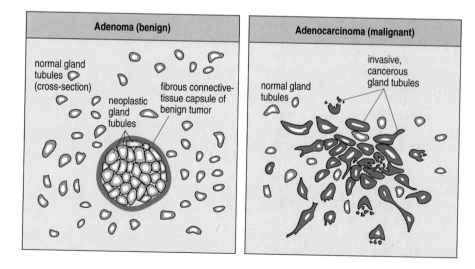

Figure 16.1 The contrast between benign and malignant tumors derived from the same tissue. The diagram illustrates tumors found in the breast. Adenoma is the general name given to a benign tumor of glandular tissue; malignant tumors of glandular tissues are called adenocarcinomas.

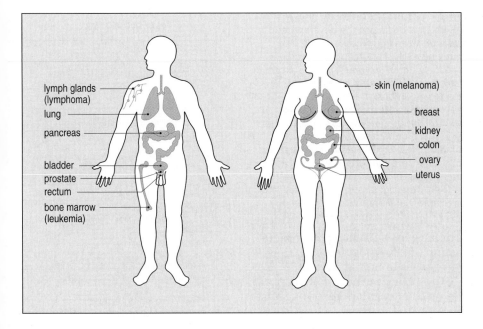

Figure 16.2 The tissues that give rise to the most common cancers in humans in the USA. For simplicity, tissues common to males and females are not all shown on both figures.

undergoing cell division and are thus most likely to accumulate mutations as a result of errors in DNA replication. These tissues include the epithelial linings of the gastrointestinal tract, urogenital tract, and mammary glands (Figure 16.2). Cancers of epithelial cells are known as **carcinomas**; cancers of other cell types as **sarcomas**. Cancers of immune system cells are known as **leukemias** when they involve circulating cells, **lymphomas** when they involve solid lymphoid tumors, and **myelomas** when they involve bone marrow (see Sections 6-16, p. 180, and 7-15, p. 204).

16-2   A cancer arises from a single cell that has accumulated multiple mutations

The best defenses against cancer lie within each cell of the human body and not with the specialized cells of the immune system. The integrity of the body is so dependent on well-controlled cell division that many mechanisms have evolved to ensure this. These include the repair of certain types of DNA damage as well as mechanisms that prevent the survival and division of cells with badly damaged DNA. Consequently, the control of cell division never depends on the function of just one protein, and a cell cannot become cancerous by mutation in just one gene. For a cell to give rise to a cancer, it must first accumulate multiple mutations, and these must occur in genes concerned with the control of cell multiplication and cell survival. When a cell becomes able to form a cancer it is said to have undergone **malignant transformation**.

There are two main classes of gene that, if mutated or misexpressed, contribute to malignant transformation. **Proto-oncogenes** are genes that normally contribute positively to the initiation and execution of cell division. More than 100 human proto-oncogenes have been identified; they encode growth factors and their receptors, and also proteins involved in signal transduction and gene transcription. The mutant forms of proto-oncogenes that contribute to malignant transformation are called **oncogenes**.

The second class of genes involved in cellular transformation are called **tumor suppressor genes** because they encode proteins that prevent the unwanted proliferation of mutant cells. A typical tumor suppressor gene is that for the protein p53. This protein is expressed in response to DNA damage and causes

Figure 16.3 A typical series of mutations acquired during the development of a cancer. This illustration shows the development of colorectal cancer by successive mutations in different genes. The morphological changes accompanying each change are indicated. *RAS* is an oncogene; *APC, DCC,* and *p53* are tumor suppressor genes. Both copies of a tumor suppressor gene must be mutated to contribute to malignant transformation, so the total number of mutations that has occurred here is seven. The sequence of events shown here usually takes 10–20 years or more. After a cell has undergone the initial series of mutations that render it malignant, the tumor cells then rapidly accumulate more mutations. Some of these contribute to making the cancerous cells more invasive, but others are thought to be simply a consequence, rather than a cause, of the cell's becoming cancerous. Once a tumor has become genetically heterogeneous, selection acts to favor those cells that divide more rapidly and are more invasive.

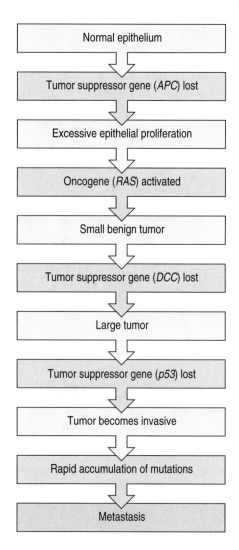

the damaged cell to die by apoptosis. Loss of the gene encoding p53 or mutations that interfere with its protective function are the most frequent mutations present in human cancers. Over 50% of cases of human cancer have a mutation in p53, revealing its importance in protecting the body from cancer. Indeed, p53 might well have evolved as an internal cellular defense against cancer.

It has been estimated that a cell must accumulate at least five or six independent mutations before it can become cancerous. The actual number depends on the cell type and the particular genes in which the mutations occur (Figure 16.3). Because of the random manner in which mutations occur, the combination of genetic lesions in each cancer is unique.

The low frequency of mutation together with the requirement for multiple mutations means that each cancer inevitably arises from a single cell that has undergone malignant transformation. This explains why in paired organs, such as the lungs, cancer initially affects only one of them. In the course of a lifetime, a person's cells accumulate mutations, and the probability that a cell somewhere in the body will have the right combination of mutations to cause a cancer increases in a nonlinear fashion. For this reason, the incidence of cancer increases with age, and cancers are largely diseases of older people.

16-3   Exposure to chemicals, radiation, and viruses can facilitate the progression to cancer

Cancer is not, however, an inevitable outcome of aging. Most people never suffer from cancer, even in old age. This reveals the existence of genetic and environmental factors that influence the risk of developing a cancer. Genetic factors include the possession of a germline mutation in one copy of a tumor suppressor gene. Such a mutation in the gene encoding p53 is the underlying cause of Li–Fraumeni syndrome, a condition in which there is an unusually strong predisposition to multiple cancers that arise at a relatively early age. This is because only one new mutation is needed, in the good copy of p53, to abolish all p53 function. In other people at least two new mutations, in both copies of p53, would be needed.

Of greater importance to the population at large are environmental insults that increase the number of mutations suffered by the body. Chemical and physical agents that damage DNA in such a way as to cause an increased rate of mutation are called **mutagens**. Many known mutagens can increase the risk of cancer and are known as **carcinogens**. People who have had heavy or prolonged exposure to carcinogenic agents, which include certain chemicals, ultraviolet light, and other forms of radiation, are more at risk of developing cancer than those who have not been exposed. Most marked is the increased frequency of lung cancer among people who smoke cigarettes.

Chemical carcinogens tend to cause mutations due to single nucleotide substitutions in DNA. Radiation, in contrast, tends to produce grosser forms of damage such as DNA breaks, cross-linked nucleotides, abnormal recombination, and chromosome translocations. The radiation released by the atomic bombs that were exploded at Hiroshima and Nagasaki in 1945 caused an increased incidence of leukemia in those who survived the acute effects of the blasts. Less marked, but more pervasive, is the increasing incidence of skin cancer caused by overexposure to ultraviolet radiation from the sun in the pursuit of work, fun, and fashion. However, despite the fear of cancer in developed societies and the well-known risk factors, people in those same societies still deliberately engage in behavior that they know increases their chances of developing cancer in later years.

Certain viruses also have the potential to transform cells, and viruses are associated with some 15% of human cancers (Figure 16.4). Such viruses are called **oncogenic viruses**. The known oncogenic viruses that affect humans are DNA viruses, except for the RNA retrovirus HTLV-1, which is associated with adult T-cell leukemia.

Human oncogenic viruses typically set up chronic infections in a cell, producing novel virally encoded proteins that override or interfere with the cell's normal mechanisms for regulating cell division. Infected cells therefore start to proliferate. For example, the Epstein–Barr virus induces infected B cells to divide repeatedly (see Section 11-4, p. 334), and if the infected cells are not cleared rapidly, a proportion of them go on to become transformed. Certain strains of papillomavirus predispose to cancer of the cervix. They encode proteins that prevent the normal tumor suppressor mechanisms from acting within the infected cells. Viral proteins bind to p53 protein and to another tumor suppressor protein called Rb, blocking their functions and enabling the virus-infected epithelial cells to proliferate. Other chronic infections can lead to cancer because the tissue damage they cause necessitates a high rate of cell renewal and, as a consequence, mutations accumulate more rapidly. The liver cancers arising from hepatitis B and C virus infections might be of

Viruses associated with human cancers		
Virus	**Associated tumors**	**Areas of high incidence**
DNA viruses		
Papillomavirus (many distinct strains)	Warts (benign) Carcinoma of uterine cervix	Worldwide Worldwide
Hepatitis B virus	Liver cancer (hepatocellular carcinoma)	Southeast Asia Tropical Africa
Epstein–Barr virus	Burkitt's lymphoma (cancer of B lymphocytes) Nasopharyngeal carcinoma B-cell lymphoproliferative disease	West Africa Papua New Guinea Southern China Greenland (Inuit) Immunosuppressed or immunodeficient patients
RNA viruses		
Human T-cell leukemia virus type 1 (HTLV-1) Human immunodeficiency virus (HIV-1) and human herpes virus 8 (HHV8)	Adult T-cell leukemia/ lymphoma Kaposi's sarcoma	Japan (Kyushu) West Indies Central Africa

Figure 16.4 **Viruses associated with human cancers.** For all the viruses listed here, the number of people infected is much larger than the number who develop cancer; the viruses must act in conjunction with other factors. Some of the viruses probably contribute to cancer only indirectly. For example, an increased incidence of Kaposi's sarcoma due to transformation of endothelial cells by HHV8 is seen not only in HIV-infected patients but also in other immunosuppressed patients. HIV-1, by obliterating cell-mediated immune defenses, is probably simply allowing the HHV8-transformed endothelial cells to thrive as a tumor instead of being destroyed by the immune system.

this type, as might the stomach cancers associated with ulcers caused by infection with the bacterium *Helicobacter pylori*.

16-4   Certain common features distinguish cancer cells from normal cells

Cancer comprises a highly heterogeneous set of diseases distinguished by their tissue of origin and their state of differentiation. In Chapter 6, for example, we saw how there are cancers representing all stages of B-cell differentiation. Despite their heterogeneity, all successful and life-threatening cancers share a core set of seven characteristics. Six of these properties have already been alluded to: cancer cells stimulate their own growth, ignore growth-inhibiting signals from other cells, avoid death by apoptosis, become well connected to the blood, leave their site of origin to invade other tissues, and inexorably expand the size of their population. The seventh requirement of a successful cancer is for it to outwit the immune system (Figure 16.5).

An increased incidence of cancer among transplant patients maintained on immunosuppressive drugs shows that the immune system can control or eliminate cancer cells. A study of around 6000 kidney transplant patients in Scandinavia revealed higher incidences of colon, lung, bladder, kidney, ureteric, and endocrine cancer than in the general population. In these immunosuppressed patients, latent Epstein–Barr virus, which is no threat to normal individuals, can reactivate and become malignant. Similarly, people with immunodeficiencies and mutant laboratory mice lacking key components of the immune system are more likely to get cancer than are immunocompetent individuals. On the positive side, the survival of cancer patients directly correlates with the extent to which there is an ongoing adaptive immune response within the tumor mass. Together, these observations point to the immune system's being able to recognize cancer cells and eliminate them before they threaten to cause disease. This function of the immune system—to survey the body for cancer—is called **immunosurveillance** or **cancer immunosurveillance**. The contribution of immunosurveillance to preventing cancer has been hotly disputed, but after several decades of skepticism, evidence from both humans and mice is now turning opinion in its favor.

16-5   Immune responses to cancer have similarities with those to virus-infected cells

Like viral infections, malignant transformation alters protein expression in cancer cells in ways that make the cells appear foreign to the immune system. Notably, there are changes in the expression of MHC class I molecules that can be detected by the NK cells of innate immunity and the cytotoxic T cells of adaptive immunity. Thus the mechanisms for cancer immunosurveillance are the same as those used for detecting and responding to viral infection.

An important difference is in the tempo of the immune response to cancer compared with that to infection. At mucosal surfaces there is a chronic response to the local microbial populations that will readily activate both innate and adaptive immunity when potential pathogens breach a mucosal barrier. In other tissues, the innate and adaptive responses to an infection are started by inflammation produced at the initial site of infection. These mechanisms make innate immune mechanisms available within hours after the start of infection, and adaptive immune responses are fully operational within 2 weeks. In comparison, the malignant transformation of a single cell that marks the start of a cancer might go unnoticed by the body. For years the growth of a cancer might cause no damage of the sort that triggers inflammation and an immune response (Figure 16.6). In time, the growth of the cancer will begin to damage healthy tissue and raise the alarm of inflammation. In

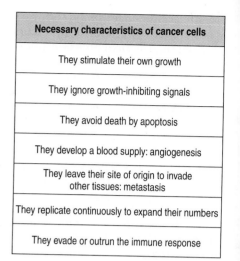

Necessary characteristics of cancer cells
They stimulate their own growth
They ignore growth-inhibiting signals
They avoid death by apoptosis
They develop a blood supply: angiogenesis
They leave their site of origin to invade other tissues: metastasis
They replicate continuously to expand their numbers
They evade or outrun the immune response

Figure 16.5 The cells that cause the many different diseases called cancer have seven critical features in common.

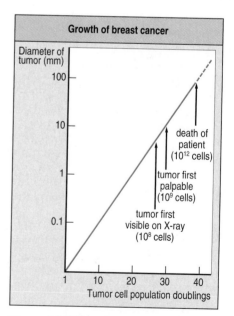

Figure 16.6 The growth and lifespan of a common human tumor. The diameter of the breast cancer is plotted on a logarithmic scale. Years can elapse before the tumor becomes noticeable to the patient or her immune system.

some cases, the immune response will be able to control or even eliminate the cancer, but in others the tumor load will be so great that the immune system will be overwhelmed.

Early in the twentieth century it was noted that some cancer patients contracting bacterial infections experienced regression—reduction in mass—of their tumors. It was surmised that the inflammation induced by infection had stimulated an immune response to the tumor as well as to the pathogen. A more recent application of this phenomenon is the established treatment for superficial bladder cancer in which chronic inflammation of the bladder is achieved by introduction of the BCG vaccine through a procedure known as intravesicular BCG therapy, in which viable *Mycobacterium bovis* cells are introduced into the patient by bladder instillation. The component of the vaccine that gives it this antitumor effect is the unmethylated CpG-containing DNA of the mycobacteria, the ligand for the Toll-like receptor TLR9 (see Section 2-11, p. 45).

As with infections, it is likely that many of the potentially cancerous cells that arise in the human body are detected by immunosurveillance and eliminated at an early stage before they have had any perturbing effect on the anatomy and function of the human body. Only a small minority of such cells is able to grow, multiply, and give rise to clinical cancer.

16-6 Differences in MHC class I allow tumor cells to be attacked and eliminated by cytotoxic T cells

Tumors of laboratory mice will grow when transplanted to mice of identical MHC type but fail to take hold when transplanted to mice of different MHC type (Figure 16.7). In these animals, the tumor cells are killed by alloreactive CD8 T cells that recognize the allogeneic MHC class I molecules. Similarly, human tumor cells are easily killed by alloreactive CD8 T cells but will also survive passage to an HLA-compatible person—as has occurred in some transplant patients. In one such case, the two kidneys from a deceased woman were given to different, unrelated HLA-matched recipients. Within two years, both the transplant recipients had developed metastatic melanoma. This was not a coincidence. It turned out that 16 years before transplantation the donor had been successfully treated for primary melanoma and had been considered free of tumor cells. On the contrary, however, she seems to have retained some tumor cells that were controlled by her immune system. When these tumor cells were transferred in the donated kidneys to the immunosuppressed transplant patients they were released from immunological restraint and were free to grow uncontrollably. Some cancers are thought to be propagated by minority subpopulations of **cancer stem cells** that are self-renewing and more resistant to the toxins and radiation that are commonly used to treat cancer. Thus treatment for melanoma may have eliminated all the donor's tumor cells except for some cancer stem cells, which remained quiescent and in low numbers until after they were transplanted.

In animal populations that have limited MHC polymorphism, tumors can be passed from one individual to another and become essentially infective. Passage of venereal sarcoma between domestic dogs occurs in the course of copulation. In the wild, a fatal facial tumor is passed between Tasmanian devils, when these aggressive marsupials fight and bite each other's faces (Figure 16.8). The cells of all these tumors are characterized by distinctive chromosomal aberrations showing they are all part of the same clone which originally arose in one individual. Immunological acceptance of this infectious, malignant allograft occurs because the population of Tasmanian devils has very limited MHC diversity. The lethality of this infectious tumor, which was first observed as recently as 1996, now threatens the Tasmanian devil with extinction, vividly illustrating the survival value of a highly polymorphic MHC.

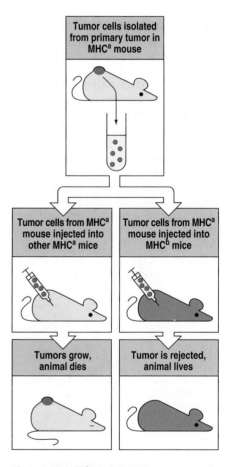

Figure 16.7 When tumors are transplanted between MHC-incompatible mice they are rejected by the alloreactive immune response to MHC differences. The left panels show the transplantation of a tumor between two mice of the same MHC type. The tumor grows in the recipient. In the right panels, the tumor is transplanted to a mouse of a different MHC type and is rejected. Experiments of precisely this type led to the discovery of the MHC, although their intent was to discover tumor-specific antigens.

Although human individuals and populations are prone to equally aggressive behavior, the high polymorphism of their HLA class I and II genes provides a formidable barrier that prevents the growth of any tumor cells that are passed from one person to another. Situations in which such transfer could potentially occur are the same as those that serve to spread HIV: intimate contact in love and war, blood transfusion, and sharing of syringes and needles in the non-medical use of drugs.

16-7   Mutations in cellular genes acquired during oncogenesis provide tumor-specific antigens

Cells that become malignantly transformed have genomic differences that distinguish them from every other cell in the body. Cells transformed by oncogenic viruses are genetically the most distinctive, whereas at the other end of the spectrum are tumors arising from a transformed cell with point mutations in some half-dozen oncogenes or tumor suppressor genes. As a tumor grows, further mutations occur, introducing genetic heterogeneity into the tumor-cell population. Selection for variant cells with faster growth or increased metastatic potential causes the genomes of the tumor and host to diverge even further.

Almost all cancer patients make an adaptive immune response against their tumor. The antigens to which they respond are called **tumor antigens**. By studying the tumor-reactive antibodies and CD8 T cells made by patients with different types of cancer, more than 1000 different tumor antigens have been defined. Antigens present on tumor cells but not on normal cells are called **tumor-specific antigens**; antigens expressed on tumor cells but also found on certain normal cells, often in smaller amounts, are called **tumor-associated antigens** (Figure 16.9). Tumor-specific antigens have structures that are not present in any normal cell. They can derive from viral proteins, from the mutated parts of mutant cellular proteins, or from amino acid sequences spanning tumor-specific recombination sites between genes. Some examples

Figure 16.8 The 'infectious' facial tumor of the Tasmanian devil.
A native of Tasmania, the smaller of the two main islands that make up Australia, the Tasmanian devil in an aggressive carnivorous marsupial, hence its name. The tumor that disfigures the right eye of this individual is caused by a clone of tumor cells, which is passed from one Tasmanian devil to another when they fight and bite their faces. Transferred tumor cells grow because of the limited MHC diversity in the population of Tasmanian devils. Thus the tumor cells are MHC compatible with many individuals and recognized as self by their immune systems. Photograph courtesy of Menna Jones.

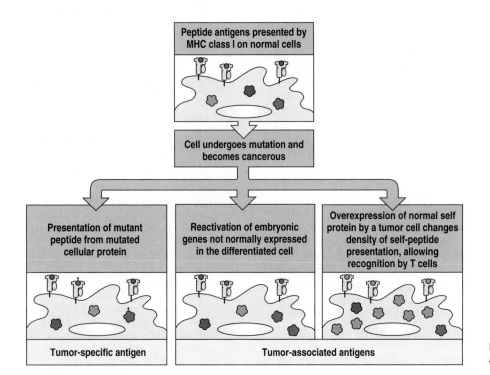

Figure 16.9 Sources of tumor-specific and tumor-associated antigens.

Mutated tumor antigens			
Source of peptide antigen	Disease	HLA restriction	Peptide antigen
MART2	Melanoma	A1	FLEGNEVGKTY
ME1	Non-small cell lung carcinoma	A2	FLDEFMEGV
p53	Head and neck squamous cell carcinoma	A2	VVCEPPEV
KIAA0205	Bladder tumor	B44	AEPINIQTW
Triosephosphate isomerase	Melanoma	DR1	GELIGILNAKVPAD
BCR-ABL fusion protein	Chronic myelogenous leukemia	DR4 B8 A2	ATGFKQSS∣KALQRPVAS GFKQSS∣KAL SS∣KALQRPV ▲ point of fusion

Figure 16.10 Examples of tumor antigens that contain tumor-specific mutations. Shown here are tumor antigens discovered in the course of studying the T-cell response to different types of cancer: five are peptide antigens containing a point mutation (in magenta), and three are peptides derived from the novel protein produced by the fusion of the *BCR* and *ABL* genes that characterizes chronic myelogenous leukemia. *ABL* is an oncogene and *BCR* encodes a kinase (not the B-cell receptor). The HLA molecule that presents each peptide antigen, and thus 'restricts' the immune response, is shown. These data are taken from the Cancer Immunity Peptide Database website [http://www.cancerimmunity.org/peptidedatabase/mutation.htm], which contains many more examples of mutant tumor antigens.

of this class of mutant tumor antigens are shown in Figure 16.10. Peptides containing the mutations in cellular proteins such as p53 are bound by HLA class I and II molecules and presented to CD8 and CD4 T cells. Associated with 95% of the cases of chronic myelogenous leukemia is a chromosomal rearrangement that involves a fusion of the kinase-encoding *BCR* gene on chromosome 22 with the *ABL* proto-oncogene on chromosome 9. The protein encoded by the fused genes is instrumental in the progression to cancer, but also provides novel peptide antigens that span the point of fusion and can be presented by both HLA class I and II molecules. Tumor antigens can also derive from abnormal patterns of protein modification, including glycosylation and phosphorylation, or translation from unusually spliced mRNAs.

Another way in which 'foreign' peptide antigens can be produced from self proteins is by peptide splicing. Several peptide antigens recognized by tumor-specific T cells do not correspond to a continuous linear sequence of amino acid residues in the protein from which the peptide is derived. One clone of CD8 T cells obtained from a melanoma patient recognizes a nonamer peptide derived from a melanocyte glycoprotein that is presented by HLA-A32. This peptide corresponds to residues 40–42 and 47–52 of the melanocyte glycoprotein. In the course of the glycoprotein's degradation by the proteasome, cleavage after residues 42 and 46 removes residues 43–46, which is followed by the formation of a peptide bond between residues 42 and 47. This peptide does not correspond to a sequence in the native protein (Figure 16.11). In other tumor-specific antigens, the peptides are stitched together in an order that is different from that in the protein providing the epitope.

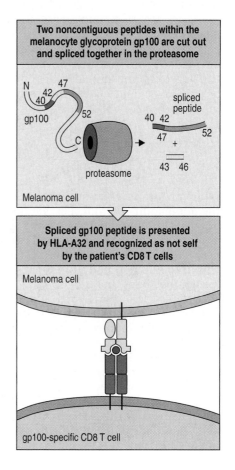

Figure 16.11 Some tumor antigens are produced by cutting and pasting peptides from self proteins. The proteasome cleaves the melanocyte glycoprotein gp100 after residues 39, 42, 46, and 52. It then splices together the peptides corresponding to resides 40-42 and residues 47–52 to produce a nonamer peptide that does not represent a contiguous sequence in the native protein. This peptide is bound by HLA-A*3201 and is presented to CD8 T cells. In splicing the peptides together, the protease activity of the proteasome runs 'backwards' to form, rather than break, a peptide bond. Several tumor antigens of this kind have been described, and their formation can involve the splicing of peptides that are much further apart in the protein than the example shown.

Figure 16.12

Figure 16.12 Genes encoding cancer/testis antigens are concentrated on the X chromosome. Thirty-eight different types of cancer/testis antigen have been defined and numbered in a CT series analogous to the CD series of differentiation antigens. Shown here are the first 10, and best-characterized, CT antigens, the number of genes that encode them, and the chromosomal location of the genes. Genes encoding 17 of the 38 CT antigens are on the X chromosome; the others are distributed between 11 autosomes, none of which has genes for more than three CT antigens. This information and more can be found on the Cancer Immunity CT Gene Database website [http://www.cancerimmunity.org/CTdatabase].

Cancer/testis antigens			
Antigen	Alternative name	Chromosome	Number of genes
CT1	MAGEA	X	11
CT2	BAGE	13	5
CT3	MAGEB	X	4
CT4	GAGE1	X	8
CT5	SSX	X	4
CT6	NY-ESO-1 LAGE	X	2
CT7	MAGEC	X	2
CT8	SYPC1	1	1
CT9	BRDT	1	1
CT10	MAGEC2	X	1

16-8 Cancer/testis antigens are a prominent type of tumor-associated antigen

Tumor-associated antigens derive from normal cellular proteins to which the immune system is not tolerant and that become immunogenic when expressed by the tumor. These can be proteins that are normally made in immunologically privileged sites or proteins that are normally not produced in an amount sufficient to be seen by T cells. Many tumor-associated antigens correspond to human proteins normally expressed only by immature sperm in the testis or by the trophoblast (the cells of the very early embryo that contribute to the placenta) and to which the immune system is not tolerant because neither cell type expresses HLA-A, HLA-B or HLA class II. The genes encoding these tumor antigens are not expressed by any somatic cells but become turned on in cancer cells. This subset of **tumor-associated antigens** is called the **cancer/testis antigens** or **CT antigens**. Some 90 different genes, including several gene families, encode CT antigens and about half of them are on the X chromosome (Figure 16.12). This bias, which is unlikely to be fortuitous, indicates that genes whose functions are normally restricted to reproduction and fetal development can be hijacked by somatic cells to produce cancers. During the course of pregnancy the mother's energy and resources become increasingly devoted to promoting the growth and development of the fetus. Similarly, as a cancer grows and differentiates it too consumes an increasing part of the body's energy and nourishment—but unlike in pregnancy the natural endpoint is not birth but death.

16-9 Successful tumors evade and manipulate the immune response

When an immune response is made against a tumor it imposes selection on the population of tumor cells. Variant cells that have low expression of tumor antigens or mutant epitopes that are no longer recognized by effector T cells or antibodies may be able to evade the immune response. The longer a cancer grows, expands its population, and colonizes different sites and environments within the human body, the more genetic variation it acquires and the less likely it becomes that the immune response can delay or terminate the disease. Thus the immune system has its best chance of eliminating cancer in the early stages, when the population of cancer cells is small and has yet to adapt to the immune response.

When epithelial cells become transformed they increase the amount of MIC proteins on their surface. The MIC proteins are ligands for the activating NKG2D receptor, which is expressed on NK cells, $\gamma{:}\delta$ T cells, and cytotoxic CD8 T cells (see Section 10-22, p. 316). Expression of MIC by the nascent tumor cells enables these three types of lymphocyte to attack and kill the tumor cells, and if it is done efficiently the tumor is driven to extinction. If only a fraction of the tumor cells are killed, then the remainder has the opportunity to

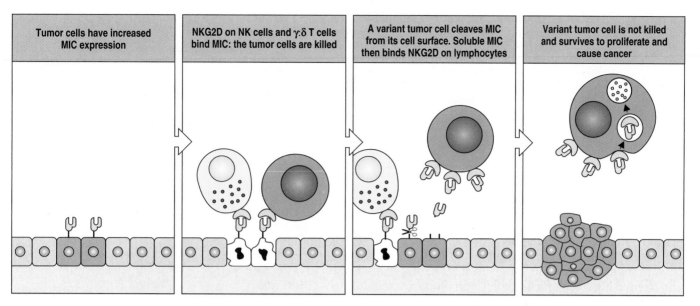

| Tumor cells have increased MIC expression | NKG2D on NK cells and γ:δ T cells bind MIC: the tumor cells are killed | A variant tumor cell cleaves MIC from its cell surface. Soluble MIC then binds NKG2D on lymphocytes | Variant tumor cell is not killed and survives to proliferate and cause cancer |

Figure 16.13 Human epithelial tumors can inhibit the response of lymphocytes expressing NKG2D. Malignant transformation of epithelial cells induces the expression of MIC proteins. These are recognized by the NKG2D receptors of NK cells, γ:δ cells, and CD8 T cells, which enables these cells to kill the tumor cells. Variant tumor cells can evade this response by making a protease that cleaves MIC from the cell surface. This has two effects to the tumor's advantage. First, the variant tumor now lacks the ligand for NKG2D; second, the binding of soluble MIC to NKG2D on the lymphocyte surface causes endocytosis and degradation of the receptor:MIC complex.

proliferate and acquire mutations and changes in gene expression that allow them to evade the immune response. Some successful epithelial tumors make proteases that cleave MIC from their own cell surfaces, producing a soluble form that binds to the NKG2D receptor of infiltrating lymphocytes. Binding of MIC to NKG2D induces receptor-mediated endocytosis that removes NKG2D from the surface and speeds its degradation. By removing MIC from their own surfaces and removing NKG2D from lymphocyte surfaces, the tumor cells evade attack by NK cells, γ:δ T cells, and cytotoxic CD8 T cells (Figure 16.13).

Cytotoxic CD8 T cells are the best effector cells for killing tumor cells. One way in which tumor cells escape a cytotoxic T cell is to stop expressing the autologous HLA class I molecule that presents the tumor antigen to the T cell. Between one-third and one-half of all human tumors have defective expression of one or more of their HLA class I allotypes (Figure 16.14). What this tells us is that many patients have made a CD8 T-cell response against their cancer but variant cells lacking particular HLA class I allotypes have escaped that immune response and expanded to keep the cancer going.

Loss of HLA class I expression can in some circumstances make a tumor susceptible to attack by NK cells. The killing of acute myelogenous leukemia cells by alloreactive NK cells in an HLA-haploidentical bone marrow transplant (see Section 15-24, p. 482) occurs because the recipient's tumor cells lack an HLA class I allotype that the donor-derived NK cells see as self. This is an example of NK cells attacking a cell that is 'missing self' (see Section 10-26, p. 319).

As well as evading the immune response, tumors can manipulate the immune response in their favor. In the absence of inflammation, tumor-cell antigens can be processed and presented to T cells by dendritic cells that lack B7 co-stimulators. As a result, the tumor-specific T cells will become anergic (see Section 7-12, p. 202). Some tumors secrete cytokines such as TGF-β that create an immunosuppressive environment in the tumor, which can be reinforced by the recruitment of regulatory cells (Figure 16.15).

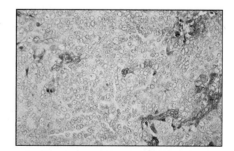

Figure 16.14 Loss of HLA class I expression in a cancer of the prostate gland. The section is of a human prostate cancer that has been stained with a monoclonal antibody specific for HLA class I molecules. The antibody is conjugated to horseradish peroxidase, which produces a brown stain wherever the antibody binds. The stain and HLA class I molecules are not seen on the tumor mass but are restricted to lymphocytes infiltrating the tumor and tissue stromal cells. Photograph courtesy of G. Stamp.

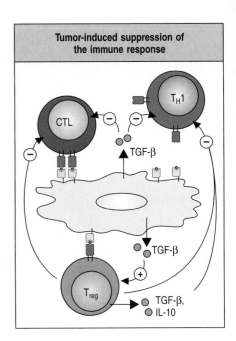

Figure 16.15 Manipulation of the immune response by a tumor. Tumors can protect themselves from the immune response by secreting immunomodulatory cytokines, such as TGF-β, which suppress the inflammatory response and recruit regulatory T cells (T$_{reg}$) into the tumor tissue. The combined effect of TGF-β and IL-10 made by the regulatory T cells is to suppress the actions of effector CD8 T cells and CD4 T$_H$1 helper cells that are specific for tumor antigens.

16-10 By preventing infection, vaccination protects against cancers caused by viruses

Several types of cancer are caused by chronic viral infection. In principle, the cancers caused by oncogenic viruses should be easier to prevent than those caused by other agents. Viruses provide epitopes to which the immune system can respond, and vaccination can provide protective immunity that will prevent infection. Cervical cancer in women and genital warts in both men and women are strongly associated with human papillomavirus (HPV) infections and are the most common sexually transmitted diseases. Approximately 50% of sexually active individuals experience an HPV infection, and it is estimated that 4000 women will die of cervical cancer in 2008 in the United States. Out of more than 100 different types of papillomavirus, 40 infect the genitalia, and types HPV16 and HPV18 account for 72% of cervical cancer. The infection localizes to the outer epithelium at the mucosal surface. In the infected cells, viral proteins bind to the human tumor suppressors p53 and retinoblastoma protein (Rb), which induces the epithelial cells to divide abnormally and increases the probability of malignant transformation. Recently completed trials have shown that immunization with vaccines containing the LI capsid proteins of the HPV16, HPV18, and other papillomaviruses induces an antibody response that completely prevents infection by HPV and the occurrence of precancerous lesions in the cervix. To be effective, girls need to be vaccinated before they become sexually active. Current recommendations are for three immunizations over a 6-month period for females aged 11–26 years. The actual effect of vaccination on the incidence of cancer will only be known as cohorts of vaccinated and unvaccinated women are followed over time. Chronic infection with hepatitis B virus (HBV) is associated with hepatocellular carcinoma, a type of liver cancer. A universal program to vaccinate newborns against HBV was started in Taiwan in 1986 and is now seen to have reduced the incidence of hepatocellular carcinoma by two-thirds.

16-11 Vaccination with tumor antigens can cause cancer to regress

Knowing the immune system has potential to make a response that kills tumor cells but is poorly triggered to make such response, tumor immunologists are exploring ways to stimulate and enhance human immune responses to cancer. They hope to devise cancer vaccines that can be used for both treatment and prophylaxis. Several studies and clinical trials have focused on patients with malignant melanoma, an aggressive disease for which there is no reliable treatment. The first CT antigens were discovered as targets for cytotoxic T cells obtained from melanoma patients and were called MAGEA1 and MAGEA3 (*melanoma antigen encoding*). Figure 16.16 shows the effects of vaccination with an epitope of the CT1 antigen MAGEA3, which is presented to CD8 T cells by HLA-A1. Before vaccination the patient had had surgery to remove a cutaneous melanoma and 2 months later developed metastases that expressed the MAGEA3 antigen. Over a period of 2 years the patient was vaccinated 11 times with either a recombinant virus encoding the epitope or a synthetic peptide of the same sequence. As a result of the vaccination, the frequency of MAGEA3-specific cytotoxic CD8 T cells was increased 30-fold, commensurate

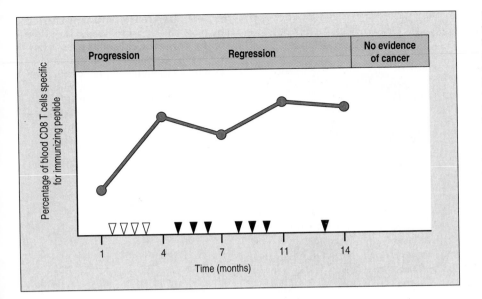

Figure 16.16 The results of vaccination of a cancer patient with a tumor antigen. A patient with metastatic melanoma was vaccinated with an antigenic peptide from the MAGEA3 protein that is presented by the patient's HLA-A1 allotype. The first four immunizations used a recombinant viral vaccine (open arrowheads); the last seven immunizations were with synthetic peptides. The red line shows the percentage of total CD8 T cells in the peripheral blood that are specific for the peptide. The status of the cancer's growth is indicated in the blue box above the graph. Data courtesy of Pierre Coulie.

with a steady regression of the tumor to produce a complete remission that lasted for more than 2 years (see Figure 16.16). For reasons that are not understood, tumor regression was seen in only 20% of the vaccinated patients and remission was achieved only in 10%. Several other ongoing clinical trials are focused on vaccination with the CT6 antigen NY-ESO-1, which is well expressed by various types of cancer and stimulates strong T-cell and B-cell responses.

16-12 Increasing co-stimulation can boost the T-cell response to tumor cells

Although most cancer patients make an adaptive immune response against their tumor cells, that response is usually too weak to control or eliminate the tumor. One approach for improving the antitumor response is to increase the strength and persistence of T-cell co-stimulation by treating the patients with a fully human anti-CTLA4 monoclonal antibody. Because CTLA4 is an inhibitory regulator of co-stimulation that competes with CD28 for the binding to B7 ligands (see Section 8-5, p. 220), the general effect of anti-CTLA4 is to elevate T-cell responses. During chronic immune responses of the type that develop against tumor cells, the expression of CTLA4 by T cells increases and reduces the effectiveness of the immune response. The first clinical trials of the effect of anti-CTLA4 in patients with melanoma and renal carcinoma have shown some regression of tumor but also symptoms of autoimmunity.

16-13 Heat-shock proteins can provide natural adjuvants of tumor immunity

The strategy for cancer vaccination described in previous sections is to target a few well-characterized antigens that are shared by many tumors, such as the CT antigens, and not antigens that are specific to a particular tumor. A drawback to this approach is that tumors can more easily escape the immune response produced by such vaccines, as has been observed in patients in whom vaccine-induced regression was temporary. A different approach to vaccine design seeks to embrace a wider range of tumor antigens, including those that have not been defined. It employs a group of ubiquitous cellular proteins called the heat-shock proteins.

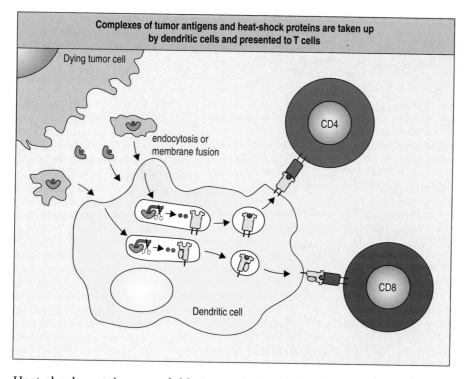

Complexes of tumor antigens and heat-shock proteins are taken up by dendritic cells and presented to T cells

Dying tumor cell

endocytosis or membrane fusion

CD4

Dendritic cell

CD8

Figure 16.17 Tumor antigens carried by heat-shock proteins can be taken up by dendritic cells and presented to T cells. Complexes of heat-shock protein 70 (green) and peptides derived from tumor antigens (red and purple circles) are taken up by dendritic cells, either as soluble complexes or associated with membrane vesicles. The bound peptides are delivered to endosomal and lysosomal pathways for antigen processing. The peptides are presented by MHC class II proteins to CD4 T cells or are cross-presented by MHC class I proteins to CD8 T cells.

Heat-shock proteins are soluble intracellular chaperone proteins that organize the folding, assembly, and degradation of cellular proteins. They were first discovered as proteins that were produced in cells in response to the stress of raised temperature—hence the name 'heat shock.' Heat-shock proteins bind peptides and disorganized parts of proteins in a manner akin to MHC molecules, but they bind with more promiscuous specificity. Some heat-shock proteins, for example calnexin and calreticulin, are familiar for their contribution to antigen processing and presentation (see Section 5-11, p. 138), but there are many others. Animal models of cancer vaccination have shown that immunization with heat-shock proteins extracted from cancer cells produces a powerful antitumor response. The heat-shock proteins bind and carry a potent cargo of protein and peptide tumor antigens. These they specifically deliver to dendritic cells for presentation to T cells (Figure 16.17). When taken up by dendritic cells, heat-shock proteins can deliver tumor antigens directly to the antigen-processing machinery of the dendritic cell. They may also activate dendritic cells through signaling receptors. Because of these functions, heat-shock proteins have been described as natural or internal adjuvants of the human immune system.

That some fraction of cancers escape the immune system is largely because of failures in starting the response early enough and making it strong enough. Manipulation of dendritic cells and heat-shock proteins might help to overcome these limitations. The basic idea is to isolate dendritic cells from a patient's blood and load them with heat-shock proteins isolated from the patient's tumor or with defined tumor antigens. Thus charged, the dendritic cells will be replaced in the patient's circulation, where they will home to the secondary lymphoid tissues and initiate a tumor-specific immune response.

16-14 Monoclonal antibodies against cell-surface tumor antigens can be used for diagnosis and immunotherapy

Monoclonal antibodies specific for tumor antigens are used in both tumor analysis and therapy. The location of tumor cells within the body can be revealed by giving patients purified monoclonal antibody specific for a tumor

Figure 16.18 Detection of colon cancer by using a radiolabeled antibody against carcinoembryonic antigen. Anterior projection of the pelvis showing uptake of anti-carcinoembryonic antigen (CEA) antibody in a pelvic tumor. A patient with a suspected recurrence of colon cancer was injected intravenously with a radioactive indium-111-labeled monoclonal antibody against CEA. The tumor is seen as two dark red spots in the pelvic region. The blood vessels are faintly outlined by circulating antibody that has not bound to the tumor but may have bound to soluble CEA released from the tumor and is therefore circulating in the blood. R, right side; L, left side. Photograph courtesy of A.M. Peters.

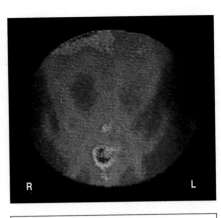

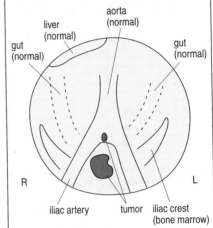

antigen and covalently coupled to a radioactive isotope, for example iodine-131. The binding of the antibodies to tumor cells concentrates the radioactivity at sites in the body containing tumors (Figure 16.18). In this manner, the size and location of a primary tumor can be determined, as can the extent of its metastasis. Such information can help to determine the type of therapy required.

Humanized monoclonal antibodies are seeing increased use as treatments for cancer. Five of the six antibodies shown in Figure 16.19 are specific for cell-surface molecules that target the tumor cells for killing either by antibody-dependent cell-mediated cytotoxicity (ADCC) or by opsonization and phagocytosis. An antibody specific for vascular endothelial growth factor (VEGF) has a different effect: it neutralizes this cytokine and prevents the angiogenesis that is necessary for the tumor to grow.

A second use of monoclonal antibodies in cancer therapy does not rely on the natural effector functions of the antibodies. Instead, the antibodies are used to deliver a toxic agent that kills the cells to which the antibody binds. Three such antibodies have been approved for clinical use (Figure 16.20). The antibody gemtuzumab is specific for CD33, a marker of immature myeloid cells, and is used as a treatment for acute myelogenous leukemia when conjugated to ozogamicin, a cytotoxic drug. On binding to CD33 on the surface of leukemia cells, the conjugate is endocytosed. In the lysosomes the drug is cleaved from the antibody and activated by glutathione-mediated reduction. The activated drug causes double-strand breaks in the cell's DNA and induces apoptosis (Figure 16.21). Conjugates of a biological toxin and an antibody are called **immunotoxins**. The advantage of immunotoxins over conventional chemotherapy is that the destructive power of the toxin is more specifically targeted toward the tumor cells and away from proliferating normal tissues.

Monoclonal antibodies used in the treatment of cancer				
Antibody	Antigen	Function of antigen	Cancers treated	Year approved
Rituximab	CD20	B-cell signaling receptor	Non-Hodgkin's lymphoma	1997
Trastuzumab	HER2/neu protein	Growth factor receptor	Breast cancer	1998
Alemtuzumab	CD52	Differentiation antigen	Chronic lymphocytic leukemia	2001
Cetruximab	Epidermal growth factor receptor (EGFR)	Growth factor receptor	Colorectal cancer	2004
			Head and neck cancer	2006
Panitumumab	EGFR	Growth factor receptor	Colorectal cancer	2004
Bevacizumab	Vascular endothelial growth factor (VEGF)	Growth factor promoting angiogenesis	Colorectal cancer	2004
			Non-small cell lung cancer	2006

Figure 16.19 Humanized monoclonal antibodies used in the treatment of patients with cancer. Shown here are antibodies that function purely as antibodies, not as delivery systems for chemotherapy or radiation therapy. 'Year approved' refers to their approval for medical use in the United States by the US Food and Drug Administration.

Conjugated monoclonal antibodies used in the treatment of cancer				
Antibody	Antigen	Disease	Small molecule conjugate	Year approved
Gemtuzumab	CD33	Acute myelogenous leukemia	Conjugated to ozogamicin, a cytotoxic antibiotic derivative	2000
Ibritumomab	CD20	Non-Hodgkin's lymphoma	Conjugated through the chelator tiuexetan to indium-111 for imaging and yttrium-90 for treatment	2002
Tositumomab	CD20	Non-Hodgkin's lymphoma	Conjugated to iodine-131	2003

Figure 16.20 Conjugated monoclonal antibodies used in the treatment of cancer.

Ibritumomab and tositumomab are both anti-CD20 antibodies that are conjugated to radioactive isotopes and used to treat non-Hodgkin's lymphoma. Ibritumomab conjugated to indium-111 is first used to target and image the tumor by detecting the indium-111 γ-radiation; it is then followed up with the yttrium-90 conjugate, which kills the tumor cells with its β-radiation. Tositumomab is directly conjugated to iodine-131. The advantage of these radioactive antibody conjugates over conventional radiation therapy is their greater precision. The radiation does not have to pass through and damage other tissues to reach the tumor but is generated directly at the surface of the tumor cells (Figure 16.22).

Summary to Chapter 16

Cancer, a disease that can affect any tissue in the body, results from the uncontrolled growth of human cells. Each case of cancer derives from a single cell that has undergone malignant transformation, a process requiring mutations in several genes important for cell survival and cell division. Cancer cells become divorced from the mechanisms that maintain tissue integrity, and they gradually compete with normal cells for food and space within the human body. In time, the effect of the competition becomes manifested as disease and ultimately death. Every cancer is unique and dies with the person in which it arose.

The principal defenses against cancer are intrinsic to every cell of the body. These are the normal cell processes that monitor DNA integrity, and control DNA replication and cell division; they kill cells growing abnormally and ensure that multiple mutations are necessary for malignant transformation to occur. Environmental factors such as chemicals, radiation, and viruses that mutate, rearrange, and change the expression of human genes all increase the probability of cancer.

The second line of defense against cancer is the immune system, which can detect the tumor antigens due to the abnormalities in the transformed cells with the same set of mechanisms that are used to respond to virus-infected cells. The immune system surveys the body for nascent cancer and probably eliminates most transformed cells that escape the intrinsic defenses at an early stage. The successful cancers that eventually manifest themselves as disease and are diagnosed by oncologists are those that have defeated the

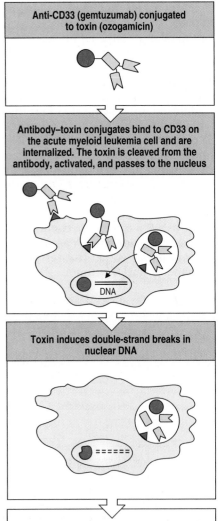

Figure 16.21 Antibodies can target toxins to the tumor-cell surface. In this example, the anti-CD33 antibody gemtuzumab is coupled to the inactive toxin ozogamicin. This conjugate is used to treat acute myelogenous leukemia. The toxin is activated after the conjugate has bound to the tumor cell surface and has been internalized into the lysosomes. The toxin is cleaved from the antibody and passes to the cell's nucleus, where it damages DNA.

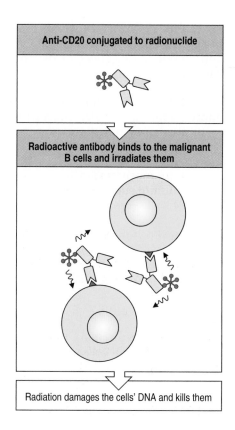

Figure 16.22 Antibodies can target radioactive isotypes to the tumor-cell surface. Conjugates of anti-CD20 antibodies with radioactive isotopes are used to irradiate and kill non-Hodgkin's B-cell lymphomas.

immune response. After transformation, the expanding population of cancer cells continues to accumulate new mutations, which can yield new tumor cell variants that evade or suppress the immune response. Although tumor-specific cytotoxic CD8 T cells and antibodies can be detected in patients with cancer, these immune responses cannot control or eliminate the disease.

Although cancer immunotherapy is still an emerging and experimental field, a variety of approaches show some promise. Monoclonal antibodies specific for surface antigens of tumor cells, or growth factors on which they depend, lead to the selective destruction of some cancers. To enhance the autologous immune response to cancer, patients are being immunized with peptides or proteins corresponding to their own tumor antigens. Other experimental therapies, such as the administration of anti-CTLA4, aim at releasing the natural restraints on the immune response and creating a more inflammatory environment in which both the innate and the adaptive response to the tumor will be more forceful. For the past 50 years, the success and application of organ transplantation has gradually increased through combining many different approaches and methods, each with its own advantages and limitations. Over the next 50 years a similar strategy should lead to the more general application of immunotherapy in the treatment of cancer.

Questions

16–1
 A. What are tumor-specific antigens?
 B. How do they originate?
 C. Give two examples, together with the type of tumor they come from.

16–2
 A. What are tumor-associated antigens?
 B. How do they originate?
 C. Give two examples together with the type of tumor they come from.

16–3 Indicate which of the following statements is true (T) or false (F).
 a. A mutation in only one copy of a tumor suppressor gene can cause malignant transformation.
 b. A neoplasm is characterized by abnormal cell division, which can cause disruption to organ function.
 c. Malignant tumors are often encapsulated and are limited in size.
 d. Most cancers develop in tissues that are actively undergoing division.
 e. Cancers of immune system cells that form solid tumors are known as sarcomas.
 f. Individuals with Li–Fraumeni syndrome who have been treated successfully for cancer are likely to develop another primary malignancy at a later time.

16–4 Identify two *in vitro* strategies by which an antitumor immune response could be boosted in cancer patients without the use of modified tumor cells.

16–5 How are mouse monoclonal antibodies (MoAbs) used by oncologists for the diagnosis and immunotherapy of cancer?

16–6
 A. What are the two main classes of gene that, if not expressed correctly, can lead to malignant transformation?
 B. What are the gene products made by each class?

16–7 Explain why the incidence of cancer is higher in the elderly than in young people.

16–8
 A. What is Li–Fraumeni syndrome?
 B. What is the underlying cause?

16–9 What are the seven critical features of cancer cells that contribute to their capacity to establish disease?

16–10 Explain how the immune mechanisms for the detection of cancer cells are (A) similar to and (B) different from those that detect virus-infected cells.

16–11 Provide examples where tumor cells transferred between a donor and a recipient are either (A) rejected and do not cause cancer, or (B) not rejected and cause cancer.

16–12 When the immune system is involved in _____, an appraisal for cancer cells is occurring.
   a. malignant transformation
   b. apoptosis
   c. antineoplasia
   d. immunosurveillance
   e. immunosuppression.

16–13
   A. How is it possible for a tumor-specific antigen to be composed of a unique sequence of amino acids not encoded in the genome of the tumor cell?
   B. Give a specific example.

16–14 Epithelial tumors express MIC proteins on their cell surfaces.
   A. Why does this make the tumor cells susceptible to attack by NK cells, γ:δ T cells, and cytotoxic CD8 T cells?
   B. How do the tumor cells evade such attack?

16–15 Describe four potential ways in which T-cell responses to tumor cells can be enhanced *in vivo* and lead to tumor regression.

16–16 Explain two ways in which monoclonal antibodies can be used in tumor therapy to deliver a particular molecule directly to the tumor, and give an example for each.

16–17 At 63 years old, Lauren Brooks was successfully treated with chemotherapy and radiation for cancer of the urinary bladder epithelium. At a later follow-up appointment with her oncologist, however, tests revealed that the cancer had returned. Her physician opted to try a different strategy aimed at inducing an *in situ* state of chronic inflammation intended to stimulate an antitumor immune response. Which of the following is the most likely course of action?
   a. intramuscular vaccination with the BCG vaccine
   b. intradermal vaccination with tumor antigens derived from the patient's tumor cells from a bladder biopsy
   c. intravenous infusion of a monoclonal antibody specific for IL-10, an anti-inflammatory cytokine
   d. intravenous infusion of the patient's tumor cells transfected *ex vivo* with a gene encoding IL-13, an anti-inflammatory cytokine
   e. intravesical adjuvant therapy with BCG vaccine via bladder instillation.

Answers

Chapter 1

1–1   The four classes of pathogen are: bacteria, viruses, fungi, and parasites (protozoa and worms). Examples of these pathogens are given in Figure 1.4 (pp. 6–7).

1–2   d

1–3
A. Skin; mucosal epithelium of the gastrointestinal tract; mucosal epithelium of the respiratory tract; mucosal epithelium of the urogenital tract.
B. (i) Mechanical (physical) barriers: tight junctions between the epithelial cells prevent the penetration of pathogens between the cells to underlying tissues. In addition, there is a flow of air and fluid over epithelial surfaces, which oxygenates and flushes the surface, preventing anaerobic bacterial growth and transient adhesion. On ciliated epithelial surfaces, such as those of the respiratory tract, the formation of a layer of mucus that is kept in continual movement by the beating cilia inhibits colonization and invasion by microorganisms. (ii) Chemical barriers: the epithelium produces a variety of chemical substances that interfere with the adherence of microorganisms to epithelium and with their replication. The skin produces fatty acids in sebaceous glands, which helps to create an acid environment inhibitory to the growth of many bacteria. Lysozyme, an enzyme that inhibits cell-wall formation in bacteria, is secreted in tears, saliva, and sweat. The stomach produces strong hydrochloric acid, creating a highly acidic and formidable environment, which when combined with the stomach enzyme pepsin (an acid protease) poses one of the most inhospitable environments for microbial growth in our bodies. Defensins are antimicrobial peptides secreted by all the protective epithelia. (iii) Microbiological barriers: a flora of non-pathogenic commensal microorganisms colonizes many epithelial surfaces and provides an additional barrier to infection. These microorganisms compete with pathogenic microbes for space and nutrients, and sometimes produce antibacterial proteins that further inhibit attachment to epithelium. For example, *Escherichia coli* in the large intestine produce colicins, which prevent colonization by other bacteria.

1–4   b

1–5   Antibiotics attack the microbiological barriers of intestinal epithelia. The normal flora sensitive to the antibiotics are killed off and the intestine can then be re-colonized and overgrown by microorganisms that in normal circumstances are present in very small numbers and thus do not cause a problem. An example is a condition called pseudomembranous colitis caused by the overgrowth of *Clostridium difficile*. A membrane-like substance is produced in the large intestine, causing an obstruction that can block intestinal flow and usually requires surgical removal.

1–6   The hallmarks of inflammation are heat, redness, pain, and swelling (edema). These are caused by a combination of vasodilation (causing redness and heat), increased vascular permeability and the consequent infiltration of fluid and leukocytes from the blood into the infected site (causing swelling, and also pain as a result of the increased pressure on local nerve endings).

1–7   a, c, e

1–8   b

1–9   Innate immune responses are initiated almost immediately after infection, whereas adaptive immunity takes longer to develop. Innate immunity uses generalized and invariant mechanisms to recognize pathogens. Examples of these are the receptors on phagocytes that recognize surface molecules shared by many different pathogens and stimulate phagocytosis, and serum proteins such as complement. Innate immunity is often unable to eradicate the pathogen completely, and even when it does, it does not produce immunity to reinfection. An adaptive immune response, in contrast, involves specific recognition of the particular pathogen by a highly specific receptor on a subset of lymphocytes, which are selected from a pool of millions of lymphocytes each bearing receptors specific for different molecules. Adaptive immunity is often powerful enough to eradicate the infection and provides long-term protective immunity through immunological memory.

1–10

A. The two major progenitor subsets of leukocytes are the common lymphoid progenitor and the myeloid progenitor.

B. In adults all leukocytes originate in the bone marrow and are derived from pluripotent hematopoietic stem cells.

C. The common lymphoid progenitor differentiates into three cell types: B cells, T cells, and natural killer (NK) cells. The myeloid progenitor differentiates into six main cell types: basophils, eosinophils, neutrophils, mast cells, dendritic cells, and monocytes. Monocytes are circulating leukocytes that enter tissues, where they then differentiate into macrophages.

1–11 c

1–12 e

1–13 In an adaptive immune response to a pathogen, the term clonal selection describes the fact that only those lymphocytes that can recognize the particular pathogen and respond to it are selected to participate in the immune response. Clonal expansion describes the proliferation and subsequent differentiation of these few original lymphocytes to provide large numbers of effector lymphocytes. Clonal selection ensures that the adaptive response will be tailored specifically for the particular type of pathogen involved in the infection. Clonal expansion ensures that the few original lymphocytes specific for the pathogen produce a large population of effector lymphocytes that can make an effective immune response against the pathogen.

1–14 The last case of smallpox was reported in the 1970s. As a result, children are no longer vaccinated routinely as was the case before smallpox was eradicated. A large proportion of any given population today would be unvaccinated, and thus susceptible, to smallpox infection. The mortality rate would be high (30–50%) among those not protected by vaccination.

1–15 **Rationale**: The correct answer is c. This patient needs to be protected from encapsulated bacteria by vaccination because the spleen is important in protection against this type of bacteria. For example, virulent *Streptococcus pneumoniae* resists innate immune responses such as phagocytosis because of its capsular polysaccharide. Effective clearance of encapsulated bacteria is therefore dependent on the presence of anti-capsular opsonizing antibodies. These are normally first produced in the spleen, which filters these bacteria out of the bloodstream. In the absence of a spleen, an antibody response will be much delayed, allowing the bacteria to multiply in the bloodstream. Vaccination and regular boosters with pneumococcal capsular polysaccharides will induce protective antibodies against pathogenic *S. pneumoniae*.

1–16 **Rationale**: The correct answer is c. This is most likely to be a case of antibiotic-associated colitis.

Enterohemorrhagic *Escherichia coli* (EHEC) causes bloody diarrhea, but Mrs Ratamacher does not have blood in her stool. Her symptoms are probably a consequence of taking long-term antibiotics, which have affected the composition of her normal gut flora. The numbers of antibiotic-susceptible *E. coli* will have decreased over time, and the concentration of antibacterial colicins made by these *E. coli* will consequently also decrease. Recolonization with *Clostridium difficile* leads to toxin production, causing diarrhea and sometimes a condition known as pseudomembranous colitis.

Chapter 2

2–1 d

2–2

A. The cleavage of C3 into C3a and C3b and the covalent bonding of C3b to the pathogen surface is called complement fixation, and is the reaction on which the alternative, lectin, and classical pathways of complement activation converge.

B. The enzyme responsible for cleaving C3 into C3a and C3b is called C3 convertase and it differs in composition depending on the particular complement pathway. The classical and lectin pathways use the classical C3 convertase (C4b2a), whereas the alternative pathway uses the alternative convertase (C3bBb).

C. C3 is the most abundant complement component in the plasma and circulates as a zymogen, an inactive enzyme. When cleaved into C3a and C3b, three different effector mechanisms are armed: (1) C3b binds to and tags pathogens for destruction by phagocytes through binding to a C3b receptor, CR1; (2) C3b contributes to a multicomponent enzyme, C5 convertase, that catalyzes the assembly of the terminal complement components and the formation of the membrane-attack complex; and (3) C3a is an inflammatory mediator that serves as a chemoattractant and recruits inflammatory cells to the infection site.

2–3 d

2–4 The CR1 on the macrophage can bind to C3b that is coating a bacterial surface after complement activation, and the macrophage then engulfs the bacterium through receptor-mediated endocytosis. The macrophage membrane invaginates and forms an intracellular vesicle called a phagosome. The phagosome fuses with a lysosome forming a phagolysosome, where toxic mediators and degradative enzymes are localized. The bacterium is destroyed.

2–5

i. The soluble proteins include S protein, clusterin, and factor J, which all inhibit C5b, C6, and C7 from binding to cell membranes.

ii. The cell surface-associated proteins include homologous restriction factor (HRF) and CD59 (protectin), which both prevent the recruitment of C9 and thus block C9 polymerization.

2–6

A. (1) The classical pathway is activated in two ways, either by the presence of antibody bound to the surface of the microorganism (for example, IgM bound to lipopolysaccharide of Gram-negative bacteria) or by the presence of C-reactive protein bound to a bacterium. (2) The lectin pathway requires the presence of mannose-binding lectin, an acute-phase protein made by the liver in response to IL-6 (secreted by activated macrophages) and which accumulates in plasma during infection. (3) The alternative pathway requires an activating surface of a pathogen, which stabilizes complement components.

B. Only the classical pathway is considered part of the adaptive immune response because of the requirement for antibody. However, the classical pathway is also considered part of innate immunity because of the ability of C-reactive protein, an acute-phase protein, to activate it. The other two pathways are considered part of innate immunity because they are initiated independently of antibody.

2–7   a: 3; b: 6; c: 1, 2, 4, 5; d: 1, 2, 4, 5; e: 7; f: 5; g: 8; h: 9.

2–8   TLR-5, TLR-4, TLR1:TLR2, and TLR2:TLR6 are transmembrane receptors anchored on the plasma membrane surface of human cells and interact with pathogens located in extracellular locations. In contrast, TLRs 3, 7, 8, and 9 are anchored in endosomal membranes located in the cytosol, where the intracellular degradation of pathogens takes place.

2–9   Because many pathogens possess features that are common to different groups of pathogens, for example LPS in Gram-negative bacteria, only a small number of TLRs are required to act as sensors of molecular patterns shared by pathogens.

2–10   NFκB is a transcription factor found in the cytoplasm, in an inactive form in macrophages before activation. Signaling through TLRs results in a phosphorylation cascade that converts NFκB to its active form, which then gains the ability to translocate into the nucleus and direct the transcription of specific genes enabling the effector capacity of the macrophage.

2–11   d

2–12

A. (i) Both are phagocytic white blood cells (leukocytes) produced by the bone marrow. Both ingest extracellular pathogens and destroy them intracellularly. They both carry receptors on their surface that recognize pathogens and their components and facilitate pathogen uptake. They both produce defensins (antimicrobial peptides). (ii)

Macrophages are resident in the tissues; neutrophils circulate in the blood and enter tissues only after an infection has become established. Macrophages are long-lived and have important functions other than the uptake and killing of pathogens: once activated by the presence of pathogens, they produce cytokines that induce an inflammatory response, help attract neutrophils to the site of infection, and help initiate an adaptive immune response. Also, once activated, they produce additional cell-surface molecules that enable them to act as professional antigen-presenting cells. In contrast, once in the tissues, neutrophils are short-lived cells. Their only function is to take up and kill extracellular microorganisms. In addition, macrophages carry Toll-like receptors (TLR4), whereas neutrophils do not.

B. Neutrophils and macrophages ingest microorganisms by phagocytosis, taking them up into phagolysosomes, where they are destroyed by bactericidal substances produced by the phagocyte. These include toxic oxygen derivatives such as superoxide, hydrogen peroxide, singlet oxygen, hydroxyl radical, hypohalite and nitric oxide generated by NADPH-dependent oxidases, nitric oxide synthase, and myeloperoxidase. Peptides called defensins and cationic proteins also inhibit microbial growth, whereas lysozyme inhibits cell wall formation. Phagolysosomes have a pH of about 3.5–4.0, a hostile environment for bacteria, fungi, and some enveloped viruses, and one that favors the activity of acid proteases and hydrolases, which degrade the phagocytosed material. Extracellular secretion of lactoferrin serves to compete for essential iron otherwise bound by bacterial siderophores.

2–13   a

2–14

A. Type I interferon genes (encoding interferon-α and -β) are transcribed as a result of the presence of double-stranded RNA.

B. Normal cells not infected with virus do not contain double-stranded RNA; however, cells infected with virus often do. Some viruses either have double-stranded RNA genomes or use double-stranded RNA as an intermediate in the replication cycle.

C. Type I interferons (IFN-α and -β) block virus replication in infected cells and protect uninfected cells nearby from becoming infected. This is accomplished by: (1) inducing cellular genes that destroy viral RNA through endonuclease attack; and (2) inhibiting protein synthesis of viral mRNA by modifying initiation factors required for protein synthesis. In addition, IFN-α and -β activate natural killer (NK) cells. NK cells kill virus-infected cells by releasing cytotoxic granules and cytokines induced by a mechanism involving the recognition of altered cell-surface proteins elevated during viral infection, for example stress-related proteins MIC-A and MIC-B, which are recognized by NKG2D.

2–15   b

2–16   **Rationale**: The correct answer is b. In the absence of factor I, the alternative C3 convertase C3bBb is forming continuously, even in the absence of infection. This results in depletion of the C3 pool in the blood, lymph, and other extracellular fluids, lowering its concentration to levels below the threshold needed to mount effective innate immune responses that rely on the alternative pathway of complement activation. Encapsulated bacteria pose a particular threat because their eradication is dependent on opsonization, in which the opsonin C3b has a key role.

2–17   **Rationale**: The correct answer is a. Two important clues are (1) that Mary has granulomas and (2) that the microorganism cultured from her groin, *Staphylococcus aureus*, is catalase-positive—both suggestive of the condition chronic granulomatous disease (CGD). One of the reactive oxygen intermediates made by phagocytes is hydrogen peroxide, which requires both NADPH oxidase and superoxide dismutase activities. NADPH is therefore essential for neutrophils to produce superoxide and hydrogen peroxide as part of the respiratory burst after phagocytosis. If hydrogen peroxide is not synthesized, then the pH of the phagosome cannot be elevated to the level necessary for activation of the antimicrobial peptides and proteins required to kill phagocytosed bacteria and fungi. This leads to chronic infections because intracellular killing of ingested microbes is ineffective.

Chapter 3

3–1
  i.   Unique differences distinguishing particular pathogens from other pathogens and from the vertebrate host are targeted specifically.
  ii. Pathogen-specific memory cells generated during first exposure are available during a secondary encounter and mediate a more rapid and effective clearance during a recall response.
  iii. Even as pathogens mutate at a rapid rate, adaptive immune responses retain the capacity to recognize molecular changes in evolving pathogens.

3–2   c

3–3   Because it takes several days for the immune system to generate adequate numbers of pathogen-specific lymphocytes after clonal selection, the host is most vulnerable to disease during this time.

3–4   c

3–5   b

3–6   d

3–7   (A) CD4 T cells, also known as helper T cells, function by secreting cytokines that instruct other cells to acquire effector function. CD8 T cells differentiate into cytotoxic effector cells and kill the target cells that they recognize. (B) CD4 T cells recognize only antigens presented by MHC class II molecules. CD8 T cells recognize only antigens presented by MHC class I molecules.

3–8
  A. CD8 T cells recognize antigen that was derived from intracellular sites in the host such as viruses and some bacteria that live and replicate within cells. CD4 T cells recognize antigen derived from extracellular fluid and interstitial spaces between cells and tissues where all types of pathogen can be found, and where some toxins secreted by pathogens are located.
  B. The host's immune response needs to be able to identify all types of pathogen in the various anatomical locations in which they exist. This is essential so that effective clearance mechanisms can be tailored to combat infections regardless of which strategy the pathogen has evolved for its survival and dissemination in the host.

3–9   (A) MHC class I molecules do not bind to peptides derived from pathogen-derived proteins until the peptides have been transported into the endoplasmic reticulum. Transport to the endoplasmic reticulum does not occur until after proteolytic cleavage of the pathogen proteins has occurred in the cytoplasm. Once the peptide has bound to an MHC class I molecule, the peptide:MHC class I complex is transported to the cell surface for presentation to CD8 T cells. (B) MHC class II molecules bind pathogen-derived peptides in a location distinct from MHC class I molecules inside endocytic vesicles, the same location where proteolytic cleavage of pathogen proteins occurs after engulfment of pathogen-laden material from infected tissue. Once the peptide has bound to an MHC class II molecule, the peptide:MHC class II complex is transported to the cell surface for presentation to CD4 T cells.

3–10   e

3–11   In fact there are vaccines made solely of polysaccharide constituents of encapsulated bacteria that stimulate antibody responses (for example, Pneumovax is used in humans older than 2 years of age). The most successful vaccines of this nature, however, are conjugated; this means that the bacterial polysaccharide is covalently linked to an antigenic protein by a chemical process (for example, the Hib vaccine is a conjugate of *Haemophilus influenzae* polysaccharide and tetanus toxoid and is administered to infants). For CD4 T helper cells to participate in B cell activation, protein sequences must be included in the vaccine; CD4 T cells are activated only when pathogen peptides are presented on the surface of antigen-presenting cells together with MHC class II molecules. Vaccines devoid of protein constituents will therefore not provoke T cell responses, and may be limited to weaker T-independent B-cell responses.

3–12   (A) Neutralization and opsonization are similar in that once antibody is bound to antigen it is targeted for destruction by phagocytes after ingestion. This process is mediated by receptors for antibodies. (B) Neutralization differs from opsonization in that by binding to antigen (such as a toxin or a virus), the host cell receptor is unable to bind to native antigen, uptake is blocked, and the antigen is thus inhibited from damaging the host cell.

3–13   When bound to the surface of a pathogen, IgM induces the classical pathway of complement activation. This in turn leads to the deposition of C3b on the pathogen's surface. Phagocytes contain complement receptor CR1, which binds to C3b fixed on the pathogen's surface; receptor-mediated endocytosis—that is, opsonization—ensues.

3–14   (i) Somatic hypermutation is a mutational mechanism that results in nucleotide substitutions affecting coding regions of the variable loci of heavy- and light-chain genes in immunoglobulins. Three possible consequences arise as a result of a nucleotide substitution in the variable region: (1) no change to the affected amino acid; (2) a change reducing the affinity of antibody:antigen interaction; and (3) a change increasing the affinity of antibody:antigen interaction. The last of these interacts with antigen most effectively, and cells bearing them will be selected for clonal expansion and differentiation into plasma and memory cells. (ii) Isotype switching is the second mutational mechanism affecting antibody quality and memory B cells. Here the constant region is involved; the variable region, which binds to antigen, is preserved. By switching the constant region from IgM to IgG, IgA, or IgE, the secreted antibodies are more effectively transported to the appropriate anatomical locations where antigen is located, and subsequent elimination of the antigen by antibody-receptor-bearing leukocytes occurs.

3–15   (i) Allergy. IgE antibodies made against normally innocuous environmental antigens trigger widespread mast-cell activation. This can lead to allergic diseases such as asthma or to a potentially fatal anaphylactic reaction. (ii) Autoimmune disease. Chronic immune responses by B cells or T cells to self antigens can cause tissue damage and chronic illnesses such as diabetes, multiple sclerosis, and myasthenia gravis. Autoimmunity is sometimes provoked as a consequence of an immune response to pathogen-derived antigen that cross-reacts on healthy host cells or tissue. (iii) Transplant rejection. A person's immune system will make an immune response against the foreign MHC molecules on transplanted tissue that is MHC-incompatible.

3–16   **Rationale**: The correct answer is c. One mechanism for achieving immunological tolerance to self proteins is negative selection in the thymus. Developing T cells bearing T-cell receptors that bind too strongly to self-MHC or to self-MHC:self-peptide complexes are signaled to die by apoptosis, which eliminates self-reactive T cells. Self proteins not presented to developing T cells in the thymus, such as proteins sequestered in the central nervous system, would not be scrutinized during negative selection, and self-reactive T cells would result.

Chapter 4

4–1
   A. Immunoglobulins are the membrane-bound form of the antigen receptors of B cells. Antibodies are the secreted form of the same immunoglobulins.
   B. Immature, mature, and memory B cells have membrane-bound immunoglobulin. Antibodies are produced by plasma cells.

4–2   An antibody molecule is made of four polypeptide chains—two identical heavy chains and two identical and smaller light chains, with a total molecular weight of approximately 150 kDa. Each chain is made up of a series of structurally similar domains known as immunoglobulin domains. The amino-terminal portion of each H chain combines with one L chain, and the two carboxy-terminal portions of the H chains combine with each other, forming a Y-shaped quaternary structure. Disulfide bonds hold the H and L chains together, hold the two H chains together (interchain disulfide bonds), and stabilize the domain structure of the chains (intrachain disulfide bonds). The arms of the antibody molecule are called Fab (fragment antigen binding) and interact with antigen. The stalk is called Fc (fragment crystallizable) and is made up of H chains only. The amino-terminal domains of an H and an L chain together make up a site that binds directly to antigen and varies greatly between different antibodies. These domains are referred to as the variable region, and each antibody has two identical antigen-binding sites. The remaining domains of both H and L chains are the same in all antibodies of a given class (isotype). These domains are referred to as the constant region. The variable region of each chain includes hypervariable regions of amino acid sequences that differ the most between different antibodies. These are nested within less variable sequences known as the framework regions. The hypervariable regions make loops at one end of the domain structure and are also known as complementarity-determining regions because they confer specificity on the antigen-binding site.

4–3   d

4–4
   A. An epitope is the specific part of the antigen that is recognized by an antibody and binds to the

complementarity-determining regions in the antibody variable domains. Epitopes are sometimes referred to as antigenic determinants. Epitopes can be part of a protein or can be carbohydrate or lipid structures present in the glycoproteins, polysaccharides, glycolipids, and proteoglycans of pathogens.

B. Multivalent antigens are complex macromolecules that contain more than one epitope.

C. Linear epitopes are epitopes in proteins that comprise a contiguous amino acid sequence. They are also called continuous epitopes. In contrast, a conformational epitope is formed by amino acids that are brought together as a result of protein folding and are not adjacent to each other in the protein sequence. Conformational epitopes are also known as discontinuous epitopes.

D. Antibodies bind antigens via noncovalent bonding such as hydrogen bonds, hydrophobic interactions, van der Waals forces, and electrostatic attraction.

4–5   c

4–6
A. In developing B cells, gene rearrangements within the genetic loci for immunoglobulin light and heavy chains can produce an almost unlimited variety of different variable regions, and thus produce the huge repertoire of antibodies with different specificities for many types of antigens. This gene rearrangement mechanism is called somatic recombination. In the germline configuration, before gene rearrangement, the immunoglobulin loci in progenitor B cells are composed of sequences encoding the constant regions and families of gene segments encoding different portions of the variable region. Heavy-chain loci contain a series of gene segments called variable (V), diversity (D), and joining (J). Light-chain loci contain only V and J gene segments. In somatic recombination in developing B cells, one of each family of gene segments is randomly selected and joined together to give a complete variable-region sequence, which is subsequently expressed as an immunoglobulin heavy or light chain. Immunoglobulin gene rearrangement is irreversible, leading to permanent alteration of the chromosome; it occurs exclusively in B cells.

B. A D gene segment first joins to a J to form DJ, followed by a V becoming joined to DJ to form VDJ, which encodes a complete variable region.

C. The heavy-chain locus rearranges before the light-chain loci. For light chains in humans, the κ locus rearranges first and is followed by the λ locus only if both κ loci fail to produce a successful rearrangement.

4–7   c, d

4–8   b

4–9   An individual with this genetic defect would be unable to rearrange either immunoglobulin or T-cell receptor genes somatically. There would be a severe combined immunodeficiency (SCID) owing to the absence of mature B cells and T cells.

4–10   a

4–11   a = F; b = T; c = F; d = T; e = F

4–12   b

4–13   a

4–14   a

4–15   **Rationale**: The correct answer is f. Aliya has IgM antibodies, so this is clearly not a case of X-linked agammaglobulinemia, in which a complete absence of immunoglobulin is expected. The father has normal immunoglobulin levels and the patient is a girl, so X-linked hyper IgM syndrome is highly unlikely. IgA deficiency can be ruled out because she also does not make IgG. Patients with acute lymphoblastic leukemia do not have an isotype-switching defect, and children with severe combined immunodeficiency die in their first year without treatment. A defect in both copies of the gene encoding activation-induced cytidine deaminase (AID) is most likely; AID is required for both isotype switching and somatic hypermutation, which accounts for Aliya's ability to make IgM but her inability to make any other antibody isotype. Both of her parents would be heterozygous carriers of the AID defect but would produce sufficient amounts of AID to escape immunodeficiency.

Chapter 5

5–1
A. Similarities. (1) The T-cell receptor has a similar overall structure to the membrane-bound Fab fragment of immunoglobulin, containing an antigen-binding site, two variable domains and two constant domains. (2) T-cell receptors and immunoglobulins are both generated through somatic recombination of sets of gene segments. (3) The variable region of the T-cell receptor contains three complementarity-determining regions (CDRs) encoded by the V_α domain and three CDRs encoded by the V_β domain, analogous to the CDRs encoded by the V_H and V_L domains. (4) There is huge diversity in the T-cell receptor repertoire and it is generated in the same way as that in the B-cell repertoire (by combination of different gene segments, junctional diversity due to P- and N-nucleotides, and combination of two different chains). (5) T-cell receptors are not expressed at the cell surface by themselves but require association with the CD3 γ, δ, ε, and ζ chains for stabilization and signal transduction, analogous to the Igα and Igβ chains required for

immunoglobulin cell-surface expression and signal transduction.

B. Differences. (1) A T-cell receptor has one antigen-binding site; an immunoglobulin has at least two. (2) T-cell receptors are never secreted. (3) T-cell receptors are generated in the thymus, not the bone marrow. (4) The constant region of the T-cell receptor has no effector function and it does not switch isotype. (5) T-cell receptors do not undergo somatic hypermutation.

5–2   The organization of the TCRα locus resembles that of an immunoglobulin light-chain locus, in that both contain V and J gene segments and no D gene segments. The TCRα locus on chromosome 14 contains about 80 V gene segments, 61 J gene segments, and one C gene. The immunoglobulin light-chain loci, λ and κ, are encoded on chromosomes 22 and 2, respectively. The λ locus contains about 30 V gene segments and four J gene segments, each paired with a C gene. The κ locus contains about 35 V gene segments, five J segments, and one C gene segment. The arrangement of the κ locus more closely resembles that of the TCRα locus except that there are more J segments in the T-cell receptor locus.

The organization of the TCRβ locus resembles that of the immunoglobulin heavy-chain locus; both contain V, D, and J gene segments. The TCRβ locus contains about 52 V gene segments, two D gene segments, 13 J gene segments, and two C genes, encoded on chromosome 7. Each C gene is associated with a set of D and J gene segments. The immunoglobulin heavy-chain locus on chromosome 14 contains about 40 V segments, 23 D segments, and six J segments, followed by nine C genes, each specifying a different immunoglobulin isotype. The heavy-chain C genes determine the effector function of the antibody.

5–3   T-cell receptors are not made in a secreted form and their constant regions do not contribute to T-cell effector function. Other molecules secreted by T cells are used for effector functions. There is therefore no need for isotype switching in T cells, and the T-cell receptor loci do not contain numerous alternative C genes.

5–4   a

5–5   a, c, d

5–6   a

5–7
A. (i) The complete MHC class I molecule is a heterodimer made up of one α chain and a smaller chain called β2-microglobulin. The α chain consists of three extracellular domains—α_1, α_2, and α_3—a transmembrane region and a cytoplasmic tail. β2-Microglobulin is a single-domain protein noncovalently associated with the extracellular portion of the α chain, providing support and stability. (ii) The polymorphic class I molecules in humans are called HLA-A, HLA-B, and HLA-C. The α chain is encoded in the MHC region by an MHC class I gene. The gene for β2-microglobulin is elsewhere in the genome. (iii) The antigen-binding site is formed by the α_1 and α_2 domains, the ones farthest from the membrane, which create a peptide-binding groove. The region of the MHC molecule that binds to the T-cell receptor encompasses the α helices of the α_1 and α_2 domains that make up the outer surfaces of the peptide-binding groove. The α_3 domain binds to the T-cell co-receptor CD8. (iv) The most polymorphic parts of the α chain are the regions of the α_1 and α_2 domains that bind antigen and the T-cell receptor. β2-Microglobulin is invariant; that is, it is the same in all individuals.

B. (i) MHC class II molecules are heterodimers made up of an α chain and a β chain. The α chain consists of α_1 and α_2 extracellular domains, a transmembrane region, and a cytoplasmic tail. The β chain contains β_1 and β_2 extracellular domains, a transmembrane region, and a cytoplasmic tail. (ii) In humans there are three polymorphic MHC class II molecules called HLA-DP, HLA-DQ, and HLA-DR. Both chains of an MHC class II molecule are encoded by genes in the MHC region. (iii) Antigen binds in the peptide-binding groove formed by the α_1 and β_1 domains. The α helices of the α_1 and β_1 domains interact with the T-cell receptor. The β_2 domain binds to the T-cell co-receptor CD4. (iv) With the exception of HLA-DRα, which is dimorphic, both the α and β chains of MHC class II molecules are highly polymorphic. Polymorphism is concentrated around the regions that bind antigen and the T-cell receptor in the α_1 and β_1 domains.

5–8   c

5–9
A. Antigen processing is the intracellular breakdown of pathogen-derived proteins into peptide fragments that are of the appropriate size and specificity required to bind to MHC molecules.

B. Antigen presentation is the assembly of peptides with MHC molecules and the display of these complexes on the surface of antigen-presenting cells.

C. Antigen processing and presentation must occur for T cells to be activated because (1) T-cell receptors cannot bind to intact protein, only to peptides, and (2) T-cell receptors do not bind antigen directly, but rather must recognize antigen bound to MHC molecules on the surface of antigen-presenting cells.

5–10
A. Proteins from pathogens growing in the cytosol are broken down into small peptide fragments in proteasomes. The peptides are transported into the lumen of the endoplasmic reticulum (ER) using TAP (transporter associated with antigen processing), which is a heterodimer of TAP-1 and TAP-2 proteins located in the ER membrane.

Peptides bearing the appropriate peptide-binding motif bind to MHC class I molecules already delivered into the ER. MHC class I α chains are bound to the chaperone calnexin until β_2-microglobulin binds, and then are bound by the chaperones calreticulin and tapasin until peptide binds. Tapasin binds to TAP-1, positioning the MHC class I molecule near the peptide source. MHC class I molecules bound to peptide dissociate from the chaperone molecules and progress to the Golgi apparatus for completion of glycosylation and transport to the cell surface in membrane-bound vesicles.

B. (i) If an MHC class I α chain is unable to bind β_2-microglobulin, it will be retained in the ER and will not be transported to the cell surface. It will remain bound to calnexin and will not fold into the conformation needed to bind to peptide. Thus, antigens will not be presented using that particular MHC class I molecule. (ii) If TAP-1 or TAP-2 proteins are mutated and not expressed, then peptides will not be transported into the lumen of the ER. Without peptide, an MHC class I molecule cannot complete its assembly and will not leave the ER. A rare immunodeficiency disease called bare lymphocyte syndrome (MHC class I immunodeficiency) is characterized by a defective TAP protein, causing less than 1% of MHC class I molecules to be expressed on the cell surface in comparison with normal. Thus, T-cell responses to all pathogen antigens that would normally be recognized on MHC class I molecules will be impaired.

5–11 a

5–12

A. Extracellular pathogens are taken up by endocytosis or phagocytosis and degraded by enzymes into smaller peptide fragments inside acidified intracellular vesicles called phagolysosomes. MHC class II molecules delivered into the ER and being transported to the cell surface intersect with the phagolysosomes, where these peptides are encountered and loaded into the antigen-binding groove. To prevent MHC class II molecules from binding to peptides prematurely, invariant chain (Ii) binds to the MHC class II antigen-binding site in the ER. Ii is also involved in transporting MHC class II molecules to the phagolysosomes via the Golgi as part of the interconnected vesicle system. Ii chain is removed from MHC class II molecules once the phagolysosome is reached. Removal is achieved in two steps: (1) proteolysis cleaves Ii into smaller fragments, leaving a small peptide called CLIP (class II-associated invariant chain peptide) in the antigen-binding groove of the MHC class II molecule; and (2) CLIP is then released by HLA-DM catalysis. Once CLIP is removed, peptides derived from endocytosed material will bind (if the correct

peptide-binding motif is possessed) and the peptide:MHC class II molecule complex will progress to the cell surface.

B. (i) Defects in the invariant chain would impair normal MHC class II function because invariant chain not only protects the peptide-binding groove from binding prematurely to peptides present in the ER but is also required for transport of MHC class II molecules to the phagolysosome. (ii) If HLA-DM were not expressed, most MHC class II molecules on the cell surface would be occupied by CLIP rather than endocytosed material. This would compromise the presentation of extracellular antigens at the threshold levels required for T-cell activation.

5–13

A. Multigene family refers to the presence of multiple genes for MHC class I and MHC class II molecules in the genome, encoding a set of structurally similar proteins with similar functions. MHC polymorphism is the presence of multiple alleles (in some cases several hundreds) of most of the MHC class I and class II genes in the human population.

B. T cells recognize peptide antigens in the form of peptide:MHC complexes, which they bind using their T-cell receptors. To bind specifically, the T-cell receptor must fit both the peptide and the part of the MHC molecule surrounding it in the peptide-binding groove. (i) Because each individual produces a number of different MHC molecules from their MHC class I and class II multigene family, the T-cell receptor repertoire is not restricted to recognizing peptides that bind to just one MHC molecule (and thus all must have the same peptide-binding motif). Instead, the T-cell receptor repertoire can recognize peptides with different peptide-binding motifs during an immune response, increasing the likelihood of antigen recognition and, hence, T-cell activation. (ii) The polymorphism in MHC molecules is localized in the regions affecting T-cell receptor and peptide binding. Thus, a T-cell receptor that recognizes a given peptide bound to variant 'a' of a particular MHC molecule is likely not to recognize the same peptide bound to variant 'b' of the same MHC molecule. Polymorphism also means that the MHC molecules of one person will bind a different set of peptides from those in another person. Taken together, these outcomes mean that because of MHC polymorphism, each individual recognizes a somewhat different range of peptide antigens using a different repertoire of T-cell receptors.

5–14 MHC polymorphisms are non-randomly localized predominantly in the region of the molecule that makes contact with peptide and T-cell receptors. Random DNA mutations, in contrast, would be scattered through the gene, giving rise to amino acid changes throughout MHC molecules and not just in those areas important for peptide binding and presentation.

5–15 **Rationale:** The correct answer is d. A deficiency in either TAP-1 or TAP-2, whose function is to transport peptide fragments to the lumen of the endoplasmic reticulum, would inhibit the expression of MHC class I on the cell surface because peptide must be bound to an MHC class I molecule to permit its transport from the endoplasmic reticulum to the cell membrane. Low levels of MHC class I account for the low levels of CD8 T cells; MHC class I-restricted CD8 T cells would fail to undergo positive selection in the thymus if MHC class I levels were abnormally low. Deficiencies of HLA-DM, invariant chain (which gives rise to CLIP), or CIITA would affect the MHC class II pathway of antigen presentation, but would have no effect on the MHC class I pathway.

Chapter 6

6–1   e → a → f → d → c → b

6–2   d → a → e → b → f → c

6–3
A. Bone marrow stromal cells provide the necessary environment for B-cell development by expressing secreted products and membrane-bound adhesion molecules. For example, VCAM-1 adhesion molecule binds to the integrin VLA-4 on early B-cell progenitors. Cytokines such as IL-7 have an important role in later stages of B-cell development, serving to stimulate the growth and cell division of late pro-B and pre-B cells.
B. If anti-IL-7 antibody was introduced into this environment, developing B cells would be arrested at the late pro-B or pre-B cell stage and would not be able to progress normally to the immature B-cell stage. Interestingly, in transgenic mice overexpressing IL-7, significant increases in pre-B cells are observed in the bone marrow and secondary lymphoid organs, while in IL-7 knockout mice (where the gene locus encoding IL-7 is interrupted and no IL-7 is produced) early B-cell expansion is significantly impaired. These experiments in mice demonstrate clearly the importance of IL-7 in B-cell maturation.

6–4
A. Checkpoint 1 is marked by the formation of a complex of a μ heavy-chain complexed with the surrogate light-chain VpreBλ5, Igα, and Igβ. Checkpoint 2 is when a complete B-cell receptor, comprising μ heavy chains, κ or λ light chains, and Igα and Igβ chains, is expressed on the B-cell surface.
B. At checkpoint 1, if the V(D)J rearrangement gives rise to a functional pre-B-cell receptor the late pro-B cell will be permitted to survive and undergo clonal proliferation. If V(D)J rearrange-

ment produces a nonfunctional heavy chain and no pre-B-cell receptor is assembled, the pro-B cell undergoes apoptosis and dies. Similarly, at checkpoint 2, production of a functional light chain results in the assembly of a functional surface immunoglobulin and the survival and maturation of the B cell. Non-production of a light chain results eventually in apoptosis.
C. Checkpoint 1 delivers an important signal to the cell, verifying that a functional heavy chain has been made. This triggers the cessation of heavy-chain gene rearrangement followed by the inactivation of surrogate light-chain synthesis. Thus, only one heavy-chain locus ends up producing a product. As surrogate light chain becomes unavailable, μ accumulates and is retained in the endoplasmic reticulum, ready to bind to functional light chain when that is synthesized after successful light-chain gene rearrangement. Checkpoint 2 signals the cessation of light-chain rearrangement. This ensures that only one light-chain locus out of the possible four produces a functional product.

6–5   N nucleotides would be added at the VJ joints of all rearranged light-chain genes during gene rearrangement (instead of about half), resulting in an increase in immunoglobulin diversity. It is interesting to note that because TdT is not expressed until after birth, B-1 cells that are generated prenatally lack N nucleotides in the VD and DJ junctions of their rearranged heavy-chain genes as well as in the VJ junctions of all light-chain genes.

6–6   e

6–7
A. Unlike conventional B-2 B cells, B-1 B cells express the cell-surface protein CD5, possess few N nucleotides at VDJ junctions, and have a restricted range of antigen specificities. They produce IgM antibodies of low affinity and respond mainly to carbohydrate, rather than protein, epitopes. Individual B-1 cells are polyspecific for antigen; that is, their immunoglobulins bind several different antigens.
B. B-1 cells are probably best associated with innate immune responses because of their rapid response to antigen, their limited diversity and their polyspecificity.

6–8   a, c

6–9   e

6–10   e

6–11   To survive, circulating B cells must enter primary follicles where survival signals are delivered by cells in the follicles, including follicular dendritic cells (which are the stromal cells of primary lymphoid follicles).

Circulating B cells that fail to enter follicles in secondary lymphoid tissues will die in the peripheral circulation with a half-life of about 3 days. B cells with antigen receptors specific for soluble self antigen are generally rendered anergic in the bone marrow or the circulation. Anergic B cells that enter secondary lymphoid organs are held in the T-cell areas adjacent to primary follicles and are not permitted to penetrate the follicle. As a result, they do not receive the necessary stimulatory signal for survival. Instead, anergic B cells will undergo apoptosis in the T-cell zone. This is an efficient cleansing mechanism and serves to delete potentially autoreactive B cells from the circulation.

6–12   a, b, e

6–13

A. Memory enables faster, more efficient recall responses when antigen is encountered subsequently. This enables the body to get rid of a pathogen before it has time to cause disease.

B. Immunoglobulin produced during a primary immune response is mainly IgM, in low concentration (titer) and of low affinity for the antigen. Immunoglobulin expressed during a secondary immune response has undergone isotype switching and is often of the IgG isotype. It also has a higher titer and, through the process of somatic hypermutation, will have a higher affinity for its corresponding antigen.

6–14

A. A B-cell tumor comprises cells deriving from a single cell that has undergone transformation, resulting in uncontrolled growth. No further maturation of the B cell occurs after transformation. If the B cell has rearranged the heavy-chain and light-chain genes before transformation, then immunoglobulin will be expressed at the cell surface. Because all of the cells in the tumor belong to the same clone, the immunoglobulin on all of them will be made of the same heavy and light chains.

B. Pre-B-cell leukemia is characterized by transformation before the rearrangement of light-chain genes. If transformation occurs at the large pre-B-cell stage, the immunoglobulin on the cell surface will be composed of μ:VpreBλ5. If transformation occurs at the small pre-B-cell stage, there would be little or no immunoglobulin on the cell surface because surrogate light chain expression is turned off at this stage and the μ heavy chains are retained in the endoplasmic reticulum. Normal immature B cells that have not undergone transformation express IgM containing μ plus a κ or λ light chain.

6–15   **Rationale:** The correct answer is b. Multiple myeloma results from the outgrowth of a single plasma cell that has undergone malignant transformation in the bone marrow. Tumor masses expand to the point at which the relatively limited amount of space available in the bone becomes filled with tumor cells, constraining the development of erythrocytes and neutrophils and causing anemia and neutropenia, respectively. Serum IgG would be predominantly monoclonal, because the tumor was derived from a single plasma cell. Likewise, the Bence-Jones protein would be of the λ type, not κ, because this is an IgG λ multiple myeloma. Patients with multiple myeloma would be susceptible to pyogenic infections because the diversity of immunoglobulins will be limited, making these patients immunocompromised.

Chapter 7

7–1   The analog of VpreB:λ5 in developing T cells is the protein preTα (pTα), which combines with the T-cell receptor β chain, the first of the two T-cell receptor chains to be expressed, to form the pre-T-cell receptor. The β chain, like the immunoglobulin heavy chain, contains V, D, and J segments. pTα also binds CD3 and ζ components to this complex, and the assembly of the complete complex induces T-cell proliferation and the cessation of rearrangement at the TCRβ loci (leading to allelic exclusion). Formation of the analogous pre-B-cell receptor complex of VpreB:λ5 and heavy chain with Igα and Igβ in B cells similarly prevents further rearrangement of the heavy-chain loci.

7–2   e

7–3   d

7–4   The first checkpoint occurs after the rearrangement of the β-chain locus, which tests for the ability of the β-chain to associate with pTα, the surrogate light chain, and form the pre-T-cell receptor on the cell surface. The second checkpoint occurs after the rearrangement of the α-chain locus, which tests for the ability of the α-chain to associate with the β-chain and form the T-cell receptor on the cell surface.

7–5   e

7–6

A. Cells that express MHC class II are the professional antigen-presenting cells (B cells, macrophages, and dendritic cells), thymic epithelial cells, neural microglia, and activated T cells (in humans).

B. Macrophages, dendritic cells, and thymic epithelial cells either populate the thymus or circulate through it. Cortical thymic epithelial cells participate in positive selection by presenting MHC class I and class II molecules with self peptides to double-positive (CD4 CD8) thymocytes. These are developing T cells that have successfully rearranged the TCRα and TCRβ genes. Only T cells that have T-cell receptors that can interact with self MHC are positively selected, thus shaping a

T-cell repertoire that is specific for self-MHC molecules. If the affinity for self-peptide:self-MHC is too weak, the T cells die by neglect, through apoptosis. Cortical and medullary thymic epithelium may also participate in negative selection by inducing the apoptosis of thymocytes that bear a T-cell receptor with high affinity for self MHC, self peptide or a combination of the two. Thymic epithelium, circulating macrophages, and dendritic cells found primarily at the cortico-medullary junction participate in negative selection and aid in the elimination of potentially self-reactive T cells bearing high-affinity T-cell receptors for complexes of self peptides:self MHC. Because these cells are circulating between tissues, organs, and the thymus, a heterogeneous array of self peptides will be transported to the thymus and displayed. This is important because there are some self peptides that are expressed in locations other than the thymus and may be encountered in secondary lymphoid organs.

C. Although circulating dendritic cells and macrophages are effective at endocytosing cellular debris in extrathymic locations and then presenting self peptides derived from this debris to thymocytes, there are some self proteins that are excluded from this housekeeping function. Some self proteins are located in immunologically privileged sites where leukocytes do not usually circulate. These sites are normally MHC class II negative, and this is important because this avoids the presentation of self peptides with MHC class II that were not presented in the thymus during negative selection. It should be pointed out, however, that some of these tissues can be induced to express MHC class II molecules under the influence of certain cytokines, for example interferon-γ, during an inflammatory response. This is believed to be one mechanism by which tolerance can be broken, resulting in autoimmunity.

7–7

A. Only one of the receptors will have to be positively selected for the cell to get the survival signals necessary for it to pass on to the next stage. Even if the other receptor does not react with self MHC this will have no effect on the cell.

B. In contrast, both receptors will have to pass the negative selection test for the T cell to survive, because if only one of them fails it, the cell will die.

C. Yes. Imagine this situation. The T cell with dual specificity could be activated appropriately during a genuine infection by a professional antigen-presenting cell plus foreign antigen 1 using T-cell receptor 1. But that same T cell, because it is now an activated effector T cell, would also be able to respond to a second peptide, which might be a self peptide, using T-cell receptor 2, without requiring the co-stimulatory signals that only

professional antigen-presenting cells deliver. Thus it could cause a reaction against a self tissue, either directly, if it is a CD8 cytotoxic T cell, or indirectly, if it is a CD4 T cell, by activating potentially autoreactive B cells.

Furthermore, interferon-γ produced in the response against foreign antigen 1 could activate nonprofessional antigen-presenting cells nearby, inducing the expression of MHC class II with presentation of the self peptide above. Effector T cells with T-cell receptor 2 could make an autoimmune response against it.

7–8   Because T cells drive almost all immune responses, once they have been activated their receptors must continue to recognize the exact complex of foreign antigen and MHC molecule (which does not itself change) that activated them. Because of this requirement for dual recognition (MHC restriction), somatic hypermutation would be more likely than not to change the T-cell receptor to make it unable to recognize either the peptide or the MHC molecule, or the combination of both, thus rendering it unable to give help to B cells or to attack infected cells. This would destroy both the primary immune response and the development of immunity. Even changes that simply increased the affinity of the T cell for its antigen would have no real advantage, because it would not make the immune response any stronger or improve immunological memory in the same way that affinity maturation of B cells does. Also, if somatic hypermutation changed the specificity of the T-cell receptor so that it now recognized a self peptide, this could result in an autoimmune reaction. These considerations do not apply to B cells, because they require T-cell help to produce antibody and will only receive it if their B-cell receptor still recognizes the original antigen.

7–9

A. MHC class II deficiency affects the development of CD4 T cells in the thymus. If the thymic epithelium lacks MHC class II, positive selection of CD4 T cells will not take place. CD8 T cells are not affected because MHC class I expression is unaffected by this defect.

B. To produce antibody, B cells require T-cell help in the form of cytokines produced by CD4 T_H2 cells. Low immunoglobulin levels (hypogammaglobulinemia) in MHC class II deficiency are attributed to the inability of B cells to proliferate and differentiate into plasma cells in the absence of T_H2 cytokines.

7–10

A. After thymic atrophy or thymectomy, T cells in the periphery self-renew by cell division and are long lived.

B. B cells are short lived and replenish from immature precursors derived from the bone marrow.

7–11

A. (i) As well as expressing their own thymus-specific self antigens, medullary epithelial cells in the thymus produce a transcription factor called autoimmune regulator (AIRE), which causes several hundred genes normally expressed in other tissues to be expressed in these cells. The proteins can then be processed to form self peptides that will be presented by MHC class I molecules. (ii) The thymic epithelium uses different proteases for self-protein degradation; cathepsin L is used for peptide production instead of cathepsin S, which is used by other cell types.

B. Generating a more comprehensive repertoire of self peptides in the thymus increases the types of potentially autoreactive T cell that are removed from the peripheral T-cell repertoire during negative selection.

7–12

A. T_{reg} cells suppress the proliferation of naive autoreactive CD4 T cells by secreting inhibitory cytokines. This inhibitory action requires that both the T_{reg} and the other CD4 T cell are interacting with the same antigen-presenting cell.

B. Unlike non-regulatory CD4 T cells, T_{reg} express CD25 on the cell surface and the FoxP3 transcriptional repressor protein.

7–13 a

7–14 b

7–15 **Rationale**: The correct answer is b. Depletion of T cells, but not of B cells, and the absence of a thymic shadow on the X-ray are critical clues. Development of both CD8 and CD4 T cells is affected in this patient because the thymus is the primary lymphoid organ required for T-cell development. Bare lymphocyte syndrome affects either CD8 (type I) or CD4 (type II), but not both. Although patents without a thymus succumb to infections also common in individuals with AIDS, this option can be ruled out because Giulia lacks CD8 T cells, a condition not seen in AIDS patients. Chronic granulomatous disease is a defect of neutrophil, not T-cell, function.

Chapter 8

8–1

A. Naive T cells encounter antigen, and start the primary immune response, in a secondary lymphoid tissue (for example, lymph nodes, spleen, Peyer's patches, tonsils).

B. (i) Lymph nodes. The pathogen, and dendritic cells that have ingested the pathogen, are carried to the nearest lymph node in the afferent lymph. (ii) Gut-associated lymphoid tissues (GALT) such as Peyer's patches. Pathogens enter GALT via specialized cells (M cells) in the gut epithelium. (iii) Spleen. Pathogens circulating in the blood enter the spleen directly from the blood vessels that feed it.

C. Naive T cells are delivered to all secondary lymphoid organs from the blood. They can also pass from one lymph node to another via a lymphatic.

D. After activation by antigen only CD8 T cells and T_H1 cells exit the lymphoid tissue (via efferent lymph, which delivers them eventually into the blood) in search of infected tissues. Antigen-activated T_H2 cells remain in the lymphoid tissue, where they provide help to antigen-specific B cells.

8–2   First, antigen needs to be transported to a nearby secondary lymphoid tissue, processed, and presented by antigen-presenting cells to naive cytotoxic T cells or helper T cells so as to activate them. Second, the number of T cells specific for a given pathogen will be only around 1 in 10,000 to 1 in 100,000 (10^{-4} to 10^{-6}) of the circulating T-cell repertoire; thus it can take some time before the relevant T cells circulating through the secondary lymphoid tissues reach the tissue containing the antigen that will activate them. Finally, it takes several days for an activated T cell to proliferate and differentiate into a large clone of fully functional effector T cells.

8–3

A. T cells (and B cells) express L-selectin, which binds to sulfated carbohydrates of mucin-like vascular addressins in HEVs. Three types of mucin-like vascular addressin are involved: GlyCAM-1 and CD34 expressed on lymph-node HEVs, and MAdCAM-1, expressed on mucosal endothelium.

B. Chemokines made by endothelium and bound to its extracellular matrix induce T cells to express LFA-1, an integrin. LFA-1 binds with high affinity to an intercellular adhesion molecule, ICAM-1, expressed on the endothelium. Finally, the T cell squeezes between the endothelium junctions, a process called diapedesis, and hence gains entry into the lymph node.

8–4   b

8–5

A. Expression of B7, a co-stimulator molecule, distinguishes professional antigen-presenting cells from other cells.

B. When B7 binds to CD28, the B7 receptor expressed earliest on T cells, an activation signal is delivered and T cells undergo clonal expansion and differentiation. This interaction requires, of course, that the T-cell receptor and CD4 co-receptor are engaged specifically with a peptide:MHC class II complex. The second B7 receptor, CTLA-4, binds B7 with about 20-fold higher affinity than does CD28. An inhibitory signal is delivered to the T cell when B7 on the professional antigen-presenting cell binds to CTLA-4. This mechanism serves to regulate T-cell proliferation and to suppress T-cell activation after an immune response.

C. If they engage antigen in the absence of B7 expression and, hence, co-stimulation, T cells will become irreversibly non-responsive (anergic) instead of activated. This is one mechanism by which T-cell tolerance may be achieved.

8–6   a

8–7

A. Cytotoxic T cells, T_H1 cells and T_H2 cells

B. Cytotoxic T cells recognize antigen on the surface of a cell that is presenting antigen bound to MHC class I molecules. The cytotoxic T cell responds by killing these target cells by inducing the apoptotic pathway. T_H1 cells recognize antigen bound to MHC class II molecules on the surface of an antigen-presenting cell such as a macrophage. The T_H1 cell responds by activating the macrophage to destroy intravesicular bacteria and increase the phagocytosis of extracellular bacteria it has taken up. T_H2 cells recognize antigen bound to MHC class II molecules on the surface of B cells. The T_H2 cell responds by secreting cytokines that will activate B cells to differentiate into antibody-producing plasma cells.

C. Cytotoxic T-cell antigens are derived from proteins from pathogens (such as viruses) replicating in the cytosol of the target cell. Typical T_H1-cell antigens are proteins from bacteria (such as *Mycobacterium tuberculosis*) that live in the vesicles of macrophages. Typical T_H2-cell antigens are derived from protein components of pathogens circulating in the blood or tissue fluids (such as bacterial toxins such as the diphtheria toxin produced by *Corynebacterium diphtheriae* or proteins of circulating virus particles).

8–8

A. Cytotoxic T cells focus their killing machinery on target cells through a process called polarization. The cytoskeleton and the cytoplasmic vesicles containing lytic granules are oriented toward the area on the target cell where MHC class I:peptide complexes are engaging T-cell receptors. In the T cell, the microtubule-organizing center, Golgi apparatus, and lytic granules, which contain cytotoxins, align toward the target cells. The lytic granules then fuse with the cell membrane, releasing their contents into the small gap between the T cell and the target cell, resulting in the deposition of cytotoxins on the surface of the target cell. The cytotoxic T cell is not killed in this process and will continue to make cytotoxins for release onto other target cells, thereby killing numerous target cells in a localized area in succession.

B. The cytotoxins include perforin, granzymes, and granulysin, molecules that induce apoptosis (programmed cell death) of the target cell.

8–9   c, d, e

8–10

(i) The CD3 subunits γ, δ, and ε associated with the antigen-binding T-cell receptor help transmit the signal from the T-cell receptor–peptide–MHC interaction at the cell surface into the interior of the cell through immunoreceptor tyrosine-based activation motifs (ITAMs) present on their cytoplasmic tails. These are phosphorylated by associated protein tyrosine kinases, such as Fyn, when the antigen receptor is activated, and in turn activate further molecules of the signaling pathway. (ii) Lck associates with the tails of the CD4 and CD8 co-receptors. When these participate in binding to peptide:MHC complexes, Lck is activated and phosphorylates ZAP-70, a cytoplasmic protein tyrosine kinase. (iii) CD45 is a cell-surface protein phosphatase that helps activate Lck and other kinases by removing inhibitory phosphate groups from their tails. (iv) When ZAP-70 is phosphorylated it binds to the phosphorylated ITAMs of (v) the ζ chain, which initiates the signal transduction cascade by activating phospholipase C-γ (PLC-γ) and guanine-exchange factors. (vi) IP_3, which is produced by the action of PLC-γ on membrane inositol phospholipids, causes an increase in intracellular Ca^{2+} levels, which leads to the activation of the protein calcineurin. (vii) Calcineurin activates the transcription factor NFAT by removing an inhibitory phosphate group. Activated NFAT enters the nucleus, and together with the transcription factors NFκB and AP-1 will initiate the transcription of genes that lead to T-cell proliferation and differentiation.

8–11   c

8–12

A. A granuloma contains at its center macrophages and giant multinucleated cells created by macrophage fusion, infected with bacteria that are replicating intracellularly. Epithelioid cells surround the core, which consists of non-fused macrophages. Epithelioid cells are surrounded by activated CD4 T cells.

B. Chronic infections resistant to the killing mechanisms of macrophages can lead to the formation of granulomas, for example in tuberculosis, caused by *Mycobacterium tuberculosis* growing inside intracellular vesicles.

C. Granulomas surrounded by CD4 T cells cut off the infection from the blood supply, and the cells in the core of the granuloma will die from oxygen deprivation and toxic products of the macrophages. In tuberculosis, the dead tissue is referred to as caseation necrosis because it has a cheesy consistency. If the infection were not localized in this fashion, it could disseminate systemically to other anatomical locations.

8–13   In an immune response, the binding of IL-2 to a high-affinity IL-2 receptor composed of α, β, and γ chains drives the proliferation and differentiation of T cells that occurs after they have encountered their specific antigen. The activated T cell will divide two to three times daily for about a week, producing a clone of thousands of identical antigen-specific, effector T cells. The α chain is required to form a complex with preexisting γ and β chains to make the high-affinity receptor. The receptor formed by the γ and β chains on their own has a low affinity for IL-2. In the absence of IL-2 and its high-affinity receptor, T cells will not be fully activated, will not differentiate and will not undergo clonal expansion. Thus, by preventing the production of IL-2 and its receptor, cyclosporin prevents the clonal expansion of T cells specific for the foreign antigens on the graft and their differentiation into effector T cells, and thus suppresses the immune response directed against the graft.

8–14

A. Many bacteria are surrounded by a polysaccharide capsule. In some cases, antibodies against the capsular polysaccharides give protective immunity against the pathogen. Antibodies produced against polysaccharide antigens are generally restricted to the IgM isotype, because the help needed to switch isotypes to IgG is provided by T cells, which recognize only peptide antigens. Adult humans make effective immune responses to polysaccharides alone and thus can be protected by subunit vaccines made from the capsular polysaccharides of encapsulated bacteria. Their antibody responses are polysaccharide-specific, T-cell independent, and involve antibodies of the IgM isotype. In contrast, children do not make effective immune responses to polysaccharides alone and thus cannot be immunized with such vaccines.

However, if the polysaccharide is conjugated to a protein, peptides from the protein part of the molecule can activate specific T$_H$2 cells. B cells specific for polysaccharide will bind and internalize the whole antigen via their antigen receptors, process it, and then present peptides from the protein part on their surface. T cells specific for these peptides will interact with the B cell, delivering the necessary cytokines (such as IL-4) and the CD40–CD40-ligand signal required for isotype switching. The B cell will then produce IgG anti-polysaccharide antibodies. This type of vaccine can be used to immunize children so as to induce protective anti-polysaccharide antigens.

B. A vaccine of this type has been produced against *Haemophilus influenzae* B (HiBC), which can cause pneumonia and meningitis. The conjugate vaccine is composed of a capsular polysaccharide of *H. influenzae* conjugated to tetanus or diphtheria toxoid (a protein). The antibody response is polysaccharide-specific, T-cell dependent, and comprises IgG that protects children from the meningitis caused by this microorganism.

8–15   Rationale: The correct answer is c. This is a case of autoimmune lymphoproliferative syndrome (ALPS). Splenomegaly and hyperplasia of lymphoid follicles suggest a defect in the regulation of cell division or survival of lymphoid cells. Angelina's low level of platelets may be the result of autoimmune attack. The interaction between Fas and Fas ligand is an important mechanism for eliminating lymphocytes when an infection has been terminated and during lymphocyte development. In the absence of functional Fas, the size of the lymphocyte pool can no longer be regulated, and removal of autoimmune cells is ineffective. Secondary lymphoid tissues such as the spleen and lymph nodes remain enlarged even in the absence of infection, and autoimmune responses are common—both these conditions are seen here. ALPS affects individuals who are heterozygous or homozygous for Fas mutations, because Fas acts as a homotrimer, which will not signal if one or more of the subunits is a mutant.

8–16   Rationale: The correct answer is b. Leprosy, caused by *Mycobacterium leprae*, can come in two forms. Lepromatous leprosy is characterized by a bias toward T$_H$2 cells, which produce the cytokines IL-4, IL-5, and IL-10. In contrast, tuberculoid leprosy is associated with a bias toward T$_H$1 cells, which characteristically produce IL-2, IFN-γ, and lymphotoxin (LT). The location of the lesions on Vijay's extremities is also consistent with an infection with *M. leprae* rather than with *M. tuberculosis* (which causes tuberculosis), because *M. leprae* grows optimally at 30°C whereas *M. tuberculosis* grows best at 37°C and primarily causes lung disease. Vijay most probably contracted *M. leprae* in an area of South India where leprosy is endemic before his emigration to the United States.

Chapter 9

9–1   The B-cell co-receptor, made up of CD21 (complement receptor 2), CD19, and CD81 (TAPA-1), cooperates with the B-cell receptor in B-cell activation and increases the sensitivity of the B cell to antigen 1000–10,000-fold. This becomes important when antigen concentrations are low. CD21 binds to complement components that have been deposited on the surface of pathogens, for example C3d. CD19 provides the long cytoplasmic tail involved in signaling. When the co-receptor and B-cell receptor and co-receptor are both ligated by antigen and C3d, respectively, Lyn and the cytoplasmic tail of CD19 come close together. Lyn is a protein tyrosine kinase bound to the immunoreceptor tyrosine-based activation motifs (ITAMs) of Igα, which phosphorylates CD19 when it is nearby. The phosphorylated CD19 tail will initiate activation signals that complement those generated by the B-cell receptor complex.

Depending on the characteristics of the antigen, an additional signal provided by CD4 T$_H$2 cells may be needed for B-cell activation. T$_H$2 cells express CD40 ligand on their surface, which binds to CD40 on the B cell and delivers a stimulatory signal. They also secrete the cytokines IL-4, IL-5, and IL-6. By binding to specific receptors these help stimulate B cells to proliferate and differentiate into plasma cells.

9–2
    A. Thymus-dependent antigens can only induce the generation of antibodies by B cells when T-cell help is available. They induce immunological memory as well as affinity maturation after somatic hypermutation. Thymus-independent (TI) antigens are able to induce antibodies in the absence of T-cell help, even in patients with DiGeorge syndrome, who lack a thymus. Because of the absence of T-cell cytokines in the response to TI antigens, these antigens do not induce iso-type switching or somatic hypermutation to any great extent, and thus a TI antibody response does not undergo affinity maturation. Immunological memory is also not produced.

    B. Examples of TI-1 antigens are the bacterial lipopolysaccharide (LPS) of Gram-negative bacterial cell walls and bacterial DNA. TI-1 antigens not only engage the B-cell receptor and co-receptor but also bind to other activating cell-surface receptors, such as Toll-like receptors, and thus provide additional activating signals that drive B cells to proliferate and differentiate into plasma cells. In the absence of the T-cell cytokines needed for isotype switching, only IgM antibodies are made.

    C. TI-2 antigens are highly repetitive protein or carbohydrate epitopes on the surface of pathogens. An example is the polysaccharide in the capsules of *Streptococcus pneumoniae*. The B cells responding to TI-2 antigen are often B-1 cells, and both IgM and IgG antibodies can be produced; however, predominantly IgM is made. TI-2 antigens by-pass the requirement for T-cell help because the number of B-cell receptors occupied by these antigens is very high and cross-linking is efficient, thereby generating strong signals from the B-cell receptor.

    D. A B-cell might bind one component of the bacterial cell surface (antigen A) with its B-cell receptor and another component (such as LPS) with a Toll-like receptor. Together, the signals from these receptors activate the B cell to differentiate into plasma cells producing antibody against antigen A. However, if antigen A is administered on its own, the signal from the Toll-like receptor will be absent, and T-cell help would be needed to provoke the production of anti-A antibody.

9–3
    A. False. A plasma cell is a terminally differentiated B cell that is not dividing and cannot change its antibody specificity. Somatic hypermutation and selection of B cells with higher-affinity immunoglobulin receptors occur in activated B cells before they differentiate into plasma cells.

    B. True.

    C. True.

    D. False. TI-2 polysaccharide antigens activate only mature B cells. Often these B cells are B-1 cells, which do not acquire full function until a child is about 5 years old. TI-2 antigens therefore do not stimulate efficient antibody responses in infants. Recall from Chapter 8 that vaccines administered to children to prevent meningitis caused by *Haemophilus influenzae* type b are conjugate vaccines composed of polysaccharide coupled to diphtheria or tetanus toxoid. The toxoid component stimulates T cells, which then provide the necessary help to activate T-dependent antigen responses to polysaccharide in B cells of infants and young children.

9–4   CD40 ligand on T cells binds to CD40 on B cells, signaling B cells to activate NFκB. NFκB is a transcription factor that upregulates the expression of ICAM-1, an adhesion molecule, on the surface of B cells. ICAM-1 prolongs interaction between the B cell and T cell, trapping the B cell in the T-cell zone and causing a primary focus of proliferating B cells (B lymphoblasts) to form.

9–5   b, c

9–6
    A. The main effector function of IgM is complement activation; it can also neutralize pathogens and toxins.

    B. (i) IgM is the first antibody to be produced by plasma cells during a primary antibody response and is secreted as a pentamer that circulates in the blood. Because of the large size of pentameric IgM, it does not penetrate effectively into infected tissues and is most effective against pathogens in the bloodstream. (ii) In the classical pathway of complement activation, at least two Fc regions are needed to bind C1, the first complement component in the pathway. A single pentameric molecule of IgM can thus initiate complement activation. In contrast, two IgG antibodies in close proximity to each other are needed to bind C1. (iii) Phagocytic cells carry both complement receptors and Fc receptors for IgG (FcγR) and IgA (FcαR), but there are no Fc receptors for IgM. Thus, immune complexes of IgM and antigen alone cannot be taken up by macrophages through Fc receptor-mediated endocytosis. An IgM:antigen:C3b complex can be phagocytosed by a macrophage after binding to complement receptors, but this is not as efficient as having both complement receptors and Fc receptors cooperating in inducing phagocytosis.

9–7
    A. Dimeric IgA is made in mucosa-associated lymphoid tissue (MALT) and is transported across the barrier of the mucosal epithelium. First, dimeric IgA binds to the poly-Ig receptor on the basolateral surface of an epithelial cell, followed

by uptake through receptor-mediated endocytosis into an endocytic vesicle. On reaching the opposite face of the cell, the apical surface, the vesicle fuses with the membrane. Here the poly-Ig receptor is cleaved proteolytically between the membrane-anchoring and the IgA-binding regions, thus releasing IgA into the mucous layer on the surface of the epithelium. Dimeric IgA remains attached to a small piece of the poly-Ig receptor, called the secretory component, which holds the IgA at the epithelial surface through interactions with molecules in the mucus. The rest of the poly-Ig receptor is degraded and serves no further purpose.

B. Dimeric IgA is released into the lumen of the gastrointestinal, urogenital, and respiratory tracts, onto the surface of the eyes, into the nose and throat, and into breast milk (which is the route by which newborn babies receive protective maternal IgA).

9–8

A. IgG antibody binds to FcRn in the membranes of endothelial cells in blood capillary walls and is transported by receptor-mediated transcytosis from the blood into the extracellular spaces in tissues. IgG binds to two FcRn molecules on the apical side of the endothelium in the lumen of the capillary, followed by receptor-mediated endocytosis into an endocytic vesicle. On reaching the basal surface of the endothelial cell, the vesicle fuses with the membrane, and IgG is released into the extracellular space.

B. IgG is transported into infected tissues and is also transported across the placenta into the fetal circulation during pregnancy.

9–9

A. Similarities: (1) Activation of both mast cells and NK cells occurs only when their Fc receptors are bound to antigen:antibody complexes. (2) When cross-linking occurs, both mast cells and NK cells release the contents of granules through exocytosis, which involves the fusion of vesicles containing preformed proteins with the cell membrane.

B. Differences: (1) Mast cells bind IgE, whereas NK cells bind IgG. (2) Exocytosis of granules from mast cells occurs at random around the cell membrane. Exocytosis of granules from NK cells is highly polarized, focusing only on the target cell to minimize damage to neighboring cells. (3) IgE binds to FcεRI with high affinity in the absence of antigen; mast cells become activated when antigen becomes available and binds to the receptor-bound IgE. NK cells bind IgG with low affinity, and bind IgG effectively only when it is already bound to multivalent antigen. (4) Activated mast cells release inflammatory mediators (histamine and serotonin) that affect other cells, for example endothelium, causing increased vascular permeability and vasodilation. Activated NK cells release

apoptosis-inducing compounds (perforin and granzyme/fragmentin) that kill target cells directly. (5) Antibody-dependent cell-mediated cytotoxicity (ADCC) carried out by NK cells could be induced in newborn infants by maternal IgG acquired transplacentally. IgE cannot be transferred across the placenta, and so newborn babies cannot activate mast cells via maternal IgE.

9–10 B lymphoblasts that have bound specific antigen and encountered their cognate T cells in the T-cell areas of a lymph node are activated and start to proliferate, forming a primary focus. The B cells move from the primary foci into primary follicles, which are primarily B-cell areas, where they become centroblasts—large, metabolically active, dividing cells. As centroblasts accumulate and proliferate, the primary follicle enlarges and changes morphologically into a germinal center. Centroblasts undergo somatic hypermutation while dividing in the germinal center, producing centrocytes with mutated surface immunoglobulin. Only cells with mutated surface immunoglobulin that can take up antigen efficiently through receptor-mediated endocytosis and present it to helper T cells (T_H2) will be selected to differentiate into plasma cells or memory cells. Antigen will be encountered at the surface of follicular dendritic cells as an immune complex. If B cells do not encounter their specific antigen, they will undergo apoptosis and then be ingested and cleared by tingible body macrophages. This process takes around 7 days after an infection begins, and the increase in cell numbers due to lymphocyte proliferation accounts for the swollen lymph nodes.

9–11

A. Passive transfer of immunity refers to the process of transferring preformed immunity from an immune subject to a nonimmune subject. This can be achieved by transferring whole serum (antiserum), purified antibody, monoclonal antibody, or intact effector or memory lymphocytes (adoptive transfer).

B. (i) IgG antibodies transported transplacentally provide passive protection in the bloodstream and extracellular spaces of tissues until the baby can begin making its own antibodies, after which time maternal IgG levels decrease. IgG1 is the isotype that is transported most effectively. (ii) IgA is transferred into the infant's gastrointestinal tract in breast milk and protects the gastrointestinal epithelia from colonization and invasion by ingested microorganisms.

C. It is possible for autoreactive antibodies to be transferred passively to a fetus via the placenta if the isotype is IgG. Any reaction will persist only for as long as the antibodies are present. Maternal antibodies can be removed by plasmapheresis, a procedure that involves replacing blood plasma, and consequently the removal of maternal immunoglobulins, or they will eventually be degraded by serum proteases.

9–12   During transcytosis the poly-Ig receptor is cleaved by a protease, leaving a small piece of the original receptor, called the secretory component, still bound to the J chain via disulfide bonds. Once dimeric IgA is released at the apical face, the carbohydrate moieties of the secretory component anchor the antibody to mucins of the mucus, enabling the antibodies to bind subsequently to microbes in the mucus and inhibit the ability of microbes to bind to and invade the mucosal epithelium of the gut. Instead, the microbe is expelled from the body via mucosal secretions in the feces.

9–13   When IgE binds to antigen, leading to cross-linking of Fcε RI on mast cells in connective and mucosal tissues, the mast cells rapidly release chemicals that activate smooth muscle to contract. Muscle activity leads to vomiting and diarrhea in the gastrointestinal tract, and sneezing and coughing in the respiratory tract, helping to expel the offending pathogen or toxic material.

9–14   Newborn infants are afforded passive immunity to the pathogens in their environment through IgG and dimeric IgA. IgG is transferred transplacentally, and dimeric IgA is acquired through breast milk. If an endemic infection develops in the newborn infant, the IgG antibodies in the infant's bloodstream may not have the appropriate specificity for the foreign antigens because the mother would not have encountered these antigens previously in their homeland, and therefore the newborn infant would not have acquired them passively during fetal development. Furthermore, without maternal IgG, infants are particularly susceptible to infection for the first 6 months of their life, when their immune systems are unable to produce significant levels of IgG. In addition, infections that breach mucosal surfaces may be more likely to develop during this time because dimeric IgA against such a pathogen will not be formed in the breast milk until about a week after the mother has been exposed to the same pathogen.

9–15
   A.   A module composed of one heavy chain and one light chain of one IgG4 molecule can exchange with a similar module from a different IgG4 molecule, creating an antibody that has two different antigen-binding sites and thus two antigen specificities (bispecificity).
   B.   This antibody is monovalent for its specific antigens, and so functions that require two identical antigen-binding sites, for example complement activation and the consequent promotion of inflammation, are compromised. Although IgG4 is able to neutralize pathogens to some extent, its function also seems to be to interfere with certain immune responses, such as allergic reactions mediated by IgE, and to help reduce their severity.

9–16   **Rationale**: The answer is c. This is a case of systemic lupus erythematosus (SLE). The vascular rash on the face, and the painful finger joints and hips, suggest that Amanda has developed a systemic inflammatory condition. The decreased levels of C3 are consistent with increased complement fixation by the classical pathway mediated by the anti-nuclear antibodies typical of SLE. Elevated levels of protein in the urine are suggestive of glomerulonephritis, another common complication of SLE. When immune complexes of antibody, complement, and antigen are deposited in the synovia of joints, in blood vessel walls, and in kidney glomeruli rather than being cleared from the circulation, inflammation results.

Chapter 10

10–1
   A.   Mucosal secondary lymphoid tissues have the same general microanatomy and organization as secondary lymphoid tissues found at other anatomical locations, with distinct compartmentalization of B-cell and T-cell zones. Both mucosal and systemic secondary lymphoid tissues function as sites where naive lymphocytes are activated by antigen and adaptive immune responses are initiated.
   B.   In the systemic immune system, adaptive immune responses are activated in secondary lymphoid tissues that are quite distinct from and often distant from the site of infection. In contrast, in the mucosal immune system, the adaptive immune response is initiated at secondary lymphoid tissues at the site of infection.

10–2   Tonsils and adenoids are located in the oral cavity and are composed of extensive secondary lymphoid tissue making up Waldeyer's ring. They are responsible for the production of secretory IgA specific for infectious material entering the gut and airways. The oral polio vaccine elicits the most effect protective immunity through the production of secretory IgA, including the secondary lymphoid tissues of Waldeyer's ring.

10–3
   (i)   M-cell dependent capture relies on transcytosis of microorganisms across the gut epithelium into pockets containing dendritic cells. The dendritic cells then process and present antigens to T lymphocytes in either the T-cell areas of the Peyer's patch or the mesenteric lymph node.
   (ii) M-cell independent capture occurs in the lamina propria as a result of dendritic cells extending cytoplasmic processes between intracellular enterocyte junctions. Antigen is taken up by these processes and processed, and the dendritic cell then presents antigens to T cells in gut-associated lymphoid tissue or mesenteric lymph nodes.

10–4
   A.   Intestinal macrophages are able to phagocytose and kill bacterial pathogens.
   B.   They do not, however, stimulate inflammatory

responses because they lack TLRs and the signaling receptors required for activation pathways leading to inflammatory cytokine production.

10–5   The chemokines CCL21 and CCL19 synthesized in Peyer's patches bind to CCR7 on naive T cells and recruit them into the lymphoid tissue across a high endothelial venule. The T cells bearing appropriate antigen-specific receptors are stimulated by dendritic cells and undergo proliferation and differentiation within the Peyer's patches. Activated T cells then leave the Peyer's patch and enter the lymph, and travel through mesenteric lymph nodes before arriving via the lymph at the thoracic duct and then enter the bloodstream. From the bloodstream, activated T cells home to the same type of mucosa as that in which they were activated, a homing process mediated by expression of appropriate integrin and chemokine receptors. They cross the vascular endothelium into the lamina propria (some subsequently entering the epithelium), where they secrete cytokines and mediate killing.

10–6
   (i)   lamina propria; transcytosis to mucosal layer
   (ii) endosomal compartment; neutralization of endocytosed antigen
   (iii) mucosal layer; neutralization of antigen at mucosal surfaces
   (iv) M-cell surface; transport to secondary lymphoid tissue.

10–7   In nursing mothers, B cells activated in the gut or other mucosal tissue can home to the lactating mammary gland and secrete their dimeric IgA antibodies into the breast milk. This is due to a general property of the mucosal immune system in that lymphocytes activated in one mucosal tissue can recirculate to another one as well as to the tissue in which they were activated. This is because lymphocytes activated in a mucosal tissue carry integrins that bind to the vascular addressin MAdCAM-1, which is expressed on the walls of blood vessels serving different types of mucosal tissue.

10–8   Individuals with selective IgA deficiency possess compensatory mechanisms for combating infections at mucosal surfaces, most notably by making increased levels of IgM, which can be secreted as a pentamer across mucosal epithelium.

10–9
   (i)   IL-13 secreted by T_H2 cells enhances epithelial cell turnover and the sloughing of parasitized epithelial cells. IL-13 also stimulates goblet cells to produce mucus, which then interferes with adherence of helminths to mucosal surfaces (see Figure 10.16).
   (ii) IL-5 secreted by T_H2 cells attracts eosinophils and activates them to secrete major basic protein, which is cytotoxic to helminths. Eosinophils can also attack parasites directly when their Fc receptors are cross-linked by parasites that have

become coated with antibody, especially IgE. In this situation the eosinophil pours out its toxic granule contents directly onto the parasite surface.
   (iii) The IL-4 produced by T_H2 cells when they interact with their target B cells preferentially stimulates the B cells to switch to the production of immunoglobulin of isotype IgE. The anti-parasite IgE antibodies subsequently produced bind to high-affinity Fcε receptors on mast cells. Cross-linking of the bound IgE by parasite antigens activates the mast cells to release chemical mediators that help to eliminate the parasite. These mediators include histamine, which causes the contraction of smooth muscle, TNF-α, which activates epithelium and thus enables the recruitment of additional effector leukocytes, and matrix metalloproteinase, which helps remodel the mucosa.
   (iv) IL-3 and IL-9 secreted by T_H2 cells attract mast cells to the site of parasite infection.

10–10
   A.   In a secondary immune response, there are 10–100 times more antigen-specific memory B cells participating in the immune response. These B cells have undergone isotype switching and somatic hypermutation and thus have higher-affinity B-cell receptors that can respond to lower levels of pathogen. Activated memory B cells also differentiate into plasma cells more quickly than naive B cells do, producing antibody as soon as 4 days after antigen is encountered. Memory B cells also make stronger cognate interactions with helper T cells as a result of higher levels of MHC class II and B7 co-stimulatory molecules on the memory B-cell surface.
   B.   Memory T cells have a different recirculation pattern from that of naive T cells: they can enter peripheral tissues without the need for previous activation in secondary lymphoid organs and can therefore be activated directly at sites of infection in peripheral tissues. In addition, memory T cells do not require co-stimulation via CD28–B7 interaction to differentiate into effector T cells, and so do not need to wait for activation of the antigen-presenting cell and production of co-stimulatory molecules before becoming reactivated.

10–11
   A.   Naive B cells carry the inhibitory Fc receptor FcγRIIB1. Complexes composed of antigen and IgG produced in the primary response, or by reactivated memory cells, cross-link FcγRIIB1 and the B-cell receptor, which suppresses naive B-cell activation. In contrast, memory B cells do not carry this receptor, and so are not inhibited in this way.
   B.   The suppression of naive B cells means that only reactivated memory B cells (which have already undergone isotype switching and somatic hypermutation) make antibodies. Thus all the anti-

bodies made are of high affinity and are primarily of the IgG, IgA, or IgE isotype. Suppression of naive B cells eliminates repetition of the events that took place in the primary immune response, which would, if not inhibited, lead to the production of low-affinity IgM antibodies rather than high-affinity, isotype-switched antibodies that are more effective at removing the pathogen.

10–12 The measles virus is a relatively invariant pathogen that has little if any antigenic change. Antibodies made by memory B cells will be just as effective in a recall response compared with a primary challenge. In fact, antibodies made in secondary immune responses by memory B cells will be more effective because of isotype switching and somatic hypermutation. In contrast, the influenza virus is highly mutable; as a result, new strains emerge each year bearing new epitopes that have not previously stimulated a primary response. Memory response and the suppression of naive B cells restrict antibody production to only those epitopes shared by the infecting strain and the original strain. Over time the influenza virus will express only a limited number of epitopes that are able to activate memory B cells, and the new epitopes will lack the capacity to stimulate naive B cells.

10–13
A. Short-term immunological memory operates shortly after an adaptive immune response has cleared the infection in an individual and while the pathogen is still present in the community. If the individual is reexposed and reinfected, antibodies generated in the first round of infection can bind immediately to the pathogen, blocking its action by neutralization and mediating its removal and destruction by complement fixation and phagocytosis. In addition, any remaining effector T cells or activated B cells can respond straight away to the presence of antigen. These activities ensure that the infection does not reestablish itself and also generate a fresh supply of antibodies and effector cells.
B. Long-term immunological memory is mediated through long-lived memory lymphocytes that are generated in the primary immune response. These are cells that can be rapidly stimulated by reexposure to the same antigen to produce a strong and effective immune response that rapidly clears the pathogen.

10–14
A. The killing activity of the NK cell. Like cytotoxic T cells, NK cells can kill other cells by releasing molecules that induce apoptosis. When an activating receptor on an NK cell recognizes its ligand on the surface of a target cell, this tends to activate the killing function of the NK cell. But when an inhibitory receptor also recognizes its ligand on the target cell, this tends to inhibit the killing activity of the NK cell, even if activating receptors

are also engaged. Whether the NK cell kills the target cell depends on the balance between the activating and inhibiting signals. The known ligands for the inhibitory receptors are MHC class I molecules, and if there are normal levels of these molecules on the target-cell surface, the cell is not killed.
B. Virus-infected cells often have lower levels of MHC class I molecules on their surface. This feature is thought to be exploited by NK cells, which are continually monitoring levels of MHC class I molecules on host cells. When an NK cell finds a cell that lacks or has decreased MHC class I on its surface, the signals from the activating receptors predominate over those from the inhibitory receptors and the target cell is killed. Some viruses encode proteins that mimic MHC class I molecules and interfere with NK-cell attack by ligation of the NK cells' inhibitory receptors.
C. The actions of NK cells are considered part of innate immunity because NK cells can, in principle, act against any virus-infected cell: their killing activity is not dependent on the recognition of viral protein epitopes. In addition, NK cells are already present and ready to act immediately after they encounter an infected cell. Although there are large numbers of different inhibitory and activating receptors in the NK-cell repertoire, none of these receptors is encoded by a rearranging gene, and in most cases the receptors specifically recognize HLA allotypes and are relatively insensitive to the peptides bound.
D. Individual inhibitory NK-cell receptors are specific for particular allotypes of a given MHC class I molecule. The NK-cell repertoire of a person seems to be tailored to their own MHC tissue type, so that all NK cells in that person will carry at least one inhibitory receptor recognizing one of the person's own HLA class I molecules. This ensures that NK cells do not attack the healthy tissues of their own body. However, because MHC class I molecules are highly polymorphic, one person may have HLA class I allotypes that are not recognized by all the NK cells from another individual. If a tissue transplant is not matched exactly for HLA class I, therefore, some of the recipient's NK cells may not recognize the HLA class I molecules on the transplanted tissue and will attack it.

10–15
A. The ligand for CD94:NKG2A is the non-classical MHC class I molecule HLA-E bound to conserved peptides derived from the leader sequence of the heavy chains of classical MHC class I molecules HLA-A, -B, and -C.
B. This ligand will be produced only if there is a steady supply of HLA-A, -B, and -C heavy chains in the cell. If the supply of these proteins is interrupted, for example during a viral infection in which the cell's ribosomes are used primarily for

viral protein synthesis, then the leader peptides will not be provided in the lumen of the endoplasmic reticulum (ER) for binding to HLA-E. HLA-E will be retained in the ER and its levels on the cell surface will decrease.

C. The mechanism works effectively despite the high degree of classical MHC class I polymorphism because HLA-E itself is essentially monomorphic and because only peptides from the leader sequences of the classical MHC molecules are needed, and these are relatively well conserved between the different isoforms.

10–16 Rationale: The correct answer is b. This case involves hemolytic disease of the newborn, which is only a problem in families where the mother is negative for the Rhesus (Rh) antigen and the father is positive. If Fatima were Rh⁺ there would be no risk of hemolytic disease of the newborn and no need to administer RhoGAM. If Fatima's baby were Rh⁻ there would also be no concern, because even if fetal blood did enter the maternal circulation there would be no alloimmunization against the Rhesus antigen. If Samir were Rh⁻ and the baby Rh⁺, then, assuming marital fidelity, Fatima must be Rh⁺ and thus tolerant to Rhesus antigen. If Fatima is Rh⁻ and the baby Rh⁺, however, alloimmunization could occur, putting a second pregnancy with a Rh⁺ baby at greater risk of hemolytic disease. Maternal anti-Rh IgG alloantibodies would cross the placenta during pregnancy, enter the fetal circulation and cause hemolysis of fetal Rh⁺ erythrocytes, with resulting severe anemia in the newborn baby.

Chapter 11

11–1

A. Some pathogens, such as *Streptococcus pneumoniae*, exist in several antigenically different strains known as serotypes. Antibodies and memory B cells generated during an infection with one serotype will protect the host from reinfection with the same serotype but will not protect against a first infection with a different serotype of the same pathogen, because different epitopes are expressed. Thus, the immunity generated is serotype-specific.

B. *Streptococcus pneumoniae* has evolved at least 90 different serotypes that differ in their capsular polysaccharide antigens. A new infection with a different serotype provokes a new primary response rather than the more effective secondary immune response. This is advantageous to the pathogen because it prolongs the period of survival in the host and hence increases the likelihood of transmission to a new host.

11–2

A. Influenza hemagglutinin (HA) and neuraminidase (NA) bear the main epitopes against which protective antibodies are made.

B. Antigenic drift is due to the frequent point mutations in the RNA genome of influenza virus. Mutants that acquire changes in the HA and NA epitopes are selected because they are not subject to the serotype-specific immunity generated against the original strain. Because point mutation causes only relatively small changes at a time, this process is known as antigenic drift.

Antigenic shift is due to recombination between a human influenza virus and an avian influenza virus, leading to replacement of the human NA and/or HA with the avian type. The genome of the influenza virus is composed of eight separate pieces of RNA. When a human influenza virus and an avian influenza virus infect the same cell (usually in domestic livestock such as pigs, ducks, or chickens), reassortment of the RNAs can lead to the generation of a new human virus encoding the avian HA and/or NA that has never before been encountered by the human population.

C. Worldwide pandemics are caused by influenza viruses that have undergone antigenic shift, because no human population has any immunity to them. In antigenic drift, where the changes are more gradual, a population will contain some people who are immune to some epitopes and some to others. Epidemics due to antigenic drift are usually relatively mild and limited.

D. The phenomenon of 'original antigenic sin' means that when a person is reinfected with a new strain of influenza virus that shares some epitopes with the virus they were first infected by, they will make an immune response (which will be a secondary immune response) only to the shared epitopes and not to the new epitopes. This is because when the virus binds via a new epitope (say HA*) to the B-cell receptor of a HA*-specific naive B cell, preexisting IgG antibodies against other epitopes on the virus can bind via their Fc regions to FcγRIIB1 receptors on the surface of the same cell. Cross-linking of the B-cell receptor and Fc receptor delivers an inhibitory signal to the B cell, thus blocking the production of a HA*-specific antibody response. As an influenza virus undergoes antigenic drift, some HA and NA epitopes change while others remain the same, so the person retains some immunity. However, when a completely new HA or NA is produced as a result of antigenic shift, and it shares no epitopes with the original antigens, the body sees the virus as a completely new infection, and disease results while the body is making a new primary response against the virus.

11–3 This strategy has been selected by evolution to maintain the long-term survival of the parasite in its host. The surface glycoproteins, known as variable surface glycoproteins (VSGs) are involved in enabling the parasite to evade the immune response of its host. The trypanosome can change the VSG that is expressed, thus rendering antibodies made against the previously expressed VSG ineffective at controlling the parasite. Switching occurs at random during the life cycle of the trypanosome by a process called gene conversion, in which a different VSG gene is rearranged to the single 'expression site,' replacing the original gene, and is expressed instead. This process is called antigenic variation and leads to the periodic rise and fall in the number of parasites that is characteristic of trypanosome infections. Antibodies made by the host against the VSG expressed on the infecting trypanosomes will start to suppress the infection and parasite numbers will fall. But if some parasites switch VSG, these antibodies will be ineffective, and parasites bearing the new VSG will multiply. Parasite numbers will rise for a time until the new immune response begins to suppress them in turn. By then a third VSG gene is likely to have been expressed, and parasites bearing that one will begin to predominate in the host, and so the cycle continues.

11–4   a: 7; b: 10; c: 6; d: 3; e: 4; f: 9; g: 2; h: 8; i: 5; j: 1

11–5   d

11–6
A. 1. X-linked agammaglobulinemia. (i) No antibody at all. (ii) It is caused by a defect in the tyrosine kinase Btk, which is necessary for B-cell development and is encoded on the X chromosome. No mature B cells develop. 2. X-linked hyper IgM syndrome. (i) Large amounts of IgM antibody, but no antibodies of isotypes other than IgM, are produced. Virtually no antibodies are made against T-cell dependent antigens. (ii) It is caused by a defect in the protein CD40 ligand, which is encoded on the X chromosome and is expressed on T cells. T cells lacking CD40 ligand cannot give help to B cells, which thus cannot respond to most protein antigens and cannot switch isotype.

Other diseases in which defects in antibodies seem to be the main deficiency are: common variable immunodeficiency (defective antibody production; cause unknown); selective IgA and/or IgG deficiency (no IgA or IgG synthesis; cause unknown).
B. Defective antibody responses lead to increased susceptibility to extracellular bacteria and to some viruses.

11–7   Diseases with this pattern of inheritance are caused by defective alleles carried on the X chromosome and are called X-linked diseases. Women have two X chromosomes, one inherited from each parent. A man only has one X chromosome, which he always inherits from his mother. A woman with one X chromosome carrying a recessive disease allele and one X chromosome with a normal allele will show no symptoms of the disease because the normal allele will compensate. However, if she passes on the defective X chromosome to a son, he will show the disease because he has no compensating normal allele. Unless her husband also has the defective allele, which is extremely unlikely as most of these immune defects are very rare, her daughter will also not show the disease even if she inherits a defective X chromosome from her mother. Diseases that show this pattern of inheritance are always recessive. A dominant disease allele carried on an X chromosome will show up as disease equally in women (even if they are heterozygous for the gene) and in men.

11–8
A. A lack of C3 or C4 function means that immune complexes of antibody and antigen do not bind C3 or C4 and thus cannot bind to the complement receptors on phagocytes that facilitate their clearance from the circulation.
B. Immune complexes accumulate in the circulation and are deposited in tissues. They damage tissues directly and also activate phagocytes (via the binding of antibody Fc regions to Fc receptors), causing inflammation and further tissue damage.
C. Increased susceptibility to bacteria of the genus *Neisseria*. C5–C9 are required to form the membrane-attack complex on the bacterial membrane that leads to bacterial lysis. An absence of any of these components means that the complex cannot be formed. Complement-mediated lysis seems to be an important means of protection against *Neisseria*.

11–9
A. Chronic granulomatous disease. Chédiak–Higashi syndrome. Leukocyte adhesion deficiency. Others are: glucose-6-phosphate dehydrogenase deficiency and myeloperoxidase deficiency.
B. Chronic granulomatous disease is caused by defects in phagocyte NADPH oxidase. Phagocytes with this defect cannot produce superoxide radicals and are less effective at intracellular killing of ingested bacteria. Macrophages infected with bacteria that they cannot kill form the granulomas characteristic of this condition.
C. Persistent infections with bacteria, especially encapsulated bacteria, and fungi.

11–10
A. Severe combined immune deficiency (SCID), characterized by an almost complete absence of T-cell and B-cell function and the inability to make any effective immune responses. Because the child cannot make functional receptors for these important cytokines, their lymphocytes cannot respond to them. The inability of naive T cells to respond to IL-2 in particular blocks the proliferation and differentiation of T cells and thus all cell-mediated immune responses and T-cell dependent antibody responses. Without treatment, children with SCID die early in infancy from common bacterial or viral infections.
B. Jak3 kinase is part of the intracellular signaling pathway that leads from activated cytokine receptors. Thus, a lack of Jak3 kinase also leads to SCID.

C. Reconstitution of a functioning immune system by bone marrow transplantation from a healthy donor can treat those types of SCID in which the genetic defect is intrinsic to lymphocytes or other bone marrow derived cells.

11–11

A. Human immunodeficiency virus (HIV) infects cells bearing CD4 molecules on their surface. These include T_H1 and T_H2 cells, macrophages, and dendritic cells. CD4 acts as a receptor involved in binding the gp120 envelope glycoprotein of HIV. A co-receptor is also required for virus entry. Two are used by HIV—the chemokine receptors CCR5 and CXCR4. CCR5 is expressed on all CD4$^+$ cells, whereas CXCR4 is restricted to T cells. After binding to the co-receptor, another viral envelope glycoprotein, gp41, facilitates fusion between the host-cell plasma membrane and viral envelope with the release of viral components into the cytoplasm.

B. The cell tropism of HIV depends on which co-receptor is used for entry. Macrophage-tropic HIV, associated with early infection, uses the CCR5 co-receptor. Lymphocyte-tropic HIV, associated with a phenotypic change in late infection in about 50% of cases, uses CXCR4.

C. About 1% of the caucasoid population is homozogous for a mutant form of CCR5 that cannot be used by HIV as a co-receptor. The mutation involves a 32-nucleotide deletion called *CCR5-Δ32*, which alters the reading frame and results in a nonfunctional CCR5 protein. This renders such individuals resistant to primary infection by macrophage-tropic variants of HIV but still susceptible to lymphocyte-tropic strains that use the CXCR4 co-receptor.

11–12

A. Seroconversion is the point during an infection at which antibodies specific for the pathogen can first be detected in the blood serum. In an HIV infection, antibody is produced when virions are being released from infected cells and are thus accessible to B-cell antigen receptors, a period referred to as acute viremia. During clinical latency the level of infectious virus in plasma decreases markedly but does not disappear.

B. Immediately after seroconversion the levels of free virus decline, T-cell numbers rise, and for a time the infection is held in check. Eventually, however, virus levels start to rise again and T-cell numbers decrease, leading eventually to AIDS. The level of virus in the blood after seroconversion is positively correlated with the severity and progression of the disease: the more virus that remains at this time, the more rapid is the progress to AIDS.

11–13 HIV is a retrovirus that uses the enzyme reverse transcriptase to complete its life cycle. This enzyme uses the single-stranded RNA genome of HIV as a template to make a double-stranded DNA (complementary DNA or cDNA) that integrates into the host genome as a provirus. Reverse transcriptase is an error-prone DNA polymerase that lacks proofreading capabilities. Mutations are therefore introduced into the HIV genome each time it is replicated. Such rapid mutation leads to variant viruses with modified antigens, which thus escape detection by antibodies or cytotoxic T cells specific for the original epitopes.

The high rate of mutation in HIV also causes the emergence of resistance to the antiviral drugs used to treat HIV infection, which are mainly drugs that inhibit viral reverse transcriptase and protease. Mutation produces viruses with mutant enzymes that are not blocked by the drugs. Multidrug regimes are used to try to eliminate the virus before the multiple mutations needed to resist all of the drugs have accumulated.

11–14 **Rationale:** The correct answer is c. The key finding here is the unresponsiveness of Christiana's PBMCs to interferon-γ (IFN-γ) plus LPS. When IFN-γ binds the IFN-γ receptors of macrophages, these cells become stimulated and very effective at killing intravesicular bacteria such as mycobacteria. Defects in the IFN-γ receptor are associated with persistent mycobacterial infections. Because of Christiana's age, she is likely to be homozygous for a recessive mutation in IFNγR1 that leads to the lack of receptor. Chronic granulomatous disease and leukocyte adhesion disease result from defects in NADPH oxidase and CD18, respectively, but these defects would not hinder the ability of PBMCs from affected individuals to produce TNF-α in response to IFN-γ. Because Christiana has normal levels of tetanus antitoxoid antibody, she does not have X-linked agammaglobulinemia or severe combined immune deficiency.

11–15 **Rationale:** The correct answer is a. This is a case of streptococcal toxic shock syndrome. Toxin produced by Group A β-hemolytic streptococci (*Streptococcus pyogenes*) acts as a superantigen, stimulating very large numbers of monocytes and CD4 T cells (but not B cells) and causing massive and unregulated production of inflammatory cytokines. The superantigen cross-links MHC class II (but not MHC class I) molecules and any T-cell receptor with a particular V_β segment. Cytokines produced include IL-1 and TNF-α (by monocytes), and IL-2 and IFN-γ (by T_H1 cells). IL-1 and TNF-α not only induce fever but also activate vascular endothelium, causing increased vascular permeability that leads to hypotension and systemic shock. The decreased white cell count is caused by apoptosis of the large numbers of T cells activated by the superantigen.

Chapter 12

12–1 (1) Inhalation: house dust mite feces, animal dander. (2) Injection: drugs administered intravenously, wasp venom. (3) Ingestion: peanuts, drugs administered orally. (4) Contact with skin: poison ivy, nickel in jewelry.

12–2 a: 3 or 5. b: 1, 7, 8. c: 1, 2, 5, 6. d: 1, 4, 5, 7.

12–3

A. If an individual becomes sensitized to antigen by making antibodies of the IgE isotype during first exposure, then a type I hypersensitivity reaction may result if antigen is encountered again. The IgE made initially binds stably via its Fc region to very high-affinity FcεRI receptors on mast-cell surfaces. When antigen binds to this IgE, cross-linking of FcεRI occurs, delivering an intracellular signal that activates the mast cell.

B. Mast cells contain preformed granules containing a wide range of inflammatory mediators that are triggered to be released extracellularly through an exocytic mechanism called degranulation. The inflammatory mediators contained in the granules and released immediately include histamine, heparin, TNF-α, and proteases involved in the remodeling of connective tissue matrix. The proteases include tryptase and chymotryptase (expressed by mucosal and connective mast cells, respectively), cathepsin G, and carboxypeptidase. Additional inflammatory mediators are generated after mast-cell activation, including IL-3, IL-4, IL-5 and IL-13, GM-CSF, CCL3, leukotrienes C4 and D4, and platelet-activating factor.

12–4 Mast cells activated by inhaled allergens in the type I hypersensitivity reaction that causes allergic rhinitis or asthma release histamine. Binding of histamine to its receptors on smooth muscle causes the bronchial constriction typical of asthma and the difficulty in breathing. Binding of histamine to receptors on vascular endothelium causes increased permeability of the epithelium and inflammation of nearby tissue, causing the runny nose and swollen eyes typical of allergic rhinitis, and an accumulation of mucus and fluid in the bronchi typical of allergic asthma. By blocking histamine action, antihistamines help alleviate these symptoms.

12–5

A. (i) The immunoglobulin on a mast cell is an IgE antibody that has become bound to the mast cell's FcεRI receptor, whereas the immunoglobulin on a B cell is a transmembrane form made by the cell itself. (ii) A B cell might have any class of immunoglobulin on its surface, whereas mast cells have only IgE. (iii) IgE molecules of many different antigen specificities can be bound to the surface of an individual mast cell, whereas an individual B cell carries immunoglobulin molecules of only one specificity.

B. After binding to antigen, mast cells become operational as effector cells without the need to undergo proliferation or differentiation. In contrast, after antigen has bound to a surface immunoglobulin on a B cell, the cell must proliferate and differentiate to produce effector cells (antibody-secreting plasma cells).

12–6

A. (i) One approach to desensitization is the subcutaneous injection of the allergen itself into sensitized individuals with the aim of skewing the immune response from an IgE to an IgG4 isotype. This is achieved by gradually increasing the subcutaneous allergen concentration over time, which favors the production of IgG over that of IgE. When antigen is encountered subsequently, IgG will compete with IgE for binding and inhibit IgE cross-linking on mast-cell surfaces. (ii) The second approach involves vaccination with allergen-derived peptides designed to be bound by HLA class II molecules and presented to allergen-specific TH2 cells with the aim of inducing anergy in these T cells. This would prevent them from giving help to allergen-specific naive B cells, thus preventing the production of more IgE antibody on repeated exposures to the environmental allergen.

B. (i) A risk of the first approach is the possibility of activating a systemic anaphylactic response after mast-cell activation. Because the patient was previously sensitized, IgE antibodies against allergen are present and, if bound to mast cells, will induce mast-cell degranulation. In the event of an anaphylactic response to an allergy injection, the patient will have epinephrine administered immediately by the attending physician or nurse practitioner. (ii) With the peptide vaccination approach, allergen-specific T cells are rendered anergic and there is no risk of triggering an anaphylactic reaction. However, the disadvantage of this approach is that because the HLA class II genes are highly polymorphic, the vaccine would have to include sufficient peptides able to bind to most HLA class II allotypes, or it would need to be custom made for each individual on the basis of their HLA class II type.

12–7

A. In allergic individuals, penicillin typically causes a type II hypersensitivity reaction. Such reactions are mediated by IgG antibodies directed towards cell- or matrix-associated antigens. Production of IgG antibody requires isotype switching, which is dependent on help from T_H2 cells. T_H2 cells cannot be activated by non-protein molecules such as penicillin. They can, however, become activated by cell-surface proteins that have been modified by penicillin and thus generate 'foreign' epitopes to which the host is not tolerant.

B. Penicillin binds to surface proteins of erythrocytes (red blood cells) by forming a covalent bond through its highly reactive β-lactam ring. Meanwhile, the bacterial infection for which the drug was administered has activated the alternative pathway of complement. As a side-effect, erythrocytes become coated with C3b through the action of alternative C3b convertase (C3bBb). Erythrocytes coated with penicillin and the opsonin C3b are ingested by macrophages bearing receptors (CR1) for C3b. Macrophages process the penicillin-bound erythrocyte proteins and present penicillin-modified peptide epitopes to specific CD4 T cells, which differentiate into T_H2 cells. T_H2 effector cells are now available to provide help to penicillin-specific B cells. Differentiation of penicillin-specific B cells to plasma cells producing penicillin-specific antibodies results in the uptake of penicillin-coated erythrocytes by B cells through receptor-mediated endocytosis, antigen processing of penicillin-modified erythrocyte surface proteins, and presentation of antigen by the B cell to effector T_H2 cells. T_H2 cytokines and CD40 ligand will direct isotype switching and the subsequent production of anti-penicillin IgG.

IgG antibodies bind to penicillin-coated erythrocytes and activate the classical pathway of complement, leading to destruction of the penicillin-coated red cells through opsonization (via FcγR and CR1 on macrophages) and through lysis mediated by formation of the membrane-attack complex.

C. If administered in large doses, penicillin can cause a form of serum sickness, a type III hypersensitivity reaction in which large amounts of immune complexes of penicillin-modified proteins and antibodies are deposited in tissues, causing a widespread inflammatory response that leads, for example, to rashes, fever, chills, vasculitis, and sometimes glomerulonephritis. Penicillin can cause serum sickness even in a non-allergic individual.

12–8   Farmer's lung is a type III hypersensitivity reaction caused by the repeated inhalation of large quantities of hay dust or mold spores to which an IgG response is mounted. Small immune complexes form in the presence of excess antigen and limiting antibody. Instead of being cleared from the circulation, as are large immune complexes, the small immune complexes become deposited at the alveolar capillary interface. These deposits result in complement activation and inflammation leading to the accumulation of exudates; this compromises gas exchange and causes difficulty in breathing and long-term irreversible lung damage.

12–9   Type III hypersensitivity reactions differ from type I in several important ways. First, IgG, not IgE, antibody is secreted by activated B cells after isotype switching, a process regulated by T-cell cytokines; for example, IFN-γ favors IgG whereas IL-4 favors IgE. Second, large amounts of antigen (allergen) are required to form the small immune complexes that become deposited at various anatomical sites such as blood vessel walls, renal glomeruli, joint spaces, perivascular sites, and the walls of the alveoli. Type I reactions, in contrast, can be provoked by

small amounts of allergen. Third, unlike type I reactions, type III reactions do not cause systemic anaphylaxis. Finally, type III reactions are often self-limiting and dissipate as antibody titer increases, leading to the formation of large immune complexes, which are cleared efficiently from the circulation. Type I reactions, however, are exacerbated by increasing IgE levels, which sensitize the individual for subsequent exposure to allergen.

12–10

A. The tuberculin test is used to determine whether someone has been infected with *Mycobacterium tuberculosis*. A mixture of proteins (tuberculin) derived from *M. tuberculosis* is injected subcutaneously. Dendritic cells and macrophages process and present the mycobacterial proteins. If memory or effector T cells against *M. tuberculosis* are present in the circulation from a previous or existing infection, these will be activated if they enter the site of injection. Activated T cells produce cytokines that activate vascular endothelium and initiate a local inflammatory reaction that eventually produces a raised red lesion at the site of antigen injection. The lesion may take 1–3 days to develop because the number of preexisting antigen-specific memory or effector T cells will be small, and because they will not be specifically attracted to the site of injection by a preexisting inflammatory reaction. This is why the reaction is referred to as delayed-type hypersensitivity. Once effector T cells are generated, these produce cytokines, which induce the activation of vascular endothelium at the site of injection. This results in increased vascular permeability and blood cell chemotaxis, culminating in edema and infiltration of cells into tissues, and, finally, the formation of a visible skin lesion.

B. The tuberculin test is a type IV delayed-type hypersensitivity reaction induced in sensitized individuals.

C. If an individual has been immunized against *M. tuberculosis*, with the BCG vaccine used commonly in the United Kingdom, for example, then a positive skin lesion will form after the injection of mycobacterial protein. The memory T cells cannot discriminate between vaccine proteins and proteins encountered during a genuine infection, and in these circumstances the tuberculin test is of no value in determining whether the individual has been naturally infected or not. In the United States the diagnostic value of the tuberculin skin test for existing infection currently outweighs the benefits of vaccination against tuberculosis and hence vaccination is not routine.

12–11

A. Pentadecacatechol is a low-molecular-weight lipid-like molecule found in the leaves and roots of poison ivy. Its lipid-like nature enables it to penetrate the epidermal layer of the skin readily and to cross the lipid bilayer of the plasma membrane into cells, where it forms covalent bonds with intracellular proteins in the cytosol. Chemically modified intracellular proteins are processed into peptides via the cytosolic pathway involving proteasomes.

B. These antigenic peptides are presented to cytotoxic T cells via MHC class I molecules. Activated cytotoxic T cells kill all cells presenting these peptides on their cell surface, causing the characteristic skin lesions of poison ivy rash.

12–12

    (i) Avoid contact with allergens as much as possible by modifying their behavior and their home environment.

    (ii) Use pharmacological agents that inhibit allergic responses.

    (iii) Undergo desensitization therapy to divert immune responses from IgE to IgG4 isotype.

12–13 **Rationale:** The correct answer is a. This is a case of anaphylactic shock caused by a food allergy. Anita's sudden onset of shortness of breath, swelling of mucosal tissues causing difficulty in breathing, and rash are characteristic of a type I hypersensitivity reaction after rapid absorption of allergen into the bloodstream and systemic activation of mast cells in the connective tissue of blood vessels. Given the life-threatening nature of this medical emergency, in which asphyxiation due to constriction of the airways and swelling of the epiglottis may occur, immediate suppression of Anita's hyperactive response is warranted. Subcutaneous injection of epinephrine will bring the most immediate effect, not corticosteroids, antihistamines, antibiotics, or nonsteroidal anti-inflammatory agents. Epinephrine will relax bronchial constriction, stimulate the heart, and induce the reformation of tight junctions between vascular endothelial cells, which will alleviate swelling and restore blood pressure.

12–14 Anita was suffering from acute anaphylactic shock almost certainly caused by an allergic reaction to something she had just eaten at the buffet. Given that it was a dessert buffet, nuts or peanuts would be one of the most likely suspects, as they or their oils are ingredients or contaminants in many prepared foods, and they are well known to provoke severe anaphylactic reactions in susceptible people.

12–15 **Rationale:** The correct answer is c. This is an example of serum sickness, a type III hypersensitivity reaction. George has made antibodies against infliximab, which is a chimeric monoclonal antibody made of human and mouse (foreign) components. Because this is an adaptive immune response, it takes time before sufficient levels of anti-infliximab antibody are made and cause the formation of immune complexes. Deposition of the immune complexes in blood vessels, joints and glomeruli are causing George's symptoms. These symp-toms are typically self-limiting, because as anti-infliximab antibody levels increase into the zone of antibody excess, the size of immune complexes will increase and they will then be cleared effectively by splenic macrophages, Kupffer cells of the liver and mesangial cells of the kidney.

12–16 **Rationale:** The correct answer is c. This is a case of celiac disease, also known as gluten-sensitive enteropathy. An inflammatory response involving CD4 T cells is made against gluten proteins (including gliadins and glutenins) in gut-associated lymphoid tissues. Anti-gliadin and anti-glutenin IgA antibodies are secreted into the lumen of the gut. Atrophy of intestinal villi compromises absorption and, as seen in this case, affected children fail to thrive.

Chapter 13

13–1

A. Autoimmune diseases are distinguished by the type of immune reaction that is responsible for the disease. Three types of autoimmune response parallel the effector mechanisms described in Chapter 12 for hypersensitivity reactions of types II, III, and IV. The same system used for differentiating hypersensitivity reactions is also used for autoimmune diseases.

    Specifically, type II autoimmunity is caused by antibodies directed against cell-surface or extracellular matrix self antigens. Type III autoimmunity is the result of the deposition of small, soluble immune complexes in tissues. The third type of autoimmune disease that corresponds to a hypersensitivity reaction is type IV, which is mediated by effector T cells.

B. There is no type I autoimmune disease in this categorization because IgE is not involved in autoimmunity and hence no autoimmune disease corresponds to a type I hypersensitivity reaction.

13–2   a, b, d, e, f, j, k

13–3

    a. Rheumatoid arthritis: 7, B IV

    b. Subacute bacterial endocarditis: 4, D III

    c. Autoimmune hemolytic anemia: 8, A II

    d. Mixed essential cryoglobulinemia: 5, H III

    e. Multiple sclerosis: 1, I IV

    f. Systemic lupus erythematosus: 2, G III

    g. Type 1 diabetes: 9, C IV

    h. Graves' disease: 3, E II

    i. Pemphigus foliaceus: 6, F II

13–4   (1) Red blood cells coated with antibodies against red blood cell-surface components bind to splenic macrophages via Fc receptors, inducing phagocytosis of the red blood cell by receptor-mediated endocytosis. (2) Red blood cells bound by antibodies against red blood cells fix complement, which causes deposition of C3b on the surface of the red blood cell. C3b then binds to the receptor CR1 on splenic macrophages, inducing phagocytosis of red blood cells. (3) Antibody against red blood cells triggers the complement cascade and the formation of membrane-attack complexes on the red blood cell, leading to cell lysis.

13–5   First, endocrine glands synthesize tissue-specific proteins unique to that gland. These proteins are not normally found in primary lymphoid organs where lymphocyte maturation occurs. Hence, the population of T and B lymphocytes is not tolerant to some endocrine gland-specific proteins, and these self proteins are thus recognized as foreign antigens. Second, endocrine glands are highly vascularized because their products need to gain access to the circulation. This feature gives leukocytes relatively easy access to endocrine tissue.

13–6   Both Hashimoto's and Graves' diseases disrupt the normal production of the thyroid hormones tri-iodothyronine (T_3) and thyroxine (T_4), which are derived from thyroglobulin in thyroid follicles. Formation of T_3 and T_4 requires engagement of the thyroid-stimulating hormone receptor (TSHR) with thyroid-stimulating hormone (TSH) secreted from the pituitary gland, a key step in the regulation of thyroid hormone production. This step malfunctions for these two diseases.

In Hashimoto's disease, anti-thyroid antigen antibodies and T_H1 effector cells are involved. Large numbers of lymphocytes take up residence in the gland tissue, establishing germinal centers that resemble those in lymph nodes. Eventually the thyroid tissue is destroyed and thyroid follicles are no longer able to respond to TSH and make T_3 or T_4, a condition called hypothyroidism.

Graves' disease, in contrast, results in hyperthyroidism. Anti-TSHR antibodies act agonistically, mimicking TSH even in its absence. The thyroid follicle is chronically overstimulated by these antibodies and overproduces T_3 and T_4. The effector T cells are of the T_H2 type, and the absence of lymphocyte infiltration retains the thyroid gland in operable condition. Therefore T_3 and T_4, no longer regulated by TSH, are secreted continuously in excess of concentrations required by the body.

13–7   a = F; b = T; c = T; d = F; e = T

13–8
A. The negative selection of developing autoreactive T cells in the thymus.
B. The underlying genetic defect is in a gene encoding a protein called AIRE (autoimmune regulator). This is a transcription factor that, when working normally, causes several hundreds of proteins otherwise expressed only in particular peripheral tissues to be expressed by medullary epithelial cells in the thymus. This induces the negative selection of T cells specific for these proteins and their deletion from the T-cell repertoire. The T cells emerging from the thymus are therefore tolerant to a large number of antigens found primarily on organs and tissues elsewhere in the body. When AIRE is defective and these proteins are not expressed in the thymus, the population of naive T cells that develops will contain T cells reactive against these antigens of peripheral tissues.

13–9   These cells are called regulatory T cells (T_{reg} cells). When activated by encounter with their corresponding self antigen they become able to suppress the activation of naive autoreactive T cells. This active suppression of autoreactive T cells in the periphery is now thought to be an important method of preventing autoimmune reactions.

13–10
A. Genes of the HLA complex, specifically the HLA class I and class II genes. Associations with HLA class II genes are most common. Various alleles of these genes are associated with a higher or lower susceptibility to particular autoimmune diseases compared with the incidence of the diseases in the population as a whole.
B. The polymorphic HLA genes encode the proteins that present peptide antigens to T cells. It has been proposed that particular alleles of these genes are associated with particular autoimmune diseases because of their ability to present the required peptide epitope(s) to autoreactive T cells. Associations with HLA class II genes are more common than those with HLA class I because CD4 T cells rather than CD8 T cells are most commonly involved in autoimmunity.

13–11
A. Increased risk of developing type 1 diabetes is associated with the formation of a heterozygote-specific heterodimer composed of the HLA-DQ8α chain DQA1*03 and the HLA-DQ2β chain DQB1*0201.
B. In northern Europeans this combination of α and β chains is never encoded in the same haplotype, so it can be produced only in heterozygotes. Haplotypes containing both susceptibility alleles can, however, be present in Africans, and similarly confer susceptibility to type 1 diabetes.

13–12
A. Habitual cigarette smokers with Goodpasture's syndrome develop not only glomerulonephritis but also pulmonary hemorrhage.
B. Unlike non-smokers, whose basement membranes are in general not compromised in pulmonary alveoli, smokers have alveoli that are damaged as a result of chronic exposure to cigarette smoke. Damage provides a gateway by which autoantibodies can gain access to basement membranes, where they become deposited, activate complement, and cause rupture of alveolar blood vessels.

13–13   Peptidyl arginine deaminase (PAD) is an enzyme

induced in the respiratory tract after smoke-induced damage. PAD converts arginine residues in self proteins to citrulline residues, thus creating epitopes to which the T-cell repertoire is not tolerant. Citrullinated proteins are subject to proteolysis, and the resulting peptides bind to HLA-DRB1*04 and stimulate autoreactive CD4 T cells and antibodies against citrullinated self proteins. If a joint becomes infected or is wounded, inflammation of joint tissue activates PAD, which generates the same citrullinated epitopes. HLA-DRB1*04-restricted effector and memory T cells are recruited and activated and an immune response follows, initiating the development of rheumatoid arthritis.

13–14
A. Molecular mimicry refers to the phenomenon in which a pathogen expresses an antigen that bears a chemical similarity to a host-cell antigen. Once pathogen-specific antibodies or effector T cells are generated, they have the potential to cross-react with self antigen.
B. An autoimmune disease involving molecular mimicry is rheumatic fever. Infection with *Streptococcus pyogenes* (for example 'strep throat') results in the production of antibodies specific for the bacterial cell-wall proteins. These antibodies cross-react with chemically similar (but not identical) self antigen expressed on heart tissue. This is followed by complement activation and the production of inflammatory mediators, which cause damage to heart tissue and valves and the formation of scar tissue, which can lead to cardiovascular complications later in life. This type of immunological aftermath can be avoided if antibiotics are administered early during infection.

13–15 a, c

13–16 **Rationale:** The correct answer is c. This is a case of myasthenia gravis, a type II autoimmune disease caused by antagonist antibodies against acetylcholine receptors which impede the binding of the neurotransmitter acetylcholine. Lisa experienced muscle weakness and impairment of neuromuscular signaling, which affected the strength in oral, ocular, and upper extremity muscles, hallmarks of this disease. Pyridostigmine is an inhibitor of cholinesterase, an enzyme that normally degrades acetylcholine after neuromuscular transmission. That Lisa demonstrated rapid recovery when treated with this drug shows that acetylcholine was associated with her symptoms, and by increasing the acetylcholine concentration her symptoms were alleviated.

Chapter 14

14–1
A. Inactivated virus vaccines are made of virus particles that are not able to replicate because they have been chemically or physically treated (for example, by heat) in a way that inactivates the nucleic acid. Examples are Salk polio vaccine, rabies vaccine, and influenza vaccine.
B. Live-attenuated virus vaccines are made of viruses that have lost their pathogenicity and ability to reproduce efficiently in human cells through mutations accumulated as a result of growing the virus in non-human cells. Examples are Sabin polio vaccine (oral), measles vaccine, mumps vaccine, yellow fever vaccine, rubella vaccine, and varicella vaccine.
C. Subunit vaccines are composed only of particular antigenic pathogen components known to induce protective immune responses. Recombinant DNA technology enables the production of antigenic proteins in the absence of other pathogen gene products. Examples are hepatitis A vaccine, hepatitis B vaccine, and pertussis vaccine.
D. Toxoid vaccines are made from chemically inactivated toxins purified from pathogenic bacteria. Toxin activity is eliminated but not antigenic activity, so an immune response is generated in the absence of pathological damage. Examples are diphtheria vaccine and tetanus vaccine.
E. Conjugate vaccines are made by covalently coupling antigenic polysaccharide found in bacterial capsules to a carrier protein (often a toxoid). This converts the otherwise T-independent bacterial polysaccharide antigen into a T-dependent antigen. T cells respond to an epitope on the protein carrier, whereas B cells respond to epitopes on the polysaccharide portion of the conjugate. This ensures that T-cell help is provided to B cells making anti-capsule antibodies. An example is *Haemophilus influenzae* type B vaccine (HIB).

14–2
A. Some potential antigens will evoke a very weak or no immune response when injected into an animal or human in their pure form. Immunogens, in contrast, are antigens that are able to evoke a strong immune response when introduced into an animal or human.
B. Adjuvants are substances that when mixed with any antigen increase its immunogenicity. They are commonly used in experimental immunology and in vaccines. Adjuvants work by inducing a nonspecific antigen-independent inflammation, which helps to drive the immune response forward. Adjuvants delay the release of antigen at the injection site and prevent its rapid clearance from the body by converting the soluble antigen into particulate material. The most effective adjuvants also activate Toll-like receptors on macrophages and tissue dendritic cells, which then produce inflammatory cytokines, chemokines, and co-stimulatory molecules (B7) that help drive the immune response to the antigen itself.
C. The following adjuvants are used in humans: alum, a form of aluminum hydroxide; MF59, a squalene–oil–water emulsion; and bacterial

components included as part of some vaccines, for example whole *Bordetella pertussis* as part of the DTP (diphtheria, tetanus, pertussis) vaccine.

14–3 Live-attenuated virus vaccines are mutant viruses that can replicate, albeit inefficiently, in human cells, thus simulating conditions of a normal viral infection. The attenuated vaccine strains of virus have been obtained by growing the virus over many generations in non-human cells (for example, monkey cells), so that it acquires multiple mutations that allow it to replicate but prevent it from spreading in the human body and causing disease. When introduced into humans as a vaccine, there is small chance that some or all of the mutations may revert to the original nucleotide sequence, restoring the properties of the virulent strain of the virus. This occurs very rarely with the poliovirus used in the trivalent oral polio vaccine (TVOP) and, now that polio is very rare in the United States, this vaccine is no longer recommended and an inactivated polio virus vaccine is used instead.

The more rounds of replication the vaccine virus undergoes in the human host before being contained by the immune response, the greater is the potential for genetic reversion. This is why individuals who suffer from inherited or acquired immunodeficiencies should never receive live-attenuated virus vaccines.

14–4 b, d, e

14–5 a, b, c, d, e

14–6
A. As the number of susceptible individuals increases to a particular threshold, herd immunity is no longer effective in protecting individuals who have never been vaccinated. The outcome is the resurgence of the disease and an epidemic.
B. (i) A resurgence of whooping cough was documented in Japan between 1975–1980. Distrust in the vaccine followed the death of two children who had recently been vaccinated with DPT. (ii) A resurgence of measles was documented in the UK at the turn of the twenty-first century. Distrust was linked to unsubstantiated claims that the MMR vaccine induced autism in children.

14–7
A. The Rotarix vaccine is a live, attenuated single human rotavirus strain that expresses the G1 variant of V7 (VP7G1) and the P8 variant of VP4 (VP4P8). Both variable proteins are common in disease-causing strains. The RotaTeq vaccine is a mixture of five different non-pathogenic cattle rotaviruses engineered to express VP4P8 and VP7G1, VP7G2, VP7G3, and VP7G4 variants from common human pathogenic strains, in addition to cattle-specific variable proteins.
B. The RotaTeq vaccine protects against a broader range of variable proteins than Rotarix does. Both

vaccines stimulate neutralizing antibodies against VP4P8 and VP7G1, but RotaTeq also stimulates antibodies against VP7G2, VP7G3, and VP7G4, providing broader protection against the different naturally occurring rotavirus variants.
C. Broader protection is important because there are five naturally occurring variants of rotavirus that cause disease. In addition, rotavirus has the potential of RNA reassortment, as seen in influenza, because the genome is made up of 11 separate double-stranded RNA molecules, providing opportunities for the generation of additional diversity.

14–8 Determination of genome composition enables researchers to understand the life cycle and pathophysiology of the pathogen. This type of information assists with the identification of the types of immune responses that are evoked in the host, including NK-cell, T-cell, and B-cell responses. In addition, armed with specific sequence knowledge, recombinant DNA methodology can be used to engineer attenuated strains, either through site-directed mutagenesis or the deletion of virulence genes.

14–9 **Rationale:** The correct answer is d. This is an example of herd immunity, a type of immunity that is conferred in which immunization of a large portion of the population (or herd) provides protection to the minority who are unvaccinated. The number of children in her daycare facility who have been immunized with MMR (attenuated measles, mumps, and rubella viruses) is likely to be at the level needed to achieve herd immunity, thereby minimizing the probability of contracting the disease. Tolerance to measles antigen does not explain her lack of infection because tolerance is a state of immunological unresponsiveness, which would prove lethal if Madison became infected with measles. 'Dormancy' can occur with the measles virus. leading to a condition known as subacute sclerosing panencephalitis, but this occurs 2–10 years after the original measles illness. The DPT vaccine does not contain any determinants that provide cross-protection against measles. Finally, measles, mumps, and rubella vaccine viruses are not transmitted from vaccinated persons, and in fact pose no risk even during pregnancy, so it is not possible that Madison was infected through contact with children recently vaccinated.

14–10 **Rationale:** The correct answer is c. This is a case of passive immunity provided to the fetus during pregnancy through transplacental transfer of IgG. Only IgG antibodies cross the placenta, with the aid of FcRn, and enter the fetal circulation during pregnancy, providing passive immunity for the newborn for the first 3–6 months, after which antibody levels diminish as a result of catabolism, and IgG must then be made by the infant. The antibody specificity will be against the tetanus toxoid, not cell-wall entities, because the booster vaccine is a subunit vaccine made of toxoid from *Clostridium tetani*, not whole bacterial cells. Sophie will make IgG anti-toxoid antibodies, not IgA, because of the route of immunization; intramuscular injection would stimulate IgG antibody production. IgA antibody production would require mucosal delivery. Therefore, even though the newborn is breast-feeding and passively receiving IgA antibodies from Sophie, the IgA antibodies will not have specificity for tetanus toxoid.

14–11 **Rationale:** The correct answer is d. The most serious threat would be posed by the biological agent to which there is no immediate defense or antidote. Antibiotics would be effective against bacterial agents such as *Bacillus anthracis* and *Yersinia pestis*, and anti-toxin against diphtheria and botulism toxins could be administered to those affected. If the powder contained the smallpox virus, however, the population would be particularly susceptible, because there are no antibiotics available to protect the non-immune population and vaccination would take time to take effect. Those infected in the initial attack could spread this highly infectious virus, which has severe morbidity and a mortality rate of about 30%.

Chapter 15

15–1 Vaccination is used to stimulate a very specific, long-lasting immune response against a pathogen that provides protection in the event of subsequent encounter. The objective of transplantation, however, is aimed at suppressing the immune response to eliminate the rejection of a graft that bears foreign epitopes.

15–2 Acute rejection is due chiefly to immune responses made by the recipient's T cells against HLA class I and II molecules of the graft that are different from those of the recipient and that the recipient's immune system perceives as 'foreign.' The differences can be due to the HLA molecules, the self peptides they bind, or both. Transplantation between identical twins and transplantation of autografts are the only situations in which the graft and the recipient are genetically identical and there are no differences in either the HLA molecules or the bound peptides. In these situations graft rejection does not occur. Transplantation between donors and recipients who have identical HLA class I and II molecules, usually HLA-identical siblings, almost always involves differences in the peptides that are bound by the HLA molecules. These differences trigger peptide-specific alloreactive T cells to cause graft rejection through the direct pathway of allorecognition. Although it is possible to match donor and recipient for many HLA class I and II allotypes, in practice most clinical transplants involve one or more mismatched HLA loci. For these differences in HLA type, alloreactive T-cell clones activated by either the direct or indirect pathway of allorecognition cause graft rejection. Destruction of the grafted organ is effected through a type IV delayed-type hypersensitivity response.

15–3 c

15–4 Acute rejection occurs within a few days of transplantation and is mediated by an alloreactive CD4 and CD8 T cell-mediated adaptive immune response by the recipient against the 'foreign' HLA molecules on the graft and involves the direct pathway of recognition. In contrast, chronic rejection occurs months to years after transplantation and is mediated by anti-HLA class I and anti-HLA class II alloantibodies and involves the indirect pathway of recognition. CD4 T cells are first activated by recipient dendritic cells presenting donor-derived HLA class II allotypes of the recipient. These activated CD4 T cells in turn activate B cells, which are also presenting donor-derived allogeneic HLA peptides. This cognate interaction results in the production of anti-HLA class I (and also anti-HLA class II) antibodies.

15–5

A. The organ would be rejected immediately by the process of hyperacute rejection as a result of the presence in the recipient's blood of preformed antibodies against the A and B blood group antigens present on the tissues of the graft. Such antibodies are made early in life as a result of exposure to common bacteria that carry surface carbohydrates similar to those on human cells. A person of blood group O would have made antibodies against bacterial 'A' and 'B' antigens, because the person does not have these antigens on their own cells and is thus not tolerant to them. These preexisting anti-A and anti-B antibodies in the recipient's blood will immediately attack the endothelium of blood vessels throughout the transplant, which expresses the A and B blood group antigens. Blood vessels become occluded through thrombus formation. The graft is deprived of oxygen and becomes engorged with blood hemorrhaging from leaky blood vessels. Hyperacute rejection occurs almost immediately after transplantation and cannot be treated once it has started.

B. Other examples of combinations that will induce hyperacute rejection: O recipient, A donor; O recipient, B donor; A recipient, B donor; B recipient, A donor; A recipient, AB donor; and B recipient, AB donor.

C. A recipient's preformed antibodies against an HLA class I antigen expressed on the endothelial cells of the transplant can also cause hyperacute rejection. Such antibodies can be generated in pregnancies in which the fetus expresses a paternal HLA allotype different from the maternal HLA allotype. These antibodies can also arise from HLA-incompatible blood transfusions or previous transplants.

D. Hyperacute rejection can be prevented by typing and cross-matching donor and recipient for the A, B, and O blood groups and HLA antigens. The recipient's serum antibodies are assayed *in vitro* for their ability to bind to donor white blood cells.

15–6 Beyond a certain number of isotypes, the T-cell repertoire will be decreased, owing to the disproportionate increase in negative selection events as the number of different isotypes increases. Each additional isotype will decrease the number of T cells exported to the periphery, compromising the diversity of the T-cell population.

15–7

A. Class 1: corticosteroids. Examples: hydrocortisone and prednisone.
Class 2: cytotoxic drugs. Examples: azathioprine, cyclophosphamide, and methotrexate.
Class 3: T-cell activation inhibitors. Examples: cyclosporin A, tacrolimus (FK506), and rapamycin.

B. Class 1: side-effects/toxic effects: fluid retention, weight gain, diabetes, bone demineralization, and thinning of the skin.
Class 2: side-effects/toxic effects: nonspecifically prevent DNA replication in all mitotic cells causing, for example, diarrhea or hair loss. More specific effects are liver damage caused by azathioprine, and bladder damage caused by cyclophosphamide.
Class 3: side-effects/toxic effects: nephrotoxicity and suppression of B-cell and granulocyte activation.

15–8 Cyclosporin A prevents the production of IL-2 and its high-affinity receptor, and thus prevents the activation of T cells and their proliferation and differentiation. It acts by inactivating the protein calcineurin. Calcineurin is a serine/threonine protein phosphatase that is activated by the first part of the T-cell receptor signaling pathway and dephosphorylates the transcription factor NFAT. This modification is necessary for NFAT, which normally resides in the cytoplasm, to enter the nucleus and stimulate transcription of the genes for IL-2 and the IL-2 receptor α chain. Inactivation of calcineurin by cyclosporin thus prevents the production of IL-2 and its high-affinity receptor.

15–9 a

15–10

A. Anti-CD3 MoAbs are often administered to patients to suppress T-cell activity when signs of graft rejection are observed. Because CD3 is expressed only on T lymphocytes, this therapy is extremely specific. Anti-CD3 antibodies cross-link CD3:T-cell receptor complexes, leading to a reduction in the number of these complexes on the cell surface and a reduction in the number of T cells in the circulation. Suppression of effector T-cell activity protects the graft.

B. Mouse MoAbs are antigenic in species other than mice and stimulate anti-MoAb responses.

Repeated doses will exacerbate this situation and lead to the formation and clearance of MoAb:anti-MoAb immune complexes before the antibody can bind to the T cells, thus rendering the mouse antibody ineffective. A type III hypersensitivity reaction resembling serum sickness can also result when small immune complexes are formed; that is, when MoAb levels exceed anti-MoAb levels. Hence, repeated doses are discouraged, and physicians must restrict this form of immunosuppressive therapy to one episode of rejection.

15–11 b

15–12

A. Graft-versus-host disease is caused by T cells in the transplanted bone marrow making an immune response against antigens on the recipient's tissues. This can happen even though donor and recipient are HLA matched, because there are proteins other than HLA antigens that can differ between people and provoke an immune response. Such antigens are known as minor histocompatibility antigens. In a bone marrow transplant from a female to a male, the minor histocompatibility antigens most likely to cause a problem are male-specific proteins (which are encoded on the Y chromosome) that a female's T cells will not be tolerant to and will see as 'foreign' or nonself.

B. CD8 cytotoxic T cells. The proteins that act as minor histocompatibility antigens are mainly intracellular proteins. Intracellular proteins of the recipient's cells are processed into peptides by proteasomes as part of normal protein degradation and turnover. These peptides are transported into the endoplasmic reticulum and thus are eventually presented on the surface of the recipient's cells by HLA class I molecules. Any peptides that are different from those in the donor may be recognized as nonself by the donor's cytotoxic T cells, which recognize peptides bound to HLA class I molecules. The naive CD8 T cells in the bone marrow can be activated to effector status by the presentation of minor histocompatibility peptides by dendritic cells in secondary lymphoid organs.
Because the brother and sister share HLA class I type, the sister's T cells will be able to recognize nonself peptides presented by her brother's HLA molecules.

15–13

A. No.

B. Unlike bone marrow, in which a successful outcome is compromised by HLA mismatches, the liver can be transplanted even if there are major differences in HLA class I and class II between donor and recipient. The liver is more refractory to hyperacute or acute rejection than other vascularized organs such as the kidney; low expression levels of HLA class I and the absence of HLA class II contribute to this refractory state. Donor and recipient still need to be matched for ABO, and transplant patients receive immunosuppressive therapy to control chronic rejection.

15–14 First, the medical specialists involved in carrying out the transplant and overseeing the post-transplant regimen are different for organ transplantation and for bone marrow transplantation. Patients receiving organ transplants will have transplant surgeons and physicians, whereas bone marrow recipients will consult with hematologists, oncologists, and radiologists. Second, the requirement for HLA matching and immunosuppressive therapy in organ transplantation depends on which organ is being transplanted. The success of a bone marrow transplant is much more sensitive to HLA mismatches, and the recipient's immune system is not simply immunosuppressed, rather it is destroyed by myeloablative therapy involving chemotherapy and irradiation. Third, the pool of potential bone marrow donors is larger than the demand, unlike organ donations, for which thousands of patients are on waiting lists. Finally, the bone marrow donor is alive and healthy, whereas organ donors are on life support or have experienced a fatal accident.

15–15 a = F; b = T; c = F; d = F; e = F

15–16 **Rationale:** The correct answer is b. This is a case of graft-versus-host disease (GVHD) involving differences in minor histocompatibility antigens between Carter and his sister. The presence of a maculopapular rash and jaundice are characteristic of tissue reactions involved in GVHD, not of type II hypersensitivity reactions or cytomegalovirus infection. The myeloablative conditioning regimen would have destroyed Carter's immunocompetent cells, abolishing the possibility of graft rejection. However, mature T cells in the transplanted bone marrow may react to differences in major and minor histocompatibility antigens. In this case, because the HLA was identical, and the donor is a female, the donor would have mature T cells with specificity toward male-specific or polymorphic protein antigens to which she is not tolerant, for example, H-Y antigens presented by HLA class I or class II on the brother's cells. Because this is a bone marrow transplant, and not an organ transplant, acute rejection of the graft and host-versus-graft reactions are not considerations.

15–17 **Rationale:** The correct answer is a. An HLA haploidentical donor has one HLA haplotype in common with the recipient and one that is different. A bone marrow sample from a haploidentical donor contains numerous alloreactive T cells that can respond to the HLA class I and class II molecules encoded by the HLA class I and II genes of the recipient's HLA haplotype that is not shared with the donor. These alloreactive T cells have the potential to cause a severe and life-threatening graft-versus-host-disease. To prevent such an outcome, bone marrow grafts from haploidentical donors are purged of T cells

before being infused into the recipient. Haploidentical transplants are only given to patients who are unable to find an HLA-matched donor. Family members are good candidate donors because all parents and 50% of siblings (on average) have one HLA haplotype in common and one that is different from a patient such as Richard. Unlike Don, who is HLA-haploidentical to Richard, his other sibling Margaret could not be a donor because neither of her HLA haplotypes is shared with Richard.

15–18 **Rationale:** The correct answer is d. The better the match in HLA class I and class II allotypes between the donor and the recipient, the better the outcome of the transplantation. Donor 'd' has five out of six alleles in common for HLA-A, HLA-B, and DRB. Donors 'b' and 'e' share only four, and donors 'a' and 'c' only three.

Chapter 16

16–1
A. Tumor-specific antigens are antigens found exclusively on tumor cells and are not expressed by normal cells.
B. They can originate from the following: (1) as the result of mutations in normal genes in the tumor cell that cause changes in amino acid sequence that generate new epitopes; (2) by the generation of hybrid genes through genetic recombination, with the consequent production of a new protein unique to the tumor cell; or (3) from viral proteins expressed as a result of viral infection or integration into the host-cell genome.
C. Examples are MARTZ (melanoma) and BCR-ABL fusion protein (chronic myeloid leukemia) and human papillomavirus (cervical carcinoma).

16–2
A. Tumor-associated antigens are antigens expressed in tumor cells as well as some normal cells, but often at higher levels in tumor cells.
B. They can originate from the following: (1) proteins involved in mitosis that are produced at higher concentrations because the cells are continually dividing; (2) proteins expressed during embryogenesis that have become deregulated and reactivated transcriptionally; or (3) proteins expressed continuously at high levels in tumor cells that are normally expressed at low levels or transiently.
C. Examples are MAGEA1 and MAGEA3 (melanoma) and cyclin-dependent kinase 4 (melanoma).

16–3 a = F; b = T; c = F; d = T; e = F; f = T

16–4 First, the patient's tumor-specific T cells could be expanded *in vitro* to generate a large number of tumor-specific effector T cells. To achieve this, the patient's tumor-specific T cells would be co-cultured with tumor antigen, antigen-presenting cells, and cytokines. Second, the patient's purified dendritic cells could be loaded with tumor antigen *in vitro* and then be placed back in the patient, where they would stimulate tumor-specific T cells *in vivo*. Both of these techniques require the tumor antigen to have been characterized and purified so that it can be used in the co-culture *in vitro*.

16–5   Tumor-specific MoAbs are useful tools for the oncologist because of their ability to home in on tumor cells *in vivo*. The size, location, and extent of metastasis can be evaluated. The MoAb can be coupled covalently to a radioactive tag (for example, indium-111) that is detectable   by   diagnostic   imaging   equipment.

   MoAbs can also be used to target the tumor for destruction while minimizing damage to other, healthy, tissue. This can be achieved in several different ways. First, MoAbs can be coupled to radioisotopes (iodine-131 or yttrium-99); the radiation damages the tumor cell's DNA and the cell dies. Coupling MoAbs to cytotoxic drugs is another way of destroying tumor cells. Drugs coupled to antibodies are cleaved from their MoAb transporter only after they have bound to and been internalized by the tumor cell. Finally, alerting the host immune system to tumors coated with MoAbs can lead to the destruction of tumors. MoAb bound to tumor-cell surfaces will activate the classical pathway of complement and lead to complement coating the cell surface. This facilitates uptake by phagocytes, leading to the production of inflammatory mediators, which will recruit additional phagocytes. Activated complement can also form membrane-attack complexes on tumor-cell surfaces. NK cells can also participate in tumor eradication through antibody-dependent cell-mediated cytotoxicity (ADCC).

16–6
   A.   Proto-oncogenes and tumor suppressor genes.
   B.   Proto-oncogenes encode proteins that participate in cell division, such as growth factors, growth factor receptors, signal transduction proteins, and activators of gene expression. Tumor suppressor genes encode proteins that inhibit the division of mutant cells; examples are p53, APC, and DCC.

16–7   For a cell to become cancerous and undergo malignant transformation, at least five or six mutations (depending on the cell type) must accumulate independently. Because of the low frequency of mutation, significant time is needed for a single cell to accrue this number of mutations. As a person ages, the prevalence of cells bearing the required mutations increases.

16–8
   A.   Li–Fraumeni syndrome is a condition in which there is a strong predisposition to developing multiple types of cancer, even at a young age.
   B.   Progression to cancer is related to the inheritance of a mutation in one of the copies of p53, a tumor suppressor gene, which facilitates the apoptosis of cells with DNA damage. Individuals with Li–Fraumeni syndrome need accrue only one

new mutation in the functional copy of the p53 gene to develop a cell with loss of p53 activity, in contrast with other people, who must acquire mutations on both copies.

16–9   Cancer cells:
   1.   activate their own proliferation
   2.   do not yield to growth-inhibiting signals
   3.   escape apoptosis-inducing signals
   4.   become vascularized by the formation of new blood vessels
   5.   spread to other tissues at distant sites
   6.   undergo clonal expansion through repeated cell division
   7.   evade eradication by immune cells

16–10
   A.   NK cells and cytotoxic T cells detect cancer cells and virus-infected cells through alteration of MHC class I .
   B.   Unlike virus detection, which involves inflammatory responses that stimulate immediate innate immune responses, cancer cells can often grow for long periods before they induce an inflammatory state and a consequent immune response.

16–11
   A.   If primary tumor cells are donated between MHC-disparate mice, the cytotoxic T cells of the recipient recognize differences in MHC class I and kill the tumor cells.
   B.   (i) The marsupial known as the Tasmanian devil can pass tumor cells between individuals when fighting and biting each other's faces. Because of limited MHC diversity in the species, there is a lack of allograft rejection, and the tumor cells establish themselves at the site of facial injury. (ii) In humans, residual tumor cells in an HLA-matched transplanted organ will not be rejected because the recipient is immunosuppressed and unable   to   mount   an   antitumor   immune response.

16–12 d

16–13
   A.   The proteasome not only degrades proteins into peptide fragments, but it also has the potential to splice peptides together to form a novel peptide. The new peptide is derived from short sequences from different parts of the polypeptide that is undergoing proteolysis.
   B.   An example of proteasome splicing has been demonstrated   for   a   glycoprotein   (gp100) expressed in melanocytes. A unique fused peptide sequence made up of amino acids 40–42 and 47–52 of gp100 was shown to be formed and to be presented by HLA-A32, stimulating cytotoxic T-cell responses.

16–14

A. MIC proteins are ligands for the activating receptor NKG2D, which is expressed by NK cells, γ:δ T cells, and cytotoxic CD8 T cells and activates these cells to kill the tumor cells.

B. Some of the tumor cells evade being killed by expressing a protease that cleaves MIC from the cell surface. When the soluble MIC product binds to the NKG2D receptors on NK cells, γ:δ T cells, and cytotoxic T cells it induces receptor-mediated endocytosis, thus reducing the level of NKG2D on the surface of these cells. Together with the removal of MIC from tumor-cell surfaces, this helps the tumor cells evade cell-mediated killing.

16–15

(i) Increase the number of virus-specific cytotoxic CD8 T cells by vaccinating with either a recombinant virus expressing viral epitopes, or with synthetic peptides.

(ii) Administer anti-CTLA4 monoclonal antibody to maintain co-stimulatory B7 ligands on professional antigen-presenting cells, thus maintaining the activation of naive T cells.

(iii) Immunize with cancer cell-derived heat-shock proteins, which bind to tumor antigens and enable dendritic cells to become activated and present tumor antigens to CD4 and CD8 T cells.

(iv) Manipulate the patient's dendritic cells by culturing them *ex vivo* with tumor-derived heat-shock proteins or purified tumor antigens, and then return them to the patient's circulation. In the patient they will home to secondary lymphoid tissue and stimulate a T-cell response.

16–16

(i) Monoclonal antibodies can be covalently coupled to radioisotopes to concentrate radioactivity at tumor sites to track the location, size, and degree of metastasis, or to kill the tumor. One example of such an antibody is tositumomab, which is specific for CD20 and is conjugated to iodine-131. It is used to treat non-Hodgkin's lymphoma.

(ii) Monoclonal antibodies can be conjugated to biological toxins to deliver cytotoxic agents with specificity for tumor cells and to protect normal tissue. Some cytotoxic agents used in this way are not active until internalized and metabolized by the tumor cell; for example, ozogamicin is not active until it is cleaved from the antibody and activated by glutathione-mediated reduction in lysosomes.

16–17 **Rationale:** The answer is e. The *Mycobacterium bovis* BCG vaccine is an attenuated producer of unmethylated CpG-containing DNA, the bacterial-specific ligand for intracellular Toll-like receptor 9 (TLR9). TLR9 is expressed by macrophages and dendritic cells and transduces signals for inflammatory cytokines; this is integral to the *in situ* inflammatory response sought for stimulation of an antitumor immune response as well as an antimycobacterial response. So it must be applied directly to the site of the tumor to have this effect.

Glossary

α when referring to an immunoglobulin heavy chain, the isotype that is present in immunoglobulin IgA.

α:β T cells T cells that express an antigen receptor made up of α and β chains. This majority population of T cells includes all T cells that recognize peptide antigen presented by MHC class I and II molecules.

α:β T-cell receptor one of two classes of **T-cell receptor**.

α-defensins a subset of the secreted antimicrobial peptides called **defensins**.

ABO system set of blood group antigens on red blood cells that have to be appropriately matched between donor and recipient for successful blood transfusion or organ transplantation.

ACAID acronym for **anterior chamber-associated immune deviation**.

acquired immune deficiency syndrome (AIDS) disease caused by infection with the human immunodeficiency virus (HIV). It involves a gradual destruction of the CD4 T-cell population and increasing susceptibility to infection.

acquired immunity alternative term for **adaptive immunity**, pathogen-specific immunity acquired as a consequence of infection or vaccination.

activated dendritic cells dendritic cells that take up pathogens at a site of infection and are stimulated to travel to secondary lymphoid tissue and present antigens to T cells.

activation-induced cytidine deaminase (AID) enzyme that deaminates DNA at cytosine resides, converting them to uracil. The activity of this enzyme and conseqent repair of the damaged DNA is the basis of somatic hypermutation and isotype switching in activated B cells.

acute-phase proteins plasma proteins made by the liver whose synthesis is rapidly increased in response to infection. They include mannose-binding lectin (MBL), C-reactive protein (CRP), and fibrinogen.

acute-phase response response of innate immunity that occurs soon after the start of an infection and involves the synthesis of acute-phase proteins by the liver and their secretion into the blood.

acute rejection rejection of transplanted cells, tissues, or organs that is due to a T-cell response stimulated by the transplant.

ADA acronym for the enzyme **adenosine deaminase**.

adaptive immune response the response of antigen-specific B and T lymphocytes to antigen, including the development of immunological memory.

adaptive immunity the state of resistance to infection that is produced by the adaptive immune response.

ADCC acronym for **antibody-dependent cell-mediated cytotoxicity**.

adenoids mucosa-associated secondary lymphoid tissues located in the nasal cavity.

adenosine deaminase (ADA) enzyme involved in purine breakdown. Its absence leads to the accumulation of toxic purine nucleosides and nucleotides, resulting in the death of most developing lymphocytes within the thymus. The disease that results from a genetic lack of adenosine deaminase is called **adenosine deaminase deficiency**. Its symptoms are severe combined immunodeficiency.

adhesins molecules on the surfaces of bacteria that bind to epithelial cells, enabling the bacteria to colonize epithelial surfaces.

adhesion molecules cell-surface proteins that enable cells to bind to each other. *See also* **selectins**; **integrins**; **vascular addressins**; **immunoglobulin superfamily**.

adjuvants substances used in experimental immunology and in vaccines to enhance the adaptive immune response to an antigen. An adjuvant must be injected along with the antigen to have an effect. Most protein antigens are non-immunogenic without an adjuvant.

afferent lymphatic vessels the various vessels that bring lymph draining from connective tissue into a lymph node en route to the blood.

affinity measure of the strength with which one molecule binds to another at a single binding site.

affinity maturation the increase in affinity of the antigen-binding sites of antibodies for antigen that occurs during the course of an adaptive immune response. It occurs as the result of somatic hypermutation of the rearranged immunoglobulin region and the consequent selection of mutated B cells that make antigen receptors of higher affinity for their antigen.

agglutination clumping together of particles, usually caused by antibody or some other multivalent binding molecule interacting with antigens on the surfaces of adjacent particles. Such particles are said to be **agglutinated**. When the particles are red blood cells, the phenomenon is called hemagglutination. *See also* **influenza hemagglutinin**.

agonist any molecule that binds to a receptor and activates it.

AID acronym for **activation-induced cytidine deaminase**.

AIDS acronym for **acquired immune deficiency syndrome**.

AIRE acronym for **autoimmune regulator**.

ALG acronym for **antilymphocyte globulin**.

alleles the natural variants of a gene. Some genes occur in many allelic forms within a population.

allelic exclusion in reference to antibody production, the case that, in a heterozygous individual, only one of the two C-region alleles at the immunoglobulin heavy-chain or light-chain loci is expressed in each B cell. In the B-cell population, for each locus roughly half of the cells express one allele and the remaining cells express the other.

allergen any antigen that elicits hypersensitivity or allergic reactions. Allergens are usually innocuous proteins that do not inherently threaten the integrity of the body.

allergic asthma disease caused by an allergic reaction to inhaled antigen, in which the bronchi constrict and the patient has difficulty breathing.

allergic conjunctivitis allergic reactions in the conjunctiva of the eye, usually caused by airborne antigens.

allergic reaction the result of a secondary immune response to an otherwise innocuous environmental antigen, or allergen, causing a wide variety of unpleasant and sometimes life-threatening symptoms. Allergic reactions can involve either antibodies or effector T cells.

allergic rhinitis allergic reaction manifested in the nasal mucosa that causes runny nose, sneezing, and tears. It is also known as hay fever.

allergy state of hypersensitivity to a normally innocuous environmental antigen. It results from the interaction between the antigen and antibodies or T cells produced by earlier exposure to the same antigen and causes a variety of unpleasant and sometimes life-threatening symptoms depending on the route of entry of the antigen and the type of effector mechanisms involved.

alloantibody antibody that is made by immunization of one member of a species with antigen derived from another member of the same species. Alloantibodies recognize antigens that are the result of allelic variation at polymorphic genes. Common types of alloantibody are those recognizing blood group antigens and HLA class I and class II molecules.

alloantigen antigen that differs between members of the same species, such as HLA molecules and blood group antigens. Alloantigens are determined by the different alleles of polymorphic genes.

allogeneic describes two members of the same species who are genetically different.

allograft tissue graft made between genetically non-identical members of the same species.

alloreaction adaptive immune response made by one member of a species to an allogeneic antigen from another member of the same species.

alloreactive T cell T cell in one member of a species that responds to an allogeneic antigen from another member of the same species.

allotypes natural variants of a protein that are encoded by the alleles of a given gene.

alternative C3 convertase the C3 convertase of the alternative pathway of complement activation. It is composed of C3 or C3(H$_2$O) bound to proteolytically active Bb (C3Bb or C3(H$_2$O)Bb) and cleaves C3 into C3a and C3b.

alternative C5 convertase the C5 convertase of the alternative pathway of complement activation. It is composed of two molecules of C3b bound to Bb (C3b$_2$Bb) and cleaves C5 into C5a and C5b.

alternative pathway of complement activation one of three pathways of complement activation. It is triggered by the presence of infection but does not involve antibody. *See also* **classical pathway of complement activation; lectin pathway of complement activation.**

anaphylactic shock IgE-mediated allergic reaction to a systemically administered antigen (e.g., insect venom entering the bloodstream

via a sting) that causes collapse of the circulation and suffocation due to tracheal swelling. Also called **systemic anaphylaxis**.

anaphylactoid reactions reactions that are clinically similar to anaphylactic shock but do not involve IgE.

anaphylatoxins complement fragments C3a and C5a, which are produced during complement activation. They act to induce inflammation, recruiting fluid and inflammatory cells to sites of antigen deposition. In some circumstances C3a and C5a can induce anaphylactic shock.

anchor residues amino acid residues in MHC-binding peptides that interact with pockets in the peptide-binding groove on the MHC molecule. Peptides that bind to a given MHC allotype have the same or very similar anchor residues.

anergy state of nonresponsiveness to an antigen. People are said to be anergic when they cannot mount delayed-type hypersensitivity reactions on challenge with an antigen. T and B cells are said to be anergic when they cannot respond to their specific antigen.

angioedema swelling of the skin as a result of IgE-mediated allergic reactions, which cause increased permeability of subcutaneous blood vessels with consequent leakage of fluid into the skin.

antagonist any molecule that binds to a receptor and prevents its function.

anterior chamber-associated immune deviation (**ACAID**) active and systemic state of tolerance to foreign antigens that exists in the eye and which allows it to tolerate corneal grafts of all HLA types. It is due to immunomodulatory factors in the aqueous humor of the anterior chamber of the eye.

antibody the secreted form of the immunoglobulin made by a B cell.

antibody repertoire the total variety of antibodies that a person can make.

antibody-dependent cell-mediated cytotoxicity (**ADCC**) the killing of antibody-coated target cells by NK cells having the receptor FcγRIII (CD16), which recognizes the Fc region of the bound antibody.

antigen originally defined as any molecule that binds specifically to an antibody, the term now also refers to any molecule or molecular fragment that can be bound by an MHC molecule and presented to a T-cell receptor. *See also* **epitope**.

antigen-binding site the site on an immunoglobulin or T-cell receptor molecule that binds specific antigen.

antigen presentation the display of antigen as peptide fragments bound to MHC molecules on the surface of cells. This is the form in which antigen is recognized by most T cells.

antigen-presenting cells cells that express either MHC class I and/or MHC class II molecules and thus display complexes of MHC molecule and peptide antigen on their surfaces.

antigen processing the intracellular degradation of proteins into peptides that bind to MHC molecules for presentation to T cells.

antigen receptor the cell-surface receptor on lymphocytes that recognizes antigen. For a B cell, the antigen receptor is its cell-surface immunoglobulin; for a T cell the antigen receptor is a rather similar molecule called the T-cell receptor. All the antigen receptors on an individual lymphocyte are identical and recognize the same epitope.

antigenic determinant the portion of an antigenic molecule that is bound by an antibody or gives rise to the MHC-binding peptide that is recognized by a T-cell receptor. Also called an epitope.

antigenic drift process by which point mutations in influenza virus genes cause alterations in the structure of viral surface antigens. This causes year-to-year antigenic differences in strains of influenza virus.

antigenic shift process by which influenza viruses reassort their segmented genomes and change their surface antigens radically. New viruses arising by antigenic shift are the usual cause of influenza pandemics.

antilymphocyte globulin (**ALG**) preparation of antibodies made by immunizing animals with human lymphocytes. It is used clinically to prevent rejection in organ transplantation.

antiserum (*plural* **antisera**) fluid component of clotted blood from an immune individual that contains antibodies against a given antigen. Antiserum contains a heterogeneous collection of antibodies that bind the antigen.

antithymocyte globulin (**ATG**) preparation of antibodies made by immunizing animals with human thymocytes. It is used clinically to prevent rejection in organ transplantation.

AP-1 family of transcription factors, some of which participate in lymphocyte activation.

APD, APECED acronyms for **autoimmune polyendocrinopathy–candidiasis–ectodermal dystrophy** the genetic autoimmune disease affecting many different tissues that is caused by the absence of the autoimmune regulator.

apoptosis mechanism of cell death in which the cells to be killed are induced to degrade themselves from within, in a tidy manner. Also called **programmed cell death**.

appendix gut-associated secondary lymphoid tissue located at the beginning of the colon.

Arthus reaction immune reaction in the skin caused by the injection of antigen into the dermis. The antigen reacts with specific IgG antibodies in the extracellular spaces, activating complement and phagocytic cells to produce a local inflammatory response.

ATG acronym for **antithymocyte globulin**.

atopic dermatitis an IgE-mediated allergic disease that affects the skin and runs in families. Also called eczema.

atopy the genetically determined tendency of some people to produce IgE-mediated hypersensitivity reactions against innocuous substances.

autoantibody antibody produced by an individual against an antigen produced by their own body.

autoantigen self antigen, an antigenic component of the body that provokes an immune response by the individual's own immune system.

autocrine describes a cytokine or other secreted molecule that acts on the same type of cell as the one that secreted it.

autograft graft of tissue taken from one anatomical site and transplanted to another in the same individual.

autoimmune disease disease in which the pathology is caused by an immune response to normal components of healthy tissue.

autoimmune hemolytic anemia disease characterized by reduced numbers of red blood cells (anemia). The reduction is caused by autoantibodies that bind to surface antigens of red blood cells, resulting in destruction of the cells.

autoimmune polyendocrinopathy–candidiasis–ectodermal dystrophy (**APECED**) autoimmune disease caused by a lack of a protein called autoimmune regulator (AIRE), which results in the production of T cells reactive against a number of tissues in the body. Also known as **autoimmune polyglandular disease (APD)**.

autoimmune polyglandular disease (**APD**) alternative name for **autoimmune polyendocrinopathy–candidiasis–ectodermal dystrophy (APECED)**.

autoimmune regulator (**AIRE**) transcription factor that causes several hundred tissue-specific genes to be transcribed by a subpopulation of epithelial cells in the medulla of the thymus, and which thus enables the developing T-cell population to become tolerant of antigens that normally function only outside the thymus.

autoimmune response adaptive immune response directed at an antigenic component of the individual's own body.

autoimmune T cells T cells that recognize self antigens (antigens of an individual's own body) and react against them.

autoimmunity adaptive immunity specific for an antigenic component of the individual's own body.

autologous describes cells, HLA molecules, and so on that derive from the individual in question.

autologous bone marrow transplantation bone marrow transplantation in which the donor and recipient are the same person. In such cases, bone marrow is removed from the patient, treated in some way to remove diseased or harmful cells, and then reinfused.

autoreactive describes lymphocytes or their antigen receptors that bind to self antigens.

avidity the overall strength of binding of an antibody to its antigen with multiple sites, in contrast to the affinity, which is the strength of binding at a single site.

azathioprine immunosuppressive drug that kills dividing cells. It is used in transplantation to help suppress rejection reactions.

azurophilic granules preformed cytoplasmic granules in neutrophils that contain proteins and peptides that can disrupt and digest ingested microbes. Also known as **primary granules** but distinct from the specific granules of the neutrophil.

β-defensins a subset of the secreted antimicrobial peptides called **defensins**.

β₂-microglobulin (β_2m) invariant polypeptide that is common to all MHC class I molecules. Also called the **light chain** of **MHC class I molecules**.

B-1 cells the minority population of B cells. They express the CD5 glycoprotein and make antibodies of broad specificities. They are also known as **CD5 B cells**.

B-2 cells the majority population of B cells. They do not express the CD5 glycoprotein and make antibodies of narrow specificities.

B7 molecules the **B7.1** and **B7.2** proteins, which are co-stimulatory molecules present on the surface of professional antigen-presenting cells such as dendritic cells.

B cell one of the two main classes of lymphocyte (the other is the T cell). B cells are dedicated to making immunoglobulins and antibodies. Also known as **B lymphocyte**.

B-cell co-receptor complex of the polypeptides CD19, TAPA-1, and CR2 that associates with the B-cell receptor and augments its response to specific antigen.

B-cell receptor the antigen receptor on B cells, which is a membrane-bound immunoglobulin molecule. Each B cell is programmed to make a single type of immunoglobulin. The cell-surface form of this immunoglobulin serves as the B-cell receptor for specific antigen. Associated in the membrane with the immunoglobulin are the signal transduction molecules Igα and Igβ.

B lymphocyte an alternative name for **B cell**, the lymphocytes of adaptive immunity that have immunoglobulin receptors for antigen.

bacteria (*singular* **bacterium**) diverse prokaryotic microorganisms that are responsible for many infectious diseases of humans and other animals.

bacterial lipopolysaccharide (**LPS**) component of the surface of Gram-negative bacteria that activates Toll receptors on macrophages and other leukocytes as part of the innate immune response.

BAFF acronym for B-cell activating factor in the TNF family. A cytokine that promotes the survival of B cells.

balancing selection type of evolutionary selection that acts to maintain a variety of phenotypes (such as different variants of MHC molecules) in a population.

BALT acronym for **bronchial-associated lymphoid tissues**.

bare lymphocyte syndromes genetically determined diseases in which either MHC class I or MHC class II molecules are not expressed on cells. They can be caused by various different regulatory gene defects, and their effect is severe immunodeficiency. On its own, the term **bare lymphocyte syndrome** often refers to a lack of

MHC class II molecules, whereas a lack of MHC class I molecules is designated **bare lymphocyte syndrome (MHC class I)**.

basophil white blood cell present in small numbers in the blood; it is one of the three types of granulocyte. Basophils contain granules that stain with basic dyes, hence their name.

benign tumor cellular growth, such as a wart, which is caused by abnormal proliferation of cells but is localized and contained by epithelial barriers.

bone marrow tissue in the center of certain bones that is the major site of the generation of all the cellular elements of blood (hematopoiesis).

bone marrow transplant the replacement of a person's diseased or functionally deficient bone marrow with healthy bone marrow from a donor.

bronchial-associated lymphoid tissues (BALT) the lymphoid cells and organized lymphoid tissues of the respiratory tract.

bronchiectasis chronic inflammation of the bronchioles of the lung.

C1 complement protein C1, the first complement protein activated in the classical pathway of complement fixation. It is a complex of proteins C1q, C1s and C1r and binds to C-reactive protein or antibody coating an antigen surface. Activated C1 has serine protease activity and cleaves complement component **C4** to produce C4b, and complement component **C2** to produce C2a, which together form the classical C3 convertase C4b2a.

C1 inhibitor (C1INH) regulatory protein in plasma that inhibits the enzyme activity of activated complement component **C1**. C1INH deficiency causes the disease hereditary angioneurotic edema, in which spontaneous complement activation causes episodes of epiglottal swelling and other symptoms.

C2 complement protein C2, *for details see* **C1**.

C3 complement protein C3, which becomes covalently bound, as the cleavage product C3b, to the surface of pathogens in all three pathways of complement fixation, thus facilitating phagocytosis of the pathogen. The other cleavage product of C3, C3a, is an anaphylatoxin.

C3 convertases proteolytic enzymes that are formed during complement activation and cleave complement component C3 to C3b and C3a, thereby enabling C3b to bond covalently to antigens. *See also* **alternative C3 convertase; classical C3 convertase**.

C3(H$_2$O) form of the complement protein C3 produced in the first step in the alternative pathway of complement activation, in which a thioester bond is hydrolyzed but the molecule is not cleaved. Also called **iC3**.

C4 complement protein C4, *for details see* **C1**.

C4-binding protein (C4BP) a regulatory protein in plasma that inactivates the classical C3 convertase by binding to C4b and displacing C2a.

C5 complement protein C5, which is cleaved during complement activation to produce C5b (which binds complement proteins C6 and C7 to form the nucleus of the membrane-attack complex) and the anaphylatoxin C5a.

C5 convertases proteolytic enzymes that are formed during complement activation and cleave complement component C5 to form C5a and C5b.

C6, C7, C8, C9 complement proteins that form the membrane-attack complex on the surface of cells.

C domain protein domains that make up the constant regions of immunoglobulin and T-cell receptor chains. Immunoglobulin light chains and T-cell receptor chains have one C domain, immunoglobulin heavy chains have 3 or 4 C domains.

C-reactive protein (CRP) soluble acute-phase protein that binds to phosphocholine, a surface constituent of various bacteria. CRP binds to bacteria, opsonizing them for uptake by phagocytes. It can also activate the classical pathway of complement fixation.

C region acronym for **constant region** the part of an immunoglobulin or T-cell receptor chain that is not involved in binding to antigen and is identical in sequence in antibodies or T-cell receptor with different antigenic specificities.

calcineurin cytosolic serine/threonine phosphatase that contributes to T-cell activation. The immunosuppressive drugs cyclosporin A and tacrolimus act by inhibiting calcineurin.

calnexin membrane protein of the endoplasmic reticulum that facilitates the folding of MHC molecules and other glycoproteins.

CAM acronym for **cell adhesion molecule**.

cancer diseases caused by abnormal and invasive cell proliferation.

cancer immunosurveillance this describes the capacity of the immune system to identify cancerous cells at an early stage in their development and to kill them before they cause disease. Also called immunosurveillance.

cancer stem cell minority population of cells in a tumor that are self-renewing and more resistant to the toxins and radiation that are commonly used to treat cancer.

cancer/testis antigens (CT antigens) a subset of tumor-associated antigens that normally are expressed only in immature sperm in the testis.

carcinogen chemical or physical agent that increases the risk of cancer.

carcinoma cancer of epithelial cells.

carrier person who carries one copy of a recessive allele for a hereditary disease and does not show symptoms. Such a person can pass the allele on to successive generations, where in a different genetic context disease may result.

caseation necrosis form of necrosis seen in the center of some large granulomas. The term comes from the white cheesy appearance of the central necrotic area.

catalytic antibody antibody that binds an antigen, chemically changes it, and then releases it.

CCL18 chemokine secreted by mature dendritic cells in secondary lymphoid tissues, which specifically attracts naive T cells.

CCL19 and **CCL21** chemokines that are secreted by stromal cells and dendritic cells in secondary lymphoid tissues and attract leukocytes and dendritic cells to the tissue.

CCP acronym for **complement control proteins**.

CCR7 receptor on leukocytes for chemokines CCL19 and CCL21.

CD2 adhesion molecule on T cells that binds to the LFA-3 adhesion molecule on antigen-presenting cells.

CD3 complex complex of signaling proteins that associates with T-cell receptors. It consists of CD3γ, δ, and ε chains, and ζ chains.

CD4 cell-surface glycoprotein that distinguishes a subset of T cells that recognize antigens presented by MHC class II molecules. CD4 binds to MHC class II molecules on the antigen-presenting cell and acts as a co-receptor to augment the T cell's response to antigen.

CD4 T cells the subset of T cells that express the CD4 co-receptor and recognize peptide antigens presented by MHC class II molecules.

CD5 B cells an alternative name for human B-1 cells, the minority population that of B cells that do not require T-cell help, do not undergo affinity maturation and participate in early defense. *See also* **B-1 cells**.

CD8 cell-surface glycoprotein that distinguishes a subset of T cells that recognize antigens presented by MHC class I molecules. CD8 binds to MHC class I molecules on the antigen-presenting cell and acts as a co-receptor to augment the T cell's response to antigen.

CD8 T cells the subset of T cells that express the CD8 co-receptor and recognize peptide antigens presented by MHC class I molecules.

CD19 component of the B-cell co-receptor.

CD21 another name for complement receptor 2 (CR2). It is a component of the B-cell receptor.

CD28 the low-affinity receptor on T cells that interacts with B7 costimulatory molecules to promote T-cell activation.

CD34 vascular addressin that is expressed on high endothelial venules in lymph nodes and is involved in the extravasation of white blood cells.

CD40 cell-surface glycoprotein on B cells whose interaction with **CD40 ligand** on T cells triggers B-cell proliferation.

CD59 alternative term for **protectin**, a complement control protein.

CD81 component of the B-cell receptor that is also a cellular receptor for hepatitis C virus. Also called TAPA-1.

CDR acronym for **complementarity-determining regions**, the antigen binding loops of immunoglobulins, antibodies and T-cell receptors.

celiac disease inflammatory hypersensitivity disease of the gut mucosa caused by an immune response to the gluten proteins present in some cereals such as wheat and barley but not rice.

cell-mediated immunity (**cellular immunity**) any adaptive immune response in which antigen-specific effector T cells dominate. It is defined operationally as all adaptive immunity that cannot be transferred to a naive recipient with serum antibody.

central lymphoid tissues the tissues where lymphocytes develop from progenitor cells. Principally the bone marrow and the thymus. Also called primary lymphoid tissues.

central memory cells one of two subsets of memory T cells (the other being effector memory T cells) that are distinguished by different activation requirements. Central memory T cells have a preference for the T-cell zones of secondary lymphoid tissues and take longer than effector memory cells to mature into functioning effector T cells after encounter with their specific antigen.

central MHC the class III region of the major histocompatibility complex, located between the class I and II regions.

central tolerance tolerance to self antigens that is generated in B and T cell populations during their development in the primary (central) lymphoid organs (bone marrow and thymus).

centroblast large dividing B cell present in germinal centers. Somatic hypermutation occurs in centroblasts, and antibody-secreting and memory B cells derive from them.

centrocyte nondividing B cell in germinal centers. Centrocytes have undergone isotype switching and somatic hypermutation.

CGD acronym for **chronic granulomatous disease**.

checkpoint stage in lymphocyte development when the production of potentially functional immunoglobulin chains and T-cell receptor chains from rearranged genes is tested by the cell. There are several such checkpoints in lymphocyte development.

Chédiak–Higashi syndrome genetic disease in which phagocytes malfunction. Their lysosomes fail to fuse properly with phagosomes, and killing of ingested bacteria is impaired.

chemokines large group of small proteins involved in guiding white blood cells to sites where their functions are needed. They have a central role in inflammatory responses.

chimeric monoclonal antibodies monoclonal antibodies that combine mouse variable regions with human constant regions.

chronic asthma disease characterized by chronic inflammation of the airways and difficulty in breathing. Although probably initiated by exposure to an allergen, chronic asthma can be perpetuated in its absence and is exacerbated by smoking.

chronic granulomatous disease (**CGD**) immunodeficiency disease in which multiple granulomas form as a result of defective elimination of bacteria by phagocytic cells. It is caused by a defect in the

NADPH oxidase system of enzymes, which generates the superoxide radical involved in bacterial killing.

chronic rejection rejection of organ grafts that occurs years after transplantation and is characterized by degeneration and occlusion of the blood vessels in the graft. It is caused by an antibody response to the HLA class I alloantigens of the graft.

chronic thyroiditis autoimmune disease that causes progressive destruction of the thyroid gland. Also called Hashimoto's thyroiditis or Hashimoto's disease.

CIITA acronym for the **MHC class II transactivator**, a key transcription factor in the expression of MHC class II genes.

class an alternative term for **isotype**, particularly as it pertains to immunoglobulins.

class I region the part of the major histocompatibility complex that contains the MHC class I heavy-chain genes.

class II region the part of the major histocompatibility complex that contains the MHC class II α- and β-chain genes.

class II-associated invariant-chain peptide (**CLIP**) peptide of variable length cleaved from the invariant chain protein by proteases in the endosomal pathway. CLIP remains unstably bound in the peptide-binding cleft of an MHC class II molecule until it is removed by the HLA-DM protein and replaced by an antigen peptide.

class III region the region of the major histocompatibility complex between the class I and II regions. Also known as the **central MHC**, it contains no genes for class I or II MHC molecules.

class switching the mechanism by which B cells can change the heavy-chain isotype of the immunoglobulin they make. This produces immunoglobulin of a different class. Also called isotype switching.

classical C3 convertase surface-associated serine protease of the classical pathway of complement activation. It is composed of the complement components C4b2a and cleaves C3 into C3a and C3b.

classical C5 convertase surface-associated serine protease of the classical pathway of complement activation. It is composed of the complement components C3b4b2a and cleaves C5 into C5b and C5a.

classical pathway of complement activation one of three pathways of complement activation. It is activated by antibody bound to antigen, and involves complement components C1, C4, and C2 in the generation of the classical C3 and C5 convertases. *See also* **alternative pathway of complement activation; lectin pathway of complement activation**.

CLIP acronym for **class II-associated invariant-chain peptide**.

clonal deletion the elimination of immature lymphocytes that bind to self antigens. Clonal deletion is the main mechanism that produces self-tolerance.

clonal selection the central principle of adaptive immunity. It is the mechanism by which adaptive immune responses derive only from individual antigen-specific lymphocytes, which are stimulated by the antigen to proliferate and differentiate into antigen-specific effector cells.

coagulation system collection of enzymes in blood plasma whose activity forms blood clots. The coagulation system is activated by damage to blood vessels.

coding joint the joint between the ends of two rearranged gene segments in the chromosome.

cognate interactions cell–cell interactions between B and T lymphocytes specific for the same antigen, in which the T cell recognizes peptide antigens processed and presented by the B cell.

collectins family of calcium-dependent sugar-binding proteins (lectins) containing collagen-like sequences. An example is mannose-binding lectin.

combination therapy antiviral therapy (for HIV, for example) in which several antiviral drugs are used together to try and avoid the rapid generation of mutant viruses resistant to one of the drugs alone.

combination vaccine vaccine that contains antigens derived from more than one pathogen and is designed to provide protection against more than one disease.

commensal describes a microorganism that habitually lives on or in the human body, and which normally causes no disease or harm and can even be beneficial.

common gamma chain (γ_c) protein chain that is the signaling component of several different cytokine receptors, including those for IL-2, IL-4, IL-7, IL-9, and IL-15.

complement set of plasma proteins that act in a cascade of reactions to attack extracellular forms of pathogens. Many of them are serine proteases. As a result of complement activation, pathogens become coated with complement components, which can either kill the pathogen directly or facilitate its engulfment and destruction by phagocytes.

complement activation the initiation by pathogens of a series of reactions involving the complement components of plasma, leading to the death and elimination of the pathogen. *See also* **alternative pathway of complement activation; classical pathway of complement activation; lectin pathway of complement activation**.

complement component C3 the central and most important component of the complement system. Also called C3 for short.

complement control protein any of a diverse group of proteins that inhibit complement activation at various stages and by different mechanisms. *See* **C1 inhibitor; C4-binding protein; decay-accelerating factor; factor I; membrane cofactor protein; protectin**.

complement control protein (CCP) modules family of structurally similar protein modules found in many of the proteins that regulate complement activity.

complement fixation the covalent attachment of C3b or C4b to pathogen surfaces, which is a central feature of the action of complement because it facilitates phagocytosis of the pathogen.

complement receptors (CRs) cell-surface proteins on various cell types that recognize and bind complement proteins bound to antigens. Complement receptors on phagocytes facilitate the phagocytic engulfment of pathogens coated with complement. Complement receptors include **CR1, CR2, CR3, CR4**, and the receptor for C1q.

complement system a system of some 30 soluble and cell-surface proteins that is a major mechanism in innate and adaptive immunity for identifying and eliminating pathogens and their products. Also called complement for short.

complementarity-determining regions (CDRs) the localized regions of immunoglobulin and T-cell receptor chains that determine the antigenic specificity and bind to the antigen. The CDRs are the most variable parts of the variable domains and are also called **hypervariable regions**.

conformational epitope epitope on a protein antigen that is formed from several separate regions in the primary sequence of a protein brought together by protein folding. Antibodies that bind conformational epitopes bind only to native folded proteins. Also called a **discontinuous epitope**.

conjugate pair a CD4 effector T cell bound via its antigen receptor to its target cell (either a macrophage or a B cell).

conjugate vaccine vaccine made from capsular polysaccharides bound to an immunogenic protein such as tetanus toxoid. The protein provides peptide epitopes that stimulate CD4 T cells to help B cells specific for the polysaccharide.

connective tissue mast cell one of two types of mast cell in humans (the other is the mucosal mast cell). It is found in connective tissues throughout the body.

constant domains (C domains) the constituent domains of the constant regions of immunoglobulin and T-cell receptor polypeptides.

constant region (C region) the part of an immunoglobulin or T-cell receptor (or of its constituent polypeptide chains) that is of identical amino acid sequence in molecules of the same isotype but different antigen-binding specificities.

contact sensitivity form of delayed-type hypersensitivity in which T cells respond to antigens that are introduced into the body by contact with the skin.

co-receptor cell-surface protein that increases the sensitivity of an antigen receptor to its antigen. A co-receptor can accomplish this by increasing adhesive interactions between the interacting cells, and/or by enhancing signal transduction from the antigen receptor.

corticosteroids compounds related to the steroid hormones produced in the adrenal cortex, such as cortisone. They suppress immune responses and are widely used as anti-inflammatory and immunosuppressive agents in medicine.

co-stimulator molecule molecule on an antigen-presenting cell that delivers signals to an interacting naive lymphocyte that are required in addition to the antigen-binding signal for the lymphocyte to respond. Co-stimulator molecules include the B7.1 and B7.2 proteins on professional antigen-presenting cells, which engage molecules CD28 and CTLA-4 on T cells. CD40 ligand serves a co-stimulatory role when it binds to CD40 on B cells.

co-stimulatory describes any signal that is required for the activation of a naive lymphocyte in addition to the signal delivered via the antigen receptor. *See also* **co-stimulator molecule**.

CR acronym for the group of four complement receptor proteins.

CR1, CR2, CR3, CR4 acronyms for the individual **complement receptors**.

cross-match test test used in blood typing and histocompatibility testing to determine whether donor and recipient have antibodies against each other's cells that might interfere with successful transfusion or transplantation.

cross-presentation process whereby antigen of extracellular origin can be presented by class I MHC molecules.

cross-priming initiation of an adaptive immune response by cross-presentation of antigen.

CRP acronym for **C-reactive protein**, a major serum protein of innate immunity.

cryptdins α-defensins HD5 and HD6, which are antibacterial proteins secreted by the Paneth cells of the small intestine.

cryptic epitope antigenic determinant on a molecule that is normally hidden from the immune system but becomes revealed under conditions of infection or inflammation.

CT antigens alternative name for **cancer/testis antigens**.

CTLA4 high-affinity inhibitory receptor on T cells that interacts with B7 co-stimulatory molecules.

CXCL2 chemokine produced by T_H1 cells that guides other leukocytes into infected tissues.

CXCL8 chemokine produced by activated macrophages and involved in the extravasation of neutrophils into infected tissue.

CXCL13 chemokine secreted by follicular dendritic cells that attracts B cells into lymphoid follicles.

cyclophilins family of cytoplasmic proteins that bind to the immunosuppressive drugs cyclosporin A and tacrolimus. The complex of drug and cyclophilin binds calcineurin, and this prevents T-cell activation.

cyclophosphamide alkylating agent used as an immunosuppressive drug. It acts by killing rapidly dividing cells, including lymphocytes proliferating in response to antigen.

cyclosporin A immunosuppressive drug that specifically prevents T-cell activation and effector function. Also called cyclosporine.

cytokines proteins made by cells that affect the behavior of other cells. Cytokines made by lymphocytes are often called lymphokines or interleukins (abbreviated IL). Cytokines bind to specific receptors on their target cells.

cytotoxic T cells subset of effector T cells that kill their target cells. They express the CD8 co-receptor and recognize peptide antigen presented by MHC class I molecules. They are important in host defense against viruses and other cytosolic pathogens because they can recognize and kill the infected cells.

cytotoxins proteins made by cytotoxic T cells that participate in the destruction of target cells. Perforins, granzymes, and granulysin are examples of cytotoxins.

δ when referring to an immunoglobulin heavy chain, the isotype that is present in immunoglobulin IgD.

D gene segments short DNA sequences present in multiple versions in immunoglobulin heavy-chain loci and in T-cell receptor β- and δ-chain loci. In the rearranged functional genes at these loci, a D gene segment connects the V and J gene segments. The D stands for 'diversity', because the D gene segments provide additional diversity in these receptor chains.

DAF acronym for **decay-accelerating factor**, a complement control protein.

dark zone the part of a germinal center in secondary lymphoid tissue that contains dividing centroblasts.

DC-SIGN adhesion molecule unique to activated dendritic cells that binds ICAM-3 on T-cell surfaces. Also called CD209.

decay-accelerating factor (**DAF**) cell-surface protein that prevents complement activation on human cells. DAF binds to C3 convertases of both the alternative and classical pathways of complement activation and, by displacing Bb and C2a, respectively, prevents their action.

defensins family of antimicrobial peptides 35–40 amino acids long that can penetrate microbial membranes and disrupt their integrity.

delayed-type hypersensitivity reaction (**DTH**) form of cell-mediated immunity elicited by antigen in the skin and mediated by CD4 T_H1 cells. It is called delayed-type hypersensitivity because the reaction appears hours to days after antigen is injected.

dendritic cells professional antigen-presenting cells with a branched, dendritic morphology. They are the most potent stimulators of T-cell responses. Also known as interdigitating reticular cells, they are derived from the bone marrow and are distinct from the follicular dendritic cell that presents antigen to B cells. **Immature dendritic cells** take up and process antigens but cannot yet stimulate T cells. **Mature** or **activated dendritic cells** are present in secondary lymphoid tissues and are able to stimulate T cells.

desensitization therapeutic procedure in which an allergic individual is exposed to increasing doses of allergen with the goal of inhibiting their allergic reactions. It probably works by shifting the balance between CD4 T_H1 and T_H2 cells, thus changing the antibody produced from IgE to IgG.

diapedesis movement of white blood cells from the blood across blood vessel walls into tissues.

DiGeorge syndrome recessive genetic immunodeficiency disease in which thymic epithelium fails to develop.

diphtheria toxin cytotoxic protein secreted by the bacterium *Corynebacterium diphtheriae*, the cause of diphtheria, and which causes the disease symptoms. The diphtheria vaccine consists of an inactive form of the toxin called diphtheria toxoid.

direct pathway of allorecognition type of alloreactive response in which T cells of the recipient of a transplant are stimulated by direct interaction of their receptors with the allogeneic HLA molecules expressed by dendritic cells from the donor, present in the transplant.

directional selection type of natural selection that replaces older alleles with newer variants (for example in the MHC). Its characteristic outcome is change.

discontinuous epitope term given to an epitope recognized by an immunoglobulin or antibody that is made up of amino-acid residues that come from different parts of the sequence of the protein antigen and are brought together by the protein's folding. Also referred to as a conformational epitope.

diversity gene segments one of three types of gene segments that are brought together by rearrangement to form the V region gene of an expressed immunoglobulin heavy chain gene and of expressed T-cell receptor β-chain and δ-chain genes. The diversity segment is place between the variable segment and the joining segment. Also called D segments.

dominant describes an allele that influences the phenotype in homozygous and heterozygous individuals. Usually refers to disease-causing alleles that encode proteins with functional defects.

double-negative thymocyte (**DN thymocyte**) immature T cell in the thymus that expresses neither CD4 nor CD8.

double-positive thymocyte (**DP thymocyte**) T cell at an intermediate stage of development in the thymus. It expresses both CD4 and CD8.

draining lymph node the lymph node to which extracellular fluid collected at a site of infection first travels.

DTH acronym for **delayed-type hypensensitivity**, which is mediated by effector T cells and is characterized by a **delayed-type hypersensitivity reaction**.

ε when referring to an immunoglobulin heavy chain, the isotype that is present in immunoglobulin IgE.

E-selectin a cell adhesion molecule that is one of several **selectins**. Also called CD62E.

early pro-B cell an early stage of differentiation in the development of B cells.

ectopic lymphoid tissue tissue resembling secondary lymphoid organs that forms in diseased and inflamed organs that do not normally contain lymphoid tissues, for example in the thyroid in Hashimoto's disease.

eczema common allergic skin disease of children, appearing as scaly, reddened and itchy patches on the skin. Also called atopic dermatitis.

edema abnormal accumulation of fluid in connective tissue, leading to swelling.

effector cells lymphocytes that can act to remove pathogens from the body without the need for further differentiation.

effector mechanisms the physiological and cellular processes used by the immune system to destroy pathogens and remove them from the body.

effector memory cells one of two subsets of memory T cells (the other being central memory T cells) that are distinguished by different activation requirements. Effector memory T cells have a preference for inflamed tissues and are activated more quickly than central memory cells to mature into functioning effector T cells after encounter with their specific antigen.

efferent lymphatic vessel the single vessel in which lymph and lymphocytes leave a lymph node en route to the blood.

encapsulated bacteria bacteria that possess thick carbohydrate coats that protect them from phagocytosis. Encapsulated bacteria cause extracellular infections and can be dealt with by phagocytes only if the bacteria are first coated with antibody and complement.

endocytosis the uptake of extracellular material into cells by endosomes that form by pinching off pieces of plasma membrane. *See also* **receptor-mediated endocytosis**.

endoplasmic reticulum aminopeptidase (**ERAP**) enzyme in the endoplasmic reticulum that removes amino acids from the amino-terminal end of peptides bound to MHC class I molecules to improve their fit.

endosome intracellular membrane-bounded vesicle that is formed by the invagination and pinching-off of a portion of plasma membrane. It contains extracellular material.

endothelium epithelium lining the interior of blood vessels.

engraftment the time at which a bone marrow transplant is making new blood cells.

eosinophil white blood cell that is one of the three types of granulocyte. It contains granules that stain with eosin (hence their name) and whose contents are secreted when the cell is stimulated. Eosinophils contribute chiefly to defense against parasitic infections.

eotaxin chemokine that attracts eosinophils. Also called CXCL11.

epidemic outbreak of infectious disease that affects many individuals within a population.

epithelium general name for the layers of cells that line the outer surface and the inner cavities of the body.

epitope the portion of an antigenic molecule that is bound by an antibody or gives rise to the MHC-binding peptide that is recognized by a T-cell receptor. Also called an antigenic determinant.

ERAP acronym for **endoplasmic reticulum aminopeptidase**.

erythrocyte red blood cell.

extravasation the movement of cells or fluid from within blood vessels to the surrounding tissues.

Fab fragment a proteolytic fragment of IgG that consists of the light chain and the amino-terminal half of the heavy chain held together by an interchain disulfide bond. It is called Fab because it is the Fragment with **a**ntigen **b**inding specificity. In the intact IgG molecule the parts corresponding to the Fab fragment are often called Fab or Fab arms.

factor B plasma protein that binds to C3(H$_2$O) or C3b and is cleaved to form part of the C3 convertase (C3(H$_2$O)Bb and C3bBb) in the alternative pathway of complement activation.

factor D plasma protease that cleaves factor B to Bb and Ba in the alternative pathway of complement activation.

factor H complement regulatory protein of plasma that inactivates the C3 convertase of the alternative pathway and C5 convertases by binding to C3b and rendering it susceptible to cleavage by factor I to produce inactive iC3b.

factor I protease that regulates complement action by cleaving C3b and C4b into inactive forms.

factor P an alternative name for **properdin**, the complement factor that stabilizes the alternative C3 convertase.

farmer's lung hypersensitivity disease caused by the interaction of IgG antibodies with large particles of inhaled allergen in the alveolar wall of the lung. The resulting inflammation compromises gas exchange by the lungs. The disease is caused by chronic exposure to large quantities of the allergen.

Fas member of the TNF receptor family that is expressed on certain cells and makes them susceptible to killing by cells expressing **Fas ligand**, a cell-surface member of the TNF family of proteins. Binding of Fas ligand to Fas triggers apoptosis in the Fas-bearing cell.

Fc fragment fragment of an antibody that consists of the carboxy-terminal halves of the two heavy chains disulfide-bonded to each other by the residual hinge region. It is produced by proteolytic cleavage of antibody. It is called Fc because it was the fragment that was most readily crystallized in early studies of IgG antibody structure. In an intact antibody the part corresponding to the Fc fragment is called **Fc**, **Fc region**, or **Fc piece**.

Fc receptors cell-surface receptors for the Fc portion of some immunoglobulin isotypes. They include, for example, the Fcγ receptor FcγRI and the Fcε receptor **FcεRI**.

Fc region an alternative term for the **Fc fragment** of an antibody.

Fcε receptor (FcεRI) receptor present on the surface of mast cells, basophils and activated eosinophils that binds free IgE with very high affinity. When antigen binds to IgE and cross-links FcεRI it causes cellular activation and degranulation.

Fcγ receptors receptors present on various cell types that are specific for the Fc regions of IgG antibodies. There are different receptors for different subclasses of IgG.

FcRn an Fc receptor that transports IgG across epithelia and has a structure resembling an MHC class I molecule.

FDC acronym for the **follicular dendritic cells** that nurture developing B-cells.

fever rise of body temperature above the normal. It is caused by cytokines produced in response to infection.

FK506 an alternative and earlier name for the immunosuppressive drug **tacrolimus**.

FK-binding proteins another name for the cyclophilins that bind tacrolimus (FK506). *See also* **cyclophilins**.

flora the community of microbial species that inhabits a particular niche in the human body, such as skin, mouth, gut or vagina.

follicle substructure in secondary lymphoid cells where naive B cells congregate and are nurtured by follicular dendritic cells. Also called lymphoid follicle and primary follicle.

follicular center cell lymphoma a malignancy of mature B cells.

follicular dendritic cells (**FDCs**) characteristic nonlymphoid cells of follicles in secondary lymphoid tissues. They have long branching processes that make intimate contact with B cells and have Fc and complement receptors that hold antigen:antibody:complement complexes on their surfaces for long periods. These cells are crucial in selecting antigen-binding B cells during antibody responses.

follicular helper cells subset of central memory cells that reside in the follicles of secondary lymphoid tissues, produce IL-2, and provide help to B cells.

framework regions relatively invariant regions within the variable domains of immunoglobulins and T-cell receptors that provide a protein scaffold for the hypervariable regions.

Freund's complete adjuvant emulsion of mineral oil and water containing killed mycobacteria that is mixed with purified antigens before injection to render them immunogenic. *See also* **adjuvants**.

fungi single-celled and multicellular eukaryotic organisms, including the yeasts and molds, that can cause a variety of diseases. Immunity to fungi involves both humoral and cell-mediated responses.

γ when referring to an immunoglobulin heavy chain, the isotype that is present in immunoglobulin IgG.

γc an alternative designation for the **common gamma chain** of Fc receptors.

γ:δ T cells minority population of T cells expressing receptors made up of γ and δ chains. The antigen specificities and functions of these cells are uncertain.

GALT acronym for **gut-associated lymphoid tissues**, the most extensive secondary lymphoid tissues in the human body.

gamma globulin antibody-containing preparation made from the plasma of healthy blood donors. Such preparations contain antibodies against a range of common pathogens.

gene conversion process whereby one copy of a gene, or part of a gene, is replaced by a different version of that gene, or part of a gene.

gene family set of genes encoding proteins of similar structure, and often of similar function, such as the MHC class I genes.

gene rearrangement in immunology the term refers to the somatic recombination process that occurs at immunoglobulin and T-cell receptor loci in developing B cells and T cells and assembles the sequence encoding a functional variable region of an immunoglobulin or T-cell receptor chain.

gene segments multiple short DNA sequences in the immunoglobulin and T-cell receptor genes. These can be rearranged in many different combinations to produce the vast diversity of immunoglobulin or T-cell receptor polypeptide chains. *See also* **D gene segments; J gene segments; V gene segment**.

genetic polymorphism variation in a population owing to the existence of two or more alleles of a given gene.

germinal center area in secondary lymphoid tissue that is a site of intense B-cell proliferation, selection, maturation, and death. Germinal centers form around follicular dendritic cell networks when activated B cells migrate into lymphoid follicles. The cellular and morphological events that form the germinal center and take place there are called the **germinal center reaction**.

germline configuration, **germline form** the original unrearranged organization of the immunoglobulin and T-cell receptor genes in the DNA of germ cells and in somatic cells other than T cells and B cells.

glucan receptor receptor on the surface of macrophages and neutrophils that recognizes microbial carbohydrates.

GlyCAM-1 vascular addressin present on the high endothelial venules of secondary lymphoid tissues. It is an important ligand for the L-selectin adhesion molecule on naive lymphocytes and directs these cells to leave the blood and enter the lymphoid tissues.

GM-CSF acronym for **granulocyte–macrophage colony-stimulating factor**.

Goodpasture's syndrome autoimmune disease in which autoantibodies against type IV collagen of the basement membrane of blood vessel endothelium cause extensive vasculitis.

graft-versus-host disease (GVHD) pathological condition caused by the **graft-versus-host reaction (GVHR)**, which is the response of mature donor-derived T cells in transplanted bone marrow to the alloantigens of the recipient's tissues.

graft-versus-leukemia (GVL) effect in bone marrow transplantation as therapy for leukemia, some degree of genetic incompatibility between donor and recipient is thought to help T cells or NK cells from the transplant to eliminate residual leukemia cells in the recipient. Also known as the **graft-versus-tumor effect**.

granulocyte–macrophage colony-stimulating factor a cytokine that stimulates the development of white blood cells from hematopoietic stem cells and is given to bone marrow transplant patients to stimulate the reconstitution of their immune system.

granulocytes white blood cells with irregularly shaped, multilobed nuclei and cytoplasmic granules. There are three types of granulocyte: neutrophils, eosinophils, and basophils. They are also known as polymorphonuclear leukocytes.

granuloma site of chronic inflammation usually triggered by persistent infectious agents such as mycobacteria, or by a non-degradable foreign body. Granulomas have a central area of macrophages, often fused into multinucleate giant cells, surrounded by T lymphocytes.

granulysin membrane-perturbing protein present in the granules of cytotoxic T cells and NK cells. With perforin and serglycin it is thought to make pores in the target cell's membrane.

granzymes serine esterases present in the granules of cytotoxic T cells and NK cells. On entering the cytosol of a target cell, granzymes induce apoptosis of the target. Also called fragmentins.

Graves' disease autoimmune disease in which antibodies against the thyroid-stimulating hormone receptor cause the overproduction of thyroid hormone and the symptoms of hyperthyroidism.

gut-associated lymphoid tissues (GALT) lymphoid tissues closely associated with the gastrointestinal tract, including the palatine tonsils, Peyer's patches in the intestine, and layers of intraepithelial lymphocytes.

GVHD acronym for **graft-versus-host disease**, a complication of bone-marrow transplantation.

GVHR acronym for **graft-versus-host reaction**, the response of donor-derived alloreactive T cells that causes **graft-versus-host disease**.

GVL acronym for **graft-versus-leukemia reaction**, the beneficial effect of donor-derived alloreactive T cells or NK cells in patients receiving bone marrow transplants for leukemia that leads to the elimination of leukemia cells that escaped chemotherapy and irradiation.

H chain alternative name for the **heavy chain** of immunoglobulins and antibodies.

HAART acronym for the **highly active anti-retroviral therapy** used to treat HIV-infected patients.

half-life in reference to cellular life span, the period of time during which a population of cells reduces to half its original size.

HANE acronym for **hereditary angioneurotic edema**, an immunodeficiency disease resulting from deficiency of the C1 inhibitor (C1INH) of complement activation.

haploidentical transplant tissue or organ transplant from a donor who shares one HLA haplotype with the patient but differs in the second.

haplotype in respect of a linked cluster of polymorphic genes, the set of alleles carried on a single chromosome is called a haplotype. Every person inherits two haplotypes, one from each parent. The term was first used in connection with the genes of the major histocompatibility complex.

Hashimoto's disease autoimmune disease characterized by persistent high levels of antibodies against thyroid-specific antigens. These antibodies recruit NK cells to the thyroid, leading to damage and inflammation; secondary lymphoid tissue forms in the thyroid, replacing functional thyroid tissue. Also known as **Hashimoto's thyroiditis**.

heavy chain (H chain) the larger of the two types of polypeptide in an immunoglobulin molecule. It consists of one variable domain and a number of constant domains. Immunoglobulin heavy chains come in a variety of heavy-chain isotypes, or classes, each of which confers a distinctive effector function on the antibody molecule.

helper T cells CD4 T cells are sometimes generally known as helper cells because their function is to help other cell types to perform their roles. The term helper T cell sometimes refers to T_H2 cells only, the cells that help B cells to produce antibody.

hematopoiesis the generation of the cellular elements of blood, including the red blood cells, white blood cells, and platelets. These cells all originate from pluripotent **hematopoietic stem cells** whose differentiated progeny divide under the influence of various **hematopoietic growth factors**.

hematopoietic cell any blood cell or its precursor cell types.

hematopoietic cell transplantation, **hematopoietic stem cell transplantation** any transplantation in which the role of the graft is to replace the hematopoietic system. Sources of hematopoietic stem cells include bone marrow, peripheral blood and umbilical cord blood.

hemolytic anemia of the newborn, **hemolytic disease of the newborn** a potentially fatal disease caused by maternal IgG antibodies directed toward paternal antigens expressed on fetal red blood cells. The usual target of this response is the Rh blood group antigen. Maternal anti-Rh IgG antibodies cross the placenta to attack the fetal red blood cells. Also called erythroblastosis fetalis.

herd immunity the phenomenon whereby those people in a population who have no protective immunity against a pathogen are largely protected from infection when the majority of the population has protective immunity and is resistant to the pathogen.

hereditary angioneurotic edema (HANE) genetic disease that results from deficiency of the C1 inhibitor of the complement system (C1INH). In the absence of C1 inhibitor, spontaneous activation of the complement system causes diffuse fluid leakage from blood vessels, the most serious consequence of which is epiglottal swelling leading to suffocation.

heterozygous describes an individual who has inherited different forms (alleles) of a given gene from their two parents.

highly active anti-retroviral therapy (HAART) combination therapy for HIV infection, in which several antiviral drugs are used together to try and avoid the rapid generation of drug-resistant mutant viruses that occurs when one of the drugs is used alone.

highly polymorphic describes genes that have many alleles and for which most individuals in a population are heterozygotes.

histamine vasoactive amine stored in mast-cell granules. Histamine is released when antigen binds to IgE molecules on mast cells. It causes dilation of local blood vessels and contraction of smooth muscle, producing some of the symptoms of immediate hypersensitivity reactions. Antihistamines are drugs that counter histamine action.

histocompatibility literally, the ability of tissues (Greek *histo*) to get along with each other. The term is used in immunology to describe the genetic systems that determine the rejection of tissue and organ grafts as a result of the immunological recognition of alloantigens (sometimes called in this context histocompatibility antigens).

HIV acronym for **human immunodeficiency virus**, the cause of acquired immunodeficiency syndrome (AIDS).

hives itchy red swellings in the skin caused by IgE-mediated reactions. Also called **urticaria** or nettle rash.

HLA the acronym for **Human Leukocyte Antigen**. It is the genetic designation for the human MHC. Individual loci are designated by capital letters, as in HLA-A, and alleles are designated by numbers, as in HLA-A*0201.

HLA class I molecules the name for the human version of the **MHC class I molecules**.

HLA class II molecules the name for the human version of the **MHC class II molecules**.

HLA-A, **HLA-B**, and **HLA-C** the highly polymorphic human MHC class I genes.

HLA-DM an invariant MHC class II molecule in humans that is involved in the intracellular loading of MHC class II molecules with peptides.

HLA-DO a relatively invariant human MHC class II molecule. Although not precisely defined, the function of HLA-DO is thought to modify that of HLA-DM.

HLA-DP, **HLA-DQ**, and **HLA-DR** the highly polymorphic human MHC class II molecules. Each class II molecule is made from α and β chains encoded by A and B genes, respectively. For example, the HLA-DPα and HLA-DPβ chains are encoded by the HLA-DPA and HLA-DPB genes, respectively. All the genes are in the MHC.

HLA-E and **HLA-G** relatively invariant human MHC class I molecules that form ligands for NK-cell receptors.

HLA-F monomorphic human MHC class I molecule of unknown function that is expressed intracellularly.

HLA type the combination of HLA class I and class II allotypes that a person expresses.

HMCV acronym for **human cytomegalovirus**, a common herpes virus.

Hodgkin's disease malignant disease caused by transformed germinal center B cells.

homing the movement of naive T cells into secondary lymphoid tissues, or of effector T cells to a site of infection.

homozygous describes an individual who has inherited the same form (allele) of a given gene from both parents.

human cytomegalovirus (HCMV) very common herpesvirus that often causes no or few symptoms in young healthy people but can cause life-threatening infections in immunocompromised patients.

human immunodeficiency virus (HIV) causative agent of the acquired immune deficiency syndrome (AIDS). HIV is a retrovirus of the lentivirus family that infects CD4 T cells, leading to their slow depletion, which eventually results in immunodeficiency.

human leukocyte antigen (HLA) complex the human major histocompatibility complex (MHC).

humanize to replace, by means of genetic engineering, the CDR loops in a human antibody with the corresponding CDR sequences of a desired specificity from a mouse antibody.

humoral immunity immunity that is mediated by antibodies and can therefore be transferred to a non-immune recipient by serum.

HV regions an alternative name for **hypervariable regions**.

hybridomas hybrid cell lines that make monoclonal antibodies of defined specificity. They are formed by fusing a specific antibody-producing B lymphocyte with a myeloma cell that grows in tissue culture and does not make any immunoglobulin chains of its own.

hygiene hypothesis hypothesis advanced to explain the increasing incidence over the past 50 years of hypersensitivity and autoimmune diseases in developed countries, in which the increase is proposed to be due to widespread hygiene, vaccination, and antibiotic therapy, preventing children's immune systems from becoming used to dealing correctly with either natural infections or innocuous environmental antigens.

hyperacute rejection rejection of an allogeneic tissue graft as a result of preformed antibodies that react against ABO blood group antigens or HLA class I antigens on the graft. The antibodies bind to endothelium and trigger the blood clotting cascade, leading to ischemia and death of the transplanted organ or tissue.

hypereosinophila abnormally high numbers of eosinophils in the blood.

hyper IgM immunodeficiency genetically determined immunodeficiency disease in which B cells cannot switch their immunoglobulin heavy-chain isotype. It can be due to several different underlying mutations. Also called **hyper IgM syndrome** and X-linked hyper IgM syndrome. hypersensitivity reactions immune responses to innocuous antigens that lead to symptomatic reactions on reexposure. These can cause **hypersensitivity diseases** if they occur repetitively. This state of heightened reactivity to an antigen is called **hypersensitivity**. Hypersensitivity reactions are classified by mechanism: type I hypersensitivity reactions involve the triggering of mast cells by IgE antibodies; type II hypersensitivity reactions involve IgG antibodies against cell-surface or matrix antigens; type III hypersensitivity reactions involve antigen:antibody complexes; and type IV hypersensitivity reactions are mediated by effector T cells.

hyperthyroid describes an abnormally high production of thyroid hormones by the thyroid gland.

hypervariable regions (HV regions) small regions of high amino acid sequence diversity within the variable regions of immunoglobulin and T-cell receptors. They correspond to the complementarity-determining regions.

hypothyroid describes an abnormally low production of thyroid hormone by the thyroid gland.

iC3 the product formed when the thioester bond of complement component C3 is hydrolyzed by water. Alternatively named $C3(H_2O)$.

ICAM-1, **ICAM-2**, and **ICAM-3** are the names of three different types of intercellular adhesion molecule (ICAM).

iccosomes immune-complex coated bodies. Small fragments of membrane coated with immune complexes that bud off from the processes of follicular dendritic cells in lymphoid follicles.

IDDM insulin-dependent diabetes mellitus; *see* **type 1 diabetes**.

IFN-α, **IFN-β**, **IFN-γ** the names for individual human cytokines called **interferons** that interfere with the viral infection of cells.

IFN-γ receptor deficiency general name for a genetically determined immunodeficiency caused by lack of or low levels of the IFN-γ receptor on macrophages and monocytes, and thus a deficit in their function. The disease is characterized by an inability to clear intracellular bacteria, especially mycobacteria.

Ig acronym for **immunoglobulin**.

Igα, Igβ components of the B-cell receptor that transduce signals to the interior of the B cell when the B-cell receptor binds antigen.

Ig domain acronym for **immunoglobulin domain**.

IgA the class of immunoglobulin having α heavy chains. Dimeric IgA antibodies are the main antibodies present in mucosal secretions. Monomeric IgA is present in the blood.

IgD the class of immunoglobulin having δ heavy chains. It appears as surface immunoglobulin on mature naive B cells but its function is unknown. Its transcription is coordinated with that of IgM.

IgE the class of immunoglobulin having ε heavy chains. IgE is involved in allergic reactions.

IgG the class of immunoglobulin having γ heavy chains. IgG is the most abundant class of immunoglobulin in plasma.

IgM the class of immunoglobulin having μ heavy chains. IgM is the first immunoglobulin to appear on the surface of B cells and the first antibody secreted during an immune response. It is secreted in pentameric form.

Ii acronym for the **invariant chain** of the MHC class II molecule.

IL-1 (interleukin-1) cytokine released by macrophages, which with IL-6 and TNF-α induces a wide range of inflammatory responses at early times in infection.

IL-2 (interleukin-2) cytokine produced by activated T cells that is essential for the proliferation of activated T cells and the development of an adaptive immune response.

IL-4 (interleukin-4) cytokine secreted by CD4 T_H2 cells that helps to initiate the proliferation and clonal expansion of B cells.

IL-6 (interleukin-6) cytokine released by macrophages, which with IL-1 and TNF-α induces a wide range of inflammatory responses at early times in infection.

IL-10 (interleukin-10) cytokine released by CD4 T_H2 cells and regulatory T cells, which promotes the development of T_H2 cells and inhibits macrophage activation.

IL-12 (interleukin-12) cytokine released by macrophages that activates NK cells.

IL-12 receptor deficiency genetically determined immunodeficiency caused by a lack of the IL-12 receptor on macrophages and monocytes, and thus a deficit in their function. The disease is characterized by an inability to clear intracellular bacteria, especially mycobacteria.

IL-13 (interleukin-13) cytokine secreted by CD4 T_H2 cells and regulatory T cells that inhibits macrophage activation.

immature B cell B cell that has rearranged a heavy-chain gene and a light-chain gene and expresses surface IgM but not IgD.

immature dendritic cell dendritic cell present in tissues, which takes up antigen but does not express co-stimulatory molecules and cannot yet act as a professional antigen-presenting cell to naive T cells.

immediate hypersensitivity hypersensitivity reactions occurring within minutes of exposure to antigen. They are caused by preexisting antibodies in the circulation. Also called **immediate reactions**.

immune resistant to infection.

immune complex protein complex formed by the binding of antibodies to soluble antigens. The size of immune complexes depends on the relative concentrations of antigen and antibody. Large immune complexes are cleared by phagocytes bearing Fc and complement receptors. Small soluble immune complexes tend to be deposited on the walls of small blood vessels, where they can activate complement and cause damage.

immune-complex coated bodies an alternative name for iccosomes, the membrane blebs arising from the dendrites of follicular dendritic cells and which are coated with antigen-antibody complexes.

immune dysregulation, polyendocrinopathy, enteropathy, and **X-linked syndrome** an immunodeficiency disease, for which the acronym is **IPEX**, caused by lack of FoxP3, a transcription factor necessary for the development of regulatory T cells.

immune stimulatory complexes (ISCOMs) lipid complexed with antigens that has adjuvant properties and potential application in vaccination.

immune system the tissues, cells, and molecules involved in the defense of the body against infectious agents.

immunity the ability to resist infection.

immunization the deliberate provocation of an adaptive immune response by introducing antigen into the body.

immunodeficiency diseases group of inherited or acquired disorders in which some part or parts of the immune system are either absent or defective, resulting in failure to mount an effective immune response to pathogens.

immunogen substance that is able to provoke an adaptive immune response if injected on its own. Such substances are described as **immunogenic**.

immunogenetics subfield of immunology that was originally concerned with the analysis of genetic traits by means of antibodies against genetically polymorphic molecules such as blood group antigens and MHC proteins. Immunogenetics now encompasses the genetic analysis, by any technique, of molecules that are of specific importance to the immune system.

immunoglobulin (Ig) the antigen-binding molecules of B cells.

immunoglobulin A full name for **IgA**.

immunoglobulin D full name for **IgD**.

immunoglobulin domain (Ig domain) protein structure module consisting of about 100 amino acids that fold into a sandwich of two β sheets held together by a disulfide bond. Immunoglobulin heavy and light chains are made up of a series of immunoglobulin domains. Similar domains are present in many other proteins.

immunoglobulin E full name for **IgE**.

immunoglobulin G full name for **IgG**.

immunoglobulin-like domains protein domains that resemble the immunoglobulin domain in structure but are present in a variety of other proteins including MHC class I and II, CD4 and CD8.

immunoglobulin M full name for **IgM**.

immunoglobulin superfamily (Ig superfamily) the name given to all the proteins that contain one or more immunoglobulin or immunoglobulin-like domains.

immunological memory the capacity of the immune system to make quicker and stronger adaptive immune responses to successive encounters with an antigen. Immunological memory is specific for a particular antigen and is long-lived.

immunological synapse localized region of contact between cell adhesion molecules and other cell-surface receptor–ligand pairs that is formed when a lymphocyte binds to its target cell via cell-surface receptors. T cells, B cells, and NK cells all form immunological synapses known respectively as the T-cell synapse, the B-cell synapse, and the NK-cell synapse.

immunological tolerance situation where the immune system does not make a response to a given antigen. *See also* **self-tolerance**.

immunophilins intracellular proteins with peptidyl–prolyl **cis–trans** isomerase activity that bind the immunosuppressive drugs cyclosporin A, tacrolimus, and sirolimus (rapamycin).

immunoproteasome type of proteasome that specializes in making peptides having a hydrophobic or a basic residue at the carboxy terminus, features that enable them to bind to MHC class I molecules.

immunoreceptor tyrosine-based activation motifs (ITAMs) sequences in the cytoplasmic domains of membrane receptors that are sites of tyrosine phosphorylation and of association with tyrosine kinases and phosphotyrosine-binding proteins involved in signal transduction. Related motifs with opposing effects are **immunoreceptor tyrosine-based inhibitory motifs (ITIMs)**, which recruit phosphatases that remove the phosphate groups added by tyrosine kinases.

immunosuppressive inhibiting the immune system and preventing immune responses.

immunosurveillance the presumed ability of the immune system to recognize cancer cells at an early stage and eliminate them before they cause disease.

immunotoxin conjugate composed of a specific antibody chemically coupled to a toxic protein usually derived from a plant or microbe. The antibody is designed to specifically bind to target cells, such as cancer cells, and deliver the toxin to kill them.

inactivated virus vaccine an anti-viral vaccine that contains virus that has been killed by treatment with heat, chemicals or irradiation. Also called killed virus vaccine.

indirect pathway of allorecognition one means by which alloreactive T cells in a transplant recipient can be stimulated to react against the transplant. The alloreactive T cells do not directly recognize the transplanted cells but recognize subcellular material that has been processed and presented by autologous antigen-presenting cells.

inflammation general term for the local accumulation of fluid, plasma proteins, and white blood cells that is initiated by physical injury, infection, or a local immune response. This is also known as an **inflammatory response**. The cells that invade tissues undergoing inflammatory responses are often called **inflammatory cells** or an inflammatory infiltrate. Those cytokines that promote inflammation are known as **inflammatory cytokines**.

inflammatory mediators variety of substances released by various cell types that contribute to the production of inflammation at a site of infection or trauma.

influenza hemagglutinin glycoprotein of the influenza virus coat that binds to certain carbohydrates on human cells, the first step in infection of cells with the virus. Changes in the hemagglutinin are the major source of antigenic shift. The protein is called a hemagglutinin because it can agglutinate red bood cells.

innate immunity the host defense mechanisms that act from the start of an infection and do not adapt to a particular pathogen. Also called the **innate immune response**.

insulin-dependent diabetes mellitus (IDDM) the autoimmune disease caused by the destruction of the insulin-producing β-cells of the pancreas. Also called type 1 diabetes and juvenile onset diabetes.

insulitis infiltration of the pancreatic islets of Langerhans with lymphocytes and other leukocytes. This is a symptom of incipient diabetes.

integrins class of cell-surface glycoproteins that mediate adhesive interactions between cells and the extracellular matrix.

interallelic conversion mechanism of genetic recombination between two alleles of a locus in which a segment of one allele is replaced with the homologous segment from the other. This mechanism is used to generate new HLA class I and II alleles. Also called segmental exchange, a term that can also include exchange of a homologous segment between two different genes in a gene family, for example between HLA-C and HLA-B.

interdigitating reticular cells name given to interdigitating dendritic cells seen in sections of secondary lymphoid tissue.

interferons cytokines that help cells to resist viral infection. **Interferon-α (IFN-α)** and **interferon-β (IFN-β)** are called type I interferons and are produced by leukocytes and fibroblasts, respectively, as well as by other cells. They act specifically to induce cells to resist viral infection. The structurally and functionally unrelated **interferon-γ (IFN-γ)** is a product of CD4 T$_H$1 cells, CD8 T cells, and NK cells, and has more general functions in immune responses, acting principally to activate macrophages.

interferon response changes in the expression of a variety of human genes in cells exposed to interferon.

interferon-producing cells (IPCs) specialized lymphocyte-like cells in the blood that secrete up to 1000-fold more type I interferon (interferons-α and -β) than other cells. *See also* **plasmacytoid dendritic cells**.

interleukin (IL) generic term used for many of the cytokines produced by leukocytes. *See also* **IL-1, IL-2, IL-3**, etc.

intermolecular epitope spreading process by which the immune response initially reacts against an epitope of one antigenic molecule and then progresses to epitopes on different proteins.

intramolecular epitope spreading process by which the immune response initially reacts against epitopes in one part of an antigenic molecule and then progresses to respond to other, non-cross-reactive, epitopes in the same molecule.

invariant chain (Ii) polypeptide that associates with major histocompatibility complex (MHC) class II proteins in the endoplasmic reticulum and prevents them from binding peptides there. It guides the MHC class II molecules to endosomes, where Ii is degraded, enabling MHC class II molecules to bind peptides present in the endosomes.

IPC acronym for the **interferon-producing cells** that make large quantities of type-1 interferons in the innate immune response.

IPEX (immune dysregulation, polyendocrinopathy, enteropathy, and X-linked syndrome) genetic disease caused by a deficiency of FoxP3 and a consequent lack of regulatory T cells.

ischemia deficient blood supply due to obstructed blood vessels.

ISCOMs laboratory-designed lipid mixtures that can carry antigens and act as adjuvants. Also called **immune stimulatory complexes**.

islets of Langerhans the endocrine hormone-producing tissue of the pancreas, which includes the β cells that produce insulin.

isoforms the different forms of a protein that are encoded by the alleles of a gene or by different but closely related genes.

isograft tissue or organ graft from one genetically identical individual to another.

isolated lymphoid follicle secondary lymphoid tissue in the gut wall that resembles a lymphoid follicle and is composed mainly of B cells.

isotype the class of an immunoglobulin—that is, IgM, IgG, IgD, IgA, and IgE—each of which has a distinct heavy-chain constant region encoded by a different constant-region gene. The heavy-chain constant region determines the effector properties of each antibody class.

isotype switching the process by which a B cell changes the class of immunoglobulin it makes while preserving the antigenic specificity of the immunoglobulin. Isotype switching involves a somatic recombination process that attaches a different heavy-chain constant-region gene to the existing variable-region exon.

ITAMs acronym for **immunoreceptor tyrosine-based activation motifs**.

ITIMs immunoreceptor tyrosine-based inhibitory motifs; *see entry for* **immunoreceptor tyrosine-based activation motifs**.

J gene segments relatively short DNA sequences present in multiple different copies at all immunoglobulin and T-cell receptor loci. At gene rearrangement, a J and a V gene segment are joined directly in immunoglobulin light-chain genes and T-cell receptor α- and γ-chain genes, and via a D gene segment in the immunoglobulin heavy-chain gene and T-cell receptor β- and δ-chain genes. The J stands for 'joining.'

Janus kinases (JAKs) family of tyrosine kinases that transduce activating signals from cytokine receptors.

junctional diversity diversity present in immunoglobulin and T-cell receptor polypeptides that is created during the process of gene rearrangement by the addition or removal of nucleotides at the junctions between gene segments.

junctional gene segment a gene segment common to the rearranged variable region genes encoding immunoglobulin and T-cell receptor chains. Also called joining gene segment and J segment.

κ kappa, one of the two types of immunoglobulin light chain; the other is λ.

killed virus vaccines vaccines that contain viral particles that have been deliberately killed by heat, chemicals, or radiation.

killer-cell immunoglobulin-like receptors (KIRs) family of receptors on NK cells that bind to MHC class I molecules on target cells and can send either activating or inhibitory signals to the NK cell.

kinin system enzymatic cascade of plasma proteins that is triggered by tissue damage and helps facilitate wound healing.

Kupffer cells phagocytic cells of the liver that line the hepatic sinusoids.

λ lambda, one of the two types of immunoglobulin light chain; the other is κ.

λ5 the surrogate light chain of the pre-B-cell receptor.

L chain an abbreviated form of the immunoglobulin light chain.

L-selectin an adhesion molecule of the selectin family found on lymphocytes. It binds to CD34 and GlyCAM-1 on high endothelial venules to initiate the migration of naive lymphocytes into secondary lymphoid tissue.

Langerhans cells phagocytic dendritic cells found in the epidermis. They can migrate in lymph from the epidermis to lymph nodes, where they differentiate into dendritic cells.

large granular lymphocyte an alternative and earlier name for the natural killer cell. Microscopic examination revealed two types of blood lymphocyte: the small lymphocytes (naïve B and T cells) and the large granular lymphocytes (NK cells).

large pre-B cells immature B cells that have a cell-surface pre-B-cell receptor.

late pro-B cell an early stage in the development of B cells in the bone marrow.

latency state adopted by some viruses in which they have entered cells but do not replicate.

late-phase reaction that part of a type 1 immediate hypersensitivity reaction that occurs 7–12 hours after contact with antigen and is resistant to treatment with antihistamine.

Lck intracellular protein tyrosine kinase associated with the CD4 and CD8 co-receptors of T cells.

LD acronym for **linkage disequilibrium**.

lectins receptors and plasma proteins that recognize carbohydrates.

lectin pathway of complement activation one of the three pathways of complement activation. It is activated by the binding of a mannose-binding lectin present in blood plasma to mannose-containing peptidoglycans on bacterial surfaces. *See also* **alternative pathway of complement activation; classical pathway of complement activation**.

lentiviruses group of slow retroviruses that includes the human immunodeficiency virus (HIV). They have a long incubation period, and disease can take years to become apparent.

leukemia a disease resulting from the unrestrained proliferation of a malignant white blood cell and characterized by abnormally high numbers of white cells in the blood. A leukemia can be lymphocytic, myelocytic, or monocytic, depending on the type of white blood cell that became malignant.

leukocyte general term for a white blood cell. Lymphocytes, granulocytes, and monocytes are all leukocytes.

leukocyte adhesion deficiency immunodeficiency disease in which the common β chain of the leukocyte integrins is not made. The resulting deficiency of integrins mainly affects the ability of leukocytes to enter sites infected with extracellular pathogens, and so such infections cannot be effectively eradicated.

leukocyte function-associated antigen-1 (LFA-1) one of the leukocyte integrins that mediate the adhesion of lymphocytes, especially T cells, to endothelial cells and antigen-presenting cells.

leukocyte function-associated antigen-3 (LFA-3) cell adhesion molecule of the immunoglobulin superfamily that is expressed on antigen-presenting cells (and other types of cell) and mediates their adhesion to T cells.

leukocytosis increased numbers of leukocytes in the blood. It is commonly seen in acute infection.

LFA-1 acronym for **leukocyte function-associated antigen-1**.

LFA-3 acronym for **leukocyte function-associated antigen-3**.

light chain (L chain) the smaller of the two types of polypeptide chain in an immunoglobulin molecule. It consists of one variable and one constant domain, and is disulfide-bonded to a heavy chain in the immunoglobulin molecule. There are two classes of light chain, known as κ and λ.

light zone the part of a germinal center in secondary lymphoid tissue that contains non-dividing centrocytes interacting with follicular dendritic cells.

linear epitope epitope of a protein recognized by antibody that consists of a linear sequence of amino acids within the protein's primary structure.

lingual tonsils aggregates of secondary lymphoid tissue at the back of the tongue.

linkage disequilibrium (LD) the situation when particular alleles of two or more polymorphic genes (for example those comprising an HLA haplotype) are inherited together at frequencies higher than expected by chance.

lipopolysaccharide an alternative term for **bacterial lipopolysaccharide**.

live-attenuated virus vaccines vaccines composed of live viruses that have an accumulation of mutations that impedes their growth in human cells and their capacity to cause disease.

LPS acronym for **bacterial lipopolysaccharide**.

LPS-binding protein plasma protein that binds bacterial lipopolysaccharide (LPS) and delivers it to LPS receptors on neutrophils and macrophages.

LT acronym for **lymphotoxin**.

lymph mixture of extracellular fluid and cells that is carried by the lymphatic system.

lymph nodes a type of secondary lymphoid tissue found at many sites in the body where lymphatic vessels converge. Antigens are delivered by the lymph and presented to lymphocytes within the lymph node where adaptive immune responses are initiated.

lymphatic vessels (lymphatics) thin-walled vessels that carry lymph from tissues to secondary lymphoid tissues (with the exception of the spleen) and from secondary lymphoid tissues to the thoracic duct.

lymphocytes class of white blood cells that consists of small and large lymphocytes. The small lymphocytes bear variable cell-surface receptors for antigen and are responsible for adaptive immune responses. There are two main classes of small lymphocyte—B lymphocytes (B cells) and T lymphocytes (T cells). Large granular lymphocytes are natural killer (NK) cells, the lymphocytes of innate immunity.

lymphocyte recirculation of lymphocytes, their continual migration from blood to secondary lymphoid tissues to lymph and back to the blood. An exception to this pattern is traffic to the spleen; lymphocytes both enter and leave the spleen in the blood.

lymphoid containing lymphocytes, or pertaining to lymphocytes.

lymphoid follicle more or less spherical aggregation of mainly B cells in secondary lymphoid tissues. Naive B cells must pass through follicles for their survival, and after B cells are activated by antigen they enter the follicles where they proliferate to form a germinal center and undergo somatic hypermutation and isotype switching.

lymphoid lineage all types of lymphocyte, and the bone marrow cells that give rise to them.

lymphoid organs (**lymphoid tissues**) organized tissues that contain very large numbers of lymphocytes held in a non-lymphoid stroma. The primary lymphoid organs, where lymphocytes are generated, are the thymus and bone marrow. The main secondary lymphoid tissues, in which adaptive immune responses are initiated, are the lymph nodes, spleen, and mucosa-associated lymphoid tissues such as tonsils, Peyer's patches, and the appendix.

lymphoid progenitor stem cell in bone marrow cell that gives rise to all lymphocytes.

lymphokines cytokines produced by lymphocytes.

lymphoma tumor of lymphocytes that grows in lymphoid and other tissues but in which the malignant lymphocytes do not enter the blood in large numbers.

lymphotoxin (**LT**) subset of TNF-family cytokines involved in development and maintenance of the architecture of secondary lymphoid tissues and organs. Also called TNF-β.

lysosome digestive intracellular organelle that contains degradative enzymes and breaks down macromolecules.

lytic granules intracellular storage granules of cytotoxic T cells and NK cells that contain the cytotoxins perforin, granulysin, and granzymes. These proteins are released when the lymphocytes interact with their target cells and kills them.

μ when referring to an immunoglobulin heavy chain, the isotype that is present in immunoglobulin IgM.

M cell specialized cell type in intestinal epithelium through which antigens and pathogens enter gut-associated lymphoid tissue from the intestines. Short for 'microfold cell.'

macrophage large mononuclear phagocytic cell resident in most tissues. Macrophages are derived from blood monocytes and contribute to innate immunity and early nonadaptive phases of host defense. They function as professional antigen-presenting cells and as effector cells in humoral and cell-mediated immunity.

macrophage activation stimulation of macrophages, which increases their phagocytic, antigen-presenting, and bacterial killing functions. It occurs in the course of infection.

macropinocytosis the nonspecific uptake of large amounts of extracellular fluid by endocytosis, a characteristic of dendritic cells.

MAdCAM-1 mucosal cell adhesion molecule-1, a mucosal addressin that is recognized by the lymphocyte surface proteins L-selectin and VLA-4. This interaction mediates the specific homing of lymphocytes to mucosal tissues.

major basic protein constituent of eosinophil granules that is released on eosinophil activation. It acts on mast cells to cause their degranulation.

major histocompatibility complex (**MHC**) large cluster of mainly immune system genes on the short arm of human chromosome 6 that encodes, among other proteins, a set of polymorphic membrane glycoproteins called the MHC molecules, which are involved in presenting peptide antigens to T cells.

major histocompatibility complex (**MHC**) **molecules** the highly polymorphic proteins that bind peptide antigens and present them to T cells. Also called **MHC molecules** and transplantation antigens.

malignant transformation the changes that occur in a cell to make it cancerous.

malignant tumors tumors that are capable of uncontrolled and invasive growth.

MALT acronym for **mucosa-associated lymphoid tissue**.

mannose-binding lectin (**MBL**) soluble acute-phase protein in the blood that binds to mannose residues on pathogen surfaces and, when bound, activates the complement system by the lectin pathway.

mannose receptor cell-surface receptor on dendritic cells, macrophages, and other leukocytes that binds to mannose residues on the surfaces of pathogens.

mast cell large bone marrow derived cell resident in connective tissues throughout the body. Mast cells contain large granules that store a variety of chemical mediators including histamine. Mast cells have high-affinity Fcε receptors (FcεRI) that bind free IgE. Antigen binding to IgE associated with mast cells triggers mast-cell activation and degranulation, producing a local or systemic immediate hypersensitivity reaction. Mast cells have a crucial role in allergic reactions.

mature B cell in B-cell development, B cell that has IgM and IgD on its surface and is able to respond to antigen.

mature dendritic cell dendritic cell in secondary lymphoid tissues that expresses co-stimulatory molecules and other cell-surface molecules that enables it to present antigen to naive T cells and activate them.

MBL acronym for **mannose-binding lectin**, a soluble pathogen-binding and complement-fixing protein of the innate immune response.

MCP acronym for **membrane cofactor protein**, a membrane-associated complement control protein.

megakaryocyte large cell of the erythroid lineage, produced in the bone marrow and resident there. Megakaryocytes produce platelets.

membrane cofactor protein (**MCP**) complement regulatory protein on human cells that promotes the inactivation of C3b and C4b by factor I.

membrane-attack complex the complex of terminal complement components that forms a pore in the membrane of the target cell, damaging the membrane and leading to cell lysis.

memory B cells long-lived antigen-specific B cells that are produced from activated B cells during the primary immune response to an antigen. On subsequent exposure to their specific antigen they are reactivated to differentiate into plasma cells as part of the secondary and subsequent immune responses that antigen.

memory cells general term for lymphocytes that are responsible for the phenomenon of immunological memory.

memory response a person's adaptive immune response to a pathogen or antigen to which he has previously been exposed and made a primary immune response. Also called the secondary immune response or the secondary adaptive immune response.

memory T cells long-lived antigen-specific T cells that are produced from activated T cells during the primary response to an antigen. On subsequent exposure to their specific antigen they are activated to differentiate into effector T cells as part of the secondary and subsequent immune responses to that antigen.

metastasis the spread of a tumor from its site of origin to other tissues. Some cells from the primary tumor invade other tissues and grow and divide to become secondary tumors.

methotrexate cytotoxic drug used to inhibit graft-versus-host reactions in bone marrow transplant recipients.

MHC acronym for **major histocompatibility complex**, the genetic region on chromosome 6 that contains the MHC class I and II genes and many other immune-system genes.

MHC class I molecules the class of MHC molecules that present peptides generated in the cytosol to CD8 T cells. They consist of a heterodimer of a class I heavy chain associated with β$_2$-microglobulin. Sometimes shortened to **MHC class I** or MHCI.

MHC class II compartment (**MIIC**) intracellular endocytic vesicles where MHC class II molecules bind peptides derived from extracellular sources.

MHC class II molecules the class of MHC molecules that present peptides generated in intracellular vesicles to CD4 T cells. They consist of a heterodimer of class II α and β chains. Sometimes shortened to **MHC class II** or MHCII.

MHC class II transactivator (**CIITA**) a transcriptional activator of MHC class II genes. When defective it causes a type of bare lymphocyte syndrome.

MHC molecules major histocompatibility complex molecules. Highly polymorphic glycoproteins encoded by the major histocompatibility complex (MHC). They form complexes with peptides and present peptide antigens to T cells. There are two classes—MHC class I and MHC class II molecules—with different roles in the immune response. They are also known as major histocompatibility antigens because they are the main alloantigens involved in the rejection of transplanted tissues.

MHC restriction the fact that a given T-cell receptor will recognize its peptide antigen only when the peptide is bound to a particular form of MHC molecule.

MIC proteins (**MICs**) MIC-A and MIC-B, stress-induced proteins that appear on the surface of epithelial cells in response to infection, damage, or other stresses, and which are recognized by the NKG2D receptor of NK cells.

microfold cell cell in the gut mucosa that facilitate the transport of pathogens and antigens from the gut lumen to secondary lymphoid tissues underlying the gut epithelium. Often abbreviated to M cell.

MIIC acronym for **MHC class II compartment** and pronounced 'em-two-see'. An endosomal compartment in professional antigen-presenting cells where MHC class II molecules load with peptides derived from pathogens and antigens that have been taken up from the extracellular environment.

minor histocompatibility antigens (**minor H antigens**) peptides of polymorphic cellular proteins that can lead to graft rejection when they are bound by MHC molecules and recognized by T cells.

minor histocompatibility loci the genes encoding proteins that can act as minor histocompatibility antigens.

mixed lymphocyte reaction cellular assay for detecting MHC differences between two individuals. The T cells from one individual proliferate in response to allogeneic MHC molecules on the cells of the other individual.

molecular mimicry antigenic similarity between a pathogen antigen and a cellular antigen, which results in the induction of antibodies or T cells that act against the pathogen but also cross-react with the self antigen.

monoclonal antibodies antibodies produced by a single clone of B lymphocytes and which are therefore all identical in structure and antigen specificity.

monocyte white blood cell with a bean-shaped nucleus. It is the precursor of the tissue macrophage.

monomorphic having only one form. The term is used, for example, for genes that have only one allele.

mucosa (*plural* **mucosae**) a mucus-secreting epithelium such as those that line the respiratory, intestinal, and urogenital tracts. The conjunctiva of the eye and the mammary glands are also in this category.

mucosa-associated lymphoid tissue (**MALT**) aggregations of lymphoid cells in mucosal epithelia and in the lamina propria beneath. The main mucosa-associated lymphoid tissues are the gut-associated lymphoid tissues (GALT) and the bronchial-associated lymphoid tissues (BALT).

mucosal mast cell one of two types of mast cell in humans (the other is the connective tissue mast cell). It is found in mucosal tissues throughout the body.

mucosal surfaces the mucus-coated outer surfaces of tissues, such as the gut, lungs, eyes and vagina, which communicate with the external environment to provide the body with material and information. The surfaces are delicate and they are protected by the mucus.

mucus slimy protective secretion composed of glycoproteins, proteoglycans, peptides, and enzymes that is produced by the goblet cells in many internal epithelia.

multiple sclerosis chronic progressive neurological disease characterized by patches of demyelination in the central nervous system and lymphocyte infiltration into the brain. It is believed to be an autoimmune disease.

multivalent having more than one binding site for the same or different ligands.

mutagen any agent, such as a chemical or radiation, that can cause a mutation.

mutation an alteration in the DNA sequence of a gene.

myasthenia gravis autoimmune disease in which autoantibodies against the acetylcholine receptor on skeletal muscle cells cause a block of signal transmission from nerve to muscle at neuromuscular junctions, leading to progressive muscle weakness and eventually to death.

mycophenolic acid the active derivative of the immunosuppressive and cytotoxic drug mycophenolate mofetil, which is metabolized to mycophenolic acid in the liver. It acts by inhibiting guanine synthesis.

myeloablative therapy conditioning regime used before a bone marrow transplant. The recipient's immune system is destroyed by a combination of cytotoxic drugs and irradiation.

myeloid lineage a subset of bone marrow derived cells comprising granulocytes, monocytes, and macrophages.

myeloid progenitor stem cell in bone marrow that gives rise to granulocytes, monocytes, and macrophages.

myeloma tumor of plasma cells resident in bone marrow.

myelopoiesis production of monocytes and granulocytes in the bone marrow.

N nucleotides nucleotides added at the junctions between gene segments of T-cell receptor and immunoglobulin heavy-chain variable-region sequences during somatic recombination, which contribute to the diversity of these molecules. They are not encoded in the gene segments but are inserted by the enzyme terminal deoxynucleotidyltransferase (TdT).

NADPH oxidase multisubunit enzyme that produces superoxide radicals and contributes to the killing of internalized pathogens in neutrophils. A deficiency in any of the NADPH subunits can be a cause of chronic granulomatous disease, an immunodeficiency disease.

naive B cell mature B cell that has left the bone marrow but has not yet encountered its specific antigen.

naive T cell a mature T cell that has left the thymus but not yet encountered its specific antigen.

natural interferon-producing cells (**NIPCs**) plasmacytoid dendritic cells that are highly specialized to synthesize and secret interferon-α and interferon-β in the innate immune response. Also shortened to interferon-producing cells.

natural killer cell (**NK cell**) large, granular, cytotoxic lymphocyte that circulates in the blood and is important in innate immunity to intracellular pathogens such as viruses. NK cells do not have variable receptors for antigen, but have numerous other receptors by which they can recognize and kill virus-infected cells and certain tumor cells. Also known as **natural killer lymphocyte** or **large granular lymphocyte**.

negative selection process in the thymus whereby developing T cells that recognize self antigens are induced to die by apoptosis.

NEMO deficiency acronym for **nuclear factor-κB essential modulator**. The *NEMO* gene is on the X chromosome and patients (mostly boys) who lack NEMO function suffer from **X-linked hypohidrotic ectodermal dysplasia and immunodeficiency**.

neoplasm a tumor, which may be benign or malignant.

neutralization the mechanism by which antibodies binding to sites on pathogens prevent growth of the pathogen and/or its entry into cells. The toxicity of bacterial toxins can similarly be **neutralized** by bound antibody.

neutralizing antibodies high-affinity IgA and IgG antibodies that bind to pathogens and prevent their growth or entry into cells.

neutropenia abnormally low numbers of neutrophils in the blood.

neutrophil phagocytic white blood cell that enters infected tissues in large numbers and engulfs and kills extracellular pathogens. Neutrophils are a type of granulocyte and contain granules that stain with neutral dyes; hence their name. They are the most abundant white blood cell.

NFAT (nuclear factor of activated T cells) transcription factor that is activated as a result of signaling from the T-cell receptor. It functions as a complex of the NFAT protein with the dimer of Fos and Jun proteins known as AP-1.

NFκB (nuclear factor κB), a transcription factor activated by signaling from the Toll-like receptors and many other receptors, and that helps turn on the expression of many immune-system genes.

NK cell acronym for **natural killer cell**.

NK-cell immunoglobulin-like receptors one of the two main structural classes of NK-cell receptors. They bind their ligands via immunoglobulin-like domains.

NK-cell lectin-like receptors one of the two main structural classes of NK-cell receptors. They bind both protein and carbohydrate ligands via lectin-like domains.

NOD proteins intracellular proteins structurally related to the Toll-like receptors (TLRs) and which bind components of intracellular bacteria in infected cells, activating similar pathways to the TLRs. NOD proteins contain a *n*ucleotide *o*ligomerization *d*omain, hence their name.

nonproductive rearrangements in developing lymphocytes, gene rearrangements at the immunoglobulin and T-cell receptor loci that that do not translate into a useful protein chain.

nuclear factor κB a transcription factor centrally involved in inflammatory and immune responses. Often shortened to NFκB.

oligomorphic describes genes that have only a small number of different alleles in the population.

Omenn syndrome genetically determined severe immunodeficiency caused by missense mutations that produce RAG proteins with only partial enzymatic activity, leading to a failure of lymphocyte development.

oncogenes genes involved in the control of cell growth. When these genes are defective in either structure or expression, they can cause cells to proliferate abnormally and form a tumor.

oncogenic viruses viruses that are involved in causing cancer.

oncology the clinical discipline that deals with the study, diagnosis, and treatment of cancer.

opportunistic pathogen microorganism that causes disease only in individuals whose immune systems are in some way compromised.

opsonins antibodies and complement components that bind to pathogens and facilitate their phagocytosis by neutrophils or macrophages.

opsonization the coating of the surface of a pathogen or other particle with any molecule that makes it more readily ingested by phagocytes. Antibody and complement opsonize extracellular bacteria for phagocytosis by neutrophils and macrophages because the phagocytic cells carry receptors for these molecules.

organ-specific autoimmune diseases autoimmune diseases targeted at a particular organ, such as the thyroid in Graves' disease.

original antigenic sin a bias seen in successive immune responses to structurally related antigens such as those on different strains of influenza virus. On infection for the second time with influenza, the antibody response is restricted to epitopes that the second strain shares with the first strain to which the person was exposed. Other highly immunogenic epitopes on the second and subsequent viruses are ignored.

P nucleotides nucleotides added into the junctions between gene segments during the somatic recombination that generates a rearranged variable-region sequence. They are an inverse repeat (a palindrome) of the nucleotide sequence at the end of the adjacent gene segment, which gives them their name of palindromic or P nucleotides.

P-selectin an adhesion molecule on endothelial cells that guides leukocyte from the blood into inflamed tissues. One of several selectins.

palatine tonsils aggregates of secondary lymphoid tissue at the sides of the throat.

pandemic outbreak of an infectious disease that spreads worldwide.

panel reactive antibody (PRA) a measure of the likelihood that a patient seeking a transplant is sensitized to potential donors. The patient's serum is tested against tissue from a representative panel of individuals for antibodies that would cause immediate hyperacute rejection of a graft. The PRA is the percentage of individuals in the panel whose cells react with the patient's antibodies.

paracrine term applied to a cytokine that is released from one type of cell and acts on other cells nearby.

parasites the unicellular protozoa and multicellular worms that infect animals and humans and live within them.

paroxysmal nocturnal hemoglobinuria (PNH) a disease in which the complement regulatory proteins CD59 and DAF are defective, so that complement activation leads to episodes of spontaneous hemolysis. The defect is in the attachment of CD59 and DAF to cell membranes via a glycolipid anchor.

passive immunity immunity to a given pathogen that has been acquired by injection of preformed antibodies, antiserum or T cells.

passive immunization the injection of specific antibodies to provide protection against a pathogen or toxin. The administered antibodies may derive from human blood donors, immunized animals, or hybridoma cell lines.

passive transfer of immunity transfer of immunity to a non-immune individual by the injection of specific antibody, immune serum, or T cells.

pathogen organism, most commonly a microorganism, that can cause disease.

pathology the detectable damage to tissue caused by disease. The term is also used for the study of such damage.

pemphigus foliaceus, **pemphigus vulgaris** autoimmune diseases involving blistering of the skin.

pentraxins family of pentameric acute-phase proteins to which C-reactive protein belongs. Pentraxins are formed of five identical subunits.

peptide-binding motif of an MHC isoform, the combination of anchor residues that are common to the amino acid sequences of peptides that bind to the isoform.

peptide-loading complex a protein complex in the endoplasmic reticulum membrane that loads peptides from the TAP transporter onto MHC class I molecules.

perforin one of the proteins released by cytotoxic T cells on contact with their target cells. It forms pores in the target cell membrane that contribute to cell killing.

peripheral lymphoid tissues all lymphoid tissues except the bone marrow and thymus. Also called **secondary lymphoid tissues**.

peripheral tolerance tolerance to self antigens that is acquired by the lymphocyte population outside the primary lymphoid organs (thymus and bone marrow).

Peyer's patches organized gut-associated lymphoid tissue present in the wall of the small intestine, especially the ileum.

phagocyte cell specialized to perform phagocytosis. The principal phagocytic cells in mammals are neutrophils and macrophages.

phagocytosis cellular internalization of particulate matter, such as bacteria, by means of endocytosis. Cells specialized for phagocytosis are known as **phagocytes**.

phagolysosome intracellular vesicle formed by fusion of a phagosome with a lysosome, in which the phagocytosed material is broken down by degradative lysosomal enzymes.

phagosome intracellular vesicle containing material taken up by phagocytosis.

phosphoantigens phosphorylated small molecules produced in the bacterial isoprenoid biosynthesis pathway that are recognized by the T-cell receptors of γ:δ T cells.

plasma cell terminally differentiated B lymphocyte that secretes antibody.

plasmacytoid dendritic cells dendritic-like cells derived from interferon-producing cells cultured with microbial products and inflammatory cytokines.

pluripotent hematopoietic stem cell stem cell in bone marrow that gives rise to all the cellular elements of the blood.

PNH acronym for **paroxysmal nocturnal hemoglobinuria**, the immunodeficiency disease caused by defective attachment to cell membranes of complement control proteins CD59 and DAF.

PNP acronym for **purine nucleoside phosphorylase**.

polarized describes cells in which one end (pole) is recognizably and functionally different from the other.

poly-Ig receptor receptor present on the basolateral membrane of epithelial cells that binds polymeric immunoglobulins, especially dimeric IgA but also IgM, and transports them across the epithelium by transcytosis.

polymorphism the existence of different variants of a gene or trait in the population. Genetic polymorphism is defined as the existence of two or more forms (alleles) of a given gene within the population, with the variant alleles each occurring at a frequency greater than 1%.

polymorphonuclear leukocytes an alternative name given to **granulocytes**—neutrophils, eosinophils and basophils—because of the varied morphology of their nuclei.

polyspecificity the ability to bind to many different antigens, a property shown by some antibodies. It is also known as polyreactivity.

positive selection process occurring in the thymus during T-cell development that selects immature T cells with receptors recognizing peptide antigens presented by self MHC molecules. Only cells that are positively selected are allowed to continue their maturation.

PRA acronym for **panel reactive antibody**.

pre-B cell stage in B-cell development at which a B cell has rearranged its heavy-chain genes but not its light-chain genes.

pre-B-cell receptor immunoglobulin-like receptor that is expressed on the surface of pre-B cells. It consists of μ heavy chains in association with surrogate light chains formed from **λ5** and **VpreB**. Its appearance signals the cessation of heavy-chain gene rearrangement.

prednisolone the biologically active compound to which the immunosuppressive drug prednisone is converted *in vivo*.

prednisone a synthetic steroid drug with potent anti-inflammatory and immunosuppressive activity. Prescribed to patients who have had organ transplants.

present cells carrying cell-surface complexes of peptide antigens and MHC molecules are said to present these antigens to T lymphocytes. The process is called presentation.

pre-T cell developing T cell that has productively rearranged a T-cell receptor β-chain gene and has produced a β chain that can form a complex with the surrogate α chain, pTα.

pre-T-cell receptor receptor that is present on the surface of some immature thymocytes. It consists of a T-cell receptor β chain associated with a surrogate α chain called pTα.

primary focus temporary aggregate of proliferating activated antigen-specific B cells and T cells that forms in a secondary lymphoid tissue at the beginning of an adaptive immune response.

primary immune response or **primary response** the adaptive immune response that follows a person's first exposure to an antigen.

primary immunodeficiency diseases diseases in which there is a failure of immunological function as a result of defects in genes for components of the immune system.

primary lymphoid follicles the B-cell areas of secondary lymphoid tissues in the absence of an immune response. They contain resting B lymphocytes.

primary granules an alternative name for the **azurophilic granules** of neutrophils. They have similarities to lysosomes in the types of hydrolytic and destructive enzymes that they store. Distinct from the **specific granules** of the neutrophil.

primary lymphoid tissues anatomical sites of lymphocyte development. In humans, B lymphocytes develop in bone marrow, whereas T lymphocytes develop in the thymus from precursor cells that have migrated there from the bone marrow.

pro-B cell B cell early in its development when it first expresses B-cell marker proteins and rearranges its heavy-chain genes. D_H to J_H joining occurs at the **early pro-B cell** stage, followed by V_H to DJ_H joining at the **late pro-B cell** stage.

pro-drug biologically inactive compound that is metabolized in the body to produce an active drug.

productive rearrangements DNA rearrangements at immunoglobulin and T-cell receptor loci that lead to a gene that can direct the synthesis of a functional polypeptide chain.

professional antigen-presenting cell cell that can present antigen to naive T cells and activate them. A professional antigen-presenting cell not only displays peptide antigens bound to appropriate MHC molecules but also has co-stimulatory molecules on its surface that are needed to activate the T cell. Only dendritic cells, macrophages, and B cells can be professional antigen-presenting cells.

programmed cell death an alternative term for **apoptosis**.

promiscuous binding specificity the ability to bind a number of different ligands; an example of this is the MHC molecules, which can each bind numerous peptides of differing sequence.

properdin blood protein that helps to activate the alternative pathway of complement activation. It binds to and stabilizes the alternative pathway C3 and C5 convertases on the surfaces of bacterial cells. It is also known as factor P.

protease inhibitors small proteins such as the serpins that can bind to proteases and inhibit their enzymatic activity.

proteasome large multisubunit protease present in the cytosol of all cells that degrades cytoplasmic proteins. It generates the peptides presented by MHC class I molecules.

protectin protein on the surface of human cells that prevents the assembly of the complement membrane-attack complex on the cell surface, thus protecting human cells against complement-mediated lysis. Also called CD59.

protective immunity the specific immunological resistance to a pathogen that is present in the environment during months after either vaccination or recovery from an infection with the pathogen, and which is due to pathogen-specific antibodies and effector T cells produced during the primary response.

protein tyrosine kinases enzymes that phosphorylate tyrosine residues on proteins. They are involved in signal transduction from many types of cell-surface receptor, including the antigen receptors on B cells and T cells.

proto-oncogenes cellular genes that regulate the control of cell growth and division. When mutated or aberrantly expressed, they contribute to the malignant transformation of cells, which leads to cancer.

provirus the DNA form of a retrovirus when it is integrated into the host cell genome. In this state it can remain transcriptionally inactive for a long time.

pTα the surrogate α chain that combines with the T-cell receptor β chain to form the **pre-T-cell receptor**.

purine nucleoside phosphorylase (PNP) enzyme involved in purine metabolism. Its deficiency results in the accumulation of purine nucleosides, which are toxic for developing T cells. This leads to severe combined immunodeficiency.

pus thick yellowish-white fluid that is formed in infected wounds. It is composed of dead and dying white blood cells (principally neutrophils), tissue debris, and dead microorganisms.

pyogenic causing the formation of pus.

pyogenic bacteria extracellular encapsulated bacteria that cause the formation of pus at sites of infection.

pyrogens molecules that induce fever, such as some bacterial products and certain cytokines.

RAG-1, RAG-2 acronyms for **recombination-activating gene 1** and **recombination-activating gene 2**, respectively. The proteins they encode, are essential for the mechanism of receptor-gene rearrangement in B cells and T cells.

rapamycin alternative and earlier name for the immunosuppressive drug **sirolimus** used to prevent rejection of organ transplants.

RCA acronym for **regulators of complement activation**.

rearranging genes in immunology, the immunoglobulin and T-cell receptor loci in developing B cells and T cells, in which somatic recombination brings together the V(D) and J gene segments and assembles the sequence encoding a functional variable region for an immunoglobulin or T-cell receptor chain. Further rearrangement of the immunoglobulin heavy-chain gene is used to switch the isotype from μ to γ, α or ε. *See also* **somatic recombination**.

receptor editing a process in developing B cells that have generated a self-reactive receptor by which further rounds of light-chain gene rearrangement can produce a new light chain to replace the self-reactive one.

receptor-mediated endocytosis internalization of extracellular material that is bound to a receptor at the cell surface.

recessive describes an allele that determines the phenotype only when present in two copies (that is, on both homologous chromosomes) or when no other copy of the allele is present (for example, for genes on the X chromosome in males). Usually refers to disease-causing alleles that encode proteins with functional defects.

recombination-activating genes (*RAG-1, RAG-2*) two genes whose expression is required for the rearrangement of immunoglobulin and T-cell receptor genes in B cells and T cells, respectively.

recombination signal sequences (**RSSs**) short stretches of DNA flanking the gene segments that are rearranged to generate V-region exons. They are the sites at which somatic recombination occurs.

regulators of complement activation (**RCAs**) proteins that regulate the activity of complement and contain one or more copies of a particular structural motif of ~60 amino acid residues called the CCP (complement control protein) motif. The CCP motif is also called a sushi domain because of its similarity in shape to a slice of a sushi roll.

regulatory CD4 T cell, regulatory T cell (T$_{reg}$) antigen-specific CD4 T cell whose actions can suppress immune responses. Also called suppressor T cells.

respiratory burst metabolic change accompanied by a transient increase in oxygen consumption that occurs in neutrophils and macrophages when they have taken up opsonized particles. It leads to the generation of toxic oxygen metabolites and other antibacterial substances that attack the phagocytosed material.

respiratory syncytial virus (**RSV**) paramyxovirus that is a common cause of severe chest infection in young children. Often associated with wheezing.

retrovirus RNA virus that replicates via a DNA intermediate that inserts into the host cell chromosome. The human immune deficiency virus (HIV) is a retrovirus.

rhesus (**Rh**) human blood group antigens that must be matched for successful blood transfusion. Rh mismatch between fetus and mother is the cause of hemolytic disease of the newborn. The Rh antigens were originally named rhesus because they were incorrectly believed to be identical to antigens detected on the surface of erythrocytes of the rhesus monkey, a type of macaque.

rheumatic fever autoimmune disease involving inflammation of the heart, joints, and kidneys, which can follow 2–3 weeks after a throat infection with certain strains of *Streptococcus pyogenes*. It is due to antibodies made against bacterial antigens cross-reacting with components of heart tissue, and the continued deposition of immune complexes.

rheumatoid arthritis common inflammatory disease of joints that is due to an autoimmune response.

rheumatoid factor IgM antibody with specificity for human IgG that is produced in some people with rheumatoid arthritis.

RSS acronym for the **recombination signal sequences** flanking the V, D, and J segments in immunoglobulin and T-cell receptor genes.

RSV acronym for **respiratory syncytial virus**.

sarcoma tumor that arises from a cell of connective tissue.

scavenger receptor phagocytic receptor of macrophages that binds to an assortment of negatively charged ligands, including sulfated polysaccharides, nucleic acids, and the phosphate-containing lipoteichoic acids in the cell walls of Gram-positive bacteria.

SCID, scid acronym for **severe combined immunodeficiency**.

SE acronym for **staphylococcal enterotoxins**.

secondary adaptive immune response, secondary immune response, secondary response the adaptive immune response provoked by a second exposure to an antigen. It differs from the primary response by starting sooner and building more quickly and is due to the presence of long-lived memory B cells and T cells.

secondary immunodeficiency disease diseases in which there is a failure of immunological function as a result of infection or the use of immunosuppressive drugs, rather than as a result of defects in genes for components of the immune system.

secondary lymphoid follicle the name given to the B-cell areas of secondary lymphoid tissues that are responding to antigen. They contain proliferating B cells.

secondary lymphoid tissues the lymph nodes, spleen and mucosa-associated lymphoid tissues. These are the tissues in which immune responses are initiated. The more highly organized tissues such as lymph nodes and spleen are also often known as **secondary lymphoid organs**.

secretory component, secretory piece fragment of the poly-Ig receptor left attached to dimeric IgA after its transport across epithelial cells.

secretory IgA dimeric IgA molecules that are produced by plasma cells in mucosal tissues and secreted across the mucosal surface.

segmental exchange a mutational mechanism that occurs during meiosis by which a segment of one gene is replaced by the homologous segment of another related gene. It can occur between different alleles of the same locus, when it is called **interallelic conversion**, or between different genes within a gene family, when it is called gene conversion.

selectins family of adhesion molecules present on the surfaces of leukocytes and endothelial cells. They are lectins and bind to sugar moieties on certain mucin-like glycoproteins.

selective IgA deficiency condition in which little or no IgA is made but the production of all other isotypes is normal. It usually causes no apparent symptoms of immunodeficiency.

self, self antigens describes all the normal constituents of the body to which the immune system would respond were it not for the mechanisms of tolerance that destroy or inactivate self-reactive B and T cells.

self MHC a person's own MHC class I and class II molecules.

self peptides peptides produced from the body's own proteins. In the absence of infection these peptides occupy the peptide-binding sites of MHC molecules on cell surfaces.

self-reactive responding to self antigens.

self-renewal the ability of a population of cells to renew itself.

self-tolerance the normal situation whereby a person's immune system does not respond to constituents of that person's body. The circulating lymphocyte population in an individual is thus said to be **self-tolerant**.

sensitized describes the first exposure to an allergen that elicits an IgE response and sets up the conditions for an allergic reaction to subsequent exposure to the allergen.

sepsis the toxic effects of infection of the bloodstream. Usually caused by Gram-negative bacteria. *See also* **septic shock**.

septic shock shock syndrome that is frequently fatal, caused by the systemic release of the cytokine TNF-α after bacterial infection of the bloodstream, usually with Gram-negative bacteria.

septicemia bacterial invasion of the blood.

serglycin proteoglycan that forms a complex with granulysin and perforin in the lytic granules of T cells and NK cells.

seroconversion phase of an infection when antibodies against the infecting agent are first detectable in the blood.

serotypes antigenically different strains of a bacterium or other pathogen that can be distinguished by immunological means, for example by antibody-based detection tests. Also used to describe different types of human alloantigen such as HLA and blood group antigens.

serpins class of protease inhibitor proteins that inhibit serine and cystine proteases. The C1 inhibitor is an example of a serpin.

serum sickness the collection of symptoms that follow the injection of large amounts of a foreign molecule (such as a serum protein) into a person. It is caused by the immune complexes formed between the injected protein and the antibodies that are made against it. Characterized by fever, arthralgias, and nephritis.

severe combined immunodeficiency disease (**SCID**) immune deficiency disease in which neither antibody nor T-cell responses are made. It is usually the result of genetic defects that lead to T-cell deficiencies, and is fatal in childhood if not treated.

signal joint the joint made in the discarded circle of DNA that is formed as a result of V(D)J recombination.

single-positive thymocytes a late stage of T-cell development in the thymus characterized by the expression of either the CD4 or the CD8 co-receptor on the cell surface.

sirolimus the proprietary name for the immunosuppressive drug rapamycin. Both names are in common use.

SLE acronym for the systemic autoimmune disease **systemic lupus erythematosus**.

small lymphocytes the general name for recirculating resting T and B lymphocytes.

small pre-B cells pre-B cells in which light-chain gene rearrangements are being made.

somatic gene therapy potential therapy for curing inherited immunodeficiency diseases. Hematopoietic stem cells isolated from the patient are transfected with a normal copy of the gene that is defective in the patient. The stem cells, which now express a good copy of the formerly defective gene, are reinfused into the patient's circulation.

somatic hypermutation mutation that occurs at high frequency in the rearranged variable-region DNA of immunoglobulin genes in activated B cells, resulting in the production of variant antibodies, some of which have a higher affinity for the antigen.

somatic recombination DNA recombination that occurs between gene segments in the immunoglobulin genes and T-cell receptor genes in developing B cells and T cells, respectively. It generates a complete exon composed of a V gene segment and a J gene segment (and a D gene segment at the immunoglobulin and T-cell receptor heavy-chain loci) that encodes the variable region of an immunoglobulin or T-cell receptor polypeptide chain.

specific describes the property of antibodies and other antigen-binding molecules and immune system receptors that interact selectively with only one type or a few types of molecule or cell.

specific granules membrane-bound storage granules in the cytoplasm of neutrophils that contain antimicrobial agents including lactoferrin, lysozyme, and defensins. They are termed specific because they are only found in neutrophils and are distinct from the **primary** or **azurophilic granules** that have similarity to the lysosomes present in other types of cell.

spleen organ situated adjacent to the cardiac end of the stomach. One function of the spleen is to remove old or damaged red blood cells from the circulation; the other is as a secondary lymphoid organ that responds to blood-borne pathogens and antigens.

staphylococcal enterotoxins (**SEs**) toxins secreted by strains of staphylococcus that act on the gut and cause the symptoms of food poisoning. They are also superantigens.

STATs acronym for **signal transducers and activators of transcription**. Proteins that are activated to become transcription factors through the Janus kinase signaling pathway from cytokine receptors.

stromal cells cells that provide the supporting framework of a tissue or organ.

subunit vaccines vaccines composed only of isolated antigenic components of a pathogen and not the pathogen itself, either alive or dead.

superantigens molecules that, by binding nonspecifically to MHC class II molecules and T-cell receptors, stimulate the polyclonal activation of T cells.

surrogate light chain protein that mimics an immunoglobulin light chain. It is made up of two subunits, VpreB and λ5, and is produced by pro-B cells. Together with the μ heavy chain it forms the pre-B-cell receptor.

survival signals in immunology, generally refers to the signals that developing and naive mature B cells and T cells must receive if they are to survive.

switch regions, switch sequences short DNA sequences preceding heavy-chain constant-region genes at which somatic recombination occurs when B cells switch from producing one immunoglobulin isotype to another.

sympathetic ophthalmia autoimmune response that sometimes follows damage to an eye and affects both the damaged eye and the healthy eye.

syngeneic genetically identical. A syngeneic tissue graft is one made between two genetically identical individuals. It does not provoke an immune response or a rejection reaction.

systemic anaphylaxis rapid-onset and potentially fatal IgE-mediated allergic reaction, in which antigen in the bloodstream triggers the activation of mast cells throughout the body, causing circulatory collapse and suffocation due to tracheal swelling.

systemic autoimmune disease autoimmunity that involves reactions to common components of the body and whose effects are not confined to one particular organ.

systemic lupus erythematosus (**SLE**) systemic autoimmune disease in which autoantibodies made against DNA, RNA, and nucleoprotein particles form immune complexes that damage small blood vessels.

T$_{reg}$ abbreviated form of **regulatory T cell**.

T cell one of the two main classes of lymphocyte (the other is the B cell). T cells develop in the thymus and are responsible for cell-mediated immunity. Their cell-surface antigen receptor is called the T-cell receptor. Also known as **T lymphocyte**.

T-cell activation the stimulation of mature naive T cells by antigen presented to them by professional antigen-presenting cells. It leads to their proliferation and differentiation into effector T cells.

T-cell area that part of a secondary lymphoid tissue where the lymphocytes are predominantly T cells.

T-cell precursor undifferentiated cell that gives rise to α:β T cells and γ:δ T cells in the thymus.

T-cell priming the activation of mature naive T cells by antigen presented to them by professional antigen-presenting cells.

T-cell receptor (TCR) the highly variable antigen receptor of T lymphocytes. On most T cells it is composed of a variable α chain and a variable β chain and is known as the α:β T-cell receptor. This receptor recognizes peptide antigens derived from the breakdown of proteins. On a minority of T cells, the variable chains are γ and δ chains, and this receptor is known as the γ:δ T-cell receptor. The type of antigen it recognizes is less well defined. Both types of receptor are present at the cell surface in association with the complex of invariant CD3 chains and ζ chains, which have a signaling function.

T-cell receptor α chain (TCRα) one of the two polypeptide chains that make up the most common form of T-cell receptor, the α:β T-cell receptor.

T-cell receptor β chain (TCRβ) one of the two polypeptide chains that make up the most common form of T-cell receptor, the α:β T-cell receptor.

T-cell receptor complex the complex of T-cell receptor and the invariant CD3 and ζ chains that makes up a functional antigen receptor on the T-cell surface.

T-cell synapse the localized area of contact between T cell and an interacting B cell or macrophage where receptors and their ligands congregate, signals are transduced and cytokines are exchanged. The T cell version of the **immunological synapse**.

TCR acronym for **T-cell receptor**.

T lymphocyte an alternative name for **T cell**, the lymphocytes of adaptive immunity that have T-cell receptors for antigen.

tacrolimus immunosuppressive polypeptide drug that inactivates T cells by inhibiting signal transduction from the T-cell receptor. It is widely used to suppress transplant rejection. It is also known as FK506.

TAP acronym for the **transporter associated with antigen processing**.

tapasin TAP-associated protein, a chaperone protein involved in the assembly of peptide:MHC class I molecule complexes in the endoplasmic reticulum.

target cell any cell that is acted on directly by effector T cells, effector cells or molecules. For example, virus-infected cells are the targets of cytotoxic T cells, which kill them, and naive B cells are the targets of effector CD4 T cells, which help stimulate them to produce antibodies.

TdT acronym for **terminal deoxynucleotidyltransferase**.

terminal deoxynucleotidyltransferase (TdT) enzyme that inserts nontemplated nucleotides (N nucleotides) into the junctions between gene segments during the rearrangement of the heavy-chain genes for T-cell receptor and immunoglobulin.

tetanus toxin protein neurotoxin produced by the bacterium *Clostridium tetani*. The cause of the disease tetanus.

TGF-β abbreviated name for the immunosuppressive cytokine **transforming growth factor-β**.

T$_H$1 cells subset of effector CD4 T cells that are characterized by the inflammatory cytokines they produce. They are involved mainly in activating macrophages. Also called **inflammatory T cells**.

T$_H$2 cells subset of effector CD4 T cells that are characterized by the non-inflammatory cytokines they produce. They are involved mainly in stimulating B cells to produce antibody. Also called **helper T cells**.

T$_H$17 cells recently defined subset of effector CD4 T cells that are characterized by the production of IL-17 and are thought to be involved in promoting inflammatory responses. They are associated with autoimmune conditions.

thymectomy the surgical removal of the thymus gland.

thymic anlage the tissue from which the thymic stroma develops during embryogenesis.

thymic stroma the reticular epithelial cells and connective tissue of the thymus, which form the essential microenvironment for T-cell development.

thymocytes developing T cells in the thymus.

thymus a lymphoepithelial organ in the upper part of the middle of the chest, just behind the breastbone. It is the site of T-cell development.

thymus-dependent lymphocyte lymphocytes of the adaptive immune response that differentiate in the thymus gland from progenitor cells coming from the bone marrow. Often shortened to **T cells**.

thymus-independent antigens (TI antigens) antigens that can elicit antibody production in the absence of T cells. There are two types of TI antigen: **TI-1 antigens**, which have intrinsic B-cell activating activity, and **TI-2 antigens**, which have multiple identical epitopes that cross-link B-cell receptors.

tingible body macrophages phagocytic cells that are seen in histological sections engulfing apoptotic B cells in germinal centers.

tissue type the combination of HLA molecules expressed by an individual.

TLR acronym for the **Toll-like receptors** of innate immunity.

TNF-α acronym for the inflammatory cytokine **tumor necrosis factor-α**.

tolerance when the immune system of a person does not or cannot respond to an antigen. In such cases the individual is said to be **tolerant** of the antigen.

Toll-like receptors (TLRs) class of receptors of innate immunity that are present in many types of leukocyte, especially macrophages, dendritic cells, and neutrophils. Each type of TLR is specific for a different type of common pathogen component, such as bacterial lipopolysaccharide or other cell wall components. The responses of macrophages and dendritic cells to stimulation via TLRs are also essential to enabling the initiation of an adaptive immune response.

tonsils large aggregates of lymphoid cells lying on each side of the pharynx.

toxic shock syndrome a systemic toxic reaction caused by the overproduction of cytokines by CD4 T cells activated by the bacterial superantigen **toxic shock syndrome toxin-1 (TSST-1)**, which is secreted by *Staphylococcus aureus*.

toxoids toxins that have been deliberately inactivated by heat or chemical treatment so that they are no longer toxic but can still provoke a protective immune response on vaccination.

transcytosis the transport of molecules from one side of an epithelium to the other by endocytosis into vesicles within the epithelial cells at one face of the epithelium and release of the vesicles at the other.

transforming growth factor-β (TGF-β) multifunctional cytokine produced by T$_H$2 cells and other cell types.

transfusion effect improved outcome of transplantation if the recipient has previously been given blood transfusions from people who share an HLA-DR allotype with the organ with which the patient is subsequently transplanted.

translocation chromosomal abnormality in which a piece of one chromosome has become joined to another chromosome. The rearranging genes of B and T cells are often the sites of translocations in B-cell and T-cell tumors.

transplant rejection immune reaction that is directed toward transplanted tissue and leads to death of the graft.

transplantation the grafting of organs or tissues from one individual to another.

transplantation antigen name historically given to the MHC molecules because they are the main antigens that provoke the rejection of transplanted organs.

transporter associated with antigen processing (TAP) ATP-binding protein in the endoplasmic reticulum membrane that transports peptides from the cytosol to the lumen of the endoplasmic reticulum. It is composed of two subunits, TAP-1 and TAP-2. It supplies MHC class I molecules with peptides.

tuberculin test clinical test used to detect exposure to *Mycobacterium tuberculosis*, the causal agent of tuberculosis. Subcutaneous injection of purified protein derivative (PPD) of *M. tuberculosis* elicits a delayed-type hypersensitivity reaction in individuals who have had tuberculosis or have been immunized against it.

tumor a growth arising from uncontrolled cell proliferation. It may be benign and self-limiting, or malignant and invasive.

tumor antigens cell-surface components of tumor cells that can elicit an immune response in the affected individuals. *See also* **tumor-associated antigens; tumor-specific antigens.**

tumor-associated antigens antigens characteristic of certain tumor cells that are also present on some types of normal cell.

tumor necrosis factor-α (TNF-α) cytokine produced by macrophages and T cells that has several functions in the immune response and is the prototype of the TNF family of cytokines. These cytokines function as cell-associated or secreted proteins that interact with receptors of the tumor necrosis factor receptor (TNFR) family.

tumor-specific antigens antigens that are characteristic of tumor cells and are not expressed by normal healthy cells.

tumor suppressor genes genes for cellular proteins that function to prevent cells from becoming cancerous.

TSST-1 acronym for the **toxic shock syndrome toxin-1,** a **superantigen** secreted by *Staphylococcus aureus.*

type 1 diabetes an autoimmune disease in which insulin-secreting β cells of the pancreatic islets of Langerhans are gradually destroyed.

type I hypersensitivity reaction tissue-damaging allergic reaction caused by exposure to an allergen against which IgE antibodies were made during the immune response to an earlier exposure.

type II hypersensitivity reaction tissue-damaging immune reaction caused by the secondary response to small chemically reactive molecules that modify cell surface components and stimulate production of specific IgG antibodies.

type III hypersensitivity reaction tissue-damaging immune reaction caused by immune complexes formed during the secondary immune response to soluble proteins of non-human origin.

type IV hypersensitivity reaction a delayed-type tissue-damaging immune reaction caused by the secondary response of T cells specific for peptides derived from human proteins that have been modified by small chemically reactive molecules from the environment.

type I interferon interferons-α and -β, which are cytokines produced by virus-infected cells that interfere with viral replication by the infected cell, and also signal neighboring uninfected cells to prepare for resisting infection.

type II interferon interferon-γ, which is a cytokine with multiple pro-inflammatory functions in an immune response.

unproductive rearrangements DNA rearrangements of immunoglobulin and T-cell receptor genes that produce a gene unable to make a functional polypeptide chain.

urticaria the technical term for hives, which are red, itchy skin welts usually brought on by an allergic reaction.

V domain abbreviated form of **variable domain.**

V gene segment abbreviated form of **variable gene segment.**

V region abbreviated form of **variable region.**

vaccination the deliberate induction of protective immunity to a pathogen by the administration of killed or non-pathogenic forms of the pathogen, or its antigens, to induce an immune response.

vaccine any preparation made from a pathogen that is used for vaccination and provides protective immunity against infection with the pathogen.

vaccinia the cowpox virus. It causes a limited infection in humans that leads to immunity to the human smallpox virus. Was used as a vaccine against smallpox.

variable domain (V domain) the amino-terminal domain of immunoglobulin and T-cell receptor polypeptide chains. Paired variable domains make up the antigen-binding site.

variable gene segment (V gene segment) DNA sequence in the immunoglobulin or T-cell receptor genes that encodes the first 95 or so amino acids of the V domain. There are multiple different V gene segments in the germline genome. To produce a complete exon encoding a V domain, one V gene segment must be rearranged to join up with a J or a rearranged DJ gene segment.

variable region (V region) the part of an immunoglobulin or T-cell receptor, or of its constituent polypeptide chains, that varies in amino acid sequence between isoforms having different antigenic specificities. It is responsible for determining antigen specificity.

variable surface glycoproteins (VSGs) glycoproteins that form the surface coat of African trypanosomes. The trypanosome can repeatedly change its glycoprotein coat by expressing different glycoprotein genes using a process akin to gene conversion.

variola name given to both the smallpox virus and the disease it causes: smallpox.

variolation historical procedure for immunization against smallpox in which a small amount of live smallpox virus was introduced through scarification of the skin.

vascular addressins mucin-like cell adhesion molecules on endothelial cells to which some leukocyte adhesion molecules bind. They determine the selective homing of leukocytes to particular sites in the body.

V(D)J recombinase the set of enzymes needed to recombine V, D, and J segments in the rearrangement of immunoglobulin and T-cell receptor genes.

V(D)J recombination the mechanism of gene rearrangement by which V, D, and J segments of immunoglobulin and T-cell receptor genes are brought into register, with the elimination of the intervening segments of DNA, to make function variable-region genes. *See also* **somatic recombination.**

viruses submicroscopic pathogens composed of a nucleic acid genome enclosed in a protein coat. They replicate only in a living cell because they do not possess all the metabolic machinery required for independent life. A viral particle is called a virion.

VLA-4 integrin present on the surface of T lymphocytes that binds to mucosal cell adhesion molecules.

VpreB abbreviated form for **pre-B-cell receptor.**

VSG acronym for the **variable surface glycoproteins** of African trypanosomes.

WAS acronym for **Wiskott–Aldrich syndrome** a genetic immunodeficiency disease.

Weibel–Palade bodies granules containing P-selectin, which are present inside endothelial cells.

wheal and flare reaction observed when small amounts of allergen are injected into the dermis of an individual who is allergic to the antigen. The reaction consists of a raised area of skin containing fluid (the wheal) with a spreading, red, itchy reaction around it (the flare).

Wiskott–Aldrich syndrome (WAS) genetic immunodeficiency disease in which interactions between T cells and B cells are defective, owing to a defective protein (WASP) that is normally required for cytoskeletal reorganization. Antibody responses in particular are deficient.

X-linked agammaglobulinemia (XLA) a genetic disorder in which B-cell development is arrested at the pre-B-cell stage and neither mature B cells or antibodies are formed. The disease is due to a defect in the gene encoding the protein tyrosine kinase Btk.

X-linked diseases heritable diseases caused by recessive mutations in genes carried on the X chromosome. Because males have only one copy of the X chromosome, such diseases show up much more frequently in males than in females.

X-linked hyper IgM syndrome alternative name for **hyper IgM immunodeficiency**.

X-linked hypohidrotic ectodermal dysplasia and immunodeficiency condition characterized by developmental defects and immunodeficiency, caused by a defect in a component of the NFκB signaling pathway. Also called NEMO deficiency.

X-linked lymphoproliferative syndrome immunodeficiency due to a defect in the SH2D1A gene on the X chromosome, resulting in an inability to control EBV infections.

xenoantibodies antibodies produced in one species of animal against the antigens of another animal species.

xenoantigens antigens on the tissues of one species of animal that provoke an immune response in members of another animal species.

xenogeneic coming from a different species of animal.

xenograft an organ from one animal species transplanted into another animal species.

xenotransplantation the transplantation of organs from one animal species into another. It is being explored as a possible solution to the shortage of human organs for clinical transplantation.

XLA acronym for the immunodeficiency disease **X-linked agammaglobulinemia**.

ZAP-70 acronym for Zeta-chain-associated protein kinase of 70 kDa molecular weight. It is a cytoplasmic tyrosine kinase in T cells that forms part of the signal transduction pathway from T-cell receptors.

Figure Acknowledgments

Chapter 1

Opener image: Copyright 2008 The Wellcome Trust Medical Photographic Library and Stephanie Schuller.

Figure 1.1 lower panel from World Health Organization: WHO slide set on the diagnosis of smallpox.

Figure 1.3 panel a © Omikron/Science Photo Library; panel b, c, d, e, i, j, k, © Eye of Science/Science Photo Library; panel f © A. Ryter/Institut Pasteur; panel g © Philipe Gontier/Science Photo Library; panel h © A.B. Dowsett/Science Photo Library; panel l © Sinclair Stammers/Science Photo Library.

Figure 1.23 upper photograph reprinted from P.R. Wheater et al., *Functional Histology* 2nd Edition. © (1987) with permission from Elsevier.

Chapter 2

Opener image: Copyright 2006 The Wellcome Trust Medical Photographic Library and Derek Penman.

Figure 2.13 from S. Bhadki et al., *Blut*, **60**(6):309–318. © (1990) Springer-Verlag.

Figure 2.42 reprinted by permission of Federation of the European Biochemical Societies from Structure and functions associated with the group of mammalian lectins containing collagen-like sequences, by S. Thiel and K.B.M. Reid, *FEBS Letters*, **250**:78–84. © (1989).

Figure 2.46 from F.P. Siegal et al., *Science*, **284**(5421):1835–1837, 1999. Reprinted with permission from AAAS.

Chapter 3

Opener image: Copyright 2008 The Wellcome Trust Medical Photographic Library and Stephen Fuller.

Chapter 4

Opener image: Copyright 2008 SPL – Custom Medical Stock Photo, All Rights Reserved.

Figure 4.6 panel a reprinted by permission from L. Harris et al., *Nature*, **360**:369–372. © (1992) Macmillan Magazines Ltd.

Figure 4.11 panel a from R.L. Stanfield et al., *Science*, **248**:712–719. © (1990) American Association for the Advancement of Science. Reprinted with permission of AAAS.

Figure 4.29 from A.C. Davis et al., *The European Journal of Immunology*, **18**:1001–1008. © (1988) Wiley-VCH.

Chapter 5

Opener image: Copyright 2006 The Wellcome Trust Medical Photographic Library and Paul Duprex.

Figures 5.2 and 5.22 from K.C. Garcia et al., *Science*, **274**:209–219. © (1996) American Association for the Advancement of Science. Reprinted with permission of AAAS.

Figure 5.4 panel a reprinted from I. Roitt et al., *Immunology* 5th Edition © (1998) with permission from Elsevier.

Chapter 6

Opener image: Copyright Dennis Kunkel Microscopy, Inc.

Figure 6.5 panel b from P.L. Witte et al., *The European Journal of Immunology*, **17**:1473–1484. © (1987) Wiley-VCH.

Chapter 7

Opener image: Copyright 2008 Joseph R. Siebert, Ph.D. – Custom Medical Stock Photo, All Rights Reserved.

Figure 7.8 reprinted by permission from C.D. Surh and J. Sprent, *Nature*, **372**:100–103. © (1994) Macmillan Magazines Ltd.

Chapter 8

Opener image: Courtesy of Yasodha Natkunam.

Figure 8.2 reprinted by permission from Macmillan Publishers Ltd: P. Pierre et al., *Nature*, **388**:787–792 © 1997.

Figure 8.21 from G. Kaplan and Z.A. Cohn, *International Review of Experimental Pathology*, **28**:45–78. © (1986) with permission from Elsevier.

Figure 8.29 panel c from P.A. Henkart and E. Martz (eds), *Second International Workshop on Cell Mediated Cytotoxicity* © (1985) Kluwer/Plenum Publishers. With kind permission of Springer Science and Business Media.

Figure 8.33 from H.D. Oche et al, *Primary Immunodeficiency Diseases: A Molecular and Genetic Approach.* © (1998) by permission Oxford University Press, Inc. Also Chapter 28, J. Puck et. al., *Inherited Disorders with Autoimmunity and Defective Lyphocyte Regulation.* (1999) Oxford University Press.

Chapter 9

Opener image: Copyright 2008 SPL – Custom Medical Stock Photo, All Rights Reserved.

Figure 9.11 from D. Zucker-Franklin et al., *Atlas of Blood Cells: Function and Pathology* 2nd Edition. Milan, Italy (1988) Lippincott, Williams and Wilkins.

Figure 9.12 reprinted with permission from A.K. Szakal et al., *The Journal of Immunology*, **134**:1349–1359. © (1985) The American Association of Immunologists, Inc.

Chapter 10

Opener image: Copyright Dennis Kunkel Microscopy, Inc.

Figure 10.7 from J.H. Niess, *Science*, **307**:254–257, 2005. Reprinted with permission from AAAS.

Figures 10.13 and 10.14 from P. Brandtzaeg and F.E. Johansen, *Immunological Reviews*, **206**:32–63, 2005, Wiley.

Chapter 11

Opener image: Copyright 2006 The Wellcome Trust Medical Photographic Library and Hugh Sturrock.

Figure 11.10 from S. Dorman et al. Clinical features of dominant and recessive interferon γ receptor 1 deficiencies. *Lancet*, **364**:2113–2121, 2004. © (2004) with permission from Elsevier.

Figure 11.25 adapted with permission from Macmillan Publishers Ltd: Martin, M.P. et al., *Nature Genetics*, **39**:733–740 © 2008.

Chapter 12

Opener image: Copyright 2005 The Wellcome Trust Medical Photographic Library and University of Edinburgh.

Figure 12.1 pollen © E. Gueho/CNRI/Science Photo Library; house dust mite © K.H. Kjeldsen/Science Photo Library; wasp © Claude Nuridsany and Marie Perennon/Science Photo Library; vaccine © Chris Priest and Mark Clarke/Science Photo Library; peanuts © Adrienne Hart-Davis/Science Photo Library; shellfish © Sheila Terry/Science Photo Library; poison ivy © Ken Samuelson/Tony Stone Images, Images ® Copyright 1999 Photodisc, Inc.

Figure 12.39 (left photograph) from A. Mowat, J.L. Viney, *Immunological Reviews*, **156**:145–166, 1997. By permission of Blackwell Publishing Ltd.

Chapter 13

Opener image: Copyright 2006 The Wellcome Trust Medical Photographic Library and Stephanie Schuller.

Figure 13.30 adapted from L. Klareskog et al, *Annual Review of Immunology*, **26**:651–675, 2008. By permission of Annual Reviews.

Figure 13.37 reprinted from C.M. Weyland et al., *Experimental Gerontology*, **38**:833–841. © Elsevier (2003).

Chapter 14

Opener image: from D.J. Smith et al, *Science*, **305**:372–376, 2004. Reprinted with permission from AAAS.

Figure 14.6 from V.A.A. Jansen et al., *Science*, **301**:804, 2003. Reprinted with permission from AAAS.

Figure 14.11 from R.I. Glass, New hope for defeating rotavirus. *Scientific American* April 2006:47–55.

Chapter 15

Opener image: Copyright 2008 The Wellcome Trust Medical Photographic Library and Dave McCarthy and Annie Cavanagh.

Figures 15.6 and 15.11 from B.D. Kahan and C. Ponticelli, *Principles and Practice of Renal Transplantation* 1st edition (2000). Reproduced with permission Thomson Publishing.

Figure 15.13 from G. Opelz et al., *Review of Immunogenetics*, **1**(3):334–42 (1999). By permission of Wiley-Blackwell Publishing Ltd.

Figure 15.22 from E.J. Johnson and D. Goldstein, *Science*, **302**:1338–1339, 2003. Reprinted with permission from AAAS.

Figure 15.23 from J.P. McCulley, J.Y. Niederkorn, Immune mechanisms of corneal allograft rejection. *Current Eye Research*, **32**:1005–1016, 2007. Reprinted by permission of the publisher (Taylor & Francis Ltd, http://www.tandf.co.uk/journals)

Figure 15.28 This research was originally published in *Blood*. E.W. Petersdorf et. al., *Blood*, 1998; **92**:3515–3520. © American Society of Hematology, and P. Parham and K. McQueen, *Nature Reviews Immunology*, **3**,(2) 108–122 (2003). By permission of Macmillan Magazines Ltd.

Figure 15.34 from F. Claas, Immunological tolerance in an HLA non-identical chimeric twin, *Human Immunology*, **61**:190–192, 2000. © American Society for Histocompatibility and Immunogenetics (2000).

Chapter 16

Opener image: Courtesy of the Skin Cancer Foundation.

Figure 16.16 from P. Coulie et al., *Immunological Reviews*, **188**:22–32 (2002). By permission of Blackwell Publishing Ltd.

Index